Human Body Composition

SECOND EDITION

Steven B. Heymsfield, MD
Columbia University

Timothy G. Lohman, PhD
University of Arizona

ZiMian Wang, PhD
Columbia University

Scott B. Going, PhD
University of Arizona

Editors

Human Kinetics

Library of Congress Cataloging-in-Publication Data

Human body composition / Steven B. Heymsfield ... [et al.], editors.-- 2nd ed.
  p. ; cm.
  Includes bibliographical references and index.
  ISBN 0-7360-4655-0 (hard cover)
  1. Body composition. 2. Body composition--Measurement. 3. Anthropometry.
  [DNLM: 1. Body Composition. 2. Anthropometry--methods. QU 100 H9175 2005] I. Heymsfield, Steven, 1944-
QP33.5.H85 2005
  612--dc22

                                                                    2004026523

ISBN: 0-7360-4655-0

Acquisitions Editor: Loarn D. Robertson, PhD
Managing Editor: Amanda S. Ewing
Copyeditor: Julie Anderson
Proofreader: Erin Cler
Indexer: Michael Ferreira
Permission Manager: Dalene Reeder
Graphic Designer: Fred Starbird
Graphic Artist: Angela K. Snyder
Photo Manager: Kelly Huff
Cover Designer: Keith Blomberg
Photographer (cover): © Zhuang Mu
Art Manager: Kelly Hendren
Illustrator: Keri Evans
Printer: Edwards Brothers

Printed in the United States of America

10   9   8   7   6   5   4   3   2   1

Human Kinetics
Web site: www.HumanKinetics.com

United States: Human Kinetics
P.O. Box 5076
Champaign, IL 61825-5076
800-747-4457
e-mail: humank@hkusa.com

Canada: Human Kinetics
475 Devonshire Road Unit 100
Windsor, ON N8Y 2L5
800-465-7301 (in Canada only)
e-mail: orders@hkcanada.com

Europe: Human Kinetics
107 Bradford Road
Stanningley
Leeds LS28 6AT, United Kingdom
+44 (0) 113 255 5665
e-mail: hk@hkeurope.com

Australia: Human Kinetics
57A Price Avenue
Lower Mitcham, South Australia 5062
08 8277 1555
e-mail: liaw@hkaustralia.com

New Zealand: Human Kinetics
Division of Sports Distributors NZ Ltd.
P.O. Box 300 226 Albany
North Shore City
Auckland
0064 9 448 1207
e-mail: blairc@hknewz.com

Contents

Chapter 8 Anthropometry and Ultrasound

Anna Bellisari and Alex F. Roche

109

Chapter 9 Pediatric Body Composition Methods

Aviva Sopher, Wei Shen, and Angelo Pietrobelli

129

Chapter 10 Animal Body Composition Methods

Maria S. Johnson and Tim R. Nagy

141

Chapter 11 Statistical Methods

Shumei S. Sun and Wm. Cameron Chumlea

151

Part III BODY COMPOSITION MODELS AND COMPONENTS   161

Chapter 12 Multicomponent Molecular-Level Models of Body Composition Analysis   163

ZiMian Wang, Wei Shen, Robert T. Withers, and Steven B. Heymsfield

Chapter 13 Measuring Adiposity and Fat Distribution in Relation to Health   177

Luís B. Sardinha and Pedro J. Teixeira

Chapter 14 Assessing Muscle Mass   203

Henry C. Lukaski

Chapter 15 Body Composition, Organ Mass, and Resting Energy Expenditure   219

Dympna Gallagher and Marinos Elia

Part IV BODY COMPOSITION AND BIOLOGICAL INFLUENCES 241

Chapter 16 Genetic Influences on Human Body Composition 243

Peter Katzmarzyk and Claude Bouchard

Chapter 17 Age 259

Richard N. Baumgartner

Chapter 18 Variation in Body Composition Associated With Sex and Ethnicity 271

Robert M. Malina

Chapter 19 Pregnancy 299

Sally Ann Lederman

Chapter **25**   **Body Composition Studies in People With HIV**   **377**
Donald P. Kotler and Ellen S. Engelson

Chapter **26**   **Inflammatory Diseases and Body Composition**   **389**
Ian Janssen and Ronenn Roubenoff

Preface

The study of body composition in humans and animals is in transition. Whereas body composition was of interest to only a few scientists around the world several decades ago, body composition measurements are now routinely carried out in basic science laboratories, in clinical settings, and even by the public in their homes. The study of body composition is now sufficiently prominent that dedicated sections are provided for research articles in journals and for presentations at national and international meetings. Every 3 years, since 1986, an International Symposia on In Vivo Body Composition Studies has been convened solely for the purpose of discussing new developments in body composition research. *The International Journal of Body Composition Research* was recently launched as the first peer-reviewed publication dedicated solely to the science of body composition research. With so much activity in different areas, the possibility exists for evolution of discordant nomenclature, varying measurement protocols, and overall limited diffusion of new relevant knowledge between scientific disciplines.

The intent of the first edition of this book was to bring together in a single treatise definitive information on each of the prevailing methods and to thus serve as the standard for educating students and new investigators entering the field. This second edition maintains this focus and provides new information on recently introduced methods and topics of biological importance.

The book includes five parts and an appendix that encompass the field of body composition research. The first section presents an overview of the field and provides a foundation for the chapters that follow. The second section provides an extensive review of the available body composition methods. Each chapter provides a historical overview of the topic, the method's scientific underpinnings, and the key methodological details. Other reviewed topics in each chapter include measurement cost, safety, and practicality. The field of molecular genetics is increasingly focusing on animal phenotypes, and body composition applications abound. Accordingly, we have added a new chapter on animal body composition to fill a large void on this topic.

In the third section, reviews are provided on specific body composition components of widespread interest, adipose tissue and skeletal muscle, along with a review of body composition models. Obesity and sarcopenia are increasing in prevalence, and these chapters are designed to provide the reader with updated knowledge on the measurement and clinical implications of components linked with adiposity and muscularity. Because body composition is a determinant of energy expenditure, particularly at rest, an important new chapter is provided that reviews this area in detail. New and inexpensive systems for measuring energy expenditure are appearing on the market, and we anticipate greatly increased interest in evaluation of human energy requirements.

The fourth section covers biologically fundamental areas including growth, development, genetics, aging, sex, race, exercise, and hormones, and there is a new chapter on pregnancy. The six chapters in this portion of the book provide a broad overview of body composition across the life span and associated influencing factors.

In the fifth section, several currently important pathological states involving body composition are reviewed. In addition to the earlier topics covering morbidity–mortality linkages with body composition, obesity, and diabetes, new chapters on HIV-associated illnesses, cancer, and inflammatory diseases review developments in these rapidly evolving areas of global interest.

The appendix presents a summary of reference body composition data from fetus to adult.

We thank the authors for their outstanding efforts and the staff of Human Kinetics, particularly Loarn Robertson, for their help and encouragement. Thanks are due also to many others, including Jon Hutchens and Michele Graves for their administrative assistance. Editing this book was an enjoyable and instructive experience for us, and we hope that the readers will enjoy the result.

Contributors

Richard N. Baumgartner, PhD
Aging and Genetic Epidemiology Program
University of New Mexico
Albuquerque, New Mexico, United States

Anna Bellisari, PhD
Departments of Sociology/Anthropology and
    Community Health
Lifespan Health Research Center
Wright State University School of Medicine
Dayton, Ohio, United States

Per Bjorntorp, MD, PhD (deceased)
The Cardiovascular Institute
University of Goteborg
Goteborg, Sweden

Claude Bouchard, PhD
Human Genomics Laboratory
Pennington Biomedical Research Center
Baton Rouge, Louisiana, United States

Zhao Chen, PhD, MPH
Division of Epidemiology and Biostatistics
Mel and Enid Zuckerman College of
    Public Health
University of Arizona
Tuscon, Arizona, United States

Wm. Cameron Chumlea, PhD
Fels Professor, Departments of Community
    Health and Pediatrics
Lifespan Health Research Center
Wright State University School of Medicine
Dayton, Ohio, United States

Marinos Elia, MD
Professor of Clinical Nutrition and Metabolism
Institute of Human Nutrition
University of Southampton
Southampton, England, United Kingdom

Kenneth J. Ellis, PhD
Department of Pediatrics
Baylor College of Medicine
Houston, Texas, United States

Ellen S. Engelson, EdD
Gastrointestinal Division, Department of Medicine
St. Luke's–Roosevelt Hospital
Columbia University College of Physicians
    and Surgeons
New York, New York, United States

Dympna Gallagher, EdD
Assistant Professor of Nutritional Medicine
Department of Medicine and Institute of
    Human Nutrition
Columbia University
New York, New York, United States
and
Director Body Composition Unit,
    Obesity Research Center
St. Luke's–Roosevelt Hospital
New York, New York, United States

Bret H. Goodpaster, PhD
Division of Endocrinology and Metabolism
Department of Medicine
University of Pittsburgh School of Medicine
Pittsburgh, Pennsylvania, United States

Ian Janssen, PhD
School of Physical and Health Education
Department of Community Health
    and Epidemiology
Queen's University
Kingston, Ontario, Canada
and
Nutrition, Exercise Physiology, and
    Sarcopenia Laboratory
Jean Mayer USDA Human Nutrition Research
    Center on Aging at Tufts University
Boston, Massachusetts, United States

Maria S. Johnson, PhD
Division of Physiology and Metabolism
Department of Nutrition Sciences
University of Alabama at Birmingham
Birmingham, Alabama, United States

Peter Katzmarzyk, PhD
School of Physical and Health Education
Queen's University
Kingston, Ontario, Canada

David E. Kelley, MD
Division of Endocrinology and Metabolism,
    Department of Medicine
University of Pittsburgh School of Medicine
Pittsburgh, Pennsylvania, United States

Donald P. Kotler, MD
Gastrointestinal Division, Department of Medicine
St. Luke's–Roosevelt Hospital
Columbia University College of Physicians
    and Surgeons
New York, New York, United States

Sally Ann Lederman, PhD
Obesity Research Center
St. Luke's–Roosevelt Hospital
New York, New York, United States

Henry C. Lukaski, PhD, FACSM
U.S. Department of Agriculture
Agricultural Research Center
Grand Forks Human Nutrition Research Center
Grand Forks, North Dakota, United States

Robert M. Malina, PhD, FACSM
Research Professor
Tarleton State University
Stephenville, Texas, United States

Tim R. Nagy, PhD
Division of Physiology and Metabolism
Department of Nutrition Sciences
University of Alabama at Birmingham
Birmingham, Alabama, United States

Richard N. Pierson, Jr., MD
St. Luke's–Roosevelt Hospital
College of Physicians & Surgeons
Columbia University
New York, New York, United States

Angelo Pietrobelli, MD
Pediatric Unit
Policlinic "G.B. Rossi"
Verona University Medical School
Verona, Italy

Alex F. Roche, MD, PhD, DSc, FRACP
Professor Emeritus
Department of Community Health and Pediatrics
Lifespan Health Research Center
Wright State University School of Medicine
Dayton, Ohio, United States

Robert Ross, PhD
School of Physical and Health Education, Medicine
Queen's University
Kingston, Ontario, Canada

Ronenn Roubenoff, MD, MHS
Nutrition, Exercise Physiology, and Sarcopenia Laboratory
Jean Mayer USDA Human Nutrition Research Center on Aging at Tufts University
Boston, Massachusetts, United States
and
Millennium Pharmaceuticals, Inc.
Cambridge, Massachusetts, United States

Luís B. Sardinha, PhD
Faculty of Human Movement
Technical University of Lisbon
Estrada da Costa
Cruz-Quebraba, Portugal

Dale A. Schoeller, PhD
Nutritional Sciences
University of Wisconson–Madison
Madison, Wisconsin, United States

Jacob C. Seidell, PhD
Department for Nutrition and Health
Free University and VU Medical Center
Amsterdam, The Netherlands

Wei Shen, MD
Institute of Human Nutrition
Obesity Research Center
Columbia University
New York, New York, United States

Aviva Sopher, MD
Department of Pediatrics
Jacobi Medical Center
A. Einstein College of Medicine and
    Institute of Human Nutrition
Bronx, New York, United States

Marie-Pierre St-Onge, PhD
Division of Physiology and Metabolism
Department of Nutrition Science
University of Alabama at Birmingham
Birmingham, Alabama, United States

Shumei S. Sun, PhD
Lifespan Health Research Center
Department of Community Health
Wright State University School of Medicine
Dayton, Ohio, United States

Pedro J. Teixeira, PhD
Faculty of Human Movement
Technical University of Lisbon
Estrada da Costa
Cruz-Quebraba, Portugal

Daniel P. Williams, PhD
College of Health
Exercise and Sport Science Department
University of Utah
Salt Lake City, Utah, United States

Robert T. Withers, PhD, FACSM
Exercise Physiology Laboratory
School of Education
The Flinders University of South Australia
Adelaide, Australia

Part I
The Science of Body Composition Research

The area of scientific inquiry that we now recognize as body composition research had its formal beginning several centuries ago with roots in antiquity. Scientists from many disciplines participate in this research.

Chapter 1 presents an overview of the body composition research field as we practice it today. The central five-level model, comprising the atomic, molecular, cellular, tissue–organ, and whole-body levels, creates a formal structure for developing appropriate body composition methods and models. This chapter also defines the scope of body composition research as three interconnected areas: body composition rules and models, body composition methodology, and body composition variation. The authors examine the historical development of concepts in each of the areas and clearly define terms applied throughout the book.

Having traced the long and distinguished history of body composition research, the authors of chapter 1 then look forward to the exciting new developments rapidly transforming the field: increasing availability of highly sophisticated measuring systems, an expanding focus beyond static "mass" measurements to mass-function associations, and an influx of imaginative investigators from other research areas to the field of body composition research. This first chapter gives those with little background in the field an important basis for understanding concepts and methods reviewed in the remaining parts of the book.

Chapter 1

Study of Body Composition: An Overview

Wei Shen, Marie-Pierre St-Onge, ZiMian Wang,
and Steven B. Heymsfield

An organism's composition reflects net lifetime accumulation of nutrients and other substrates acquired from the environment and retained by the body. These components empower the organism with life. Components ranging from elements to tissues and organs are the building blocks that give mass, shape, and function to all living things. Body composition analysis techniques allow scientists to study how these building blocks function and change with age and metabolic state. Scientists from many different disciplines, as well as health care workers, rely on body composition measurements for research involving both animals and humans as well as for diagnostic purposes.

The study of body composition is organized into three interconnected areas (figure 1.1). The first of these areas includes body composition rules and models, which involves the components themselves, definitions, and links between them. There are about 30 to 40 major components, including those that represent component combinations, at different levels; when combined in some mathematical fashion by investigators, they are referred to as models. A classic model is the prediction of fat-free mass (FFM, kg) from total body water (TBW, kg) as FFM = TBW/0.732 (Pace

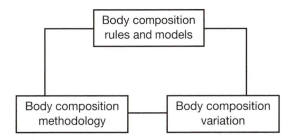

Figure 1.1 The study of human body composition as defined by three interacting research areas.

From Z. M. Wang et al., 1992, "The five-level model: a new approach to organizing body-composition research," *American Journal of Clinical Nutrition* 56: 19-28. Adapted with permission by the American Journal of Clinical Nutrition. © Am J Clin Nutr. American Society for Clinical Nutrition.

and Rathbun 1945). An assumption of this model is that FFM is 73.2% water.

The central model in body composition research is the five-level model in which body mass is considered as the sum of all components at each of the five levels—atomic, molecular, cellular, tissue–organ, and whole body (Wang et al. 1992; figure 1.2). Certain rules are inherent in the five-level model, and, ultimately, all body composition models follow these rules. We explore body

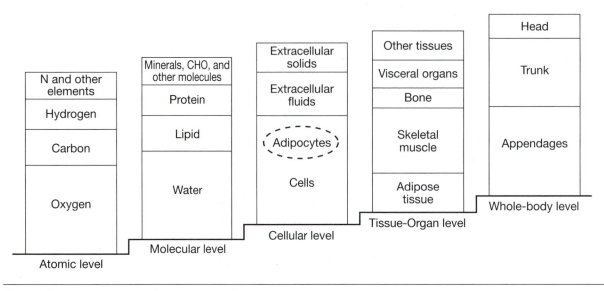

Figure 1.2 The five body composition levels. N = nitrogen; CHO = carbohydrates.

From Z.M. Wang et al., 1992, "The five-level model: a new approach to organizing body-composition research," *American Journal of Clinical Nutrition* 56: 19-28. Adapted with permission by the American Journal of Clinical Nutrition. © Am J Clin Nutr. American Society for Clinical Nutrition.

composition models in this chapter, and the developed concepts recur in many of the chapters that follow.

The second area of body composition research involves body composition methodology. Different methods are available to measure the major body components of the five levels in vivo and in vitro. Many of the available methods rely on stable between-subject relationships that exist among components within and between levels. Some methods are also based on these relationships and are often formulated as specific models.

The third area of body composition research is body composition variation, and it involves the changes in body composition related to physiological or pathological conditions. Areas investigated include growth, development, aging, race, nutrition, hormonal effects, and physical activity, as well as some diseases and medications that influence one's body composition.

Interactions occur among the three areas of body composition research. For example, an ever-increasing number of adult human and mammal studies support a stable ratio of TBW to FFM of approximately 0.732 (Pace and Rathbun 1945; Sheng and Huggins 1979; Wang et al. 1999). The "hydration law," that TBW/FFM = 0.732, is applied in measuring FFM with labeled water. Fat mass (FM) is then calculated as the difference between body mass and FFM. This method has been used to study body composition and associated physiological processes in humans and mammals.

Some of these studies, in turn, provided feedback to support or refute the hydration law in various conditions and metabolic states. Studying these relations also provides the opportunity to formulate and discover new rules and body composition laws.

The three interacting areas of body composition research, including each of the reviewed levels and components, are the basis of this book. Part I of the book examines each major body composition method in detail. Part II expands the measurement area to body composition models and specific components. Parts III and IV then examine biological and pathological variation in body composition, respectively.

History

Many contemporary scientists contributed to the field of body composition research as it exists today even though interest in the topic extends back several thousand years. The first body composition concepts can be traced to the Greeks around 400 b.c. (Schultz 2002). Events such as rain or disease were considered to have natural causes. Accordingly, the Greeks believed humans were made of the same basic elements that make up the cosmos: fire, water, air, and earth (figure 1.3). The qualities of these elements could be hot, cold, dry, or moist. Ingested food consisted of these elements, and digestion was thought to convert them

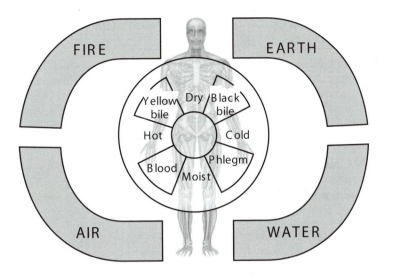

Figure 1.3 The relations among the elements that make up the cosmos, their qualities, and the humors present in humans according to the ancient Greek model.

NEWS IN PHYSIOLOGICAL SCIENCES by S.G. Schultz. Copyright 2002 by AMERICAN PHYSIOLOGICAL SOCIETY. Adapted with permission of the AMERICAN PHYSIOLOGICAL SOCIETY in the format Textbook via Copyright Clearance Center.

to four body juices, or humors—blood, phlegm, yellow bile, and black bile.

The modern era of body composition research can be traced back about 1 century, and some research highlights are presented in the Chronology of Body Composition Research sidebar on pages 6-9 (Wang et al. 1999). Substantial groundwork for the field was prepared by scientists, such as Justus von Liebig, working during the mid-19th and early 20th centuries. Although there were some important conceptual advances during this period, available methods were not very practical or accurate. Important advances were made in the early 20th century by investigators such as J. Matiegka, who reported an anthropometric model to estimate total body muscle mass in 1921. The field rapidly advanced, entering a golden era, beginning in the 1930s, with many new ideas and notions fueled by the introduction of metabolic concepts and stable and radioactive labeled isotopes. Behnke's introduction of the underwater weighing method and two-component model in the early 1940s is a benchmark because it provided a simple and practical method for investigators throughout the world to measure FM and FFM in humans (Behnke et al. 1942). Many see the culmination of this golden era of body composition research as the heralded New York Academy of Science's First Body Composition Symposium in 1963 (Brozek et al. 1963). Important methods such as whole-body ^{40}K counting and in vivo neutron activation analysis were introduced and refined during this period by investigators such as Gilbert Forbes and Stanton Cohn and their colleagues. The modern era traces back to the origins of dual-energy X-ray absorptiometry (DXA) in the early 1970s and the development of and refinement in computed axial tomography (CT) and magnetic resonance imaging (MRI) within the next two decades (Foster et al. 1984; Hounsfield 1973; Mazess et al. 1970; Tokunaga et al. 1983).

Chronology of Body Composition Research

1850s	Justus von Liebig (1803-1873) found that the human body contains many substances present in food and that body fluids contain more sodium and less potassium than tissues.
1857	A. von Bezold (1857) discovered that animal growth was accompanied by a decrease in water and an increase in ash content.
1859	J. Moleschott (1859) first reported values for the amounts of protein, fat, extractives, salts, and water per 1,000 parts of the human body. J.B. Lawes and J.H. Gilbert (1859) documented that animal body water varied inversely with fat content.
1863	E. Bischoff (1863) analyzed adult human cadavers for water content.
1871	L.A.J. Quetelet (1871) first observed that among adults, weight seemed to increase in proportion to the square of stature and established Quetelet's index, which was renamed body mass index by A. Keys and colleagues (1972).
1876	H. Fehling (1876) analyzed fetuses and newborns for water content.
1887	L. Pfeiffer (1887) noted that the variation in water content of animal bodies could be reduced if the data were expressed on a fat-free basis.
1896	J. Katz (1896) reported detailed chemical analysis of muscle.
1900	W. Camerer and C. Söldner (1900) estimated the chemical composition of fetuses including water, fat, nitrogen, and major minerals.
1901	E. Voit (1901) and M. Rubner (1902) announced the concept of "active protoplasmic mass" related to certain physiological functions.
1905	O. Folin (1905) hypothesized that urinary creatinine was a qualitative indicator of body composition.
1906	A. Magnus-Levy (1906) announced for the first time the concept of "fat-free body mass."
1907	E.P. Cathcart (1907) found that nitrogen was lost from the body during fasting.
1909	P.A. Shaffer and W. Coleman (1909) used urinary creatinine excretion as an index of muscle mass.
1914	E. Benjamin (1914) reported that infants retained nitrogen as they grew.
1915	N.M. Keith, L.G. Rowntree, and J.T. Geraghty (1915) first determined plasma and blood volume by dilution of dies (Vital Red and Congo Red).
1916	D. Du Bois and E.F. Du Bois (1916) proposed the height–weight equation to estimate whole-body surface area.
1921	J. Matiegka (1921) derived an anthropometric model to estimate total body muscle mass.
1923	C.R. Moulton (1923) advanced the concept of "chemical maturity of body composition." W.M. Marriott (1923) advanced the concept of "anhydremia" and initiated the modern study of body fluids.

Justus von Liebig (1803-1873)

From the Perry-Castañeda Library, University of Texas at Austin, with permission

Jindřich Matiegka (1862-1941)

Courtesy of the Hrdlièka Museum of Man, Czech Republic

1934	G. von Hevesy and E. Hofer (1934) used deuterium to estimate total body water volume.
1938	N.B. Talbot (1938) estimated that 1 g of creatinine excreted during a 24-hr period was derived from approximately 18 kg of muscle mass. V. Iob and W.W. Swanson (1938) reported chemical composition in fetuses and newborns.
1940	H.C. Stuart, P. Hill, and C. Shaw (1940) first used two-dimensional standard radiography to estimate bone, adipose tissue, and skeletal muscle shadows.

Albert Behnke
(1903-1996)
U.S. Navy Photo

1942	A.R. Behnke, Jr., B.G. Feen, and W.C. Welham (1942) estimated the relative proportion of lean and fat in the human body based on Archimedes' principle.
1943	J. Nyboer developed and applied tetrapolar bioimpedance analysis (BIA) to evaluate fluid compartments.
1945	N. Pace and E.N. Rathbun (1945) found the relatively constant ratio of total body water to fat-free body mass and suggested a method for estimating body fat from total body water. H.H. Mitchell, T.S. Hamilton, F.R. Steggerda, and H.W. Bean (1945) first reported whole-body composition analysis on the molecular level (water, fat, protein, ash, Ca, and P) for an adult human cadaver.

Francis Moore
(1913-2001)
Courtesy of the Harvard Medical School, Public Affairs

1946	F.D. Moore (1946) introduced the concept of total body exchangeable sodium and potassium.
1950	E.L. Reynolds (1950) discussed fat distribution in children.
1951	R.A. McCance and E.M. Widdowson (1951) determined extracellular fluid volume by using dilution techniques with thiocyanate. E.M. Widdowson, R.A. McCance, and C.M. Spray (1951) first reported whole-body composition analysis on the atomic level (Ca, P, K, Na, Mg, Fe, Cu, and Zn) for adult human cadavers.
1953	A. Keys and J. Brozek (1953) provided a detailed analysis of the densitometric technique. A. Keys and his colleagues carried out the classic Minnesota Experiment and traced the effects of semistarvation and refeeding on body components in young male volunteers (see figure 1.4). F. Fidanza, A. Keys, and J.T. Anderson (1953) reported remarkable uniformity in the density (0.90 g/cm^3) of human fat across individuals.

Ancel Keys (1904-)
Courtesy of the Division of Epidemiology, University of Minnesota

1954	S.J. Fomon begins his pioneering pediatric body composition research and later introduces reference values such as those for percent fat shown in figure 1.5.
1955	N. Lifson, G.B. Gordon, and R. McClintock (1955) measured total body water and total body carbon dioxide production by using $D_2^{18}O$ dilution method.
1958	I.S. Edelman, J. Leibman, M.P. O'Meara, and L.W. Birkenfeld (1958) first observed that in normal extracellular fluid the total cation forms a linear relationship with total osmotic solute. R. Kulwich, L. Feinstein, and E.C. Anderson (1958) and E.C. Anderson and W. Langham (1959) reported the existence of a correlation between natural ^{40}K concentration and fat-free body mass.

Samuel J. Fomon
(1923-)
Photo courtesy of Samuel J. Fomon

1960	J.M. Foy and H. Schneider (1960) determined total body water by using tritium dilution method.
1961	W.E. Siri (1961) developed a three-component model to estimate total body fat mass. G.B. Forbes, J. Hursh, and J. Gallup (1961) estimated fat and lean contents by using whole-body ^{40}K counting.
1962	A. Thomasset (1962) first introduced the BIA method.
1963	J. Brozek organized the First Body Composition Symposium, held at the New York Academy of Sciences. J. Brozek, F. Grande, J.T. Anderson, and A. Keys (1963) developed a two-component model to estimate total body fat mass. F.D. Moore and colleagues (1963) proposed the concept of body cell mass (BCM). E.C. Anderson proposed a three-component system of muscle, muscle-free lean, and adipose tissue from potassium and water using Van Dobeln's original tissue model. J.A. Sorenson and J.R. Cameron developed the theoretical basis of dual-photon absorptiometry (DPA) for body composition.
1964	J. Anderson and colleagues (1964) and S.H. Cohn and C.S. Dombrowski (1971) developed in vivo neutron activation method for body composition analysis.
1966	J. Rundo and L.J. Bunce (1966) established the first in vivo prompt-γ neutron activation facility for measuring body hydrogen mass.
1968	D.B. Cheek's (1968) book on human growth and body composition was published.
1970	R.B. Mazess, J.R. Cameron, and J.A. Sorenson (1970) developed DPA method for peripheral body composition in vivo.
1972	A.F. Roche and colleagues (1972) reported their growth velocity studies that later evolved to encompass body composition research. G.N. Hounsfield (1973) reports the first computerized tomographic imaging system that revolutionizes clinical medicine and body composition research. Hounsfield's discovery culminates the efforts of many scientists over two centuries, starting with the mathematical models developed by J. Fourier (1768-1830), which formed the basis of magnetic resonance imaging. Modern development of magnetic resonance methods began with the discovery of nuclear magnetic resonance by F. Bloch and E. Mills Purcell and was followed by the development of magnetic resonance imaging concepts and systems by P. Lauterbur and R. Damadian.
1973	W. Harker first introduced the total body electrical conductivity (TOBEC) method.
1975	W.S. Snyder and colleagues (1975) summarized data on Reference Man, a concept developed in 1952 by J. Brožek.
1978	Selinger developed a four-component model and equation.
1979	S.B. Heymsfield, R.P. Olafson, M.H. Kutner, and D.W. Nixon (1979) first used computed axial tomography (CT) for body composition analysis.

William Siri (1919-)
Courtesy of the Berkeley Lab Image Library

Gilbert Forbes (1915-2003)
Courtesy of University of Rochester

Stanton Cohn (1929-)
Courtesy of Stanton Cohn

Godfrey N. Hounsfield (1919-2004)
Courtesy of the Nobel Foundation. © The Nobel Foundation.

Alex F. Roche (1921-)
Courtesy of Anna Bellisari

| 1981 | H.C. Lukaski, J. Mendez, E.R. Buskirk, and S.H. Cohn (1981) developed urinary 3-methylhistidine method to estimate total body skeletal muscle mass. |

| 1982 | S.J. Fomon (Fomon, Haschke, Ziegler and Nelson 1982) proposed Reference Children from birth to age 10 years. |

| 1983 | CT was used in whole-body composition analysis (G.A. Borkan et al., 1983; K. Tokunaga et al., 1983; and L. Sjöström et al., 1986). |

Felix Bloch (1905-1983)

Courtesy of the Nobel Foundation. © The Nobel Foundation.

| 1984 | The Brussels cadaver study (12 men and 13 women) on tissue–system level was published by J.P. Clarys, A.D. Martin, and D.T. Drinkwater (1985). M.A. Foster, J.M.S. Hutchison, J.R. Mallard, and M. Fuller (1984) were among the first to demonstrate that magnetic resonance imaging (MRI) could accurately measure body composition. R.B. Mazess, W.W. Peppler, and M. Gibbons developed DPA for total body composition measurements (see figure 1.6). |

| 1986 | The First International Symposium on In Vivo Body Composition Studies was held in New York. T.G. Lohman and colleagues established chemical maturity of the prepubescent child from a four-component model and empirical data. |

| 1987 | J.J. Kehayias, K.J. Ellis, S.H. Cohn, and J.H. Weinlein (1987) established the first inelastic scattering facility for estimating total body carbon and oxygen. |

Edward Mills Purcell (1912-1997)

Courtesy of the Nobel Foundation. © The Nobel Foundation.

| 1990 | S.B. Heymsfield and colleagues (1990) estimated appendicular skeletal muscle by dual-energy X-ray absorptiometry (DXA). K.J. Ellis (1990) introduced the Reference Woman concept. |

| 1991 | J.J. Kehayias and colleagues (1991) developed a method for assessing total body fat mass from total body carbon mass by in vivo neutron activation analysis. |

| 1992 | Z.M. Wang, R.N. Pierson, Jr., and S.B. Heymsfield (1992) proposed the five-level model of human body composition. |

| 1992-2003 | DXA and BIA systems proliferated worldwide and were incorporated into many ongoing research and clinical programs. Air displacement plethysmography (Bod Pod) was developed and commercialized, advancing the older and less practical underwater weighing method. Multislice CT and MRI became increasingly practical, even in the clinical setting. New uses for the classical whole-body ^{40}K counting method were reported, although centers with in vivo neutron activation analysis remained limited throughout the world. Clinical and research applications in the area of body composition variation abounded, driving model and method development to ever-higher resolution and practicality. Growth in the merger of classical body composition measurements with corresponding in vivo functional and metabolic processes was heralded by the creation of a new National Institutes of Health program in biomedical imaging. The International Symposium on In Vivo Body Composition Studies was held in Toronto (the second symposium, 1989), Houston, TX (the third symposium, 1992), Malmo, Sweden (the fourth symposium, 1996), New York (the fifth symposium, 1999), and Rome (the sixth symposium, 2002). *The International Journal of Body Composition Research* published its first volume in 2003 (chief editor P. Deunenberg). |

Paul Lauterbur (1929-)

Photo courtesy of Paul Lauterbur

Raymond Damadian (1936-)

Photo courtesy of Raymond Damadian

This list highlights major advances in the study of human body composition and was prepared by the authors with advice from colleagues who work in the field. The list should be considered a work in progress and is not meant to be an exhaustive compilation of research in the field.

Adapted from Z.M. Wang, Z.M. Wang, and S.B. Heymsfield, 1999, "History of the study of human body composition: A brief review," *American Journal of Human Biology* 11:157-165.

Pediatric body composition research traces its modern origins to C.R. Moulton, who introduced the concept of chemical maturity in 1923. Samuel J. Fomon and his colleagues reported values for reference children in 1982 and new pediatric body composition methods continue to be introduced (figure 1.5). Donald B. Cheek and Alex F. Roche reported new information of the kinetics of component growth, including cells and the skeleton.

A continuing interest among body composition researchers is the associations among body mass, adiposity, and heat production. Lambert Adolphe Jacques Quetelet developed his equation for body mass index (weight/height^2) in 1871 that is now used as a proxy measure of body adiposity. Max

Rubner reported the surface area law of heat production in the early 20th century (Rubner 1902), and Max Kleiber reported Kleiber's law, the proportionality of mammal heat production to body mass$^{0.75}$ in 1932.

Alongside the technological measurement advances are parallel important discoveries in biological variation and model development. A new phase is emerging, or at least additional advances are appearing at a rapid rate, and we review these in the future development section to follow. We now turn our attention to what may be considered the core of body composition research, the recognized components that collectively comprise the field, as we now know it.

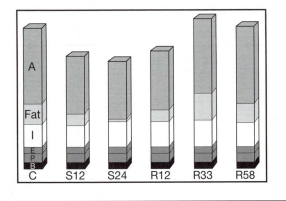

Figure 1.4 The classic Minnesota Experiment. A multicomponent body composition model was applied by the investigators to track component changes over the weeks of semistarvation (S) and refeeding (R). A, active tissue; I, interstitial fluid; E, erythrocytes; P, blood plasma; B, bone mineral.

Courtesy, Division of Epidemiology, University of Minnesota.

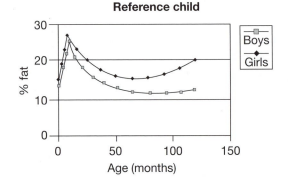

Figure 1.5 Samuel J. Fomon pioneered pediatric body composition research and introduced reference values such as those for percent fat shown in this figure.

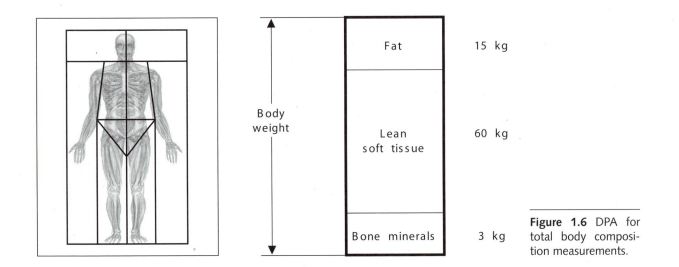

Figure 1.6 DPA for total body composition measurements.

Body Composition Rules and Models

Throughout this book, authors review topics related to body composition components. Over the past century, many different components were established and authors varied widely in their use of terminology and measurement methods. In this section we set forth currently accepted terminology and body composition classification as they apply throughout this book.

Body mass can be viewed as five distinct and separate but integrated levels, beginning with the atomic level and moving to molecular, cellular, tissue–organ, and whole-body levels (Wang et al. 1992). The specific components of the five levels are presented in figure 1.2 (Wang et al. 1992). The sum of all components at each of the five levels is equivalent to body mass. Some of the common models at each level are presented in table 1.1.

Atomic Level

At the atomic level (figure 1.2), body mass includes 11 major elements. More than 96% of body mass is accounted for by four elements: oxygen, carbon, hydrogen, and nitrogen. Other important elements are calcium, potassium, phosphorus, sulfur, sodium, chlorine, and magnesium. Most of these elements can be measured in vivo by neutron activation analysis (Cohn and Dombrowski 1971) and whole-body counting (Cohn et al.

1969; see chapter 4). Elements are important to investigators in fields such as radiobiology and nuclear medicine. The major elements can also be linked to higher-level components. For example, total body carbon, nitrogen, and potassium can be used with appropriate models to derive total body fat (Sutcliffe et al. 1993), protein (Sutcliffe et al. 1993), and body cell mass (BCM; Moore et al., 1963), respectively.

Molecular Level

The molecular level consists of six major components: water, lipid, protein, carbohydrates, bone minerals, and soft tissue minerals (figure 1.7). Various models of the molecular level can be created that range from two to six components (table 1.1). The two-component model, which includes FM and FFM, is presently one of the most widely applied models in body composition research. FFM is typically considered the actively metabolizing component at the molecular level of body composition. Models including three or more components are referred to as multicomponent models. These models further divide the FFM into additional components that are quantifiable in vivo. The figure also shows the three-component DXA model in which FFM consists of lean soft tissue and bone mineral. The widely used four-component model partitions FFM into water, protein, and mineral components. Models are available that include up to six molecular-level components.

Table 1.1 Representative Multicomponent Models at the Five Body Composition Levels

Level	Body composition model	Number of components
Atomic	BM = H + O + N + C + Na + K + Cl + P + Ca + Mg + S	11
Molecular	BM = FM + TBW + TBPro + Mo + Ms + CHO	6
	BM = FM + TBW + TBPro + M	4
	BM = FM + TBW + nonfat solids	3
	BM = FM + Mo + residual	3
	BM = FM + FFM	2
Cellular	BM = cells + ECF + ECS	3
	BM = FM + BCM + ECF + ECS	4
Tissue–organ	BW = AT + SM + bone + visceral organs + other tissues	5
Whole-body	BW = head + trunk + appendages	3

Note. AT = adipose tissue; BCM = body cell mass; BM = body mass; CHO = carbohydrates; ECF = extracellular fluid; ECS = extracellular solids; FFM = fat-free mass; FM = fat mass; M = mineral; Mo = bone mineral; Ms = soft tissue mineral; SM = skeletal muscle; TBPro = total body protein; TBW = total body water.

Many different definitions have been applied over the years by investigators studying molecular-level components. The terms *lipid* and *fat* may be used interchangeably, but their meanings in the body composition research field differ (figure 1.8; Gurr and Harwood 1991). Lipid includes all of the biological matter extracted with lipid solvents such as ether and chloroform. These extracted lipids include triglycerides, phospholipids, and structural lipids that occur in relatively small

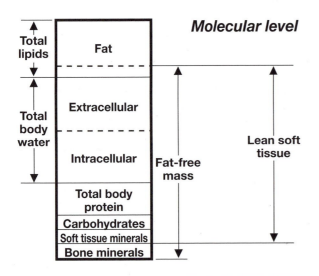

Figure 1.7   Main components of the molecular level of body composition.

From Z. M. Wang et al., 1992, "The five-level model: a new approach to organizing body-composition research," *American Journal of Clinical Nutrition* 56: 19-28. Adapted with permission by the American Journal of Clinical Nutrition. © Am J Clin Nutr. American Society for Clinical Nutrition.

quantities in vivo (Gurr and Harwood 1991). In contrast, fats, in body composition research, refer to the specific family of lipids consisting of triglycerides (Wang et al. 1992). Although this terminology is not uniform, in this book we use the term *lipid* for all ether and chloroform extractable lipids from tissues and the term *fat* for triglycerides. Based on the reference man (Snyder et al. 1975), approximately 90% of total body lipid in healthy adults is triglyceride, although this proportion differs with dietary intake and health conditions (Comizio et al. 1998). The rest, approximately 10% of total body lipid (i.e., nonfat lipids), are mainly composed of glycerophosphatides and sphingolipids.

Although widely used in early body composition research, the term *lean body mass* is rarely used today. This is because modern body composition models are increasingly specific in the composition and physical characteristics of included components, and lean body mass is not a precisely defined and uniformly agreed upon entity. A similar term, *lean soft tissue,* is used in reference to DXA body composition models as outlined later in chapters 5 and 12.

Cellular Level

The cellular level includes three components: extracellular solids, extracellular fluid, and cells, as shown in figure 1.2. The cells can be additionally partitioned into two components, fat and BCM, the latter of which is the actively metabolizing component at the cellular level of body composition (Moore et al. 1963).

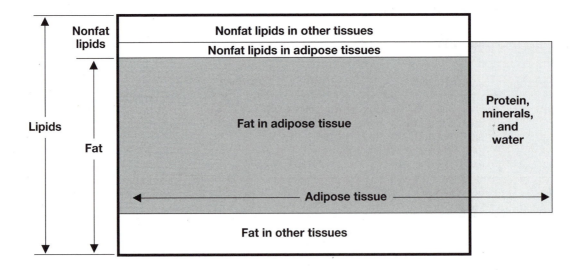

Figure 1.8   The relationships between molecular-level components lipid and fat and the tissue–organ-level component adipose tissue.

Tissue–Organ Level

The tissue–organ level of body composition, as shown in figure 1.2, consists of major components including adipose tissue, skeletal muscle, visceral organs, and bone. Some tissue–organ-level components are single solid organs such as brain, heart, liver, and spleen. Others, such as skeletal muscle and adipose tissue, are dispersed throughout the body.

Fat and *adipose tissue* are often interchanged in common usage, although in the body composition field, fat and adipose tissue are distinct, and different components at different levels and their taxonomic separation is important when measuring their mass and metabolic characteristics (figure 1.8). Although fat is found primarily in adipose tissue, intracellular triglyceride pools are also observed in liver, skeletal muscle (Anderwald et al. 2002), and other organs, particularly in pathological conditions such as hepatic steatosis and various forms of lipidosis. There also exist small circulating extracellular pools of triglyceride, mainly in the form of lipoproteins. Adipose tissue consists of adipocytes, extracellular fluid, nerves, and blood vessels. Adipose tissue components are present throughout the body, and the metabolic properties of these components vary across anatomic locations (Björntorp 2000; Enevoldsen et al. 2001). Adipose tissue components are closely linked with health-related conditions. For example, visceral adipose tissue is associated with insulin sensitivity, metabolic syndrome (Wajchenberg 2000), and type 2 diabetes (Bermudez and Tucker 2001). Figure 1.8 shows the relations among molecular level fat and lipid and tissue–organ-level adipose tissue.

Whole-Body Level

The fifth level of body composition, the whole-body level, can be divided into regions such as appendages, trunk, and head. Rather than discrete components, as used in other levels, trunk and appendages are usually described by anthropometric measures (chapter 8) such as circumference, skinfolds, and length. Waist circumference, an important anthropometric variable, is closely connected to obesity-related morbidity and mortality (Zhu et al. 2002). The development of body composition prediction equations from anthropometry is an ongoing field with many statistical concerns highlighted by Lohman (1981). In addition to age- and sex-related variations in body composition at the whole-body level, differences in these regions exist between different ethnic groups (Raji et al. 2001).

Each of the components at the five levels is distinct, and there is no overlap between any of the components at the same level. This discrete nature of each component is important because it avoids confusion when developing body composition models. This lack of overlap between components also prevents redundancy in multicomponent body composition models as developed in chapter 12 and summarized in table 1.1.

Although major body components are defined and identified at each of the five levels, some established component studies in body composition research are done across levels. For example, intracellular and extracellular portions of TBW, trunk adipose tissue, and limb muscles are combined-level terms and components. The liver fat component combines molecular and tissue–organ-level components. Some terms, such as *intramyocellular lipid,* combine components at three levels: molecular, cellular, and tissue–organ. Thus, although the five-level model provides the basic structure of human body composition and creates a framework for explaining the relationships between major body components, cross-level components can be investigated according to specific research and clinical interests and are a promising area for future research.

An important concept in body composition research is that when body mass and energy balance are stable, the major components remain stable and thus maintain predictable relationships with each other. These stable relationships are the basis of some of the body composition models. In addition to the FFM hydration factor mentioned earlier (Pace and Rathbun 1945; Sheng and Huggins 1979), the density of FFM is another well-recognized physical characteristic of steady-state body composition (Lohman 1986). These widely recognized "constants" are used in developing body composition models, and their use is described in later chapters.

Future Directions

Body composition research has a long and distinguished history, but a transition is now taking place that will ultimately transform the field. First, practical and highly sophisticated measuring systems are becoming available on a widespread scale. Accordingly, body composition research and clinical applications, previously used by a few specialized laboratories, are being used by a wide array of biological scientists and health practitioners. For example, underwater weighing (chapter 2), a traditional method of assessing body composition, requires a reasonable amount of technical skill and substantial subject participation, including underwater submersion and exhalation with breath hold. Elderly or very obese subjects, as well as some children and nonswimmers, are often unable to complete the underwater weighing procedure. In contrast, the newer air displacement plethysmography method (i.e., Bod Pod, chapter 2) requires less operator technical skill and system maintenance along with minimal subject participation. Similarly, both DXA (chapter 5) and MRI (chapter 7) are advanced measurement methods that are widely available, are relatively easy to carry out, and require minimal subject participation. These newer tools for body composition assessment and the smaller subject burden associated with them allow body composition research to extend from reliance on the two-component model to multicomponent models and from the hands of a few to broad scientific and consumer communities. The availability and practicality of the newer methods extend the possibility of interfacing body composition research with many other fields.

Second, investigators in traditional body composition research concentrated on measuring the major body components. Mass, volume, and distribution of the different components (mainly fat and fat-free mass) are the primary current measures of the field. Another emerging area examines in vivo metabolism and includes measurement methods such as positron emission tomography, functional MRI, and magnetic resonance spectroscopy. These newer approaches enable investigators to assess the quality of tissues and organs, their composition, fuel utilization, oxygen uptake, blood flow, and other functional properties. A clear pathway of current and future development merges traditional body composition research with these newer in vivo measurement methods.

In the past 2 decades, the relationships between obesity and diabetes, for example, and major body components have been carefully investigated (Bermudez and Tucker 2001; Colberg et al. 1995; Ross et al. 1996). In the past 5 years, through a combination of traditional body composition methods and the newer magnetic resonance spectroscopy, investigators have not only observed the effects of obesity and diabetes on body components but have gained an improved understanding of the mechanisms of these pathological conditions. In the future, advancing technology and collaboration between investigators from the field of body composition and other fields promise to engage body composition research in studies of normal metabolism and pathological mechanisms of diseases.

Part II

Body Composition Measurement Methods

The measurement of body components is central to the study of body composition in humans and animals. In part II, 10 chapters examine the recognized methods and statistical approaches for quantifying each of the major components at the five body composition levels. Part II is organized by method, not by measurable components, because some methods are capable of measuring more than one component.

Body volume is a fundamental physical property that can be used in developing body composition models and can be measured in human subjects through two methods: hydrodensitometry (underwater weighing) and air displacement plethysmography. These methods are reviewed in chapter 2 along with the main body composition models that rely on estimates of body volume. This chapter provides an extensive theoretical and practical overview of the hydrodensitometry method and the recent development of an accurate air displacement plethysmograph (Bod Pod), which is another approach for measuring body volume.

A large proportion of body mass consists of water, and total body water is relatively easy to measure in both animals and humans. Chapter 3 reviews the available methods for measuring total body water and also examines the body composition models that rely on estimates of total body water. This chapter describes the dilution concept (used for quantifying body fluid spaces including total body and extracellular water) and reviews methods for estimating fluid spaces.

Two related methods, whole-body counting and in vivo neutron activation analysis, are important for measuring the major body elements. These methods, which provide us with information at the atomic body composition level, are reviewed in chapter 4. These elements form the basis of models for estimating body compartments such as total body protein and bone mineral. Chapter 4 also describes the evolution and current status of these methods.

Two additional physical properties measured by body composition methods are X-ray attenuation and impedance. Two methods for estimating body compartments that exploit these tissue properties are dual-energy X-ray absorptiometry (chapter 5) and bioimpedance analysis (chapter 6).

Dual-energy X-ray absorptiometry is now one of the most widely applied techniques used in body composition studies. Chapter 5 deals with the key issues of underlying assumptions and their validity, applicability, hardware and software, measurement procedures, calibration, and precision in the estimation of both total and regional body composition, including the estimation of bone mineral content, lean tissue mass, fat-free mass, and fat mass.

Bioimpedance analysis (BIA), the group of methods reviewed in chapter 6, is now regarded as either a substitute or supplement to conventional anthropometry in field studies. The BIA method is also appropriate for quick bedside assessments in clinical settings in which limited accuracy is acceptable. Chapter 6 discusses the current widely used single-frequency BIA as well as newly developed segmental and multifrequency BIA techniques and recent results with estimating changes in body fluid.

One of the most important recent advances in body composition research, the availability of imaging methods, is reviewed in chapter 7. Although imaging methods were used in body composition research more than 5 decades ago, the major advances in this field came in the mid-1970s with the introduction of computed tomography. Both computed tomography and the more recent magnetic resonance imaging (MRI) are some of the most important advances in the field of body composition assessment. Chapter 7 describes the technical aspects of imaging methods and measurement of two major components, adipose tissue and skeletal muscle, as well as their distribution.

A group of classic and still widely applied body composition methods, classified collectively as anthropometry, is reviewed in chapter 8. Anthropometry has considerable appeal because it can be applied in the laboratory and in rural or urban field situations. The instruments are portable and relatively inexpensive. Chapter 8 describes the relationships between anthropometric and body composition variables. The use of ultrasound and its application to the study of body composition are also discussed.

The field of body composition research encompasses studies of both humans and animals. Methods applicable in adult humans are not always appropriate for use in children, and pediatric methods are specifically reviewed in chapter 9. Body composition methods rely on several assumptions that are often age dependent. Body composition analysis is of increasing importance in pediatric research and clinical practice. Chapter 9 highlights body composition methods for use in children and provides an overview of body composition changes with growth and development.

Chapter 10 presents an overview of animal body composition research. As with children, not all body composition methods applied in adult humans are appropriate for use in animals. Animals varying widely in body size are increasingly the subjects of in vivo body composition analysis. The varied available methods, their accuracy and precision, and strengths and limitations are the subject of chapter 10.

An important branch of body composition research, statistical methods, is presented in chapter 11. In epidemiological and clinical settings, it is frequently necessary to predict body composition for groups or individuals because the application of sophisticated direct methods is not practical. Chapter 11 discusses the statistical methods for the development and application of predictive equations for body composition.

Chapter 2

Hydrodensitometry and Air Displacement Plethysmography

Scott B. Going

The term *densitometry* refers to the general procedure of estimating body composition from body density. Although several methods can be used to estimate body density, densitometry has been virtually synonymous with underwater weighing, also called hydrostatic weighing or *hydrodensitometry*. More recently, *air displacement plethysmography* has become a viable alternative method for estimating body density. This chapter focuses on measurement issues related to hydrodensitometry (HD) and air displacement plethysmography (ADP), because of their widespread application. However, the discussion of the limitations and potential errors of assessing body composition applies to all densitometric techniques.

The density of the human body (D_b), like any material, is equivalent to the ratio of its mass (MA) and volume (V):

$$D_b = MA/V. \qquad (2.1)$$

Body mass estimated from body weight is relatively easy to measure, and recommendations for accurately measuring weight have been published (Lohman et al. 1988). Thus, the primary requirement for accurately estimating D_b is to obtain an accurate measure of body volume. Indeed, most methods commonly considered to be densitometric techniques are methods for estimating body volume. Once volume is known, density can be calculated from equation 2.1, and composition can then be estimated as outlined subsequently.

Long considered the "gold standard," hydrodensitometry often has been used as the criterion method in validation studies of new body composition assessment methods. The findings from many studies, however, using both anatomical (Clarys et al. 1984) and chemical (Baumgartner et al. 1991; Heymsfield et al. 1989) models of body composition, have emphasized the limitations of densitometry, especially when applied across a wide age range without adjustments for the changes that occur with growth and maturation (Lohman 1986) and aging (Going et al. 1995; Heymsfield et al. 1989). Although density can be estimated with acceptable precision and accuracy in most groups, the assumption of an invariant fat-free composition, commonly used to convert density to composition, may not be valid for many individuals. Indeed, the magnitude of the deviation from the assumed fat-free composition, more than measurement errors in body density, ultimately determines the accuracy of densitometric estimates of body composition for any individual or group.

Body Composition Models

The density of any material is a function of the proportions and densities of its components. In the classic two-component model of body composition, body weight is divided into fat (F) and fat-free (FFM) fractions. Thus,

$$1/D_b = F/D_F + FFM/D_{FFM} \qquad (2.2)$$

where $1/D_b$ equals body mass, set equal to unity, divided by body density (D_b), and F/D_F and FFM/D_{FFM} are the proportions of the fat and fat-free masses divided by their respective densities.

The fat-free mass is a heterogeneous compartment that can be further divided into its primary constituents of water (W), protein (P), and mineral (M). These constituents can then be combined to form a variety of three-component and four-com-

ponent models. On the basis of limited data from chemical analyses of animal carcasses and human cadavers (Brozek et al. 1963; Keys and Brozek 1953), D_F, W, P, and M have been estimated, and an estimate of D_{FFM} has been derived (table 2.1). Using these values and solving for F, we can derive formulas for calculating percent body fat from body density based on two-component, three-component, and four-component models (table 2.2) The Siri (1956) equation, based on equation 2.2, and the Brozek equation (Brozek et al. 1963)

Table 2.1 Composition and Density of Fat-Free Mass and Reference Body

Component	Density (g/ml)	Fat-free mass (%)	Reference body (%)
Water	0.9937	73.8	62.4
Protein	1.34	19.4	16.4
Mineral	3.038	6.8	5.9
Osseous	2.982	5.6	4.8
Nonosseous	3.317	1.2	1.1
Fat	0.9007		15.3
Fat-free mass	1.100	100	84.7
Total reference body	1.064		100

Note. Densities are at 36° C.

Adapted, by permission, from J. Brozek et al., 1963, "Densitometric analysis of body composition: revision of some quantitative assumptions," *Annals of the New York Academy of Sciences* 110: 113-140. Copyright by New York Academy of Sciences.

Table 2.2 Equations for Estimating Percent Fat Based on Two-, Three-, and Four-Component Models of Body Composition

Model	Equation	Reference
Two-component	$\% Fat = (\frac{4.95}{D_b} - 4.50) 100$ $\% Fat = (\frac{4.570}{D_b} - 4.142) 100$	Siri 1956 Brozek et al. 1963
Three-component	$\% Fat = (\frac{2.118}{D_b} - 0.78\,W - 1.354) 100$ $\% Fat = (\frac{6.386}{D_b} - 3.96\,M - 6.090) 100$	Siri 1961 Lohman 1986
Four-component	$\% Fat = (\frac{2.747}{D_b} - 0.714W + 1.146B - 2.0503) 100$	Selinger 1977

Note. D_b = body density; W = total body water as a fraction of body weight; M = mineral (osseous + nonosseous) as a fraction of body weight; B = osseous mineral as a fraction of body weight.

are the simplest and most common fat-estimating formulas. The validity and accuracy of any equation depend on whether the assumed fractions and densities of the body components represented in the underlying model are appropriate for the person being measured. A variety of studies have demonstrated considerable variation in FFM composition and density attributable to growth and maturation, aging, and specialized training. Sex and racial differences also exist, and even within a population there is considerable interindividual variation that challenges the assumption of FFM "chemical constancy." Consequently, three-component and four-component equations are generally more valid and accurate, because they require fewer assumptions, although the potential for introducing greater measurement error exists. A more thorough discussion of multicomponent models can be found in chapter 12 and elsewhere (Going 1996; Lohman and Going 1993).

Estimation of Body Volume by Underwater Weighing

Estimation of body volume by underwater weighing uses Archimedes' principle that a body immersed in a fluid is acted on by a buoyancy force, which is evidenced by a "loss" of weight equal to the weight of the displaced fluid. Thus, when a subject is submerged in water, body volume (BV) is equal to the loss of weight in water, corrected for the density of water (D_w) corresponding to the temperature of the water at the time of the submersion:

$$BV = (W_a - W_w)/D_w \qquad (2.3)$$

where W_a and W_w are the subject's weight in air and water, respectively. Air in the lungs and flatus in the gastrointestinal tract at the time of measurement are two extraneous volumes included in total body volume that must be reconciled in the final calculation. Usually, the underwater weight is measured after a maximal expiration, and a correction for the residual lung volume (RV) is made, although other volumes have been used. The residual volume makes a sizable contribution to the estimate of total body volume (1-2 L) and, because residual volume is highly variable, it is essential to obtain an accurate estimate of the individual's residual volume at the time of weighing. The volume of flatus is considerably smaller and is not measured. Buskirk (1961) proposed the use of a constant correction of 100 ml to approximate the volume of gas in the gastrointestinal tract.

With the corrections for RV and gastrointestinal tract gas volume, the calculation of body density becomes as follows:

$$Db = \cfrac{Wa}{\cfrac{(W_a - W_w)}{D_w} - (RV + 0.100)} \qquad (2.4)$$

Equipment

The accurate estimation of BV from underwater weight (UWW) requires an appropriate site and the equipment to make precise and accurate measurements of body weight, underwater weight, expired gas, water and ambient temperatures, and barometric pressure. An autopsy (spring) scale or transducer system is used to measure underwater weight, and equipment for RV depends on whether RV is measured in the water or outside the tank. In addition, a heater, water circulator, and filter appropriate for the size of the tank are also needed.

Measurement Site

The underwater weight of most subjects can be measured in almost any body of fresh water that is at least 2.3 m deep. Underwater weight can be measured in the field at poolside, although water turbulence can make poolside measurements difficult. A redwood tub (or its equivalent) or a stainless steel or Plexiglas enforced tank is recommended for laboratory use. A tank no smaller than 3 m × 3 m × 3 m is suitable to accommodate a range of subject sizes (a 4-m depth is helpful for subjects taller than 200 cm). The tank should have an easily accessible emergency water drainage system. In addition, a water heater and filtering system are recommended. For subject comfort and to promote compliance, water temperatures close to skin temperature (≈32-35° C) are desirable. Water quality must be maintained through filtering and regular chemical treatments to maintain chlorine levels at the manufacturer's recommended levels for spas. Water pH must be maintained between 7.4 and 7.6.

Underwater Weight

Underwater weight is typically measured with the subject seated on a chair assembly suspended from a spring-loaded autopsy scale (figure 2.1) or with the subject seated or kneeling on a weighing

Figure 2.1 Redwood tank and spring-loaded autopsy scales for underwater weighing. The weighing chair is constructed from 3.81-cm (inner diameter) PVC pipe.

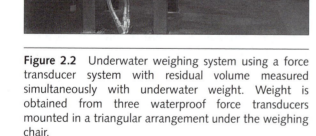

Figure 2.2 Underwater weighing system using a force transducer system with residual volume measured simultaneously with underwater weight. Weight is obtained from three waterproof force transducers mounted in a triangular arrangement under the weighing chair.

Reprinted, by permission, from L.W. Organ, A.D. Eklund, and J.D. Ledbetter, 1994, "An automated real time underwater weighing system," *Medicine and Science in Sports and Exercise* 26: 383-391.

platform supported from force transducers (figure 2.2) with digital or analog display. A 9 kg × 10 g Chatillon autopsy scale, or an equivalent scale, is adequate for most subjects, although a 15 kg × 25 g scale may be necessary for very large individu-

als with low body fat. Scale oscillations caused by underwater movements limit the accuracy of autopsy scales. Although interpolation within 20 to 60 g is often possible, for some subjects "swing averaging" can be grossly inaccurate (>100 g). A stable, comfortable chair assembly, such as can be constructed from 2-in. plastic pipe and nylon webbing, minimizes subject motion underwater.

Using strain gauges or load cells that are linked directly to a computer is a more sophisticated technique for measuring UWW. Although fewer load cells might be used, four load cells mounted at the corners of a rectangular weighing platform give maximum stability. If the extreme range of underwater weights is ±10 kg, then transducers of ±0.1% accuracy and repeatability and a linear range of 20 kg are adequate to accommodate the weight of the subject and weighing platform and to measure underwater weight with the same accuracy with which subjects can be weighed in air (±20 g; Akers and Buskirk 1969). Although the additional instrumentation (force transducers, signal conditioner for integrating the transducer outputs, and analog recorder or digital display) makes these systems more expensive, they have several advantages over autopsy scales, including greater stability, less motion underwater and fewer fluctuations in underwater weight, a permanent record of the weight, and more objective and accurate measurements. When the objective is to obtain accurate individual results or to detect small changes in underwater weight, a force transducer system should be used, whereas the autopsy scale is adequate for large-scale screening and measurements in field settings.

Residual Volume

Residual volume (RV) is commonly measured using either the closed-circuit approach, which involves dilution and eventual equilibration of an inert tracer or indicator gas such as nitrogen, oxygen, or helium, or the open-circuit approach, where nitrogen is "washed out" of the lungs during a specified period of oxygen breathing. Both approaches yield precise estimates of RV and with appropriate equipment and procedural modifications can be used to estimate RV with the subject either inside the tank (simultaneously with underwater weighing) or outside the tank.

The closed-circuit oxygen dilution technique described by Wilmore (1969a) has the advantage of being very rapid (five to eight breaths), making it suitable for multiple measurements in a reasonable amount of time. This technique entails having the subject breathe in and out of a spirometer filled with a known volume ($\approx$80-90% of vital capacity) and concentration (medical grade) of oxygen until the nitrogen (N_2) concentration in the lungs and spirometer has equilibrated. The spirometer should be modified to reduce dead space as described by Wilmore (1969a), which enables a faster clearance of N_2 from the system before testing and increases accuracy by decreasing the total volume of the system. A N_2 gas analyzer in the stream of respired air is used for continuous electronic gas analysis for N_2 concentration. A digital display or analog recorder is used to monitor and record nitrogen concentration. After a maximal exhalation, the subject breathes in and out of the spirometer at approximately two thirds of vital capacity at a rate of one respiration every 3 s until N_2 equilibrium is reached, usually within five to eight breaths. The subject then inspires maximally and expires maximally, and the breathing valve is switched back to room air when the maximal expiration is complete. Measurements of fractional N_2 concentration are made at the termination of the initial maximal expiration (A_{in2}), which is assumed to represent the initial alveolar N_2 concentration, at equilibrium (E_{n2}), and at the endpoint of the final maximal expiration, which represents the final alveolar concentration (A_{fn2}). Residual volume (ml) is calculated as follows:

$$RV = [\frac{VO_2(E_{n2} - I_{n2})}{A_{in2} - A_{fn2}} - DS] \times BTPS\ factor \quad (2.5)$$

where VO_2 is the initial volume of O_2 in the spirometer system, including the dead space between the breathing valve and spirometer bell; E_{n2} is the fractional percentage of nitrogen at the point of equilibrium; I_{n2} is the fractional percentage of nitrogen initially in VO_2 (impurity); A_{in2} is the fractional percentage of nitrogen in alveolar air initially when the subject is breathing room air; A_{fn2} is the fractional percentage of nitrogen in alveolar air at termination of the test; DS is the dead space of the mouthpiece, sensing element of the nitrogen analyzer (if one is used), and breathing valve; and BTPS factor is the correction of volume using body temperature, ambient pressure, and spirometer temperature.

Because of its simplicity and the minimal time involved, the oxygen dilution technique is particularly appealing for measuring RV at the time of underwater weighing. For this procedure, flexible plastic tubing is used to deliver respiratory gases

to the subject so that the rebreathing procedure can be done while the subject is in the tank. The mouthpiece is connected directly to the tubing, which is connected to a pneumatic three-port breathing valve with nitrogen-sensor assembly or needle valve. A 5-L rubber rebreathing (anesthesia) bag, or spirometer, is attached to the second port, and the third port is used to deliver oxygen to the bag or spirometer.

To measure RV simultaneously with UWW, the subject exhales maximally while bending forward and submerging the head and shoulders and remains submerged at the end of expiration until the underwater weight is obtained. The breathing valve is then switched from room air to the bag (spirometer) and the subject breathes in and out of the bag of oxygen until nitrogen equilibrium is reached. Constants are used because the remote location of the nitrogen sensor away from the mouth makes it impossible to measure A_{in2} and A_{fn2} accurately. Thus, equation 2.5 for the calculation of residual volume becomes

$$RV = [\frac{VO_2(E_{n2} - I_{n2})}{(0.80 - A_{fn2})} - DS] \times BTPS\ factor \quad (2.6)$$

where 0.80 is A_{in2}, assumed constant at 80%; A_{fn2} is the fractional percentage of nitrogen in alveolar air at the end of the test, assumed to be 0.2% N_2 higher than the equilibrium percentage (i.e., E_{n2} + 0.2% N_2); and DS is the dead space in the mouthpiece, flexible plastic hose, and breathing valve. Wilmore and colleagues (1980) showed good average agreement (8 ml) and a high correlation ($r = .92$) between this simplified approach and the more complex procedure originally described by Wilmore (1969a).

Direct comparisons between the closed-circuit oxygen dilution technique and the open-circuit nitrogen washout technique have shown that the two procedures are very reliable ($r \geqslant .96$, mean differences = 5 ml) and give very similar (mean difference = 26 ml) and highly correlated ($r = .96$) estimates of residual volume in healthy men and women (Cournand et al. 1941; Wilmore 1969a). In addition, the excellent reliability ($r = .99$; standard error = 28-30 ml) of the oxygen dilution technique was confirmed in additional samples of young men ($n = 195$) and young women ($n = 102$) by Wilmore (1969a). Thus, for normal healthy adults, the closed-circuit technique gives valid, reliable, and accurate estimates of residual volume and, because of its rapidity, is useful for multiple determinations of residual volume obtained simultaneously with measurements of underwater weight. Whether similar results are possible in older individuals and patients with impaired pulmonary function has not been established. Hence, in these populations, the open-circuit nitrogen washout procedure may be a better technique.

To estimate residual volume using the nitrogen washout technique, a breathing system consisting of a metered source of oxygen, a three-port breathing valve and hose assembly, and a collecting spirometer is used. In the system described by Akers and Buskirk (1969), oxygen is fed from a cylinder into a small spirometer, which acts as a demand-breathing system for the subject, and a 150-L chain-compensated Tissot spirometer is used to collect expired gases. The inspired and expired gases flow through flexible plastic tubing brought in through the side of the tank to facilitate measurements with the subject in the tank. While on the weighing platform, the subject breathes room air normally and then exhales maximally and submerges for the measurement of underwater weight. As soon as the underwater weight is obtained, the breathing valve is switched and the subject surfaces and inspires oxygen and expires into the collecting spirometer. The nitrogen in the subject's lungs is flushed into the spirometer during a 4- to 7-min "washout" period. The volume of expired nitrogen is determined from the volume of collected gas and the nitrogen concentration measured with a nitrogen gas analyzer. Residual volume is then calculated as follows:

$$RV = \frac{(V + DS)N_{2T} - DSxN_{2DS}}{N_{2i} - N_{2f}} \qquad (2.7)$$

where V is the volume of air expired into the collecting spirometer during the washout; DS is the dead space of the collecting spirometer; N_{2T} is the fraction of nitrogen in the dead space before the washout period; N_{2i} is the fraction of nitrogen in the end alveolar air at the point of maximal expiration, just before the start of the washout period; and N_{2f} is the fraction of nitrogen in the end alveolar sample at the point of maximal expiration at the completion of the washout period. Although the extended washout period makes it difficult to accomplish more than one or two trials using this technique, the nitrogen washout technique may be less affected by incomplete mixing of gases in the lungs and thus may provide more accurate estimates of residual volume, particularly in people with impaired pulmonary function.

Automated Real-Time System

Organ and colleagues (1994) described an automated, real-time underwater weighing system using a weighing chair mounted on three waterproof force transducers positioned in a triangular arrangement to measure underwater weight combined with simultaneous measurement of RV by the oxygen dilution technique. The instrumentation for data acquisition consists of a custom input–output module with an Intel 80V196KB 16-bit microcontroller that serves as a strain gauge signal and amplifier and performs 10-bit analog-to-digital signal conversion for the analog outputs from the spirometer, the nitrogen analyzer, and the summed output from the force transducers (figure 2.3). The digital signals are read by a microcomputer over serial RS232 communications lines; and a custom program for data acquisition, analysis, and display is used to enter subject information and system dead space, control calibration, record and display underwater weight and nitrogen concentration, and calculate and display body density, FFM, and percent fat. The real-time display of data and immediate results provided by this system make it possible to use percent fat as the criterion for selecting trials for averaging rather than maximum underwater weight, which depends on accurate measurements of residual volume. Accuracy of percent fat requires only that the corresponding values of underwater weight and lung volume be known, without the necessity that the weight be maximum or the lung volume truly residual. Thus, the number of trials is reduced because each trial is potentially useful regardless of the magnitude of underwater weight. Also, the graphic display of underwater weight can be reviewed with the subject to reinforce instructions and facilitate learning the procedure.

Methodological Issues

There are several methodological issues to consider when estimating body volume from underwater weight. These include subject position, residual volume, number of trials and selection criteria, alternative lung volumes, and head placement.

Subject Position

Weighing a subject underwater is usually accomplished with the subject completely submerged—sitting, kneeling, or prone—after a

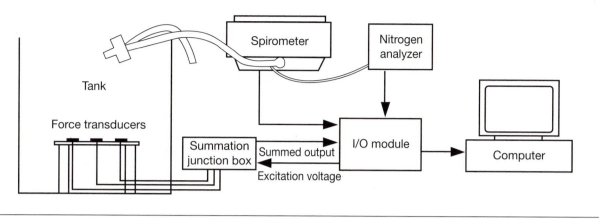

Figure 2.3 Instrumentation and data flow for a real-time underwater weighing system.

Reprinted, by permission, from L.W. Organ, A.D. Eklund, J.D. Ledbetter, 1994, "An automated real time underwater weighing system," *Medicine and Science in Sports and Exercise* 26(3): 383-391.

maximal expiration. Alternatively, other positions and lung volumes have been used, and sometimes subjects have been measured with the head above water. The specifications for tank size depend on which positions are likely to be used. Subject position should be chosen with subject comfort and flexibility and data stability and reproducibility as determining factors. If residual volume is measured with the subject outside the tank, it is important that the positions for weighing and RV measurements be as similar as possible. The seated position is generally used when weight is measured with an autopsy scale, whereas seated or kneeling positions are used with weighing platforms. For many older persons and subjects with arthritic knees, a sitting position on the weighing platform may be most comfortable. Young children and shorter adults may be more at ease in the water when in a kneeling position. Likewise, subjects with protruding abdomens and subjects with limited low back flexibility may find it easier to bend forward and submerge their head and shoulders when kneeling compared with sitting in the water.

Residual Volume

The residual volume can be measured using either the open-circuit or closed-circuit technique on land before underwater weighing or simultaneously with underwater weighing with the subject in the tank. Although it is slightly more complex to set up the equipment for simultaneous measurements, this approach is preferred because it is not necessary to assume that subjects are able to accurately match maximal exhalations on land and in the water because the actual volume of air in the lungs at the time of underwater weighing

is measured. Measurement inside the tank has two other advantages: The error attributable to the compressive force of water on the thoracic cavity is avoided (Girandola et al. 1977), and fewer trials are needed to obtain a reliable and accurate estimate of density. Although good agreement between average measurements of RV on land and in the water is possible (Wilmore 1969a), and measurements outside the tank are adequate when a group description (average) is the primary goal, comparisons have been limited to measurements in healthy young adults measured by experienced investigators, and even under optimal conditions some significant individual differences ($\geqslant$200 ml) occurred. Whether similar agreement would be found in other groups—for example, children and the elderly—is doubtful. For these groups, and when detection of individual differences or changes in serial measurements are of primary importance, measurements at the time of underwater weighing are required.

Number of Trials and Selection Criteria

Variation in D_b of 0.0015 to 0.0020 g/ml is characteristic of the expected trial-to-trial variation within a day, reflecting the combined measurement error from several sources. This degree of agreement is easily accomplished with a force transducer system for measuring UWW and with residual volume measured at the time of underwater weighing. Generally one or two practice trials are sufficient to acquaint the subject with the procedure, followed by three additional trials averaged to obtain the criterion estimate of body density. An on-line computer system programmed for immediate calculation is ideal for monitoring trial-to-trial variation in underwater weight, residual volume, and body density. In many settings, immediate calculation of body density is not feasible, making it necessary to monitor digital or analog recordings and the covariance in underwater weight and equilibrium nitrogen concentration to determine qualitatively whether the system is functioning appropriately.

The possibility of a "practice curve" associated with successive trials of underwater weighing may make it necessary to use more trials when residual volume is measured outside the tank. Katch (1969) demonstrated a progressive increase in underwater weight, suggesting that subjects learn to expire more air from their lungs with each additional trial. His data support the use of the average of the 8th, 9th, and 10th trials to obtain the best estimate of the "true" underwater weight. Behnke and Wilmore (1974) also advocated the use of 10 trials but recommended different selection criteria. Katch (1969) and Bonge and Donnelly (1989) did not show a practice curve. In their data, there was a range of only 25 g across Trials 2 through 10. Adjacent trial correlations were $r = .98$ or higher for Trials 2 through 10, and higher underwater weights were observed more frequently (64%) during Trials 1 through 5 than Trials 6 through 10. Moreover, estimates of body density and percent body fat (%BF) derived from the average of Trials 8 to 10, the average of the first three consecutive trials having a range of 100 g or less, and the average of the first three nonconsecutive trials having a range of 100 g or less were found to be virtually identical and highly correlated ($r = .99$). These results suggest that using four or five trials and using the average of three trials agreeing within 100 g represent an acceptable alternative in subjects for whom 10 trials are burdensome. For others, given the conflicting results, it would seem prudent to conduct sufficient trials to establish a plateau in underwater weight until more definitive research is done.

Alternative Lung Volumes

Residual volume has been the lung volume used most widely during underwater weighing because it is the volume least affected by hydrostatic pressure (Welch and Crisp 1958) and may be the most precise. Concern regarding the subject's ability to accurately and reliably reproduce land residual volume underwater has led some investigators to determine body density at other lung volumes, most notably, functional residual capacity (Thomas and Etheridge 1980), a constant fraction of vital capacity (Welch and Crisp 1958), and total

lung capacity (TLC; Timson and Coffman 1984; Weltman and Katch 1981). For some subjects, determining body density at lung volumes other than residual volume is more comfortable, which may improve subject compliance. In particular, measurements at TLC may be advantageous because subjects may be more at ease and able to stay submerged longer, allowing more time for the scale to steady and thus reducing scale oscillations when underwater weight is read. Other potential advantages include minimizing the problem of air trapping caused by airway closure (Thomas and Etheridge 1980), absence of a "practice curve," and easing the burden for subjects who are unable to perform the novel task of maximal exhalation while underwater.

Whether more precise and accurate estimates of body density are possible at TLC remains to be resolved. The studies that have addressed this issue have given conflicting results, most likely because in some studies TLC was measured on land, and in others TLC was measured in the water. The question of which lung volume to measure is moot, however, when lung volume is measured simultaneously with underwater weighing (Goldman and Buskirk 1961; Thomas and Etheridge 1980) as long as a stable measurement is obtained, because the actual lung volume at the time of weighing is measured.

Head Placement

In an attempt to enhance subject compliance and broaden the application of underwater weighing, some investigators have estimated body density using underwater weighing without head submersion (Donnelly et al. 1988). Although this method is potentially advantageous in some subjects who are unaccustomed to submerging their heads or unable to do so, an unknown and variable, albeit relatively small, error is introduced. Moreover, there is the potential for additional error unless care is taken to ensure that subject position in the water is consistent (Donnelly et al. 1988). Although generally not recommended, this approach may have utility in clinical work or for subjects who are obese, elderly, very young, or mentally or physically disabled. However, newer techniques, such as air displacement plethysmography and dual-energy X-ray absorptiometry, are more appropriate for these populations.

Technical Errors

Technical errors in the measurement of D_b are well documented. Measurement errors are evident in the variation in density observed from trial to trial, which reflects primarily technical errors, and in the variation observed in repeated measurements over several days, caused by both technical errors and biological variation. A third source of error relates to whether the residual volume is measured in the water simultaneously with underwater weight or with the subject outside of the water. Akers and Buskirk (1969) estimated the combined technical and biological error in the measurement of body density to be 0.0017 g/ml in an 80-kg man. Siri (1956) estimated the measurement error to be slightly higher at 0.0025 g/ml, and in the laboratory the error was found to be 0.0020 g/ml (Buskirk 1961).

Akers and Buskirk (1969) carefully analyzed the sources of error in body density and showed errors in RV to be the major source of variation. The variations in body weight, UWW, and water temperature at the time of weighing have much smaller effects, leading to a combined error of 0.006 g/ml if body weight is measured within 0.02 kg, underwater weight within 0.02 kg, and water temperature within $0.0005°$ C. The combined error from residual volume, when RV is measured within 100 ml (0.00139 g/ml variation in density), plus the previously mentioned errors is 0.0017 g/ml or approximately 0.8% fat. Variation in D_b of 0.0015 to 0.0020 g/ml is characteristic of the trial-to-trial variation within a day, reflecting the combined measurement error inherent in most underwater weighing systems. Within-subject variation greater than 0.0020 g/ml reflects larger measurement errors in one or more of the components of body density and indicates a need to improve measurement precision. Lohman (1992) suggested periodic estimates of within-subject variation using at least 10 subjects, each measured three or more times, as an important quality control procedure. The technical error for repeated estimates of body density over several days, estimated to be 0.0030 g/ml (Jackson et al. 1988) or 1.1% fat in men and 1.2% fat in women, is somewhat larger than the error associated with repeated trials within a day. The source of the additional variation is likely to be fluctuation in body water and variation in gastrointestinal flatus, which is assumed constant at 100 ml. Despite the additional variation in day-to-day measurements, it is clear that the technical errors associated with within-subject variation both within a day and between days are quite small when residual volume is measured and that

%BF can be estimated with a precision of 1% or more. The error is inflated when residual volume is estimated, for example, from age, sex, and stature, which can easily lead to errors in RV of 300 to 400 ml in a given subject.

Total Error of Estimating Fat Content From Underwater Weighing

The total error associated with the estimation of %BF from body density using the underwater weighing technique can be estimated from the combined errors attributable to biological variability in FFM composition and D_{FFM} and the measurement errors associated with estimation of body density. Given a variation in D_{FFM} of 0.0059 g/ml in a specific population and a technical error of 0.0020 g/ml, the combined error is estimated to be .0062 g/ml, or the equivalent of approximately 2% fat. This estimate represents the theoretical limit of accuracy when other techniques are used to estimate density attributable to the inherent limitations in the underwater weighing technique.

The impact of measurement errors in body weight, underwater weight, residual volume, and water temperature on estimates of body density and percent fat is illustrated in table 2.3. As indicated in the foregoing discussion, inaccuracies in residual volume represent a major source of error in body density and percent fat. As shown in the table, for every 100-ml error in residual volume or 100-g error in underwater weight, percent fat will be in error by approximately 0.7% fat units. Typically the error in underwater weight will be less than 100 g, although larger errors may occur when an autopsy scale is used to measure subjects because of subject difficulty remaining motionless underwater. In contrast, discrepancies of 100 to 200 ml (0.7-1.4% fat units) can easily occur when residual volume is measured on land rather than simultaneously with underwater weight, and larger errors (300-400 ml) are likely when residual volume is estimated rather than measured directly. Relatively large errors in water temperature and body weight in air have relatively minor effects on body density and %BF; nevertheless, all variables must be measured as accurately as possible to minimize the combined total error, and it is essential that all components of the underwater weighing system be calibrated before each testing session to avoid introducing constant errors.

Table 2.3 Effect of Errors in Residual Volume, Underwater Weight, Body Weight, and Water Temperature on Body Density and Percent Fat

Measure	Actual	ERRORS		
		1	2	3
Residual volume (L)	1.200	1.300	1.600	2.200
D_b (g/ml)	1.0645	1.0661	1.0710	1.0809
%BF	15.0	14.3	12.2	8.0
Underwater weight (kg)	3.36	3.38	3.41	3.46
D_b (g/ml)	1.0645	1.0648	1.0653	1.0661
%BF	15.0	14.9	14.6	14.3
Body weight (kg)	70.0	70.0	70.5	71.0
D_b (g/ml)	1.0645	1.0643	1.0639	1.0634
%BF	15.0	15.1	15.3	15.5
Water temperature (C°)	36.0	36.1	36.5	37.0
D_b (g/ml)	1.0645	1.0644	1.0643	1.0641
%BF	15.0	15.1	15.1	15.2

Note. D_b = body density; BF = body fat. The error attributed to each variable was calculated individually with the other variables held constant at the actual values.

Recommended Procedures for Hydrodensitometry

Given considerations of expense and the precision and accuracy of measurement, the underwater weighing technique continues to be a useful method for estimating body volume leading to the assessment of body composition. The most precise and accurate estimates of body volume and density are derived using a force transducer system to measure underwater weight, with simultaneous measurement of residual volume by either the oxygen dilution or nitrogen washout techniques. The oxygen dilution technique with its rapid equilibration times is particularly suited for multiple determinations along with underwater weight, whereas the nitrogen washout technique may be more accurate in individuals with impaired lung function. Recommended procedures for system calibration, subject preparation, and measurement of weight, underwater weight, and residual volume have been previously published (Going 1996) and will not be repeated here.

Air Displacement Plethysmography

For young children, the elderly, and infirm, disabled, and other special populations, complete submersion in water and thus hydrodensitometry are very difficult, if not impossible. An alternative approach, air displacement plethysmography (ADP), uses pressure–volume relationships to estimate volume and density. Although ADP avoids many of the challenges of HD, past applications in humans have been limited, in part by technical difficulties in adjusting for irregularities in temperature and humidity of the air next to the skin and hair (Graedinger et al. 1963; Taylor et al. 1985). A new system, the Bod Pod (Life Measurement Instruments, Inc., Concord, CA), has improved precision and accuracy compared with past techniques (Demerath et al. 2002; Dempster and Aitkens 1995; Fields et al. 2002; McCrory et al. 1995) and has been preferred by subjects in some studies over underwater weighing (figure 2.4).

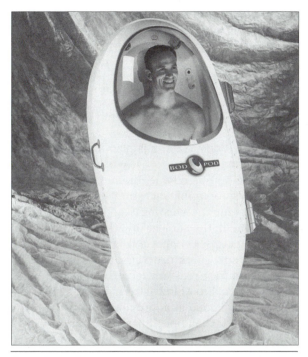

Figure 2.4 The Bod Pod Body Composition System. The plethysmograph is constructed of fiberglass with a large acrylic window. A molded seat separates the unit into a test chamber (front) and a reference chamber (rear) where instrumentation is housed (see figure 2.5).

Courtesy of Life Measurement Instruments, Inc., Concord, CA.

Theoretical Approach: Relevant Gas Laws

The pressure–volume relationship at constant temperature (i.e., isothermal conditions) is described by Boyle's law:

$$P_1/P_2 = V_2/V_1 \qquad (2.8)$$

where P_1 and V_1 represent one paired condition of pressure and volume, and P_2 and V_2 represent

a second condition (Faires 1962). According to Boyle's law, a quantity of air compressed under isothermal conditions will decrease its volume in proportion to the increasing pressure. In contrast to isothermal conditions, under adiabatic conditions (no gain or loss of heat) the temperature of air does not remain constant as its volume changes and the molecules gain or lose kinetic energy. The relationship between pressure and volume under adiabatic conditions is defined by Poisson's law,

$$P_1/P_2 = (V_2/V_1)^\gamma \qquad (2.9)$$

where γ is the ratio of the specific heat of the gas at constant pressure to that at constant volume (Sly et al. 1990). Air under isothermal conditions is easier to compress than air under adiabatic conditions; that is, for a given gas volume and change in volume, the change in pressure will be lower under isothermal conditions compared with adiabatic conditions. As estimated by Dempster and Aitkens (1995), for small changes in volume (relative to total volume), air under isothermal conditions changes its pressure by 40% less than it would under adiabatic conditions. Failure to correct for this difference introduces significant volume measurement errors (~2.5%), and, consequently, early efforts to develop ADP were abandoned (Graedinger et al. 1963).

Bod Pod: Design and Operating Principle

The Bod Pod provides a means of determining body volume by application of Poisson's law. The device consists of a single structure with two complementary chambers: a test chamber of approximately 450 L where the subject is seated and a reference chamber of approximately 300 L (figure 2.5). A molded fiberglass seat forms a

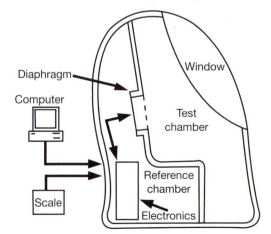

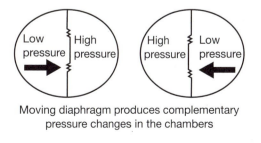

Moving diaphragm produces complementary pressure changes in the chambers

Figure 2.5 Components of the Bod Pod plethysmograph.

Courtesy of Life Measurement Instruments, Inc., Concord, CA.

common wall separating the chambers, and subject entry is permitted through a door to the front chamber that is closed by a series of electromagnets. A moving diaphragm mounted on the front wall between the chambers is made to oscillate under computer control, producing small, sinusoidal volume and pressure perturbations in both chambers, equal in magnitude and opposite in sign. Fourier coefficients are used to calculate the pressure amplitude at the frequency of oscillations. Because the perturbations are small compared with the ambient pressure of the chamber, the ratio of the volumes of the test and reference chambers is equal to the ratio of their pressure amplitudes, according to Poisson's law. To maintain equivalency of gas composition in the two chambers, and hence constancy of γ in the pressure–volume relationship (equation 2.9), an air circulation system is used to mix the air between the two chambers. Because any air added to one chamber is subtracted from the other, there is no effect on the equality of perturbation. Furthermore, the use of sinusoidal perturbation and Fourier coefficients effectively eliminates the deleterious effects of temperature change during the measurement period.

Calibration

Variations in conditions such as chamber size and transducer sensitivity are resolved by a two-point calibration process that precedes each test. The computed ratio of the pressure measurements with the test chamber empty and with a 50-L calibration cylinder inside allows computation of a calibration line, $y = b_1 (x) + b_o$, where y is the volume of an object in the test chamber and x is the ratio of the reference and test chamber pressure amplitudes.

Human Measurement

Because humans maintain themselves in an isothermal condition, conditions of the test are not adiabatic as required by Poisson's law. Air close to skin, hair, and clothing is more compressible than air under adiabatic conditions. In addition, the relatively large quantity of air in the lungs is maintained at very close to isothermal conditions. Thus, to measure body volume with adequate accuracy, it is necessary to either eliminate or account for the effects of clothing, hair, skin surface area, and lung volume. The effects of clothing and hair are adequately dealt with by wearing minimal clothing (bathing suit) and compressing the hair with a swim cap. The isothermal effects related to skin surface area are minimized by adjusting the raw body volume for the surface area artifact,

which is calculated as a constant (K) multiplied by the total body surface area from the Dubois formula (Dubois and Dubois 1916):

$$\text{Surface area artifact (L)} = K(L/cm^2) \cdot BSA\ (cm^2), \quad (2.10)$$

where BSA $(cm^2) = 71.84 \cdot$ weight $(kg)^{0.425} \cdot$ height $(cm)^{0.725}$.

The raw body volume is also adjusted for the thoracic gas volume (V_{TG}), which is measured during the test using a technique very similar to the plethysmographic measurement of V_{TG} used in pulmonary function testing (Ruppell 1994). During measurement, the subject breathes normally. The fluctuation in body volume attributable to breathing is monitored by pressure transducers in the chambers, producing a real-time record of tidal breathing. Once the tidal breathing pattern is established, the airway is occluded at approximately average lung volume (midexhalation) for a brief period (~3 s), and a gentle "puffing" maneuver (alternating contraction and relaxation of the diaphragm at about 1 Hz) is performed. Airway pressures are recorded during the maneuver to provide a measure of the volume of exhaled air (the thoracic gas volume) that comes into contact with the chamber gas volume during determination of the raw body volume. During occlusion, airflow is essentially zero, so there is little difference between pressure in the alveoli and the airway. The pressure changes result in external volume changes, which are reflected in the tidal breathing record. Time-correlated analysis of airway pressure and the tidal breathing record yields V_{TG} at the point when the airway was occluded. Because the pressure variations in the airway and chamber are produced by equal but complementary volume variations, the two pressure curves have a similar shape and may be superimposed by scaling and translation. Lung volume (V_{TG}) is related to the degree of scaling and chamber volume and is estimated as

$$V_{TG}(L) = (m/1.4) - \text{dead space} \atop \text{of breathing apparatus} \quad (2.11)$$

where m is derived from the least squares solution of $[Y_i - (mx_i + b)]$ with Y_i equal to the product of chamber pressure and chamber volume and x_i equal to the corresponding airway pressure during the 3-s occlusion. The constant 1.4 corrects for the overestimation by 40% of V_{TG} attributable to the isothermal air in the lungs compared with adiabatic air in the chamber. Similarly, during body volume measurement, the volume of air in the lungs "appears" 40% larger than it is so that body volume is underestimated by 40% of V_{TG}. Hence, 40% of the measured V_{TG} is added to the raw body volume to account for this fact.

To verify subject compliance with the breathing maneuver, Dempster and Aitkens (1995) developed a mathematical "figure of merit" to assess how well the airway and chamber records superimpose after scaling and translation. This is done by dividing the difference in the shape of the two curves by the magnitude of the excursions. The difference, D, is defined as the sum of the squares of $[Y_i - (mx_i + b)]$. D is divided by the sum of the squares, S, of the differences between adjacent points $Y_i - (Y_i - 1)$. The figure of merit is calculated as $M = 100\sqrt{D / S}$, with M = 0 indicating perfect agreement and M approaching 100 indicating little agreement between the two curves. Dempster and Aitkens (1995) reported that M values below 1.0 are easily obtainable, and M = 1.0 was established as the criterion for acceptance of the V_{TG} measured during testing. High M values indicate poor compliance, that is, leakage of air at the mouth seal or other conditions that cause the pressure signals in the airway and chamber to differ in shape.

A complete test involves measurement of the uncorrected (raw) body volume, computation of the surface area artifact (SAA), and measurement of V_{TG}. Corrected body volume is calculated as follows:

$$\text{Body volume (L)} = \text{Body volume }_{raw} \text{ (L)} - \text{SAA (L)} + 40\% \ V_{TG} \text{ (L).} \quad (2.12)$$

Reliability

Demerath and colleagues (2002) and Fields and colleagues (2002) recently reviewed studies of the reliability and accuracy of the Bod Pod. Across studies, test–retest correlation coefficients, standard deviations (SD), coefficients of variation (SD or technical error of measurement as a percent of the mean), and estimates of precision (SD/n/$\sqrt{d}$, where n = sample size and d = number of repeat measurements) have been used to assess consistency between two or more measurements. Excellent precision and accuracy have been demonstrated for estimating the volume of an inanimate object. For example, 20 successive tests on 2 consecutive days of an aluminum cylinder (50,039 ml) gave mean volumes of 50,027 ml and 50,030 ml on days 1 and 2, respectively, with a maximum deviation of 30 ml. Mean volume errors (±SD), as a percent of actual, were 0.024 ± 0.025% and 0.018 ± 0.027%. Volume errors of this magnitude correspond to percent fat errors of approximately 0.1% fat. Dempster and Aitkens (1995) also

demonstrated that the system was linear over a range of 25 to 150 L (equivalent to the expected range of human body volumes). The mean percent error was less than 0.1% volume at all levels (25-L increments) except for the smallest volume (25 L), where the error was slightly larger (0.13%).

In humans, reliability of the Bod Pod is good to excellent. Between-day, test–retest correlation coefficients in children and adults for body density and percent fat generally exceed $r = .90$, with a trend for somewhat higher coefficients in adults ($r > .95$) than children ($r = .90$; Demerath et al. 2002). In adults, within-subject coefficients of variation (CVs) for percent body fat have ranged from 1.7% to 4.5% within a day (Biaggi et al. 1999; Iwaoka et al. 1998; McCrory et al. 1995; Miyatake et al. 1999; Sardinha et al. 1998) and from 2.0% to 2.3% between days (Levenhagen et al. 1999; Miyatake et al. 1999; Nunez et al. 1999). These CVs are within the range of CVs reported for HD (Pierson et al. 1991; Van Der Ploeg et al. 2000) and dual-energy X-ray absorptiometry (DXA; Economos et al. 1997; Lohman and Chen 2004). In two studies that examined within-day repeatability of Bod Pod and HD percent fat in the same subjects, CVs for the two methods were similar: 1.7% versus 2.3% (McCrory et al. 1995) and 3.7% versus 4.3% (Iwaoka et al. 1998). In two studies, reliability of body volume measurements by HD and Bod Pod were compared. Both Dewit and colleagues (2000) and Wells and colleagues (2000) reported better precision with Bod Pod (0.07 and 0.11 L, respectively). In both studies, however, V_{TG} was predicted rather than measured, whereas lung volume at submersion was measured in conjunction with underwater weight. The use of a constant (predicted) V_{TG} would tend to bias the precision of the Bod Pod toward a more consistent body volume measurement compared with when the precision of HD is calculated using a measured and presumably varied lung volume.

Only a few studies have reported CVs and precision for percent fat and body volume in children. Demerath and colleagues (2002) reported that the technical error of measurement (TEM) for percent fat was higher in children (1.84%) compared with adults (1.63%). As a percent of the mean, TEM was 8.5% and 6.17% in children and adults, respectively. Using the precision statistic described previously, Wells and Fuller (2001) reported the precision of percent body fat to be 0.83% for boys and 0.99% for girls. Precision was not related to body size because the precision for duplicate measurements in men and women (0.99% and 0.76% body fat) was similar to children. Dewit and colleagues (2000) reported the precision of body volume measurements in children 7 to 14 years old. Precision was 0.07 L, which was as good as the precision for adults in the same study. Similar precision was reported in another study by the same group (Wells and Fuller 2001). It has been suggested that a relatively small ratio of chamber volume to subject volume would optimize the precision of body volume measurement (Graedinger et al. 1963; Petty et al. 1984). Using the data from Dewit and colleagues (2000) and assuming a Bod Pod test chamber volume of 450 L, Fields and colleagues (2002) showed that precision was similar despite a larger ratio for children (14:1 for children compared with 8:1 for adults). Thus, within the range of body sizes studied up to this time, the ratio of chamber volume to subject volume may be a negligible source of error.

Validity

Estimates of body composition by the Bod Pod have been validated against HD and DXA (Demerath et al. 2002; Fields et al. 2002). Most of the studies in adults with HD as the criterion method have been conducted in young to middle-aged adults, except for a study by Nunez and colleagues (1999) that included subjects up to 86 years of age. The body mass index in these studies varied across a relatively wide range (~17-40 kg/m^2). Average differences between Bod Pod and HD estimates of percent fat ranged from –4.0% to 1.9% fat. Several studies have reported a significant difference between Bod Pod and HD percent fat (Collins et al. 1999; Dewit et al. 2000; Fields et al. 2000; Iwaoka et al. 1998; Millard-Stafford et al. 2001; Wagner et al. 2000; Wells et al. 2000), although the direction of differences has been inconsistent. Two studies with the largest differences (–4.0% and –3.3%) had very small sample sizes ($n \leq 10$; Dewit et al. 2000; Iwaoka et al. 1998). In some studies reporting significant differences between methods, ethnicity was mixed (Collins et al. 1999; Millard-Stafford et al. 2001) and the potential effects of ethnicity were not examined. In other studies with multiple ethnicities represented, ethnicity did not contribute significantly to differences between the methods (McCrory et al. 1995; Nunez et al. 1999). In studies that have used regression analysis to describe the relationship between percent fat measured by HD and percent fat measured by Bod Pod, the slope of this relationship has ranged from 0.76 to 0.96 (Biaggi et al. 1999; Collins et al. 1999; McCrory et al. 1995; Fields et al. 2000, 2001; Iwaoka et al. 1998; Levenhagen et al. 1999; Millard-Stafford et al. 2001). In some studies, the slope was much lower (0.76-0.82) than desirable (Biaggi et al. 1999; Iwaoka et al. 1998; Levenhagen et al. 1999; Millard-Stafford et al.

2001), although only two studies reported slopes that differed significantly from 1.00 (Biaggi et al. 1999; Collins et al. 1999). Across studies reporting multiple correlation coefficients, R^2 ranged from .78 to .94, and standard errors of estimate (SEEs; reported in only four studies) ranged from 1.8 to 2.3%, in the excellent range ($\leq$ 2.5% body fat) according to Lohman (1992). However, Bland–Altman plots for limits of agreement (mean difference ±2 SD; Bland and Altman 1986) indicated wide variations in agreement between the Bod Pod and HD for individuals ($\approx$9-16% body fat), even when mean differences were small.

The results of studies comparing body composition measurements by DXA and Bod Pod in adults are generally similar to studies in which HD was the criterion method (Collins et al. 1999; Fields et al. 2001; Koda et al. 2000; Levenhagen et al. 1999; Millard-Stafford et al. 2001; Miyatake et al. 1999; Nunez et al. 1999; Sardinha et al. 1998; Wagner et al. 2000). Mean differences between the methods in percent fat varied widely (–3.0% to 1.7% fat). In one study with an overall difference of –0.1% fat, there was a significant negative mean difference for females (–1.3%) and a significant positive mean difference for males (1.2%; Koda et al. 2000). The slopes for predicting percent fat by DXA from percent fat by Bod Pod in studies reporting regression analyses were generally very close to 1.00 (ranged from 0.91 to 1.02) and SEEs ranged from 2.4% to 3.5% body fat (Collins et al. 1999; Fields et al. 2001; Levenhagen et al. 1999; Nunez et al. 1999; Wagner et al. 2000). The 95% limits of agreement ranged from 10% to 15% in three studies reporting Bland–Altman analyses (Fields et al. 2001; Levenhagen et al. 1999; Sardinha et al. 1998), indicating large differences between DXA and Bod Pod in some individuals.

Validation studies of Bod Pod in children have yielded results similar to those in adults. Average differences between percent fat from HD and Bod Pod range from –2.9% fat to 1.2% fat, with no consistent direction of difference (Dewit et al. 2000; Fields and Goran 2000; Lockner et al. 2000; Nunez et al. 1999; Wells et al. 2000). In the few studies reporting multiple correlations, R^2 values ranged from .72 to .87 (Fields and Goran 2000; Lockner et al. 2000; Nunez et al. 1999). In the only study reporting a slope (0.86), it was not significantly different from 1.00 (Fields and Goran 2000), and in the same study, the SEE was 3.3% body fat. Bland–Altman limits of agreement, calculated by Fields et al. (2002) from the data of Fields and Goran (2000), were –4.4% to 9.6% body fat, indicating large individual variations in the differences between Bod Pod and HD.

The results of comparisons of Bod Pod with DXA in children are mixed. In two studies, a significant, negative average difference between the two methods was reported (–3.9% and –2.9% body fat; Fields and Goran 2000; Lockner et al. 2000), and in a third study (Nunez et al. 1999), the average difference (–0.1% body fat) was negligible. Available evidence suggests that percent fat from Bod Pod accounts for 81% to 88% of the variance in percent fat by DXA (Fields and Goran 2000; Lockner et al. 2000; Nunez et al. 1999), and reported SEEs range from 3.4% to 4.1% fat. A wide range of individual differences between Bod Pod and DXA percent fat was demonstrated by Bland–Altman analysis (limits of agreement, –11.9% to 4.1% body fat). In addition, a nonsignificant upward trend in the Bland–Altman plot was reported by Nunez and colleagues (1999), whereas Fields and Goran (2000) found no such trend.

Because the same equation (e.g., the Siri equation) is used to convert density from Bod Pod and HD to percent fat, differences between the methods must be attributable to factors that influence estimation of body volume, including the effects of clothing; moisture on the body, in the hair, and in the swimsuit; elevated metabolism; and the use of predicted rather than measured lung volumes. As noted previously, isothermal air is 40% more compressible than is adiabatic air. Excess clothing causes a significant underestimation of body volume (and thus overestimation of D_b and underestimation of percent fat) because the more clothing that is worn, the greater the layer of isothermal air. The effect of clothing on percent fat was demonstrated by Fields and colleagues (2000), who showed no difference in percent fat in women when a one-piece or two-piece swimsuit was worn compared with a 5% lower %BF when a hospital gown was worn. Moisture on the body, in the hair, and in the swimsuit would alter the correction for the compressibility of air next to the body surface as well as artificially inflate body weight, leading to an underestimation of percent fat (Fields et al. 2002). Furthermore, if subjects are recovering from exercise or other situations that elevate metabolism, breathing patterns may vary over time, which in turn would confound estimation of V_{TG}. A key assumption in Bod Pod testing is that breathing patterns are similar during raw body volume and V_{TG} measurement, which may not be valid if subjects are recovering from a physical or thermal stress. The situation is analogous to HD when RV is measured on land and it is assumed that the subject exhales to the same end point both on

land and in the water. In both cases, the exact lung volume is not a concern but the lung volume should be the same during the HD and RV measurement procedures and, likewise, during raw body volume and V_{TG} measurement procedures.

Interlaboratory variation may be an important source of differences between Bod Pod and HD across the various studies. The extent to which different Bod Pod systems vary is unclear although the calibration that precedes measurement should minimize system differences. In contrast, there are considerable differences in underwater weighing systems and protocols, especially in weighing scales (load cells vs. spring-loaded scales) and approaches to estimating the lung volume corresponding with the underwater weight (inside the tank, simultaneous with weighing vs. outside the tank). As shown by Fields and colleagues (2002), in four pairs of studies, the results within a pair from the same laboratory were more alike than results across laboratories (Biaggi et al. 1999; Collins et al. 1999; Dewit et al. 2000; Fields et al. 2000, 2001; Levenhagen et al. 1999; Millard-Stafford et al. 2001; Wells et al. 2000). These similar findings within studies done in the same laboratory suggest that interlaboratory variation in protocols, equipment, or both may contribute to the variation in results observed among studies.

Other factors that may explain differences between Bod Pod and HD are subject sex, subject size, and errors in RV and V_{TG}. Whether subject sex systematically affects the results of comparisons between Bod Pod and HD remains to be determined. Whereas some studies have reported a significant sex effect (Biaggi et al. 1999; Levenhagen et al. 1999) and a significant upward trend in the Bland–Altman plot, others have failed to find similar effects (McCrory et al. 1995; Nunez et al. 1999). Because males tend to have less body fat than females, it is difficult to determine whether the significant effect of sex is attributable to an effect of sex per se or to differences in body size and fatness. It is also possible that the sex effect could be attributable to a greater amount of body hair in men compared with women, because excess hair may reduce apparent body volume by increasing the amount of isothermal air near the surface of the body, which is unaccounted for by the SAA correction. The effect of excess hair has not been systematically studied (Higgins et al. 2001; Taylor et al. 1985). Similarly, the effect of size on differences between Bod Pod and HD has not been fully elucidated. The notion that a smaller ratio of chamber volume to subject volume would improve precision, and findings by Lockner et al. (2000) that the difference between D_b by HD and Bod Pod in children was related to height, body mass, and surface area (with the largest differences in the smallest children), suggest that body size is an important factor. However, other studies have failed to find differences between children and adults (Fields et al. 2002), and other potential confounders, such as difficulty complying with the underwater weighing protocol and the V_{TG} protocol, cannot be easily dismissed.

Errors in estimates of lung volumes are a likely source of differences between HD and Bod Pod estimates of volume and percent fat. The largest contributor to variability in HD is error in the measurement of residual volume (Akers and Buskirk 1969; Buskirk 1961; Friedl et al. 1992), with errors of up to 4% fat depending on the measurement technique (Forsyth et al. 1988). In theory, an error in V_{TG} has less of an effect on measurements made with Bod Pod than an equal error in RV has on HD (McCrory et al. 1998), although the variability of V_{TG} relative to RV and the validity of V_{TG} measurements by Bod Pod compared with standard pulmonary plethysmography are not well established. The use of predicted rather than measured lung volume is another source of error. The use of predicted residual volume with HD is not recommended. In some studies with Bod Pod, predicted V_{TG} was used when some subjects could not adequately perform the panting maneuver to obtain measured V_{TG} (Lockner et al. 2000; Nunez et al. 1999). The utility of this approach is not certain. Whereas McCrory and colleagues (1998) reported no significant difference between mean predicted and measured V_{TG}, other studies have shown that predicted V_{TG} was significantly higher than measured V_{TG} in collegiate football players (>344 ml; Collins et al. 1999) and in children (>190 ml; Lockner et al. 2000). The reason for the different results is not clear, although software differences provide one possible explanation. The findings suggest that Bod Pod's current method for prediction of V_{TG} may not be valid for all populations. The issue may be more problematic in children, who sometimes have more difficulty performing the V_{TG} measurement procedure (Lockner et al. 2000), than in adults. In one study in children (Dewit et al. 2000), child-specific equations for functional residual capacity (FRC) (Rosenthal et al. 1993) and tidal volume (Zapletal et al. 1976) were used to calculate child-specific V_{TG} and body composition. In this study, the difference in percent fat (Bod Pod – HD) changed from 0.8% (using the Bod Pod software) to –0.9% body fat, suggesting that the use of the adult equations in Bod Pod overestimates V_{TG} in children. As noted previously, errors in prediction generally have a small effect on percent fat because only 40% of V_{TG} is incorporated into the equation to calculate body volume. Nevertheless, more work is needed to improve both the V_{TG} measurement and the accuracy of V_{TG} prediction in different populations.

The factors discussed here that contribute to Bod Pod errors and differences between Bod Pod and HD percent fat (e.g., clothing and prediction of V_{TG}) also contribute to differences between Bod Pod and DXA measurements. In addition, limitations in DXA and errors attributable to the limitations of the assumptions of the two-component models of densitometry used to derive Bod Pod estimates of percent fat also influence comparisons between Bod Pod and DXA. Because DXA does not depend on subject performance and because it is only minimally affected by the typical variations in subject hydration, it is regarded as a standard against which other methods can be validated. However, like any method, DXA is subject to errors and results may vary because of manufacturer differences, technology (pencil beam compared with fan beam), software differences, and variation in subject thickness, especially at the extremes of BMI (Lohman and Chen 2004).

Invalid assumptions of the two-component model, as noted in the initial sections of this chapter, can be a significant source of error when percent fat is estimated from body density. Thus, three-component and four-component models that adjust for variation in FFM components provide better criterion estimates of body composition for validating new methods, particularly in children, the elderly, and samples of broad age range and mixed race or ethnicity. Comparisons of Bod Pod percent fat against percent fat from multicomponent models are affected by the technical errors associated with the measurements of body water and mineral as well as by model error. Consequently, the precision and accuracy of the water and mineral measurements must be considered along with body volume when one interprets the results of studies in which multicomponent models are used. If water and mineral are measured accurately, differences between Bod Pod percent fat from three-component and four-component models reflect model error. As long as the estimates of body volume are equivalent, using Bod Pod or HD with multicomponent models will give equivalent results.

Four studies in adults (Collins et al. 1999; Fields et al. 2001; Millard-Stafford et al. 2001; Yee et al. 2001) and one study in children (Fields and Goran 2000) used a multicomponent model to validate the Bod Pod. Overall, these studies reported significant mean differences ($\approx$2-4%) between Bod Pod and three-component and four-component models, significant correlations ($R \geqslant .90$), and limits of agreement (95% confidence interval around the mean difference; $\approx$2.0 to –7.0%) that are narrower in comparison with the confidence intervals from studies comparing Bod Pod with HD or DXA. The differences between Bod Pod and four-component model estimates of percent fat were positively and negatively associated with the aqueous (W/FFM) mineral (M/FFM) fractions of the fat-free mass, respectively, reflecting the expected influence of variation in these components on estimates of percent fat from the two-component model. The mean difference (Bod Pod – four-component model) and magnitude of the correlation between the differences and W/FFM and M/FFM will necessarily vary with the characteristics of each sample (i.e., above or below the assumed W/FFM and M/FFM). Thus, it is not surprising that findings have varied across studies.

In contrast to the other studies in adults, Yee and colleagues (2001) reported no significant differences between two-component, three-component, and four-component estimates of percent fat in 70-year-old men and women, although there was a significant sex-by-model interaction and the range of estimates across models was greater in men (~19% to ~29% fat) than women (~35% to 41% fat). Use of density from Bod Pod or HD gave similar estimates of percent fat in each model. In children, Fields and Goran (2000) reported that Bod Pod percent fat, estimated using Lohman's age-adjusted formulas (Lohman 1986), agreed better than HD percent fat with percent fat from a four-component model. In addition, Bod Pod was the only method (from among Bod Pod, HD, DXA, and total body water) for which the errors were not correlated with percent fat. Thus, Bod Pod, with age-appropriate formulas, is an appropriate choice in children as judged against the four-component model.

The results of available studies support the Bod Pod as a reasonable alternate to traditional hydrodensitometry. Indeed, given the technical challenges of HD, Bod Pod may be preferred because it potentially can be used for a broader segment of the population. The differences between Bod Pod and HD preclude ADP from being used interchangeably with HD on an individual basis. Further research with Bod Pod is warranted, because the reasons for the differences among individuals within a study, as well as the differences across studies, remain largely unknown. There are a number of potential contributors to differences, including differences in laboratory equipment, study design and protocol, subject characteristics, and criterion methods. A multilaboratory validation study with a standardized protocol, including rigorous procedures for calibrating and comparing equipment across laboratories, would help determine sources of differences across laboratories. Ideally, males and females of different ages, race, size, and composition would be included, and multiple criterion methods (three-component and four-component models) would be used. Because the Bod Pod is designed to measure body volume, data on body volume should be reported. Moreover, because the critical measurement for estimating body volume is V_{TG}, a criterion measure of V_{TG} should also be obtained. A comprehensive study such as the one proposed, although logistically challenging and expensive, would provide the best strategy for validating Bod Pod and quantifying the contribution of the various sources to the overall error in ADP.

Until further research is completed to better describe the impact of the various sources of error on Bod Pod estimates of body composition, strict adherence to a standard protocol similar to the one from the manufacturer is recommended.

Summary

Hydrodensitometry, historically considered the "gold standard" for body composition, remains a useful method in laboratories with limited access to newer methods such as dual-energy X-ray absorptiometry. With the development of the Bod Pod, air displacement plethysmography is a viable alternative for estimating body volume, density, and composition, although pending further research, these two methods should not be used interchangeably. Both HD and ADP are limited by the validity of the assumptions that underlie conversion of D_b to composition, and when it is unclear whether the assumptions are met, HD and ADP are best used in combination with other methods in three-component and four-component models.

Chapter 3

Hydrometry

Dale A. Schoeller

Water is by far the most abundant constituent of the body (Forbes 1962; Keys and Brozek 1953; Moore et al. 1963). The percentage of body weight as water varies from 70% to 75% at birth to less than 40% in obese adults. Water is essential for life, serving as a solvent for biochemical reactions and as a transport media. A 15% decrease in body water attributable to dehydration is life threatening. Even a small change in total body water (TBW), however, can produce a measurable change in body weight, and thus determining TBW is central to measuring body composition.

Water is an important constituent at the molecular, cellular, and tissue levels of models describing body composition (Wang et al. 1992). Unlike the other components of the body at the molecular level, the water compartment consists of a single molecular species, hydrogen oxide. This unique molecular structure simplifies the task of measurement, and thus TBW is a common method for the assessment of body composition at the molecular level.

At the cellular level, water cannot be viewed as a single entity. It is found in two compartments: the body cell mass, which is about 73% water and 27% solids (Moore et al. 1963), and the extracellular fluid compartment, which is about 94% water and 6% solids (Wang et al. 1992). Finally, on a tissue or anatomical level, water is viewed as being in five compartments. These are intracellular water, which is found in the cytoplasm and nucleus of every tissue in the body; plasma water; interstitial water, which is the water in the lymphatic system; dense connective tissue water, which includes the water found in bone, cartilage, and other dense connective tissues; and transcellular water, which is a diverse collection of largely excretory

extracellular fluids such as bile, gastrointestinal secretions, mucuses, cerebrospinal fluids, and other minor components (Edelman and Leibman 1959). The distribution of water at each level for the reference male is summarized in table 3.1.

Water's property as a singular molecular species lends itself well to the use of the dilution principle, which in its simplest form states that the volume of the compartment is equal to the amount of tracer added to the compartment divided by the concentration of the tracer in that compartment (Edelman et al. 1952).

Total Body Water

Total body water can be measured using the dilution principle. Application of the dilution principle in vivo, however, is more complex than in vitro. This complexity arises because the tracers used in the in vivo dilution do not behave in an ideal manner. Thus, measurement of total body water in vivo requires careful attention to these deviations from the basic assumptions underlying the dilution principle. If these deviations are considered, it is possible to design a protocol that will maximize the accuracy and precision of measuring total body water.

Assumptions and Their Validity

There are four assumptions in the measurement of total body water by dilution. These assumptions are basic to all applications of the dilution principle. The assumptions are as follows:

1. The tracer is distributed only in body water.

Table 3.1   Approximate Distribution of Water in the Young Male

Model	Compartment	kg	% Total body water
Molecular	Total body water	40	100
Cellular	Intracellular	23	57
	Extracellular	17	43
Tissue	Intracellular	23	57
	Plasma	2.8	7
	Interstitial	8.0	20
	Bone	2.8	7
	Dense connective tissue	2.8	7
	Transcellular	1.6	4

Adapted, by permission, from the International Commission on Radiologic Protection, 1975, *Report of the task force on reference man,* (Oxford, UK: Pergamon Press), 28.

2. The tracer is equally distributed in all anatomical water compartments.

3. The rate of equilibration of the tracer is rapid.

4. Neither the tracer nor body water is metabolized during the time of tracer equilibration.

The validity of these assumptions depends on the tracer used for the measurement of TBW by dilution. Tracers that have been used include antipyrine, ethanol, urea, and isotopically labeled water. Although useful results can be obtained from the nonisotopic tracers, they and other water-soluble tracers are inferior to isotopically labeled water because of deviations with regard to these assumptions. Each of these nonisotopic tracers is rapidly metabolized, and thus significant elimination from the body occurs during the time for equilibration. In addition, antipyrine and urea are not equally distributed in body water, antipyrine because of a very small degree of plasma protein binding and urea because of significant two-compartment distribution. Because of this, the following discussion of these assumptions is limited to the isotopic tracers of water. These are the radioactive tracer tritium oxide and the two stable isotopic tracers deuterium oxide and oxygen-18 hydride.

Assumption 1: The Tracer Is Distributed Only in Body Water

None of the isotopic tracers is distributed only in body water. Each tracer exchanges to a small degree with nonaqueous molecules; thus, the volume of distribution or dilution space of the isotope will be slightly greater than TBW. Until recently, however, there was less than universal

agreement about the extent of the overestimate because only tritium and deuterium oxide tracers were readily available and estimates of deuterium exchange were based on comparisons of the dilution space with TBW measured by desiccation in animal models. Estimates of hydrogen isotope exchange with nonaqueous molecules range from 0% to more than 20% of TBW (Sheng and Huggins 1979), although the majority of the comparisons indicate that the hydrogen exchange is between 2% and 6% (table 3.2). Part of the controversy arises because it has been difficult to assess the fraction of the variation in these comparisons that is attributable to exchange and that attributable to measurement error. Both isotope dilution and desiccation are subject to systematic errors. Moreover, these errors tend to be in opposite directions and may lead to exaggerated estimates of exchange.

The recent increase in use of oxygen-18 hydride as a tracer for TBW provides an additional tool for the investigation of hydrogen isotope exchange. When hydrogen- and oxygen-labeled waters are administered simultaneously, the increment in the hydrogen dilution space can be estimated from the differences in the dilution spaces without error from the desiccation method or from incomplete dosing, because both spaces are equally affected when the isotopes are mixed. Simultaneous determinations using both tracers have been reported in more than 23 studies involving 270 subjects. These studies have demonstrated that the deuterium dilution space is 2.2% greater than that for oxygen in premature infants and 3.4% greater in infants and adults (Racette et al. 1994). The measured ratio is not constant between subjects, having a standard deviation of 1.5 to 2.0. It has been demonstrated, however, that more than half

Table 3.2 Comparison of In Vivo Isotope Dilution Space With Total Body Water by Dessication

Isotope	Species	No.	% Difference from desiccation (SD)	Reference
Tritium	Rat	10	6.6	Foy and Schneider 1960
		32	12.0[a]	Tisavipat et al. 1974
		21	1.7 (2.4)	Culebras and Moore 1977
		32	4.3	Rothwell and Stock 1979
		—	2.1	Nagy and Costa 1980
	Rabbit	—	3.1 (0.4)	Green and Dunsmore 1978
	Gopher	12	9.2[a]	Gettinger 1983
	Seal	4	4.0 (0.6)	Reilly and Fedak 1990
Deuterium	Rat	16	6.4 (3.1)	Lifson et al. 1955
		24	3.4	Kanto and Clawson 1980
	Pig	24	2.2	Housman et al. 1973
	Seal	4	2.8 (0.9)	Reilly and Fedak 1990
Oxygen-18	Rat	6	1.7 (1.4)	Lifson et al. 1955
		10	1.0 (2.7)	Nagy and Costa 1980
	Pig	45	2.2 (0.1)	Whyte et al. 1985
Hydrogen, mean (SD)			3.7 (1.7)	
Oxygen, mean (SD)			1.6 (0.6)	

[a]Excluded from mean.

the variance in the ratio of the dilution spaces is attributable to measurement error and that the physiological variation is therefore quite small (Racette et al. 1994; Speakman et al. 1993). Unless the deuterium and oxygen dilution spaces vary systematically, it can be inferred that the variation in deuterium exchange is only a small fraction of the TBW, and thus the variability reported previously (Sheng and Huggins 1979) is mostly attributable to measurement error.

The difference between premature infants and adults in the degree of exchange suggested by the smaller dilution space ratio in premature infants, however, appears to be physiological. Specifically, the extracellular fluid volume is expanded in premature infants, which leads to a large increase in the ratio of body water to protein. Because protein is the likely primary source of nonaqueous exchange (Culebras and Moore 1977), the increase in the ratio of water to protein from 3:1 in adults to 5:1 at birth (Fomon et al. 1982) fully explains the difference in the relative dilution spaces between adults and premature infants.

Although the use of water containing both hydrogen and oxygen tracers has provided much-needed evidence indicating that the physiological variation in hydrogen exchange is quite small in human beings, it does not provide an absolute measure of the exchange because oxygen is also subject to exchange. However, desiccation data (table 3.2) indicate that the oxygen exchange is slightly more than 1% and that the deuterium–tritium exchange is slightly less than 4%. Theoretical considerations provide only slightly different values, with oxygen exchange estimated at slightly less than 1% (Schoeller et al. 1980) and maximal deuterium–tritium exchange calculated as 5% (Culebras and Moore 1977). Together, these lines of evidence indicate that the oxygen exchange is 0.7% of TBW and that the deuterium (or tritium) exchange is 4.2% of TBW in adult human beings (Racette et al. 1994).

Assumption 2: The Tracer Is Equally Distributed in All Anatomical Water Compartments

Although isotopic tracers are almost identical to body water, differences in molecular weight can lead to isotopic fractionation, which is a change in the abundance of the isotopes in the product relative to that in the reactant, when a compound undergoes either a chemical or a physical change. The phenomenon of isotopic fractionation includes both equilibrium isotope effects, which result from differences in the relative free energies of the isotopic species in the products and reactants, and kinetic isotope effects, which result from differences in reaction or distribution rates of the isotopic species under nonequilibrium conditions. Comparisons of the isotopic concentration in various anatomical water compartments have demonstrated that there is very little isotopic fractionation within the body. Plasma, urine, and sweat do not show fractionation (Schoeller et al. 1986a; Wong et al. 1988). Older studies, although slightly less precise, also reported an absence of fractionation between water from plasma, liver, gastric fluid, and cisternal fluid (Edelman 1952).

Isotopic fractionation, however, has been noted in water leaving the body by evaporation (Schoeller et al. 1986a; Wong et al. 1988). Isotopic fractionation between water vapor and liquid water is a well-described physical property (Dansgaard, 1964). The fractionation is greater for the hydrogen isotopes because of the differences in the hydrogen-bonding energies of the three hydrogen isotopes. Water collected from exhaled breath is depleted in deuterium and oxygen-18 relative to body water. Similarly, transdermal evaporation loss, which is insensible water loss from the skin through routes other than the sweat glands, is isotopically fractionated (Schoeller et al. 1986a; Wong et al. 1988). These two data sets are in good agreement with regard to the values of the fractionation factors and, equally important, in good agreement with the values determined in vitro (Dansgaard 1964). Saliva collected by spitting or similar methods is slightly enriched in deuterium

and oxygen-18, because it is the residual after the lighter evaporative water loss is removed, but saliva collected by cannulation of the salivary gland is expected to be unfractionated. Similar in vivo measures are not available for tritium, but the fractionation factors between water and water vapor are known. These fractionation factors for all three isotopes are presented as the ratio of the heavy isotope concentration in water vapor divided by that in water (.990 for ^{18}O, .945 for ^{2}H, .92 for ^{3}H, all at 37° C). These fractionation factors were determined in vitro, but they have been confirmed in vivo for deuterium and oxygen-18 before administration of any isotope (Schoeller et al. 1986a; Wong et al. 1988).

Assumption 3: The Rate of Equilibration of the Tracer Is Rapid

The definition of equilibration time depends on the precision of the measurement method, because a more precise method requires a closer approach to equilibrium. For this discussion, it will be assumed that the precision of the measurement is 1%. Schloerb and colleagues (1950) investigated the rate of equilibration as a function of route of deuterium oxide administration and found that equilibration was reached 2 hr after intravenous administration and 3 hr after subcutaneous or oral administration. Wong and colleagues (1988) also performed an extensive investigation of the time to equilibration after oral administration of labeled water to healthy subjects. These authors demonstrated that the time to equilibration was less than 3 hr regardless of whether plasma, breath CO_2, breath water, saliva, or urine was sampled. The degree of error attributable to disequilibria did, however, differ between physiological samples during the first 1 to 2 hr after the dose. Breath CO_2 demonstrated an initial 40% overestimate for ^{18}O relative to venous plasma water, and urine demonstrated a low isotopic enrichment for both stable isotopes of water relative to venous plasma water (Wong et al. 1988). The low enrichment of urine relative to plasma probably represents a memory effect in the bladder attributable to incomplete emptying of urine produced before isotope administration. This occurs despite a modest rate of isotope exchange across the wall of the bladder (Johnson et al. 1951). Because of this memory effect, it is also important to consider the number of voids, in addition to the time after the dose, when urine is sampled for the measurement of total body water. Three voids are recommended with the third void between 3 and 6 hr after the dose.

When the data from a large number of subjects are combined, a slight disequilibrium is still detectable at 3 hr after the dose. In 63 subjects receiving either deuterium or ^{18}O, we detected a 0.3% (+0.1%) smaller TBW at 3 hr than at 4 hr after oral isotope administration (Schoeller et al. 1985). Although a third of this difference can be accounted for by water turnover (see discussion of Assumption 4), the 0.2% underestimate in measured TBW is probably attributable to a small continued mixing of the isotopes with poorly perfused body water compartments. This small underestimate, however, is negligible for most TBW measurements. Additional care is required in older subjects. Postvoid residual volume in the bladder tends to increase with aging, and it was found that 10% of a relatively healthy, high-functioning cohort of individuals in the 8th decade of life did not reach urinary equilibrium even after the third void (Blanc et al. 2002). Use of saliva or blood sampling can prevent errors caused by incomplete equilibrium.

Even a small disequilibrium in healthy subjects raises concern about equilibration time in subjects with excess water in poorly vascularized compartments. Denne and colleagues (1990) measured the time to equilibration among pregnant women and found that pregnancy delayed equilibration by about 1 hr. McCullough and colleagues (1991) determined the equilibration time in patients with ascites and also noted a delay in equilibration, with the plateau not occurring until 4 hr after the dose. Because of these observations, equilibration times of 4 or 5 hr after oral isotope administration are suggested in patients with expanded extracellular water compartments.

Assumption 4: Neither the Tracer Nor Body Water Undergoes Metabolism During the Time of Tracer Equilibration

As indicated earlier, the inaccuracy of this assumption causes water-soluble compounds like ethanol and antipyrine to be poor tracers for measuring TBW, but even isotopically labeled water is not free of metabolic complications. Body water is in a constant state of flux. In temperate climates, the average fractional turnover rate in adults is 8% each day (Schoeller 1988). In the reference male, this includes inputs of 1.5 L/day from beverages, 1 L/day from water of hydration in food, 0.2 L/day from metabolic water produced during the oxidation of fuels, and 0.2 L/day of exchange with atmospheric moisture. This is balanced by an output of 1.6 L/day in urine; 1.2 L/day of insensible losses as sweat, breath water, or transdermal evaporation; and 0.1 L/day in stool (National Research Council 1989). Water turnover is 50% to 100% greater in tropical climates (Singh et al. 1989) because of increased insensible water loss.

This constant turnover has led to two approaches to the measurement of total body water. The first is the plateau method, which is the method discussed in the preceding paragraphs. When the plateau method is used, labeled water is administered, samples are collected for 3 to 5 hr, and TBW is calculated from the enrichment of the samples collected after the enrichment has reached a plateau or constant value.

Because this plateau is not perfectly constant, given the metabolism of water, a back-extrapolation or slope-intercept method has also been used (Coward 1988). For this method, samples are collected for up to 14 days after the dose, and the zero time intercept is calculated by back-extrapolation to the time of the dose. Because of the continuous turnover of body water, one would expect the back-extrapolation method of calculating TBW to give a smaller and more accurate estimation. In general, however, both methods give very similar results. These methods require different approaches to the treatment of water flux. When the plateau method is used, efforts are made to reduce water intake during the equilibration period to minimize water turnover and thus flatten the plateau. With the back-extrapolation method, efforts are made during the first day and subsequent days to maintain constancy of the fractional turnover rate and thus ensure that the assumption of a log-linear elimination rate is met. The difficulty of maintaining constant turnover is the likely cause of the discrepancy between the plateau method and back-extrapolation method reported by Wong and colleagues (1989a). They found that TBW calculated from the 6-hr postdose data in 10 lactating women averaged 0.6 kg more than the back-extrapolation values. Although the authors attributed this to failure of the 6-hr sample to equilibrate, it is more likely that it is attributable to an overestimation of TBW by back-extrapolation. Failure to equilibrate results in high enrichment values and thus underestimates TBW in most fluids with the possible exception of urine. In contrast, a low turnover on the first day after the dose, attributable to a 6-hr period without water intake, results in a slightly high value of the intercept and hence an underestimate of TBW by back-extrapolation, which is what Wong and colleagues observed.

Equipment

Labeled water for the measurement of TBW can be assayed with one of several methods. The choice of method is often one of convenience and depends on the availability of instruments and expertise. Although deuterium was the first tracer available to investigators for the measurement of TBW, its use was soon displaced by tritium because of the greater availability of scintillation counters and the simplicity of the assay, especially relative to the falling drop method (Schloerb et al. 1951), a technique that can best be described as an art form. The use of tritium, however, involves a small but finite radiation hazard. Because of this, the use of stable isotopic labels has again become common, especially for studies involving children and women of childbearing age. Of the two, deuterium is used more often than ^{18}O. Oxygen-18 has the advantage that its dilution space more closely approximates TBW, but, as indicated previously, the reproducibility of the differences between the dilution spaces of deuterium and ^{18}O is such that the spaces can be mathematically interconverted with little error. The disadvantage of ^{18}O is that it can be adequately measured only by isotope ratio mass spectrometry, and, even at the precisions offered by this instrumentation, the cost of ^{18}O-labeled water for measurement of TBW in an adult is $100 to $300 compared with $5 to $10 for deuterium.

Measurement Procedures

Attaining a precision of 1% in the measurement of TBW requires careful attention to detail. Each aspect of the measurement, including subject preparation, dosing, sample collection, and isotope analysis, must be controlled such that systematic and random errors are less than 0.5%.

Subject preparation is somewhat dependent on the goals of the investigation. The most rigorous preparation is required for the extrapolation from TBW to body composition with additional measures of other body components. For this, the subject must be euvolemic. Overhydration or dehydration reduces the accuracy of the extrapolation. To establish euvolemia, the subject should have normal fluid and food intake the day before the measurement and avoid vigorous exercise after the final meal of the previous day to avoid dehydration or depletion of glycogen stores. Similarly, ambient conditions should be such that the subject does not sweat excessively after the last meal of the previous day. The final meal should be eaten between 12 and 15 hr before the dose to minimize the water content of the intestine. Last, the subject should not drink for several hours before the test to avoid overhydration. The measurement can be performed in the morning to minimize the discomfort of fasting.

The dose should be aliquoted by weight using a balance with a precision and accuracy such that the relative uncertainty in the dose of labeled water is less than 0.3%. Furthermore, the dose must be weighed and transported in a screw-capped container to minimize evaporation. Although the dose can be given intravenously, which will reduce the equilibration time by about 1 hr, oral dosing is quite effective and is less invasive. Even for oral dosing, Millipore filtering is recommended. Syringe cap filters are generally adequate, but where extensive precautions are indicated or where the minor losses of costly ^{18}O are to be avoided, an online system can be used (Wong et al. 1991). Intravenous or subcutaneous dosing requires that the dose be sterile and pyrogen-free.

The dose should be given with the subject fasting to maximize the rate of absorption. Although isotopically labeled water will equilibrate with body water under conditions of reduced gastric emptying (Jones et al. 1987), studies have shown that absorption will be delayed (Scholer and Code 1954).

Usually, when the plateau method is used, subjects are prevented from eating or drinking during the equilibration period. This minimizes the changes in the body pool size. In some subject groups, such as young children, the continued fast is stressful for all involved and a light meal can be given 1 hr after the dose. One hour is chosen on the assumption that the dose will have emptied from the stomach and yet there is still time for the water in the meal to mix with the body water pool during the equilibration period. The ingestion of large boluses (2-3% of the pool size) of fluid disturbs the isotopic enrichment of plasma water for up to an hour after the bolus (Drews and Stein 1992). The volumes of water in food and beverages should be carefully recorded so that they can be subtracted from the dilution space.

When the back-extrapolation method is used, meals should be delayed for 1 hr after the dose to permit isotope absorption, but then normal intakes should be instituted so that the isotopic elimination approximates the habitual value. The research group at the Dunn Nutrition Centre (Cambridge, UK) suggests that the subject be awakened at 4 a.m. to administer the labeled water so that a normal meal pattern can be established even on the day of dose administration (personal communication).

The dose should be administered with great care to avoid losses (Roberts et al. 1990). After a dose is given orally, the container used for the dose should be washed with 20 to 50 ml of water and this should be drunk by the subject. Similarly, if the dose is administered via a nasogastric tube, the tube should be flushed with water after the dose to rinse all of the dose from the tube. If the undiluted volume is less than 10 ml, it is advisable to dilute the enriched dose waters before administration to minimize the effects of small losses. It is also advisable to have on hand a preweighed tissue and a sealed plastic bag. Should a small amount of the dose be spilled, the spill can be absorbed onto the tissue, the tissue resealed in the bag, and the tissue and bag reweighed to measure the loss.

Physiological samples must be collected over a period long enough to ensure equilibration. As indicated previously, equilibration for the plateau method, defined as 99.5% of the final enrichment, is reached within 3 hr for subjects with normal water compartmentalization and 4 hr for subjects with expanded extracellular water volumes. Measurements of urine, however, require that a third void be produced to minimize the effects of incomplete emptying of the bladder. There is less certainty about recommendations for sample collection in the back-extrapolation method. Again, however, it appears prudent that the first sample for analysis should be collected within the 6-hr period after the dose and that it be the third void after the dose. At least three additional samples should be collected on subsequent days at the same time of day to minimize the effects of diurnal variation on the elimination rate (Schoeller et al. 1986b).

Physiological samples must also be collected before the dose, because each of the isotopes under consideration occurs naturally in the body. When isotope ratio mass spectrometry is used, the baseline sample should be collected within 24 hr of the dose to minimize any day-to-day natural variation. This period must be shortened if the subjects are undergoing serial body composition measurements because isotope may still be present from the previous study. If the intervals between isotope doses are less than three biological half-lives (circa 3 weeks), then it may be advisable to collect a series of samples for 3 to 6 hr before the dose to estimate the rate of isotope elimination from the previous dose.

Sample size depends on the need for the assay, but samples smaller than 1 ml are generally not recommended. Small samples are subject to systematic error because of contamination with the moisture in air. All samples should be stored in airtight containers with minimal dead space within seconds after collection. When smaller samples are required, storage in flame-sealed glass capillaries are recommended. Although these samples are quite stable during storage, refrigeration or freezing at $-10°C$ is recommended to minimize bacterial growth.

Generalized procedures for the isotopic analyses cannot be provided here because of the many types of assays that can be used (infrared spectrometry, Lukaski and Johnson 1985; nuclear magnetic resonance, Khaled et al. 1987; mass spectrometry, Wentzel et al. 1958; isotope ratio mass spectrometry, Horvitz and Schoeller 2001). It is, however, imperative that a sample of the dose be saved and analyzed for isotope enrichment using the same procedure used to analyze the physiological samples. This aliquot of the dose should be gravimetrically diluted with tap water so that the enrichment approximates that of the physiological samples to determine the exact concentration of the tracer. Use of the manufacturer's specifications regarding enrichment is not recommended because there may be systematic differences in the analytical procedures. It is recommended that the sample of diluted dose and diluting water be analyzed in the same batch as the physiological samples for maximal precision (Coward 1990).

Calculation of Dilution Spaces

Under ideal conditions, the calculation of the isotope dilution space (N) in moles involves the simple application of the dilution principle. Thus

$$N = d \cdot f \cdot E_{dose}/E_{bw} \qquad (3.1)$$

where d is the moles of water given in the dose, f is the fractionation factor for the physiological sample relative to body water, and E_{dose} and E_{bw} are the enrichments of the dose and body water, respectively. A limitation of this equation is that it often involves interconversion of the units with respect to both the dose (i.e., grams to moles) and the various enrichment values to atom percent excess. A more user-friendly method is to measure both a sample of the diluted dose and the physiological samples during the same analytical run and to calculate dilution space from mass and instrumental units directly (Coward 1990):

$$N = (WA/a)(S_a - S_t)f/(S_s - S_p) \qquad (3.2)$$

where N is expressed in grams, W is the mass of water used to dilute the dose, A is the dose administered to the subject, a is the mass of dose used in preparing the diluted dose, f is the fractionation factor for the physiological sample relative to body water, S_a is the measured value for the diluted dose, S_t is the value for the tap water used in the dilution, S_s is the value for the physiological sample, and S_p is the value for the pre-dose physiological sample.

The value of S_s, the isotopic abundance, can be obtained by the plateau method or back-extrapolation to the time of the dose. For the back-extrapolation technique, the isotope enrichment of each postdose sample should be calculated relative to the isotope abundance in a sample collected before isotope administration, and the zero time enrichments should be calculated by linear regression using the natural logarithms. If samples are collected for more than two biological half-lives after the dose, this back-extrapolation may be inordinately affected by imprecise measurements at the latter time points (Schoeller et al. 1985). If excess isotope is not present in the subject from a previous measurement, then the enrichment may be used directly for calculating dilution spaces. If excess isotope is present, the enrichment relative to a physiological sample collected just before the second dosing must be calculated and used in the equations just given.

When the back-extrapolation method is used, the dilution space does not require correction for isotope loss in urine or insensible loss (Spears et al. 1974) or for the addition of water to the pool because these factors are fully compensated for by the back-extrapolation, as long as the subject is in water balance and the turnover is constant. The plateau method, however, requires correction for water flux during the equilibration period. When the goal of the protocol is to measure dilution space at the time of the dose, any new water added to the pool from the dose, dietary intake, beverages, metabolic water, and atmospheric exchange in humid environments should be subtracted from the isotope dilution space. Estimates of these rates are presented in table 3.3, and detailed equations are presented elsewhere (Fjeld et al. 1988; Schoeller 1991).

If the goal is to measure the dilution space at the time of sample collection (e.g., in the morning after an isotope dose given the night before), the dose must be corrected for all isotope losses from the body water pool. These routes of loss include urine and insensible water loss, assuming that fecal losses are negligible (Spears et al. 1974). Although these corrections will be small and often negligible in studies of adults, where the subjects are not allowed to eat or drink during the equilibration period, the errors can be quite large in infants and children. The influences of these small isotope losses are generally small, and the comparability of back-extrapolation and the plateau methods has been demonstrated (LaForgia and Withers 2002).

The final step in determining total body water from isotope dilution is to correct the previously calculated isotope dilution space for exchange with the nonaqueous compartment. As indicated, the bulk of the data suggest that tritium and deuterium overestimate the body water pool by 4.2% in adults and children and that deuterium overestimates the body water pool by 2% in premature infants. The overestimate for [18]O is smaller and estimated to be 0.7% in adults and 0.5% in premature infants.

Use of Body Water to Estimate Body Composition

As indicated at the beginning of the chapter, body water can be used to estimate body composition on three levels: molecular, cellular, and tissue. Because the cellular level requires an additional measurement to partition body water into an

Table 3.3   Estimates of Average Daily Total Body Water (TBW) Flux in Young Adult Subjects Living in a Temperate Climate

Route	TBW turnover %/hr	Absolute g/hr
Urine	0.23	82
Breath	0.10	34
Sweat	0.02	6
Transdermal	0.02	8
Total	0.37	130

Calculated from data recorded in Chicago, Illinois (Schoeller et al. 1986a).

intracellular compartment, it is discussed later in this chapter. In the absence of measures other than weight, the most common and useful body composition model is the two-component model of fat and fat-free mass (FFM). This model is based on the knowledge that lipids are hydrophobic and thus free of water, which is therefore restricted to the fat-free compartment. The calculation of FFM from body water depends on an assumption of constant hydration of FFM, that is, that the ratio of water to solids in FFM is the same in all subjects. Clearly, this assumption is incorrect in subjects who either are dehydrated or have abnormal water metabolism leading to edema. Among healthy subjects, however, hydration is relatively constant.

The most commonly used hydration constant is 0.73, which was first recommended by Pace and Rathbun (1945), who reviewed chemical analytic data from several small mammal species. The literature on the subject has expanded extensively since then, and this constant has been reinvestigated in a great number of animal species. Many of the chemical analyses of the hydration of FFM were performed on eviscerated carcasses, including those reported by Pace and Rathbun, and thus do not necessarily apply to in vivo models. However, a review of the literature restricted to whole-animal analyses (table 3.4) confirms the constant of 0.73 recommended by Pace and Rathbun (1945).

Table 3.4 Hydration of Fat-Free Mass (FFM) by Chemical Analysis of Whole Nonruminant Mammals

Species	No.	Sex	% Hydration FFM (SD)	Reference
Mouse	14	F	74.0 (1.4)	Annegers 1954
	128		74.0	Dawson et al. 1972
	17		70.7 (1.0)	Holleman and Dieterich 1975
Rat	16	F	72.7 (0.8)	Annegers 1954
	7	M	73.0 (0.7)	Annegers 1954
	112		71.4 (0.9)[a]	Babineau and Page 1955
	16	F	72.3 (1.0)	Rothwell and Stock 1979
	16	M	73.1 (1.0)	Rothwell and Stock 1979
	72		73.0 (2.2)	Lesser et al. 1980
Lemming	5		73.7 (0.3)	Holleman and Dieterich 1975
Vole	4		72.2 (0.3)	Holleman and Dieterich 1975
Rabbit	3		76.3 (1.4)[a]	Harrison et al. 1936
Dog	2		74.4 (0.7)[a]	Harrison et al. 1936
Seal	4		72.2 (0.8)	Reilly and Fedak 1990
Monkey	2		73.2 (0.3)[a]	Harrison et al. 1936
Human	4		72.9 (3.8)	Keys and Brozek 1953
	1		73.7[b]	Moore et al. 1968
	2		72.8 (0.2)	Knight et al. 1986
Mean (SD)			73.1 (1.3)	

[a]Excludes intestines or intestinal contents; [b]composition of viscera estimated from literature.

Although few adult human cadaver studies have been performed to determine the hydration constant (73.0 ± 2.7%, *n* = 7), results are in excellent agreement with animal studies (table 3.4), which is impressive considering that most of the subjects were quite ill before their death and analysis. In vivo studies in adults indicate that there is no effect of aging on the constant through age 70 years (Ritz 2000; Schoeller 1989). An exception to constancy of hydration occurs in infancy, when the hydration of FFM is elevated relative to adult values in animals and humans. With regard to humans, Fomon and colleagues (1982) estimated the hydration constant for ages ranging from birth to 10 years; however, they assumed that deuterium oxide overestimated total body water by only 1.3%. These values have been recalculated assuming a 4% overestimate (table 3.5). Lohman (1986, 1989) and Boileau and colleagues (1984), in a large sample of children from prepubescent to

postpubescent, found a higher water content in the FFM for prepubescent children. Similar values for the hydration of 8- to 12-year-old children were reported by Wells and colleagues (1999). See chapter 9 for an additional discussion of this topic.

The theoretical basis of this hydration constant and factors that can result in individual variations have been described in detail (Wang et al. 1999). The model provides support for the pediatric values provided by Foman and colleagues (1982) and Lohman (1989) and suggests that among healthy adults, any hydration measurement below 0.69 or above 0.77 are considered unphysiological and hence inaccurate. Using multicompartment models, investigators have reported that individual hydration constants vary with a standard deviation between 1.5% and 3% (Hewitt et al. 1993; Lohman et al. 2000; Visser et al. 1999). If these results include both physiological and measurement variation, the typical physiological

Table 3.5 Hydration of Fat-Free Mass in Children

Age	Girls			Boys		
	Wt (g)	TBW (g)	% of FFM	Wt (g)	TBW (g)	% of FFM
Birth	3,325	2,280	80.6	3,545	2,467	80.6
1 m	4,131	2,716	80.1	4,452	2,966	80.1
2 m	4,989	3,071	79.7	5,509	3,450	79.8
3 m	5,743	3,407	79.5	6,435	3,848	79.6
6 m	7,250	4,124	78.9	8,030	4,646	79.2
9 m	8,270	4,777	78.6	9,180	5,392	78.9
1 y	9,180	5,374	78.3	10,150	6,050	78.6
2 y	11,910	7,215	77.7	12,590	7,713	77.7
3 y	14,100	8,721	77.4	14,675	9,134	77.0
4 y	15,960	9,995	77.3	16,690	10,534	76.6
5 y	17,660	11,112	77.1	18,670	11,893	76.1
6 y	19,520	12,301	77.0	20,690	13,300	75.8
7 y	21,840	13,699	76.9	22,850	14,733	75.5
8 y	24,840	15,436	76.8	25,300	16,215	75.2
9 y	28,460	17,464	76.6	28,130	17,919	74.9
10 y	32,550	19,656	76.5	31,440	19,843	74.6

Note. Wt = weight; TBW = total body water; FFM = fat-free mass; m = months; y = years.

From S.J. Foman et al., 1982, "Body composition of reference children from birth to age 10 years," *American Journal of Clinical Nutrition* 35: 1169-1175. Adapted with permission by the American Journal of Clinical Nutrition. © Am J Clin Nutr. American Society for Clinical Nutrition.

range among healthy adults is probably even less than 0.69 to 0.77. Lohman and colleagues (2000) discussed the variation in water and FFM among investigators and suggested a standard deviation of three percent or less for appropriate variation in the hydration level of normal population.

Despite the general agreement between various desiccation studies, the modest variation between investigations has raised some concern about the validity of the use of a hydration constant (Sheng and Huggins 1979). However, chemical analysis is a very difficult procedure that is prone to errors. For example, underestimates of hydration can result from insensible water loss between the time of death and the time of analysis or from incomplete desiccation. Errors that can lead to overestimates of the hydration constant include loss of tissues during dissection and loss of volatile solids during drying (Culebras et al. 1977). Without estimates of these measurement errors, it is not possible to estimate the true physiological variation in the hydration constant of healthy adults. However, if we take 1% as the average measurement error and 1.1% as the average within-laboratory standard deviation of the in vivo measure of hydration, then the physiological variation in the hydration constant can be estimated to be 0.5%, which is quite small. In addition to children, other groups have hydration factors that differ from 73%. Malnourished patients with severe protein depletion have been reported to have average hydration factors as high as 75% (Beddoe et al. 1985). Disease states that alter water metabolism producing edema also result in higher hydration constants (Keys and Brozek 1953). Bodybuilders with expanded skeletal muscle compartments have hydration factors that are elevated by 2% to 3% (Modelsky et al. 1996), but other athletes display no difference (Penn et al. 1994). This comes about not because of greater hydration of FFM but because of a larger skeletal muscle as a proportion of FFM. Pregnancy also results in an increase in hydration that is dependent on the trimester (Sohlstrom and Forsum 1997).

Precision

The precision of the total body water measurement depends on the analytical method as well as the dose of tracer administered to the subject. In general, mass spectrometric methods have been the most precise. These methods, especially high-precision isotope ratio mass spectrometry, can detect very small excesses of deuterium or ^{18}O, and thus the investigator can administer a dose in which the increase in enrichment of body water exceeds the random measurement error by a factor of 200 to 500. Under these conditions, the precision of the measurement as estimated from either repeat measurements (Schoeller et al. 1985; Speakman et al. 1993) or simultaneous measurements using two isotopes (Racette et al. 1994) is between 1% and 2%. Estimates of repeatability using other analytical methods are between 2% and 4% (Bartoli et al. 1993; Lukaski 1987; Mendez et al. 1970; Wang et al. 1973).

Accuracy

Unless there is bias caused by failure to reach equilibrium, the accuracy of the isotope dilution method for the measurement of TBW is excellent. The accuracy depends only on the uncertainty of the estimate of nonaqueous exchange, which is about 1%. There is a further loss of accuracy in estimating FFM attributable to uncertainty of the hydration constant, which is roughly estimated to be 2% in healthy adult subjects (Wang et al. 1999).

Recommended Standard Procedure for Measuring TBW

There is more than one standard procedure for measuring TBW, and these cannot all be detailed because of space limitations. What follows is one recommended procedure for measuring TBW by the plateau method in adults.

- The subject should fast overnight and not drink fluids after midnight. The subject should also refrain from exercise after the previous meal and avoid excessive insensible water loss attributable to high ambient temperatures.
- Collect a baseline physiological sample of saliva, plasma, urine, or breath water (with fractionation correction).
- Weigh the subject in a hospital gown or some other minimal weight clothing.
- Administer a weighed dose of isotope by mouth. Rinse the capped container with 50 ml of water and administer this water to the subject.
- Subject should not take anything by mouth during the sample collection period.
- If saliva, plasma, or breath water is sampled, postdose samples should be collected at 3 and 4 hr after the dose. If there is excess extracellular water, samples should be collected at 4 and 5 hr after the dose.

- If urine is sampled, the subject should void once before the previously mentioned times and this specimen should be discarded. Two specimens should then be collected at the prescribed times.
- Samples should be stored in airtight vessels until analysis.
- Enrichments of the two postdose samples should agree within two standard deviations of the particular assay.

Summary

Water is the most abundant compound in the human body. The volume of water in the body can readily be measured by isotope dilution using tritium, deuterium, or ^{18}O-labeled water. These tracers are distributed rapidly within the body, but they are not perfect tracers. Equilibration after an oral dose requires 3 to 4 hr, and corrections are required for exchange with nonaqueous hydrogen or oxygen. In addition, physiological samples that undergo a chemical or physical change require correction for isotope fractionation. With careful attention to detail, however, total body water can be measured with a precision and accuracy of 1% to 2%.

Intracellular Water and Extracellular Water

The volume of extracellular water (ECW) can be measured in vivo using the dilution principle, a method similar to the measurement of total body water. The anatomy of extracellular water, however, is less defined than total body water. Furthermore, the common tracers used in analysis of extracellular water by dilution are less ideal than those used for total body water. Because of this, the deviations from ideal dilution behavior can be significant, and they must be carefully considered in the design of the protocol to measure extracellular water volume.

Assumptions and Their Validity

Total body water can be divided into intracellular and extracellular water. Although ICW is quite difficult to measure directly, extracellular water (ECW) can be measured by dilution and ICW calculated as the difference from TBW. The assumptions underlying the measurement of TBW by dilution are the same four assumptions underlying the measurement of ECW:

1. The tracer is distributed only in ECW.
2. The tracer is evenly distributed in ECW.
3. The rate of equilibration of the tracer is rapid.
4. Neither the tracer nor ECW is metabolized during the time of tracer equilibration.

Investigators have proposed a number of tracers for the measurement of ECW including bromide, chloride, thiocyanate, thiosulfate, sulfate, insulin, sucrose, and mannitol. Each of these behaves differently with respect to the four assumptions, and a complete discussion of each would be quite lengthy. Previous reviews have indicated that the disaccharides fail to penetrate dense connective tissue and transcellular water and that thiosulfate and sulfate fails to penetrate transcellular water (Bell 1985). Bromide and isotopic chloride dilution come the closest to approximating the extracellular space (Edelman and Leibman 1959), and, with the advent of improved analytical techniques, bromine has become the most commonly used tracer. Neither bromine nor any other tracer used to date provides an exact measure of ECW because the physiological properties of the various compartments of ECW (i.e., plasma, interstitial, dense connective tissue, bone, and transcellular water) differ from one another. As such, dilution spaces may differ significantly between the various tracers, and comparisons must be made with caution.

Assumption 1: The Tracer Is Distributed Only in Extracellular Water

The bromide ion behaves similarly to the chloride ion. Thus, bromine will distribute within all the ECW compartments because all contain chloride, but bromide will overestimate ECW because of penetration into the intracellular space of erythrocytes and leukocytes as well as some cells within the testes and gastric mucosa (Edelman and Leibman 1959). Chemical analyses and radiochlorine dilution studies indicate that this intracellular penetration accounts for 10% of the fully equilibrated bromide dilution space, of which about half is attributable to erythrocytes (Edelman and Leibman 1959). This assumption appears to break down even more so during disease, where the bromide space appears enlarged relative to the expectation for ECW, possibly attributable to bromide penetrating the ICW (Schober et al. 1982).

Assumption 2: The Tracer Is Equally Distributed in All Extracellular Water Compartments

This assumption is examined on the basis of chloride distribution because the bromide concentrations in the various extracellular compartments mimic those of chloride. As such, partial failure of this assumption is anticipated because of differences in the concentration of chloride in plasma and filtrates of plasma. The differences, which are referred to as the Gibbs–Donnan effect (Donnan and Allmand 1914), arise when a membrane-insoluble ionic material is present on one side of a membrane. Interstitial lymph fluids contain 1% to 3% more chloride per liter than plasma, whereas dense connective tissue contains 8% to 14% more chloride, an excess so great that it may also involve binding of chloride to extracellular proteins (Scatchard et al. 1950). The chloride concentration in bone water is very similar to that in plasma, whereas the transcellular chloride concentrations are less than those of plasma but quite variable (Edelman and Leibman 1959). The variation in concentration of chloride in these extracellular fluids, relative to plasma, means that the isotopic chloride and bromide dilution spaces will appear about 5% greater than the extracellular water space. As such, apparent bromide dilution must be corrected by 5%, which is the so-called Donnan correction.

Assumption 3: The Rate of Equilibration of the Tracer Is Very Rapid

The equilibration of bromide has both a fast and a slow component. This is demonstrated by a low ratio of bromide to chloride in brain and cerebrospinal fluid relative to that in plasma (Dunning et al. 1951; Gamble et al. 1953; Wallace and Brodie 1939) for up to 24 hr after the administration of bromide.

Although the distribution kinetics of bromide and other halides do not reach a complete equilibrium plateau until 1 day after the dose, the changes are less than a few percent between 3 and 6 hr after the dose (McCullough et al. 1991), indicating that a relatively stable distribution has been reached. Furthermore, the ratio of bromide to chloride in most tissues is also stable 3 to 5 hr after the dose (Weir and Hastings 1939). Thus, excepting cranial fluid, bromide equilibration time is similar to that of isotopic water—that is, 4 hr in normal individuals (Pierson et al. 1978). The equilibration time is extended to 6 hr or longer in subjects with expanded extracellular spaces caused by ascites.

Assumption 4: Neither Bromide Nor ECW Is Metabolized During the Time of Tracer Equilibration

Bromide does not undergo biological alteration, but it is taken up by the kidney and excreted from the body. Fortunately, the rate of excretion is relatively slow, and only 0.3% is excreted during 6 hr if the subjects are not allowed either food or water (Spears et al. 1974). Longer-term losses are dependent on water and salt flux but typically average 4% or less in 12 hr (Cheek 1953).

At the same time, the ECW volume may undergo changes. Many of these changes are similar to those of body water and reflect its dynamic nature. However, evaporative losses of water will not be accompanied by a loss of bromide. As such, these losses lead to a concentration of bromide in plasma and hence a decrease in measured ECW. In addition, however, the ratio of ECW to ICW can change if fluid redistributes within the body. Changes in posture, for example, can cause a shift from ECW to ICW over a period of several hours, and thus the subject should be allowed to move about only briefly during equilibration or should be required to assume a fixed posture during the period (Thompson and Yates 1941; Thornton et al. 1987).

Equipment

Interest in the quantification of bromide has led to the development of many methods for measuring bromide. These include fluorimetry (Trapp and Bell 1989), ion chromatography (Wong et al. 1989b), neutron activation (Vaisman et al. 1987), mass spectrometry (Janghorbani et al. 1988), and beta counting for radiobromide (Pierson et al. 1978). Most of these techniques require special instrumentation, but analytical developments during the last decade have increased the availability of methods for measuring bromide and increased the confidence in the results.

Measurement Procedures

The procedures for measuring bromide dilution are less standardized than those for TBW. This probably reflects less experience with the technique as well as a realization that accuracy in calculating ECW from bromide dilution space is limited by imperfections of bromide as a tracer for ECW.

Subject preparation before the measurement involves the same considerations as for the measurement of TBW. The subject should be in a euvolemic state, unless, of course, the goals of the measurement are to document the effects of water imbalance on body composition. The subject should therefore have normal food and fluid intake the day before the measurement and should avoid vigorous exercise, sweating, or diuretics, and the measurement should be performed in the morning between 12 and 15 hr after the previous meal.

The bromide should be given as the sodium salt. Because sodium bromide is hygroscopic, the salt must be stored in a desiccator. In humid areas, it is advisable to prepare the dose as a stock solution to avoid problems with moisture. In either case, the dose is given in water. Although oral administration is the least invasive and the most common means of administration, intravenous administration and subcutaneous administration have been used without a detectable difference in distribution time. When injected, sodium bromide is given as a sterile, pyrogen-free 2% solution. The optimal dose depends on the precision of the particular assay. Sodium bromide tastes salty and is toxic at high doses, but doses up to 25 mg/kg will not cause the concentration to exceed the pharmacological level (Basalt 1980).

Sampling is limited to blood plasma, because bromide concentrations are not the same in all fluids (Brodie et al. 1939; Gamble et al. 1953). The concentration must be measured relative to pre-dose levels because bromide is normally present in plasma and because its concentration is variable depending on dietary intake.

Although the distribution kinetics of bromide include a slow terminal phase, the sampling interval in most subjects is similar to that for isotopes of water for TBW because this terminal phase represents a relatively small volume, of which a major portion may be exchanged with bound chloride. Thus, sampling at 3 to 4 hr in normal individuals and 5 to 6 hr in subjects with expanded extracellular spaces is reasonable, although McCullough and colleagues (1991) noted that equilibration was only 88% complete in the peritoneal fluid of patients with appreciable ascites 6 hr after the dose.

The requirements and recommendations for ensuring quantitative administration of the dose and evaporation-free storage of the plasma samples for bromide are identical to those for labeled water. One additional requirement is that the plasma samples be obtained without hemolysis because bromide concentrations in erythrocytes are about 25% less than those in plasma (Weir and Hastings 1939).

Calculation of the Bromide Dilution Space and ECW

The bromide dilution space (N_{Br}) is calculated from the dose of bromide and the increment of the concentration of bromide in plasma water using the dilution principle as already described for total body water.

$$N_{Br} = (WA/a)(S_a - S_t) C_1/(S_s - S_p) \qquad (3.3)$$

where W is the mass of water used to dilute the dose, A is the mass of bromide administered to the subject, a is the mass of dose used to make the diluted dose, S_a is the concentration of bromide in the diluted dose, S_t is the concentration of bromide in the water used to make the diluted dose, S_s is the concentration of bromide in the plasma sample, and S_p is the concentration of bromide in the pre-dose plasma sample. For maximal precision, concentrations are best measured per unit of specimen mass, and the correction, C_1, is the weight fraction of water in plasma. This can be determined gravimetrically by drying a sample of plasma but is usually assumed to be 0.94. This correction will differ if the plasma is ultrafiltered, because of the removal of proteins, or if the subject is severely malnourished and levels of plasma proteins are reduced.

The bromide dilution space overestimates the extracellular water space because of the Gibbs–Donnan effect on the concentration of bromide in various extracellular fluids and because of the penetration of bromide in the intracellular space in erythrocytes, leukocytes, and secretory cells (Edelman and Leibman 1959). Thus, ECW is calculated as follows:

$$ECW = N_{BR}(c_2 c_3) \qquad (3.4)$$

where c_2 is the Gibbs–Donnan correction, and c_3 is the intracellular correction. The Gibbs–Donnan correction is generally assumed to be 0.95, and the intracellular correction is estimated to be 0.90. Intracellular water is calculated as the difference between TBW and ECW:

$$ICW = TBW - ECW. \qquad (3.5)$$

Use of ECW and ICW in Estimating Body Composition

Both ECW and ICW are components of FFM; however, the relationship between ICW and the metabolic properties of the body is much stronger than that of ECW or TBW (Barac-Nieto et al. 1979; Moore et al. 1963). By its very nature, ICW is valuable for estimating body composition at the cellular level (Moore et al. 1963).

Precision

The relative precision of measuring the ECW by dilution is not as well characterized as that of TBW. Published distribution time curves of isotopes of both bromide and water show that the residuals about the curves, which are assumed to represent random error, are about twice as large for bromide as for water. As with TBW, the precision of bromide dilution depends on the dose of tracer and the analytical method. Thomas and colleagues (1991) suggested that relative precisions of 1% should be attainable, if the doses are carefully chosen for the particular analytical method.

The precision for the determination of ICW, when calculated by the difference between TBW and ECW, is worse because errors in both variables propagate through the calculation. Thus, even if TBW and ECW are determined with 1.5% relative precision (i.e., 0.6 and 0.2 kg, respectively), the precision of the ICW calculation will be 0.64 kg or about 2.5%. This propagation of error is usually the limiting factor in the usefulness of ICW measurements for analyzing body composition, particularly on an individual basis.

Accuracy

The accuracy of the determination of ECW and ICW is unknown because direct chemical methods are not available to determine criterion values for these components of the body. The estimate of exchangeable chloride by in vivo bromide dilution, however, is within 1% of the chemically determined exchangeable chloride (Edelman and Leibman 1959); thus, the determination of ECW by bromide in healthy subjects is probably accurate to 1%. The accuracy in subjects with atypical ECW spaces, however, is probably no better than 2% to 5% because of uncertainty in the correction constants for plasma water (c_1), Gibbs–Donnan equilibration (c_2), and penetration into the intracellular space (c_3), all of which are attributable to changes in plasma protein concentrations, hematocrit, and the relative distribution of the various extracellular fluids.

Recommended Standard Procedure for Measuring Bromide Dilution

There is more than one standard procedure for measuring bromide dilution. A representative procedure for adults is presented here that is purposefully similar to that for TBW so that the two components can be measured simultaneously.

- The subject should fast overnight and not drink fluids after midnight. The subject should also refrain from exercise after the previous meal and avoid excessive insensible water loss caused by high ambient temperatures.
- Collect a baseline plasma sample.
- Weigh the subject in a hospital gown or some other minimal weight clothing.
- Administer a weighed dose of sodium bromide in 50 ml of water by mouth. Larger volumes may be used if the total dose exceeds 1 g. Rinse the capped container with 50 ml of water and administer this water to the subject.
- Subject should not take anything by mouth during the sample collection period.
- Collect postdose plasma specimens at 3 and 4 hr after the dose, except when there is excess extracellular water, in which case specimens should be collected at 5 and 6 hr after the dose.
- Store specimens in airtight vessels until analysis.
- Bromide concentrations of the two postdose specimens should agree within two standard deviations of the particular assay.

Summary

Total body water can be further subdivided into intracellular and extracellular water. The ICW cannot be readily measured, but ECW can be closely approximated from bromide dilution. The bromide dilution can be performed simultaneously with isotope dilution for TBW, but plasma sampling is required. Relative precision for ECW can approach that of TBW. The accuracy of the ECW determination is less clear, and significant (15%) corrections for unequal bromide concentrations in extracellular fluids and intracellular penetration are necessary. ICW is calculated by the difference between TBW and ECW with modest precision but good accuracy in groups of healthy subjects.

Chapter 4

Whole-Body Counting and Neutron Activation Analysis

Kenneth J. Ellis

The composition of the human body can be described using a number of different models (see chapter 12). The most basic is the atomic model. This chapter describes several nuclear-based techniques that have been used to obtain direct in vivo chemical assays of the whole body of humans. In particular, the body's content of potassium, calcium, phosphorus, sodium, chlorine, nitrogen, hydrogen, and carbon can be measured with high precision and accuracy. Furthermore, organ-specific assays for cadmium, lead, mercury, manganese, and silicon also have been developed, in part to assess environmental and industrial exposure to these metals. For this chapter, we will focus on the elemental composition most related to the body's bone, muscle, protein, and fat content.

Whole-Body Counting

Whole-body counters were initially built to monitor possible contamination of workers in the nuclear-based industries and scientists at national laboratories. Health physicists wanted these instruments to examine individuals in case of a nuclear accident or accidental exposure while working in the laboratory. Fortunately, accidents were rare. These instruments were costly to build and operate, and most were located at isolated nuclear research facilities. As nuclear medicine procedures evolved, some of these counters could be used to follow excretion of radioactive compounds from the body. As part of the routine monitoring of workers, and to follow residual activity after administration of an isotope, it was noted that the natural background signal from the body had several identifiable peaks in the spectra. With the development of scintillation detectors with sufficient energy resolution, one of the natural signals emitted from the human body could be identified as an isotope of potassium. It would be difficult to credit a single individual or research group with the development of the in vivo assay of body potassium (Anderson and Langham 1959; Forbes et al. 1961; Kulwich et al. 1958). Although a number of centers developed whole-body counters for the animal sciences, this chapter focuses only on human measurements. Over the next 15 years, more than 180 whole-body counters were built around the world, with about two thirds performing body potassium measurements in humans (International Atomic Energy Agency [IAEA] 1970). Thirty years ago, there were more than 20 counter systems in the United States; today less than six remain operational (not counting those used for health physics monitoring at nuclear power plants).

Theoretical Basis

Natural potassium (K) is distributed in three isotopic states (93.1% ^{39}K, 6.9% ^{41}K, and 0.0118% ^{40}K). Only ^{40}K is radioactive. Although it is a very small percentage of natural potassium, 1 g of K contains 1.8×10^{18} radioactive atoms. One of the fundamental laws of nuclear physics is that a radioactive iso-

This work is a publication of the Children's Nutrition Research Center, Houston, TX, and was supported with funds from the USDA/ARS Cooperative Agreement No. 58-7MN1-6-100 with Baylor College of Medicine. The contents of this publication do not necessarily reflect the views or policies of the USDA, nor does mention of trade names, commercial products, or organizations imply endorsement by the U.S. government.

tope will decay at a characteristic rate governed by the isotope's half-life ($t_{1/2}$). That is, $dN/dt = \lambda N$, where dN/dt is the number of disintegrations per unit time, N is the number of atoms, and λ is the decay constant ($\lambda = \ln 2/t_{1/2}$, with $t_{1/2}$ the physical half-life [1.3×10^9 years for ^{40}K]). The initial impression is that with such a small amount of radioactive potassium and its long half-life, there would not be a very intense signal. However, when one substitutes the appropriate values into the equation for dN/dt, the results indicate that there are about 1,800 disintegrations per minute per gram of potassium. Of these decays, 11% go through a pathway that produces gamma rays of 1.46 MeV energy. Thus, for every gram of potassium in the body, about 200 gammas per minute are produced, with more than half of these exiting the body. This translates to about 30,000 gamma rays per minute for an adult male with an average total body potassium (TBK) content of 140 g (Forbes and Lewis 1956). Women tend to be smaller and have less potassium, so the count rate is about 20,000 gamma rays per minute. The smaller the subject, the lower the absolute amount of potassium (^{40}K) so that the body's gamma signal is less intense. Infants, however, still have a sufficient amount of potassium to generate about 3,500 gamma rays per minute.

Assumptions and Their Validity

The basic assumptions of in vivo detection of radioactivity in the human body are well documented (IAEA 1970). For TBK, the gamma rays have a signature energy of 1.46 MeV. The detection systems need to have sufficient energy resolution to separate this gamma energy from other potential radioactive sources, natural and man-made, that can also be present in the body. A discussion of the conversion between TBK and body composition is included in this chapter.

Applicability

The precision of a TBK measurement is determined mainly by the detector volume and the counting time. Factors such as the subject's age, fitness, or restricted mobility attributable to surgery or illness have minimal impact. Also, because the ^{40}K signal is virtually continuous, the measurement can be interrupted when needed (i.e., to assist an infant or older person) and resumed without loss of precision or accuracy until the required counting time is completed. The possibility of claustrophobia is reduced by using bright colors for the shielding and installation of intercoms

with television, radio, or taped music as needed. As could be expected, toddlers (18 months to 4 years) are the most difficult to persuade to cooperate for the usual counting times of 5 to 20 min. For this reason, children at these ages are often measured while asleep or for clinical applications after having received a mild sedative. These difficulties most likely explain why there is an almost complete lack of TBK data for this age group.

Equipment

To measure ^{40}K in humans, there are three general design criteria: (a) Gamma ray detectors must be available, placed close to the subject, (b) the detectors should have sufficient shielding to reduce the natural background levels (cosmic rays, radioactive contaminants in construction materials), and (c) a computer-based analyzer system is needed that can identify the unique gamma rays from ^{40}K. The third requirement is especially critical because the human body also contains other naturally occurring radioactive isotopes, although in extremely small amounts (Watson 1987). To discriminate against the gamma rays from these isotopes, one needs to use detectors, such as sodium iodide (NaI[Tl]) or pure germanium (HpGe), which have adequate energy resolution. The ^{40}K signal can also be detected using plastic and liquid scintillation detectors, but the poorer energy resolution will not allow for complete correction for the interference from natural background isotopes. Also, because the whole-body counter is measuring trace amounts of radioactivity in the body, it is not wise to have the counter located near any intense source of radiation (medical therapy units, cyclotrons, radio cobalt facilities). Also, personnel from a nuclear medicine department should be carefully screened before coming too close to the counter in order to reduce the risk of contamination of the counter.

A basic scanning design for a whole-body counter, called a "shadow-shield" counter, is shown in figure 4.1. This type of counter is sufficiently small that it is considered a portable counter (i.e., it can be transported between sites). The subject, in a supine position on the bed, is scanned slowly beneath the lead-shielded NaI(Tl) detector. A second detector can also be placed below the subject. Several other shadow-shield and scanning bed designs have been built (IAEA 1970).

A more common approach is to build the shielding as part of the room when a multidetector counter is used (figure 4.2). The shield can be made

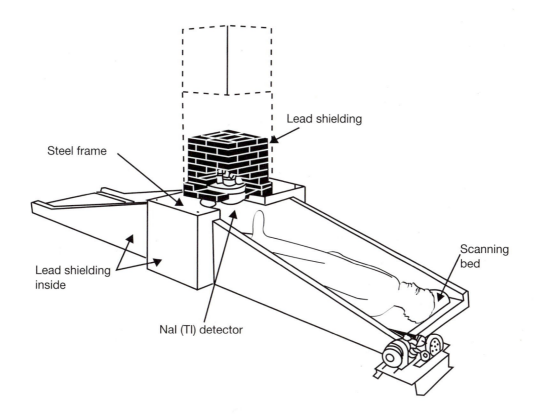

Figure 4.1 A "shadow-shield" counter. The subject is scanned under the shielded NaI(Tl) detector. Additional shielding extends out from the detector region to reduce interference from the environmental background levels.

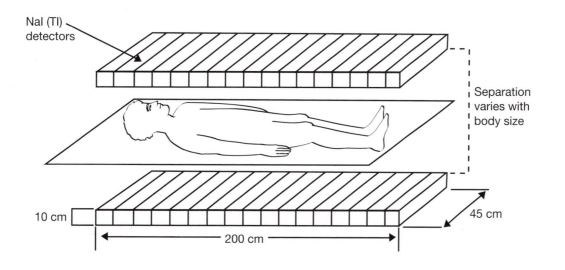

Figure 4.2 Design for a multidetector whole-body counter, housed inside a low-background shield (Fe or Pb walls).

using different materials (e.g., concrete, steel, lead, water), but the goal is to achieve a gamma attenuation that is equivalent to about 10 to 12 cm of lead. If the inside surface area of a steel-shielded room is sufficiently large, an inner liner of lead, aluminum, copper, and stainless steel is used to reduce the signal from the low-energy X rays and gammas produced by cosmic interactions in the steel. In any case, the materials for the shielding should be free of radioactive contaminants (IAEA 1970). The overall room size is determined by the counting geometry (sitting, lying, or standing) and the size and number of detectors (see figure 4.2). Adequate ventilation, lighting, and some form of voice communication between the subject and operator are needed.

Measurement Procedures

The subject should not be counted in his or her street clothes because the clothes can have dust particles that may contain environmental background contamination. The subject should also remove jewelry (which may have some trace radioactivity in the metal) before counting. There is no need for the subject to fast for several hours before the measurement, because most foods contain about the same potassium concentration as the whole body.

The ^{40}K technique is a noninvasive in vivo body composition chemical assay. The measurement procedure requires minimum cooperation from the subject, and repeat assays can be performed as frequently as needed. The counting time varies with the number of detectors, typically lasting from 5 to 20 min. For shadow-shield counters with small detectors, the time can be 40 min.

At least one center (Lykken et al. 1983) found it helpful to have the subject rest for several hours before the whole-body count to reduce the possibility of contamination caused by increased ^{214}Bi interference from the decay of inhaled radon after long-distance running or vigorous indoor exercise. However, other centers have not observed changes in the potassium measurement under similar conditions. The primary source of contamination would be unexpected events, like the Chernobyl accident, where there was a significant release of radioactivity into the environment so that transient background interference for the TBK measurement was noted for several years (Watson 1987).

Calibration of Whole-Body Counters

The detection of a ^{40}K signal from the body is rather easy to obtain. It is the conversion of this signal to a quantitative assay for TBK that is important. Phantoms or models that simulate the human body have been developed for checking the calibration of whole-body counters (Bewley 1988), but the following methods are more often used to derive a counter's calibration factor.

One of the better techniques is to count the gamma spectra for subjects who have received a tracer amount of ^{42}K, a man-made isotope of potassium ($t_{1/2}$ = 12.4 hr). The tracer can be swallowed or injected and the subject counted several times over the next 24 hr; urine needs to be collected to account for loss of tracer from the body. The counter's absolute efficiency and the body's self-absorption factor are obtained (Cohn et al. 1969). The second approach is to count a series of anthropomorphic-shaped phantoms, usually constructed of ground meat that has been chemically assayed for its K content. In either approaches, a set of calibration factors (Cf_i) for body size (weight $\times$ stature) is established.

Cross-calibration of whole-body counters has been rare. The only published comparison was performed in the United Kingdom and used a multinuclide anthropomorphic phantom (Fenwick et al. 1991). The results showed good agreement between the 10 counters using a phantom with a known amount of radioactivity. Similar efforts have been proposed in the United States, but no organized comparison of whole-body counters in the United States has been performed.

Calculation for Total Body Potassium

The calculation of a whole-body counter for the TBK assay based on the ^{40}K counts is straightforward: TBK(g) = $CF_i \times$ ^{40}K counts, where the calibration factor (CF_i) is obtained as outlined later in this chapter. For example, if the subject is administered a ^{42}K tracer dose, and a BOMAB phantom (a standardized set of 10 bottles) is used to simulate the average body size, then the calibration factor becomes

$$CF_i = (g\ K_{bottles}/^{40}K_{bottles}) \times (^{42}K\ count_{bottle}/^{42}K\ count_{subject}). \qquad (4.1)$$

An alternate approach is to define the calibration factor by using a distributed radioactive source that covers an area larger than the 2-D planar image of the body. The subject is counted with and without this source placed below the bed, and the ratio of counts provides a customized correction unique to each subject (Cohn et al. 1969). Independent of the method of calibration, most whole-body counters are not calibrated for the extremely obese subject (weight >140 kg). However, it may be possible to simulate this condition by placing potassium-free materials around the body of a normal-weight subject to establish the counter's response to very large body sizes.

Body Composition Variables, Equations, and Constants

The major application of the TBK measurement is to assess body cell mass (BCM), which is defined as "the working, energy-metabolizing portion of the human body in relation to its supporting structures" (Moore et al. 1963, p. 19). BCM consists of the cellular components of the body: muscle, visceral organs, blood, and brain. For adults, it is generally assumed that these tissues combined have an average K–N ratio of 3 mEq/g, and nitrogen is 4% based on wet tissue weight. Using these assumptions, Francis Moore and his associates (1963) defined BCM as follows: BCM (kg) = 0.00833 × K (mmol). Lean tissues for infants contain more water than adult tissues, which lowers the K concentration, such that the TBK–BCM ratio is about 92.5 mmol/kg (Burmeister 1965).

The measurement of TBK has also been used to estimate the body's fat-free mass (FFM). Based on the cadaver data for five adults (four males, one female), the TBK–FFM ratio was set at 68.1 mmol/kg FFM (Forbes and Lewis 1956). Over the years, this value became a standard of body composition research. It was used, for example, to derive reference models of body composition for infants, children, and adolescents (Fomon et al. 1982; Haschke 1989; Ziegler et al. 1976).

It is not unusual to see an original conversion factor, such as the TBK–FFM ratio, become so widely used over time that it is almost enshrined as an absolute constant. However, a number of investigators have consistently indicated that the potassium content of FFM may be lower and it may not be constant. Studies that have reported the measurement of the TBK–FFM ratio for adults are summarized in table 4.1. Without statistical weighting for differences in sample size or the method used to derive FFM, the overall average for the combined studies is 59.6 ± 4.2 mmol/kg

Table 4.1 Mean Values of TBK–FFM (mmol/kg) for Adults Based on Whole-Body Counting[a]

Age range (years)	Female	Male	Reference
CAUCASIAN (WHITE)			
30-60	64.2	68.1	Forbes et al. (1961)
17-54	59.7	66.4	Womersley et al. (1972)
20-79[b]	57.9	64.5	Cohn et al. (1980)
21-70[b]	60.8	66.8	Pierson et al. (1984)
24-56	62.0	—	Sjostrom et al. (1986)
24-56	62.0	64.7	Kvist et al. (1988)
20-37	64.0	66.0	Hassager et al. (1989)
24-78[b]	56.6	—	Ortiz et al. (1992)
26-93[b]	52.0	60.0	Heymsfield et al. (1993)
18-37	—	63.7	Penn et al. (1994)
20-50	—	63.9	Gerace et al. (1994)
54-76[b]	57.0	—	Hansen and Allen (2002)
31-61[b]	66.8[c]	68.4[c]	Larsson et al. (2003)
AFRICAN AMERICAN (BLACK)			
24-79	63.1	—	Ortiz et al. (1992)
20-60	—	62.3	Gerace et al. (1994)

Note. TBK = total body potassium; FFM = fat-free mass.

[a]Forbes and Lewis (1956) reported 66.5, 66.6, 66.8, and 72.8 mmol/kg for four human cadavers; [b]report a decline in the TBK–FFM ratio with age; [c]age range divided into five groups, values listed are for the 47- to 51-year-old group.

for females and 64.8 ± 2.1 mmol/kg for males. For studies in children, the sex-specific values are 54 to 59 mmol/kg for girls and 59 to 62 mmol/kg for boys (Cordain et al. 1989; Lohman 1986). The New York Body Composition Group has developed a model that provides a theoretical basis for the relative constancy of the TBK–FFM ratio for healthy adults (Wang et al. 2001). In particular, they have shown that variations in the TBK–FFM ratio are influenced by changes in the extracellular water to intracellular water ratio.

From a physiological or clinical perspective, the concept of body cell mass probably has more importance than that of the FFM (Pierson and Wang 1988). The BCM is a component of the FFM that is most likely to show the earliest effects of disease progression, medications, changes in nutrition, or reduced physical activity over short period of time (days or weeks). In these cases, changes in TBK will reflect changes in BCM, but not necessarily for the total FFM. Furthermore, comparing a patient's TBK value with that for an age- and size-matched healthy subject can give some measure of the patient's level of depletion (Ellis et al. 1974). As noted previously, BCM is only about 50% to 60% of FFM; thus, substantial changes can occur in FFM that are relatively independent of TBK. In fact, changes in FFM and TBK can be uncoupled for short periods of time, such that one is possibly decreasing while the other increases. For these reasons, assessment of body fatness based solely on TBK measurements can become problematic, especially for unhealthy subjects (Pierson et al. 1991).

Precision, Accuracy, and Instrument Cost

The primary error for the whole-body counting procedure is governed by Poisson statistics attributable to the random nature of radioactive decay. The precision or percent counting error is calculated as

$$\text{Precision (\%)} = ([N_i/t_i + B_r/t_i + B_r/t_b]^{0.5}) \times (100/N_i) \quad (4.2)$$

where N_i denotes the net count rate for the subject, B_r is background counts for the empty counter, and t_i and t_b are the counting times for the subject and background, respectively. Increasing the counting times or the subject's net count rate or decreasing the background rate will improve the overall precision.

The counting time used for most whole-body counters is in the range of 10 to 15 min, translating to precision in the range of 2% to 5% for adults (Cohn and Parr 1985). This includes errors caused by repositioning of the subject in the counter, which have been shown to be about 0.5% to 1% (IAEA 1970).

Whole-body counting of ^{40}K in infants and young children is mainly limited by the low count rate, so the precision for these instruments is poorer than for adults. Using a 4π plastic scintillator design, Forbes (1968) achieved a precision of 8% to 12% using 15- to 40-min count times for full-term newborn infants, whereas Ellis and Shypailo (1992b) achieved 3% precision using a shielded NaI detector array.

The TBK precision is defined by the counting statistics, but the accuracy of the TBK estimate for each whole-body counter is less firmly established. The ideal verification would be to compare the TBK value with results obtained by chemical analysis of human cadavers. The latter task, however, is difficult to perform and is not without its own substantial inaccuracies. Unfortunately, there is no record that whole-body counting was performed on the human cadavers analyzed by Forbes and Lewis (1956). The closest approximation to cadaver analysis may be the calibration based on the absolute counts using a potassium radiotracer (^{42}K) with a population representing a wide range of body sizes (Cohn et al. 1969). For our infant counter, we obtained an average agreement of ±6% between the absolute mass of TBK (via ^{40}K counting) and total carcass K analysis for 74 piglets in the weight range 1 to 40 kg (Ellis and Shypailo 1992b).

Independent of the accuracy of the TBK value, one is still faced with the decision of how best to convert this information into a more physiologically useful concept of body composition (chapter 12). Several reviews have reaffirmed the original concepts of Moore and his coworkers (1963) that TBK provided the best index for the body's metabolically active tissues or the body cell mass (Pierson 2000; Pierson and Wang 1988). There may be better methods for deriving an estimate of the total FFM than to base it solely on the measurement of TBK, especially for the very obese or extremely malnourished subject. In these cases, there can be significant alterations in extracellular–intracellular fluid ratio, which has been shown to alter the TBK–FFM ratio (Wang et al. 2001).

Recommended Standard Procedure and Cost

There is no standardized design for a whole-body counter (IAEA 1970). However, there are two basic requirements: (a) detectors sensitive to high-energy gamma rays and (b) shielding to reduce natural background interference. It is recommended that the subject change from his or her street clothing, which may be contaminated (not enough to be a hazard, but enough to interfere with the potassium measurement). For the same reasons, the subject should remove all jewelry when possible and eyeglasses during the count period. A reasonable time, no longer than 30 min, should be used. The instrument's performance should be tested frequently, using phantoms whenever possible, to ensure quality control for precision and accuracy.

Commercial whole-body counters, built for monitoring workers in the nuclear power industry, typically cost about $100,000. The cost for a research counter will vary depending on its size and the complexity of the detector array and associated electronics. In general, a 10 cm × 10 cm × 8 cm NaI(Tl) detector plus all supporting electronics and interface with a PC will cost about $5,500 (U.S.). Lead or steel shielding for a whole-body counter is about $2.75 (U.S.) per kilogram. A PC-based operating system starts at about $15,000 (U.S.), which will include all of the necessary nuclear spectroscopy software. Because there is no standardization, the final design of the counter may require the local user to develop some custom software.

Neutron Activation Analysis

The first demonstration that neutron activation could be used for the in vivo measurement of human body composition was reported 40 years ago (Anderson et al. 1964). These researchers used a cyclotron to produce the neutrons and monitored the activity induced in cadavers and living humans with a whole-body counter. The radiation exposures for these initial studies were relatively high compared with today's techniques; however, they were comparable with many diagnostic radiographic techniques that are still routinely used. Since then, several research centers have built their own instruments for neutron activation analysis (NAA). Two International Atomic Energy Agency workshops (Cohn and Parr 1985; Parr 1973) are available that reviewed the technical aspects of NAA for body composition analysis in humans. Since 1986, six international conferences have provided updates on the application of NAA in the study of human biology and physiology (Alpsten and Mattsson 1998; De Lorenzo and Mohamed 2003; Ellis and Eastman 1993; Ellis et al. 1987; Yasumura et al. 1990, 2000).

When a subject is placed in a neutron field, there is a small probability that some of the atoms in the body will undergo some type of nuclear reaction depending on the energy of the neutron.

Humans are already continuously exposed to cosmic sources of neutrons, but the intensity of the natural background neutron flux is too weak to allow us to detect the induced gamma rays coming from the body. To achieve a useful in vivo detection limit (defined by an acceptable accuracy and precision at a dose as low as possible), an external source of neutrons is needed. With this approach, meaningful measurements of the body elemental content can be achieved. The elements that have been measured in vivo in humans are calcium, sodium, chlorine, phosphorus, nitrogen, hydrogen, oxygen, carbon, cadmium, mercury, iron, iodine, aluminum, boron, lithium, and silicon (Chettle and Fremlin 1984; Cohn and Parr 1985). This chapter describes only the NAA techniques used to measure the major elements related to body composition.

Delayed Gamma Activation (DGA) Analysis

When an atom captures a neutron, the atom is immediately transformed to the next nuclear state of the same chemical element, which can be stable or radioactive. In the latter case, the new atom will decay at a fixed rate depending on its half-life. The induced activity (Act) is described by the following equation:

$$\text{Act} = k \times M \times \varepsilon_1 \times \varepsilon_2 \times \varphi \times \text{Exp} \times \text{Delay} \times \text{Counting} \qquad (4.3)$$

where M is the mass of the element. The Exp, Delay, and Counting terms account for the decay effects during the irradiation procedure, the time it takes to get to the body counter, and the counting time, respectively. The term k combines the constants of nature for a particular isotope $\{k = [S \times N \times f_1 \times f_2]/(A \times \lambda)\}$, where S = reaction cross-section, N = Avogadro's number, A = atomic number, f_1 = isotopic abundance, f_2 = gamma decay ratio, and $\lambda = \ln(2)/T_{1/2}$, where $T_{1/2}$ is the half-life of the isotope. The counting system's detection efficiency parameters are included: ε_1 = energy efficiency (detector type and volume), ε_2 = geometry efficiency (number and position of detectors). The equations for the decay effects are: Exp = 1 – exp[–$\lambda \times T_{act}$], Delay = exp[$\lambda \times T_{delay}$]$^{-1}$, and Counting = 1 – exp[–$\lambda \times T_{count}$], where the activation time (T_{act}), delay time between the end of activation and the start of counting (T_{delay}), and the total counting time (T_{count}) are known.

As with whole-body counters, there has been no design standardization among the centers that have used in vivo NAA (Chettle and Fremlin 1984; Cohn and Parr 1985). The design is often influenced by the availability of a neutron source and the gamma counting system. Each laboratory has established its own empirical calibrations, generally using human-shaped phantoms filled with tissue-equivalent fluids. Although this may appear troublesome, the recent comparison of total body calcium estimates using two independently developed systems showed very good agreement (Ma et al. 1999). Part of the success of these two facilities is that phantoms that are used are reasonable approximations of the shape of the human body.

When these phantoms are used, several of the terms in the NAA general equation can be eliminated if the procedure is standardized; that is, the activation time (T_{act}), delay time (T_{delay}), and counting time (T_{count}) are held constant. The net counts in the photopeak ($C_{i,net}$) generated under these conditions for a given element of mass, M, produce a phantom calibration factor (CF_i): CF_i = $K_{phantom} \times (M/C_{i,net})$. Once the initial calibration factors (CF_i) are established for each element (Ca, P, Na, Cl), the photopeak counts in the subject's spectra can be compared with the values for the same elemental reference phantom library:

$$MA_{i,subject} = CF_i \times (\text{Net Counts})_{i,subject}. \quad (4.4)$$

This calibration procedure is common for all neutron activation facilities. From a scientific perspective, direct chemical verification using human cadavers would be highly desirable; however, the wet chemistry techniques are not performed without their own set of difficulties, not to mention the ethical issues that may be raised. Limited carcass verification using animals has been reported (Ellis et al. 1992).

Prompt Gamma Activation (PGA) Analysis

When a neutron is captured by a nucleus, the resulting nucleus, whether stable or radioactive, is usually not in its nuclear ground state because of the added energy provided by the neutron. In a fraction of a nanosecond, the nucleus returns to its lowest nuclear state, most often with the emission of multiple gamma rays. Thus, the major technical difference between the DGA and PGA techniques is that the induced gamma signal must be measured

at the same time as the neutron exposure. The basic equation that describes the PGA technique is less complex, because the delay and counting terms used for the DGA technique are eliminated. The induced activity is directly proportional to the neutron fluence and activation time:

$$\text{Activity} = k \times \Phi \times T_{act}, \quad (4.5)$$

where Φ is the neutron fluence, T_{act} is the exposure time, and the physical constants are combined into the k term along with the system's efficiency factors (ε_1 and ε_2), which are determined by the detector type, volume, and geometry.

The measurement of body hydrogen was the first successful demonstration of the in vivo PGA technique (Rundo and Bunce 1966). However, the major application of PGA in terms of body composition analysis is the measurement of body nitrogen (Biggin et al. 1972; Ellis 1992). For this application, about 15% of the $^{14}N(n, \gamma)^{15}N$ reactions produces a characteristic gamma ray at 10.83 MeV. Because there is no environmental background signal at this energy level, it serves as a unique marker for body nitrogen, an index of body protein.

A hydrogen peak at 2.2 MeV is also present in the PGA spectra and is used as an "internal normalization" for differences in body size. The advantage of this approach is that it reduces the need for precise uniformity of the thermal neutron flux within the body. Instead, the ratio of nitrogen to hydrogen counts is assumed to be proportional to the body's mass ratio of nitrogen to hydrogen. If a phantom with known amounts of nitrogen and hydrogen is used, the system's calibration factor (CF_i) can be derived. For the subject, the equation for the measurement of total body nitrogen (TBN) becomes

$$TBN = CF_i \times (N_{net}/H_{net})_{Counts} \times TBH \quad (4.6)$$

where body hydrogen (TBH) can be approximated as 11% of body weight for most subjects (Ellis 1992; Stamatelatos et al. 1992, 1993; Sutcliffe et al. 1993a). The primary application of the TBN measurement is that it should be directly proportional to body protein mass (protein = 6.25 × TBN).

A number of medical centers worldwide have developed clinical facilities for the measurement of body nitrogen (Ahlgren et al. 1999; Baur et al. 1991; Borovnicar et al. 1996; Ellis et al. 1992; Garrett and Mitra 1991; Stamatelatos et al. 1993). Figure 4.3 shows the components for a PGA instrument. For the TBN measurement, precision is reported at 2% to 5% for scanning times of 15 to 30 min and

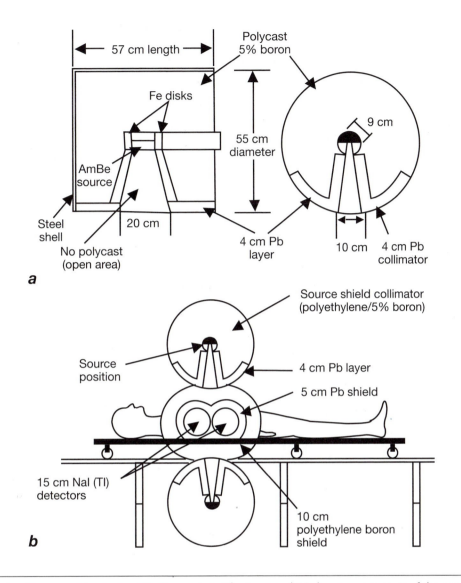

Figure 4.3 Design for a prompt gamma activation system. The top panel (*a*) shows components of the neutron shielding with the Pb-lined collimator opening. The bottom panel (*b*) shows the scanning geometry with the relative positions of the neutron sources (with shielding and collimators) and the shielded NaI(Tl) detectors at the sides of the bed.

an average dose of about 0.3 mSv. Before the September 11, 2001, terrorist attacks, most facilities used radioactive sources (^{252}Cf, ^{238}PuBe, ^{241}AmBe). With the increasing international risk of terrorist attacks and the concerns related to the use of "dirty" bombs, there is a renewed interest in the use of accelerator-produced neutron sources (Ellis and Shypailo 2005).

Inelastic Scattering Activation (ISA) Analysis

Most neutron generators can be pulsed to provide a noncontinuous neutron beam. This has allowed for the detection of gammas produced during the burst of the neuron pulse. One center has performed most of the investigation of this technique, mainly for the in vivo measurements of total body carbon (TBC) and total body oxygen (TBO) in humans (Kehayias et al. 1987, 1991; Kyere et al. 1982). It has been proposed that monitoring the TBC–TBO ratio for an individual may provide an index for total energy expenditure (Kehayias and Zhuang 1993).

The ISA technique requires a pulsed source of high-energy neutrons, which can be typically produced by a (D,T) neutron generator, operating at a 4- to 10-kHz cycle. The primary gamma rays that need to be detected are 4.43 MeV for carbon and 6.13 MeV for oxygen. The counting system uses large-volume bismuth germanate oxide (BGO) detectors positioned to the side of the bed, with the neutron generator below the bed. The shielding components used around the generator and near the detectors need to be constructed of carbon- and oxygen-free materials as much as possible. The electronics for ISA are more complex than those used for PGA, because the counting time must be set precisely with the time of the neutron pulse. It has also been shown that BGO detectors provide a significant improvement over NaI(Tl) detectors for this application (Kehayias and Zhuang 1993). As with the DGA and PGA applications, there is no standardized ISA instrument design.

Assumptions and Their Validity

Body composition assessment based on the NAA procedures is solidly based on the chemical model. For example, about 99% of body calcium is in bone, and body nitrogen is a direct assay for protein, whereas at least 95% of body carbon is in body fat (International Commission on Radiological Protection 1984). In vivo NAA can be used to determine the total body water, protein, fat, and bone calcium (Sutcliffe et al. 1993a, 1993b). Alternatively, in vivo NAA data can be used to estimate energy expenditure (Kehayias and Zhuang 1993).

Several concerns have limited the use of NAA techniques. The first is that no commercial "turnkey" instruments are available. An experienced investigator (often a medical physicist) has been needed to establish and calibrate a new facility.

Second, using radioactive sources raises licensing and safety issues. The limitations of radiation exposure for the subject have played less of a role, because the doses associated with in vivo NAA are well within the range used for routine diagnostic radiographic procedures. Some routine life experiences with risks that are comparable to that for a PGA measurement of body nitrogen are given in table 4.2. For example, a commercial round-trip airline flight across the United States, flying at about 10,000 m, results in increased exposure to cosmic radiation. Third, the accuracy of NAA for body composition estimates may not be as high as previously reported (Wang et al. 1996). When NAA is compared with body composition using computed tomography scans, the standard error of estimate is large (4.4 kg) compared with DXA (1.6 kg).

System Components

To establish a new NAA, PGA, or ISA system, several aspects of the design of these facilities should be considered first. These relate to the choice of neutron sources that are available, the selection of the gamma-ray detectors and their shielding, and the computer-based system needed to operate the system.

Neutron Sources and Geometry

There are a number of choices for neutron sources, both radioactive and machine-produced. The selection of a neutron source is based, in part, on the elements to be measured, the acceptable

Table 4.2   Life Experiences With a Risk Comparable to That for the In Vivo Nitrogen Measurement

Activity (duration)	Outcome[a]

Travel options
    Air (150 miles)
    Air (3,000-mile)
    Car (15 miles)
	Accident
	Cancer (cosmic rays)
	Accident

Housing and geographic location
    5,000 ft above sea level (3-4 months)
    Living in a stone building (2-3 weeks)
	Cancer (cosmic rays)
	Cancer (radon)

Employment conditions
    Working in average U.S. factory (3-4 days)
    Working in a U.S. coal mine (30-40 min)
	Accident
	Accident

Smoking (1-2 cigarettes)	Cancer

[a]For each activity, the probability of death has been estimated at 1 in 8 to 10 million.

level of accuracy, and the allowable radiation dose (Chettle and Fremlin 1984; Ellis 1991). Three radioactive neutron sources (continuous output) have been used: ^{238}PuBe, ^{241}AmBe, and ^{252}Cf. The (D,T) generator has been used, in both continuous and pulsed mode. The clear advantage of the neutron generator is that it can be turned off (no neutron output) when not in use, which eliminates the need for extra shielding when using radioactive sources. A number of different geometries have been used for whole-body DGA neutron activation systems (Chettle and Fremlin 1984); the most recent system is illustrated in figure 4.4 (Ellis and Shypailo 1992a).

Because the PGA technique for body nitrogen uses a thermal neutron reaction, ^{252}Cf tends to provide the best sensitivity to dose ratio (Stamatelatos et al. 1992). Most facilities have used two neutron sources in a bilateral scanning geometry in which the subject is moved past the collimated beam. The typical arrangement between the source, bed, and detectors is shown in figure 4.3. For some systems only a single source has been used, with the subject scanned in both a prone and supine position, thus increasing the total time of the measurement. The advantage, however, is that with one source the background interference is reduced.

Detectors

For DGA technique, the gammas of interest are in the 1.0 to 3.3 MeV energy range; thus, large-volume NaI(Tl) detectors are needed. A whole-body counter designed to measure TBK can usually be used. If only total body calcium is to be measured, then a low-background room is not needed. The five most recently built whole-body counting facilities in the United States for ^{40}K analysis have used at least 32 NaI(Tl) detectors, providing a quasi-2π counting geometry. These also would work for DGA analysis.

The energy of the gammas for the PGA and ISA techniques requires that NaI(Tl) or BGO detectors be used. The technical concern with these detectors is that "pile-up" events (multiple neutron events occurring within the detectors themselves) will increase the background under the nitrogen, hydrogen, carbon, and oxygen peaks, reducing precision. Monte Carlo simulations have demonstrated that using a larger number of smaller detectors, as compared with several large-volume detectors, appears more suited to this application (Chung et al. 1993; Stamatelatos et al. 1992). For the pulsed ISA technique, BGO detectors appear to be the best choice (Kehayias and Zhuang 1993).

Computer-Based Analyzer

Complete "turnkey" multichannel nuclear spectroscopy analyzers, including software, are available from several commercial suppliers. The in vivo NAA user will probably have to add some custom-designed software to reflect the design of his or her facility.

Measurement Procedures

Once the choice of activation techniques has been selected and the facility calibrated, measuring a subject is rather simple. In most cases, the

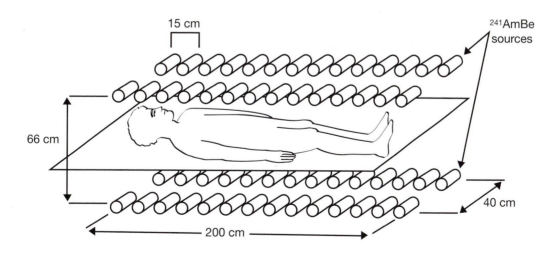

15 cm

^{241}AmBe sources

66 cm

40 cm

200 cm

Figure 4.4 Design of the Children's Nutrition Research Center's multisource neutron irradiator for delayed gamma activation analysis. The ^{241}AmBe neutron sources are arranged in two 2 × 14 arrays above and below the subject.

subject will lie in a supine position on a bed that is moved into the neutron beam area for a preset time. The DGA procedures have used activation times of 2 to 5 min, followed by a delay of 1 to 3 min to transport the subject to the whole-body counter, followed by a counting time of 10 to 15 min. With the PGA and ISA techniques, the subject is slowly moved into the collimated neutron beam area, typically taking 10 to 25 min to complete the whole-body scan.

Precision and Accuracy

Each technique's inherent precision is defined by the basic physics equations presented earlier in this chapter. It is clear that the precision is dependent on the total exposure time and the efficiency of the detection system. Repeated activation measurements in the same persons over a short time period have not been performed mainly because of the cumulative dose that would result. However, precision (reproducibility) has been determined using multiple measurements with anthropometric-shaped phantoms. For the DGA technique, the precision has been reported at ±1% for calcium and ±2% to 3% for body sodium and chlorine (Cohn and Dombrowski 1971; Cohn and Parr 1985). The precision for nitrogen using the PGA technique is in the 2.5% to 5.0% range, whereas ISA for body carbon and oxygen is about ±3% (Baur et al. 1991; Ellis et al. 1992, Kehayias et al. 1991; Stamatelatos et al. 1993).

As noted earlier, the classic wet chemistry analysis of the human body is difficult to perform. NAA results have only been compared with human cadaver results for two adults (Knight et al. 1986). The TBN results by PGA were in good agreement (4-40 g, or 0.7-2.7%) with the chemical data. Body chlorine was also assayed, and the agreement was within 2 to 3 g (2-9%) of the chemical assay. There is no reason not to believe that for normal-sized individuals, similar results for body calcium, sodium, and carbon could be expected. When anthropometric-shaped phantoms were used, the accuracy is reported at ±5% (Cohn and Parr 1985). We have performed pig studies with similar results (Ellis et al. 1992).

Recommended Standard Procedure and Instrument Costs

Because the NAA instruments are not standardized, it makes little sense to attempt to have a standard measurement procedure. It is clear that all the instruments have been built to minimize the subject's level of physical discomfort (i.e., the subject lies on a bed). The time to complete the procedure is reasonable, typically lasting no more than 30 min. And in all cases, the lowest possible dose is used without a significant reduction in precision. Each center has established operating procedures that conform with good laboratory practices, including a quality-control component based on frequent measurements of phantoms.

At present, the design, construction, and calibration of the DGA, PGA, and ISA facilities tend to be highly technical, often requiring an experienced medical physicist. However, once facilities are installed there should be sufficient quality control procedures in place to ensure that routine measurement procedures are being performed correctly. Once the staff members have received their initial training, continued operation of the instruments is rather routine. If a facility already has a whole-body counter, the addition of a DGA irradiator system can be added at relatively low cost (Cohn and Parr 1985). Custom-built PGA and ISA systems are already in the price range of dual-energy X-ray absorptiometry instruments. As neutron generators improve and more efficient gamma detectors are produced, it is reasonable to expect more of these systems to be built, possibly at lower cost.

Chapter 5

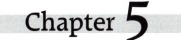

Dual-Energy X-Ray Absorptiometry

Timothy G. Lohman and Zhao Chen

This chapter focuses on the use of dual-energy X-ray absorptiometry (DXA) to measure total and regional body composition, including the estimation of lean soft tissue mass, fat-free mass (FFM), fat mass (FM), and bone mineral content. Recent reviews of DXA and body composition (Genton et al. 2002; Kohrt 1998; Lohman, 1995; Lohman et al. 2000; Pietrobelli et al. 1996) indicate theoretical and empirical validity of this method in estimating FM and FFM. Multicomponent models have shown that DXA results (pencil beam) are closely related to body composition (Evans et al. 1999; Kohrt 1998; Prior et al. 1997; Tataranni and Ravussin 1995; Withers et al. 1998). Evaluation of fan-beam DXA is under way (Salamone et al. 2000; Schoeller et al. in press; Tylavsky et al. 2000; Visser et al. 1999). Standardization has not been established in validation studies across investigators, and there is variation in criterion methods used, in the type of DXA hardware and software instrumentation selected, and in sample characteristics. This chapter discusses underlying assumptions and their validity, applicability, hardware and software, measurement procedures, calibration, and precision and accuracy in the use of DXA to estimate both total and regional body composition.

History and Development of DXA

Before the development of DXA, single-photon absorptiometry (SPA) and dual-photon absorptiometry (DPA) were used to estimate regional characteristics. The SPA technique was developed first and used iodine-125 as the photon source. This approach enabled bone mineral estimates of both distal and proximal sites for the radius and ulna. Validation, standardization, and normative data were developed by Cameron and Sorenson (1963) and Mazess and colleagues (1972).

The development, standardization, and validation of DPA, in which iodine-125 was replaced with gadolinium-153, which has gamma emissions at both 44 and 100 keV, were described by several authors (Gotfredsen et al. 1984; Mazess et al. 1972, 1981). The use of DPA allowed estimates of bone mineral density (BMD) for the lumbar vertebrae and parts of the femur. This approach was extended to provide estimates of whole-body composition (Gotfredsen et al. 1986; Heymsfield et al. 1989; Mazess et al. 1984; Wang et al. 1989).

Limitations of DPA led to the development of DXA, in which the radioactive source is replaced by an X-ray tube with a filter to convert the polychromatic X-ray beam into low- and high-energy peaks. This new technique allows greater precision of measurement and estimation of soft tissue composition to correct for regional variation in fat content and thus provides better estimates of BMD and soft tissue composition. These advances led to the widespread use of DXA in the body composition field to estimate fat, lean tissue, and bone mineral in animals and humans.

The acronyms for body composition estimated from DXA include total body bone mineral (Mo), total bone mineral density (BMD), bone mineral–free lean tissue mass (LTM), fat mass (FM), soft tissue mass (STM = LTM + FM), and fat-free mass (FFM = LTM + Mo). For consistency, we recommend that these acronyms be used in future investigations.

Assumptions of DXA and Their Validity

Several studies have described the theoretical basis for the use of DXA in estimating soft tissue composition and bone mineral (Cullum et al. 1989; Johnson and Dawson-Hughes 1991; Kelly et al. 1988; Mazess et al. 1990; Pietrobelli et al. 1996, 1998; Roubenoff et al. 1993).

Soft Tissue Composition

The fat content in soft tissue is estimated from the assumed constant attenuation of pure fat (R_f) and of bone mineral–free lean tissue (R_l). This is the first main assumption of the DXA method. R_f is 1.21 for pure fat using the X-ray energies of 40 kV and 70 kV (Mazess et al. 1990). Pietrobelli and colleagues (1996) measured six chemical elements in 11 men by neutron activation. These authors reported a value of 1.18 for R_f, similar to the theoretical R_f calculated from various triglycerides. For lean tissue, the R_l calculated from the measured elements was 1.399 ± 0.002 (Pietrobelli et al. 1996). Given the near constancy of these two values from subject to subject, it follows that the ratio of the attenuation at the lower energy relative to the higher energy soft tissue (R_{st}), for the low- and high-energy X-rays, is a function of the proportion of fat (R_f) and lean (R_l) in each pixel. From the R_{st}, the fraction of soft tissue as lean (l_f) is given by Gotfredsen et al. (1986):

$$l_f = \frac{R_{st} - R_f}{R_l - R_f} \qquad (5.1)$$

Pietrobelli and colleagues (1996) further extended theoretical validity to DXA body composition estimates by describing the physical basis of DXA, analyses of DXA as a two-component model, and initial estimates of hydration effect on DXA lean soft tissue estimates. Limitations of DXA accuracy include beam hardening, dead time losses, and crossover, all of which have been dealt with in more recent software. Subject thickness has also been addressed and continues to be a source of variation in DXA estimates.

Because DXA provides the proportion of fat and lean in each pixel, it estimates measures of fat rather than adipose tissue, unlike computed tomography (CT) and magnetic resonance imaging. Furthermore, CT differs from DXA in that pixels are allocated to adipose tissue or to lean tissue depending on levels of Hounsfield units; pixels are not graded as mixtures of lean tissue and adipose tissue.

The fat and lean content can be calculated by solving the following two equations using known values of R_f and R_l.

$$R_{st} = f_l \times R_l + f_f \times R_f \qquad (5.2)$$

$$f_l + f_f = 1 \qquad (5.3)$$

The constancy of R_f for subjects has not been contested by most researchers because the attenuation of fat varies little among individuals (Pietrobelli et al. 1996). Pietrobelli and colleagues (1996, 1998) addressed the constancy of R_l for lean soft tissue, calculating an R value of 1.369 for the lean soft tissue of reference man using elemental analysis and neutron activation analysis (Pietrobelli et al. 1996). Pietrobelli and colleagues (1998) focused on hydration and DXA estimates of lean soft tissue effect, using both empirical and theoretical analysis.

A second assumption underlying the use of DXA is that the measurements are not affected by the anteroposterior thickness of the body. Although this appears to be the case for subject thicknesses less than 20 cm, thicknesses greater than 25 cm may have an effect. Subject thickness was investigated by Laskey and colleagues (1992) and Jebb and colleagues (1993). Both studies used in vitro simulations to document systematic effects of increasing tissue thicknesses by increasing the thickness of water (to represent fat-free soft tissue) at varying levels of fatness (oil or lard was used to represent fat). Laskey and colleagues (1992) concluded that for body thicknesses less than 20 cm, fat is overestimated by only 4% or less and fat-free soft tissue is overestimated by 2% or less. Milliken and colleagues (1996) found greater effects in the abdomen than thigh when adding a layer of lard over the subjects (DXA, pencil beam). Kohrt (1998) found a 96% detection of added fat for both thigh and trunk using pencil-beam technology (Hologic QDR 1000/w). Salamone and colleagues (2000) added 1- and 2-kg lard packets to both trunk and leg areas (4 kg in all) and found a 91% detection of added fat over the leg and a 62% detection of added fat over the trunk using fan-beam DXA technology (Hologic 4500A).

A third assumption of DXA is related to the area of the body analyzed to obtain composition data and the degree to which the fat content of the area analyzed is associated with the fat content of the area that is not analyzed. It has been estimated that 40% to 45% of the 21,000 pixels in a typical whole-body scan contain bone in addition to soft tissue, and these pixels therefore are excluded from the calculation of values for soft tissues. Thus, the extent to which the composition in the excluded area differs from the considered area is a source of systematic error for that individual. In addition, individual variation in composition between the measured and nonmeasured area is a further source of error in the estimation of total composition. Related to this problem in the estimation of total body composition by DXA is the assumption that the composition of each body region is equally represented, per unit volume, in the calculated total body values. The influence of the arm and thorax on total body composition estimates may be underrepresented because of the relatively large areas of bone in these regions (Roubenoff et al. 1993). Because the software approach to this problem is proprietary information, it is difficult to evaluate how well the composition is estimated for the unanalyzed portions.

Bone Mineral Assessment

Estimation of bone mineral mass (g), bone mineral content (g), and bone mineral area density (g/cm^2) can be obtained from DXA. In these calculations, the measurements expressed in centimeters or square centimeters are adjusted for the widths or areas, respectively, of the parts of the skeleton that are scanned. The theoretical basis for the dual-energy assessment of bone mineral measurements has been described (Heymsfield et al. 1989; Mazess et al. 1984; Peppler and Mazess 1981). Investigations have established a higher precision and accuracy for DXA than for DPA (Kelly et al. 1988; Mazess et al. 1990). Early work on the validation of the dual-energy approach using in vivo neutron activation analysis for total body calcium established a close relationship between the two methods. The calibration of one DXA instrument (Lunar DPX) was described by Mazess and colleagues (1991) using samples of calcium hydroxyapatite and an aluminum spine phantom within a water bath covered with different thicknesses of lard, paraffin, and plastic.

The effect of subject thickness on BMD estimates was investigated by Laskey and colleagues (1992) and Jebb and colleagues (1993) using phantoms as previously described. Both studies showed an increase of about 2% in BMD with an increase in subject thickness up to 28 cm. For Lunar DPX, a manual adjustment of bone edges was required to maintain accuracy when local tissue thicknesses were greater than 22 cm. Tothill and Avenell (1994) studied the effect of soft tissue thickness in four spinal models with Lunar DPX (software version 3.4), Hologic QDR (4.47p), and Stratec (2.23). The BMD values were affected by thickness, machine, and the spinal phantom used for calibration, but the magnitude of the effects of increasing thickness generally was less than 3%. Going and colleagues (1993) showed that small changes in the hydration of the FFM do not affect BMD estimates. Tothill and colleagues (1999), Tothill and Hannan (2000), and Genton and colleagues (2002) showed further discrepancies between fan-beam technologies in regional and total BMD and bone mineral content (BMC).

Applicability of DXA

DXA can be used in human populations of all ages because of the low radiation exposure. The exposure for a whole-body scan ranges from 0.02 to 1.5 mrem depending on the instrument and the scan speed. Because this exposure is less than that during one transcontinental flight across the United States (4-6 mrem), less than the 10 to 15 mrem from DPA, and much less than the typical radiation exposure with conventional X rays of 25 to 270 mrem (chest X ray, CT scan), DXA is used widely for subjects of all ages. Because some radiation is involved, DXA is not recommended for use with pregnant women, and a pregnancy test is necessary before DXA measurements are made in women of childbearing age. Special software (version 5.61, Hologic) that assumes a higher hydration of lean tissue and, therefore, a different attenuation coefficient is used for infants. Special pediatric software is also available from Lunar. Systematic error in estimating body composition in piglets and premature infants indicates the need for further validation studies in this population (Brunton et al. 1993; Picaud et al. 1996; Svendsen et al. 1993a, 1993b). Food and fluid intake have only small effects on DXA estimates of body composition (Horber et al. 1992). Similarly, when 1 to 4 kg of salt-containing fluid was removed by hemodialysis, the estimates of lean mass decreased, as expected, with little change in bone mineral content or fat mass (Horber et al. 1992).

Large subjects may not be measured accurately by DXA. For subjects taller than 193 cm or wider than the scan area (58-65 cm), a whole-body scan cannot be obtained because part of the body will be outside the scan area. Also, for thicker subjects, for those heavier than 100 kg, or for those with a value for $\sqrt{W/S}$ greater than 0.72, where W is weight (kg) and S is stature (cm), the accuracy of body composition estimates may be reduced because the attenuation coefficients for soft tissue and bone mineral depend on subject thickness, as noted earlier.

An interesting approach to the problem of large subjects was reported by Tataranni and Ravussin (1995), who used DXA to compare the composition of right and left sides of obese subjects. Because of the high association between values from the two sides, the authors recommended that, in studies of subjects wider than the scan area, scans be made of the right half of the body and total body composition be estimated assuming bilateral symmetry.

Equipment: Hardware

There are three commercial versions of DXA in the United States. Each version is based on a different configuration of hardware and software, but all assess bone mineral and fat and lean soft tissue masses. The three manufacturers refer to their instruments as QDR (Hologic, Waltham, MA), DPX (General Electric Lunar Corporation, Madison, WI), and XR (STRATEC Biomedical Systems AG, Fort Atkinson, WI). The characteristics of the latest DXA devices were summarized by Genton and colleagues (2002). A description of each commercial version is provided in the following sections.

Hologic QDR

The Hologic QDR systems were first described by Kelly and colleagues (1988). Two X-ray beams of different energies (70 and 140 kVp) are pulsed alternately. The resulting spectra have maximum photon energies at 45 and 100 keV. The QDR uses an internal calibration system with a rotating filter wheel composed of two sections of epoxy-resin-based material. At each measurement location, the beam passes through the calibration system to provide continuous internal calibration. The QDR-1000 allows for regional BMD measurements of the spine, hip, and forearm. The QDR-1000/w adds whole-body scanning to the regional assessment and allows for both BMD and body composition assessments. Each pixel covers a 1 × 1 mm area. A total body scan requires about 4 min, and the radiation exposure is 1.5 mrem. The

QDR-2000 has both pencil-beam and fan-beam configurations. The fan beam provides better resolution and precision, shorter scanning time, and the ability to make lateral scans of the spine. Recent comparisons among the QDR-1000/w, QDR-2000 pencil, and QDR-2000 fan configurations show excellent agreement among these scanner modes with low standard error of estimate (SEE) and comparable mean BMD values (Faulkner et al. 1992; Harper et al. 1992). If a change is made from QDR-1000 to QDR-2000 pencil or to QDR-2000 fan, cross-calibration should be completed using a series of phantom and patient scans on both scanners (Faulkner et al. 1992). It was found that when one device (a single-beam monodetector QDR-1000/w) is replaced by another (a fan-beam multidetector QDR-4500/A), appropriate quality control procedures must be strictly observed to ensure precision (Barthe et al. 1997).

Lunar DPX and Prodigy

The Lunar DPX system was described by Mazess and colleagues (1990). This instrument uses a constant potential X-ray source and a K-edge filter to achieve an X-ray beam of stable energy radiation of 38 and 70 keV. The X rays are emitted from a source below the subject and pass through the subject, who lies in a supine position on a table. The total scan area is 60 × 200 cm. The attenuated X rays, after passing through the subject, are measured with an energy-discriminating detector situated above the subject on the scanning arm. The DPX instrument makes transverse scans of the body at 1-cm intervals from head to toe. For each transverse scan, about 120 pixel elements yield data on the attenuation ratio, with each pixel covering a 5 × 10 mm area for the fast and medium mode (10 and 20 min) and a 5 × 5 mm area for the detail mode (20 min). The radiation exposure varies from 0.02 mrem (fast mode) to 0.06 mrem (detail mode). System hardware and software allow for lateral spinal scans in addition to anteroposterior scans and forearm scans (DPX-L).

Recently, Lunar Corporation jointed GE and became GE Lunar Corporation. From December 1997 to May 2002, a serial of Lunar pencil-beam bone densitometer system was developed including DPX-NT, DPX-IQ, DPX-MD, DPX-MD+, DPX–, DPX+, and DPXA. An improvement for DPX-NT is that patient thickness determines the appropriate measurement mode. The program selects the appropriate mode (thick, standard, and thin) based on the patient's height and weight.

As for DXAs from other companies, calibration studies should be conducted when Lunar DXA is upgraded, because calibration may change long-term precision for follow-up measurements. DPX data files should be reanalyzed with the new software of DPX-IQ so that longitudinal changes in BMD can be accurately assessed (Wong and Griffiths 2002).

The first generation of Lunar fan-beam DXA is known as the Expert. The fan-beam angle of Expert is 12°. The Lunar Prodigy DXA machine is a new generation of fan-beam machine. It employs a cadmium–zinc–telluride detector, which is an energy-sensitive material that directly converts X rays into an electronic signal without the intermediate conversion to light as in previous fan-beam machines. Because of the direct conversion, smaller radiation exposure is used in Prodigy to produce high-precision measurements. The patient and operator radiation exposure is 10 times smaller than that of the previous generation of fan beam. It has been noticed that a wide-angle fan beam (30°) may be subject to magnification errors (Geriffiths et al. 1997). The angle of the Prodigy fan beam is narrow (4.5°), which may reduce magnification errors and is less sensitive to differences in patients' positions for repeat measurements (Mazess et al. 2000). In addition, Prodigy uses a new method of fan-beam acquisition with multiple, overlapping passes over the region of interest, which further reduces magnification errors.

Mazess and Barden (2000) recently compared body composition measurements from two fan-beam geometry DXA machines (Expert and Prodigy) and one pencil-beam geometry DXA machine (DPX). Overall, fan-beam DXA was faster than the pencil-beam DXA. Although radiation exposure was higher in fan-beam machines than in pencil-beam machines, the difference in radiation exposure was very small between Prodigy and DPX. Body composition and BMD measurements were closely correlated between fan-beam machines and the pencil-beam DXA. The means between Prodigy and DXA differences were generally small (<1% relative to the mean value) for both BMD and body composition variables.

Stratec System

The Stratec Biomedical (formerly Norland XR system), described by Clark and colleagues (1993) and by Nord and Payne (1995), uses a K-edge filter (samarium) to produce two photon peaks at 44 and 100 keV. For the XR-26 model, the pixel size is 6 ×

13 mm, and the total body scan is made in about 20 min with a radiation exposure of 0.05 mrem. Both total body and regional BMD can be obtained. The XR-36 DXA system includes a unique feature called "dynamic filtration," which varies the X-ray intensity by patient thickness. Thus, the hardware of this instrument compensates for the effects of tissue thickness on the precision and accuracy of the measurements. This approach to compensate for tissue thickness needs to be evaluated in comparison with the software of other instruments that has been updated to achieve a similar purpose. The XR-36 DXA system has a scanning area of 193 × 64 cm and allows analyses of total bone mineral and body composition, regional BMC, forearm BMD, and small subject scanning. Regional body composition analyses can also be made using software versions 2.4 and 2.5. The XR-46 and Excell Plus are the latest pencil-beam DXA machines of Stratec Biomedical Systems (Genton et al. 2002).

Equipment: Software

The software versions for each commercial unit have been upgraded many times during the past 5 years. Each software version is based on specific assumptions to estimate BMD and body composition for the total body and for body regions. It is essential that researchers identify the software versions used in their studies so readers can evaluate the findings and the performance of the upgrades. In longitudinal studies, it may be important to repeat the analyses with the most recent software versions. The following paragraphs and the sidebar on page 68 briefly describe the software updates for each company.

The Hologic software versions 5.0 to 5.64 have undergone several changes to increase the precision and accuracy of body composition analysis. In addition, Hologic has developed a standard whole-body analysis and an enhanced whole-body analysis for recent software versions (5.53-5.64). Hologic recommends using the enhanced whole-body analysis for subjects with large changes in body mass. All the Hologic fan-beam whole-body measurements use the enhanced analysis. Hologic recommends the standard analysis for scanning pigs, dogs, and certain whole-body phantoms.

Hologic software version 8.21 was designed by the manufacturer to better address magnification effects on estimations of soft tissue lean mass. A comparison study with 20 weight-stable, obese, postmenopausal, Caucasian women aged 40 to 70 years was conduced to investigate the long-term

Software Versions for DXA Instruments

Lunar		Hologic
DPX Series (DPX, DPX-L, DPX-A, DPX-alpha, DPX-SF)	**Prodigy and DPX-NT and MD+**	**(QDR-1000/w and QDR-2000)**
3.0 Original version	1.10 Original	5.39 to 5.47 Early version of body composition
3.1 Corrected for artifacts; provides BMD in thin tissue	2.00 Total body dual-image (skeletal vs. soft tissue)	5.48 to 5.54 Whole-body and enhanced whole-body; corrected for tissue thickness and large weight change
3.2 Further corrections for artifacts and anomalous points	2.10 No change in body composition measurements	5.55 Rats: whole-body
3.4 Changes for spine and femur	2.15 Forearm and lateral spine measurements	5.56 Infant: whole-body
3.4R Corrections for tissue thickness, high-density metal	2.16 Corrections for total body thickness (increased number of tissue thickness points)	5.57 Whole-body analysis 5.60 Forearm analysis 5.64 Enhanced whole-body analysis (released May 1993)
3.5 Updated version of 3.4 and 3.4R	2.17 No change in body composition	5.65 to 5.67 No changes in body composition
3.6 (1.3y) Flexible cuts for different regions, high-resolution mode, tissue thickness corrections, forearm and lateral spine analysis (released August 1992)	2.20 Female body composition reference data (a population assembled from all of the published studies on body composition we could find at that time) and measurements for small animals	5.73 Support for fan-beam and pencil-beam WB exams **QDR-4500 A/W (DDS)** 8.1a Early versions without correction for magnification 8.21 Corrections for magnification released for research studies
3.65 Year 2000 compliant; no changes in body composition measurements	2.25 No changes in body composition	8.26 Mass magnification solution implemented
DPX-IQ and MD	3.00 Male body composition reference	**QDR for Windows (Delphi and QDR 4500 A/W)** 11.2 High-power whole-body released for obesity research
4.1d Original	3.50 More tissue thickness reference points and better algorithm for cut placement	**QDR for Windows XP (Discovery A/W)** 12.0 Support for standard ROIs and user-defined ROIs for bone and body composition analysis
4.1g No changes in total body 4.3b Year 2000 compliant 4.3c No changes to total body		
4.5 Support of IQ NET 4.6b No changes to total body	3.60 Imaging of lumbar and thoracic vertebrae looking for vertebral deformities	**Stratec System** 2.2 & 2.3 Whole-body composition; regional BMD
4.6c Additional non-English languages supported	4.00 No change in body composition	2.4 & 2.5 Recalibrated fat–lean system; regional body composition added with flexible cuts (released March 1993)
4.6d Enhanced reference graphs	5.00 No change in body composition	
4.6e Ability to choose different reference populations for different scan types	6.00 No change in body composition	
4.6f Additional non-English languages supported	6.10 No change in body composition	
4.7 Dual-femur scan type 4.7a No changes in total body	6.50/6.60 Allows user to enter custom reference population values into software	
4.7b New pediatric reference graph		
4.7c WHO graphs for females all regions	7.00 No change in body composition	
4.7d Support of nonstandard English characters in file names		

Note. BMD = bone mineral density; WHO = World Health Organization; WB = whole body; ROI = region of interest; NET = network.

and short-term precision of a fan-beam Hologic QDR-4500A absorptiometer and the comparability of software versions 8.1a and 8.21. The results suggested that reproducibility of all variables was comparable between software versions. However, software version 8.21 yielded smaller percentage mean differences between scale and DXA-estimated weights and higher fat and lean weights than version 8.1a (Cordero-MacIntyre et al. 2000b).

Lunar software versions 3.1 through 3.6 provide similar mean body composition values (see the sidebar on page 68). The more recent software versions (3.4-3.6), however, provide different body composition estimates for amenorrheic subjects and obese subjects with body thickness greater than 20 cm. Version 3.6 allows for more flexible boundaries for regional analysis of body composition. The DXA series final software version is 3.65; the major change was to make it Year 2000 compliant. The DXA series is still supported by the company, but no further development is occurring. The DPX-IQ contains software versions 4.1 to 4.7. The manufacturer recommends that all users of the DPX-IQ update to version 4.7 to take advantage of all corrections and enhancements. The DPX-IQ series is still supported, but no further development is occurring (see the sidebar on page 68).

No new functionality has been added since the introduction of Prodigy in late 1999. The current software platform for Prodigy (Prodigy Oracle and Prodigy Vision) and DPX-NT/MD+ systems is enCORE. The software version and enhancements in each version are listed in the sidebar on page 68. The major change for version 2.00 was adding dual-femur and total body dual image. Version 2.10 was released corresponding to the introduction of DPX-NT. Version 2.15 has forearm and lateral scan features. Version 2.16 added correction for total body thickness. There are no changes for body composition measurement in version 2.17, and version 2.20 has total body composition reference for female and features for small animal measurement. Male body composition references were added in version 3.00 (see the sidebar on page 68).

Stratec's earlier software versions 2.2 and 2.3 allowed for whole-body composition and total regional BMD measurements (see the sidebar on page 68). Software versions 2.4 and 2.5 were recalibrated for fat and lean with a new set of standards (Nord and Payne 1995). A new fat distribution model was used and options were introduced to allow regional analysis of body composition.

Measurement Procedures

The measurement procedures for assessing bone mineral and body composition are similar for each instrument (Hologic, Lunar, Stratec). In general, all body composition measurements are made in the anteroposterior position. A series of transverse scans is made from head to toe of the subject at 0.6- to 1.0-cm intervals over the entire scan area. The whole-body scan takes 5 to 30 min depending on the instrument. Special subject preparation and requirements for measurements in the fasting state are not needed to obtain accurate results.

Careful positioning is essential to achieve reproducible regional bone mineral assessments. For precise measurements of regional BMD, separate scans of the spine (anteroposterior or lateral), femur, and forearm are conducted. BMD of the distal or ultradistal radius shaft can be measured by DXA with Hologic QDR, Lunar DPX, or Stratec XR.

Training in DXA operation, positioning of subjects, and data management are essential to obtain precise and accurate data from DXA, and this training is provided by each company along with a standard operating manual. In some states in the United States, DXA measurements must be made by a licensed X-ray technician.

Calibration Procedures

The calibration of DXA involves the use of different standards and assumptions depending on the company and software version that is applied. Mazess and colleagues (1990) used water, lard, Delrin, and mixtures of water and isopropyl alcohol to simulate various mixtures of fat and lean tissues. Also comparisons were made with DPA, which had been calibrated with a series of grade mixtures of lean beef and fat (Wang et al. 1989). The calibration procedure for bone mineral was described by Mazess and colleagues (1991). The calibration resulted in the following formula: % fat DXA = 500 ($1.40 - R_{st}$) (Hansen et al. 1993).

The equations used by Hologic and Norland have not been published. Both companies used a mixture of 0.6% NaCl in water solution and stearic acid to simulate pure lean tissue and fat, respectively (Nord and Payne 1995). With mixtures of these substances, any DXA instrument can be calibrated to derive an equation relating R_{st} to percent fat. The authors (Nord and Payne 1995) found equivalent results for the three instruments when 0.8% NaCl was used. More recent calibration procedures comparing various hardware and software combinations were outlined by Genton and colleagues (2002).

Precision of Total and Regional Composition Estimates

The precision of total and regional body composition estimates from DXA can be evaluated over the short term (within and between days) and long term (during months within individuals) for bone mineral, body fat, and lean soft tissue. Precision is less for regional body composition assessment than for whole-body assessments.

The short-term precision of total body composition assessments by DXA is usually evaluated by repeatedly scanning subjects and calculating an intraindividual standard deviation for each subject. The standard deviations are then pooled to obtain a more reliable estimate. In general, the standard deviation is about 1.0% for percent fat, which is comparable to the precision of other methods (e.g., body density and bioelectric impedance).

In general, it is recommended that percent fat be included in future studies of the precision of DXA hardware and software and that both the standard deviation and coefficient of variation be reported. For BMC, BMD, and LTM, coefficients of variation for short-term precision may vary from 0.6% to 1.6%, with about 1% being a representative value for each of these variables.

In a short-term precision study, 20 volunteers were scanned once each day for 4 consecutive days with a Lunar DPX-L DXA scanner and manufacturer-supplied software (version 1.3z). Coefficients of variation (CV) derived from data using the (preferred) extended research mode of analysis were 0.62%, 1.89%, 0.63%, 2.0%, 1.11%, 1.10%, and 1.09% for total body bone mineral density (BMD), total percentage fat, total body tissue mass, fat mass, lean mass, bone mineral content, and total bone calcium, respectively. Small but statistically significant differences in mean values for most body composition variables were found when data were compared between extended and standard modes of analysis. The investigators suggested that inconsistent use of analysis mode in a cohort or when a patient is followed longitudinally may negatively affect precision (Kiebzak et al. 2000).

More recent studies of reliability with fan-beam DXA technology show a similar level of precision (Cordero-MacIntyre et al. 2000b, 2002).

Long-term precision of posteroanterior lumbar spine, femoral neck, and total hip BMD measurements in 40 postmenopausal women was evaluated. Despite the good long-term precision (SEE/mean × 100%, where SEE = standard error of estimate) for lumbar spine BMD (1.1-1.6%), femoral neck BMD (2.2-2.5%), and total hip BMD (1.3-1.6%), the study showed that obesity may have an adverse effect on precision errors in individual patients. Hence, particular care is necessary to ensure reproducible patient positioning for femur scans (Patel et al. 2000).

Regional body composition assessments are somewhat less precise than whole-body assessments. Although there is a good agreement between these reports in the precision found for fat, lean mass, and BMD in most regions, the fat CV found by Fuller and colleagues (1992) was considerably less than that reported by Mazess and colleagues (1990) for legs and trunk but not for the arms (both used Lunar DPX). Kiebzak and colleagues (2000) also found that regional measurements (arm, leg, trunk, pelvis, and spine) were less precise than total body measurements, with CVs of 1% to 3% for total body, 4.3% for fat mass for arms, and 3.1% for trunk. When regional body composition changes are estimated, it is recommended that DXA be assessed two times within each measurement period (within 7-10 days) to decrease measurement error.

Accuracy of DXA for the Measurement of Fatness

The prediction of body fatness from DXA can be evaluated in several ways. First, there may be systematic errors in comparing DXA estimates with those from another reference method, and the magnitude of these systematic errors may vary with different populations and with different DXA instruments. The systematic error may be attributable to inaccuracy of the reference method, inaccuracies of the DXA estimates, or a combination of these. Second, the SEE for percent fat should be less than 3% for a new method to be accepted as an accurate reference method. Errors exceeding 4% show too much variability, and errors between 3% and 4% show limited validity (Lohman 1992). Many methods tend to overestimate fatness in lean populations and underestimate fatness in the obese. Animal studies allow validation of a new method against chemical analyses of carcasses. Third, new software has to be tested to determine if the prediction of fatness is improved. Multicomponent models are essential to validate DXA.

Human Validation Studies

If the two methods have theoretical validity and are calibrated correctly to estimate percent fat, then the regression line relating the values from the two methods should have a slope equal to 1.0. A slope that deviates markedly from unity shows that a unit change in the DXA value does not correspond to a unit change in the reference value. In addition, a SEE between 2% and 3% and a systematic bias between criterion and DXA methods less than 2% are both essential. Finally, the error between methods must remain uncorrelated by the mean value (Altman and Bland 1983). A summary of recent validation studies by Lohman and colleagues (2000) indicated that DXA and multicomponent models are in agreement in terms of mean values (table 5.1). In general, there is agreement with pencil-beam DXA and percent fat within 1% to 3% (Evans et al. 1999; Kohrt 1998; Withers et al. 1998). An underestimation of FM using fan-beam DXA (Hologic 4500) was found by Salamone et al. (2000).

One of the most important validation studies compared DXA with CT and neutron activation analysis for measurement of skeletal muscle mass (Wang et al. 1996). The CT procedure required 22 cross-sectional images at specific locations and could be predicted with a SEE of 1.6 kg (r^2 = .91) from DXA in contrast to a SEE of 4.4 kg with neutron activation analysis used to predict CT skeletal muscle mass. Further developments in the area of skeletal muscle assessment by DXA were published by Wang and colleagues (1999) and Kim and colleagues (2002) and show greater promise as a measure of muscle mass as well as percent fat.

With the development of fan-beam DXA technology (Hologic 4500A and Delphi Lunar, Expert, and Prodigy), whole-body composition and bone density scans can be completed in much less time with equal or improved precision (Kelly et al. 1988; Nord et al. 2000).

Only a limited number of validation studies have been conducted with fan-beam DXA and a multicomponent body composition criterion method. Because a national probability sample has used DXA fan-beam technology (National Health and Nutrition Examination Survey IV) and

because fan-beam technology is being used more and more in clinical trials, fan-beam technology is preferable. Thus, validation studies are important to determine if biased estimates are obtained with fan-beam technology, to evaluate the standard error of estimate using a multicomponent criterion method in various populations from pediatric to the aged, and to establish the independence of the error from the magnitude of body composition estimates (Bland–Altman approach).

For Hologic 4500A, compared with both the 4C model and body water, research has shown that even with modification of software (Hologic software version 8.21) to minimize effects of magnification (Cordero-MacIntyre et al. 2000a, 2000b), there is a systematic overestimation of FFM (Schoeller et al. 2004; Tylavsky et al. 2003; Visser et al. 1999).

Tothill and colleagues (1999, 2000) conducted studies with fan-beam technology using the Lunar Expert versus Hologic 4500A, finding a number of differences in regional as well as total body composition. Preliminary data with Lunar Prodigy compared with DPX-IQ indicate general agreement in body composition and BMC (Nord et al. 2000). Genton and colleagues (2002) recently compared body composition studies with DXA pencil beam and fan beam; in 11 studies between 1993 and 1998, the authors found differences in BMC, FM, and FFM between various hardware and software systems ranging from 0.5% to 19.4%, with most differences between 1% and 5%. The work of Ellis and Shypailo (1998) provides evidence that differences between pencil beam and fan beam for children are smaller than the differences for adults, emphasizing the importance of separate calibration and validation studies in children versus adults. There are limited data on the use of fan-beam DXA (Prodigy and QDR- 4500A) and multicomponent models in children.

Animal Validation Studies

Several animal validation studies have been published using Hologic and Lunar instruments. Svendsen and colleagues (1993a), using Lunar DPX (3.2) in seven pig carcasses ranging in weight from 35 to 95 kg, found a mean difference in fat content of 2.2% between carcass chemical analysis and DXA. The SEE for the regression analysis was 2.9%.

Picaud and colleagues (1996) assessed the reproducibility, accuracy, and precision of DXA (QDR-2000) by scanning 13 piglets. These authors found that body weight was measured accurately,

Table 5.1 Mean Percent Fat From Dual-Energy X-Ray Absorptiometry Versus Other Reference Methods

	N	Sex	Mean age	Age range	POPULATIONS					MEAN % FAT (DXA)	
					4-comp Heyms[a]	4-comp Baumg[b]	3-comp Siri,UW[c]	2-comp Siri, UW[d]	2-comp Brozek, UW[e]	DXA	DXA hardware
Prior et al. 1997	81	F	20.7 ± 2.6	—	22.3 ±7A			22.7 ± 1.7		22.5 ± 8.4	Hologic QDR-1000
	91	M	21.2 ± 2.1		12.5 ± 5.9			14.4 ± 5.4		13.1 ± 5.7	
Kohrt 1998		F	20-39						21.4 ± 5.1	24.2 ± 6.6	Hologic QDR-1000
	225	F	40-59	21-81					28.4 ± 6.3	31.1 ± 7.0	
		F	>60						38.5 ± 5.2	40.5 ± 7.3	
		M	20-39						14.5 ± 5.6	12.9 ± 6.5	
	110	M	40-59						22.2 ± 5.9	21.2 ± 7.4	
		F-M	>60						27.5 ± 5.7	25.9 ± 7.5	
Evans et al. 1999[1] C.1[f]	9	F	34.7 ± 5.8	24-40		43.0 ± 4.1				47.2 ± 5.0	Hologic QDR-1000
D.1[g]	9	F	32.0 ± 6.9	22-40		41.1 ± 4.5				42.8 ± 4.7	
D+E.1[h]	9	F	29.2 ± 6.6	21-39		43.2 ± 3.4				44.6 ± 3.9	

					Fat mass (kg)			Fat mass (kg)	
Salamone et al. 2000	30	M	73.9 ± 2.2		22.3 ± 7.3			21.6 ± 6.6	Hologic 4500
	30	F	73.6 ± 2.3	70-79	26.3 ± 8.4			25.5 ± 8.0	
	60	Total	73.7 ± 2.2		24.2 ± 8.0			23.5 ± 7.5	
Withers et al. 1992	12	M	22.3 ± 5.1	—	12.1 ± 2.8	12.0 ± 2.8	9.8 ± 3.6	8.6 ± 2.8	Lunar (1.3Z)
	12	M	24.7 ± 4.5		21.8 ± 8.2	21.7 ± 8.1	19.5 ± 8.3	20.5 ± 9.3	
	12	F	23.8 ± 5.7		16.4 ± 2.4	16.2 ± 2.3	13.6 ± 2.3	15.1 ± 3.0	
	12	F	22.4 ± 2.8		28.9 ± 4.7	28.4 ± 4.7	26.6 ± 5.9	28.5 ± 4.9	

Note. DXA = dual-energy X-ray absorptiometry; comp = component; UW = underwater weighing; F = female; M = male.

[a]The four-component model using the Heymsfield formula (Heymsfield et al. 1990a). [b]The four-component model using the Baumgartner formula (Baumgartner et al. 1991). [c]The three-component model using underwater weighing with the Siri equation (Siri 1961). [d]The two-component model using underwater weighing with the Siri equation (Siri 1961). [e]The two-component model using underwater weighing with the Brozek equation (Brozek et al. 1963). [f]The control group at baseline. [g]Diet + exercise intervention group at baseline.

but fat content was overestimated by DXA. Nevertheless, DXA estimates of body weight, BMC, and fat content were significantly correlated with scale body weight, ash weight, chemical calcium, and chemical fat (r was in the range of .955-.999) and gave excellent reproducibility.

Brunton and colleagues (1993) used Hologic QDR-1000/w and the pediatric whole-body software (6.01) in small piglets (mean weight 1.6 kg). In comparison with chemical analyses of carcasses, DXA overestimated fat by 100%; lean estimates were within 6% agreement, and bone mineral was underestimated by 30%. In larger piglets (mean weight 6.0 kg), DXA overestimated fat by 36%, but there was excellent agreement for lean mass and bone mineral.

A recent study that used a piglet model showed that fan-beam DXA provided accurate and precise body composition measurements for piglets, suggesting that rapid scan acquisition by fan-beam DXA can be used in studies with growing humans and small animals (Winston et al. 2002). Using rats, Makan and colleagues (1997) compared DXA-derived body composition measurements with measurements from carcass analysis. These authors found good precision and high levels of agreement between body composition measurements from DXA and chemical analysis.

Accuracy of body composition measurement by DXA was compared with direct chemical analysis in 10 adult rhesus monkeys. This study confirmed the accurate measurement of fat and lean tissue mass by DXA in rhesus monkeys. DXA also accurately measured lumbar spine BMC but underestimated total body BMC (Black et al. 2001).

Lukaski and colleagues (2001) compared DXA-derived body composition with chemical composition of 49 male, weanling Sprague–Dawley rats that were fed different types of diet for weight loss. DXA significantly underestimated body mass (1-2%) and fat-free and bone-free mass (3%) and significantly overestimated fat mass and body fatness (3-25%). The greatest measurement errors of DXA were found in diet restriction groups, in which body mass was diminished and body hydration was decreased.

Accuracy of DXA Regional Body Composition Assessments

Going and colleagues (1990) also assessed abdominal fat by DPA at L2-4 combined with an abdominal skinfold thickness and found higher correlations between DPA-estimated abdominal fat and truncal skinfold thicknesses than for extremity skinfold thicknesses. These authors also reported high precision for abdominal fat from DPA.

Lohman (1992) suggested that DXA abdominal fat, in combination with skinfold thicknesses, could be used to estimate intra-abdominal fat. Svendsen and colleagues (1993c) validated DXA estimates of abdominal fat using CT as a reference method. These authors found that DXA measures of abdominal fat accounted for 80% of the variance in intra-abdominal fat by CT and that the combination of trunk skinfold thicknesses and abdominal fat by DXA accounted for 91% of the variation (CV = 14.8%) in CT intra-abdominal fat. Additional studies are greatly needed to compare DXA abdominal fat with intra-abdominal fat measured by magnetic resonance imagery as independent risk factors for chronic diseases.

Limb composition measures from DXA have been validated. Fuller and colleagues (1992) reported moderate correlations between anthropometric estimates of limb composition and volume estimates from DXA. Heymsfield and colleagues (1990b) used DPA to estimate extremity skeletal muscle mass and compared these estimates with those from anthropometry and estimates of whole-body skeletal muscle from total body potassium and nitrogen. The work of Wang and colleagues (1999) and Kim and colleagues (2002) provided additional validation of DXA as a measure of skeletal muscle mass. Baumgartner and colleagues (1998) proposed DXA FFM in relation to height as a measure of sarcopenia, and further validation studies are needed to refine this approach.

Estimating Body Composition Changes

Important to the accurate estimation of body composition changes using DXA is the comparison of body weight with DXA sum of parts at baseline and after the intervention as a quality control measure. Lohman and colleagues (2000) reviewed failure to obtain closed agreements in the mean difference and standard deviation of the difference. Mean agreement should be within 1 kg with a standard deviation less than ±2 kg at baseline and again after intervention (Houtkooper et al. 2000). It is essential to show this agreement between body scale weight and sum of DXA parts as a part of the quality control of DXA data, and all future investigations, both DXA validation studies and studies with DXA body composition changes, should include this comparison.

The accuracy of estimating fat, lean, and bone mineral changes with significant weight gain or loss is a more difficult question. In a study of in vivo weight loss in an obese sample, Cordero-MacIntyre and colleagues (2000a) found little change in BMC or BMD when using a Hologic QDR-4500A DXA scanner. Van Loan and colleagues (1998), using the Lunar DPX, found a reduction of 1.4% in BMD and no change in BMC. For changes in fatness, Evans and colleagues (1999) compared DXA with a multicomponent model in different weight loss interventions and found mean percent fat agreement to vary between DXA and a multicomponent model from 0.3% to 2.0% within the range of sampling error. Tylavsky and colleagues (2000) estimated similar changes in soft tissue using both pencil and fan beam DXA.

Hydration Status As a Confounder in DXA Body Composition Estimates

Thomsen and colleagues (1998) found that drinking water resulted in significantly increased values for tissue and lean tissue mass from DXA, which corresponded to water intake. From a theoretical standpoint, a decrease in hydration level should increase the elemental content (e.g., Na, K, Cl), leading to a higher molecular weight of the LTM and changing its attenuation coefficient. In the study by Going and colleagues (1993), body weight decreased 1.5 kg because of dehydration. The authors found that 98% of the change in weight with dehydration was measured by DXA. The measures of bone mineral and fat were unaffected by changes in hydration in this study. Thus, normal variations in hydration (1-3% of body weight) have little effect on the ability of DXA to detect body composition changes.

Reviewing the recent literature with water and multicomponent models, Lohman and colleagues (2000) concluded that the water content of FFM in healthy adults is between 72% and 74.5% with a standard deviation of 3% (2-4% for most investigations). Given this level of variation, questions have been raised about the effects of hydration level on DXA estimates of lean and fat tissue (Roubenoff et al. 1993). The theoretical work of Pietrobelli and colleagues (1996) and empirical work of Going and colleagues (1993), Prior and colleagues (1997), and Evans and colleagues (1999) all indicate that an increase or a decrease in hydration levels of 5%

biases DXA estimates of percent fat only 1% to 2.5% (Lohman et al. 2000). Thus, hydration level is not a major source of variation in DXA body composition estimates in the normal healthy population.

Application of DXA to Pediatric Populations

Because of the lower amount of radiation exposure and because of the changes in fat-free body composition during growth and development (Lohman 1986, 1989, 1992), DXA provides an ideal method to assess body composition in children and youth and to track changes in body composition over time. Unlike the assumption for the two-component model (body density, body water, bioelectric impedance analysis, and anthropometry), DXA estimates of bone mineral, soft tissue, and fat are not greatly affected by the hydration level of FFM. Theoretical analyses in adults (Pietrobelli et al. 1998) and in children from infancy to 10 years of age (Testolin et al. 2000) have shown that the natural variation in FFM hydration alters DXA estimates of percent fat less than 1% during growth and development.

Ellis and colleagues (2000) developed reference models for children (Reference Child and Adolescent Model) using DXA in a population of black, white, and Hispanic children from 5 to 19 years of age. In general, results confirm older FFM and FM estimates of Fomon and colleagues (1982) and Haschke (1989) using indirect estimates of body composition from the literature.

The use of DXA in children also provides an ideal method for confirming the previous results of Fomon and colleagues (1982), Haschke (1989), and Lohman (1986, 1992) on the changes in hydration and bone mineral content of FFM with age. Although it is generally recognized that higher FFM hydration levels and lower bone mineral levels in prepubescent children limit the accuracy of two-component models using underwater weighing and air displacement plethysmography (Fields and Goran 2000; Fields et al. 2002; Goran et al. 1996; Roemmich et al. 1997; Wells et al. 1999), use of multicomponent models in children's body composition studies is not universal (Lockner et al. 2000; Nunez et al. 1999). Using DXA in children allows for direct measurements of bone mineral content, lean soft tissue, and fat and thus offers an alternative to the four-component model in this population; furthermore, DXA can estimate FFM changes in water and mineral from prepubescence to postpubescence.

DXA Use in Multicomponent Models

Multicomponent models are discussed in chapter 12. Selinger's four-component model (1977), discussed by Lohman (1986, 1992) and Lohman and Going (1993), uses bone mineral (Mo) and estimated total body mineral (M) in the same multicomponent equation:

$$\%\text{fat} = (2.747/D_b - 0.714\ TBW/BW + 1.146\ Mo/BW - 2.0503) \times 100$$
$$\text{(Lohman 1986)} \qquad (5.4)$$

$$\%\text{fat} = (2.749/D_b - 0.714\ TBW/BW + 1.146\ M/BW - 2.0503) \times 100$$
$$\text{(Lohman and Going 1993)} \qquad (5.5)$$

where M = total body mineral.

Lohman and Going's (1993) equation is derived from the more general equation of Selinger (1977):

$$\%\text{fat} = (\frac{2.747}{D_b} - 0.714\ TBW/BW + 1.129\ Mo/BW + 1.222\ Ms/BW - 2.0503) \times 100$$

where Ms = soft tissue minerals.

Bartok-Olson and colleagues (2000) indicated that M is the correct term (total body mineral) and not Mo as in equation 5.4. Thus, equation 5.4 is not recommended and equation 5.5 is recommended.

In a thorough review of soft tissue mineral models, Wang and colleagues (2002) published a new model for estimation of Ms based on TBW and extracellular water, avoiding the assumptions made by Brozek and colleagues (1963) and Selinger (1977) and using neutron activation analysis. Results indicate less bias in soft tissue mineral estimates and thus offer an alternative multicomponent model:

$$\%\text{fat} = (2.748/D_b - 0.699\ TBW/BW + 1.129\ M/BW - 2.051) \times 100 \qquad (5.6)$$

where M is Mo + Ms. Wang and colleagues explore these and other models in detail in chapter 12.

If the multicomponent model calls for bone mineral directly such as that recommended by Withers and colleagues (1992), then no conversion is needed and DXA bone mineral content can be used directly.

$$\%\text{fat} = (\frac{2.513}{D_b} - 0.739\ TBW/BW + 0.947\ Mo/BW - 1.790) \times 100 \qquad (5.7)$$

A final problem in the use of multicomponent models is the type of DXA (hardware and software) used to estimate bone mineral content. For example, Modlesky and colleagues (1996) compared the Hologic 1000X with the Lunar DPX and found a 10% difference in BMC. Genton and colleagues (2002) compared several estimates of BMC among instruments and found considerable variation among instruments. In general, it is well established that Hologic estimates of BMC (1000X, 2000, 4500A) are lower than Lunar estimates (DPX, Expert, Prodigy) and that a difference of 10% in Mo/BW or M/BW is associated with about a 1% fat difference in a four-component model.

The conversion of DXA bone mineral to total body mineral has been expressed as 1.22 × DXA bone mineral (Lohman 1992), 1.23 (Tylavsky et al. 2003; Visser et al. 1999), 1.235 (Kopp-Hoolihan et al. 1999), 1.25 (Salamone et al. 2000), 1.274 (Fuller et al. 1992), and 1.279 (Baumgartner et al. 1991; Heymsfield et al. 1990b). There are also multicomponent models based on bone mineral and bone ash from DXA converted to bone mineral that use 1.0436 (Roemmich et al. 1997 used 1.14; Withers et al. 1998). The total body mineral models are reviewed in chapter 12.

Steps Toward Standardization of DXA

Genton and colleagues (2002) reviewed several studies comparing different DXA technologies and showed that between DXA instruments, differences can be as high as 19% for BMC, 7% for percent FM, and 4% for FM and LTM. Part of the observed difference between instruments is attributable to different boundaries being used for the same region, depending on the software version, hardware (such as fan beam vs. pencil beam), and company. Previous research with pencil-beam DXA by Van Loan and colleagues (1994), Modlesky and colleagues (1996), Korht (1998), and Tothill and colleagues (1999) indicates a lack of agreement among different hardware and software DXA results. Comparisons are needed among fan-beam technology in various populations. Documenting these differences, and validating the results against data from a multicomponent model, could lead to further standardization of software among the instruments.

Recommended Procedure for DXA Measurements

1. Always compare the sum of DXA parts (total DXA weight) and scale weight to assess quality control of DXA measurements.
2. Use the mean of two scans when measuring regional composition.
3. Use the right side of body when the subject is too large for the table.
4. Always cite software version and hardware version when publishing DXA data.
5. In multicomponent models using DXA for bone mineral, use equation 5.5, 5.6, or 5.7.

Summary

Several aspects of DXA methodology have been reviewed in this chapter. There is general agreement that DXA offers a precise method to estimate aspects of body composition, including lean soft tissue, appendicular muscle mass, FM, and BMD. Regional body composition can also be estimated with adequate precision using DXA (although the precision is somewhat less than that for total body composition), and assessment of abdominal fat offers an additional advantage of this method.

When percent body fat from DXA is used to predict percent body fat obtained by a reference method, the SEE values are generally similar to or lower than the values obtained when other traditional methods are used to predict values from a reference method. When DXA percent fat is compared against percent body fat using a multicomponent method, the SEE values typically range between 2.5% and 3.5%.

Factors that determine the accuracy of DXA include subject thickness, subject size, calibration procedures, software version, and instrument, company, and model. The recent development of fan-beam technology adds an additional source of variation. Until new approaches are validated in the same sample against data from a multicomponent model, it is difficult to recommend a standard procedure, approach, or instrument. Validation studies are especially needed to determine the accuracy of fan-beam DXA in pediatric samples and in weight loss studies. Recent studies in the elderly, such as the Health ABC study, show great promise for DXA as a reference method for studying changes in body composition. Also, studies with CT show that DXA provides an excellent measure of appendicular skeletal muscle mass.

Because of widespread use of DXA to estimate total and regional body composition changes with age, exercise, and diet, it is essential that changes in body weight and fat distribution not affect the accuracy of the estimates. Further studies are needed with fan-beam DXA to determine the effects of changes in fat distribution and body weight on whole-body and regional estimates of FM, FFM, Mo, and BMD.

DXA offers precise estimates of body composition. It can also be used in a multicomponent approach with other methods or by itself to yield estimates of bone mineral, lean soft tissue, and fat mass for regional and total body composition. DXA is becoming an accepted reference method to estimate body composition. Future studies are likely to clarify the limitations of its accuracy and lead to more standardized approaches in all populations.

Chapter **6**

Bioelectrical Impedance Analysis

Wm. Cameron Chumlea and Shumei S. Sun

The principle underlying the use of bioelectrical impedance for assessing body composition is the relationship of body composition to the water content of the body (Hoffer et al. 1969; Jenin et al. 1975; Nyboer 1959, 1970; Thomasset 1962). Like all body composition methods, bioelectrical impedance depends on several static assumptions and dynamic relationships regarding electrical properties of the body; its composition, hydration, and density; and the age, race, sex, and physical condition of those assessed (Baumgartner et al. 1990; Kushner 1992; Lohman 1986). This chapter reviews briefly the development of bioelectrical impedance and its application to research and clinical studies and identifies areas for further study. We start with the physical properties of bioelectrical impedance followed by a short historical review. The relationships of impedance to total body water (TBW), intra- and extracellular water (ICW, ECW), and fat-free mass (FFM) are discussed, followed by the present state of epidemiological, clinical, and commercial applications of bioelectrical impedance analysis.

Physical Electrical Properties

Impedance is the frequency-dependent opposition of a conductor to the flow of an alternating electric current. Impedance is determined by the vector relationship between resistance (R) and reactance (Xc) measured at a current frequency according

to the equation $Z^2 = R^2 + Xc^2$ as demonstrated in figure 6.1. Resistance is the pure opposition of the conductor to the alternating current, and reactance is the dielectric component of impedance. The values of resistance and reactance depend on the frequency of the electric current (figure 6.1). At low frequency, impedance equals resistance and reactance is zero. As the current frequency increases, reactance occurs if there are multiple current pathways within the conductor and some of these retard the current more than others. The value of reactance increases with the frequency, but it reaches a maximum at a specific frequency that depends on the composition of the conductor. Afterward, reactance falls as the frequency continues to increase so that at some high frequency the impedance for that conductor is again equal to resistance only. As can be seen

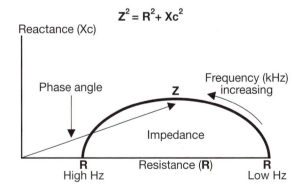

Figure 6.1 Impedance plot curve of resistance and reactance with frequency.

This work was supported by grants HD-12252 and HD-27063 from the National Institutes of Health, Bethesda, MD.

in figure 6.1, the impedance vector creates an angle with the resistance vector as the frequency changes from low to high. This is the phase angle, and it is the arctangent of the ratio of reactance and resistance, or Xc/R.

Geometrical relationships between a conductor's shape and its resistance according to Ohm's law are important in understanding the application of bioelectrical impedance to body composition assessment. Resistance is proportional to the length of a conductor and inversely proportional to its cross-sectional area. This means that a long conductor will have a greater resistance than a short one and that a conductor with a small cross-sectional area will have a greater resistance than one with a wider area. If we make algebraic substitutions in the following formula, the volume of a conductor can be estimated from the ratio of its length squared divided by it resistance.

$$\text{Conductor volume (V) =}$$
$$\text{length (L)} \times \text{area (A)} \qquad (6.1)$$

$$A = V/L$$
$$\text{Resistance (R)} = \rho\,(L/A)$$
$$R = \rho\,L\,(L/V)$$
$$V = \rho\,L^2/R$$

This volumetric relationship assumes that the conductor has a uniform shape and that the current is distributed throughout the conductor uniformly. The specific resistivity, or ρ, in this volumetric formula is an electrical property of a homogeneous conductor, independent of its size or shape, and it has units of ohm·cm (Baker 1989); it is a constant physical property similar to specific gravity. By substituting a person's stature (S) for the L in the preceding equation, then the impedance index, S^2/R, is proportional to body volume (Hoffer et al. 1969).

Bioelectrical Properties

Bioelectrical impedance analyzers use an alternating current that enters the body at a very low and safe amperage. The conductor is the water content of the body, and a bioelectrical impedance analyzer measures the impedance of this fluid conductor (Hoffer et al. 1969; Kushner and Schoeller 1986). Resistance in the body is the same as in nonbiological conductors (Kay et al. 1954; Nyboer 1959). Reactance is caused by the capacitant effect of cell membranes, tissue interfaces, and non-ionic tissues that retard a portion of the

electric current through these multiple current pathways (Barnett and Bagno 1936; Schwan and Kay 1956). The electric current flows differentially through the ECW and ICW as a function of the current frequency (Hoffer et al. 1969; Kushner and Schoeller 1986). At frequencies of 5 kHz or less, the current flows through ECW and reactance is minimal because the capacitant properties of body tissues are bypassed, but this was questioned by Settle and colleagues (1980). As the frequency increases, the current also enters the intracellular space, and the capacitant aspects of the body such as cell membranes and tissue interfaces retard the current, causing reactance. This capacitance has a maximum reactance at a specific high frequency, but as the frequency continues to increase, the effects are reduced and reactance falls. At high frequencies, above 100 kHz, the current penetrates all body tissues equally, and reactance is again minimized. Electrical circuit diagrams (figure 6.2) can describe or model these electrical characteristics of the body (Nyboer 1959; Schwan and Li 1953).

Reactance and the phase angle also describe relationships between bioelectrical impedance and the body (Baumgartner et al. 1988; Lukaski and Bolonchuk 1987). Studies have demonstrated the association of Xc and the phase angle with physiological and nutritional variables (Barnett 1937; Barnett and Bagno 1936; Kyle et al. 2001; Spence et al. 1979; Subramanyan et al. 1980; VanderJagt et al. 2002). Specific resistivity, ρ in the formula $\rho\,L^2/R$, is assumed to be a constant for the whole body; however, each tissue has a characteristic specific resistivity, and the observed specific resistivity for a body segment or the whole body is the mean specific resistivity of all conductive tissues (Geddes and Baker 1967; Rush et al. 1963;

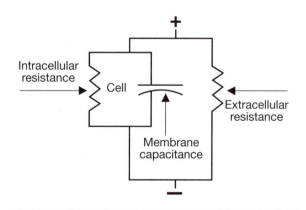

Figure 6.2 Intracellular and extracellular electrical circuit diagram.

Schwan and Li 1953). Variation in specific resistivities among body tissues and segments and between individuals occurs because of intra- and interindividual differences in tissue composition. This variation, in part, accounts for some of the interindividual differences and predictive errors in the use of impedance to estimate body composition (Chumlea et al. 1988).

Measuring Impedance

The tetrapolar method is the most common way to measure impedance. Two electrodes are attached to the body through which the alternating electric current enters the body. Two detection electrodes are also attached to the body within the linear location of these two current electrodes, and the values of resistance and reactance are measured across these two detection electrodes. The length of the conductor is technically the linear distance between the two detection electrodes. In adults, there is a minimal allowed distance between current and detection electrodes of about 4 to 5 cm to avoid electrical interference. This distance, the type and number of electrodes, and the electrode placement schemes on the body are machine- and manufacturer-specific (Bedogni et al. 2002; Deurenberg and Deurenberg-Yap 2002).

In theory, the current penetrates or energizes the entire body uniformly so that the detection electrodes can be placed at any location on the body and the distance between them is the length of the conductor. However, the human body is not a uniform conductor, nor are all commercial impedance analyzers capable of producing a constant current throughout the body. Simplistically, the body consists of limbs and a trunk, and the head is ignored. Within a limb, resistance is a function of the amount of muscle tissue because of its high water content and the high resistance of bone and fat according to the formula for resistors in parallel. The reactance value within a limb depends on the current frequency used. The trunk of the body is complex in its structure and composition with the presence of the lungs, the heart, large major blood vessels, intestinal organs, and abdominal adipose tissue. As a result, impedance estimates of body composition that use measures from or include the trunk should be looked at cautiously.

Early measures of impedance were taken with the subject in a supine position and the electrodes connected to the right hand–wrist and right ankle–foot. The rationale for a supine position was to counter the effects of gravity that tend to pool body water in the legs while the subject was standing (Slinde et al. 2003). In part because of the use of segmental impedance and multifrequency impedance analyzers, measurements are now taken from hand to hand or from foot to foot with the subject standing or supine depending on the model and manufacturer of the impedance analyzer used (Andreoli et al. 2002; Cable et al. 2001; Tyrrell et al. 2001).

Impedance History

Early studies of bioelectrical impedance and body composition focused on the relationship of impedance to TBW and to physiological variables such as thyroid function, basal metabolic rate, estrogenic activity, and blood flow (Barnett 1937; Lukaski and Bolonchuk 1987; Spence et al. 1979). Nyboer (1959) and Hoffer and colleagues (1969) first reported impedance estimates of TBW at a current frequency of 50 kHz using the index S^2/R. This is the frequency for the maximum reactance for muscle tissue (Nyboer 1970), and some studies also measured Xc at 50 kHz to differentiate amounts of TBW among individuals (Lukaski and Bolonchuk 1987; Segal et al. 1985). Thomasset (1962), Ducrot and colleagues (1970), Bolot and colleagues (1977) and Jenin and colleagues (1975) were some of the first investigators to describe the proportion of ECW in TBW using ratios of bioelectrical impedance measured at separate frequencies. Analyses of resistance and reactance data collected at several frequencies were also related to physiological characteristics of the body (Kanai et al. 1987; Rush et al. 1963), and impedance was used to study body fluid distribution in persons with congestive heart failure or end-stage renal disease on hemodialysis (Spence et al. 1979; Subramanyan et al. 1980).

In the mid-1980s, the first commercial impedance analyzers appeared along with numerous reports of the use of these machines in assessing body composition. The criterion methods available then were body density from underwater weighing and total body water based on the two-component models of Siri (1961) and Brozek and colleagues (1963). These studies looked at the relationship of impedance to FFM, TBW, and body fatness in children and adults (Deurenberg et al. 1989; Guo et al. 1987, 1989; Kushner and Schoeller 1986; Lukaski et al. 1985, 1986). In addition, studies of specific resistivity, the association of phase angle and reactance to body composition, and segmental impedance measurements were reported (Baumgartner et al. 1988; Chumlea et al. 1987, 1988). These studies and many other reports at the time established predictive relationships of S^2/R with TBW and FFM, and many of these equations are still in use.

By about 1990, several models of body composition impedance analyzers from different manufacturers were commercially available, and there were efforts to develop noncommercial impedance analyzers in research laboratories. The commercial units were marketed to research, clinical, sports, and weight reduction facilities, and there were numerous but small published studies and reports on the use of impedance to estimate body composition. The scientific and medical potential of impedance to estimate body composition and its inclusion in the third National Health and Nutrition Examination Survey (NHANES) led the National Institutes of Health (NIH) to convene a technology assessment conference (1994) in regard to possible research goals and directions for bioelectrical impedance. This conference assessed the state of knowledge and technological development of bioelectrical impedance at that time. The recommendations of this conference addressed relevant and important aspects of bioelectrical impedance: safety, standardization of measurements, bioelectrical parameters, validity, clinical use, and limitations. These are briefly summarized here and critiqued with regard to any progress that has occurred.

Safety

At that time, bioelectrical impedance measured with commercially available analyzers was considered safe, as it is today. A systematic assessment of all safety-related issues is not known to have ever been conducted or published. Before commercial analyzers are sold publicly, they receive approval from the Food and Drug Administration.

Standardization

There is little if any standardization of impedance measurement techniques outside of the accepted use of the tetrapolar configuration of electrodes. At the time of the NIH report (NIH 1996), the accuracy and precision of bioelectrical impedance measures were not well known (Roche et al. 1986) but have been reported since (Chumlea et al. 1994). Other variables that might affect the

precision of the measurements included body position, hydration status, consumption of food or beverages, ambient air and skin temperatures, recent physical activity, and bladder activity. Because these variables were known to affect criterion methods, they were reasonably assumed to affect impedance measurements also. Most investigators, when taking impedance measurements, follow certain conventions regarding the intake of food, recent physical activity, and bladder voiding, but there have been few published studies of the effects of these factors on impedance measures and subsequent estimates of body composition (Berneis and Keller 2000; Demura et al. 2002; Slinde and Rossander-Hulthen 2001).

Bioelectrical Properties

Impedance measurements do not provide any direct information as to how much current travels through extracellular versus intracellular volumes. Impedance measurements of biological and bioelectrical parameters were and still are important (Lafarque et al. 2002) because current paths in the body vary from person to person because of differences in body size, shape, electrolytes, water distribution, or other aspects of the body's composition in health and disease. Bioelectrical impedance measurements do not directly measure any biological quantity of interest on the basis of a physical or biophysical model, and additional research that links bioelectrical impedance measurements to the underlying physiological and biophysical structure will help place bioelectrical impedance technology on a much stronger scientific basis. However, the assumptions underlying bioelectrical impedance are not fulfilled in humans. Instead, the impedance index, S^2/R, is used as an independent variable in statistical regression equations. Bioelectrical impedance equations describe statistical relationships found for a particular population, and each equation is useful for subjects who are a close match to the reference population used in the development of each equation (Mast et al. 2002; Sun et al. 2003). Prediction equations using combinations of impedance and anthropometry are the norm for estimating body composition (Chumlea et al. 2002; Kyle et al. 2001; Sun et al. 2003). For commercial impedance analyzers, there is still difficulty in obtaining the computational algorithms and the data on which most, if not all, manufacturers' equations were derived. Information is needed on the validation samples used to derive the equations along with the predictive errors of the equations used to predict body composition.

Validity

The validity of impedance and its estimates of body composition remains a significant issue (Fornetti et al. 1999; Lukaski et al. 1986). The disproportionality of the body in terms of its size, shape, and composition between limbs and trunk affects impedance measurements. The ability to predict fatness in severely obese subjects remains a problem because they have a greater proportion of body mass and body water accounted for by the trunk, the hydration of FFM is greater in the obese, and the ratio of ECW to ICW is increased in the obese (Tagliabue et al. 2001). A major clinical research need was reference norms for impedance values to improve data interpretation and population-specific equations based on multicomponent criterion methods and statistically accepted cross-validation methods. The NHANES III single-frequency bioelectrical impedance data were thought useful to examine the relation of body composition estimates from this technique to clinical risk factors such as blood pressure, blood lipids, and glucose intolerance. Recently, impedance prediction equations have been published (Sun et al. 2003) that were validated and cross-validated using multicomponent methods for application to the impedance data from NHANES III. With these equations, body composition estimates for non-Hispanic whites, non-Hispanic blacks, and Mexican American males and females from 12 to 90 years of age have been published along with reference values for resistance, reactance, and S^2/R from the NHANES III impedance values (Chumlea et al. 2002).

Clinical Use

The clinical uses of bioelectrical impedance frequently concern conditions where water distribution is disturbed, such as during critical illness, and the assumptions of impedance and TBW are invalid in these cases (O'Brien et al. 2002). In most clinical settings, the role of bioelectrical impedance was not thought to be clearly defined, but it is useful for patients undergoing hemodialysis in the prescription and monitoring of dialysis based on urea kinetic modeling (Chertow et al. 1997) and may also improve interpretation of drug pharmacokinetics. The use of impedance continues to be of value in persons with cancer or HIV to assess nutritional status (Bauer et al. 2002; Horlick et al. 2002; Schwenk et al. 2001). The use of impedance in assessing TBW volume in persons with end-stage renal disease is an area of continued research (Konings et al. 2003; Lee et al. 2001).

Limitations

Disturbances of ICW are characteristic of protein calorie malnutrition, and direct or indirect measures of TBW do not reliably reflect FFM in these conditions (Barak et al. 2003; Barbosa-Silva et al. 2003). This likely invalidates bioelectrical impedance as an assessment of the response to parenteral and enteral nutrition in such patients, at least in terms of changes in FFM that reflect protein accretion. There are clinical conditions for which knowledge of TBW will be helpful in monitoring the critically ill, but the role of bioelectrical impedance in this assessment remains to be defined.

Bioelectrical impedance does not appear to be useful to assess the response to parenteral and enteral nutrition in terms of changes in FFM that reflect protein accretion. Acute weight changes attributable to dieting in the obese and by infusion and acute loss attributable to protein calorie malnutrition also do not appear to be reliably detected by bioelectrical impedance. Nutritional repletion may be assessed by bioelectrical impedance in the malnourished who are not critically ill. Bioelectrical impedance is useful in describing status and has considerable limitations in describing change when used in prediction equations.

The NIH technology assessment report provided recommendations for the direction of research and clinical investigations of bioelectrical impedance (NIH 1996). Some of these have been followed (Ellis et al. 1999) and others have not, in part because of continual changes in impedance technology and the rapid commercialization of impedance analyzers and their availability to the general public. In general, arm and leg impedance is less predictive of FFM than whole-body impedance and presents a limitation of commercial impedance analyzers.

Single- and Multiple-Frequency Impedance

The quantification of TBW from single-frequency impedance measures is reasonably accurate (Kushner and Roxe 2002; Sun et al. 2003). Single-frequency impedance analyzers almost all operate at a current frequency of 50 kHz (but different amperages), and the use of the term *single frequency* generally implies a measure of impedance at 50 kHz. At this frequency, the impedance index, S^2/R, is directly related to the volume of TBW (Simpson et al. 2001), but use of the impedance index to estimate FFM and body fatness is based on the fraction of 73% of TBW in FFM (Kushner and Schoeller 1986; Lohman 1986; Lukaski 1987; Moore 1963). This hydration fraction of FFM is not constant among individuals or groups (Lohman 1986; Moore 1963) so that S^2/R at 50 kHz is used in combination with other anthropometric data to predict body composition based on valid criterion methods (Guo et al. 1987; Sun et al. 2003).

Numerous equations have been published using single frequency to estimate body composition, and several sets of equations were developed for the commercial single-frequency impedance analyzers. A brief list of some of these equations is presented in table 6.1. Overall, these published equations (table 6.1) and those not listed here

provide reasonably accurate body composition estimates for groups, but their accuracy for individuals depends on numerous factors specific to the construction of the equations (Chumlea et al. 2002; Lohman et al. 2000; Sun et al. 2003).

Single-frequency impedance analyzers are limited in their ability to distinguish the distribution of body water into its intra- and extracellular compartments. The ability of multifrequency impedance to differentiate TBW into ICW and ECW is potentially important to describe fluid shifts and balance and to explore variations in levels of hydration (Chumlea and Guo 1994). The clinical significance of being able to measure ECW is considerable (Chamney et al. 2002).

Impedance values measured at a spectrum of frequencies, at several discrete frequencies, or some combination of frequencies can explain interindividual variations in body composition more precisely than can an impedance measurement at a single frequency. The most appropriate

Table 6.1   Comparison of Impedance Prediction Equations of Fat-Free Mass

Authors	Publication year	Age range (years)	r^2, SEE
Chumlea et al.	1990	18-63	.90, 3.65
Conlisk et al.	1992	11-25	Females: .95, 1.32 Males: .97, 1.82
Deurenberg et al.	1990	60-83	.96, 2.5
Deurenberg et al.	1991	7-83	<15 yrs: .97, 1.68 >16 yrs: .93, 2.63
Deurenberg et al.	1989	8-11	.89, 1.31
Deurenberg et al.	1991	7-25	.99, 2.39
Deurenberg et al.	1989	11-16	.97, 2.06
Dittmar and Reber	2001	60-90	.84, 1.73
Guo et al.	1987	7-25	Males: .98, 5.02 Females: .95, 5.80
Heitmann	1990	35-65	.90, 3.61
Kyle et al.	2001	22-94	.98, 1.72
Lukaski et al.	1986	18-60	.98, 2.06
Segal et al.	1985	17-59	.96, 3.06
Segal et al.	1991	17-62	Females: .89, 2.43 Males: .90, 3.61
Sun et al.	2003	12-94	Males: .90, 3.9 Females: .84, 2.9

combination of frequencies and multivariate methods of using multifrequency impedance values in estimating body composition and changes in body composition is still being examined and modeled.

Pairs and ratios of low- to high-frequency impedance values have been used to explore variations in levels of hydration and to differentiate disease conditions (Ducrot et al. 1970; Fredrix et al. 1990; Schols et al. 1991). More recently, multiple-frequency impedance has been used to estimate body composition in normal individuals and those with chronic disease (Bauer et al. 2002; Cornish et al. 1993; Dumler and Kilates 2003). A comparison of these multifrequency body composition studies is presented in table 6.2. Van Loan and Mayclin (1992) measured a spectrum of multifrequency impedance values at 25 frequencies. However, Van Loan and Mayclin did not attempt to model the spectrum of frequencies but simply used regression analysis to select a single impedance value to use in prediction equations, and Deurenberg and Schouten (1992) did the same. Others used measures of impedance at four separate frequencies to predict body composition (table 6.2), but these were used separately as independent variables in prediction equations.

From many of these multifrequency impedance studies, there is little new in the way of analyses compared with what has been reported with single-frequency 50-kHz impedance (Dittmar and Reber 2001). The reliability and validity of multiple-frequency impedance to estimate TBW, ECW, and FFM have been reported (Chumlea et al. 1994). Observer reliability with multiple-frequency impedance is good at all frequencies tested and similar to that reported for 50-kHz impedance machines. Recently, some investigators have used multiple-frequency impedance to estimate TBW, and the results of these studies have been accurate (Kyle et al. 2001).

New developments in multifrequency analysis come with the use of Cole–Cole model and Hanai model (Gudivaka et al. 1999; Ho et al. 1994; Lichtenbelt et al. 1997; Matthie and Withers 1998; Matthie et al. 1998). With these new approaches, multifrequency bioelectrical impedance has expanded the use of impedance to quantify the distributions of TBW, ECW, and ICW and body composition in clinical and nutritional studies. In general, multifrequency impedance has not improved body composition estimates over the use of single-frequency impedance (Dittmar and Reber 2001; Simpson et al. 2001), but it has been

Table 6.2 Comparison of Frequencies Used With Frequencies Measured in Published Studies of Multifrequency Bioelectrical Impedance

Author, year	Frequencies measured (kHz)	N	Frequencies used in models
Segal et al. 1991	5, 50, 100	36	Only a single frequency used with stature and weight
Van Loan and Mayclin 1992	1, 2, 3, 4, 5, 6, 8, 10, 15, 20, 27, 37, 50, 64, 90, 100, 122, 167, 224, 300, 400, 548, 740, 1,000, 1,348	60	Only a single frequency used with stature, weight, and sex
Deurenberg and Schouten 1992	1, 5, 10, 15, 20, 25, 50, 75, 100, 250, 750, 1,000, 1,250, 1,350	12	Only a single frequency used with stature
Hannan et al. 1994	5, 50, 100, 500, 1,000	43	One or two frequencies used with stature and weight
Sergi et al. 1994	1, 50	40	Only a single frequency used with stature, weight, sex, and health status
van Marken Lichtenbelt et al. 1994	1, 50, 100, 400	29	Four frequencies used with stature, sex, age, and BMI
Visser et al. 1995	1, 5, 50, 100, 250, 500, 1,000, 1,350	117	Only a single frequency used with stature, weight, sex, and age
Deurenberg et al. 1996	0, 1, 5, 50, 100, 250, 500, ∞	48	Four frequencies used with stature and weight

able to provide accurate and precise estimates of TBW and ECW, which were limited with single-frequency 50-kHz impedance. Multifrequency impedance analyzers are used in research and clinical settings especially in the area of end-stage renal disease and dialysis prescription.

Impedance and Body Composition

Estimates of body composition using bioelectrical impedance require a regression equation validated against a criterion body composition method. This is the only way to convert impedance values of resistance and reactance into estimates of body composition. Numerous studies and manufacturers have developed body composition prediction equations, but such equations and their estimated values are only as accurate as the criterion method used to determine the dependent variable in the equation. Most published impedance prediction equations use a two-component body composition model derived from densitometry, TBW, or DXA as the criterion measure, with a few exceptions (Guo et al. 1989; Houtkooper et al. 1992). The two-component and the DXA method have limitations (Withers et al. 1999) based on sex, body size, and a constant of 73% of water in FFM, which is erroneous (Siri 1961). Recently, TBW and FFM impedance prediction equations using a multicomponent body composition model with direct measures of body density from underwater weighing, bone mineral content from DXA, and TBW from isotope dilution were published for children and adults (Sun et al. 2003). These impedance prediction equations are also some of the few that have been cross-validated.

Most published impedance prediction equations are of limited general use because of a narrow age range and specificity to the racial and ethnic makeup of their sample. Among published equations, there are wide variations in the goodness-of-fit measure or root mean square error (RMSE) for TBW and for FFM; the RMSE for TBW ranges from 1.3 to 8.7 L, and for FFM the range is from 1.1 to 4.6 kg (Deurenberg et al. 1991; Guo et al. 1987; Houtkooper et al. 1992, 1996; Roubenoff 1996).

Many of the published equations are for white persons only. Body composition studies of nonwhite ethnic groups have never equaled those conducted for white samples (Kotler et al. 1996; Rising et al. 1991; Stolarczyk et al. 1994; Zillikens and Conway 1991). There have been a limited number of reports of the use of impedance with Native American and African American samples (Lohman et al. 2000, 2003; Wagner et al. 1997). Because of the specificity of these ethnic samples, the results have been anecdotal or of limited application to other corresponding samples in the United States.

The recently published impedance equations did not perform as well in non-Hispanic blacks as in non-Hispanic whites (Sun et al. 2003). These equations tended to underpredict TBW and FFM in non-Hispanic black males and females by about 1.5 to 2.0 L or kg. As a result, these equations are more valid for non-Hispanic whites than for non-Hispanic blacks. As many non-Hispanic blacks as possible were included in these equations, but there have been too few body composition studies that include black persons, Mexican Americans, and other ethnic groups of the U.S. population. However, a multicomponent body composition model was used to account for age, sex, and race differences in densities of FFM. Nevertheless, there were residual racial and ethnic differences for impedance and its weight-based predictions of TBW and FFM (Schoeller and Luke 2000).

In the obese, clinical cases, or simply those individuals with greater than normal amounts of adipose tissue, errors of prediction will be exacerbated. These recent equations for TBW and FFM provide reasonable prediction for individuals at the extremes of the distribution with only a slight tendency to overpredict at the lower end of the distribution and to underpredict at the upper end of the distribution. These problems have also been noted in previous studies and attributed to the criterion method of underwater weighing (Hodgdon and Fitzgerald 1987).

Changes in Body Composition

Impedance is able to estimate body composition at a point in time, but its ability to estimate change in body composition over time is questionable (Berneis and Keller 2000; Chanchairujira and Mehta 2001; Forbes et al. 1992). From analysis of the data reporting body composition changes, Forbes and coworkers (1992) questioned the ability of impedance to determine the validity of repeated estimates of body composition from the same individuals. Part of the problem is the limited sensitivity of the impedance index. Resistance is proportional to the number of ions in a conductor, and a change in the composition that affects the number of ions should be reflected in the value of resistance. However, when the resistance is used in the index S^2/R and related to FFM, then changes in body weight or FFM must be sufficient to produce concurrent changes in the volume of conductor before significant changes are detectable by impedance. A measure of resistance can only account for the fat-free portion or water in the change. Therefore, if a change in weight is predominantly fat, it will not be easily detected by resistance unless a sufficient change in the FFM or TBW has occurred.

The effect of concurrent changes in body volume and composition on the sensitivity of impedance is unknown. A change in the ionic composition without a concurrent change in the volume of a biological conductor will not be detected accurately without knowing the change in the value of the specific resistivity. Recently, however, there have been reports of the successful use of impedance to estimate changes in body composition accurately (Phillips et al. 2003). In contrast, Lohman and colleagues (2003) found that bioelectrical impedance analysis showed more variation than skinfolds in assessing body composition changes from a 3-year multisite intervention with Native American children.

Segmental Bioelectrical Impedance

Patterson (1989) reported that segmental impedance was sensitive to changes in body composition. Segmental bioelectrical impedance was first used by Settle and colleagues (1980), who noted that 85% of total body impedance was accounted for by the sum of impedance of the arm and the leg although these segments accounted for only 35% of the total body volume. This approach has been used to estimate total and segmental body composition with good results (Fuller and Elia 1989). The arm impedance index is reported to be highly correlated with whole-body FFM, followed by the impedance indexes for the leg and the trunk (Baumgartner et al. 1988; Pietrobelli et al. 2002; Salinari et al. 2002). The use of segmental impedance to estimate regional aspects of body composition is being expanded with the development of electrical impedance tomography.

There is also increased clinical use of segmental impedance in the assessment of diseases that affect body fluid balance (Fuller and Elia 1989; Scheltinga et al. 1991; Ward et al. 1992). The segmental imped-

ance approach has been incorporated by several manufacturers of the newer impedance analyzers where only the impedance of the legs is measured on machines that resemble bathroom scales. The utility of segmental impedance is that an arm or leg is easier to measure than the whole body, and the structure of the limb conforms to the assumptions of impedance theory and body composition better than does a measure of the whole body. However, the accuracy and precision of segmental impedance total body composition estimates in large samples are generally less than those obtained with whole-body impedance estimates.

Impedance and Blood Chemistry

There are few if any significant associations reported among physiological variables and bioelectrical impedance, and possible associations of bioelectrical impedance to blood chemistry and electrolytes have not been generally considered in detail (Azcue et al. 1993; Cha et al. 1994; Shirreffs and Maughan 1994; Shirreffs et al. 1994). Factors that affect body fluid or electrical activity potentially affect measures of impedance. Multifrequency impedance values are significantly correlated with blood chemistry variables including hemoglobin, hematocrit, serum sodium and potassium, serum creatinine and serum osmolality, and whole blood viscosity (Berneis and Keller 2000; Chumlea et al. 1994; Varlet-Marie et al. 2003). The effect of interrelationships with blood chemistry values on the variance in impedance measures and subsequent estimates of body composition needs further study.

Summary

Bioelectrical impedance has an interesting research and clinical history. Much research has been conducted and more is currently under way primarily with multiple-frequency analyzers. The NIH Technology Assessment Conference (NIH 1996) highlighted areas where impedance technology might be directed. Since that time, this technology progressed from single- to multifrequency analyzers and from a cumbersome methodology to relatively simple measurement methods. Single-frequency impedance is a useful technique for assessing TBW and FFM in groups both healthy and with disease, even though there are numerous equations for estimating these values. Impedance is a useful method of predicting body composition and is better than anthropometry alone (Baumgartner et al. 1988; Sun et al. 2003), but there has not really been any significant improvement in the research potential of single-frequency impedance and body composition beyond the current state of affairs. In many ways this is attributable to the availability of single-frequency analyzers at local food markets and pharmacies at prices of less than $50. When the general public purchases a technology that is marketed to them as complete, then future research is probably of limited value and funding difficult to obtain. At present, multifrequency impedance is the area where significant gains in knowledge will occur in the next few years. This technology is being used in the current National Health and Nutrition Examination Survey and is used widely in clinical areas where there is still a need to quantify TBW and extra- and intracellular water and the changes in these body volumes.

Chapter 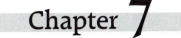7

Computed Tomography and Magnetic Resonance Imaging

Robert Ross and Ian Janssen

Imaging methods such as computed tomography (CT) and magnetic resonance imaging (MRI) are considered the most accurate means available for in vivo quantification of body composition at the tissue–organ level. Although access and cost remain obstacles to routine use, these imaging approaches are now used extensively in body composition research. CT and MRI are the methods of choice for calibration of field methods designed to measure adipose tissue (AT) and skeletal muscle in vivo and are the only methods available for measurement of internal tissues and organs. Recently, both methods have been used to measure tissue composition in skeletal muscle and liver.

CT Image Acquisition

The basic CT system consists of an X-ray tube and receiver that rotate in a perpendicular plane to the subject. The X-ray tube emits X rays (0.1-0.2 A, 60-120 kVp) that are attenuated as they pass through tissues (Sprawls 1977). A receiver detects the attenuated X rays, and the image is reconstructed with mathematical techniques based on two-dimensional Fourier analysis, filtered back projection, or a combination of these methods.

The X-ray attenuation is expressed as the linear attenuation coefficient or CT number. The CT number is a measure of attenuation relative to air and

water, with the CT numbers of air and water defined as –1,000 and 0 Hounsfield units (HU), respectively. Physical density is the main determinant of attenuation or HU. Cross-sectional CT images are composed of picture elements or pixels, usually 1 mm by 1 mm squares. Each of the image pixels has a CT number or HU, which gives contrast to the image. Each pixel is assigned an HU value on a gray scale that reflects the composition of the tissue. The lower the density of the tissue, the lower the HU values for the pixels that make up that tissue. For example, the density of AT is less than that of water, and the CT number for adipose tissue pixels ranges from about –190 to –30 HU (Chowdhury et al. 1994; Kvist et al. 1986). Conversely, the density of skeletal muscle is greater than that of water, and the CT number for AT-free skeletal muscle pixels ranges from about 30 to 100 HU (Chowdhury et al. 1994).

MRI Image Acquisition

The estimation of body composition components at the tissue level with MRI is essentially the same as for CT. The two methods primarily differ in the manner in which the images are acquired, which has a subsequent bearing on practical considerations of cost as well as technical aspects of image analysis, accuracy, and reliability.

MRI does not use ionizing radiation. Instead, it is based on the interaction between hydrogen nuclei (protons), which are abundant in all biological tissues, and the magnetic fields generated and controlled by the MRI system's instrumentation. Hydrogen protons have nonzero magnetic

Research supported by grants from the Canadian Institutes of Health Research (MT13448) and Mars Corporation to R. Ross. I. Janssen is supported by a Canadian Institutes of Health Research Postdoctoral Fellowship.

moments, which cause them to behave like tiny magnets. When a subject is placed inside the magnet of an MRI unit, where the field strength is typically 15,000 times stronger than the earth's (e.g., 1.5 Tesla), the magnetic moments of the protons align themselves with the magnetic field. After the hydrogen protons are aligned in a known direction, a pulsed radio-frequency field is applied to the body tissues causing a number of hydrogen protons to absorb energy. When the radio frequency pulse is turned off, the protons gradually return to their original positions, in the process releasing energy that is absorbed in the form of a radio frequency signal. The MRI system uses this signal to generate the cross-sectional images. Manipulating the radio frequency parameters allows one to exploit the differences in relaxation times between different tissues and organs and, in so doing, provides the necessary tissue contrast for high-quality MRI images.

Historically, an important problem with the application of MRI to body composition research has been the substantial time needed to obtain images of sufficient quality and resolution for reliable measurements. For example, in early studies the acquisition of a set of abdominal MRI images required between 8 (Ross et al. 1992) and 16 (Seidell et al. 1987) min. Recent advances in MRI technology have reduced the time needed to obtain the same quality images in the abdomen to approximately 25 s (Ross 1996; Ross et al. 1996b). During this time, the subject must hold his or her breath to reduce the effects of respiratory motion on image quality (e.g., motion artifact). A series of images for whole-body analysis can now be acquired in less than 30 min (Ross 1996; Ross et al. 1996b; Thomas et al. 1998). These advances have made MRI a much more useful instrument for body composition research. An example of a multiple-slice MRI protocol for whole-body analysis is illustrated in figure 7.1 (Ross et al. 1992, 1996b).

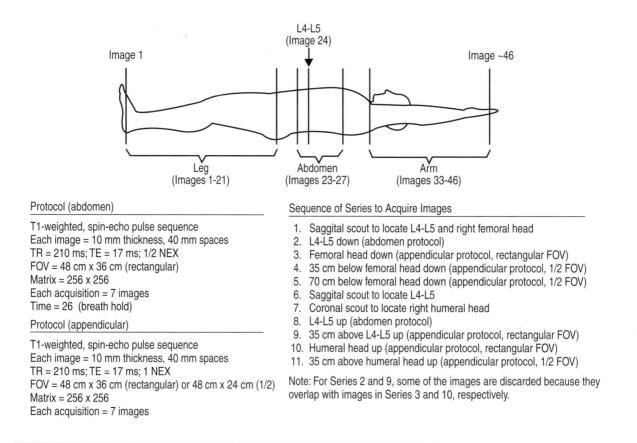

Protocol (abdomen)

T1-weighted, spin-echo pulse sequence
Each image = 10 mm thickness, 40 mm spaces
TR = 210 ms; TE = 17 ms; 1/2 NEX
FOV = 48 cm x 36 cm (rectangular)
Matrix = 256 x 256
Each acquisition = 7 images
Time = 26 (breath hold)

Protocol (appendicular)

T1-weighted, spin-echo pulse sequence
Each image = 10 mm thickness, 40 mm spaces
TR = 210 ms; TE = 17 ms; 1 NEX
FOV = 48 cm x 36 cm (rectangular) or 48 cm x 24 cm (1/2)
Matrix = 256 x 256
Each acquisition = 7 images

Sequence of Series to Acquire Images

1. Saggital scout to locate L4-L5 and right femoral head
2. L4-L5 down (abdomen protocol)
3. Femoral head down (appendicular protocol, rectangular FOV)
4. 35 cm below femoral head down (appendicular protocol, 1/2 FOV)
5. 70 cm below femoral head down (appendicular protocol, 1/2 FOV)
6. Saggital scout to locate L4-L5
7. Coronal scout to locate right humeral head
8. L4-L5 up (abdomen protocol)
9. 35 cm above L4-L5 up (appendicular protocol, rectangular FOV)
10. Humeral head up (appendicular protocol, rectangular FOV)
11. 35 cm above humeral head up (appendicular protocol, 1/2 FOV)

Note: For Series 2 and 9, some of the images are discarded because they overlap with images in Series 3 and 10, respectively.

Figure 7.1 Magnetic resonance imaging protocol for both whole-body and regional (e.g., arm, leg, abdomen) measures of adipose tissue and lean tissue. These images can be acquired in approximately 30 min and can be analyzed using specially designed computer software (Slice-O-Matic, Tomovision Inc., Montreal, Canada) in approximately 3 hr. TR = repetition time; TE = echo time; NEX = number of excitations; FOV = field of view.

CT Image Analysis

CT body composition methods are designed to quantify components at the tissue–organ level of body composition. The main components are AT, skeletal muscle, bone, visceral organs, and brain. The tissue area (cm²) for the different tissues on each cross-sectional CT image can be determined using one of two methods. In one technique, the technician traces the perimeter of the target tissue with a light pen, mouse, or track–ball-controlled cursor. The second and more commonly used technique uses a computer-automated procedure that identifies the area of the target tissue by selecting pixels within a given HU range, for example, –190 to –30 HU for AT (Kvist et al. 1986). After the pixels for a given tissue have been identified, the area (cm²) of the tissue is calculated by multiplying the number of pixels for a given tissue by the surface area of the individual pixels (figure 7.2). CT image analysis is typically performed with software installed on the CT scanner console; however, specially designed image analysis software is also available.

MRI Image Analysis

In contrast to CT, MRI image analysis software is not included on most MRI scanner consoles. As a result, MRI data must be transferred to a separate workstation equipped with image analysis software. Once this is accomplished, the approach for analyzing MRI images is similar to that used for CT images. The perimeter of the tissue of interest can be traced using a light pen or a track–ball- or mouse-controlled pointer (Abate et al. 1994); the area within the perimeter can be calculated by multiplying the number of pixels in the high- lighted region by their known area. Alternatively,

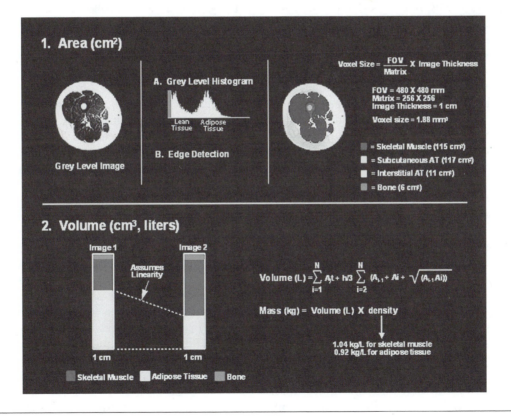

Figure 7.2 In the first step, tissue area for a given gray level image is determined by subjecting the image to automated techniques such as a histogram (based on voxel attenuation thresholds) or edge detection (groups together voxels of similar intensity) procedure. In the next step, the tissue area (cm²) in a given image is determined by multiplying the number of voxels by the individual voxel surface area. In the third step, whole-body or regional volumes are calculated using a three-dimensional formula. This formula adds the volumes of truncated cones as defined by pairs of consecutive images where i is the image number, A is the tissue area for a given image, t is the slice thickness, h is the distance between consecutive images, and N is the number of images. Skeletal muscle and adipose tissue mass (kg) can be determined by multiplying the volume (liters) by 1.04 and 0.92, respectively, the assumed constant density values for skeletal muscle and adipose tissue. FOV = field of view.

image segmentation algorithms can be used that highlight all pixels within a selected range of intensities believed to be representative of a specific tissue (Mourier et al. 1997). The latter approach, however, is considered more problematic when applied to MRI than to CT images for three reasons: The distributions of pixel intensity (gray scale) values for different tissues overlap more for MRI than for CT images, noise caused by respiratory motion blurs the borders between tissues in the abdomen to a greater extent in MRI than in CT, and inhomogeneity in the magnetic field can produce "shading" at the peripheries of MRI images (Ross et al. 1992).

Mitsiopoulos and colleagues (1998) described an example of special, interactive image analysis software that allows the analyst to analyze CT or MRI images, as illustrated in figure 7.2. Initially, an automated filter is used to distinguish between different gray-level regions (e.g., a group of pixels) on the MRI image. After the software has determined the edges of the regions, it draws lines on the images that reflect the edges. Once the lines are drawn, the observer uses a mouse pointer and color codes (e.g., different color for each tissue) to identify what tissue each region is within. The image is then reviewed with an interactive slice-editor program that allows for verification and, when necessary, correction of the segmented results. Using a transparency feature that allows the color-coded image to be superimposed on top of the original gray-level image facilitates these steps. The advantage of this technique is that it uses both computer-automated and manual procedures. This may result in smaller intra- and interobserver errors than would be obtained with the tracer methodology (see preceding paragraph) while at the same time allowing for correction of the pixels that were misclassified with the automated procedures.

The three techniques that are routinely used to measure tissue size on CT and MRI images vary considerably in terms of computer automation, manual editing, and the anatomical expertise required by the individual analyzing the images. However, whether the method used to analyze CT and MRI images influences the tissue area (cm^2) being measured is unknown. No published studies have directly compared the validity (accuracy) and reproducibility (intra- and interobserver error) of the different analysis methods.

Determination of Tissue Volume

If multiple CT or MRI images are obtained, tissue volumes can be calculated by integrating the cross-sectional area data from consecutive slices (figure 7.2). Because acquisition and analysis of contiguous images over the whole body or a given region are very time-consuming and expensive, axial images are typically collected with interslice gaps (e.g., space between the top of one image and the bottom of the next image), usually ranging from 20 to 40 mm. Volumes are then calculated using geometrical models based on the tissue areas in the images and the distance between adjacent images. To date, three models have been used. The first, originally proposed by Kvist and colleagues (1986), suggested that tissue volume can be calculated according to a "parallel trapezium" model as follows:

$$V = \sum^{N} h(A_1 + A_2)/2 \qquad (7.1)$$

where V is the volume (cm^3) of the tissue, h is the distance between the centers (not the gaps) of two consecutive images, A_1 is the tissue area (cm^2) in the first image, and A_2 is the tissue area in the second image. As an example, if there were two 10-mm thick images separated by a 40-mm gap, and the tissue areas for the first and second images were 100 cm^2 and 115 cm^2, respectively, the total volume for the 50-mm region (center of first image to center of second image) would be 537.5 cm^3. This model assumes that the change in cross-sectional area between two consecutive images in linear and that the irregular shapes of the tissues in the axial images are parallel trapeziums.

More recently Ross (Ross 1996; Ross et al. 1996b) defined a "truncated cone" geometrical model for deriving tissue volumes from measurements of a series of axial images as follows:

$$V = t \sum^{N} A_1 + \sum^{N} h/3[(A_1 + A_2) + (\sqrt{A_1} * A_2)] \quad (7.2)$$

where V is the volume (cm^3) of the tissue, t is the thickness of each slice, h is the distance (gap) between consecutive images, A_1 is the tissue area (cm^2) in the first image, A_2 is the tissue area in the second image, and N is the total number of images. As an example, if there were two 10-mm thick images separated by a 40-mm gap, and the tissue areas for the first and second images were 100 cm^2 and 115 cm^2, respectively, the total volume for the 50-mm region (top of first image to top of second image, thus the volume of the second image is not included) would be 529.7 cm^3. This model recognizes that each image has a slice thickness (e.g., 10 mm) and that the pixels are actually volume elements or voxels (figure 7.2). Thus, the truncated cone model, in theory, is a better representation of the three-dimensional nature of CT and MRI images than is the parallel trapezium model proposed by Kvist and colleagues (1986). The underlying assumptions of the truncated cone model are that the change in cross-sectional area between two images is linear and that the irregular shapes of the tissue in the axial images are circular. Although no studies have compared the truncated cone and parallel trapezium models in humans, it has been reported that the truncated cone and parallel trapezium equations provide similar and precise volume calculations in comparison to phantoms (Mitsiopoulos et al. 1998). Shen and colleagues (2003) recently defined a new geometrical model for estimating tissue volumes with CT or MRI images, which they referred to as the "two-column" model, as follows:

$$V = (t + h) \sum_{1}^{N} A_1 \qquad (7.3)$$

where t is the thickness of each image, h is the distance (gap) between consecutive images, A_1 is the tissue area (cm^2) in the first image, and N is the total number of images. As an example, if there were two 10-mm thick images separated by a 40-mm gap, and the tissue areas for the first and second images were 100 cm^2 and 115 cm^2, respectively, the total volume for the 50-mm region (top of first image to top of second image) would be 500 cm^3. In this model, the geometry of the tissue of interest is assumed to be represented by two columns, both having the original shape of the tissue in the cross-sectional image and a height of one half of the distance between the image centers.

Shen and colleagues (2003) compared volume estimates with the truncated cone and two-column models, using the Visible Women (U.S. National Library of Medicine 1995) as the reference. Actual volumes were calculated from 1,730 contiguous 1-mm thick Visible Women images that were segmented for major tissues, including subcutaneous AT, visceral AT, skeletal muscle, and lung. Between-slice intervals (gaps) were varied (10, 20, 30, 40,

50, 60, 70, 80 mm) using both geometrical models in four different regions (head, trunk, arms, legs). Independent of tissue, image spacing, and region, the means of the two-column model were the same as the corresponding reference volume. However, the volumes derived from the truncated cone model tended to be slightly smaller than the reference volumes. Similarly, the coefficient of variations for the two-column model was smaller than for the truncated cone mode. These findings suggest that the equation based on the two-column model is more accurate in estimating tissue volumes than the equation based on the truncated cone model, and the findings have important implications for future CT- and MRI-based research wherein volume calculations are made from multiple images. These findings obtained from a single cadaver require duplication in larger samples of human subjects varying in adiposity, age, race, and sex.

Because tissue densities for AT, skeletal muscle, and organs are fairly constant from person to person, CT and MRI volume measures for these tissues can be converted to mass units by multiplying the volume by the assumed density values for that tissue (figure 7.2). For example, the constant densities for AT and skeletal muscle are 0.92 g/cm^3 and 1.04 g/cm^3, respectively (Snyder et al. 1975). Density values are also available for the brain and visceral organs, although these vary from organ to organ (Gallagher et al. 1998).

CT Measurement of Skeletal Muscle Tissue Quality

Although CT was originally used in body composition research to assess tissue size, more recently it has been applied to measure tissue composition. The attenuation or HU value for each CT voxel depends on the molecular composition of that voxel. Because lipid has a lower density than water and protein, the intensity value of an AT voxel is lower than the intensity value of a lean skeletal muscle voxel. Typically, voxels that range from –190 to –30 HU denote AT, whereas voxels that range from 0 to 100 HU denote skeletal muscle (Kvist et al. 1986; Sjöström 1991). Thus, the marbled or intermuscular AT can be separated from lean skeletal muscle based on differences in attenuation characteristics (Goodpaster et al. 1997, 2000b).

In addition, the average HU for AT-free skeletal muscle (i.e., free of AT visible to the naked eye) voxels can be used as an index of skeletal muscle lipid content (figure 7.3). The lower the average

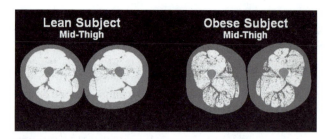

Figure 7.3 CT images obtained in the midthigh for lean (left) and obese (right) subjects. The darker appearing pixels within the skeletal muscle tissue of both subjects have attenuation values between 0 and 29 HU and represent low-density (e.g., high-fat) muscle. The lighter appearing pixels within the skeletal muscle tissue of both subjects have attenuation values between 30 and 100 HU and represent high-density (e.g., low-fat) muscle. The greater the proportion of dark (low-density) skeletal muscle pixels, the greater the infiltration of lipid (both intra- and extramyocellular lipid) within the muscle. Thus, the obese subject has a greater skeletal muscle lipid content than the lean subject. The mean attenuation value of all skeletal muscle pixels within the range of 0 to 100 HU can also be used as an index of muscle composition, with lower mean attenuation values indicating a greater infiltration of lipid within the muscle.

skeletal muscle HU value (i.e., mean attenuation value), or the greater the number of low-density skeletal muscle pixels (e.g., 0-30 HU), the higher the skeletal muscle lipid content. It is important to note that muscle attenuation values by CT are not analogous to intramyocellular lipid values obtained by skeletal muscle biopsy or proton magnetic resonance spectroscopy. Muscle attenuation determined by CT is a reflection of both intramyocellular and extramyocellular lipid content.

CT Measurement of Liver Tissue Quality

In a manner similar to that used to determine skeletal muscle density, CT has also been used to determine the density of liver tissue (Banerji et al. 1995; Goto et al. 1995; Ricci et al. 1997). In body composition studies, this approach has been used to identify fatty liver, another component of fat distribution that reflects the utility of body composition assessment in clinical medicine.

As with skeletal muscle, the lower the mean liver HU value, the lower the density and the greater the fat content (figure 7.4). Therefore, liver density is inversely related to liver fat and thus is a surrogate for it (Ricci et al. 1997). However, although extremely low HU values have been measured in

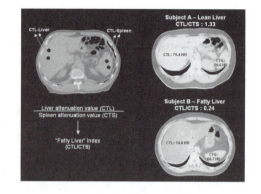

Figure 7.4 Computed tomography (CT) image and derivation of "fatty liver index" are shown on the left. The attenuation values for the liver and spleen are determined by identifying one or more regions of interest within each organ and then calculating the average attenuation value for those regions. A ratio of mean liver to spleen attenuation values (CTL–CTS) is used as an index of liver fat. Lower-density values for the liver relative to the spleen indicated excess infiltration of lipid or fatty liver. On the right (top), the CT image for Subject A provides an example of a person with a lean liver (CTL/CTS 1.33). Subject B (bottom) is an example of a person with a fatty liver (CTL/CTS 0.24). When CT is used, fatty liver is defined as a CTL–CTS ratio less than 1. Note that despite large variation in liver attenuation, the attenuation values (Hounsfield units, or HU) for the spleen in the two subjects are similar.

livers infiltrated with fat, an overlap exists between normal and abnormal liver HU values (Piekarski et al. 1980). Therefore, the absolute liver density determined by CT may not be sensitive for predicting abnormal liver chemical content. Because a constant relationship exists between liver and spleen attenuation in individuals with normal livers, the ratio of mean liver to spleen attenuation values is used as an index of liver fat, as originally described by Piekarski and colleagues in 1980 and depicted in figure 7.4.

MRI Measurement of Skeletal Muscle Tissue Quality

As with CT, evidence suggests that MRI may also be used to measure the quality of skeletal muscle. However, because proton MRI integrates, rather than separates, the signals from distinct protons (e.g, protons for water, lipids) within the image voxel, conventional MRI is not useful for determining the concentration of lipid or water in skeletal muscle. To obtain basic information within the tissue volume of interest requires application of "chemical shift" imaging techniques. Several chemical shift methods have been developed that separate the water and fat signals from the region of interest, creating the potential to determine water and fat contents of skeletal muscle. Tsubahara and colleagues (1995) were among the first to establish a proton chemical shift imaging technique (Dixon method) to measure the lipid and water content of skeletal muscle in men and women. Schick and colleagues (2002) developed a spoiled gradient-echo sequence for fat selective imaging of skeletel muscle. Their results showed a high correlation between the total muscle fat assessed by fat imaging and by magnetic resonance spectroscopy ($r = .91$ in soleus, $r = .55$ in tibialis anterior). Similarly, Boada and colleagues (1999) reported preliminary findings using a water-spoiled spiral MRI imaging scheme for direct quantification of the lipid MRI signal in skeletal muscle. The lipid contents from various regions of interest within skeletal muscle were obtained by comparing the signal intensity to that of phantoms of known lipid concentrations that are contained within the field of view during image acquisition. Although preliminary, the results from this study suggest that this method may be employed to characterize and distinguish the lipid concentration in lean and obese muscle (Boada et al. 1999).

With the MRI methods just discussed (Boada et al. 1999; Schick et al. 2002; Tsubahara et al. 1995), it is not possible to separate the lipid measurement into intra- and extramyocellular compartments. Partitioning the lipid signal into separate compartments can be accomplished using proton magnetic resonance spectroscopy, details of which are reported elsewhere (Boesch and Kreis 2000; Boesch et al. 1997). Proton magnetic resonance spectroscopy is now being used extensively to describe lipid distribution (e.g., intra- and extramyocellular lipid) in human skeletal muscle and its relation to insulin resistance (Jacob et al. 1999; Perseghin et al. 1999).

Constandinides and colleagues (2000) have demonstrated the feasibility of using sodium (^{23}NA) MRI to quantify sodium concentration in lower-limb skeletal muscles in humans. In this study, ^{23}NA MRI measured variations in sodium content that are characteristic of normal and diseased muscles. Furthermore, it was demonstrated that ^{23}NA MRI may be used to characterize the selective recruitment of muscles during exercise, in cases of knee cartilage degeneration in osteoarthritis and for the diagnosis of muscle dystrophy in aging muscle.

Although these observations for lipid and sodium MRI are restricted to skeletal muscle of the lower limb, they represent unique applications of MRI that can be used to obtain novel insights into the composition of skeletal muscle, insights that are highly relevant to the adverse health consequences of obesity, type 2 diabetes, and aging.

MRI Measurement of Liver Tissue Quality

Marks and colleagues (1997) have also described a method for estimating liver fat by MRI. They suggest that a liver fat index can be created by examining the signal intensity in a region of interest within the right lobe of the liver and comparing this to the signal intensity of an identically sized region of interest within the adjacent subcutaneous AT (i.e., liver fat index = signal intensity of liver expressed as a percentage of the intensity of subcutaneous AT).

Validity of Tissue Quantity by CT

The validity and accuracy of area and volume measurements from CT have been established by comparing CT measures in human cadavers with direct measures (e.g., dissection, planimetry; table 7.1). Heymsfield and colleagues (1979) reported that organ masses estimated using CT were highly reproducible and agreed with actual mass to within 5% to 6%. Rössner et al. (1990) found good agreement between CT and cadaver AT areas (correlations ranged from .77 to .94). Engstrom and colleagues (1991) compared the cross-sectional area measurements of skeletal muscle determined from the proximal thigh in cadavers to the corresponding CT-measured cross-sectional areas and reported that the correlation coefficient between the two approached unity. Mitsiopoulos and colleagues (1998) also reported a strong correlation between cadaver and CT area measures for lean skeletal muscle and subcutaneous AT in the arms and legs (figure 7.5).

Validity of Tissue Quantity by MRI

Several studies support the validity of MRI estimates of AT and skeletal muscle by comparison

Table 7.1   Cadaver Validation Studies and Studies Comparing CT and MRI Measures

Reference	Subjects	Scan location	CORRELATION (SEE, %) [CV, %]		
			Subcutaneous AT	Visceral AT	Skeletal muscle
CT					
Rössner et al. 1990	2 Cadavers	Abdomen	.89	.88	–
Mitsiopoulos et al. 1998	2 Cadavers	Arm and leg	.99 (9)	–	.97 (10)
Engstrom et al. 1991	3 Cadavers	Thigh	–	–	.99 (10-20)
Hudash et al. 1985	1 Cadaver	Leg	–	–	(0.5-17)
MRI					
Abate and Garg 1995	3 Cadavers	Abdomen	[2]	[6]	–
Engstrom et al. 1991	3 Cadavers	Thigh	–	–	.99
Mitsiopoulos et al. 1998	2 Cadavers	Arm and leg	.99 (8)	–	.97 (10)
CT AND MRI COMPARISONS					
Seidell et al. 1990	7 Men	Abdomen	.99 [4]	.79 [5]	–
Mitsiopoulos et al. 1998	2 Cadavers	Arm and leg	.99 (5.5)	–	.97 (10)
Engstrom et al. 1991	3 Cadavers	Thigh	–	–	.99 (10-20)
Sobol et al. 1991	11 Adults	Abdomen	.98	.93	–

Note. CT = computed tomography, MRI = magnetic resonance imaging, SEE = standard error of the estimate, CV = coefficient of variation, AT = adipose tissue.

with dissection in human cadavers (table 7.1). Abate and colleagues (1994) compared MRI measures of abdominal subcutaneous and visceral AT to measures obtained by direct weighing of the same AT compartments after dissection in three human cadavers. For the various compartments, the mean difference between the two methods was about 6%. Engstrom and colleagues (1991) compared the cross-sectional area measurements of skeletal muscle determined from the proximal thigh in cadavers with the corresponding MRI-measured cross-sectional areas and reported that the correlation coefficient between the two approached unity. Mitsiopoulos and colleagues (1998) performed a similar analysis to that of Engstrom and colleagues but used images covering the entire arm and leg regions of the cadavers. Mitsiopoulos and colleagues reported a strong correlation between cadaver and MRI area measures for lean skeletal muscle and subcutaneous AT (figure 7.5).

Comparison of CT and MRI

Only a few studies have directly compared CT and MRI (table 7.1). Two studies compared CT and MRI measures of subcutaneous AT or skeletal muscle in the appendicular regions. Engstrom and colleagues (1991) compared CT skeletal muscle cross-sectional area measurements in the midthigh of cadavers to corresponding MRI images. Within each of the three cadavers, the CT and MRI measures were highly correlated for each muscle (r = .81-1.00). Mitsiopoulos and colleagues (1998) reported a strong correlation between CT and MRI area measures for AT-free skeletal muscle and subcutaneous AT obtained from an arm and leg from each of two cadavers (figure 7.5).

Two studies compared CT and MRI measures of abdominal AT, and combined these studies had a total of 18 subjects (table 7.1; Seidell et al. 1987; Sobol et al. 1991). In both studies, CT and MRI measures of abdominal subcutaneous AT were highly

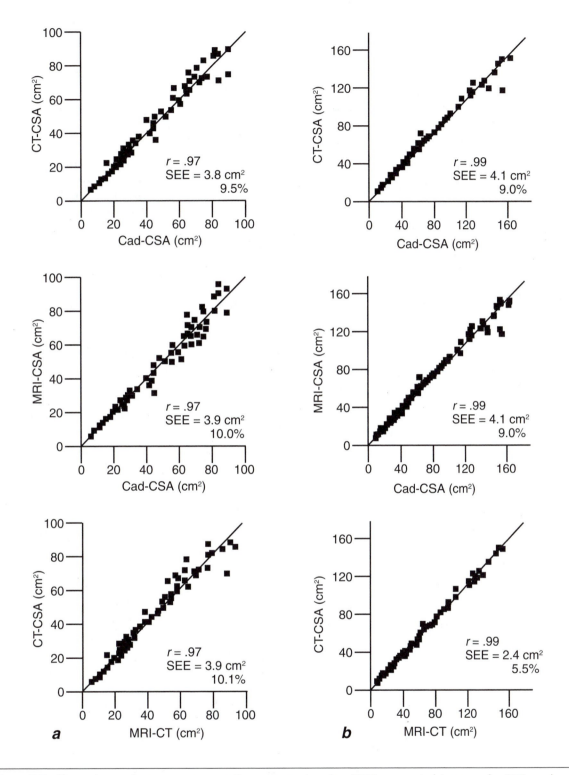

Figure 7.5 Regression analyses among magnetic resonance imaging (MRI), computed tomography (CT), and cadaver (Cad) measures of adipose tissue-free skeletal muscle (*a*) and subcutaneous adipose tissue (*b*). The cross-sectional area (CSA) values were obtained from 119 slices obtained in two cadaver limbs. SEE = standard error of estimate.

Adapted, by permission, from N. Mitsiopoulos et al., 1998, "Cadaver validation of skeletal muscle measurement by magnetic resonance imaging and computerized tomography," *Journal of Applied Physiology* 85: 115-22.

correlated and comparable. CT and MRI measures of visceral AT were also well correlated in both studies; however, the MRI estimates of visceral AT were about 15% smaller than the CT estimates of visceral AT. It is possible that methodological differences between in vivo CT and MRI measures influenced the quantification of visceral AT in these two studies. In particular, these studies were conducted before the development of MRI breath-hold sequences (Ross et al. 1996b), and motion artifact may have influenced the ability to accurately measure visceral AT. Additional studies are required wherein CT and MRI abdominal AT measures are compared in a larger and more heterogeneous sample and newer breath-hold sequences are used for acquiring the MRI data.

Validity of Tissue Quality by CT and MRI

No comparisons have been made between CT and MRI measures of tissue composition and direct chemical extraction from human cadavers or animal carcasses. Thus, for tissue quality, the validity (accuracy) of CT and MRI has been determined by comparing the imaging methods with tissue biopsy samples in humans.

CT

A recent validation study showed that CT-determined skeletal muscle attenuation characteristics are well correlated with skeletal muscle lipid levels determined in muscle biopsy samples (Goodpaster et al. 2000a). The correlations between mean muscle attenuation (e.g., average HU) in the midthigh and muscle triglyceride content and oil red O staining of lipid in muscle fibers measured in percutaneous biopsy samples from the vastus lateralis were .58 and .43, respectively. These results provide strong evidence that attenuation of skeletal muscle in vivo determined by CT is related to muscle lipid content.

MRI

Limited findings suggest that MRI can be used as a valid noninvasive tool for measuring tissue composition. Schick and colleagues (2002) found a strong correlation between the total muscle fat assessed by fat imaging with total muscle fat (intra- + extramyocellular lipid) determined by magnetic resonance spectroscopy ($r = .91$ in soleus, $r = .55$ in tibialis anterior). Marks and colleagues (1997) found a strong correlation ($r = .87$) between the liver fat index, as determined by MRI, and histo-

logical lipid assessment of liver biopsies in 10 men with type 2 diabetes. Clearly there is a need for additional studies that would confirm the accuracy of MRI measures of tissue composition.

Reproducibility of Tissue Quantity by CT and MRI

Reproducibility is a poorly understood concept in CT and MRI research. Most CT and MRI researchers have interpreted reproducibility to mean the ability to duplicate tissue area measurements on the same CT or MRI images. For example, if one observer analyzes the same set of images twice (e.g., intraobserver error), do his or her results for the first analysis compare with the results for the second analysis? When this type of repeated analysis is performed, most researchers report almost nonexistent intraobserver errors (e.g., 1%). This is expected given that the same images were analyzed and that much of the image analysis is performed using automated methods inherent to the image analysis software (e.g., based on range of HU values of, for example, −190 to −30 for AT). Ideally, intraobserver error should be determined by acquiring images on two separate occasions (e.g., separated by 1 day) in each subject and then comparing a single observer's analysis of the two sets of images. This approach will take into consideration variation in both image acquisition and image analysis. This approach was used for all of the studies listed in table 7.2.

Intraobserver Error

Inspection of table 7.2 reveals that, in general, CT tissue area and volume measurements are highly reproducible. Hudash and colleagues (1985) reported that the coefficient of variation (CV) for repeat CT skeletal muscle area measurements in the thigh was 1.4%. Kvist and colleagues (1986) reported an average error of 0.6% for whole-body AT volume for duplicate measures in four subjects. For abdominal AT measures in adults, Thaete and colleagues (1995) reported CVs of 1.9% and 3.9% for subcutaneous and visceral AT, respectively. For abdominal AT measures in children, who have much smaller abdominal AT depots, Figueroa-Colon and colleagues (1998) reported CVs of 10.7% for subcutaneous AT and 21.5% for visceral AT when repeated measurements were made up to 6 weeks apart. Without exception, the CVs presented for CT represent repeated analysis using a single HU range.

Table 7.2   Reproducibility (Intraobserver Error) of CT and MRI Measures

Reference	Subjects	Scan location	CORRELATION (SEE, %) [CV, %]		
			Subcutaneous AT	Visceral AT	Skeletal muscle
CT					
Thaete et al. 1995	16 Women	Abdomen	.99 (1.9)	.99 (3.9)	–
Figueroa-Colon et al. 1998	61 Girls	Abdomen	.96 (2.5) [10.7]	.67 (7.3) [21.5]	–
Sjöström et al. 1986	4 Women	Whole body	.99 [0.6][a]	–	–
MRI					
Mitsiopoulos et al. 1998	6 Adults	Thigh	.99 (2.5)	–	.99 (2.9)
Staten et al. 1989	6 Adults	Abdomen	.99 (1.2)	.99 (5.2)	–
Staten et al. 1989	6 Adults	Thigh	.99 (4)	–	–
Sohlstrom et al. 1993	25 Women	Whole body	[1.7]	[5.3][b]	–
Abate et al. 1994	3 Cadavers	Abdomen	[2.2]	[6.0]	–
Ross et al. 1992	3 Men	Abdomen	[1.1]	[5.5]	–
Gerard et al. 1991	4 Adults	Abdomen	[3.0]	[9.0]	–
Seidell et al. 1990	7 Men	Abdomen	.93 [10.1]	.96 [10.6]	–

Note. CT = computed tomography, MRI = magnetic resonance imaging, SEE = standard error of the estimate, CV = coefficient of variation, AT = adipose tissue.

[a]Reported as nonsubcutaneous AT; [b]total AT.

Overall, the reported reproducibilities of body composition estimates for MRI are slightly higher than those reported for CT (table 7.2). For example, Seidell and colleagues (1987) reported CVs for repeated measurements of visceral and abdominal subcutaneous AT in men of 10.6 and 10.1%, respectively. Again, this study was conducted before the advent of breath-hold sequences, which may have led to more motion artifact and increased measurement error than would be expected with current MRI acquisition protocols. Ross and colleagues (1992) reported much smaller CVs for repeated MRI measures of abdominal subcutaneous and visceral AT in three subjects of 1.1% and 5.5%, respectively. Mitsiopoulos and colleagues (1998) reported CVs for repeated in vivo MRI measures of appendicular subcutaneous AT and skeletal muscle in six subjects of 2.5% and 2.9%, respectively.

Interobserver Error

Another important concept when determining the reproducibility of CT and MRI measures is the interobserver error. This is the error introduced when two or more individuals are analyzing images for a given study. An early report indicated interobserver errors of 0.7%, 0.4%, and 2.1% for CT measures of total AT, skeletal muscle, and visceral organs, respectively (Brummer et al. 1993). For in vivo MRI measures of skeletal muscle and subcutaneous AT obtained at the midthigh level, Mitsiopoulos and colleagues (1998) reported that the correlations between two observers approached unity (r = .99 and standard error of estimate ~3% for both tissues) and that the intra- and interobserver errors were similar in magnitude. For whole-body MRI measures of skeletal muscle, total AT, subcutaneous AT, and visceral AT in six subjects, Janssen and colleagues (2000b, 2002b) reported that the mean difference between two observers was 2% or less. Together, these results suggest that minimal interobserver error is introduced for CT and MRI measures of tissue quantity.

Reproducibility of the Measurement of Tissue Quality by CT and MRI

Very little is known about the reproducibility of CT and MRI measures of tissue composition. Goodpaster and colleagues (2000a) reported that the test–retest CV for skeletal muscle attenuation in two CT scans performed in six volunteers was less than 1% for the midthigh and midcalf. Marks and colleagues (1997) reported that the CV for repeated measures of liver fat by MRI in 10 people with type 2 diabetes was 4.6%. As the application of CT and MRI for measuring tissue composition continues to increase, it will be important to determine the limits of these methods so that true biological variations in tissue quality are properly interpreted.

Applications of CT and MRI on Tissue Quantity and Quality

Heymsfield and colleagues (1979) were among the first to explore the use of CT in body composition research. They initially used CT to quantify arm muscle cross-sectional area and visceral organ volumes (Heymsfield et al. 1979) in 1979. Borkan and colleagues (1982) were the first to evaluate CT as a method for estimating visceral AT in 1982. Sjöström and colleagues (1986) introduced whole-body CT imaging and multiple-component analysis to quantify total body and regional AT, skeletal muscle, bone, and other organ–tissue volumes. Since these original studies, CT has been used to measure AT and skeletal muscle in children (Cruz et al. 2002; Figueroa-Colon et al. 1998; Goran et al. 1995; Gower et al. 1999; Herd et al. 2001), normal-weight men and women (Goodpaster et al. 1997, 2000b; Toth et al. 2001), obese men and women (Brochu et al. 2000; Goodpaster et al., 1997, 1999, 2000b), people with diabetes (Goodpaster et al. 1997, 1999), elderly persons (Brochu et al. 2000; Overend et al. 1992a, 1992b; Rice et al. 1989), and patients with muscular dystrophy, HIV, and AIDS (Brummer et al. 1993; Liu et al. 1993; Nordal et al. 1988; Saint-Marc et al. 2000). These and other studies have examined the effect of obesity, weight loss, exercise, inactivity, disease, and aging on abdominal AT distribution and associated metabolic risk and on skeletal muscle size and associated function. The vast majority of these studies have based their observations on a single CT image obtained in the abdomen, typically at the level between the fourth and fifth lumbar vertebrae (L4-L5 level), or, in the case of skeletal muscle, at the midthigh level.

The principal application of MRI in human body composition research has been to characterize the quantity and distribution of AT and skeletal muscle. Foster and colleagues (1984) were among the first to illustrate the applicability of MRI in body composition research. Hayes and colleagues (1988) first used MRI to demonstrate the quantification of subcutaneous adipose tissue distribution in human subjects. Ross and colleagues (1992) developed an MRI protocol for measuring whole-body and regional AT and lean tissue. Since these original studies, MRI has been used to measure AT and lean tissue in fetuses (Deans et al. 1989), children (de Ridder et al. 1992; Eliakim et al. 2001), normal-weight males and females (Ross et al. 1992; Thomas et al. 1998), obese males and females (Fowler et al. 1991; Gray et al. 1991; Ross and Rissanen 1994; Ross et al. 1995, 1996a, 1996b, 2000, 2002), people with diabetes (Abate et al. 1996; Gray et al. 1991), and elderly persons (Ivey et al. 2000; Janssen et al. 2000b; Klein et al. 2002; Roth et al. 2001; Tracy et al. 1999). Although the aforementioned studies based their observations in large measure on a single MRI image, it is also possible to safely acquire whole-body MRI because, unlike CT, MRI does not use ionizing radiation.

The acquisition of whole-body MRI data offers distinct advantages in assessing changes in body composition. For example, weight reduction may induce regional changes in AT or muscle. Thus, if an increase in skeletal muscle in one anatomical region is masked by a loss of skeletal muscle in another, only MRI studies could discover it. Indeed, whole-body MRI protocols have been used to make important observations concerning the effects of various diet and exercise perturbations on total and regional AT (Janssen and Ross 1999; Ross and Rissanen 1994; Ross et al. 1995, 1996b, 2000) and skeletal muscle distribution (Janssen and Ross 1999; Ross et al. 1996b, 2000). Whole-body MRI has also been used to describe age-related muscle loss or sarcopenia (Janssen et al. 2000b). Using fewer images, investigators have used MRI to assess the influence of inactivity (Abe et al. 1997), space travel (LeBlanc et al. 1995, 2000), and resistance training (Conley et al. 1997; Ivey et al. 2000; Roth et al. 2001; Tracy et al. 1999) on region-specific changes in skeletal muscle in, for example, the appendicular regions.

Abdominal AT

One of the most common applications of CT and MRI imaging methods in body composition research is the measurement of abdominal AT distribution. CT and MRI are uniquely capable of distinguishing between the subcutaneous and visceral AT depots that comprise abdominal obesity. Clearly, the ability to measure abdominal subcutaneous and visceral AT using these methods represents a major advance in our understanding of the relationships between obesity phenotype and health risk. Numerous studies have clearly identified that the division of abdominal AT into subcutaneous and visceral depots provides novel insights into the relationship between abdominal obesity and related comorbid conditions (Abate et al. 1995, 1996; Brochu et al. 2000; DeNino et al. 2001; Despres et al. 1989, 1990; Goodpaster et al. 1997, 1999; Janssen et al. 2002a; Kelley et al. 2000; Rice et al. 1999; Ross et al. 1996a, 2000, 2002; Toth et al. 2001); however, the independent contribution of these AT depots to metabolic risk remains a topic of debate (Frayne 2000; Seidell and Bouchard 1997).

Abdominal subcutaneous and visceral AT depots can be further subdivided according to differences in anatomic and metabolic characteristics. Abdominal subcutaneous AT measured by CT can be subdivided into superficial and deep compartments using the fascia superficialis (figure 7.6). The rationale for such a division comes from animal studies indicating that lipids are depleted and deposited at a faster rate in the deep layer compared with the superficial layer, suggesting that deep abdominal subcutaneous AT functions as a metabolically active tissue whereas superficial abdominal subcutaneous AT functions primarily as a thermoinsulatory or storage layer (Carey 1997; Mersmann and Leymaster 1984). These earlier findings from animal studies were confirmed by Monzon and colleagues (2002), who reported that lipolytic activity was higher in adipocytes isolated from deep subcutaneous AT compared with adipocytes isolated from superficial subcutaneous AT. If we assume that the mobilization of free fatty acids adversely affects insulin action (Randle et al. 1963; Shulman 2000), it follows that the deep AT depot would be the stronger predictor of insulin resistance.

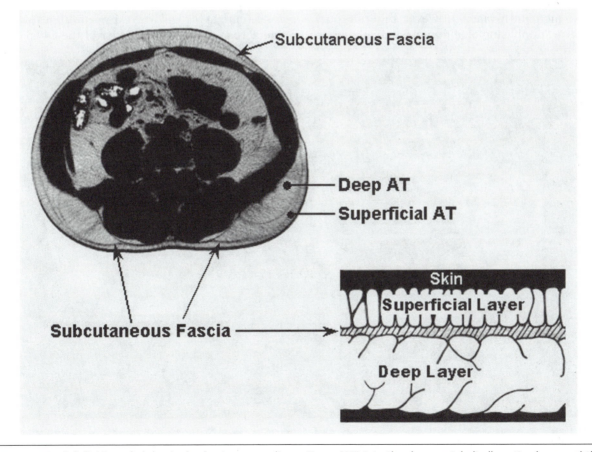

Figure 7.6 Subdivision of abdominal subcutaneous adipose tissue (AT) into the deep metabolically active layer and the superficial storage layer with the fascia superficialis.

Because the fascia superficialis is not visible on MR images, it is not possible to determine deep and superficial AT depots in a manner similar to CT. However, because the majority of the deep layer is located in the posterior half of the abdomen, a line can be drawn dissecting the abdomen into posterior and anterior depots (figure 7.7). In this way it is assumed that posterior abdominal subcutaneous AT measured by MRI is analogous to deep subcutaneous abdominal AT measured by CT (Misra et al. 1997). This is reasonable given that approximately three fourths of the deep compartment is located in the posterior abdomen (Smith et al. 2001).

Whether subdivision of abdominal subcutaneous AT provides additional insight into the relationships between abdominal AT distribution and metabolic risk per se remains unclear. Some report that deep and posterior abdominal subcutaneous ATs are stronger correlates of insulin resistance than are superficial and anterior abdominal subcutaneous ATs, respectively. For example, Kelly and colleagues (2000) report that deep, but not superficial, subcutaneous AT is strongly related to insulin resistance in a cohort of lean and obese men and women. However, others report that the subdivision of abdomnial subcutaneous

AT provides no additional insight. For example, in a group of 89 obese men Ross and colleagues (2002) observed similar correlations between fasting and oral glucose tolerance test insulin area with total abdominal subcutaneous AT as well as the anterior and posterior compartments. These discrepant observations highlight the need for additional research to clarify whether or not there are metabolic differences between the deep and superficial layers and whether or not deep subcutaneous AT is an independent predictor of insulin resistance.

The subdivision of abdominal subcutaneous AT based on metabolic characteristics is analogous to the partitioning of visceral AT into intraperitoneal and extraperitoneal depots on the premise that free fatty acids from the intraperitoneal depot alone (omental and mesenteric adipocytes) are delivered directly to the liver, the so-called portal theory (Bjorntorp 1990, 1991). Subdivision of visceral AT into intraperitoneal and extraperitoneal depots on CT or MRI images is not straightforward because the peritoneum is not visible with either method (figure 7.7). Thus, visceral AT is subdivided into intraperitoneal and extraperitoneal AT areas (cm^2) at the L4-L5 level by drawing a straight line across the anterior border of the L4-L5 disc

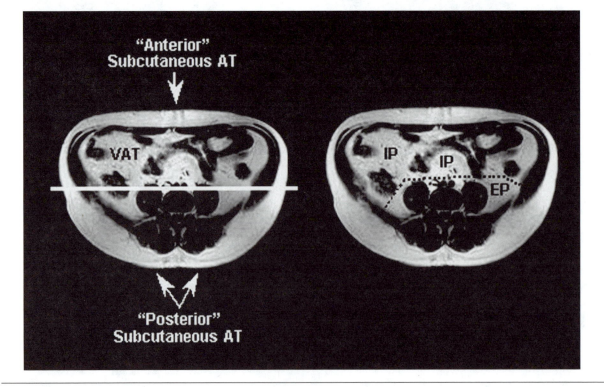

Figure 7.7   Subdivision of abdominal subcutaneous adipose tissue (AT) into anterior and posterior depots (left panel) and visceral AT into intraperitoneal and extraperitoneal depots (right panel). VAT, visceral AT; IP, intraperitoneal AT; EP, extraperitoneal AT.

and the psoas muscles, continuing on a tangent toward the inferior borders of the ascending and descending colons, and extending to the abdominal wall. For images in which the kidneys appear, an oblique line is drawn from the anterior border of the aorta and inferior vena cava to the anterior border of the kidney extending to the abdominal wall. On all images, extraperitoneal adipose tissue is defined as the AT located posterior to the lines drawn (figure 7.7).

Notwithstanding the uncertainties surrounding the accuracy of segmenting visceral AT into its component depots, whether segmentation improves on the relationships observed between visceral AT per se and metabolic risk factors is unclear. Although Abate and colleagues (1995, 1996) twice reported that intraperitoneal AT is a stronger correlate of insulin resistance compared with extraperitoneal AT, these authors did not confirm whether intraperitoneal AT was a better predictor of insulin resistance than visceral AT per se. Indeed, others reported that the subdivision of visceral AT does not improve on the relationship observed between visceral AT per se and insulin action (Rice et al. 1999; Ross et al. 1996a, 2002). Clearly there is a need for further study to clarify whether the segmentation of the intraperitoneal depot from visceral AT provides additional insight.

Skeletal Muscle

Accurate assessment of skeletal muscle has important applications in physiology (Evans 1997; Tracy et al. 1999), nutrition (Heymsfield et al. 1982), and clinical medicine (Baumgartner et al. 1998; Evans 1997; Janssen et al. 2002c). The effects of aging on muscle wasting, more commonly referred to as sarcopenia, has become a topic of interest in recent years. Janssen and colleagues (2000b) examined the effects of age on skeletal muscle mass, as determined by a whole-body MRI protocol, in a cross-sectional study of 468 men and women. In this cohort, skeletal muscle relative to body mass started to decline in the third decade whereas the absolute quantity of skeletal muscle was preserved until the fifth decade, with noticeable losses thereafter. This was true for both men and women. These results confirmed those of prior studies wherein skeletal muscle size was estimated by elemental analysis (Cohn et al. 1980), urinary creatinine excretion (Tzankoff and Norris 1977), duel-energy X-ray absorptiometry (DXA; Gallagher and Heymsfield 1998; Gallagher et al. 1997), and

single CT images (Borkan et al. 1983; Overend et al. 1992b; Rice et al. 1989). However, the acquisition of whole-body MRI data allowed these authors to make novel insights into the pattern of age-related muscle loss or sarcopenia (Janssen et al. 2000b). When examining regional changes, they found that the loss of skeletal muscle mass with age was considerably greater in the lower body than in the upper body for both sexes. This observation has important implications for the development of the optimal intervention strategies designed to prevent and treat sarcopenia. For example, these findings suggest that resistance training programs in older adults should place an extra emphasis on lower-body muscles (Janssen et al. 2000b).

Earlier studies examining changes in muscle quality (force per unit muscle mass) with imprecise estimates of muscle mass (e.g., creatinine) found that muscle quality did not change with advancing age. For example, in a study of more than 200 men and women aged 45 to 78 years, Frontera and colleagues (1991) found that isokinetic and isometric strength decreased with aging. However, when data were corrected for whole-body muscle mass estimated by creatinine excretion, the age-related strength changes disappeared. More recent findings with more precise CT and MRI measures of muscle size disagree with these earlier observations (Brooks and Faulkner 1994; Metter et al. 1999). For example, in a longitudinal cohort of 12 older men, Frontera and colleagues (2000) reported that the loss in quadriceps muscle area, as assessed by a CT scan in the midthigh, was a major predictor of the loss in quadriceps muscle strength. However, although the average reduction in muscle size was 1.4% per year, the average reduction in strength was considerably larger (2.0-2.5%, depending on the contraction velocity). These findings, and those of others who used CT or MRI to measure muscle size (Grimby 1995), suggest that there is a loss of specific force as well as muscle size with advancing age. The finding that muscle force per unit mass decreases is consistent with the observation that there are age-related neuromuscular changes including a decline in the number (Doherty et al. 1993; Tomlinson and Irving 1977) and firing rates (Roos et al. 1999) of motor units.

Organ Volume

An important area of research in the field of energy metabolism and obesity has been establishing the factors that account for individual differences in resting energy expenditure (REE). Early workers in the field found that it was necessary to adjust for differences in body size (e.g., body mass) when expressing REE (Kleiber 1932). Subsequent studies discovered that estimates of fat-free mass made for an improved measure of metabolically active body mass versus body size alone (Cunningham 1980). Nonetheless, individuals with a large fat-free mass have a lower REE to fat-free mass ratio than do individuals with a small fat-free mass (Weinsier et al. 1992). It has been hypothesized that with increasing fat-free mass, the proportion of fat-free mass as high-metabolic-rate organs (e.g., brain, liver) decreases, whereas the proportion of fat-free mass as low-metabolic-rate organs and tissues (e.g., bone and muscle) increases. However, the means of exploring this hypothesis in humans was unavailable until the advent of MRI for the study of human body composition. With MRI, investigators can safely and accurately quantify the volume of all major energy-producing tissues and organs in human subjects.

Using advanced imaging methods, researchers have recently tested the aforementioned hypothesis in cohorts of healthy adults (Gallagher et al. 1998; Heymsfield et al. 2002; Illner et al. 2000). Using MRI-derived organ and tissue volumes combined with previously reported organ–tissue metabolic rates and chemical composition, these investigators developed in vivo models to estimate whole-body REE. For example, Gallagher and colleagues (1998) estimated REE from organ and tissue volumes in 13 subjects and found that estimated REE was highly correlated ($r = .92$) with actual REE measured by indirect calorimetry. Furthermore, there were no significant differences between the mean measured (7045 kJ/d) and estimated (6926 kJ/d) REE values in these 13 subjects (Gallagher et al. 1998). The results of these three studies (Gallagher et al. 1998; Heymsfield et al. 2002; Illner et al. 2000) support the hypthesis that the proportion of high and low metabolic components varies as a function of lean body mass and that the mass of internal organs contributes significantly to the variance in REE. These findings, as well as the body composition approach employed in these studies, has the potential to greatly expand our knowledge of the energy expenditure–body size relationships in humans.

Skeletal Muscle Quality

Early studies that used CT to measure muscle composition found associations between reduced skeletal muscle attenuation (e.g., decreased attenuation = increased fat) with aging (Overend et al. 1992b; Rice et al. 1989) and Duchenne muscular dystrophy and other myopathies (Nordal et al. 1988). Goodpaster and colleagues (1997, 1999) observed that the mean muscle attenuation within skeletal muscle was reduced in obesity and type 2 diabetes mellitus and that weight loss increased the mean attenuation value of muscle.

The use of CT for examining altered muscle composition in vivo also has applications in clinical medicine given that an increase in skeletal muscle lipid content (both intramyocellular lipid and total lipid content) is implicated in insulin resistance. As shown in table 7.3, six studies have examined the relationship between CT-derived skeletal muscle attenuation characteristics and insulin resistance assessed with the hyperinsulinemic euglycemic clamp technique (Brochu et al. 2000; Goodpaster et al. 1997, 2000b; Ross et al. 2002; Simoneau et al. 1995; Toth et al. 2001). Of these six studies, three found a significant relationship between muscle composition and insulin sensitivity (Goodpaster et al. 1997, 2000b; Simoneau et al. 1995), whereas three did not (Brochu et al. 2000; Ross et al. 2002; Toth et al. 2001). It is possible that these discrepant findings are explained by differences in the cohorts examined. In the studies wherein muscle attenuation predicted insulin resistance, the investigators studied heterogeneous subjects who varied widely in body composition (e.g., lean to obese) and insulin sensitivity (Goodpaster et al. 1997, 2000b; Simoneau et al. 1995). Conversely, in the studies wherein muscle attenuation did not predict insulin resistance, the investigators studied homogeneous groups of obese (Brochu et al. 2000; Ross et al. 2002) or lean (Toth et al. 2001) subjects. Limiting the range of the dependent variable in these studies may have reduced the correlation between CT muscle attenuation characteristics and insulin sensitivity. It is also important to note that muscle attenuation values by CT are not analogous to intramyocellular lipid values. This is relevant, as intramyocellular lipid, but not extramyocellular lipid, is related to insulin resistance (Jacob et al. 1999). This is not a trivial concern and raises questions as to the utility of CT to determine the relationship between skeletal muscle lipid and insulin resistance.

In addition to its potential impact on insulin sensitivity, recent evidence suggests that skeletal muscle lipid content influences muscle strength and function. Goodpaster and colleagues (2001) report that lower mean CT attenuation values of

Table 7.3 Relationship Between CT-Measured Muscle Composition and Insulin Sensitivity

		SUBJECT CHARACTERISTICS				
Reference	N	Sex	Age (years)[a]	BMI (kg/m²)[a]	CT method for measuring muscle lipid content	Correlation with insulin sensitivity[b]
Brochu et al. 2000	44	Obese postmenopausal women	56 ± 5	35 ± 5	Mean muscle attenuation	NS
Simoneau et al. 1995	17	Lean + obese premenopausal women	31 ± 4	29 ± 6	Low-density muscle area	-.63
Goodpaster et al. 1997	54 15 39	Men + women (wide range) Lean men + women Obese men + women	37 ± 6 NR NR	31 ± 6 23 ± 2 34 ± 6	Mean muscle attenuation Mean muscle attenuation Mean muscle attenuation	.48 NS .47
Goodpaster et al. 2000b	66	15 Lean + 40 obese + 11 type 2 diabetic men and women	NR	NR	Mean muscle attenuation Low-density muscle area Intramuscular fat area	.41 −.50 −.45
Toth et al. 2001	45	Lean premenopausal women	47 ± 3	~23	Mean muscle attenuation	NS
Ross et al. 2002	40	Obese premenopausal women	42 ± 7	32 ± 3	Mean muscle attenuation Low-density muscle	NS NS

Note. CT = computed tomography, BMI = body mass index, NS = nonsignificant correlation ($p > .05$), NR = not reported.

[a]Group mean ± standard deviation; [b]insulin sensitivity was measured by the hyperinsulinemic euglycemic clamp in all studies. All listed correlations are significant ($p < .05$).

skeletal muscle in the midthigh (e.g., increased skeletal muscle lipid) were associated with lower strength values in 2,627 older men and women. As shown in figure 7.8, this relationship persisted after adjusting for skeletal muscle area, suggesting that skeletal muscle composition has an independent effect on muscle strength (Goodpaster et al. 2001). Recent observations from this group indicate that lower skeletal muscle attenuation values are associated with an increase in functional impairment in the elderly, providing further evidence that skeletal muscle lipid content has a negative impact on muscle function (Visser et al. 2002).

Liver

Several investigators have used CT to measure liver composition. A focus of this research has been to determine possible links between fatty liver and insulin resistance (Banerji et al. 1995;

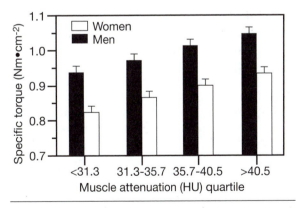

Figure 7.8 Specific torque (maximal torque per unit muscle cross-sectional area) of the knee extensors according to quintiles of skeletal muscle attenuation values. For both older men (n = 1,285) and women (n = 1,342) there was a significant increase in torque with increasing muscle attenuation (e.g., lower muscle lipid content).

Reprinted, by permission, from B.H. Goodpaster et al., 2001, "Attenuation of skeletal muscle and strength in the elderly: The health ABC study," *Journal of Applied Physiology* 90: 2157-2165.

Goto et al. 1995). Determination of liver fat may provide a mechanistic link between abdominal obesity, in particular visceral AT, and increased metabolic risk. It is hypothesized that the metabolic importance of visceral AT may be attributable to the delivery of free fatty acids into the portal system, exerting potent and direct effects on the liver (Bjorntorp 1990, 1991). Although it is not possible to readily access the portal circulation in vivo, sustained delivery of free fatty acids to the liver may increase the infiltration of lipid within the liver. Consistent with this position, liver fat determined indirectly as a diminished attenuation in CT scans is correlated with visceral AT (Banerji et al. 1995; Busetto et al. 2002; Goto et al. 1995; Mahmood et al. 1998; Nguyen-Duy et al. 2003). Furthermore, weight loss–mediated reductions in visceral AT are associated with corresponding reductions in liver fat (Busetto et al. 2002).

Few studies have considered whether a relationship exists between visceral fat, liver fat assessed by CT, and metabolic risk factors. Banerji and colleagues (1995) reported that visceral AT was associated with CT-measured liver fat and that both visceral AT and liver fat were correlates of plasma triglycerides and insulin resistance in a small cohort of black men. Nguyen-Duy and colleagues (2003) reported that visceral AT was associated with CT-measured liver fat in 161 white men and that both visceral AT and liver fat were significant independent correlates of metabolic risk factors. However, visceral AT remained a significant correlate of plasma triglycerides and HDL-cholesterol independent of liver fat, and liver fat remained a significant correlate of plasma glucose and triglycerides independent of visceral AT, suggesting that visceral AT and liver fat carry independent health risks.

Using the MRI approach, Marks and colleagues (1997) found a strong correlation between the liver fat index and MRI-measured visceral AT area ($r = .79$) in 19 patients. In addition, they reported that a 12-week dexfenfluranime treatment in 10 patients was associated with a signifcant reduction in the liver fat index that was correlated with the reduction in visceral AT ($r = .84$) as well as the improvement in insulin sensitivity ($r = .62$). These findings highlight a potential new application of MRI in body composition research.

Future Directions

At least three groups have used CT to evaluate whole-body AT and lean tissue distribution in

humans (Sjöström 1991; Tokunaga et al. 1983; Wang et al. 1996). Although whole-body CT data can be obtained within minutes, obtaining multiple CT images over the whole body exposes the subjects to high levels of radiation, which, by some standards, is considered unethical for research purposes. Several groups have used MRI to evaluate whole-body AT or lean tissue distribution (Baumgartner et al. 1993; Busetto et al. 2002; Lee et al. 2000; Ross et al. 1992; Selberg et al. 1993; Sohlstrom et al. 1993; Thomas et al. 1998). Among studies using whole-body protocols, there is considerable variation in both the number and location of the CT or MRI images used to determine either AT or skeletal muscle volume. For example, Kvist and colleagues (1986) suggested that nine carefully placed (landmarked) images can provide accurate estimates of regional and whole-body AT. By comparison, Thomas and colleagues (1998) developed a whole-body MRI protocol wherein there are only 10-mm gaps between consecutive images in the torso and 30-mm gaps between consecutive images in the arms and legs. The extent to which variation in image number, placement, or interimage gap influences the derivation of tissue volume is unknown and requires further study.

Abdominal AT

It is not yet firmly established whether a single axial image or multiple images are needed to accurately estimate abdominal AT and, if multiple images are required, how many images should be obtained. Using a contiguous multiple-slice MRI protocol that covered the majority of the abdominal region, Abate and colleagues (1997) examined the relationship between subcutaneous and intraperitoneal AT volume and area at levels corresponding to the intervertebral discs beginning at T12-L1 and extending to L5-S1 (e.g., T12-L1, L1-L2, . . ., L5-S1). In general, for both subcutaneous and intraperitoneal AT, all six images showed high predictive values for estimating AT volume. Abdominal subcutaneous AT area at the L5-S1 space was the strongest correlate ($r = .96$), whereas abdominal subcutaneous AT area at the commonly used L4-L5 space was the poorest correlate ($r = .87$) of the volume measure. Intraperitoneal AT area at the L3-L4 space was the strongest correlate ($r = .93$) and intraperitoneal AT area at the L4-L5 space was also a strong correlate ($r = .87$) of the volume measure. Others have also reported that the areas of subcutaneous and visceral AT from single images at different levels of the abdomen are highly correlated with the volumes of the respective tissues derived from a series of images covering the entire abdominal region (Abate et al. 1997; Chowdhury et al. 1994; Han et al. 1997; Ross et al. 1992, 2002).Consistent with these observations, it is reported that abdominal subcutaneous and visceral AT area determined from a single slice appears to correlate as strongly with metabolic variables as do the respective volumes determined from multiple images (Rissanen et al. 1994; Ross et al. 2002).

Another issue that remains unresolved is whether a single abdominal image can be used to classify or rank individuals according to visceral AT deposition. In other words, if individuals were ranked according to the amount of visceral AT they had at the L4-L5 level, would the ranking be the same if L3-L4 were used to determine visceral AT? This question was considered by Greenfield and colleagues (2002), who ranked 19 premenopausal women for CT-measured visceral AT deposition at four separate levels. The authors noted a marked level-dependent intrasubject variation. For example, the two subjects ranked 8th and 9th for visceral AT at the L3 level (10.7 vs. 10.8 cm^2) had a twofold difference in visceral AT at the L4 level (12.9 vs. 24.8 cm^2), where they were ranked 7th and 12th, respectively. Drawing on these findings, the authors concluded that the use of a single CT image to determine the quantity of visceral AT in premenopausal women should be interpreted with caution. A limitation of this study was the extremely low values for visceral AT. Thus, a small variation in visceral AT (e.g., 3 cm^2) had a profound influence on the ranking.

Consistent with the findings reported by Greenfield and colleagues (2002), Thomas and Bell (2003) reported that the ranking of 59 women according to their visceral AT content was significantly altered by the level of the abdominal MRI image. The authors concluded that the level of visceral AT measurement must be considered when ranking individuals. In the same study, the authors also examined the effects of a 6-month exercise intervention on changes in visceral AT at different levels of the abdomen in 19 women. The findings suggested that single-image MRI was suitable for measuring changes in visceral AT, most likely because each subject served as her own control. Prior studies have also demonstrated that the observations for diet- and exercise-induced changes in visceral and abdominal subcutaneous AT area at the L4-L5 level are consistent with those for volume changes (Janssen and Ross 1999; Ross et al. 1996b, 2000).

Apart from the discrepanices in the use of single or multiple abdominal images for estimating abdominal AT, there is considerable variation among studies using multiple-image protocols for both the number and location of the CT or MRI images used to determine either abdominal subcutaneous or visceral AT volume (Abate et al. 1997; Han et al. 1997; Ross et al. 1992, 2002). This may partly explain the large differences among studies for reported abdominal subcutaneous and visceral AT volumes in apparently similar populations.

To be consistent with the definition of visceral AT as a depot that is portally drained, it has been suggested that the T10-T11 and L5-S1 intervertebral spaces be used to define the upper and lower visceral AT landmarks (Ross 1997). These two landmarks define a region within which the majority of intra-abdominal AT is within the visceral peritoneum (e.g., intraperitoneal AT). The visceral peritoneum is the membrane that covers the abdominal organs of the gastrointestinal tract, and all AT within the visceral peritoneum is portally drained. Intra-abdominal AT depots that are not contained within the visceral peritoneum include the extraperitoneal AT that surrounds the kidneys as well as internal AT deposits in the pelvic region and those surrounding the heart and lungs. Unlike intraperitoneal AT, these AT depots are drained into the systemic circulation.

We are unaware of a textbook definition of abdominal subcutaneous AT. It is well established, however, that important regional differences in lipolysis exist between abdominal subcutaneous adipocytes and subcutaneous adipocytes in the appendicular regions (Jensen 1997; Marin et al. 1992; Rebuffe-Scrive et al. 1990). To our knowledge, it is unknown whether the metabolism of adipocytes varies across the abdominal region. For example, is the lipolytic rate of adipocytes in a region proximal to L1-L2 different from that of adipocytes in a region proximal to L5-S1? Without this information it is difficult to obtain consensus regarding a metabolically based definition for abdominal subcutaneous AT.

Regional Skeletal Muscle

It is not yet firmly established whether a single axial image or multiple images are needed to accurately estimate appendicular skeletal muscle and, if multiple images are required, how many images should be obtained. Levine and colleagues (2000) reported that CT-measured muscle cross-sectional area (cm^2) in the midthigh was significantly related to DXA-measured thigh lean soft tissue mass (the DXA equivalent of muscle mass), with a correlation of .74. However, in that study the correlation between CT muscle volume (liters) measures in the thigh and DXA-measured thigh lean soft tissue mass was much stronger at .96. These results suggest that a single CT or MRI image at the midthigh level may not accurately represent total muscle volume. The use of a single appendicular image in longitudinal or interventional studies may be associated with an even greater measurement error. This is demonstrated by the observation that resistance training results in a greater hypertrophic response in the proximal region of the quadriceps muscles (Tracy et al. 1999). Thus, obtaining a single image in the midthigh region may not accurately reflect the changes that occur in the size of the entire muscle (e.g., volume).

Potential of CT and MRI in Large-Scale Studies

Although primarily used in smaller laboratory-based studies, in recent years CT and MRI have also been used in larger-scale studies. Laboratories from Queen's University in Canada and Columbia University in New York have published a series of papers that include whole-body and regional MRI measures of skeletal muscle and AT for close to 500 subjects (Janssen et al. 2000a, 2000b, 2002b; Lee et al. 2000). The Health ABC study, a longitudinal study of approximately 3,000 elderly men and women, has acquired both abdominal and midthigh CT images (Goodpaster et al. 2001, 2003; Visser et al. 2002). These large-scale imaging studies should improve our understanding of the health implications of abdominal AT distribution and skeletal muscle size and composition.

Summary

Both CT and MRI produce high-resolution images of all major body composition components at the tissue–system level. Although the cost of CT and MRI scanning is variable, both methods are expensive and require technical skill. Furthermore, access during peak patient hours may be limited. Last, some very obese patients may be too large to fit within the scanners, and our experience is that CT and MRI study is limited to patients below a BMI of about 35 kg/m^2.

In general, CT has been shown to provide slightly more reliable and repeatable data than MRI, although this is changing rapidly with the introduction of new imaging techniques and image analysis software. A major advantage of MRI is the lack of ionizing radiation. No known health risks are associated with MRI at the current magnet field strengths of about 1.5 Tesla. Newer scanning protocols are rapid and cost is reduced. An important advantage of CT is that instruments are more widely available. A second advantage of CT is the high resolution of images and the consistency of tissue attenuation values from image to image. Within reasonable limits, skeletal muscle, AT, and other components have similar HU distributions within and between scans. The consistency of attenuation data allows development of standardized protocols for reading images and separating various tissues from each other. The major disadvantage of CT is the associated radiation exposure. Regional studies with appropriate scanner settings substantially lower radiation dose compared to whole-body studies. Radiation exposure is still a concern in long-term longitudinal studies with repeated measurements over time and in the study of children and women in childbearing years. Accordingly, application of multiple-image protocols using MRI is the method of choice for whole-body and serial measurements.

Chapter 8

Anthropometry and Ultrasound

Anna Bellisari and Alex F. Roche

The scope of this chapter is restricted to the postnatal period; comprehensive data for fetal body composition were reported by Singer and colleagues (1991), but little is known of the relationships between anthropometric variables and body composition during the fetal period. The chapter describes the relationships between anthropometric and body composition variables and the place of ultrasound in the measurement of body composition.

Anthropometry

Anthropometry can be applied in clinics and laboratories and in both urban and rural field situations. The instruments are portable and relatively inexpensive, procedures are noninvasive, and training can be provided "on the job" without prerequisite courses. Consequently, anthropometry is applicable to large samples and can provide national estimates and data for the analysis of secular changes in representative samples. However, anthropometry requires adequate training by an experienced professional and quality control, including analyses of reliability data and calibration of equipment during a study and during the course of clinical work. Calibration requires the use of standard weights, rods, or rules. Procedures for the calibration of skinfold calipers have been somewhat inadequate, but effective instruments

for this purpose were described recently (Carlyon et al. 1998; Hewitt et al. 2002).

Assumptions and Their Validity

The term *validity* refers to comparisons between observed measurements and the true values, and the term *precision* refers to repeatability judged from inter- or intraobserver differences. In the interpretation of an anthropometric value, it is commonly assumed that the tissues included in the measurement are in a "standard" state, for example, that muscles are fully relaxed and soft tissues are normally hydrated. If these conditions are not met, interpretation of the data may be invalid. Almost all anthropometric variables include a variety of tissues, and their separate influences on recorded values are not always clear. For example, variations in skin thickness among individuals affect the validity of skinfold thicknesses as measures of subcutaneous adipose tissue (SAT).

Most lengths and breadths are interpreted as skeletal dimensions because they are made between bony landmarks. These distances are influenced by soft tissues over the bony landmarks. Therefore, although skeletal breadths are highly correlated with corresponding radiographic breadths, the latter tend to be smaller (Young and Tensuan 1963). The effects of soft tissues on recorded lengths and breadths can be reduced by the use of recommended calipers and the application of firm pressure, but these effects remain considerable, especially for bi-iliac breadths. Data

This work was supported by the National Institutes of Health (Grant HD-12252).

from computed tomography (CT) studies suggest that anthropometric measures of abdominal breadth and depth are related to visceral adipose tissue (VAT) area at the same level (Kvist et al. 1988). These relationships are unlikely to be close enough to allow the estimation of VAT area from anthropometric data.

Head circumference is highly correlated with brain size during infancy (Cooke et al. 1977). Although few data are available, this relationship is probably less close after infancy because of variable increases in the thickness of the scalp and the cranium. Limb circumferences are difficult to interpret in terms of body composition because they include skin, SAT, muscle, bone, blood vessels, nerves, and deep adipose tissue (DAT). It is even more difficult to interpret trunk circumferences, which include organs in addition to the various tissues included in limb circumferences. Interpretation of buttocks (hip) circumference is uncertain because it includes large amounts of adipose tissue and muscle and is affected by pelvic size and shape. Limb and trunk circumferences are measured with a tape while minimal tension is applied so that the soft tissues will not be compressed; therefore, enlargement of muscle or SAT caused by edema increases the recorded measurements. Even standing for 1 to 2 hr, or prolonged sitting, causes an accumulation of extracellular fluid in the lower limbs leading to increases in ankle and calf circumferences.

Trunk circumferences, particularly those of the abdomen, are difficult to measure in those who are markedly overweight, but high precision may be attained (Bray et al. 1978; Rasmussen et al. 1993). In such individuals a natural waist may be lacking, and the umbilicus and the maximum anterior extension of the abdomen may be displaced from their usual vertebral levels. Consequently, measurements at the same nominal levels on the anterior surface of the abdomen (e.g., umbilicus) may be at different vertebral levels in the lean and the overweight and may differ in the organs included. Trunk circumferences are measured with the subjects standing; the values obtained are larger than those measured supine by 4.1 ± 1.7 cm (Kvist et al. 1988). The locations of most abdominal organs are relatively independent of subject positioning, but sagging of the anterior wall during standing in overweight persons is accompanied by movement of the small intestine and transverse colon to lower vertebral levels. Despite this, waist circumferences measured standing and those measured supine have the same relationship with visceral adipose tissue areas at L4-L5 (Ross et al. 1992).

As noted earlier, skinfolds include skin and SAT, the latter consisting of adipocytes that contain triglycerides and connective tissue that includes blood vessels, nerves, and tissue fluids. The thickness of a double layer of skin is about 1.8 mm, but this varies among individuals and systematically by site and with age (Bliznak and Staple 1975; Edwards et al. 1955). The paucity of SAT in lean persons can make it difficult to elevate a fold, and it is not easy to elevate skinfolds with parallel sides in those with large amounts of SAT. Consequently, skinfold thicknesses are less precise than circumferences in overweight individuals than in general populations (Bray et al. 1978). Skinfolds are, however, less affected by edema than circumferences because caliper pressure reduces the fluid content of the SAT.

Skinfold thicknesses are affected by the compressibility of SAT, which varies by site, age, sex, and recent weight loss (Kuczmarski et al. 1987; Weiss and Clark 1987). When a skinfold thickness is measured, the pressure exerted by the calipers displaces some extracellular fluid. This displacement is marked when the extracellular fluid content of SAT is high leading to increased compressibility, as occurs in preterm infants soon after birth and in malnutrition (Brans et al. 1974; Martin et al. 1985). In addition, pressure from skinfold calipers may force some adipose tissue lobules to slide into areas of lesser pressure; this sliding may be more marked for thick skinfolds in which the adipose tissue contains little connective tissue. The conformist view is that intersite and intersubject differences in skinfold compressibility reduce the utility of skinfold thicknesses. If, however, variations in compressibility reflect differences in the fluid and connective tissue content of SAT, the reduction of these differences by compression might increase the validity of skinfold thicknesses as measures of regional fatness.

The factors to be considered in the selection of skinfold sites for evaluation of body fatness, or for possible inclusion in predictive equations, include accessibility in relationship to undressing, precision, the availability of reference data, and the thickness of the fold, which is important in overweight subjects. Consideration of these criteria affirms the usual conclusion that the most useful skinfolds are the triceps and subscapular sites (Roche et al. 1985).

Anthropometry, when used in relation to body composition, is based on the assumption that tissue composition is independent of tissue size. This assumption may be violated. For example, the relative fat content of adipose tissue is positively related to SAT thicknesses within age groups (Pawan and Clode 1960; Thomas 1962), and the fat content becomes larger as SAT thicknesses increase during growth (Baker 1969; Kabir and Forsum 1993). Despite this limitation, and those noted earlier, anthropometry is an essential component of many in vivo methods for the measurement of body composition and of equations that predict body composition values. Additionally, anthropometric values, and the body composition values to which they are related, are significantly related to health.

Applicability

The choice of measures and procedures differs among some groups (Lohman et al. 1988). For example, the precise measurement of infants and preschool children requires that they be content; one cannot obtain precise measurements of hungry or thirsty infants or young children. Disabled and elderly subjects who cannot stand erect must be measured recumbent to obtain valid and precise data (Chumlea et al. 1985a, 1985b). The two-handed technique that was described by Damon (1965) increases the precision of skinfold measurements in overweight persons (Pham et al. 1995), but the maximum jaw openings of the calipers (Harpenden, 55 mm; Holtain, 50 mm; Lange, 65 mm) may make it impossible to measure skinfold thicknesses at some sites in overweight subjects (Himes 2001). As an alternative, measurements can be made at sites where there is little SAT (e.g., biceps), or ultrasound can be used to measure SAT thickness, but there are few reference data for unusual sites or for ultrasound values and few predictive equations are based on them.

The utility and interpretation of anthropometric variables are affected by short-term variations that may not be associated with changes in composition but may alter the values predicted from equations. There is a loss of stature and an increase in abdominal and calf circumferences with prolonged standing, and there is considerable day-to-day variability of weight attributable mainly to the intake and elimination of food and water (Edholm et al. 1974). These fluctuations in weight reflect alterations in extracellular water but

are not closely related to other aspects of body composition. Consequently, relationships between anthropometric variables and body composition should be determined from data recorded in the morning from subjects who are fasting and after they have eliminated. Corresponding restrictions must be applied to the collection of data to which the equations will be applied. Data should not be collected from 1 week before a menstrual period begins until the end of the period to avoid possible variation attributable to increases in the fluid content of fat-free mass (FFM) at this time (Bunt et al. 1989).

Measurement Procedures and Instruments

Detailed instructions for instrument selection and measurement procedures for anthropometric variables are provided by Lohman and colleagues (1988) and outlined on page 112. The instructions for the measurement of weights do not describe the use of electronic scales, which are now in wide use. Electronic scales are more accurate than beam scales and allow greatly improved measurements of infants. The procedures for their use are similar to those for beam scales. Interobserver coefficients of reliability (%) for most variables fall in the 95% to 99% range, including weight (~100%). However, there is a biological variation in weight that is not captured by measurements repeated after brief intervals, and coefficients of the reliability for skinfold thicknesses during infancy are 66.13 to 83.19.

Sources of reference data for most of the relevant anthropometric measures are provided by Lohman and colleagues (1988), and other valuable data are presented by Gerver and deBruin (2001). Lohman and colleagues (1988) did not describe the measurement of knee height and abdominal depth. Knee height is measured on the left side with large sliding calipers while the subject is supine with the knee and ankle each flexed to 90°. One blade of the caliper is placed under the heel and the other blade is placed on the anterior surface of the thigh just proximal to the patella with the caliper shaft parallel to the long axis of the tibia. Pressure is applied to compress the soft tissues, and the measurement is recorded to the nearest 0.1 cm. Knee height can be used to predict stature in those unable to assume the standard position for the measurement of stature (Chumlea et al. 1985a). Alternatively, arm span, which is little affected by

Selected Anthropometric Variables Relevant to Body Composition

Lengths

Stature
Recumbent length
Arm span
Knee height

Breadths

Biacromial
Bi-iliac
Knee
Ankle
Elbow
Wrist

Circumferences

Waist
Hip
Thigh
Calf
Arm
Wrist

Skinfold Thicknesses

Subscapular
Midaxillary
Paraumbilical
Suprailiac
Anterior thigh
Medial calf
Lateral calf
Triceps
Biceps

Area

Body surface area
Cross-sectional area

Volume

Body volume

Weight

Body weight

aging, can be used in place of stature for elderly individuals who are unable to stand (Kwok and Whitelaw 1991). Measured or predicted statures are used to calculate some indexes of body composition, such as the body mass index.

Abdominal depth is of interest because it is related to the amount of DAT in the abdomen (Kvist et al. 1988). The anthropometrist stands to the left of the standing subject and positions the stationary blade of a pair of spreading calipers in the midline posteriorly at the level of the maximum depth of the lumbar concavity. While the shaft of the caliper is held parallel to the floor, the moveable blade is brought into contact with the midline of the anterior abdominal wall and the measurement is read to the nearest millimeter.

Relationships Between Anthropometric Variables and Total Body Composition

In this discussion of the relationships between anthropometric variables and total body composition, some anthropometric variables or constructs, such as weight/stature^2, are referred to as indexes because they can be used to categorize individuals (e.g., lean, obese), but they do not provide metric values for aspects of body composition. Other combinations, such as triceps skinfold thickness or subscapular skinfold thickness, are used as indexes of regional body composition or adipose tissue distribution. All reported relationships between anthropometric variables and total body composition understate the actual relationships because neither the anthropometric variables nor the body composition variables are measured with exact precision. In this discussion, relationships of anthropometric values to percent body fat (%BF) and body density will be considered jointly because of their conceptual similarity. Some prefer body density to %BF as the dependent variable in predictive equations because the relationships with anthropometric values are not affected by the calculation of %BF or FFM from body density. This choice only delays the uncertainties. Almost always, after body density has been predicted, a body composition variable is calculated from it and these calculations are based on assumptions that may be inaccurate (see chapter 2).

Lengths and Breadths

Stature is not an effective predictor of FFM when used alone (Slaughter and Lohman 1980), although it is related to FFM with regression slopes of about

0.9 kg/cm for men and 0.5 kg/cm for women and lesser slopes in children and the elderly (Forbes 1974). Skeletal lengths and breadths have only low correlations with %BF, but breadths have correlations with FFM of about 0.6 that are reduced to about 0.3 when the effects of stature are removed (Himes 1991; Sinning and Wilson 1984; Sinning et al. 1985). Selected measures of the length, breadth, and depth of the skeleton can be combined conceptually as "frame size." For example, Katch and Freedson (1982) developed a frame size model that includes stature and the sum of biacromial and bitrochanteric breadths; stature is the main determinant of the frame size scores from this model. These scores are positively related to FFM and are nearly independent of %BF.

Circumferences

Correlations between abdominal circumferences and body density are about −.7 (Pollock et al. 1975, 1976; Schlemmer et al. 1990), and the correlations of limb circumferences with body density are about −.4 (Wilmore and Behnke 1969, 1970). The correlations of abdominal and limb circumferences with FFM are about .6 in each sex (Wilmore and Behnke 1969, 1970).

Skinfold Thicknesses

Skinfold thicknesses have low correlations with FFM (about .2); their correlations with %BF are higher ($r = .7$-$.9$) and do not differ markedly among the common sites (Frerichs et al. 1979; Lohman et al. 1975). Despite the relatively high correlations between skinfold thicknesses at single sites and %BF, no single site is an accurate predictor of %BF (Himes and Bouchard 1989; Lohman 1991). This reflects individual variations in the distribution of SAT and in the proportion of the total adipose tissue that is subcutaneous. Chin and cheek skinfold thicknesses are unlikely to be useful in the prediction of total body fatness because they are largely independent of thicknesses at other sites (Lohman et al. 1975; Shephard et al. 1969).

There are sex- and age-related differences in the relationships of skinfold thicknesses to %BF (Lohman 1981; Pařízková 1977). At a fixed body density and presumably similar %BF, women have thinner skinfolds than men and the elderly have thinner skinfolds than young adults. This indicates that the proportion of adipose tissue that is deep, particularly within and between muscles, may be larger in women and in the elderly. Only three or four skinfold thicknesses are needed in predictive equations (Jackson and Pollock 1978; Jackson et al. 1980). In these equations, high correlations among sites can make the regression coefficients unstable. This instability can be reduced by using the log of the sum of the skinfold thicknesses (Durnin and Womersley 1974), but this gives greater weight to the thicker skinfolds. Other methods to reduce instability of regression coefficients attributable to intercorrelations (multicollinearity) among predictor variables are described in chapter 11.

In children, skinfold thicknesses are better predictors of body density than are circumferences (Boileau et al. 1981; Harsha et al. 1978). Some have found that circumferences are more effective than skinfold thicknesses as predictors of body density in adults, perhaps because of their greater precision (Mueller and Malina 1987; Pollock et al. 1975), but others report little difference between the root mean square errors (RMSE) of equations using skinfold thicknesses and those using circumferences as predictor variables (Katch and McArdle 1973; Lohman 1988). In this chapter, the RMSE is used to summarize the differences between observed and predicted values instead of the standard error of the estimate. These terms are mathematically equivalent, but the standard error of the estimate is ambiguous because it could refer to the estimation of the regression coefficients.

Pairs of Variables

There are no known pairs of anthropometric variables that are good predictors of total body composition. Equations derived from subjects aged 4 to 65 years that use weight and skinfold thicknesses as independent variables have prediction errors of 0.2 ± 2.8 kg for total body fat (TBF) and 0.2 ± 5.2% for %BF (Kashiwazaki et al. 1996). Some pairs of anthropometric variables are used to screen for unusual body composition; the most common pairing is of weight and stature^2 as body mass index (BMI). Values of BMI are moderately correlated with %BF (r about .6-.8; Fukagawa et al. 1990; Roche et al. 1981). These correlations are increased only slightly when they are maximized by the use of fractional powers of weight and stature (Abdel-Malek et al. 1985; Sjöström 1987), and other weight-stature indexes have similar correlations with %BF and body density (Roche et al. 1981; Womersley and Durnin 1973). The RMSE of the prediction of %BF from BMI is about 3.5% to 5% BF (Deurenberg et al. 1990; Womersley and Durnin 1977). Despite these errors, BMI has high specificity (recognition of true negatives) in screening for high %BF values (Himes and Bouchard 1989).

A limb circumference and a skinfold thickness at the same level can be used to calculate the cross-sectional areas of adipose tissue and of "muscle plus bone." The correlations between calculated adipose tissue areas and %BF are only slightly stronger than those of skinfold thicknesses (Borkan et al. 1983; Himes et al. 1980). The cross-sectional areas of muscle plus bone are correlated ($r = .7$) with FFM (Reid et al. 1992).

Prediction of Percent Body Fat and Body Density

The statistical procedures to develop and cross-validate predictive equations, and guidelines for the application of such equations, are described in chapter 11. Predictive equations should be applied only after they have been successfully cross-validated in a population similar to the one that will be studied. In their application, the anthropometric procedures and instruments must match those that were used when the equations were developed. The sites for the measurement of skinfolds must match those in the validation study, and the same calipers must be used (Lohman et al. 1984; Ruiz et al. 1971). The importance of accurate site location for the measurement of skinfold thicknesses has been demonstrated by B-mode ultrasound images (Bellisari 1993).

The present account is restricted to equations in which only anthropometric variables are used as predictors. Selected predictive equations for children and adolescents, for young, middle-aged, and elderly adults, and for obese persons are considered. Equations that use combinations of anthropometric and impedance variables, are discussed in chapter 11. Few, if any, predictive equations can be recommended for individuals with major pathological conditions, partly because the assumptions underlying the calculation of body composition values from body density, total body water, or total body potassium may be invalid. In theory, disease-specific equations could be developed using valid dependent variables, but it would be difficult to obtain large samples that are homogeneous in relation to the duration, severity, and treatment of the disease. The disease and its comorbidities may make it impossible to perform the procedures needed to measure the dependent variable.

Almost all the equations considered were developed from validation samples larger than 50, the equations were cross-validated, and the RMSE for the validation group and the pure error (PE) for the cross-validation group were reported. The need for cross-validation of predictive equations is demonstrated by the findings of Boileau and colleagues (1981), who reported systematic bias in predictions of body density when equations from one group of boys were applied to another group of similar age, size, and body density.

Errors in the dependent variable should be considered in evaluating the RMSE and PE of predictive equations. The random error in observed values for %BF in young adults is about 4% BF with the two-component model and about 2% BF with a multicomponent model (Bakker and Struikenkamp 1977; Lohman 1992). The two-component model assumes the human body has two components—fat and fat-free mass—and that each of these has a fixed density. This model does not take into account known age and sex differences in the density of FFM and individual variability in this density. Multicomponent models include measures of the constituents of FFM, particularly water, mineral, and protein, for individuals. Multicomponent models, which include measures of the constituents of FFM, are preferable to the two-component model despite the additional cost of data collection, because there are systematic errors in children and adolescents (Guo et al. 1989), and perhaps in elderly and obese persons, when a two-component model is used to calculate %BF from body density.

In reviewing the selected equations, the distributions of stature, weight, and skinfold thicknesses will be considered relative to U.S. national data (Kuczmarski et al. 2002; Najjar and Rowland 1987). These comparisons indicate the nature of the groups to which the equations may be applicable. Considerable increases in weight have occurred during recent decades in the United States and in many other countries. These increases are probably associated with increases in skinfold thicknesses and may reduce the applicability of equations developed several decades ago. Most of the validation samples for the selected equations tended to be taller and lighter and to have thinner skinfolds than U.S. national samples, suggesting the applicability of these equations is not well established for general versus population groups. The selection of U.S. groups for validation is common because many predictive equations have been developed in studies of U.S. groups.

The use of skinfold thicknesses to predict %BF or body density is based on implicit assumptions that (a) measurements of skinfold thicknesses at a few sites provide an adequate description of SAT and (b) there is a fixed relationship between SAT and DAT. The first assumption appears correct because there are generally high correlations between skinfold thicknesses at commonly measured sites, but the second assumption may be inaccurate. There are age and sex differences in the relationships between SAT and DAT (Baumgartner et al. 1991; Durnin and Womersley 1974). Equations developed by stepwise regression to predict %BF or body density commonly overestimate the low values and underestimate the high values, but this tendency is reduced if power terms are included (Jackson and Pollock 1978). The inclusion of power terms in predictive equations that include skinfold thicknesses as predictors is conceptually desirable because there is a curvilinear relationship between skinfold thicknesses and body density (Boileau et al. 1981; Durnin and Womersley 1974; Mayhew et al. 1985).

Children and adolescents

Specific predictive equations are needed for children and adolescents because the distributions of adipose tissue and body proportions (e.g., leg length or stature) differ from those of adults. These differences can alter the relationships between anthropometric variables and body composition. In addition, when the dependent variable is obtained from densitometry, a four-component model should be applied as described in chapters 2 and 12 because the composition and the density of FFM change with increasing age and maturity.

Boileau and colleagues (1981) developed linear equations to predict body density in a group of boys who were slightly taller and lighter than national U.S. samples but similar in breadths and skinfold thicknesses. Nonlinear terms were not included because they did not reduce the RMSE. The RMSEs were 0.0054 to 0.0081 g/cm^3, and the PEs were about 20% to 80% larger than the RMSE. This demonstrates that the accuracy of such equations should be evaluated in another group before they are used.

Young adults

The Jackson and Pollock (1978) equation to predict body density for young men was developed from a group that was slightly taller and lighter than U.S. national data. This equation worked well in the validation group (RMSE = 0.0077 g/cm^3). It was satisfactorily cross-validated (PE = 0.0077 g/cm^3) by Thorland and colleagues (1984) in a group that tended to be slightly taller and heavier than U.S. national data but had markedly smaller skinfold

thicknesses. The Jackson and Pollock equation also was satisfactorily cross-validated by Norgan and Ferro-Luzzi (1985) in a sample that was slightly shorter but markedly lighter than U.S. national data and had markedly thinner skinfolds. These reports suggest that the Jackson–Pollock equation is widely applicable. When cross-validated by various workers, the PE for the equation of Sloan (1967) that predicts body density in young men is about 20% to 40% larger than the RMSE for the validation group (.0067 g/cm^3), except in the sample of Lohman (1981) for which the PE was approximately equal to the RSME. Haisman (1970), using data from a sample that tended to be slightly shorter and markedly lighter than U.S. national data, cross-validated several equations and reported good results for the equations of Durnin and Womersley (1974) but not for those of Chinn and Allen (1960) and Steinkamp et al. (1965). Descriptive statistics were not provided by Chinn and Allen, but the Steinkamp sample tended to be taller and lighter than U.S. national data, although similar for arm circumference.

The equation of Jackson and colleagues (1980) to predict body density for young women, which was developed from a group slightly taller and lighter than U.S. national data, had an RMSE of 0.0086 g/cm^3, which was equivalent to 3.9% BF in the validation group, and a PE equivalent to 3.7% BF on cross-validation. Withers and colleagues (1987) cross-validated this and other equations in young Australian women who were slightly taller and lighter and had markedly thinner skinfolds than U.S. national data. The best results were for the equations of Katch and Michael (1968), Katch and McArdle (1973), Pollock and colleagues (1975), Sloan and colleagues (1962), and Wilmore and Behnke (1970), each of which had a PE of 3.0% BF or less.

The log of the sum of skinfold thicknesses at the biceps, triceps, subscapular, and suprailiac sites has been used widely in age- and sex-specific equations to predict body density from which %BF can be calculated (Durnin and Womersley 1974). In the Scottish sample from which Durnin and Womersley developed their equations, the statures were similar to U.S. national data but the weights tended to be smaller. The choice of a log function reduced the RMSE slightly for men and had irregular effects on the RMSE for various age groups of women. These equations performed poorly when applied to Italian men and to Indian and Australian women (Norgan and Ferro-Luzzi 1985; Satwanti et al. 1977; Withers et al. 1987). In comparison with the Durnin and Womersley validation groups, the Italian men were similar in stature and weight, the Indian women were shorter and lighter, and the Australian women were taller and lighter, but each of these cross-validation samples had thinner skinfolds than the Durnin and Womersley validation groups. These differences in skinfold thicknesses may have been a critical factor in the failure of cross-validation.

Middle-aged and elderly subjects

Skinfold equations derived from young adults commonly underpredict %BF in middle-aged and elderly people (Pollock et al. 1976), perhaps because of changes with age in the density of FFM and in the relationships between SAT and DAT, attributable in part to increases in the fat content of muscles (Baumgartner et al. 1993; Forsberg et al. 1991; Fülöp et al. 1985). There are few, if any, anthropometric equations for middle-aged and elderly subjects that predict body composition values derived from body density using a four-component model. Therefore, although some equations have performed reasonably well on cross-validation, there may be systematic errors in both the observed and the predicted values.

The skinfold equations of Durnin and Womersley (1974) for elderly people (50-72 years), which were derived from only 24 men and 37 women, give erroneously low values for body fatness (Cohn et al. 1981; Deurenberg et al. 1989). Vu Tran and Weltman (1988) reported an equation to predict %BF in middle-aged men using data from a group that was similar in stature but heavier than U.S. national samples. The equation, based on circumferences, had an RMSE of 3.6% BF in the validation group and a PE of 4.4% BF on cross-validation.

Obese adults

There is evidence that equations to predict %BF in obese individuals should be based on circumferences rather than skinfold thicknesses (Fanelli et al. 1988). In obese people, the proportion of total body adipose tissue that is subcutaneous may be lower than in general populations, and extracellular fluid is increased (Kral et al. 1993). Therefore, equations are needed that are specific for obese people, and nonanthropometric data may be required (Segal et al. 1988; Teran et al. 1991).

An equation has been developed to predict %BF in obese women aged 18 to 50 years (Teran et al. 1991). The values of %BF were calculated from body density using a two-component model. This equation uses the log sum of circumferences, the log sum of skinfolds, and forearm circumference. Forearm circumference was kept as a separate predictor variable because, unlike the other circumferences, it was negatively correlated with %BF. The RMSE was 4.2% BF and the PE was 3.9% BF on cross-validation. This equation did not underestimate or overestimate at the lower or upper parts of the distributions of calculated %BF.

Ethnic groups

Equations to predict %BF, calculated from body density with a two-component model, may be inaccurate in some ethnic groups because of differences from the general population in the density of FFM (Boileau et al. 1984; Russell-Aulet et al. 1991, 1993) and in body proportions (Malina et al. 1987). Ethnic differences in the density of FFM do not affect the accuracy of a four-component model (Côté and Adams 1993), but ethnic-specific equations may still be needed because of different relationships between the independent and dependent variables among various ethnic groups.

African American and European American children differ in the correlations between anthropometric variables and body density (Harsha et al. 1978). In addition, ethnic differences in the distribution of SAT indicate that skinfold equations are likely to perform poorly when applied to ethnic groups other than those from which they are derived (Hammer et al. 1991; Harsha et al. 1980). Such considerations have led to the development of equations that are specific for ethnic groups (Nagamine and Suzuki 1964; Satwanti et al. 1977). Nevertheless, differences between African Americans and European Americans were not an important factor in one study to predict FFM in boys (Lohman et al. 1975).

In overview, predicted values are less accurate than observed (calculated) values, and a predictive equation should not be applied to a group that is markedly different from the group used to develop the equation. Important group differences may relate to age, sex, ethnicity, and level of body fatness. Although some anthropometric predictive equations have been successful in some cross-validation groups, they may not perform well in all other groups.

Prediction of Fat-Free Mass

There is interest in the prediction of FFM because of its relationships to morbidity, mortality, physical performance, and caloric requirements. The major constituents of FFM are muscle, bone, vital organs, and extracellular fluid. FFM is the whole body except the mass of extractable fat, whereas lean body mass is an anatomical term that refers to the whole body other than adipose tissue that is visible to the eye. Lipids in cell membranes, the central nervous system, and bone marrow are excluded from FFM, but some or all of these are typically included with lean body mass. There is no inherent advantage in predicting FFM in preference to %BF because a predicted value for either FFM or %BF can be used to calculate the other. Nevertheless, it is logical to predict FFM if bioelectric impedance values, circumferences, breadths, and lengths have been measured and to predict %BF if skinfold thicknesses dominate among the predictor variables. In judging equations to predict FFM, it should be recalled that the error in FFM values derived from body density is about 1.9 kg for men and 1.5 kg for women (Lohman 1991). There are equations to predict FFM in boys. The observed values for FFM were calculated from total body potassium using constants derived from adults (Lohman et al. 1975). These equations, which include weight and two skinfold thicknesses as predictors, have an RMSE of 1.7 kg and a PE of about 1.2 kg. The boys in the validation and cross-validation samples closely matched U.S. national data for stature and weight.

Using a multicomponent model, Fuchs and colleagues (1978) obtained observed values for FFM from body density, total body water, and total body potassium in men from the U.S. Air Force. Fuchs and colleagues' predictive equation, which used stature and arm circumference as predictor variables, had an RMSE of 2.9 kg and a PE of 3.3 kg on cross-validation. Descriptive statistics for these samples were not reported. Crenier (1966) used data from a French sample of young men and women who were slightly shorter and considerably lighter than U.S. national data to develop equations that predicted FFM with an RMSE of 1.5 kg. On cross-validation, all the errors of the estimates for individuals were less than 1.3 kg in men and 1.4 kg in women, which is a remarkably good result. Jackson and Pollock (1976) reported predictive equations developed from a sample of young adults who tended to be taller and lighter and to have thinner skinfolds than the U.S. national data. These equations had RMSEs of 3.2 kg for men and 2.3 kg for women, but the PEs were not reported. It is reasonable to conclude that the PEs of anthropometric equations to predict FFM are about 1.2 kg in boys and 3.0 kg in young adults, although lower PEs were reported by Crenier for French samples that were lighter than U.S. national data.

Prediction of Total Body Muscle Mass

There are few equations to predict total body muscle mass because it is difficult to measure the dependent variable. Lee and colleagues (2000) recently provided a good review and set of equations on this topic. The only methods for the measurement of total body muscle mass in the living are serial whole-body CT or magnetic resonance imaging (MRI) scans, which are not applicable to large samples. Indexes of muscle mass from creatinine excretion, labeled creatinine, or the potassium–nitrogen ratio are too uncertain to be used as the dependent variable (Heymsfield et al. 1983). Despite considerable independence among radiographic muscle thicknesses at different limb sites (Malina 1986), limb cross-sectional areas of muscle plus bone from anthropometry predict limb muscle mass with an RMSE = 1.8 kg, which is about 17% of the mean (Heymsfield et al. 1982, 1990). These predicted values are useful as indexes of total body muscle mass because skeletal muscle is more abundant than other types of muscle and about 74% of the skeletal muscle mass is in the limbs (Snyder et al. 1984).

Prediction of Total Body Bone Mineral

Body density is influenced by total body bone mineral (TBBM; Bunt et al. 1990). Therefore, TBBM is included in four-component models based on density. Usually TBBM is measured by dual-energy X-ray absorptiometry (DXA) as the sum of osseous and nonosseous mineral, but the latter is a small near-constant proportion of the total. The validity of measures of TBBM from DXA and dual-photon absorptiometry has been established by comparison with total body calcium (Heymsfield et al. 1989; Mazess et al. 1981).

Anthropometric values, in combination with age, sex, and ethnicity, may be useful in predicting TBBM. During infancy and childhood, TBBM is highly correlated ($r = .9$) with weight and stature (Chan 1992; Katzman et al. 1991), but the corresponding correlations in adulthood are in the range of .3 to .7 (Rico et al. 1991; Wang et al. 1988). Adams and colleagues (1992) reported ethnic-specific equations developed from small samples of young women (26 African Americans, 26 European Americans) that predict TBBM from combinations of cross-sectional areas of muscle plus bone in the limbs, chest and head size, and joint breadths. These equations have RMSEs of about 143 g for African Americans and 187 g for European Americans, which are equivalent to about 6% of the means of the measured values.

These equations have not been cross-validated, but it was shown that a general equation for both ethnic groups combined had a larger RMSE than the ethnic-specific equations.

Estimation of Changes in Total Body Composition

The main interest in estimating changes in total body composition relates to the effects of intervention in obese persons. There are considerable errors in all body composition measures; these errors may be larger in the obese, and they are necessarily larger for predicted values than for observed values. Therefore, anthropometry applied through predictive equations is unlikely to provide accurate measures of changes in total body composition. The observed changes in total body composition with weight loss in obese persons, calculated from body density, may be inaccurate if a two-component model is used, because of changes in the density of FFM (Scherf et al. 1986; Seip et al. 1993). An alternative is to use a method that is equally valid in obese and nonobese people; serial CT scans meet this requirement but they are not readily available. It is generally agreed that efforts to estimate the changes in body composition with weight loss in obese people should be based on equations that use circumferences rather than skinfold thicknesses as predictor variables because the changes in circumferences are larger, with the possible exception of subscapular skinfold thickness (Bradfield et al. 1979; Bray et al. 1978). Studies that have evaluated predictive equations for estimating changes in composition have generally given disappointing results, although they were based on two-component models (Ballor and Katch 1989; Scherf et al. 1986; Teran et al. 1991).

The reported analyses, which are discussed later in this chapter, were based on two-component body density models. The difficulties of evaluating changes in body composition were reported by Ballor and Katch (1989) who showed that in obese women with a mean loss of 2.7% BF, none of the ten common predictive equations accurately estimated the changes. The equations that performed best were the skinfold equation of Jackson and colleagues (1980) and the circumference equation of Katch and McArdle (1973), but even those were not recommended by Ballor and Katch. Others have reported that the equations of Jackson and colleagues (1980), Jackson and Pollock (1978), and Durnin and Womersley (1974) performed poorly in estimating the total

body composition changes during weight loss in people with obesity (Scherf et al. 1986; Teran et al. 1991). The equation of Teran and colleagues (1991) for women with obesity who lost weight provided only fair predictions of the change in body fatness with loss of weight. The PE was 2.9% BF when the mean measured loss was 4.2%.

Relationships Between Anthropometric Variables and Regional Body Composition

In relation to regional body composition, anthropometric data are used to describe the distribution of SAT to predict the amount of DAT and limb composition.

Subcutaneous Adipose Tissue Distribution

The major interest in regional body composition concerns the distribution of adipose tissue. Following Bouchard (1988), we use the term *adipose tissue distribution* in reference to the absolute and relative amounts of adipose tissue in body regions. When one is interpreting the relevant literature, it is important to distinguish between fat and adipose tissue. Most methods that provide regional data measure or estimate adipose tissue. They do not measure fat, which is a chemical term. The term *fat pattern,* which is in common use, is misleading because it usually refers to adipose tissue distribution. Furthermore, the word *pattern* implies a relative configuration or design that can be described only if many variables are measured. A ratio between two skinfold thicknesses cannot define a pattern or adequately describe the distribution of SAT, although some ratios between skinfolds are related to the prevalence of diseases.

Principal component analyses of skinfold thicknesses have been used to describe the distribution of SAT. The results of such analyses depend on the skinfold thicknesses measured; these should include sites on the upper and lower limbs, the chest, and the abdomen. In principal component analysis, orthogonal components are extracted from an intercorrelation matrix (Mueller and Reid 1979). Each component is the linear weighted combination of the recorded variables that describes the maximum proportion of the variance remaining after earlier components have been extracted. These components are interpreted from the coefficients given to the individual variables. The first principal component from a set of skinfold thicknesses is usually interpreted as a measure of the mass of SAT or body fatness leading to the conclusion that latter components are independent of the mass of SAT or body fatness. This is inaccurate. Rather, the first component reflects the general level of the skinfold thicknesses that were measured and the latter components are independent of this general level but are not necessarily independent of the mass of SAT or body fatness. Additionally, the loadings on the elements within the components are related to the general level of the skinfold thicknesses that were measured (Deutsch et al. 1985; Mueller and Reid 1979). Typically, the second principal component contrasts trunk and extremity skinfold thicknesses, whereas the third component may contrast upper- and lower-body SAT (e.g., subscapular and triceps skinfold thicknesses vs. suprailiac and medial calf skinfold thicknesses). The subscapular site may, however, behave statistically like an arm site, and chin skinfold thicknesses may be the dominant factor in the second component (Norgan and Ferro-Luzzi 1986; Shephard et al. 1969).

Two studies (Mueller and Malina 1987; Mueller et al. 1989) used canonical correlations between multiple skinfold thicknesses and circumferences to select the set of circumferences that best described the distribution of skinfold thicknesses with the assumption that skinfold thicknesses provide valid measures of the distribution of SAT. There were only low canonical correlations between the skinfold thicknesses and circumferences indicating that these sets of measures are not interchangeable.

Some have used ratios of skinfold thicknesses, or of their logs, or of the thickness at one site relative to the sum of the thicknesses at several sites, as indexes of the distribution of SAT (Baumgartner et al. 1990). Commonly, these ratios are correlated with total body fatness, making it difficult to determine the independent effects of SAT distribution, but the correlations with total body fatness are low for ratios between skinfold thickness z-scores (Norgan and Ferro-Luzzi 1986; Ramirez 1993).

The terms *upper- (lower-) body obesity* and *upper- (lower-) segment obesity* are used interchangeably in the literature without clear definitions of the anatomical limits of the upper body or the upper segment. These terms have been based on various indirect criteria, including (a) the waist circumference–hip circumference ratio (WHR), (b) the ratio between triceps and subscapular skinfold thicknesses, and (c) the ratio between anthropometric cross-sectional areas of adipose tissue in the upper and lower limbs relative to the corresponding cross-sectional areas of muscle plus bone (Haffner et al. 1987; Kissebah et al. 1982). The distinction between upper- and lower-body obesity based on WHR is problematic because both waist and hip circumferences include many tissues and organs in addition to adipose tissue, and it has not been demonstrated that the circumferences at these levels predict the levels of fatness in the upper- and lower-body segments. Moreover, waist and hip circumferences have been measured at various levels in different studies, making interpretation of the literature difficult. Ratios between skinfold thicknesses at two sites or between estimates of limb composition at two levels are also used, but they are inadequate bases for the recognition or quantification of upper- or lower-body obesity.

Deep Adipose Tissue

DAT includes all the adipose tissue deep to the deep fascia; it is not synonymous with VAT, which is the intraperitoneal portion of DAT. A considerable proportion of the deep abdominal adipose tissue is extraperitoneal and is located within the muscle walls of the abdomen and between these walls and the peritoneum. The VAT of the abdomen is located within peritoneal folds that form the mesentery and omentum. The abdominal DAT that is extraperitoneal may differ functionally from abdominal intraperitoneal DAT tissue partly because the latter has direct venous drainage to the liver by the portal system. Consequently, free fatty acids from mesenteric and omental adipocytes, which are very active metabolically (Bolinder et al. 1983), pass directly to the liver, leading to increased risk of cardiovascular diseases and non-insulin-dependent diabetes mellitus (Björntorp 1990). In the discussion that follows, the mesenteric and omental adipose tissue are referred to as *portal adipose tissue,* following Björntorp.

In most studies based on CT or MRI, the extraperitoneal and portal adipose tissues have been combined and called abdominal DAT. The area and volume of this adipose tissue have been calculated as the difference between total adipose tissue and SAT in single or serial CT or MRI scans of the abdomen. The volume of abdominal DAT, obtained from serial CT scans, was 8.3 L (SD = 4.5 L) in men and 2.5 L (SD = 1.9 L) in women in the study by Kvist and colleagues (1988), but a lower value for men (4.1 L, SD = 2.2 L) was reported by Ross and colleagues (1992). A few have delineated the approximate area in which extraperitoneal adipose DAT occurs by drawing straight lines directed transversely on abdominal scans from a point in the midline between the aorta and inferior vena cava to the centers of the ascending and descending parts of the colon. About 60% to 75% of the abdominal DAT is anterior to these lines and is considered portal adipose tissue (Ashwell et al. 1987; Rössner et al. 1990).

Reports of the areas of DAT in abdominal scans vary markedly among studies (table 8.1). Much of the literature is based on scans at L4-L5, where the DAT occupies about 18% of the total cross-sectional area (Borkan et al. 1983; Bosello et al. 1993). Because this is a relatively small percentage of the total area, it is not surprising that it is difficult to

Table 8.1 Selected Data for Abdominal Adipose Tissue Areas (cm², SD) in Healthy Adults

Author	MEN		WOMEN	
	SAT	DAT	SAT	DAT
LEVEL L1				
Grauer et al. 1984	95	148	146	86
LEVEL L3				
Grauer et al. 1984	150	148	202	113
LEVEL L3-L4				
Baumgartner et al. 1988	76 (72)	105	108 (71)	49
LEVEL L4-L5				
Armellini et al. 1994	— —	— —	— —	103 (53) 139 (73)
Ashwell et al. 1985	176 (82)	91	278 (122)	86
Baumgartner et al. 1988	98 (90)	131	134 (79)	77
Baumgartner et al. 1993	199 (58)	173 (57)	256 (103)	139 (33)
DeNino et al. 2001	— — — —	— — — —	181 (87) 221 (107) 256 (88) 300 (69)	42 (21) 54 (30) 82 (41) 133 (45)
Després et al. 1991	214 (124)	101 (57)	—	—
Imbeault et al. 2000	251 (116) 266 (95)	115 (59) 175 (43)	— —	— —
Koester et al. 1992	90 (67)	79 (59)	—	—
Lemieux et al. 1994	—	120 (49)	—	104 (55)
Lemieux et al. 1996	— —	— —	448 (198) 459 (205)	103 (49) 135 (12)
Pérusse et al. 2001	229 (108)	166 (83)	338 (155)	127 (70)
Ross et al. 1992	—	118 (62)	—	—
Toth et al. 2000	—	—	240 (116)	61 (34)
van der Kooy et al. 1993	316 (78)	156 (43)	391 (100)	108 (47)
LEVEL L5				
Grauer et al. 1984	170	88	312	110

Note. L1-L5 = levels of lumbar vertebrae 1-5; SAT = subcutaneous adipose tissue; DAT = deep adipose tissue.

predict the area of DAT in a CT scan from external abdominal dimensions or skinfold thicknesses at the same level. Furthermore, for reasons given earlier, it is desirable to predict portal adipose tissue, which would be even more difficult because the portal adipose tissue area is even smaller (Ashwell et al. 1985; Baumgartner et al. 1988).

At the L4-L5 level, the reported mean SAT areas for men vary from 90 to 316 cm^2, and those for women vary from 134 to 391 cm^2. The corresponding mean areas of DAT at the same level vary from 79 to 173 cm^2 for men and from 77 to 139 cm^2 for women. The lowest values for both SAT and DAT areas were reported by Koester and colleagues (1992) for men and Baumgartner and colleagues (1988) for women. The groups studied by these workers had mean BMI values of about 24 kg/m^2. The highest mean DAT areas were reported by Baumgartner and colleagues (1993) in an elderly group (mean age about 79 years) with a mean BMI of about 24 kg/m^2. This suggests that DAT areas are markedly larger in the elderly than in young adults at the same BMI. The largest mean SAT areas were reported by van der Kooy and colleagues (1993) for each sex in groups with mean BMI values of about 31 kg/m^2. In the studies listed in table 8.1, the percentage of abdominal adipose tissue at the L4-L5 level that is deep is about 40% for men and 30% for women.

In attempts to develop equations for the prediction of abdominal DAT, the anthropometric variables have usually been obtained from CT images with the subjects supine. Supine values are affected by pressure on the posterior aspect of the trunk and buttocks, an inability to measure buttocks circumference where it is maximal, and the movement of viscera between supine and erect positions. Nevertheless, there are close relationships between abdominal depths and the circumferences of the waist and hips measured standing and those measured on CT images, although the CT measures tend to be smaller for abdominal circumferences. The latter differences are greater for circumferences just superior to the iliac crest than for those at the waist (Koester et al. 1992; Kvist et al. 1988). Investigators have attempted to develop equations to predict abdominal DAT or VAT volume from measures such as abdominal depth, waist circumference, BMI, and the ratios between waist and hip or thigh circumference (Ashwell et al. 1985; Koester et al. 1992; Kvist et al. 1988; Seidell et al. 1987, 1988b; van der Kooy et al. 1993; Weits et al. 1988).

Kvist and colleagues (1988) reported equations to predict abdominal DAT volume from abdominal depth, waist circumference, and BMI. These equations have an RMSE equal to 15% of the mean. The equations of Seidell and colleagues (1988b) and of Weits and colleagues (1988) are even less accurate. None of these equations can be recommended. Others have investigated the ratio between waist circumference and thigh circumference as an index of abdominal DAT, but this ratio is less closely related to abdominal adipose tissue area at L4-L5 than is the WHR (Ashwell et al. 1985; Seidell et al. 1987). Després and colleagues (1991) reported an equation to predict the area of DAT at L4-L5 from abdominal depth and WHR that had an RMSE equivalent to 28% of the mean with similar results on cross-validation (Koester et al. 1992; van der Kooy et al. 1993). Better results were reported by Ross and colleagues (1992), but even in their data, WHR explained only 12% of the variance in VAT volume. Among the variables examined by these workers, WHR had the largest partial value but age had the closest relationship with VAT volume. In adults, waist circumference and abdominal sagittal diameter have higher correlations with VAT than those between WHR and VAT (Pouliot et al. 1994). This conclusion is in general agreement with earlier reports (Després et al. 1991; Ferland et al. 1989; Seidell et al. 1988a). Waist circumference, in comparison with WHR, is more closely related to some risk factors for chronic cardiovascular disease in women but not in men (Pouliot et al. 1994).

Limb Composition

The composition of the leg, particularly its muscle mass, is important in elderly people because it is related to the prevalence of falls and fractures. In addition, arm muscle circumference from anthropometry, despite its inaccuracy, is related to longevity in middle-aged men (Roche 1994), perhaps because of its associations with FFM and habitual physical activity. Furthermore, judgments of total body composition in people who are sick or disabled are commonly made using anthropometric data from the arm. If accurate estimates of regional muscle areas or volumes were available, they would have widespread application, but caution is necessary. The functional capacity of skeletal muscle can vary independently of its volume, as occurs when its water content increases in severe protein-caloric malnutrition (Heymsfield et al. 1982).

Skinfold thicknesses, used alone, are likely to be effective in the prediction of limb muscle mass because the thicknesses of adipose tissue and of muscle in limbs are essentially independent (Hewitt 1958; Malina 1969). In overview, anthropometry alone does not provide accurate predictions of regional muscle mass; the more appropriate techniques are CT, MRI, or DXA. Some anthropometric indexes of limb composition assume that (a) cross-sections of the limbs are circular, (b) a skinfold thickness is equal to the thickness of SAT at the site, and (c) SAT in a cross-section of a limb is of constant thickness. These assumptions are incorrect. The anthropometric approach overestimates muscle plus bone cross-sectional areas in the limbs in comparison with values from CT or MRI, although muscle-plus-bone areas from CT or MRI include some adipose tissue and connective tissue within muscles. This overestimation is larger in elderly people, particularly women, and in obese persons (Baumgartner et al. 1992; Forbes et al. 1988) partly because of increases in intra- and intermuscular adipose tissue, which are not related to the amounts of SAT (Frantzell and Ingelmark 1951; Forsberg et al. 1991).

Measures of bone mineral content for a bone or a region of the skeleton are indexes of calcium stores and, by inference, indexes of TBBM (Chestnut et al. 1973). The validity of regional bone mineral content values from DXA has been evaluated by comparison with the ash weight of excised bones. Most reports agree that DXA values are too high by 5% to 13% and that they have an RMSE equal to about 8% of the mean (Braillon et al. 1992; Chan 1992).

Anthropometric variables are related to regional bone mineral content values. In children aged 5 to 14 years, combinations of stature, wrist breadth, biceps skinfold thickness, and limb circumferences are related to bone mineral content of the radius, lumbar spine, and regions near the hip with $r = .9$ (Miller et al. 1991). Predictive equations for adults are less effective, with r values ranging from .3 to .5 (Dawson-Hughes et al. 1987; Slemenda et al. 1990). The RMSE and PE of these equations have not been reported.

Ultrasound

Traditional skinfold caliper measurements of SAT have some limitations. Skinfolds cannot be raised for measurement at many body locations. Some skinfolds are too thick for caliper application. Although the primary function of ultrasound is to assess soft tissue structures in clinical diagnoses, ultrasound instruments can also measure tissue thicknesses such as SAT, muscle, and intra-abdominal depth (IAD) without many of these limitations. The latter is used as an index of visceral adipose tissue (abdominal DAT) by many researchers.

Technical Considerations

Ultrasound was introduced for the measurement of SAT thicknesses by Booth and colleagues (1966). B-mode (brightness-modulation) instruments use high-frequency sound waves (1-10 MHz) produced by vibrations of an electrically stimulated piezoelectric crystal within a transducer (Sprawls 1987). The ultrasound beam is propagated through the skin and partially reflected from the interfaces of dissimilar underlying tissues. While the remainder of the beam continues to travel through the interface to deeper tissues, a portion of the beam returns to the transducer as an echo. Soft tissue interfaces reflect only a small portion of the sound because biological soft tissues differ little in resistance to sound propagation (Hagen-Ansert 2001). The "reflectibility" of a tissue type is expressed as acoustic impedance, calculated from tissue density and speed of sound (Bushberg et al. 2002). For example, the impedance is $0.165 \text{ g} \cdot \text{cm}^{-1} \cdot \text{s}^{-1}$ for liver tissue, $0.170 \text{ g} \cdot \text{cm}^{-1} \cdot \text{s}^{-1}$ for muscle, and $0.138 \text{ g} \cdot \text{cm}^{-1} \cdot \text{s}^{-1}$ for fat. Air has extremely low acoustic impedance ($0.00001 \text{ g} \cdot \text{cm}^{-1} \cdot \text{s}^{-1}$) and interferes with sound transmission, completely obscuring the structures beyond air-filled lungs and bowel. Bone has relatively high acoustic impedance ($0.78 \text{ g} \cdot \text{cm}^{-1} \cdot \text{s}^{-1}$) and generates strong reflections.

The transducer converts the echoes into electric signals that are amplified to form an enlarged image on a display device such as a monitor. Image quality depends on the strength of the reflection, which in turn depends on the acoustic impedance properties of different tissues and the number of tissue interfaces that the ultrasonic beam traverses. Thus, a fat–muscle interface produces a weaker echo than a fat–bone or muscle–bone interface, and a beam that travels through several different tissue interfaces loses more energy by attenuation than one that is reflected by a single interface. The relative strength (amplitude) of echoes is represented by the relative brightness of interface images on the display monitor.

A linear-array transducer with a frequency of 5 to 7.5 MHz is usually used to measure SAT, and a 3.5 or 3.75 MHz transducer is used to measure IAD with increased focal depth but reduced resolution. In each case, dynamic cross-sectional images are displayed on a monitor and can be frozen to allow close inspection and measurement with electronic calipers to the nearest 1.0 mm. The operator can vary the brightness of the image and increase focal depth to improve tissue identification.

Ultrasonic scanning is a relatively simple procedure, but interpretation of the tissue images is another matter. Tissue characterization is a subjective, qualitative process that involves distinguishing the speckle patterns of different tissues. Interpretation improves with experience (Evens 1991). For SAT measurements, the observer must be able to distinguish between adipose and muscle tissues, which transmit sound waves at different velocities and produce images with different patterns. Generally, SAT appears as a uniformly dark tissue layer that may contain light horizontal streaks representing intralipid fascia, whereas muscle appears as a lighter, more uniformly speckled tissue. The interface between fat and muscle usually appears as a continuous, broad, bright line. After the tissue boundaries have been determined, the measurement of tissue thicknesses is objective and is accomplished with electronic calipers. The most important step is accurate placement of the two caliper points marking the boundaries of the tissue layer to be measured. This requires some practice and standardization of protocol (e.g., deciding to place the caliper on the anterior or posterior surfaces of the lines representing tissue interfaces). The ultrasound measurement of SAT layers is generally less difficult than clinical diagnosis of tissue abnormalities because only two different tissue layers of known anatomical relationship are involved. Nevertheless, ultrasonic measurement is not completely error-free.

The ultrasound beam is attenuated as it travels through tissue and is reflected to the transducer. A portion of its energy is absorbed by tissues and converted to heat. Scattering and refraction also contribute to attenuation, reducing resolution in the far field. Technical errors or anatomical irregularities can produce artifacts that may lead to misinterpretation of the image by the observer (Hagen-Ansert 2001). In SAT measurement, failure to maintain good contact between the transducer and the skin can result in reverberations that appear similar to tissue interfaces. But because these reverberations are usually equally spaced, they are relatively easy to recognize and eliminate, usually by applying a generous layer of ultrasound gel to form a close bond between transducer and skin. Bone interfaces produce such strong sound reflections that dark shadows obscure the views of deeper structures. Many of these problems can be corrected by moving the transducer slightly or changing to a transducer of a different frequency.

The transducer–skin interface is very strong and produces a "main-bang" artifact (Sanders and Maggio 1984). Because measurement of SAT requires good resolution of the skin surface for correct electronic caliper placement, a gel pad or other standoff is placed between the transducer and the skin, and both are liberally coated with ultrasound gel to prevent this artifact and allow accurate measurement of the SAT layer.

With a standardized protocol for transducer placement and adequate operator training, B-mode ultrasonic measurements are about equal in precision to skinfold caliper measurements of SAT. Ultrasound has the advantage of little or no tissue compression, although this may be less important than is generally considered. Furthermore, ultrasonic measurements can be made in obese subjects and at some sites where calipers cannot be applied (e.g., sacral, paraspinal).

Widespread clinical use for nearly four decades and extensive investigation of various exposures, at the intensities used for these measurements, suggest that ultrasound does not present a detectable health risk. Genetic damage and hyperthermia due to overexposure are not likely (Bushberg et al. 2002). The disadvantages of ultrasound, in comparison with calipers, are its greater expense and lesser portability, although sturdy, portable equipment is available.

Despite the fact that ultrasound produces excellent tissue images that have the advantage of permanent storage in paper or electronic format for reexamination, quality assessment, or additional analyses at a later time, it has not been applied widely to measure adipose tissues in human body composition studies. As is shown in table 8.2, ultrasonic measurements have high precision (Abe et al. 1994, 1995; Bellisari et al. 1993; Flygare et al. 1999; Ishida et al. 1992; Stolk et al. 2001), but the validity of ultrasound measurements of SAT and abdominal DAT is uncertain (Bellisari 1997; Bellisari et al. 1994; Moeller and Christian 1998; Orphanidou et al. 1994). Some researchers consider ultrasound measurements of adipose tissue to be valid and to assist the accurate prediction of TBF and body density (Abe et al. 1994, 1996; Armellini et al. 1993a, 1993b, 1994; Ishida et al. 1997). As with other anthropometric procedures, validity must be established and inter- and intraobserver reliability should be monitored within clinics and in research studies. Equipment performance must also be assessed periodically as part of the quality control program (Bushberg et al. 2002).

Table 8.2   Precision of Ultrasonic Measurements of Subcutaneous Adipose Tissue

Sites	Intraobserver		Interobserver	
	TEM (mm)	CR (%)	TEM (mm)	CR (%)
Suprailiac	0.14	91.07	0.15	89.77
Paraspinal	0.13	86.35	0.15	81.56
Sacral	0.08	94.34	0.09	94.06
Epigastric	0.09	95.06	0.08	96.25
Midthigh	0.08	97.69	0.13	93.66
Triceps	0.09	95.61	0.62	98.01

TEM = technical error of measurement; CR = coefficient of reliability.

Adapted, by permission, from A. Bellisari, A.F. Roche, and R.M. Siervogel, 1993, "Reliability of B-mode ultrasonic measurements of subcutaneous and intra-abdominal adipose tissue," *International Journal of Obesity* 17: 475-480. Copyright © by Stockton Press.

Ultrasonic Measurement of SAT

Ultrasound measurements of SAT thicknesses have been reported for many extremity sites (triceps, biceps, lateral forearm, anterior and posterior thigh, anterior and posterior calf, midthigh, calf) and trunk sites (suprailiac, paraspinal, sacral, epigastric, subscapular, waist). The paraspinal site is on the back 2 cm superior to the iliac crest and 2 cm to the left of the midline. The sacral site is in the midline posteriorly just superior to the lumbar fossae. See figures 8.1, 8.2, and 8.3 for examples of ultrasonic images of SAT at some of these sites.

In a study of women by Ishida and colleagues (1997), the correlation of the sum of 13 ultrasound-determined SAT thicknesses with TBF (relative to stature) was .8. The middle-aged women had the expected smaller values of muscle on the trunk and quadriceps regions and larger values of SAT over the entire body than the young women. But the authors did not establish the validity of ultrasound-determined SAT measurements by comparing them with values obtained by an accepted standard method such as CT.

On the basis of the correlation ($r = -.89$) between the sum of nine SAT thicknesses measured with ultrasound and body density determined by hydrodensitometry, Abe and coworkers (1994) concluded that summed ultrasound measurements of SAT yield more accurate predictions of body density than single-site SAT measurements or waist and hip circumferences. Abe and colleagues (1996) later used ultrasound-determined SAT thicknesses and surface areas of the trunk and extremities to calculate total body SAT volumes in 13 females. Surfaces were covered with paper on

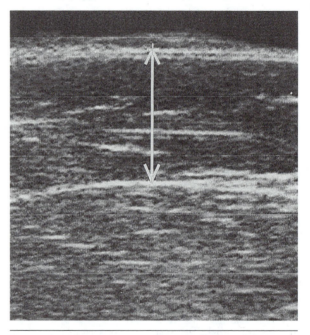

Figure 8.1 B-mode ultrasound image of subcutaneous adipose tissue (SAT) at the triceps site. The thickness of this layer (27 mm) was measured between the electronic caliper marks on the superficial surfaces of the skin and deep fascia. Lines have been added to show planes of measurement and the limits of the SAT.

A Bellisari, *Journal of Diagnostic Medical Sonography* 9(1): 11-18, copyright 1993 by Sage Publication, Inc. Reprinted by permission of Sage Publication, Inc.

which segmental boundaries were marked. The paper was subsequently removed, flattened, and measured with a planimeter. These calculated volumes did not differ significantly from MRI determinations, which had been validated by Abate and colleagues (1994) against dissection in human cadavers. Correlations between ultrasound- and

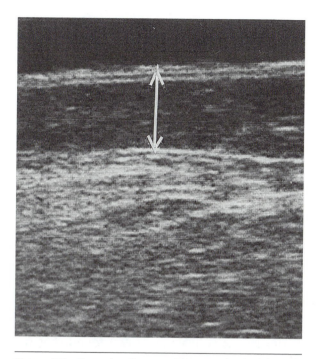

Figure 8.2 B-mode ultrasound image of subcutaneous adipose tissue (SAT) at the suprailiac site (thickness 16 mm). Lines have been added to show the planes of measurement and the limits of the SAT.

Courtesy of Anna Bellisari.

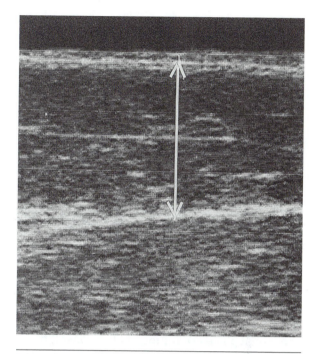

Figure 8.3 B-mode ultrasound image of subcutaneous adipose tissue (SAT) at the anterior midthigh site (thickness 32 mm). The corresponding skinfold was too thick to be measured with skinfold calipers. Note a dense connective tissue layer within the SAT that could reflect ultrasonic waves, leading to spurious results.

Courtesy of Anna Bellisari.

MRI-derived values for body segments varied (r = .79-.95). The ultrasound method underestimated SAT volume at the upper arm and thigh, resulting in 5.5% lower total SAT volumes than those obtained with MRI. Although the researchers calculated an average SAT thickness for each body segment, the moderate correlation between ultrasound- and MRI-derived values for some segments and the underestimation of upper arm and thigh SAT volume may be attributable to large variations in SAT thicknesses within body segments and to differences in segment boundaries as delineated by planimeter and MRI technology.

In 1992, Alexander and Dugdale used ultrasound to demonstrate that SAT at the triceps and abdominal sites is separated into two distinct layers by a single intralipid fascial plane. Johnson and colleagues (1996) demonstrated a similar fascial plane that is continuous around the torso and separates the more metabolically active deep SAT (DSAT; Goodpaster et al. 1997) from the superficial SAT (SSAT). The DSAT has a strong relation with insulin resistance, similar to that of abdominal DAT (Kelley et al. 2000). Kelley and coworkers found that insulin-stimulated glucose utilization was strongly correlated with both

abdominal DAT and DSAT but not SSAT in 47 lean and obese glucose-tolerant men and women.

CT is considered the gold standard for SAT measurement, but radiation exposure and high cost prevent its use in population-based studies. Comparison of SAT measurements at three abdominal sites by skinfold calipers, ultrasound, and CT showed that caliper–CT correlations were higher at all three sites (r = .60, .70, and .73) than ultrasound–CT correlations. The latter were significant at only one abdominal site (r = .54; Orphanidou et al. 1994). Eston and colleagues (1994) found that ultrasound-determined SAT measurements at the quadriceps site had the best correlation with skinfold measurements (r = .96 for European men and r = .90 for Chinese men) compared with subscapular, pectoral, calf, and abdominal sites, which ranged from .64 to .93 for European men and .75 to .76 for Chinese men. The best ultrasound predictor of %BF was the percent SAT volume in the upper leg calculated from SAT thicknesses and segmental radii (r = .79 and r = .83 for European and Chinese men, respectively), whereas the corresponding percent SAT volume of the upper arm was not a good predictor in either group (r = .55 and r = .48, respectively). Because women were not included in this study and the men were all athletes, these results may not be applicable to the general population.

Bellisari and colleagues (Bellisari et al. 1994; Bellisari 1997) compared ultrasound and MRI measurements of SAT thicknesses at various sites and found considerable variation in coefficients of reliability and technical errors by site, sex, and BMI category (coefficient of reliability [CR] = 0-.92; technical error of measurement [TEM] = 0.9-20.9 mm). For example, ultrasound midthigh site measurements were more precise than sacral site measurements. Ultrasound measurements at the triceps site were more precise in men with BMI greater than 27 kg/m^2 than in men with smaller BMI values and were more precise in women than in men in both BMI categories. In these studies, MRI images of the abdominal region clearly showed the fascial plane separating SSAT and DSAT. Many ultrasound measurements of SAT equaled the corresponding MRI measurements of SSAT thickness. It is likely that the intralipid fascial plane attenuates ultrasonic waves enough to obscure DSAT and to result in underestimation of total SAT thickness, especially when the tissues are very thick (Moeller and Christian 1998). When Stevens-Simon and colleagues (2001) compared serial ultrasound and skinfold caliper measurements of SAT in pregnant adolescents, they found high correlations between data from the two types of measurements. Nevertheless, serial skinfold caliper measurements indicated that more adipose tissue was stored on the trunk than the extremities during advancing pregnancy, whereas serial ultrasound measurements indicated the reverse. This discrepancy may have been attributable to the use of ultrasound data limited to SSAT.

Ultrasonic Measurement of Abdominal DAT

Researchers have developed three somewhat different ultrasound measuring techniques for assessing abdominal DAT. All three methods use IAD, differently defined, as an indirect measure of abdominal DAT. Armellini and colleagues (1993a, 1993b, 1994), as well as others (Cignarelli et al. 1996; Ribiero-Filho et al. 2001; Targher et al. 1996; Wirth and Steinmetz 1998), used the method developed by Armellini and colleagues (1990) or a slightly modified version. The Armellini method involves a precisely defined scan of the abdomen at the L4 level and a measurement

of the distance between the posterior surface of the linea alba and either the anterior or the posterior wall of the aorta. These studies tested correlations between changes in IAD and visceral fat decreases during weight loss programs. Correlations of ultrasound- and CT-defined abdominal DAT are about .7 (Armellini et al. 1993a, 1993b, 1994; Ribiero-Filho et al. 2001). In another study, Armellini and colleagues (1997) found that CT-determined sagittal diameter minus ultrasound-determined abdominal SAT predicts CT-determined abdominal DAT (r = .72) nearly as well as ultrasound-determined IAD (r = .78). Although the authors indicated that manual measurement of sagittal diameter can be used to simplify the procedure without loss of accuracy, the moderate relationship between abdominal DAT measurements determined by CT and by ultrasound indicates that this method for measuring abdominal DAT does not have great utility.

In contrast to Armellini and colleagues (1990), Bellisari and colleagues (1993) found that the interobserver reliability of ultrasound measurements was lower (CR = .64) for IAD than for SAT thicknesses at various sites (CR = .82-.98). As a result, Bellisari and colleagues recommended that ultrasound not be used to estimate abdominal DAT. Tornaghi and colleagues (1994), however, using a modified Armellini method, measured IAD from the linea alba to the anterior face of L4-L5 vertebrae and obtained great precision (CV = 6.3% with compression, 7.9% without compression). This is higher than that by Bellisari and colleagues (1993), who reported a CR of 64.3% and TEM of 0.18 mm, but less than the reproducibility of 4.5% reported by Armellini and colleagues (1990). Tornaghi obtained a high correlation (about .9) between ultrasound-determined IAD measured from the linea alba to vertebrae or to the anterior wall of the aorta with CT-determined abdominal DAT area.

Stolk and colleagues (2001) also modified the Armellini method. They calculated IAD as the mean of five different distance measurements from the posterior surface of the anterior abdominal muscle wall to the lumbar spine or psoas muscle at the midpoint between lower ribs and iliac crest and obtained a .83 correlation with abdominal DAT area measured by CT at the L4-L5 level. Applying this method to obese patients in a weight loss program, they found moderate correlations between changes in abdominal DAT from ultrasound and those from CT (r = .74) and MRI (r = .62).

Abe and colleagues (1995, 1996) reported an ultrasound technique for indirectly assessing abdominal DAT by subtracting SAT from total body fat determined by hydrodensitometry. These workers calculated SAT mass by ultrasonically measuring SAT thicknesses of various body segments and multiplying by the corresponding surface areas that were measured by the planimeter method described earlier. The repeated measurement difference for abdominal DAT mass was 5% and abdominal DAT mass was correlated significantly with MRI-determined abdominal DAT area at the umbilical level (r = .75 to .78). Suzuki and colleagues (1993) obtained a .7 correlation between abdominal DAT area measured by CT at the umbilical level and the maximum preperitoneal adipose tissue thickness measured with ultrasound. Their ultrasound measurement was the maximum distance from the linea alba to the visceral peritoneum along a line from the xiphoid process to the umbilicus.

Ultrasonic Assessment of Bone

Although bone mineral density is assessed using other techniques (see chapter 5), ultrasound is used to measure bone strength in an effort to determine the probability of bone fracture (Kaufman and Einhorn 1993). The calcaneus is most commonly selected for ultrasound assessment because it is accessible to the transducer with minimal surrounding soft tissue. Unlike most soft tissue, bone has a complex structure, consisting of a compact cortical shell, trabecular framework, and soft marrow. Ultrasonic speed of sound (SOS) and broadband ultrasonic attenuation (BUA) are indicators of variations in these components, and stiffness index (SI) is an ultrasound measure of bone strength derived from a combination of SOS and BUA.

Because bone strength is determined by biomechanical properties such as elasticity, compressive strength, and trabecular separation in addition to bone mineral density (Baran 1995), SI does not have a strong correlation with measures of bone mineral density (Aloia et al. 1998; Takeda et al. 1996). Moris and colleagues (1995) reported significant but moderate correlations (.57-.67) between DXA of the spine and calcaneus SOS, BUA, and SI. The precision and accuracy of ultrasound measures of bone are reported to be very high (Greenspan et al. 2001; Takeda et al. 1996; but see Çetin et al. 2001).

Summary

As new methods are introduced, anthropometry may be considered old-fashioned because it has been used for a long time, but it is far from obsolete. This chapter has addressed its utility when used alone in providing indexes and predicted values for total body composition and in measuring and predicting regional body composition. Although anthropometry is important in these contexts, it has many limitations when used alone. Anthropometry is particularly important in predictive equations that include values from bioelectric impedance.

B-mode ultrasound has been used to measure regional body composition for about 20 years. Its advantages over the measurement of skinfold thickness in obese individuals are now recognized. Continual improvements in ultrasound equipment are likely to lead to the increased use of ultrasound to measure SAT thicknesses in obese persons and to measure the thicknesses and cross-sectional areas of muscles in elderly people and during rehabilitation. Further studies are needed to document fully the location and timing of development of an intralipid fascial plane and the physiological differences between the layers of adipose tissue that it separates.

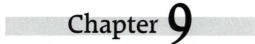

Chapter 9

Pediatric Body Composition Methods

Aviva Sopher, Wei Shen, and Angelo Pietrobelli

The importance of body composition measurements in pediatrics is gaining recognition. Significant physical changes occur during the years spanning infancy through young adulthood and are apparent both externally and internally. External changes such as body proportions, height, weight, and pubertal status are easily measured by physical examination and by simple anthropometric measurements; however, internal changes such as body composition and hormonal status require specialized testing.

Over the past several decades, body composition methods have gained acceptance in both research and clinical medicine. However, studies have demonstrated that adult body composition measurement methods and data may not be directly applicable to pediatric populations. Earlier chapters reviewed body composition model concepts (chapter 1) and measurement methods (chapters 2-8). Concepts related to growth and development are also reviewed in chapter 18. In this chapter, we focus specifically on body composition measurement methods as they apply to the pediatric population.

The distinct features of the pediatric population make the selection and application of body composition methods different from those of adults. Body composition methods requiring substantial subject cooperation cannot be easily used in infants and young children. An example is the underwater weighing method that includes submersion, breath holding, and exhalation. Additionally, exposure of pediatric subjects to high doses of radiation should be avoided when

possible. Thus, use of neutron activation analysis and computed tomography (CT) for pediatric body composition is limited, and so neutron activation is not additionally reviewed here. DXA provides a minimal amount of ionizing radiation exposure that is equivalent to 1% of the radiation exposure from a chest X ray; however, risks versus benefits of body composition assessment must be carefully considered in each case (Baur 1995).

Ideally, urine and saliva samples should be collected as alternatives to blood if possible. Appropriate topical anesthetics can be applied to reduce pain when there are no alternatives to phlebotomy. Careful oral administration of tracers is essential with equally cautious collection of urine samples, particularly in infants and young children, to ensure accuracy of results.

Compared with adults, children have much greater variation in body composition attributable to growth and development from infancy to adolescence. For example, the relative reduction in total body water after birth, especially during the first several months of life (Fomon and Nelson 2002), will influence the application of most body composition methods related to hydration (e.g., bioimpedance analysis; Lohman 1986, 1992). Some traditional reference methods used in adults, such as underwater weighing, have serious model concerns when applied in young children (Lohman 1986). These issues create limitations in the accuracy of component (e.g., total body fat) prediction methods such as anthropometry, bioimpedance analysis, and total body electrical conductivity that must be calibrated against a reference method. Methods relying on minimal age-related assumptions (e.g., magnetic resonance imaging [MRI]) thus have an advantage over methods based on adult body composition steady-state assumptions, and the applied models need to be appropriately adjusted when used in pediatric populations. The adjustment of these models is particularly problematic in longitudinal studies of dynamic changes in body components. Thus, results provided by methods that have not been carefully validated in the studied age group need to be interpreted cautiously (Butte et al. 1999).

Multicomponent Models

The two-component molecular level model in which body weight is divided into fat mass and fat-free mass (FFM) is the most widely used model in adults. The two widely used two-component models are based on the assumption that FFM has constant density and hydration (chapter 12). Although the FFM density and hydration are considered to be constant in healthy adults, they are not constant in young children and infants. Therefore, two-component models are not ideal for measuring fat mass and FFM in infants and young children. Furthermore, the two-component model provides fat and FFM estimates and does not give information regarding the nutritionally important total body protein and mineral components. These two-component models and associated measurements are relatively easy to obtain and are useful in older children in whom the focus of interest is fat mass and FFM.

In the pediatric population, the four-component molecular-level model (chapter 12) is the most widely used model to accurately describe growth of fat mass and the main FFM components—water, protein, and minerals. This model is appropriately used in pediatrics as a reference for body fat because the required model assumptions are minimal and not, according to current thinking, influenced by subject age or developmental level. The four-component model requires a combination of measurement methods, which are described in the following sections.

Butte and colleagues (2000) reported another approach specifically designed to provide reference body composition estimates in infants and young children by including more components in their model. The developed multicomponent model was as follows:

$$\text{Fat} = \text{body weight} - (\text{water} + \text{protein} + \text{glycogen} + \text{bone minerals} + \text{nonosseous minerals}) \quad (9.1)$$

Total body water (TBW) in this model was measured by deuterium dilution. Total body potassium (TBK) was then derived by ^{40}K counting, and intracellular and extracellular water was calculated from the TBW and TBK estimates (chapter 12). Nonosseous minerals were calculated by assuming there are 9.4 g of mineral per kilogram of extracellular water and 9 g of mineral per kilogram of intracellular water. Protein was calculated by assuming that protein is 16% nitrogen and that there is 461 mg of nitrogen per millimole of potassium. Bone mineral content was estimated by dual-energy X-ray absorptiometry. Finally, glycogen was estimated as 0.45% of body weight. In addition to each of the model components, Butte and colleagues (2000) derived values for the hydration, density, and potassium content of FFM (see the appendix on page 401). This pediatric modeling approach extends earlier pediatric models reported by Fomon and colleagues (1982).

Hydrodensitometry

Hydrodensitometry is designed to provide estimates of body volume or density (D_b) at body temperature (chapter 2). The body density two-component model can then be used to calculate an estimate of total body fat and FFM. The density of fat is generally accepted across all age groups as 0.9007 g/cm^3. The density of FFM in the pediatric population changes with maturation and is different from that of adults. With chemical maturation, as represented by cell hypertrophy and hyperplasia, dynamic changes in body composition have an important impact on the two-component hydrodensitometry model (Lohman 1986, 1992; Moulton 1923). Extracellular fluid has a high water content (~0.98) compared with body cell mass (BCM; ~0.70). Water has the lowest density (0.9937 at 36° C) of the major FFM components (chapter 12). At birth, when mean cell size is relatively small, the ratio of extracellular fluid to cells is large, and FFM hydration is high (Fomon and Nelson 2002). Data from the male reference infant indicate that during the first few years of life, there is a reduction in the relative size of the extracellular fluid compartment, in the ratio of extracellular to intracellular water (figure 9.1), and in FFM hydration (figure 9.2; Fomon et al. 1982).

These changes, along with a relative increase with age in high-density bone mineral, contribute to an increase in the density of FFM from birth up to the mature level of 1.100 g/cm³ (figure 9.3).

Some debate centers on appropriate model values for the density of FFM in children for age and sex. Weststrate and Deurenberg (1989) used Fomon's density of FFM (g/cm³) values to derive a set of prediction equations based on age and sex (e.g., boys and girls aged 1-1.99 years: density of FFM = 1.0635 + 0.00163 [age (months)]$^{0.5}$). Lohman (1986) also compiled multiple sources to provide estimates of FFM hydration and density (table 9.1). Much of Fomon et al.'s work (1982) is a compilation of the literature; however, both Lohman (1986) and Haschke (1983) measured large samples of children using body density and water in the same sample. In addition, Lohman (1986) estimated body mineral from forearm BMD measures as well as body density and water. All

three approaches were found to be in general agreement in estimating densities of the fat-free body with age.

Advantages of underwater weighing in children are its safety and its low cost. A number of systems have been installed in developing countries and are useful for evaluating both adults and older children. Nevertheless, underwater weighing systems are not widely available, and many children below the age of about 4 or 5 find the measurement procedure difficult. Some children are uncomfortable in water and cannot hold their breath while submerged. Also, underwater weighing is impractical to carry out in infants, very young children, and ill children. Infants and young children represent an age group in which ambiguity surrounds the two-component model coefficients. Ideally, these subjects should be studied using multicomponent models, particularly when accurate total body fat estimates are important.

Air Displacement Plethysmography

Air displacement plethysmography (ADP) by Bod Pod (Life Measurement Systems, Concord, CA) is a body composition method that measures body volume and body density (chapter 2). Air displacement plethysmography is an attractive body composition method for children because of its ease of use and safety. Because water submersion is not required, many young children find this method more acceptable than underwater weighing.

Air displacement plethysmography has limitations. ADP has not been adequately validated by a criterion method in children and adolescents for

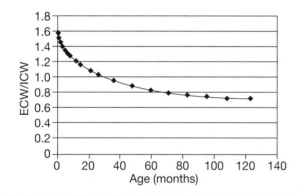

Figure 9.1 Changes in the ratio of extracellular to intracellular water (ECW and ICW) as a function of age in boys.

Data from Fomon and Nelson (2002).

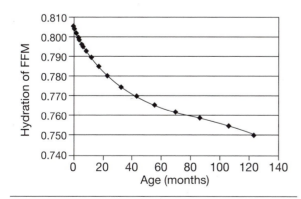

Figure 9.2 The hydration of fat-free mass (FFM) in boys from birth to the age of 10 years.

Data from Fomon and Nelson (2002).

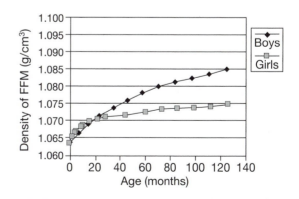

Figure 9.3 The estimated density of fat-free mass (FFM) in boys and girls from birth to the age of 10 years.

Data from Fomon and Nelson (2002).

Table 9.1 Age- and Sex-Specific Constants for Conversion of Body Density, Water, and Potassium to Percent Fat in Children and Youth

	FAT-FREE BODY COMPOSITION							
	FFM DENSITY		TBW–FFM		TBK–FFM		MINERAL–FFM	
Age (years)	Male	Female	Male	Female	Male	Female	Male	Female
7-9	1.081	1.079	76.8	77.6	61.5	59.5	5.1	4.9
9-11	1.084	1.082	76.2	77.0	62.8	60.0	5.4	5.2
11-13	1.087	1.086	75.4	76.6	64.6	60.5	5.7	5.5
13-15	1.094	1.092	74.7	75.5	65.6	61.0	6.2	5.9
15-17	1.096	1.094	74.2	75.0	66.9	61.5	6.5	6.1
17-20	1.098	1.095	74.0	74.8	67.4	61.8	6.6	6.0
20-25	1.100	1.096	73.8	74.5	68.2	62.1	6.8	6.2

Note. FFM = fat-free mass; TBW = total body water; TBK = total body potassium. Units: FFM density, g/cm^3; TBW–FFM, g/100 g; TBK–FFM, mmol/kg; mineral–FFM, g/100 g.

Data from Lohman (1986).

body composition measurements. Several studies have compared body density measurement by ADP with that by underwater weighing (Demerath et al. 2002; Dewit et al. 2000; Lockner et al. 2000; Nunez et al. 1999). The studies demonstrated relatively high correlations between the two methods (r^2 = .75-.85); however, the results were inconsistent (e.g., presence of bias, direction of bias, direction and significance of the mean body density difference; Demerath et al. 2002; Dewit et al. 2000; Lockner et al. 2000; Nunez et al. 1999). The study of Dewit and colleagues (2000) demonstrated ADP to be more precise than underwater weighing for the measurement of body density in both children and adults. One study that compared fat mass by ADP with fat mass by the four-component model in 25 children ages 9 to 14 years (Fields and Goran 2000) demonstrated no bias in the relationship between the two methods and no significant deviation of the regression line from the line of identity and the limits of agreement. The authors concluded that ADP was validated based on their analysis. However, the body density used in the four-component model was measured by ADP, and therefore this approach limits the validity of a study where the four-component model uses that same measurement in its equation.

Theoretically, ADP and underwater weighing should give similar values for body density (Fields et al. 2002). The inconsistent results in the studies that compare body density by ADP with

underwater weighing illustrate limitations that may be associated with these methods, some of which can be overcome. Factors that may explain the discrepancies and bias noted in the mentioned studies include method-specific assumptions (e.g., thoracic gas volume [V_{TG}] [ADP], residual volume [underwater weighing], intestinal gas [ADP and underwater weighing]), test conditions (e.g., clothing, degree of dryness, activity level, measured or predicted V_{TG}), and subject size, body temperature, and age (Fields et al. 2002).

Several factors intrinsic to both methodologies introduce variability and measurement error. Different weight scales are used for ADP and underwater weighing (Fields et al. 2002). Body volume measurement by ADP relies on measurement of raw body volume by the instrument itself, calculation of the surface area artifact by the instrument's software, and measurement or prediction of thoracic gas volume (Fields et al. 2002). Body volume by underwater weighing requires underwater body mass, lung, and residual volume measurement. Measured thoracic gas volume is based on functional reserve capacity predictions from height and age for subjects 17 to 91 years (Crapo et al. 1982). Predicted thoracic gas volume and functional reserve capacity both rely on equations that are based on adult values (Fields et al. 2002). Both methods are sensitive to food intake, because of the effects on intestinal gas volume. Results of one study suggest that underwater weighing is more sensitive than ADP (McCrory et al. 2000). Test conditions are extremely important for accurate ADP results. Air displacement by Bod Pod is performed under adiabatic conditions that allow for free gain and loss of heat during compression and expansion of air (Fields et al. 2002). However, clothing, body hair, and skin surface may create partially isothermal conditions, and isothermal air is 40% more compressible than adiabatic air. The surface area artifact accounts for the isothermal conditions introduced by the skin surface. The manufacturer recommends that a swim cap be worn on the head to minimize the effect of hair on creating isothermal conditions. Additionally, the manufacturer recommends that skintight clothing be worn to minimize air trapping in loose clothing. To minimize variability, all participants in Bod Pod studies should wear standardized apparel that meet the criteria described.

Currently, Bod Pod is validated for individuals who weigh more than 40 kg. The validity of Bod Pod for children below 40 kg weight remains unclear. The technical error was higher in children who weighed less than 50 kg compared with heavier children (Demerath et al. 2002). In another study of 22 children with an average weight of 33.2 kg, precision for ADP by Bod Pod was similar to that of adults (Dewit et al. 2000). Age, sex, and ability to comprehend instructions appear to be important for accurate body density measurements.

Several studies demonstrate a significant difference in the body fat obtained from measured V_{TG} when compared with predicted V_{TG}. Two pediatric studies describe higher body fat values with predicted V_{TG} than with measured V_{TG} (Demerath et al. 2002; Lockner et al. 2000). Because children cannot undergo thoracic gas volume measurements, it may be preferable to predict thoracic gas volume measurements for subjects in pediatric studies that include young children.

Air displacement plethysmography by Bod Pod has excellent potential to become a routine and accepted body composition method in children if adequately validated. However, it is important to be aware of the sensitivity of this method to many factors.

The recently developed Pea Pod Infant Body Composition System (Life Measurement Instruments, Concord, CA) for infants 0 to 6 months was reported by Urlando and colleagues (2003). Although more validations are needed, this new ADP system could provide a solution for evaluating body density in young infants.

Dilution Methods

The total body water (TBW) component is usually measured by dilution of a stable isotope such as D_2O or $H_2{}^{18}O$. Furthermore, as with hydrodensitometry and ADP two-component models, the TBW method in young children and infants is based on an FFM hydration value greater than 0.73 (Fomon and Nelson 2002; Lohman 1986). As shown in figure 9.2, FFM hydration is a maximum at birth (~0.8) and declines rapidly during the first few years of life to ultimately reach the adult value of approximately 0.73 during the teenage years. The high hydration in early life is accompanied by a high ratio of extracellular to intracellular water (figure 9.1). Corresponding estimates, based on a number of assumptions, are summarized in table 9.1.

Because of the wide between-individual variation in the hydration of FFM, the isotope dilution method may not be the optimal method for assessing total body fat in children, especially newborns (Hashimoto et al. 2002). Nevertheless, dilution methods are the most reliable approach for estimating TBW.

The high water turnover in young children also has implications on the selection of estimation methods. Plateau or back-extrapolation methods can be used to calculate TBW. Theoretically, the former overestimates TBW whereas the latter underestimates TBW (Butte et al. 2000).

Whole-Body Counting

Potassium is mainly an intracellular ion and is measured by whole-body ^{40}K counting, a safe and relatively easy method. The reproducibility of total body potassium (TBK) measurements is good, even when children weigh 20 to 25 kg (Schneider et al. 1998). Measurement of TBK is useful for estimating BCM and total body protein using appropriate prediction models (Butte et al. 2000; Wang et al. 2003). Recently, TBK estimates have been used to accurately predict total body skeletal muscle mass in adults (Wang et al. 2004). These approaches provide new possibilities for developing similar estimation methods in children.

Total body potassium counting was also used to estimate body fat using a two-component model (Forbes 1987). This model assumes stable potassium content of FFM; however, with chemical maturation, there is an increase in the potassium content of FFM during growth (figure 9.4). These changes lead to complexities in developing

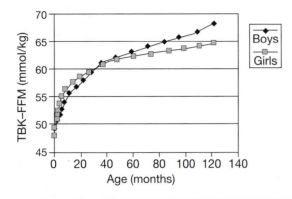

Figure 9.4 The ratio of total body potassium to fat-free mass (TBK–FFM) in boys and girls from birth to the age of 10 years.

Data from Fomon et al. (1982).

appropriate pediatric two-component model coefficients. The use of an assumed adult TBK–FFM ratio in the two-component model tends to underestimate FFM in children. Estimates of TBK/FFM as a function of age and sex are summarized in table 9.1 (Lohman 1986).

Electrical Methods

Pediatric and adult devices that measure total body electrical conductivity (Tobec; Em-Scan, Springfield, IL) can be used to provide rapid, noninvasive estimates of FFM and fat mass (Fiorotto et al. 1987; Treuth et al. 2001; Van Loan 1990). Tobec systems are limited in availability but are reported to have good accuracy and precision (Fiorotto et al. 1987; Van Loan 1990).

Tobec measurements are taken as the subject passes through a low-energy electromagnetic coil, which alters the conductance in the coil. The change in conductance is proportional to total body electrolyte content (i.e., highly conductive FFM and a minimal component of the poorly conductive fat mass). Prediction equations are used to determine fat mass and FFM (Fiorotto et al. 1987; Van Loan 1990).

The only calibration reference so far applied in children is the TBW isotope dilution method. The infant Tobec has been calibrated for FFM and TBW using data from chemical analyses of infant miniature pigs (Boileau 1988; Fiorotto et al. 1987). FFM measured by Tobec in neonates showed no significant differences with corresponding estimates by anthropometry and total body water (Fiorotto et al. 1987; Treuth et al. 2001).

The bioelectric impedance analysis (BIA) method, reviewed in chapter 6, is based on a number of assumptions related to extremity proportions and tissue composition. Application of BIA in pediatrics must therefore include prediction equations specific to age or pubertal stage. Prediction equations have been reported in the literature and are sometimes provided with commercial systems (Yanovski et al. 1996). A three-component model (Houtkooper et al. 1992) has been used to develop equations for estimating FFM for 10- to 19-year-olds, and TBW and Lohman's constants (Kushner et al. 1992) have been used to develop equations for children less than 10 years old. Additionally, race-specific prediction equations for FFM have been developed (Lewy et al. 1999). Validation studies for these equations against accepted reference methods are still needed before the value of BIA in pediatrics can be fully established.

Dual-Energy X-Ray Absorptiometry

Dual-energy X-ray absorptiometry scans are increasingly available and easily performed on children of all ages, making this method attractive for pediatric body composition measurement (chapter 5). Several studies have compared DXA with the criterion four-compartment model for body fat in pediatric subjects, and the findings of these studies are not consistent (Wong et al. 2002; Fields and Goran 2000; Roemmich et al. 1997; Wells et al. 1999).

We recently completed a cross-sectional project during which 411 healthy pediatric subjects (aged 6-18 years) performed the criterion four-component model (Sopher et al. 2004). We observed a predictable and consistent relationship between DXA and four-component measurements of percent body fat. The relationship was characterized by overestimation of percent body fat by DXA (Lunar DPX and DPX-L models) in subjects with higher percent body fat and underestimation in those with lower percent body fat. Our findings suggest that percent body fat measurements by DXA are not equivalent to those of the criterion four-component model; however, the differences between the two methods do not preclude use of DXA for clinical and research measurement of percent body fat in children and adolescents.

Our observations are similar to those of other studies that compared Lunar DXA to four-component estimates for measurement of percent body fat (Fields and Goran 2000; Gallagher et al. 2000). Pediatric body composition studies using Hologic systems demonstrated a different relationship between the two methods (Roemmich et al. 1997; Wells et al. 1999; Wong et al. 2002). One report did not find a significant difference between percent body fat by Hologic DXA (Model QDR-1000W) compared with four-component estimates (Wells et al. 1999). However, two other reports found that Hologic DXA (Models QDR-2000 and -2000W) systematically overpredicted percent body fat compared with the criterion, and the overprediction of percent body fat was independent of subject percent body fat (Roemmich et al. 1997; Wong et al. 2002). Although statistical modeling to create "translation" equations between DXA and the criterion method is possible, the variability between methods persists (Pintauro et al. 1996).

The relationships between percent body fat by DXA and four-component estimates differ by DXA manufacturer but have similar standard errors of the estimate and limits of agreement. The different pattern of the relationship to percent body fat by four-component model for each manufacturer but similar statistical characteristics suggests calibration differences associated with system-specific algorithms rather than a flaw in DXA technology.

Like all indirect in vivo body composition methods, DXA technology relies on several assumptions of tissue constancy that may not always be accurate. For example, R-values are attenuation ratios that are assumed stable for specific components such as fat and FFM. However, R-values measured by DXA for homogeneous materials may systematically change as thickness or depth varies (Pietrobelli et al. 1996; Pietrobelli et al. 1998b). In an in vitro study that used a Lunar system to measure phantoms of varying depths, all fat values were close to theoretical chemical calculations, but percent body fat of the phantoms was overestimated when phantom depth was greater and was underestimated when lower (Laskey et al. 1991). These findings are similar to the relationship we observed between our Lunar DXA system and four-component estimates. The modest but significant effect of anthropometric measures on our model may represent an effect of body thickness or fatness. In a study of healthy adults, correction for anthropometric dimensions did not improve the relationship between DXA and the criterion four-component model, emphasizing the importance of specific pediatric studies (Marcus et al. 1997).

DXA has been shown to be a precise measure of percent body fat in adults (Russell-Aulet et al.

1991). Preliminary same-day intraindividual data in our laboratory also indicate that DXA is precise for children and adolescents.

We propose that DXA has great potential as both a pediatric research tool and a clinical tool if the specific characteristics of the system being used are recognized. An example of a research use of DXA that may lead to clinical application is the prediction of comorbidity risk in obese children and adolescents (Higgins et al. 2001). Recognition that DXA differs from the criterion measure, and that not all DXA systems are the same, will lead to better interpretation of research and clinical results. Aside from calibration issues, DXA interpretation should be considered in the context of body composition changes that occur early in life and continue during childhood. The DXA method provides estimates of three main components—fat, lean soft tissue, and bone mineral. The lean soft tissue component includes two primary constituents at the cellular level of body composition—BCM and extracellular fluid. BCM is the actively metabolizing, oxygen-consuming component, whereas extracellular fluid is responsible for transfer of nutrients and waste products between cells and the environment. As shown in figure 9.5, both lean soft tissue and BCM increase with age but do so at different relative rates. The result is that the BCM component increases relative to lean soft tissue with greater age. Thus, the lean soft tissue component is not metabolically homogeneous with respect to age in children, and results should be interpreted accordingly.

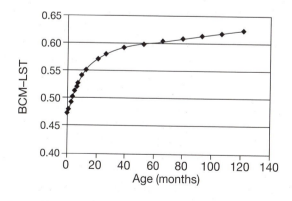

Figure 9.5 The ratio of body cell mass to lean soft tissue (BCM-LST) from birth to 10 years in boys. Intracellular water was used to estimate body cell mass by assuming cell hydration of 70%.

Data adapted from Fomon et al. (1982).

Imaging Methods

CT and MRI provide investigators with the opportunity to evaluate tissue–organ-level components in pediatric subjects (Heymsfield et al. 1997; Sjostrom 1991; chapter 7). Multiple cross-sectional images can be used to reconstruct tissue volumes in pediatric subjects including total, subcutaneous, and visceral adipose tissue, skeletal muscle, brain, heart, kidney, liver, skin, and bone (Gallagher et al. 1998; Lee et al. 2000; Shen et al. 2003).

These imaging techniques may offer new insights into the associations between intra-abdominal adipose tissue and metabolic factors (Goran 1998) because CT and MRI are more accurate than other available methods in assessing tissue–system-level body composition components. One area of particular interest is the accumulation of visceral adipose tissue (VAT) in childhood because there are significant relationships between VAT and adverse health (Caprio et al. 1996). Additionally, CT and MRI may help create an understanding of ethnic and sex differences in adipose tissue distribution (Goran 1998).

Magnetic resonance imaging for assessment of pediatric body composition is relatively new (Harrington et al. 2002; Leger et al. 1998; Olhager et al. 1998), and the lack of radiation exposure associated with its use makes it more applicable than CT. Limitations of MRI are that it is relatively costly and requires subjects to stay motionless throughout. However, because infants are short, the acquisition time for whole-body analysis can be less than 10 min (Harrington et al. 2002; Olhager et al. 1998), and the cost of image acquisition and analysis may consequently be reduced.

The interpretation of MRI in children is straightforward and does not require pediatric-specific assumptions different from those of adults. Some investigators believe that body composition research should include the fetal stage (White et al. 1991); MRI is the only in vivo method that has been used to study fetal body composition (Deans et al. 1989), and it may help provide an understanding of physiological and pathological conditions in both pregnant women and newborns.

Measurements by MRI are especially suitable for comparisons between individuals, and MRI can be used to evaluate body composition in longitudinal studies. The reproducibility of MRI measurements of subcutaneous adipose tissue in infants is 2.4% to 4% (Harrington et al. 2002; Olhager et al. 1998) and 17.6% for internal adipose tissue (Harrington et al. 2002). These values are similar to the reproducibility of similar adipose tissue compartments in adults (Elbers et al. 1997). Some investigators, however, report a large interobserver variation in adipose tissue estimates in infants ranging from 4.2% to 40.7% (Olhager et al. 1998).

An important consideration is that MRI provides tissue volume estimates, and, as noted earlier, tissue composition may vary with age. For example, the fat content of adipose tissue is 66% in neonates and gradually increases to the adult level 80% at about 13 years of age (Snyder et al. 1975; White et al. 1991). This variation in adipose tissue fat content was considered by Harrington and colleagues (2002) when they estimated total body fat from adipose tissue volume in newborns.

MRI can serve as a reference method for measuring tissue and organ volumes because estimates will be reliable independent of age. Furthermore, with the use of chemical shift imaging methods or when MRI is combined with magnetic resonance spectroscopy (MRS; Shen et al. 2003), it is likely that MRI and MRS can be used to validate measures of important molecular level components such as fat and FFM as measured by Bod Pod, BIA, DXA, and underwater weighing in different pediatric age groups.

Anthropometry

Anthropometric measurements may be used as markers of adiposity (e.g., body mass index [BMI]) or of fat distribution (e.g., waist circumference). These measurements are useful for subject groups or populations, but accuracy in individuals is limited, particularly in infants.

Weight-for-Height Measures

Many different weight-for-height indexes are reported in the pediatric literature. A list of frequently used indexes is presented in table 9.2. The existence of multiple indexes reflects the lack of a consensus on the best measure of weight-for-stature as a surrogate of adiposity in children. Presently, the most widely used weight-for-stature index is BMI, calculated as weight/height^2 (kg/m^2; Quetelet 1869).

Longitudinal studies have aided in characterizing the natural progression of the BMI curve with growth and maturation in childhood and adolescence. BMI initially increases during infancy with a peak at approximately 9 months of age and decreases until about 6 years of age. BMI rebound, originally termed adiposity rebound,

Table 9.2   Weight for Height Indexes Applied in Pediatric Populations

Type	Name	Comment	Form	References
Ratio	Relative weight	Weight as a percentage of median weight for height	W/H	Waterloo et al. 1977
Ratio	Quetelet's index	Body mass index	W/H^2	Quetelet 1869
Ratio	Rohrer's index	Ponderal index	W/H^3	Rohrer 1921
Ratio	Benn's index		W/H^p	Benn 1971
Ratio	Chinn's index	Adjusted centiles of BMI for age and sex	$(W - 9)/H^{3.7}$	Chinn et al. 1992
Ratio	Eid's index	Weight as a percentage of median weight for height and age		Eid 1970
Regression model	Tanner	Weight- and height-for-age charts	$(WFA_{SDS} - HFA_{SDS})$	Tanner 1951
Regression model	Cole	Standard deviation score	$(WFA_{SDS} - b \times HFA_{SDS})$	Cole 1993

Note. W = weight; H = height; WFA_{SDS} = standard deviation score of weight for age; HFA_{SDS} = standard deviation score of height for age; BMI = body mass index; P = constant.

occurs in childhood and represents the continued increase in BMI until adulthood is reached (Cole and Rolland-Cachera 2002). The earliest published pediatric population-based BMI reference curves were based on data obtained from French children (Rolland-Cachera et al. 1982). Currently, pediatric population-based BMI reference curves are available for American (Kuczmarski et al. 2000), British (Cole et al. 1995; He et al. 2000), Swedish, Chinese (from Hong Kong; Leung et al. 1998), Dutch (Cole and Roede 1999), and Italian (Cacciari et al. 2002) populations. Cole and colleagues (2000) proposed a BMI standard for all nationalities; however, Deurenberg and Deurenberg-Yap (2001) showed that the percent fat to BMI relationship varies with ethnicity.

In response to the rising prevalence of obesity, BMI curves have been incorporated into the routine growth assessment performed by pediatricians as a tool to prevent and detect weight problems and monitor body weight in pediatric patients in the United States. The 2000 Centers for Disease Control and Prevention growth charts represent the revised version of the 1977 National Center for Health Statistics (NCHS) growth charts. Most of the data used to construct these charts come from the National Health and Nutrition Examination Survey (NHANES), which has periodically collected height, weight, and other health information on the Ameri-

can population since the early 1960s. The growth charts were designed to document rapid growth in younger ages (0-36 months) and relatively slow growth in older ages (2-20 years). The charts for older ages can be found at http://www.cdc.gov/growthcharts. These curves help to identify children who are at risk of becoming overweight and those who are overweight. In the United States, BMI values are used to classify children as either normal weight (<85th percentile), overweight (85th-95th percentile), and obese (>95th percentile; Dietz and Bellizzi 1999; Himes and Dietz 1994). Recommendations have been made for obesity evaluation and treatment based on BMI percentile category for children and adolescents (Barlow and Dietz 1998; Himes and Dietz 1994).

Several studies delineate percent body fat levels associated with increased cardiovascular risk profiles in childhood (Dwyer and Blizzard 1996; Higgins et al. 2001; Lohman and Going 1998; Williams et al. 1992). Williams and colleagues developed equations specifically for children by using the sum of subscapular and triceps skinfolds to estimate percent fat in a biracial sample of 3,320 children and adolescents aged 5 to 18 years. The authors found that percent fat values of 25% for boys and 32% for girls are appropriate cutoff points for cardiovascular risk in children and adolescents.

Interpretation of BMI as a marker for adiposity in childhood is complex. Body mass index is a good general screening tool for pediatric obesity and is independently correlated with health risk in children and adolescents (Freedman 2002; Goran and Gower 2001; Must et al. 1992, 1999). Nevertheless, BMI reflects both fat mass and FFM and is not a direct measure of adiposity (Maynard et al. 2001). Several studies have shown good correlations between BMI and body fat in groups, but the variation is large and BMI cannot accurately predict fatness in individual subjects (Daniels et al. 1997; Goran 1998; Pietrobelli et al. 1998a). The relationship between BMI and adiposity is independently influenced by height, weight, age, sex, ethnicity, pubertal stage, body proportions, fat distribution, and medical status (Daniels et al. 1997; Ellis et al. 1999; Garn et al. 1986; Lohman and Going 1998; Mast et al. 2002; Maynard et al. 2001; Pietrobelli et al. 1998a). Although BMI may be useful as an approximate classification tool for obesity status, it lacks adequate sensitivity and specificity and may misclassify a child as obese or nonobese (Ellis et al. 1999; Mast et al. 2002; Reilly et al. 2000; Schaefer et al. 1998; Warner et al. 2001). Other confounding factors include the suggested secular increase in the ratio of fat mass to FFM in children and adolescents over the past few decades (Wells et al. 2002) and the observation that body composition in children with medical conditions differs from their equivalent age counterparts (Arpadi et al. 1998; Barera et al. 2000; Nysom et al. 1999; Sentongo et al. 2000; Stettler et al. 2000).

Future studies should continue to focus on the relationships among BMI, adiposity, and health risks in children and adolescents to further define pediatric obesity diagnosis, evaluation, and treatment.

Skinfolds

The location of specific skinfold sites and measurement methods are summarized in chapter 8. Skinfold thickness measurements in children can be used to estimate body density, FFM, fat mass, and percent body fat in conjunction with appropriate and validated pediatric prediction equations. Equations have been developed using a multicomponent method and thus provide applicability over the age range from prepubescent to postpubescent children (table 9.3). However, these

Table 9.3 Anthropometric Prediction Equations

Measurement	Ethnicity and sex	Equation
Triceps + calf (Σ)[a]	Black and white Boys Girls	%BF = 0.735 (Σ) + 1.0 %BF = 0.610 (Σ) + 5.1
Triceps + subscapular (>35 mm) (Σ)[a]	Black and white Boys Girls	%BF = 0.783 (Σ) + 1.6 %BF = 0.546 (Σ) + 9.7
Triceps + subscapular (<35 mm) (Σ)[a]	Prepubescent males—white Prepubescent males—black Pubescent males—white Pubescent males—black Postpubescent males—white Postpubescent males—black All females	%BF = 1.21 (Σ) − 0.008 $(\Sigma)^2$ − 1.7 %BF = 1.21 (Σ) − 0.008 $(\Sigma)^2$ − 3.2 %BF = 1.21 (Σ) − 0.008 $(\Sigma)^2$ − 3.4 %BF = 1.21 (Σ) − 0.008 $(\Sigma)^2$ − 5.2 %BF = 1.21 (Σ) − 0.008 $(\Sigma)^2$ − 5.5 %BF = 1.21 (Σ) − 0.008 $(\Sigma)^2$ − 6.8 %BF = 1.33 (Σ) − 0.013 $(\Sigma)^2$ − 2.5
Sum of triceps + biceps + subscapular + suprailiac	Prepubertal males[b] Prepubertal females[b] Adolescent males[c] Adolescent females[c]	BD = 1.1690 − 0.0788 log sum of 4 skinfolds BD = 1.2063 − 0.0999 log sum of 4 skinfolds %BF = ([4.95/BD] − 4.5) 100 BD = 1.1533 − 0.0643 log sum of 4 skinfolds BD = 1.1369 − 0.0598 log sum of 4 skinfolds %BF = ([4.95/BD] − 4.5) 100

Note. Σ = sum of skinfolds (mm); %BF = percent body fat; BD = body density.

[a]Slaughter et al. (1988); [b]Brook (1971); [c]Durnin et al. (1967).

equations tend to be population specific, and accuracy is variable when applied in new populations (Ellis 1998). The most useful skinfold thickness measurements in the pediatric age group are the triceps and subscapular skinfolds (Gaskin and Walker 2003; Lohman and Going 1998; Magarey et al. 2001). Additionally, skinfold measurements are subject to relatively high interobserver variability.

Percent body fat in children determined by equations that use two or more skinfolds correlates well with percent body fat determined by underwater weighing (r = .65-.90; Harsha et al. 1978). However, in infants, the use of skinfold thickness is more controversial; poor or no correlation was observed between body fat measured by water dilution methods and skinfolds (Kabir and Forsum 1993). A study by Deans and colleagues (1989) failed to show a correlation between MRI-measured fetal percentage fat and skinfold caliper measurements made after birth. Improved skinfold measurement and validation methods are needed for infants.

Circumferences and Diameters

Body circumferences and skeletal breadth are anthropometric measurements that are useful in determining body size and body proportions in children. The use of these anthropometric-based measurement methods is based on the concept that circumferences reflect fat mass and FFM and that skeletal size is associated with FFM (Wagner and Heyward 1999).

Circumferences are useful in determining fat distribution. Waist, hip, and thigh circumferences are used to predict body fat distribution in children, and waist and hip circumferences are both good predictors of intra-abdominal adipose tissue (Goran 1998).

Waist circumference and other circumferences are also useful measures of cardiovascular risk in the pediatric population. Waist circumference was found to be correlated with family history of diabetes (Giampietro et al. 2002); small, dense, low-density lipoprotein levels (Kang et al. 2002); high-density lipoprotein cholesterol levels (Teixeira et al. 2001); and blood pressure (Maffeis et al., 2001). Other studies suggest that the waist-to-hip ratio is the best predictor of cardiovascular disease risk (Hara et al. 2002; Teixeira et al. 2001).

The percentile distributions of waist circumferences among children aged 2 to 18 years in African Americans, Hispanics, and Caucasians in the U.S. population were recently reported by Fernandez and colleagues (2004). Quartile regression analysis was used to identify the linear quadratic and cubic trend of the prediction of waist circumference according to age and sex. These published values can be used to assess future trends, compare individual with population norms, and compare populations with public health recommendations.

Summary

The interest in the field of pediatric body composition is expanding rapidly. A consistent underlying theme of this field is that children are not simply little adults. Body composition methods and equations validated in adults are not necessarily appropriate for infants, children, and adolescents. For older children, standards for body composition in relation to health risks have been proposed. Despite some of these concerns, the available tools for quantifying body composition in pediatrics are well developed and useful for some age groups, and the application of these tools should yield important new information and insights. Moreover, the large gaps in available information for some topics makes pediatric research a dynamic and growing scientific area of body composition investigation.

Chapter 10

Animal Body Composition Methods

Maria S. Johnson and Tim R. Nagy

Many investigators are interested in factors that influence changes in body weight of animals. However, the composition of this weight change is potentially more important. There is a dramatic difference between a 5-g increase in body mass in a mouse during development compared with a 5-g increase of a 7-month-old mouse (middle-aged). During growth, all body components (lean, fat, and bone) are increasing, whereas an increase during middle age is likely to be mainly fat mass. If we measure only mass, the proportions of fat and lean mass gain remain unknown. Obtaining data on the relative amounts of fat and lean mass enables more accurate comparisons between treatment groups and the effects before and after interventions.

This chapter focuses on the most frequently used and most accurate methods for assessing body composition in small animals. Many of these techniques are covered in other chapters in relation to their use in human studies. Therefore, we will concentrate on the practical aspects of these techniques and discuss the accuracy and precision of each technique and in which situations each technique is most applicable.

Carcass Analysis

It is not always feasible to perform the carcass analysis immediately after death. Instead, the majority of studies involve freezing the carcasses until time permits the analysis. Freezing and thawing can result in losses of some components, especially water. Therefore, it is important to weigh the animals before freezing so losses can be assessed when the carcass is thawed and reweighed.

The amount of food consumed and the time of the last meal can have a large effect on weight and on the subsequent body composition analysis. There are several ways to address this. Animals can be fasted overnight before culling; however, in small animals as much as one third of body fat can be lost during a 14-hr fast (Bronson 1987). Alternatively, the stomach and gastrointestinal (GI) tract can be removed and the analysis performed on the "eviscerated" carcass. An alternative is to flush out the contents of the GI tract with saline to remove any food and feces. The GI tract is then returned to the carcass, which is then reweighed.

Water Content

Water content is commonly assessed by desiccation. Carcasses are weighed and placed in an oven to dry. As a general rule, carcasses should be dried at 60 to 90° C because temperatures of 100° C or more can lead to loss of organic matter, and if the oven is too cool, the carcass may decompose, thus incurring losses in addition to water. Carcasses should be dried until they reach constant weight to ensure that all the water has evaporated. Drying time is reduced by opening up the body cavity to expose the greatest surface area to the heat, which also ensures more complete drying of the carcass. In our experience, mice need to be dried for 5 to 7 days, or up to 10 days if they have large amounts of fat. Once the carcass has reached constant mass, it is cooled in a desiccator and weighed. Water content is determined as the difference between the weight before drying and the dried weight, with adjustment for any losses in water content during freezing.

An alternative to oven drying is freeze-drying. This is achieved by freezing the sample and then lowering the pressure, before heating lightly. In this way, the ice sublimes to water vapor when heated under low pressure, skipping the liquid phase. Freeze-drying has the benefit of keeping the sample intact and avoiding the problems of decomposition or volatization of components other than water associated with variations in oven temperature.

In theory, carcass analysis can be performed on any animal. The major limiting factor is the size of the oven. Small animals, such as mice, will fit easily into ovens, allowing for many to be dried simultaneously. As size increases, so does the time necessary to dry the carcass. Larger carcasses are often dried in many pieces or blended with an equal or greater weight of water to produce a homogenate. Representative samples are then analyzed (Harris and Martin, 1984; Harris et al. 1986), and percent water, fat, and ash content determined from the samples, after adjustment for the amount of water added, can be applied to the weight of the original carcass. Analyses performed on homogenate should be performed in triplicate to ensure that a representative sample has been taken. We calculated precision on a group of 60 rats analyzed in triplicate and found that the coefficients of variation (CV) were 1.91% for fat mass, 0.44% for water content, 1.37% for ash, and 1.52% for fat-free dry mass (Nagy and Johnson 2003).

Fat Content

Fat content is commonly determined using a Soxhlet apparatus with an organic solvent, such as petroleum ether or diethyl ether. The dried carcass is ground to a powder and placed in a cellulose extraction thimble, or in a glass thimble with a porous base, and weighed. The thimble is placed in the Soxhlet apparatus (figure 10.1). The flask of solvent is heated, vaporizing the solvent, which rises through the vapor tube and into the condenser. The solvent condenses, drips down into the thimble in the extractor, and dissolves the fat. When the solvent reaches the level of the reflux tube, it pours back into the flask taking the fat with it. This cycling of solvent from flask to sample and back again continues for approximately 6 to 8 hr or overnight. The thimble is removed and

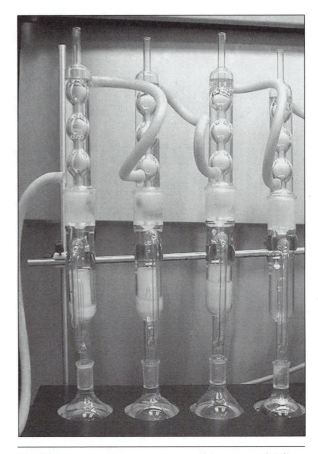

Figure 10.1 Soxhlet apparatus used to extract fat from samples.

any remaining solvent left to evaporate under the fume hood, before being dried for 30 min and then weighed. The weight lost represents the fat content of the carcass.

In theory, any organic solvent that dissolves fat could be used in this apparatus. However, it is very important that the solvent only dissolves the lipid fraction of the carcass and none of the non-lipids. Dobush and colleagues (1985) found that petroleum ether was the best solvent to use in the Soxhlet or similar Goldfisch apparatus. Although a chloroform–methanol mixture did dissolve the lipids, it also dissolved some nonlipids, and its use in these apparatuses is not recommended (Dobush et al. 1985).

The mass of the sample remaining after drying and fat extraction is commonly referred to as fat-free dry mass and contains protein, carbohydrate, and ash.

Ash Content

Ash content is determined by placing the sample remaining after drying and fat extraction in a weighed container, and burning it in a muffle furnace at 600° C for 8 hr or overnight. All the organic matter is burned off, leaving only mineral content. In a simple analysis, this is termed *ash* and taken to represent bone mineral content. However, although bones are the main source of ash, ash overestimates bone content because of cell minerals. When nonbone tissue was ashed separately from bone in a group of mice, it was found that 1.17% of the mass of the nonbone tissue was ash (Nagy and Clair 2000).

Fat Pads

When fat mass is of interest, specific fat pads can be dissected out and weighed. Frequently measured fat pads include inguinal, retroperitoneal, mesenteric, and gonadal fat pads. This process is faster than the complete carcass analysis but yields less information. West and colleagues (1994) validated the use of an adiposity index derived by dividing the sum of inguinal, epididymal, retroperitoneal, and mesenteric fat pads by the eviscerated carcass (not including fat pad weights). The adiposity index correlated strongly with carcass fat ($r = .97$).

Carcass analysis is the only technique that measures different body components directly; thus, it is the "gold standard" against which other (indirect) techniques are compared and validated. Its main limitation is its destructive nature. Thus, it allows analysis of differences between groups of animals receiving different treatments but not changes within individuals over time.

Total Body Electrical Conductivity (TOBEC)

TOBEC, based on electrical conductivity, was initially developed for the meat-packing industry to allow rapid determination of the composition of meat. In recent years, it has been used to nondestructively assess the body composition of rats (Baer et al. 1993; Bell et al. 1994; Bellinger and Williams 1993; Dickinson et al. 2001; Stenger and Bielajew 1995), rabbits (Yasui et al. 1998), fish (Barziza and Gatlin 2000; Lantry and Stewart 1999), birds (Acquarone et al. 2001, 2002; Castro et al. 1990; Golet and Irons 1999), bats (Koteja et al. 2001), ground squirrels (Frank 2002; Nunes et al. 2002), lizards (Angilletta 1999), and meadow voles (Unangst and Wunder 2001).

TOBEC provides an indirect measurement of body composition based on the greater ion concentration and hydration of lean mass, which makes it approximately 20 times more conductive than fat mass. Thus, the extent to which an animal resists the flow of current can be used as an indicator of relative lean mass. Estimates of fat mass are obtained as the difference between body weight and lean mass.

TOBEC has been validated against carcass analysis. Walsberg (1988) showed TOBEC to be closely related to lean mass in several species of mammals and birds, explaining approximately 99% of the variation in lean mass. Many validation studies have tested whether TOBEC is accurate within species and over a narrower range of body weight. A very good relationship was observed between TOBEC and lean mass in rats (Baer et al. 1993; Bell et al. 1994; Dickinson et al. 2001; Trocki et al. 1995), mice and voles (Koteja 1996; Zuercher et al. 2003), rabbits (Yasui et al. 1998), and birds (Castro et al. 1990). However, both Baer and colleagues (1993) and Trocki and colleagues (1995) found that the manufacturer's equation significantly underestimated lean mass (Baer et al. 1993; Trocki et al. 1995); they overcame this by developing their own prediction equations.

Fat mass is estimated indirectly as the difference between lean mass and body weight. The predictive ability of TOBEC to determine fat mass is generally less than that for lean mass, although a strong relationship between TOBEC and fat mass ($r^2 = .952$) was found in rats (Dickinson et al. 2001) and in rabbits ($r^2 \geq .82$). This latter estimate was associated with a high CV (24%; Fortun-Lamothe et al. 2002). The manufacturers' equations overestimated fat mass because they underestimated lean mass (Baer et al. 1993; Trocki et al. 1995). After developing a new prediction equation, Baer and colleagues (1993) found that there was no significant difference between predicted and actual fat mass; however, some values of fat mass were less than zero. For animals with low fat mass, TOBEC may not be a good method because small errors in lean mass result in greater relative errors in estimating fat mass. When animals have very little fat, body weight may be just as good as or better than TOBEC at predicting lean mass. In a validation study of red squirrels, body weight was the best predictor of lean mass, and the addition of TOBEC did not significantly improve the model (Wirsing et al. 2002). In a study of two fish species, the main predictor of total water content was body weight, although TOBEC did increase the model fit slightly (Lantry and Stewart 1999).

The conditions in which TOBEC measurements are made need to be carefully controlled to ensure accurate and reliable results (e.g., movement, temperature, and hydration). TOBEC instruments are designed to measure the conductivity of uniform cylindrical objects. Because animals vary in shape, length, and width, it is important to place the animal in the same position for each reading. Movement during the measurement should be kept to a minimum by the use of anesthesia or by physically restraining the animal. If unanesthetized animals are measured, they should be acclimated to the chamber before the measurement. Dickinson and colleagues (2001) found that acclimating rats to the chamber for 2 weeks before the measurement increased precision. Several readings are normally averaged. The number of repeat readings varies between studies and between species, with as few as three to up to seven repeats performed (Baer et al. 1993; Bell et al. 1994; Castro et al. 1990; Fortun-Lamothe et al. 2002; Lantry and Stewart 1999; Pulawa and Florant 2000; Trocki et al. 1995; Unangst and Wunder 2002; Yasui et al. 1998). The manufacturers advise that the coefficient of variation for each animal should be less than 3%.

Although anesthesia reduces the error attributable to movement in the chamber, it may create another problem by lowering the animal's body temperature, because electrical conductivity varies with temperature. Several studies have been performed to assess whether variations in ambient or body temperature significantly affect TOBEC results. In one of the first studies, Walsberg (1988) found no effect of ambient temperature (in the range of 0-40° C) on the readings of a saline model animal (300 ml of saline in a 294.5 cm^3 cylinder). However, a more recent study showed that for every degree Celsius increase in ambient temperature (2-40° C), there was an average increase of 3 TOBEC units (Robin et al. 2002). This small but significant effect of ambient temperature becomes important in field studies, especially when researchers are performing seasonal measurements on populations (e.g., summer and winter), when the machine may be exposed to large differences in ambient temperature.

Both of the previous studies also found that variations in body temperature had a much greater effect on TOBEC readings than ambient temperatures (Robin et al. 2002; Walsberg 1988). Walsberg (1988) tested samples of 35, 40, and 45° C and found that TOBEC readings significantly increased with increasing temperature. Similarly, Robin and colleagues (2002) found that when sample temperature increased from 5 to 42° C, TOBEC values increased by 64 units per degree increase. This effect was confirmed when birds were scanned alive, killed, and then rescanned (Castro et al. 1990). The dead birds, whose body temperatures had dropped to ambient (about 20° C), had significantly lower conductivities than when scanned alive. Similarly, TOBEC readings and fat-free mass (FFM) estimates of rats decreased between 4 and 44 min after anesthesia (Tobin and Finegood 1995), with an even greater effect on estimated fat mass, which increased by 63% over this time. In a separate group of rats, body temperature was found to decrease over this time by about 2° C. Although the manufacturers recognize the effect of temperature on electrical conductivity, they claim that the effect is negligible over the normal range of body temperatures (about 2° C) and is only of concern if the animals are in torpor or hibernating. However, both Tobin and Finegood (1995) and Robin and colleagues (2002) showed that variations in body temperature of 2° C can have a significant effect on the estimation of body composition.

The animals must be dry before TOBEC readings are taken. The presence of water on fur or feathers will increase the conductivity of the animal and hence falsely elevate the estimate of FFM. If the animal has urinated in the chamber, the animal and chamber must be thoroughly dried before measurements are made. For the same reason, hydration status can affect the reading. In a study of quail scanned while dehydrated and then rescanned after access to water (but no food), the TOBEC reading increased by about 15% (Walsberg 1988).

The presence of metal bands, ear tags, and wound clips may influence the TOBEC reading. Although Castro and colleagues (1990) found no effect of the U.S. Fish and Wildlife Service metal bands on the TOBEC readings in house sparrows, it is advisable to perform tests if animals are going to be routinely scanned with a metal tag, to ensure that there is no effect. TOBEC machines are relatively inexpensive and have the added benefits of being portable and battery powered. Most studies agree that whereas TOBEC may be useful for assessing population changes in body composition, it does not have the accuracy required for assessing changes within animals.

Total Body Water by Isotope Dilution

Measuring total body water (TBW) allows for the determination of FFM because of the differential hydration status of fat and FFM. FFM is generally acknowledged to contain about 73% water, whereas fat mass contains very little water. Therefore, with TBW and the hydration coefficient, FFM can be estimated.

Isotopes of hydrogen or oxygen are introduced into the body as labeled water. These molecules diffuse through the body and equilibrate with the body water. The extent to which the isotope concentration has been diluted is indicative of the amount of water in the body. The greater the body water pool, the lower the enrichment of isotope (given a standard dose).

Three main isotopes are used in animal studies: deuterium oxide (D_2O), tritiated water (TOH), and oxygen-18 labeled water. The choice of isotope depends on cost, available equipment, and issues relating to radiation hazards. Deuterium and tritium are both relatively inexpensive, whereas the cost of oxygen-18 is often prohibitive. Both deuterium and oxygen-18 are routinely measured using an isotope ratio mass spectrometer, which is very expensive and not routinely found in many departments. In contrast, tritium can be measured using scintillation counting, which is cheaper than mass spectrometry and much more accessible. Thus, if we consider cost and analysis, tritium would appear to be the best isotope. However, it is radioactive. If regulations preclude the use of a radioactive isotope, then deuterium would be the next best choice based on cost.

Before dosing, the animal is weighed and a fluid sample obtained to assess the background concentration of the isotope. The animal is then dosed with the isotope, and after allowing sufficient time for the isotope to equilibrate with the body water, a further sample is taken to assess the concentration of the isotope in body water.

The size of the dose needs to be sufficient to elevate the enrichment level in the animal a significant amount above background. The amount needed will depend on the precision of the analyzer, the background enrichment, the size of the animal, and the timing of the samples.

The closer the enrichment is to the background level, the more precise the analysis needs to be to detect the small differences. If the analysis technique is relatively imprecise, a greater dose is needed. The background enrichment of the isotopes also plays a part in determining the dose. The background levels are more important when deuterium or oxygen-18 is used, because these occur naturally within the body (approximately 150 and 2000 ppm, respectively), whereas natural levels of tritium in the body are very low. For a more detailed discussion of the importance of background isotope level, see Speakman and colleagues (2001).

The dose of the isotope depends largely on the size of the animal (table 10.1). Generally, the larger the animal, the greater the body water and hence the more isotope needed to elevate the enrichment above background levels. Some studies (e.g., Krol and Speakman 1999) appear to use a much higher dose than expected based on body size. This is probably attributable to the number and timing of samples taken. If samples are to be taken up to a day after the administration (e.g., in studies of energy expenditure or water turnover), then a much higher dose is needed compared with assessment 1 to 4 hr after administration. The dose can be administered either orally or via injection. Oral administration is only really practical for large animals, and care needs to be taken to ensure that all the isotopes are consumed. For the majority of small animals, injection is the only feasible method of administration. The dose can be administered through intravenous, intraperitoneal, intramuscular, or subcutaneous routes. Regardless of the place of injection, the dose should be injected slowly with the smallest possible needle. To avoid leakage at the injection site, the needle should remain in place for a few seconds after the injection is finished.

For calculation of TBW it is important to know the exact weight of isotope injected. This is achieved by weighing the syringe before and after injection on a four-figure balance to get the weight of isotope injected. A sample of the injectate should be kept to be analyzed with the animal samples to confirm the enrichment introduced into the animal.

The time for the isotopes to equilibrate with the body water pool also varies with the size of the animal. The greater the size of the body water pool, the longer it takes for the isotope to mix evenly. In bumblebees weighing 250 mg, for example, equilibration occurs in 10 min (Wolf et al. 1996), whereas anywhere from 1 to 6 hr has been reported for mice, hamsters, rats, cats, dogs, and kangaroos depending on their weight (Coleman et al. 1972; Culebras et al. 1977; Denny and Dawson

Table 10.1 Examples of Doses of Tritiated Water for Animals of Differing Body Masses

Species	Body mass	Dose	Reference
Mice	38 g	69 μCi	Krol and Speakman 1999
Hamster	120-140 g	1 μCi	Kodama 1971
Rats	150-250 g 227 ± 83 g	10 μCi	Culebras et al. 1977; Foy and Schnieden 1960
Rats	330-640 g	20 μCi	Barzilai et al. 1998
Chukar partridges	≤500 g	0.05 μCi/g	Degen et al. 1981
Poultry	1.9 kg	4-10 μCi/kg	Johnson and Farrell 1988
Cat	1-2.25 kg	50 μCi/kg	Foy and Schnieden 1960
Dogs	0.2-13.2 kg	10 μCi/kg	Sheng and Huggins 1971
Dogs	18.4 ± 1.5 kg	200 μCi	Coleman et al. 1972
Kangaroo	14-35 kg	25 μCi/kg	Denny and Dawson 1975
Antarctic fur seals	Adult 27-49.5 kg Pups 5.4-6.7 kg	100 μCi 50 μCi	Arnould et al. 1996

1975; Elliot et al. 2002; Ferrier et al. 2002; Kodama 1971; Krol and Speakman 1999; Moore et al. 1962; Sheng and Huggins 1971; Tisavipat et al. 1974).

Blood, urine, and saliva can all be sampled to determine isotope enrichment. Saliva should be avoided if the isotope was administered orally, because the sample may be contaminated by residual dose in the mouth. Also, for very small animals, it may not be possible to collect saliva. Urine is probably the easiest to collect, because many small mammals will urinate when handled; however, the urine present in the bladder is not available for isotope exchange. Blood is therefore most commonly collected, flame sealed in glass capillaries, and kept refrigerated until analyzed.

Total body water can be calculated from the background and equilibrium enrichments, together with the enrichment of the dose:

$$V_p = [(I_i - I_b)/(I_f - I_b)] V_i, \qquad (10.1)$$

where V_p is volume of body water pool, I_i is enrichment of the injectate, I_b is the background enrichment, I_f is the enrichment at equilibrium, and V_i is the volume of the injectate (equation 7.29 from Speakman 1997). After TBW has been calculated, fat-free mass can be estimated using the hydration coefficient of lean mass (0.73): fat-free mass = TBW/0.73.

Fat mass can then be estimated as the difference between fat-free mass and body mass.

Hydrogen isotopes generally overestimate TBW as assessed by oven or freeze-drying (table 10.2). It is generally accepted that deuterium and tritium overestimate TBW by approximately 4%; however, as seen in table 10.2, the actual error varies by study and species. Although not presented here, the error associated with TBW oxygen-18 dilution is less than for the hydrogen isotopes (Speakman 1997). The overestimation is attributed to pools of exchangeable hydrogen in the body apart from water. A more complete list of the accuracy of the technique in different species was provided by Speakman and colleagues (2001).

The precision of this technique is generally good, being shown to be 3.36% in red kangaroos (Denny and Dawson 1975) and 2.1% in dogs (Moore et al. 1962).

Dual-Energy X-Ray Absorptiometry (DXA)

The principles of DXA are given in chapter 5, and the use of DXA for small-animal research has recently been reviewed (Grier et al. 1996; Nagy 2001; Pietrobelli et al. 1996). Large clinical densitometers have proven to be useful for

Table 10.2 Accuracy of the Hydrogen Isotope Dilution Method for Estimating Total Body Water

Species	Isotope	Discrepancy (%)	Reference
Antarctic fur seals	Tritium Deuterium	1.92 ± 1.00 1.71 ± 1.74	Arnould et al. 1996
Rats	Tritium	4.25	Foy and Schnieden 1960
Rats	Tritium	1.71	Culebras et al. 1977
Rats	Tritium	8	Tisavipat et al. 1974
Hamster *M. auratus*	Tritium	1.6	Kodama 1971
Chukar partridges	Tritium	−0.2 ± 0.4	Calculated from Degen et al. 1981
Chukar partridges	Tritium	4.6 ± 2.4	Crum et al. 1985
Beagles	Tritium	13.8-15.6	Sheng and Huggins 1971
Sows (pregnant/lactating)	Deuterium	19	Shields et al. 1984
Poultry	Tritium Deuterium	10.4 8.5	Johnson and Farrell 1988

Note. Discrepancy is expressed relative to dehydration.

measuring body composition in animals ranging in mass from rats to pigs (Grier et al. 1996; Nagy 2001; Pietrobelli et al. 1996). The smaller human peripheral-DXA instruments are useful for measuring body composition in mice and lemmings and for measuring bone mineral content and density in larger animals including guinea pigs and rats (Fink et al. 2002; Hunter and Nagy 2002; Nagy and Clair 2000; Nagy et al. 2001). To our knowledge, there is currently no DXA instrument available that is capable of measuring total body composition in animals ranging in size from mice to large rats. This is a major financial burden to researchers and institutions that have to purchase two instruments to conduct research in small animals.

We have determined the precision and accuracy of two densitometers for small-animal research (Hunter and Nagy 2002; Nagy and Clair 2000; Nagy et al. 2001). The GE-Lunar PIXImus showed excellent precision when 25 mice were scanned in triplicate (total body bone mineral [TBBM], 1.6%; fat mass [FM], 2.2%; and lean tissue mass [LTM], 0.86%) and the results from DXA were highly related to carcass analysis (TBBM, r = 88; FM, r = .87; and LTM, r = .99). Precision of the Norland pDEXA Sabre was less than that of the PIXImus. Four collared lemmings were scanned five times each with repositioning resulting in CVs of 2.2%, 4.14%, and 1.63% for TBBM, FM, and LTM, respectively (Hunter and Nagy 2002). Heiman and col-

leagues (2003), working with mice, reported CVs of 8% for FM and 5.8% for LTM. Data obtained from the pDEXA Sabre correlated well with carcass analysis (TBBM, r = .92; FM, r = .99; and LTM, r = .95). Both studies (Heiman et al. 2003; Hunter and Nagy 2002) used very slow scan speeds (7 or 8 mm/s) and a high resolution (0.5 × 0.5 mm), resulting in long-duration scans (from 20 to 35 min depending on animal size). However, this is necessary to achieve good precision and accuracy with the pDEXA instrument. We have determined that increasing scan speed or decreasing the resolution adversely affects precision and accuracy of the instrument (Nagy 2001).

Using DXA to determine body composition allows noninvasive measures of fat mass and soft-lean tissue mass and the ability to determine bone mineral content and density, which cannot be obtained from TOBEC, isotope dilution, magnetic resonance spectroscopy, or magnetic resonance imaging. However, the use of DXA for small-animal research is still in its early stages, and validation studies are warranted when researchers use different species or animals of extremely low or high body fat.

Computed Tomography (CT)

The basic principles of CT are given in chapter 7. This technique, although used extensively for medical imaging and human body composition research, has received little attention for small-animal phenotyping. One of the first studies to use CT for small-animal research was conducted by Ross and colleagues (1991), who used a human CT instrument to measure body composition in rats. Seventeen rats were scanned at 21-mm intervals (3-mm thickness) over the entire length of the body (excluding the tail) for a total of 12 slices. Adipose tissue (AT) obtained by CT was compared with chemical carcass analysis to determine accuracy. Total CT AT mass was highly correlated with values from carcass analysis ($r = .99$, standard error of estimate = 6.8 g), showing the utility of this approach. However, the cost of a large clinical CT instrument and issues concerning access to the instrument may limit its usefulness for animal research.

Smaller peripheral quantitative computed tomography instruments (pQCT) may be of more use to animal researchers because of the lower cost and portability of these instruments. pQCT has been used extensively for determining volumetric bone mineral content and density in rodent models (Andersson et al. 2001; Rosen et al. 1995). However, to our knowledge these instruments have not often been used for determining relative body fat in rodents. We measured relative body fat in mice using pQCT, in a manner similar to that of Ross and colleagues (1991). Five equally spaced transverse scans were taken starting with an anterior scan distal to the head of the humerus and the most posterior scan just distal to the head of the femur. Relative fat content was averaged over the five slices, and these data were compared with carcass analysis. pQCT-derived body fat was highly related to chemically determined fat ($r = .95$, $p < .01$). Precision was determined by scanning 10 mice, five times each. The precision for determining relative body fat was 3.91%, which is similar to that of small-animal DXA instruments (Hunter and Nagy 2002; Nagy and Clair 2000). These data suggest that pQCT is useful for determining relative body composition of mice.

CT allows for the measurement of attenuation (density) of individual tissues. Thus, CT has been used in human studies to estimate the relative fat content of liver and muscle (Goto et al. 1995; Katoh et al. 2001; Simoneau et al. 1995). Because fat has a lower attenuation than lean tissue, a decrease in CT attenuation of an organ or muscle indicates an increase in fat infiltration. CT techniques have not yet trickled down to researchers using small rodents for the study of obesity and nonalcoholic fatty liver disease. Instead, animals are typically killed and liver fat content is determined histologically (Baffy et al. 2002; Brix et al. 2002) or by chemical fat extraction techniques (Brix et al. 2002). Obviously, these end-point studies do not allow for longitudinal measures of relative fat content. We validated the use of pQCT for measuring relative liver fat in a small rodent model (Nagy and Johnson 2003). Varying degrees of fatty liver were induced by fasting collared lemmings from 0 to 18 hrs. Animals were then placed in a pQCT, and a single-slice scan, 2 mm from the base of the lungs, was obtained. Attenuation of the liver slice was strongly related to percent liver fat ($r = -.98$, $p < .01$) by chemical extraction. Precision of the attenuation measure was determined by scanning eight animals, five times each with repositioning between scans, yielding a CV of 0.32%. Thus, pQCT appears to be a useful method for determining relative liver fat in small rodents.

New small-animal CT instruments using cone-beam technology are coming on the market, allowing for rapid phenotyping of mice and rats (whole-animals scans of 10-20 min). These instruments are in their infancy, and validation and software development are needed. However, the ability of CT to both image and determine density of tissues will make this technology extremely useful to small-animal researchers in the near future.

Magnetic Resonance Spectroscopy (MRS)

The basic principles of MRS are given in chapter 7. Stein and colleagues (1995) validated the use of MRS for determining relative fat content using ground meat that varied in fat content. The fat–water ratio from MRS was highly correlated ($r = .99$) with the fat–water ratio from dehydration and chemical extraction. Precision was determined by measuring the abdominal fat of rat pups. Two pups were measured five times each, and the resulting CV was 0.5%; it was not stated whether the pups were repositioned between runs.

MRS has also been used to measure body fat in mice. Mystkowski and colleagues (2000) compared MRS-derived body composition data to chemical extraction data from 20 male C57Bl/6 mice using a 4.7-T instrument. Percent body fat, total body fat, and total fat-free mass were all highly correlated between the two methods ($r > .95$). Mice were also measured in the anesthetized state and after euthanasia, and no difference was found in the data, suggesting that MRS is not sensitive to potential respiratory motion artifacts.

MRS has also been used to determine intra- versus extramyocellular triglycerides in rat skeletal muscle (Guo et al. 2001). This technique uses the fact that the methylene portion of the fatty acid chain in intramyocellular triglycerides produces a chemical shift at 1.3 ppm and extramyocellular triglyceride produces a shift at 1.5 ppm. To verify their findings, Guo and colleagues (2001) examined liver, which contains only intracellular triglycerides, and subcutaneous fat tissue. As predicted, MRS of the liver produced a peak at 1.3 ppm, whereas MRS of subcutaneous fat produced a peak at 1.5 ppm.

Magnetic Resonance Imaging (MRI)

MRI allows for the visualization of internal structures and tissues in a manner similar to CT but without harmful radiation. Therefore, MRI may be more appropriate as an imaging tool when individual animals are to be imaged multiple times.

Ross and colleagues (1991) validated MRI for measuring AT using a rat model. Seventeen rats varying in age and adiposity were anesthetized and placed in a 1.5-T whole-body scanner; 12 transverse slices, 3 mm thick, were obtained for each rat. AT volume was determined by segmenting each slice into AT and non-AT pixels, calculating the area (cm^2) by summing AT pixels and multiplying by pixel area, and then multiplying by the slice thickness to obtain volume (cm^3). The total AT volume of the rat was then interpolated by "adding the volumes of truncated pyramids defined by pairs of consecutive slices" (p. 2166). Instrument reliability was determined by scanning the same rat twice without repositioning. There were no significant differences (p = .97) between scans. Precision (test–retest, CV%) was 4.3% (range, 0.03-12.7%). To determine the accuracy of MRI for measuring total AT, the data were converted from a volume measure (cm^3) to grams and then compared with chemically extracted lipid values obtained from carcass analysis of the rats. Values obtained from MRI were not significantly different from those obtained from chemical extraction. This pioneering study showed the utility of MRI for measuring AT and gave credence to the human studies that were using MRI to measure AT but had not been previously validated.

Ross and colleagues (1991) also showed that MRI was capable of distinguishing between visceral and subcutaneous AT. MRI data obtained on visceral and subcutaneous fat correlated well with data obtained from CT, but data on fat distribution using either method were not directly compared with carcass values. Ishikawa and Koga (1998) used MRI to assess abdominal fat distribution in Otsuka Long-Evans Tokushima fatty (OLETF) and control rats. Six transverse slices (3 mm thickness) were made across the abdominal region. Intra-abdominal fat was then dissected from the animals and compared with the data obtained by MRI. The two methods were highly correlated (r^2 < .99), and the slope of the relationship did not differ significantly from 1.0. However, the intercept of the line was significantly different from zero, suggesting that the data from MRI slightly overestimated the data obtained by dissection. Together, these data suggest that MRI is a useful tool for assessing AT distribution.

MRI can also be used to determine the size of specific fat pads. Sbarbati and colleagues (1991) used MRI to determine the volume of brown adipose tissue in rats in vivo. Although the calculated volume was slightly greater than the dissected weight (140.0 mm^3 vs. 133.7 mg), this is most likely explained by the fact that the density of brown adipose tissue is significantly less than 1.0 mg/mm^3, the density used to calculate weight from volume.

MRI can also be used to determine the volume of other tissues in the body. Using a high-field (4.2-T), whole-body MRI system, Tang and colleagues (2002) determined weights (volume of tissue multiplied by the density of that tissue) of body fat, skeletal muscle, individual organs, and bone in rats. Precision of the MRI was determined by scanning three rats, three times each, with repositioning. The CVs were calculated for total body weight (0.63%), skeletal muscle (0.27%), total AT (2.15%), and intra-abdominal adipose tissue (2.96%). The precision of individual organ masses was highly variable, ranging from a high of 1.99% for brain to a low of 9.49% for heart; the low precision for heart is likely attributable to motion artifacts (Tang et al. 2002). Operator reliability was determined by segmenting three rats, three times each, on different days. The CVs ranged from a low of 0.27% for whole-body weight to a high of 5.39% for spleen. In this instance, the precision was related to the ability of the operator to distinguish one tissue from another during manual segmentation. The spleen, being surrounded by organs of similar density, is difficult to segment with precision, whereas, the brain, being surrounded by high density bone, can be reliably segmented (CV of 0.88%).

In addition to determining precision, Tang and colleagues (2002) determined the accuracy of MRI by comparing the MRI-derived data to either dissected weights, chemically extracted weights, or both. In all cases, MRI data were very highly related with either dissected or chemically extracted weights ($r^2 > .96$). Comparison of methods using the Bland–Altman technique (Bland and Altman 1986) showed that there was no bias between methods, although MRI overestimated intra-abdominal AT, kidneys, spleen, liver, and brain and underestimated heart weight. Even so, the close relationship between methods shows that MRI can be used to estimate tissue weights.

The use of MRI for phenotyping small animals has increased rapidly over the last few years (Barzilai et al. 1998; Beckmann et al. 2001; Changani et al. 2003; de Souza et al. 2001; Hockings et al. 2002; Johnson et al. 2002; Razani et al. 2002; Rudin et al. 1999). Its use in pharmaceutical research was reviewed by Rudin and colleagues (1999). Its advantages include the lack of radiation exposure as found in CT and the ability to acquire high-resolution images. At present, the main disadvantages include the lack of sophisticated software to allow segmentation of tissues, cost, and availability.

Summary

Several methods of animal body composition have been described, ranging from relatively simple, but terminal, carcass analysis to state-of-the-art computed tomography and magnetic resonance imaging. Advances in technology have allowed investigators to progress from measuring whole-body fat and lean mass, as in TOBEC and isotope dilution, to more detailed assessments of bone and fat pads. The development of DXA has allowed bone mineral content and density to be assessed in vivo, and the further development of CT has allowed for more detailed examinations of trabecular versus cortical bone and also has allowed the imaging of different organs and fat pads, as does MRI. The choice of an appropriate method depends mainly on what body component the investigator wishes to measure as well as the cost and availability of the imaging instruments.

Chapter 11

Statistical Methods

Shumei S. Sun and Wm. Cameron Chumlea

This chapter discusses the statistical methods used to develop body composition prediction equations. Body composition assessment is the quantification of fat and fat-free mass (FFM). In large studies, it is frequently necessary to predict body composition from easy-to-measure variables because direct body composition methods are impractical. The sophisticated "direct" body composition methods are time-consuming and expensive, and they require fixed dedicated equipment and support in a laboratory setting.

Epidemiological investigations of the association between body composition and cardiovascular and other chronic diseases require large sample sizes that necessitate simple and reliable procedures such as anthropometry and bioelectric impedance. Bioelectric impedance measures body resistance, which is proportional to the volume of total body water (TBW) and FFM (Sun et al. 2003). Skinfolds measure subcutaneous adipose tissue thickness, and limb circumferences measure combinations of adipose tissue and fat-free tissues (Chumlea and Guo 2000). Trunk circumferences like abdominal circumference (Chumlea and Kuczmarski 1995) provide an indirect estimate of internal adipose tissue. The body mass index or BMI (weight/stature^2; kg/m^2) is an index of the level of body fatness in adults (Guo et al. 2002) and fat and fat-free tissues in children (Maynard et al. 2001).

The prediction of body composition requires a regression equation using bioelectric impedance and selected anthropometry as predictor variables and TBW or FFM from a laboratory or direct method as a response or dependent variable (Sun et al. 2003). A prediction equation derived by a least squares method performs best in the sample from which it was derived, and its accuracy is reduced, sometimes substantially, when applied to other samples. Consequently, researchers must consider the factors that influence the accuracy of a prediction equation when applied to independent samples and how to evaluate its performance or accuracy. In addition, the precision of the predicted values should be incorporated into any subsequent analysis and interpretation of the results from the equation's application to an independent sample.

Accuracy of Predictive Equations

In accordance with common statistical usage, accuracy is the performance of a prediction equation when applied to an independent sample (cross-validation), whereas precision is the performance of the equation within the sample from which it was derived. The variable predicted by an equation is the response variable, and the variables used in the equation are the predictor variables. Some of the factors that affect the accuracy of a prediction equation include these:

- Validity of the response variable
- Precision of the predictor variables
- Biological and statistical relationships among and between the predictor variables and the response variable
- Statistical methods used to formulate the equation
- Sample size

This work was supported by Grants HD-27063 and HD-12252 from the National Institutes of Health.

Validity of the Response Variable

When developing a body composition prediction equation, a significant factor affecting precision is the validity of the response variable. If the response variable's validity is limited or its measurement contains error, the subsequent prediction equation will perform poorly when applied to independent samples.

The methods used to assess body composition can affect the validity of the measured fat and fat-free components and produce measurement errors. For example, a two-component model is commonly used to calculate percent body fat (%BF) and FFM values from body density (Conlisk et al. 1992; Deurenberg et al. 1989a; Deurenburg et al. 1989b; Deurenburg et al. 1990a; Deurenburg et al. 1990b) derived from measures of underwater weight corrected for residual volume. The precision of the underwater weight and residual volume measurements is sensitive to the compliance and performance of the subjects, and as a result the measurements contain errors. In addition, the two-component model incorrectly assumes that the density of FFM and its fraction of water are constant for all ages and both sexes (Chumlea et al. 1999; Lohman 1986), and this limits its utility. Instead, multicomponent models should be used that include measures of body density, bone mineral, and total body water (Lohman 1986; Sun et al. 2003), but there is also an accumulation of measurement errors in these models that can inflate errors in the calculated body composition values (Heymsfield et al. 1990).

Dual-energy X-ray absorptiometry (DXA) is used extensively to measure body composition. DXA measures are precise, especially for bone, and little subject compliance is required (Lukaski 1987). However, DXA estimates of soft tissue values for fat and FFM rely on computer software and pixel interpretation, and little is understood about the algorithms used, so that the errors of these estimates are unknown (Roubenoff et al. 1993). Few published prediction equations have used DXA estimates of body composition as response variables (Sun et al. 2003). DXA is used to collect body composition values in the ongoing National Health and Nutrition Examination Survey so that prediction equations can be developed from these national DXA estimates of body composition in the future. Densitometry, multicomponent models, and DXA are discussed in other chapters.

Precision of the Predictor Variables

The measurement of the predictor variables to be used in an equation should be as precise as possible. This is achieved if close attention is given to the selection and calibration of the instruments, measurement procedures, and quality control (Chumlea and Guo 2000).

Relationships Between Predictor Variables and Response Variables

Possible predictor variables include weight, stature, body circumferences, skinfold thicknesses, and measures of bioelectric impedance. These are obtained easily at low cost and with high precision (Chumlea et al. 1990). The inclusion of predictor variables depends on their biological and statistical relationships to the response variable. Predictor variables for %BF are primarily measures or indexes of subcutaneous adipose tissue easily obtained from skinfold thicknesses and body circumferences. Measures of overall body size are also possible predictor variables for %BF and FFM (Sun et al. 2003). Stature, weight, and BMI are positively correlated with FFM and with total body fat. Measures of bioelectric impedance are important predictor variables for FFM (Sun et al. 2003). The resistance index, S^2/R, where S is stature in centimeters and R is resistance in ohms, is the most significant single predictor of FFM (Guo et al. 1989, 1993; Sun et al. 2003). Consequently, when resistance is included as a predictor variable, FFM is the appropriate response variable.

Some studies have used inappropriate predictor variables to estimate body composition, relying only on statistical and not biological relationships. For example, skinfold thicknesses have been used to predict FFM, despite low correlations between these variables. Bioelectric impedance has also been used to predict %BF, although there are only weak biological and statistical relationships between these measures. The inclusion of predictor variables with strong biological and statistical relationships to the response variable in an equation increases the accuracy of that prediction equation.

Statistical Methods

Appropriate statistical methods must be selected to develop an accurate body composition prediction equation. Regression analysis is the standard analytical method used to develop a prediction equation, and it is available in numerous statistical computer packages. Regression analysis is based on several assumptions regarding the distribution of the response variable and its linear relationships with the predictor variables. If these relationships are nonlinear, the subsequent equation will have large errors of prediction, and its performance will be poor when applied to independent samples. Some problems can be avoided by transforming the predictor variables so that these relationships become linear. A scatter plot of the response variable versus each predictor variable can indicate whether the relationship is linear.

Usually, several predictor variables are included in a prediction equation, and the selection of these predictor variables is described in the next section. The relationship between the response variable and each predictor variable can be distorted if multicollinearity occurs among the predictor variables. For example, if several skinfold thicknesses and circumferences are the predictor variables in an equation and %BF is the response variable, multicollinearity will exist because the skinfolds and circumferences are all measures of subcutaneous adipose tissue and are highly interrelated. In this case, a partial regression leverage plot of the response variable versus the predictor variable, after both have been adjusted for the other predictor variables in the equation (Myers 1986), can reveal the true relationship between the response and predictor variables.

The homogeneity of the response variable should be tested also. Homogeneity assumes a constant variance of the response variable for all values of each predictor variable, but this assumption is violated if the residual plot demonstrates a pattern or trend. For example, the relationship between the response and predictor variables may be nonlinear. In such a case, other predictor variables should be included in the equation. A trend in the residuals can also be caused by influential observations or outliers in the data. When the relationship between the response and predictor variables is nonlinear, or when the variance of the response variable is heterogeneous, the data must be transformed, and several methods are available (Chatterjee and Price 1979; Myers 1986; Tukey 1977).

Regression analysis assumes that the response variable is normally distributed to allow statistical inferences about the significance of the regression parameters. Normality of the response variable can be judged by the Shapiro and Wilks test (Shapiro and Wilks 1965). However, the normality of the response variable is not as important as the absence of multicollinearity among the predictor variables and the homogeneity of the response variable.

Selection of Predictor Variables

Several statistical methods can be used to select the predictor variables in an equation; forward selection, stepwise regression, and backward elimination procedures work well. These methods are used commonly when only a few variables are considered for inclusion; however, they should not be used if multicollinearity is present.

Multicollinearity is detected by calculating the variance inflation factor (VIF). The VIF is defined as $1/(1 - R^2)$, where R^2 is derived from the regression of a particular predictor variable on the other predictor variables. Multicollinearity also affects the precision of the regression coefficients and, thus, the accuracy of the prediction when the equation is applied to other samples.

When multicollinearity occurs, a maximum R^2 or an all-possible-subsets regression procedure is the appropriate analytical choice. An all-possible-subsets regression analysis evaluates every possible combination of predictor variables. The maximum R^2 procedure finds the one-predictor variable equation that yields the highest R^2. A two-predictor variable equation is then formulated by adding another of the remaining variables that results in the largest increase in R^2 from the one-predictor variable equation. The second variable is selected by pairing each possible predictor variable with the first predictor variable until the highest R^2 is found. This process is repeated to select the best trio of predictor variables given the best pair and continues until the addition of any remaining predictor variable no longer significantly increases the R^2 values. This selects the most precise prediction equation for the possible predictor and response variables in the sample from which the equation is developed. For example, in a recent study to develop prediction equations for FFM, questions were raised as to which final equation was the best, an equation containing bioelectrical impedance and anthropometry as predictor variables or an equation containing only anthropometry (Sun et al. 2003). These prediction equations were developed by the all-possible-subsets regression procedure, and the "anthropometry-only" prediction equation was not considered further because it had smaller R^2 than the bioelectrical impedance equation (Sun et al. 2003).

In some instances there are numerous possible predictor variables, and the resulting number of equations can be large. In such cases, it is important to screen the possible predictor variables and to eliminate from consideration those that have low correlations with the response variable. This reduces the number of equations generated and the chance of multicollinearity.

Estimation Procedure

The least squares method estimates regression parameters by minimizing the sum of squares of the deviations of the predicted values from the observed values. When there is multicollinearity (i.e., the predictor variables are interrelated), the variance of the least squares estimators for the regression coefficients are inflated and the precision and accuracy of the predictions will be poor (Montgomery and Peck 1981). Ridge regression is preferable when multicollinearity occurs. Ridge regression reduces the effects of multicollinearity by adding a small constant to the diagonal of the variance–covariance matrix of the predictor variables. A ridge estimator is biased and does not provide the best fit to the data in the validation sample in the sense of minimum variance. Nevertheless, ridge regression performs better than least squares regression on cross-validation because it is less sample-specific.

Robust regression is used to circumvent influential observations (or outliers) that can cause large residuals from the regression. In robust regression, a weight function is incorporated into the analysis, and the observations with large residuals are downweighted. The two most common weight functions are Huber's weight and Tukey's biweight functions (Beaton and Tukey 1979; Huber 1963). Robust regression can improve the accuracy of the equation when applied to other samples by reducing the influence of outliers in the validation sample.

Measures of Goodness of Fit

The coefficient of determination, or R^2, represents the proportion of the total variance in the response variable that is explained by the predictor variables in an equation. The larger the R^2 value, the better the equation fits the data. The root mean square error (RMSE) is a measure of the precision of a prediction equation. When equations are compared, the one with the smallest RMSE value has the highest precision. The RMSE value can be standardized for the mean value of the response variable, and this standardized value, the coefficient of variation (CV), is useful in comparing prediction equations with different response variables and, presumably, different units.

Parsimony

As the number of predictor variables in an equation increases, there is an increase in the R^2 value, which approaches 1.0, and a decrease in the RMSE values. The rates of improvement in the R^2 and RMSE values decelerate as the number of predictor variables increases. However, if too few predictor variables are in an equation, then prediction bias can occur. If there are too many predictors in an equation, the likelihood of multicollinearity is increased. Mallows' C_p statistic (Mallows 1973) is an index of the appropriate number of predictor variables in an equation. The equation with the minimum C_p value will have the maximum R^2 value and the minimum RMSE values and, therefore, a minimum of bias and multicollinearity.

Cross-Validation

Cross-validation is the application of a prediction equation to a separate independent (cross-validation) sample from the one used to construct the equation. Ideally, cross-validation is performed using an independent sample that matches closely the circumstances in which the equation is likely to be applied. The data for the predictor and response variables in the cross-validation sample must be obtained using the same instruments and procedures that were used to record data in the sample from which the equation was derived. The pure error is used to measure the cross-validation performance of a prediction equation. The pure error is calculated as a square root of the sum of squared differences between the observed and the predicted values divided by the number of subjects in the cross-validation sample.

The pure error and the RMSE are conceptually similar but they can differ numerically. The smaller the pure error, the more accurate the equation when applied to an independent sample. A criterion value for the pure error that would denote successful cross-validation has not been set. A general rule is that the value of the pure error should be similar to the RMSE value of the same equation for the validation sample. A large pure error indicates that the prediction was poor when the equation was cross-validated, reflecting differences between the validation and cross-validation samples. These differences may be in ethnicity, age, and sex, in the predictor and response variables, or in the relationships among these variables. For example, it is likely that an equation developed from a particular ethnic group will be applicable only to other samples from the same ethnic group, as shown by Zillikens and Conway (1990). Similarly, an equation that performs well in men would perform poorly in women (Sun et al. 2003), because men and women differ in their body compositions (Chumlea et al. 2001, 2002).

Cross-validation can also be performed by the jackknife method and the prediction of the sum of squares (PRESS) procedure (Duncan 1978; Geisser 1974; Miller 1974; Stone 1974; Sun et al. 2003). In the jackknife method, the subjects are placed randomly into 10 equal-sized groups. The data for the first group are excluded when the equation is being developed, and the residuals (the differences between pairs of observed values and those predicted from the equation) are calculated for the first group. The procedure is repeated for all 10 groups, one at a time. The smaller the sums of squares of the residuals for each group, the more accurate the equation. In the PRESS procedure, each subject in the total data set is excluded, one

at a time, and regression analysis is performed. The value for each omitted subject is predicted, and the difference from the observed value is called a PRESS residual. The sum of squares of the PRESS residuals yields the PRESS statistic (Belsley et al. 1980; Mason and Gunst 1985). The jackknife and PRESS procedures are conceptually similar to data-splitting approaches in which, for example, two thirds of the available sample are used to develop an equation and the remaining one third is used to cross-validate it.

Size and Nature of the Sample

It is not possible to determine, in advance, the sample size necessary to derive an accurate prediction equation. In general, the larger the sample size, the more precise and accurate the prediction equation. The sample size required to develop an accurate equation on cross-validation depends on the relationships between the response variable and the predictor variables, the number of predictor variables, and the variance of the response variable. If the response variable and the predictor variables are highly correlated, only a small sample will be needed to develop a stable equation.

Figure 11.1 illustrates the relationship between sample size and statistical power for the signifi-

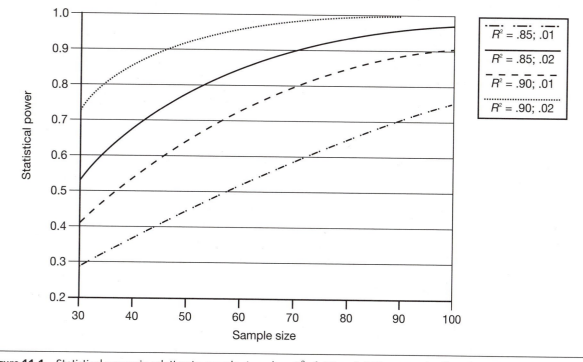

Figure 11.1 Statistical power in relation to sample size when R^2 of .85 and .90 increase significantly (.01 and .02), resulting from the addition of a predictor variable.

cant increase in R^2 attributable to the addition of another variable. For example, the lowest curve in figure 11.1 relates to an increase in the R^2 value from .85 to .86 when an additional predictor variable is added to an equation. If this occurs with a sample size of 30, the probability of significance (statistical power) is about .3, but for a sample size of 100, the probability of significance would be about .74. For increases of .02 in the R^2 value (.85-.87), the statistical power is only .53 for a sample of 30 but is .96 for a sample of 100. This illustrates one limitation that results from a small sample size: uncertainty about the significance of the predictor variables that are included in the equation. A sample of 100 is needed to achieve a significant 1% increase in R^2 precision or accuracy of a predictive equation (e.g., from R^2 = .90 to R^2 = .91) with a statistical power of 90%.

Published Predictive Equations

We have reviewed a limited selection of body composition prediction equations using bioelectric impedance or anthropometry that were published between 1985 and 2003 (table 11.1). The focus of this review is on the characteristics of the study samples, the response and the predictor variables, the statistical methods used to formulate the equations, and the performance of these equations.

In several of these studies, data from both sexes were combined, and in some the age ranges were wide, spanning a large number of years (Deurenberg et al. 1991; Sun et al. 2003). Pooling data from the two sexes and between the races and for many ages increases the sample size and simplifies the application of the equation, but these advantages are sometimes offset by the loss of accuracy attributable to increased heterogeneity. Furthermore, such equations assume that the same set of predictor variables is appropriate for each sex and for a wide range of ages. For many of the samples, ethnicity was not reported. In table 11.1, ethnicity has been assigned when the source of the subjects (e.g., Denmark) makes the ethnicity almost a certainty even though it was not reported.

Most studies used predetermined sets of predictor variables in the regression analysis to formulate the equations. The majority of the reported prediction equations are parsimonious and include no more than five predictor variables. Only a few studies have examined multicollinear-

ity among the predictor variables (Guo et al. 1989; Sun et al. 2003). Most of the equations were not cross-validated, and when cross-validation was performed, only a few investigators calculated the pure errors (Chumlea et al. 2001; Conlisk et al. 1992; Guo et al. 1989; Sun et al. 2003). Some have reported correlation coefficients between the observed and predicted values (Segal et al. 1985, 1988), but these coefficients are insufficient to evaluate the accuracy of a prediction equation.

If we examine this limited selection of prediction equations, it appears that sample specificity is a common problem. For the few reports where these equations were cross-validated, the pure errors tended to be larger than the corresponding RMSE except for males in the study by Conlisk and colleagues (Conlisk et al. 1992). In the work by Sun and colleagues (2003), the equations did not perform as well in black participants as in white participants because the equations were derived from predominantly white samples and thus were more valid for white persons than for black persons. A multicomponent body composition model can account for age, sex, and race differences in the densities of FFM as the response variable, but residual racial and ethnic differences can occur (Schoeller and Luke 2000; Sun et al. 2003). Thus, caution is always necessary in the use of body composition prediction equations in epidemiological studies. The best approach is to derive an equation from a study population and apply this equation to an independent sample where the characteristics of the samples, measuring techniques, and study protocols are minimized.

Inaccuracy is expected when a body composition prediction equation is applied to an individual such as might occur in a clinical setting (Chumlea et al. 2002). Most commercial bioelectrical impedance machines contain software that predicts body composition for an individual. The errors of the predicted values are larger for an individual than those for a group (Chumlea et al. 2001). These errors should be considered closely when predicting body composition for an individual.

A few of the equations in table 11.1 that predict FFM were applied to data from the Fels longitudinal study, and the results are summarized in table 11.2. This includes the R^2 and RMSE values from the validation samples as reported by the authors. The procedure included the following steps: In Step 1, fat-free mass was predicted for the Fels sample

Table 11.1 Summary of Selected Body Composition Equations Published Between 1985 and 2003

Author	Sex	Ethnic group	Age (years)	N	Validation response variable	RMSE	CV (%)	Predictors	Cross-validation PE	CV (%)
Chumlea et al. 2001	M	White	20-90	604	TBW	5.0 L	NA	A,W BMI	4.9	NA
	F	White	20-90	772		3.6 L		A,W,S	3.5	
	M	Black	20-80	128		3.8 L		A,W,S	3.8	
	F	Black	20-80	191		3.3 L		A,W,S	3.3	
Chumlea et al. 1990	M&F	White	18-63	63	FFM from body density (2C model)	4.10 kg	6.7	S^2/R	NA	NA
		White	18-58	72		2.80 kg	6.7	S^2/R	NA	NA
Conlisk et al. 1992	M&F	Hispanic	11-25	79	FFM from body density (2C model)	1.82 kg	4.0	a,b	1.59 kg	4.2
		Hispanic	11-25	76		1.32 kg	4.8		1.64 kg	5.0
Deurenberg et al. 1989a	M&F	Dutch	8-11	64	FFM from body density (2C model)	1.31 kg	NA	S^2/R,W	NA	NA
Deurenberg et al. 1989b	M&F	Dutch	11-16	100	FFM from body density (2C model)	2.06 kg	5.0	S^2/R,W,S	NA	NA
Deurenberg et al. 1990a	M&F	Dutch	7-25	246	FFM from body density (2C model)	2.39 kg	5.9	c	NA	NA
Deurenberg et al. 1990b	M&F	Dutch	60-83	72	FFM from body density (2C model)	3.1 kg	6.0	S^2/R,S,W thigh circumference	NA	NA
Deurenberg et al. 1991	M&F	Dutch	16-83	661	FFM from body density (2C model)	2.63 kg / 1.68 kg	5.0 / 4.9	A,W,S, S^2/R, sex	NA	NA
Deurenberg et al. 1991	M&F	Dutch	7-15	165	FFM from body density (2C model)	1.68 kg	4.9	W, S^2/R, S, sex	NA	NA
Dittmar and Reber 2001	M&F	German	60-90	160	BCM from K40	1.73 kg	NA	S^2/R,W sex	1.61 kg	6.8
Guo et al. 1989	M	White	7-25	140	FFM from body density (multi-C model)	2.31 kg	5.02	S^2/R,W, lateral calf & midaxial skinfold	2.61	7.3
	F	White	7-25	110		2.23 kg	5.80		2.46	7.0
Hassager et al. 1986	M	Danish	20-72	98	FFM from DXA	2.3 kg	2.9	W, S, A	NA	NA
	F	Danish	20-72	130		2.2 kg	3.6	W, S, A	NA	NA

(continued)

Table 11.1 *(continued)*

Author	Sex	Ethnic group	Age (years)	N	Validation response variable	RMSE	CV (%)	Predictors	Cross-validation PE	CV (%)
Heitmann 1990	M&F	Danish	36-65	72	FFM from TBK and TBW	3.61 kg	NA	S²/R, W sex, S, A	NA	NA
Heitmann 1990	M&F	Danish	35-65	139	TBW (isotope dilution)	3.47 L	NA	S²/R, W sex, S, W	NA	NA
Kwok et al. 2000	M&F	Chinese	69-82	613	%BF from DXA	4.1%	NA	Sex, BMI triceps & biceps skinfold	−0.6 to −1.8%	NA
Lukaski et al. 1986	M&F	NA	18-50	113	FFM from body density (2C model)	2.06 kg		S²/R, W, reactance	NA	NA
Morrison et al. 2001	F	White Black	6-17 6-17	65 61	FFM from DXA	1.14 1.95	3.6% 4.7%	S²/R, W, reactance or triceps skinfold	1.5 1.2	6 5
Segal et al. 1985	M&F	NA	17-59	75	FFM from body density (2C model)	3.06 kg	NA	S²/R, S, sex	NA	NA
Segal et al. 1988	M F	NA NA	17-62 17-62	1,069	FFM from body density (2C model)	3.61 kg 2.43 kg	NA NA	S²/R, W, A S²/R, W, A	NA NA	NA NA
Sun et al. 2003	M F	White & black	12-94	669 944	FFM from multi-C model	3.9 kg 2.9 kg	NA	S²/R, W, resistance	3.9 kg 2.9 kg	NA NA

Note. A = age; DXA = dual-energy X-ray absorptiometry; NA = not available; PE = pure error; S = stature; S²/R = stature²/resistance; TBK = total body potassium; TBW = total body water; 2C = two component; multi-C = multicomponent; W = weight.

[a]S²/R, bi-iliac diameter, abdominal circumference, W; [b]suprailiac skinfold thickness, S²/R, W; [c]S²/R, W, sex, S.

using the published equations. The pure errors (PE; kg) were calculated as follows:

$$\sqrt{\Sigma \frac{(\hat{y} - y)^2}{n}} \qquad (11.1)$$

where $\hat{y}$ is the predicted FFM, y is the observed FFM, and n is the number of subjects. In Step 2, multiple regression equations were calculated for the Fels sample, forcing in the variables used by each author with a new set of regression coefficients. If a predictor variable had an insignificant effect on the equation ($p > .10$), the regression was recalculated without that variable. The coefficient of variation, R^2, and the RMSE values from the Fels regression were compared with those reported by the author for the validation group.

The pure errors were consistently larger than those reported for the validation groups; the increases in these values ranged from 8.1% to 172.7% with a median value of 77.0%. This indicates that the level of accuracy was poor. For sex-specific equations derived by all possible subsets of regression, the pure errors exceeded the RMSE in an independent sample by only 12.5% for males and by 10.3% for females (Guo et al. 1989). When the Fels data were used to generate equations selected from those in table 11.1 (Step 2), with

retention of the same predictor variables but the calculation of new coefficients, the RMSE values were larger (4-104%) than those for the original equations when applied to the validation sample except for the equation reported by Deurenberg and colleagues (Deurenberg et al. 1990a). For almost all these equations, both the models and the coefficients were not well suited to Fels data. These findings reflect the effect of specificity and the size of the validation sample, and the statistical processes used to develop these equations are also important.

Summary

Body composition prediction equations have a long history and can be quite useful. Stepwise regression is the most widely used development method. Possible predictor variables should be screened to eliminate those with low statistical and biological relationships with the response variable. This will reduce the potential inaccuracy of an equation resulting from multicollinearity among the predictor variables. All possible subsets of regressions should be applied in the development of an equation because, unlike stepwise regression, this procedure evaluates every

Table 11.2   Results From the Application of Selected Equations That Predict FFM to the Fels Longitudinal Study

Author	Sex	Validation by author, R^2	RMSE (kg)	Fels cross-validation pure error (kg)	R^2	Fels validation of model RMSE (kg)	Insignificant predictors
Conlisk et al. 1992	M	.97	1.82	4.05	.97	2.62	Bicristal breadth
	F	.95	1.32	3.60	.89	2.69	—
Deurenberg et al. 1989a	M&F	.89	1.31	2.98	.82	1.97	Sex
Deurenberg et al. 1989b	M&F	.97	2.06	3.05	.94	2.76	—
Deurenberg et al. 1990a	M&F	.99	2.39	4.10	.95	2.06	Sex
Lukaski et al. 1986	M&F	.98	2.06	4.06	.89	3.68	—
Segal et al. 1985	M&F	.96	3.06	4.79	.90	3.64	—
Segal et al. 1988	M	.9	3.61	3.90	.80	3.67	—
	F	.89	2.43	3.07	.76	2.85	—

Note. FFM = fat-free mass; RMSE = root mean square error.

combination of the potential predictor variables and their relationships with the response variable. Cross-validation of prediction equations is needed to assess their generalizability. In both validation and cross-validation studies, the measures of precision (RMSE) and accuracy (PE) should be standardized for the mean of the response variable.

In recent years, several important techniques have been introduced that provide accurate measures of FFM, which is the most common response variable. These include computed tomography, magnetic resonance imaging, and neutron activation, but these are very expensive and not readily accessible except in clinical settings. DXA and bioelectric impedance are generally available. As a result, future predictive equations should be more accurate than those developed in the past, given the opportunity for more accurate measurements of the response variable using DXA and the inclusion of impedance measures with the predictor variables. Few published predictive equations have used response variables obtained from DXA, either alone or in combination with other procedures (Sun et al. 2003). This number is likely to increase because DXA provides reasonably accurate measures of body composition in special populations such as children, people with disabilities, and the elderly.

Recently, bioelectrical impedance prediction equations for TBW and FFM were developed that have several advantages over other published equations (Sun et al. 2003). Criterion body composition methods were used in a multicomponent model, and the equations were derived from data from five study sites that included subjects with a wide age range, two races, and both sexes. All possible subsets of regression were used to evaluate every possible combination of the predictor variables by allowing the simultaneous comparison among the set of possible equations in conjunction with the use of the C_p statistic, the minimum RMSE, and the significance of the regression estimates. These cross-validated prediction equations are reasonably generalizable for individuals at the extremes of the population as assessed with the PRESS procedure, and they have been applied to the bioelectrical impedance data from the Third National Health and Nutrition Examination Survey to produce body composition estimates for the U.S. population (Chumlea et al. 2002).

Measures of response and predictor variables will improve in the future, and researchers will learn more about the effects of sample characteristics (size, homogeneity) on the development of prediction equations and realize the need for independent cross-validation. Researchers and clinicians will better appreciate the sources and extent of the errors when prediction equations are applied to individuals or to groups. As a result of these changes, it is likely that in the future well-developed prediction equations that are accurate on cross-validation will be applied more commonly in epidemiological and clinical studies. It is recommended, however, that direct measures of body composition be obtained in intervention studies where it is necessary to document changes within individuals as accurately as possible. Body composition prediction equations are excellent at providing estimates of the status values for an individual or a group, but they are not reliable or accurate to account for changes in body composition.

Part III

Body Composition Models and Components

Each level of body composition consists of specific recognized components. All of the major body composition components can now be measured in vivo by one or more methods as reviewed in part II. In this section, chapter 12 contains an in-depth review of body composition models developed with the purpose of estimating specific components. The examination of components is expanded in chapters 13 and 14, which explore adipose tissue and skeletal muscle mass, respectively. One of the most important links involving body composition and other research areas is the association between components and energy expenditure. These relations of body composition and energy expenditure are reviewed in chapter 14 with a focus on new and evolving concepts.

One of the primary aims of body composition research is to study models that describe the quantitative relationships between components, and chapter 12 examines this topic in detail. The study of body composition models not only provides insight into basic biological processes but also enhances the understanding and application of body composition methods. Chapter 12 reviews representative multicomponent molecular-level models as well as recently published cellular-level models.

The two largest components at the tissue–organ level of body composition are adipose tissue and skeletal muscle, and expanded discussions of these important components are provided. Chapter 13 reviews the available methods for measuring adipose tissue mass and distribution with a concentration on special regional components and the links between these components and health. Chapter 13 also examines the topic of adipose tissue mass and distribution changes with weight loss.

Skeletal muscle, reviewed in chapter 14, represents approximately 40% and 30% of body weight in adult men and women, respectively. The need for methods of estimating either whole-body or regional skeletal muscle is compelling and reflects multidisciplinary interests. Chapter 14 reviews the available methods for the estimation of human skeletal muscle mass in vivo, with emphasis on the biological and physical basis of each approach and the application and the limitations of the methods.

The mitochondria within each cell produce heat that is referred to collectively as energy expenditure. The topic of sources of heat and energy expenditure in body composition has fascinated scientists for several centuries. Chapter 15 contains a comprehensive review of the models and methods used for defining the relation between energy exchange and the sources of heat production in body composition.

Chapter 12

Multicomponent Molecular-Level Models of Body Composition Analysis

ZiMian Wang, Wei Shen, Robert T. Withers, and Steven B. Heymsfield

Early body composition method partitions body mass into two components, usually at the molecular level, such as fat and fat-free mass. The equations used to solve for the mass of each component have a measured physical property (e.g., body density) and body mass.

In this chapter, we examine multicomponent models in which body mass is partitioned into three or more components. The equations used to solve for multiple components may have three or more unknown terms, one of which is usually body mass. Three-, four-, and even six-component models have been reported and are widely used in the research setting. Since these methods include more and different measured properties or other components than two-component models, they usually account for more biological variability. Accordingly, multicomponent methods are often selected as the reference against which other methods are compared. This chapter reviews multicomponent models and methods with an emphasis on model development concepts.

General Concepts of Molecular-Level Body Composition Methods

Over the past several decades, most body composition research was performed using the two-

component molecular-level model as outlined in chapter 1. The molecular level of body composition is important for studying the anabolic actions of drugs, the catabolic effects of disease, energy metabolism, and many other topics of biological interest.

The classic two-component model describes the human body as the sum of fat mass (FM) and fat-free mass (FFM; Brožek et al. 1963; Heymsfield and Waki 1991; Siri 1961). In this chapter we review in detail representative, published, molecular-level multicomponent models. Our focus is on multicomponent models in which the human body is divided into three or more molecular-level components. The summary presents an overview of areas in need of further investigation.

There are more than 30 major components at the five levels of body composition (Wang et al. 1992; table 12.1). Methods of quantifying these components in vivo can be organized according to the following general formula:

$$C = f(Q), \qquad (12.1)$$

where C represents an unknown component, Q a measurable quantity, and f a mathematical function relating Q to C (Wang et al. 1995). This descriptive formula states that the magnitude of an estimated component's mass is equal to the product of some mathematical function and a measurable quantity. All multicomponent molecular-level models are based on this general formula.

Supported by National Institutes of Health Grant POI-DK 42618.

Table 12.1   Main Components at the Five Levels of Human Body Composition

	Atomic level	Molecular level	Cellular level	Tissue–organ level	Whole-body level
Components	O, C, H, N, Ca, P, S, K, Na, Cl, Mg	Fat, essential lipids, water, protein, bone mineral, soft tissue minerals, carbohydrates, fat-free mass, fat-free solids	Fat cells, cells, intracellular fluid, extracellular fluid, extracellular solids, body cell mass	Adipose tissue (AT), subcutaneous AT, visceral AT, bone, skeletal muscle, skeleton, visceral organ, brain, liver	Head, neck, upper limbs, trunk, lower limbs
Component no.	11	9	6	9	5

The quantity portion (Q) of equation 12.1 can be classified into two categories. The first is a measurable *property*, such as body volume, decay profiles of specific isotopes, or electrical resistance. For example, total body water (TBW) mass can be measured by the dilution of tritium water as described in chapter 3. The measurable property in this example is the 0.018 MeV β-decay of tritium. About one half of the body composition components outlined in table 12.1 can now be quantified using property-based methods.

The second category of measurable quantities that can be used with general formula 12.1 is a known *component*. With component-based methods, the known component must be quantified first by using a property-based method. An example is the calculation of FFM from TBW. A property-based method, such as triated water dilution, must first be used to quantify TBW.

The mathematical function (i.e., *f*) in equation 12.1 can be one of two types. The first, which we refer to as *empirical*, is developed using regression analysis of experimental data to derive a prediction equation (chapter 11). Typically, a reference method is used to measure the "unknown" component (C) in a group of subjects with well-defined characteristics. The measurable quantity (Q; i.e., property or known component), as defined in equation 12.1, is also estimated in the subjects. Regression analysis is then used to establish the mathematical function (*f*) and thus develop the equation that predicts the unknown component from the measurable property or the known component. An example is the prediction of TBW from measured electrical impedance. Subjects first have TBW measured by isotope dilution and resistance measured by bioelectrical impedance analysis at 50 kHz. Regression analysis is used to develop a prediction formula, *f*, linking TBW to resistance,

stature, age, and other covariates. Thereafter, the equation can be used in new subjects to predict TBW from measured resistance.

The second type of mathematical function is based on a well-established *model*. These models usually represent ratios or proportions of measurable quantities to components that are assumed constant both within and between subjects. For example, the ratio of TBW to FFM is assumed constant at 0.73 in healthy adult mammals and can be used to develop a method for estimating FFM by multiplying TBW by 1.37 (Pace and Rathbun 1945; Sheng and Huggins 1979; Wang et al. 1999). An important feature of methods that are based on ratio or proportion models is that assumptions are needed in their development. Although ratio or proportion models are usually applicable in physiologically stable and healthy subjects, their use under certain pathological conditions may be limited. For example, the hydration value of FFM (i.e., 0.73) may not apply in patients with edema.

A useful step in understanding the multicomponent methods that follow is to analyze a classic two-component hydrodensitometry approach for measuring total body fat. Our aim here is to examine the basis of two-component methods as an introduction to the more complex multicomponent methods that are described later.

The two-component hydrodensitometry method was originally proposed by Behnke and colleagues in 1942. This method was derived from two models at the molecular level, a body mass (BM) model (BM = FM + FFM) and a body volume (BV) model (BV = FM/0.9007 + FFM/1.100). The body mass model does not involve any assumptions. The body volume model is based on two assumptions, that the density of fat is constant at 0.9007 g/cm^3 at 36° C and that the density of FFM is constant at 1.100 g/cm^3 at 36° C. If we combine the body mass model with the body volume model, there are two unknown components (FM and FFM, both in kg) and two measurable quantities (BM in kg and BV in L). The body mass and body volume simultaneous equations can then be solved for the unknown components as follows:

$$FM = 4.971 \times BV - 4.519 \times BM, \text{ and}$$

$$FFM = 5.519 \times BM - 4.971 \times BV. \quad (12.2)$$

The FM equation can also be rewritten to express FM as a fraction of body mass,

$$FM/BM = 4.971 \times BV/BM - 4.519, \text{ or}$$

$$FM/BM = 4.971/D_b - 4.519, \quad (12.3)$$

where D_b is body density (i.e., BM/BV in kg/L). These equations differ slightly from those reported by Siri (1961), who assumed a fat density in his model of 0.9000 g/cm^3 at 37° C.

According to this approach, the hydrodensitometry method is a property-based method and the mathematical function is derived from the body mass and body volume models. The main assumptions are that fat and FFM have constant densities within and between subjects. Body mass is measured using a scale and, in the original Behnke method (Behnke et al. 1942), body volume was measured using hydrodensitometry. Body volume can also now be measured using air plethysmography (chapter 2), an advance that potentially extends the use of multicomponent models to centers unable to install hydrodensitometry systems.

The purpose of developing multicomponent model methods to study body composition is twofold. First, these methods allow analysis of three or more molecular-level components rather than the two traditional fat and FFM components. There are many biological conditions in which the study of these multiple components is of interest. Second, measuring multiple components often reduces the errors of the assumptions in model methods. The two-component hydrodensitometry method is formulated on the assumption that the proportions of FFM as water, protein, and mineral are constant. Adding a measurement of TBW to the two-component method reduces the error in fat and FFM estimates attributable to individual differences in hydration. Multicomponent model methods are therefore important in biological investigations and in body composition research.

The limitations of two-component methods are particularly evident in infants, young children, and elderly people, in whom systematic and random deviations from assumed FFM component proportions are recognized (Lohman 1992). Early in life, the extracellular fluid compartment is expanded relative to FFM with a corresponding elevation in FFM hydration and reduction in FFM density (chapter 9). The proportions of FFM as water, protein, and minerals are more variable in older subjects compared with young adults, also leading to two-component model concerns.

In the sections that follow we describe some representative multicomponent models and their basic assumptions and measurements involved. Our primary intent is to describe the process and underlying theoretical basis on which multicomponent model methods are developed.

Three-, Four, and Six-Component Hydrodensitometry Methods

Behnke's two-component hydrodensitometry method assumes known and constant proportions of FFM as water, protein, and mineral (Behnke et al. 1942). It was from these assumed proportions, and the density of each chemical component, that the density of FFM was derived as 1.10 g/cm^3 (Brožek et al. 1963). This led ultimately to the development of various hydrodensitometry-based two-component methods (Brožek et al. 1963; Siri 1961).

It is obvious that deviations from the assumed proportions of the molecular-level components are possible with conditions that alter body composition such as aging, pregnancy, weight reduction in obese people, and various disease states. A more desirable approach in these conditions would be to make fewer assumptions by measuring more quantities in addition to body mass and body volume. Over the next several decades after Behnke's milestone research, the two-component model was expanded first to a three-component model by adding an estimate of water and more recently to a four-component model by adding water and mineral estimates.

Siri (1961) suggested a three-component (i.e., fat, water, and residual) method in which there are three measurable quantities: body mass, body volume, and TBW (table 12.2). The measurement of TBW reduces the errors in Behnke's two-component model (Behnke et al. 1942) related to individual differences in hydration. Siri's three-component

Table 12.2   Some Multicomponent Methods for Measuring Total Body Fat

Author	Method
Siri 1961	FM = 2.057 × BV – 0.786 × TBW – 1.286 × BM
Selinger 1977	FM = 2.747 × BV – 0.714 × TBW + 1.129 × Mo – 2.037 × BM
Lohman 1986	FM = 6.386 × BV + 3.961 × M – 6.09 × BM
Lohman 1992	FM = 2.747 × BV – 0.714 × TBW + 1.146 × Mo – 2.053 × BM
Heymsfield et al. 1990	FM = 2.748 × BV – 0.6744 × TBW + 1.4746 × TBBA – 2.051 × BM
Baumgartner et al. 1991	FM = 2.747 × BV – 0.7175 × TBW + 1.148 × M – 2.05 × BM
Fuller et al. 1992	FM = 2.747 × BV – 0.710 × TBW + 1.460 × TBBA – 2.05 × BM
Withers et al. 1992	FM = 2.513 × BV – 0.739 × TBW + 0.947 × Mo – 1.790 × BM
Friedl et al. 1992	FM = 2.559 × BV – 0.734 × TBW + 0.983 × Mo – 1.841 × BM
Siconolfi et al. 1995	FM = 2.7474 × BV – 0.7145 × TBW + 1.1457 × Mo – 2.0503 × BM
Heymsfield et al. 1996	FM = 2.513 × BV – 0.739 × TBW + 0.947 × Mo – 1.79 × BM
Forslund et al. 1996	FM = 2.559 × BV – 0.734 × TBW + 0.983 × Mo – 1.841 × BM
Wang et al. 2002	FM = 2.748 × BV – 0.699 × TBW + 1.129 × Mo – 2.051 × BM

Note. BM = body mass (kg); BV = body volume (L); FM = fat mass (kg); M = mineral (kg); Mo = bone mineral (kg); TBBA = total body bone ash (kg); TBW = total body water (kg).

model assumes a constant ratio of protein to mineral even though this ratio is known to vary between subjects.

Later investigators subsequently expanded Siri's classic three-component model by adding bone mineral (Mo) measurements to eliminate errors related to individual differences in the bone mineral content of FFM (Baumgartner et al. 1991; Friedl et al. 1992; Fuller et al. 1992; Heymsfield et al. 1990; Selinger 1977). In these models, four quantities are measured: body volume, TBW, bone mineral, and body mass. Detailed reviews of the densitometry, TBW, and bone mineral measurement methods are presented in chapters 2, 3, and 5, respectively. Representative four-compartment models are shown in table 12.2.

Although multicomponent models share assumed constant densities for fat, water, and bone mineral, two main strategies are applied in developing these methods. In one approach, the remainder of body mass after the subtraction of fat, water, and bone mineral is assumed to be protein and soft tissue minerals of known densities. In the other approach, the remainder is assumed to represent a combined residual mass (i.e., protein, soft tissue minerals, and other) of known density.

A useful exercise is to examine the simultaneous equations upon which four-component hydrodensitometry methods are developed. The body mass model for this method is

$$BM = FM + TBW + Mo + residual, \quad (12.4)$$

where all units are in kilograms. Residual mass includes protein, soft tissue minerals, and glycogen. The body volume model for this method is,

$$BV = FM/0.9007 + TBW/0.99371 + Mo/2.982 + residual/1.404. \quad (12.5)$$

This body volume model assumes constant densities for each of the four components: fat, 0.9007 g/cm^3; water, 0.99371 g/cm^3 at 36° C; bone mineral, 2.982 g/cm^3; and residual mass, 1.404 g/cm^3 for humans and other mammals at 37° C (Allen et al. 1959). The two largest components of residual mass are protein with a density of 1.34 g/cm^3 and glycogen with a density of 1.52 g/cm^3. Soft tissue minerals with a density of 3.317g/cm^3 and other miscellaneous chemical components constitute the remainder. If we combine the body mass and body volume models, body fat mass can be derived as follows:

$$FM = 2.513 \times BV - 0.739 \times TBW + 0.947$$
$$\times Mo - 1.79 \times BM. \qquad (12.6)$$

Most four-component hydrodensitometry methods give similar estimates for total body fat because they are based on closely related models and assumptions. We now review these assumptions in detail.

Density of Triglyceride

Lipids can be divided into fat and nonfat according to distribution, function, and solubility characteristics (Gurr and Harwood 1991; Wang et al. 1992). Hydrodensitometry models usually consider only fat, which consists almost entirely of triglycerides; the no-fat lipids are included in the FFM component.

Fidanza and colleagues (1953) reported that the density of 20 ether-extractable lipid samples from the intra-abdominal and subcutaneous adipose tissue of five subjects was (mean ± SD) 0.9007 ± 0.00068 g/cm^3 at 36° C with a coefficient of thermal expansion of 0.00074 g/(cm$^3 \cdot$ °C) over the range of 15 to 37° C. The small between-subject coefficient of variation of 0.08% supports the use of a "constant" fat density (0.9007 g/cm^3) in body composition models.

Although the preceding analyses were conducted on ether extracts of adipose tissue, this should not introduce major errors into the assumed density of triglyceride. Cholesterol with a density of 1.067 g/cm^3 and phospholipids with a density of 1.035 g/cm^3 comprise only 1% of the ether-extractable lipid from rabbit adipose tissue (Méndez et al. 1960). Triglyceride or fat is therefore assumed to account for 99% of the ether-extractable lipid (Wang et al. 1992). These considerations therefore lead us to conclude that the assumed density of fat (0.9007 g/cm^3 at 36° C) is reasonably accurate and stable between subjects.

Density of Water

The densities of water at either 36° C (0.9937 g/cm^3) or 37° C (0.9934 g/cm^3) are used in three- and four-component models and to convert water volume to water mass (Diem 1962).

Density of Protein

The densities of most proteins in the dry crystalline state are close to 1.27 g/cm^3 (Haurowitz 1963). However, proteins are the principal water-binding substances in humans, and the resultant hydration is accompanied by a volume contraction of both the solute and the solvent. The specific volume of hydrated protein therefore decreases until the apparent density is 1.34 g/cm^3, which appears to be the best available estimate for hydrated protein in the living cell. However, specific proteins differ in density. For example, collagen, which is found mainly in bone and skin, has an average dry density of 1.36 g/cm^3 and comprises 25% to 30% of total body protein (Hulmes and Miller 1979). The density used for total body protein is thus more tenuous than the densities used for fat and water.

Density of Bone Mineral

The density of bone mineral, 2.982 g/cm^3, is based on the mean of only four samples (Méndez et al. 1960) isolated from the long bones of animals (cow tibia, 2.9930 and 3.0066 g/cm^3 at 36° C; dog femur and tibia, 2.9624 and 2.9667 g/cm^3 at 36.7° C). Bone minerals are similar in composition across mammals and consist mainly of calcium hydroxyapatite (Biltz and Pellegrino 1969). Brožek and colleagues (1963) further verified this value against that of the stoichiometry for hydroxyapatite after allowance for water of crystallization and CO_2. Méndez and colleagues (1960) also cited earlier work by Dallemagne and Melon (1945), who reported the densities of bone and dental mineral to be 2.99 and 3.01 g/cm^3, respectively.

Density of Soft Tissue Mineral

The density of soft tissue minerals was estimated by early workers as follows. The classical cadaver analyses assumed that all calcium was contained within bone. This is a reasonable assumption because extraskeletal calcium represents less than 0.4% of total body calcium (Snyder et al. 1975). Calcium in bone was accompanied by phosphorus, sodium, and magnesium as observed in bone ash. The excess body mineral was then assigned to nonbone mineral or soft tissue mineral, the density of which was determined by measuring the apparent specific volume occupied by 1 g of each major salt represented in the nonbone mineral component. The apparent densities at 40° C (i.e., the reciprocal of apparent specific volume), which ranged from 3.07 g/cm^3 for potassium bicarbonate to 4.99 g/cm^3 for magnesium chloride, were then multiplied by their relative contributions to nonbone mineral to yield an overall density of 3.317 g/cm^3.

Density of Carbohydrates

Most of the carbohydrate component is in the form of glycogen with a density of 1.52 g/cm³ at 37° C. Four-component hydrodensitometry methods either include glycogen in residual mass or pool the small glycogen mass with the protein component. In either case, the estimates of glycogen in the various models are so approximate that glycogen must be considered a small source of model error.

Body Temperature

The density of components at the molecular level of body composition is closely related to body temperature, and this has been a source of much confusion in the body composition literature. Obviously, body temperature varies from core to skin in a predictable manner. It is usual to assume a single representative average body temperature, although structures near the skin, such as subcutaneous adipose tissue, are at a lower temperature than those located deep in the viscera.

The two-component models of Brožek and colleagues (1963) and Siri (1961) assume respective mean body temperatures of 36 and 37° C. The temperature of 37° C is representative of core temperature, which is measured in the rectum or esophagus or at the tympanic membrane. The average body temperature under basal resting conditions and in a comfortable environment is likely to be 1 to 2° C lower than the core temperature of approximately 37° C. If it is assumed that the mean skin and rectal temperatures are 34 and 37° C, respectively, in a thermoneutral environment, then the formula of Burton (1935) yields the theoretical mean body temperature of 36° C, which was used by Brožek and colleagues (1963). The average body temperature would also probably be 36° C during underwater weighing when water temperature is maintained at 35° C.

Although 36° C appears to be the best estimate of mean body temperature, it is interesting to speculate whether fat and other molecular-level components have different mean temperatures. For example, a case could be made for a lower temperature for fat because a high proportion of this component is subcutaneous and it has low levels of vascular perfusion. Conversely, it can be argued that water has a higher mean temperature because skin and subcutaneous adipose tissue contain only a small proportion of TBW. However, mathematical modeling of the human thermal system by Werner and Buse (1988) emphasizes that specific tissues have a range of temperatures in vivo. It is therefore virtually impossible to specify a mean temperature for fat and each of the molecular-level components. The best estimate of average body temperature under controlled laboratory conditions thus appears to be about 36° C.

Measurement Error

All widely applied multicomponent methods are based on well-established ratio or proportion models. These models are founded on experimental findings and thus provide reasonable component approximations with acceptable errors. The organization of body composition methods according to equation 12.1 provides an opportunity to systematically evaluate these errors. Our intent here is not to be exhaustive but rather to highlight the main types of error involved in assessing the mass of a component. Other reviews (Lohman 1992; Lohman and Going 1993; Mueller and Martorell 1988) and chapters 2 to 11 of this book provide a comprehensive assessment of errors in body composition methodology.

The first category involves errors in measuring the quantity (Q) term of equation 12.1. Each of the measurement errors can be evaluated separately in complex methods such as the use of TBW and total body potassium for estimating body fat. Additionally, measurement errors occur between observers, instruments, and laboratories and over time. Only rarely have body composition studies included a comprehensive evaluation of the many sources of measurement error.

The second group of errors involves the mathematical function term (f) of equation 12.1. With empirical methods, the prediction equations have characteristic function errors (chapter 11). There are also errors related to the reference method for assessing the unknown component. The errors in methods that include ratio or proportion errors are associated with the various assumptions that ultimately determine the model's "stability" within and between subjects.

The measurement errors for the multicomponent hydrodensitometry models are presented in table 12.3.

As can be gathered from the preceding discussion, the calculation of errors in three- and four-component hydrodensitometry methods is complex and involves many uncertainties related to assumption and measurement errors. Although greater validity should be associated

Table 12.3   Representative Errors and Radiation Exposure for Multicomponent Molecular-Level Methods

Quantity	Measurement method	Errors (CV, %)[a]	Radiation exposure (mSv)	Reference
Total body nitrogen	PGNA	3.6 2.7 4.1 3.0	0.26 0.26	Dutton 1991 Pierson et al. 1990 Beddoe et al. 1984 Mernagh et al. 1977
Total body hydrogen	PGNA	0.4	0.26	Dutton 1991
Total body potassium	WBC	1.5	0	Pierson et al. 1990
Total body carbon	INS	4.2 3.0	0.16	Dutton 1991 Pierson et al. 1990
Total body calcium	DGNA	2.6 0.8	2.50	Dutton 1991 Pierson et al. 1990
Total body phosphorus	DGNA	5.1 3.0	2.50	Dutton 1991 Pierson et al. 1990
Total body sodium	DGNA	2.5	2.50	Pierson et al. 1990
Total body chlorine	DGNA	1.5 2.5	2.50	Dutton 1991 Pierson et al. 1990
Total body water	^3H_2O dilution	1.5	0.12	Wang et al. 1973
	^2H_2O dilution	1.5 4.0	0	Pierson et al. 1990 Bartoli et al. 1993
	$H_2^{18}O$ dilution	2.0	0	Schoeller et al. 1980
Total body fat	TBC	—	3.06	Kehayias et al. 1991 Heymsfield et al. 1991
	3C UWW	—	0	
	4C IVNA	—	2.88	Cohn et al. 1984
	4C UWW	—	< 0.01	
	6C IVNA	—	3.04	Heymsfield et al. 1991
Bone mineral	DXA	1.3	< 0.01	Heymsfield et al. 1990
Body mass Body volume Stature	Scale Hydrodensitometry Stadiometer	<1% 0.6% BF 0.2	0 0 0	Heymsfield et al. 1991 Withers et al. 1998, 1999 Heymsfield et al. 1991

Note. BF = body fat; C = component; CV = coefficient of variation; DXA = dual-energy X-ray absorptiometry; DGNA = delayed-γ neutron activation analysis; INS = inelastic neutron scattering; IVNA = in vivo neutron activation analysis; PGNA = prompt-γ neutron activation analysis; TBC = total body carbon, UWW = underwater weighing; WBC = whole-body counting ^{40}K.

[a]CVs represent between-measurement variation in phantoms or humans on the same or different days. Measurement errors for total body fat estimates are discussed in the text.

with the measurement of more components, there is some concern that the resultant greater control over the biological variability of FFM may be offset somewhat by the propagation of measurement errors associated with the determination of body volume, TBW, and bone mineral (Mo). A worst-case scenario for this propagation of errors can be calculated by assuming that the squared standard errors of measurement (SEM^2) or the squared technical errors of measurement (TEM^2) are independent and additive:

$$SD \text{ of total error} = [(SEM^2 \text{ or } TEM^2 \text{ for effect of underwater weighing on } \% \text{ fat}) + (SEM^2 \text{ or } TEM^2 \text{ for effect of TBW on } \% \text{ fat}) + (SEM^2 \text{ or } TEM^2 \text{ for effect of Mo on } \% \text{ fat})]^{0.5} \quad (12.7)$$

Precision information (Withers et al. 1999) is given in table 12.4 from which results

$$SD \text{ of total error} = (0.4^2 + 0.6^2 + 0.1^2)^{0.5} = 0.7\% \text{ body fat.} \quad (12.8)$$

The test–retest reliability data thus yield a value of about 0.7% body fat units.

Friedl and colleagues (1992) investigated the reliability of fat estimates from multicomponent methods in 10 soldiers who were each tested three times. These investigators found that the greatest source of error in their four-component method was in the underwater weighing procedure (~1% of body mass) followed by TBW estimates (~0.5 L). They observed reliability coefficients of 0.991 and 0.994 and within-subject standard deviations of ±1.0 and ±1.1 for percent body fat estimates using two- and four-component models, respectively.

Siri (1961) suggested that variability in the density of FFM (~0.0084 g/cm³) resulted in an error with a standard deviation of 3.8% body fat units when percent body fat was estimated from body density alone. Lohman (1981) suggested that this error decreases to 2.7% body fat units for specific populations. Neither estimate includes the technical error associated with body density

measurement. Siri further suggested (1961) that the error could be reduced to 1.5% body fat units with a three-component method, if the error in measuring TBW could be maintained at about 1% of body mass. Friedl and colleagues (1992) found an error in TBW of about 1% of body mass. These observations therefore suggest that the additive measurement errors in three- and four-component methods do not offset the improved accuracy in estimating fat over that of the traditional two-component underwater weighing method.

Four- and Six-Component Neutron Activation Methods

Cohn and his colleagues (1984) at Brookhaven National Laboratory originally proposed a four-component total body fat method based on measurement of tritiated water dilution volume (in L), TBN (in kg), total body calcium (TBCa, in kg), and body mass (in kg). The method is designed to provide measurements of four components (in kg): fat, water, protein, and total body bone ash (TBA). The equations are as follows:

$$FM = BM - (TBW + TBPro + TBA)$$
$$TBW = 0.95 \times 0.994 \times \text{tritium dilution volume}$$
$$TBPro = 6.25 \times TBN$$
$$TBA = TBCa/0.34, \quad (12.9)$$

where TBPro is total body protein.

Our group, also working at Brookhaven National Laboratory, expanded Cohn's model (Heymsfield et al. 1991; Wang et al. 1993) to a six-component model: fat, water, protein, bone mineral, soft tissue mineral, and glycogen:

$$FM = BM - (TBW + TBPro + Mo + Ms + \text{glycogen})$$
$$TBW = 0.95 \times 0.994 \times \text{tritium dilution volume}$$
$$TBPro = 6.25 \times TBN$$
$$Mo = TBCa/0.364$$
$$Ms = 2.76 \times TBK + 1.00 \times TBNa + 1.43 \times TBCl - 0.038 \times TBCa$$
$$\text{glycogen} = 0.275 \times TBN, \quad (12.10)$$

Table 12.4 Precision Information for Multicomponent Models

Test	SEM/TEM (kg FM)	SEM/TEM (% FM)
Underwater weighing	0.28	0.4
Total body water	0.57	0.6
Bone mineral (effect on fat)	0.08	0.1

Note. SEM = standard error of measurement; TEM = technical error of measurement; FM = fat mass.

where Ms represents soft tissue minerals and TBNa and TBCl are total body sodium and chlorine (both in kg), respectively.

The Swansea (UK) and Auckland (NZ) groups have each developed models based on the analysis of total body elements (Beddoe et al. 1984; Ryde et al. 1993). All of the neutron activation whole-body ^{40}K counting models share several characteristics in common. The propagated errors for estimating total body fat by the four- and six-component methods are 2.7% and 3.4%, respectively. The main assumptions related to a multicomponent neutron activation model are as follows.

Total Body Water

Total body water is measured using tritium- or deuterium-labeled water dilution and in some laboratories by ^{18}O-labeled water (chapter 3). Each isotope measures a specific dilution volume, and the volume measured for all three isotopes is larger than the actual TBW volume. For tritium- and deuterium-labeled water, the dilution space is larger than TBW volume because the tritium and deuterium atoms exchange with hydrogen atoms associated with carboxyl, hydroxyl, and amino groups (Culebras and Moore 1977; Heymsfield and Matthews 1994). Similarly, ^{18}O atoms exchange with labile oxygen atoms in carboxyl and phosphate groups (Schoeller et al. 1980; Wong et al. 1988). The exchange of labeled atoms with unlabeled atoms from labile molecular components probably varies with a number of factors (Goran et al. 1992; Heymsfield and Matthews 1994). Most workers assume a constant exchange rate that approximates 4% to 5% for tritium and deuterium and less than 1% for ^{18}O, although some workers assume that ^{18}O exchange is negligible (Culebras and Moore 1977; Forbes 1987; Schoeller and Jones 1987). To calculate TBW, the isotope dilution space is multiplied, to account for isotope exchange, by a correction factor such as 0.95 to 0.96 for ^3H_2O and ^2H_2O dilution space and 0.98 to 0.99 for $H_2^{18}O$ dilution space. The density of water (0.9937 g/cm^3 at 37° C) is assumed constant. These corrections are extremely important because water is the largest component of the FFM and even small relative errors in its estimation affect the various multicomponent models as a whole. Nevertheless, at present, there does not appear to be an improved method of estimating isotope exchange.

Protein

At present, protein is typically measured in vivo by first estimating total body nitrogen (TBN) with neutron activation analysis (Dilmanian et al. 1990; Vartsky et al. 1979) and then converting TBN to protein as TBPro = 6.25 × TBN. As noted earlier, this approach assumes a constant nitrogen–protein ratio (0.16) and complete incorporation of TBN into total body protein.

Nitrogen and protein exist in a stable relationship within and between subjects. Mulder, who first introduced the term protein in 1838, is credited with suggesting two chemical formulas for protein ($C_{40}H_{31}N_5O_{12}$ and $C_{48}H_{36}N_6O_{14}$; Cunningham 1994). Today the general formula for meat protein is assumed to be $C_{100}H_{159}N_{26}O_{32}S_{0.7}$ with nitrogen protein equal to 0.16 (Kleiber 1975; Snyder et al. 1975). It is clear, however, that the nitrogen–protein ratio of specific proteins differs from 0.16. Knight and colleagues (1986) reported nitrogen–protein ratios of 0.172, 0.159, and 0.137 for collagen, actinomyosin, and albumin, respectively. Nevertheless, the chemical analyses of two persons who died of cancer yielded whole-body nitrogen–protein ratios of 0.158 and 0.156. The authors concluded that these data do not justify changing the assumed nitrogen–protein ratio of 0.16.

Under normal circumstances, more than 99% of TBN is incorporated into protein (Diem 1962; Snyder et al. 1975). Nonprotein sources of nitrogen include urea, creatine, creatinine, uric acid, free amino acids, and several other nitrogenous compounds. When serious disease is present, such as renal failure or congestive heart failure, total body levels of nitrogenous nonprotein compounds can increase substantially, although the impact on the TBN–TBPro ratio is not large even in these extreme circumstances. For example, in renal failure, blood urea nitrogen concentrations can increase from a normal level of 0.1 to 1.0 g/L. Assuming a TBW of 42 L, total body urea nitrogen would increase with renal failure from 4.2 to 42 g. TBN is approximately 1,800 g in the Reference Man (Snyder et al. 1975), so urea nitrogen would increase from 0.23% to 2.3% of total body nitrogen. If we assume that 99.77% of TBN is in protein (i.e., [TBN – urea N]/TBN) in healthy subjects, our total body protein estimate using 6.25 × TBN would be 11.2 kg. If instead 2.3% of TBN were urea nitrogen, then our total body protein estimate would be 11.0 kg. Hence, the classical assumption that all body nitrogen is in the form of protein is subject to a systematic error of 2% to 3% with serious renal disease and other states in which nonprotein nitrogen accumulates. Alternatively, corrections in TBN could be made for urea nitrogen and other nitrogenous nonprotein compounds using blood levels of these compounds along with TBW estimates to evaluate total body amounts as in our approximate calculations.

Bone Mineral

Bone mineral ash is the remaining noncombustible matter after bone is heated to a high temperature (>500° C; Woodard 1962). The organic matter, mainly proteins such as collagen, are volatilized during the ashing process (Snyder et al. 1975; Woodard 1964). Calcium, oxygen, and phosphorus are the abundant elements in bone mineral, with smaller amounts of sodium, potassium, and magnesium. The main bone mineral is calcium hydroxyapatite, $[Ca_3(PO_4)_2]_3Ca(OH)_2$. The ratio of calcium to bone mineral is assumed constant at 0.364 (Heymsfield et al. 1991), which is close to the proportion of calcium in calcium hydroxyapatite (0.398; Woodard 1964). The ratio of TBCa to bone mineral appears relatively stable across men and women, in normal and osteoporotic women (Burnell et al. 1982), and may even be similar across different species (Biltz and Pellegrino 1969).

Calcium has a concentration of 3 mmol/kg and 3 to 5 mmol/kg in intracellular and extracellular fluid, respectively. Only small deviations from these concentrations are observed in various disease states. Total nonosseous calcium is less than 14 g, whereas there is 1,000 g of bone calcium in the Reference Man (Snyder et al. 1975).

Bone mineral is now widely measured by dual-energy X-ray absorptiometry (DXA). The bone mineral content (BMC) measured by DXA represents ashed bone (Friedl et al. 1992). One gram of bone mineral yields 0.9582 g of ash, because labile components such as bound water and CO_2 are lost during heating (Heymsfield et al. 1989). The BMC therefore needs to be converted to bone mineral as Mo = BMC × 1.0436 (i.e., 1/0.9582). Significant differences in Mo were found between DXA commercial systems (Modlesky et al. 1996). Even though this difference is relatively large, using Mo estimates from different DXA systems has only a minimal impact on multicomponent body fat estimates (Modlesky et al. 1999).

Soft Tissue Minerals

Soft tissue minerals (Ms) consist largely of soluble minerals and electrolytes found in the soft tissue extracellular and intracellular compartments. Although soft tissue mineral mass is only approximately 0.4 kg in the Reference Man (Snyder et al. 1975), the contribution of this component to body density should be considered because Ms collectively has a higher density (3.317 g/cm³) than each of the other components including fat (0.900 g/cm³), water (0.994 g/cm³), protein (1.34 g/cm³), and bone mineral (2.982 g/cm³). Brožek and colleagues (1963) were the first to attempt a systematic analysis of soft tissue minerals and to estimate their combined density in vivo. According to three cadaver studies, Brožek and colleagues assumed Ms to be present in a constant amount relative to bone mineral (i.e., Mo). A corresponding model was thus developed, Ms (kg) = 0.221 × Mo (kg) (Brožek et al. 1963). Later, Selinger (1977) assumed that 1.05% of body mass is Ms with the corresponding equation, Ms (kg) = 0.0105 × body mass (kg). However, neither the Brožek nor the Selinger model derives theoretical support because there is no plausible reason why the relationships between Ms and Mo and between Ms and body mass should exist as stable ratios.

Wang and colleagues (2002) recently derived a soft tissue mineral model (Ms = 0.016168 × TBW – 0.006625 × ECW, where ECW is extracellular water) along with a simplified version, Ms (kg) = 0.0129 × TBW (kg). The new model has a firm physiological basis: All soft tissue minerals are distributed within the intracellular fluid and extracellular fluid compartments, and the applied coefficients are based on well-established experimental data. Both intracellular and extracellular soft tissue mineral concentrations are maintained relatively stable at 16.168 g/kg H_2O and 9.543 g/kg H_2O, respectively (Maffy 1976). The mineral–electrolyte concentrations and osmolarity of intra- and extracellular fluids are highly regulated, and these models are thus reasonable for use in healthy subjects and in people without fluid and electrolyte disturbances.

The other alternative is to estimate each soft tissue mineral component separately. For example, in our six-component model described previously, we measure total body sodium and calcium with delayed-γ neutron activation analysis. Sodium is present as a cation in several soft tissue minerals and is also bound to the crystalline matrix of bone mineral (Woodard 1962, 1964). The ratio of sodium to calcium in bone mineral is known (Woodard 1962) and can therefore be used to estimate bone mineral sodium from total body calcium: Na in bone mineral (g) = 0.038 × TBCa (g). Sodium in soft tissues can then be calculated as the difference between total body sodium and sodium in bone mineral. Similar approaches can be used to estimate six main soft tissue minerals and electrolytes (Na^+, K^+, Mg^{2+}, Cl^-, $H_2PO_4^-$, and HCO_3^-) from four measurable total body elements (TBNa, TBK, TBCl, and TBCa), which are then summed for total body soft tissue mineral mass:

$$Ms = 2.76 \times TBK + TBNa + 1.43 \times TBCl - 0.038 \times TBCa, \qquad (12.11)$$

where Ms and all elements are in grams (Heymsfield et al. 1990; Wang et al. 1993). The measurement of total body elements by in vivo neutron activation analysis is discussed in the next section.

Glycogen

The highly variable and rapidly changing total body glycogen component is estimated in most methods or is ignored because of its relatively small amount. In the six-component model, the glycogen–protein ratio is assumed constant at 0.044. Beddoe and colleagues (1984) assumed that glycogen is 0.91% of FFM. Of importance to multicomponent methods, glycogen can now be measured in vivo with ^{13}C nuclear magnetic resonance spectroscopy, and the results agree well with chemical analysis of skeletal muscle biopsies (Jue et al. 1989; Taylor et al. 1992). This ability to quantify glycogen in vivo should allow development of improved glycogen models.

Six-Component Total Body Carbon Method

Almost all body carbon is incorporated into three molecular level components: fat, protein, and glycogen (Snyder et al. 1975). The carbon fractions of fat, protein, and glycogen are 0.774, 0.532, and 0.444, respectively. A small amount of carbon is also found in bone mineral. Kyere and colleagues (1982) first proposed a model in which fat could be calculated from total body carbon (TBC) measured by in vivo neutron activation analysis. Kehayias and Heymsfield and their colleagues at Brookhaven National Laboratory (Kehayias et al.

1991; Heymsfield et al. 1991) later refined the six-component TBC model, which is as follows:

$$TBC = 0.774 \times FM + 0.532 \times$$
$$protein + 0.444 \times glycogen + 0.20 \times$$
$$carbonate\ in\ bone\ mineral + 0.197$$
$$\times\ HCO_3^-\ in\ extracellular\ fluid, \qquad (12.12)$$

where all units are in kilograms. The four remaining unknown components can be measured (protein from TBN, bone mineral from TBCa, and HCO_3^- from TBCl) or approximated (glycogen from TBN). Equation 12.12 can be rearranged to solve for body fat mass:

$$FM = 1.29 \times (TBC - 3.37 \times TBN - 0.052 \times TBCa - 0.085 \times TBCl). \qquad (12.13)$$

The main assumptions of the method are twofold: that TBC is distributed in fat, protein, glycogen, bone minerals, and electrolytes, and that the proportion of carbon in each of these molecular level components is known and constant. These assumed constant proportions are described as follows.

Carbon in Triglyceride

The method assumes that fat is 77% carbon. Triglycerides extracted from human tissues vary in fatty acid composition, and the carbon–triglyceride ratio of 0.77 represents an average based on the Reference Man (Snyder et al. 1975).

The carbon–triglyceride ratio is relatively stable because carbon only varies slightly as a fraction of these representative triglycerides (0.76-0.77).

Carbon in Minerals

There is a small amount of carbon in the carbonate component of bone mineral (Brožek et al. 1963) and in the anion HCO_3^-, which is distributed mainly in extracellular fluid. The TBC method of estimating fat assumes that all body calcium is in bone mineral and that the carbon–calcium ratio in bone is constant at 0.05 (Kehayias et al. 1991). Under most conditions, almost all body calcium is incorporated into bone mineral, and this relationship is assumed to be very stable in vivo. The carbon–calcium ratio is an approximation (Biltz and Pellegrino 1969; Brožek et al. 1963), but the errors have only small impacts on fat estimates. The small amount of carbon in extracellular fluid as HCO_3^- is estimated by assuming a constant ratio of bicarbonate to chloride; the latter is found almost entirely in extracellular fluid.

Carbon in Glycogen

Glycogen, which is present mainly in liver, skeletal muscle, and heart, comprises about 1% of FFM in weight-stable adults. The concentration of glycogen in these tissues varies with fasting and feeding so that the amount present is at a minimum of less than 0.3 to 0.5 kg for the total body early in the morning (Diem 1962; Snyder et al. 1975). This is the time of day that body composition measurements are usually made, and, because of its small amount, glycogen is not considered in most body composition models. In the six-component TBC method for estimating fat, it is assumed that the glycogen–protein ratio is 0.044 (Kehayias et al. 1991), but this must be considered an approximation. The total amount of carbon in glycogen is thus small (~0.2 kg), and measurement errors in this portion of the model have only a small impact on fat estimates.

Thus, many assumptions of varying quality form the basis of the six-component TBC method of estimating total body fat. Nevertheless, the main assumptions are highly stable and the method is thus valuable as a research tool for measuring total body fat mass independent of other conventional approaches.

Measurement of Total Body Elements

The six-component TBC method requires measurement of TBC, TBN, TBCa, and TBCl. Total body carbon is measured by neutron inelastic scattering, which is based on the reaction $^{12}C + n \rightarrow {}^{12}C^{\bullet} + n' \rightarrow {}^{12}C + n' + \gamma$ (4.44 MeV). The incoming fast neutrons (n) that interact with matter by inelastic collisions result in prompt nuclear de-excitation with γ-ray release. The source of 14 MeV neutrons is a small sealed deuterium–tritium generator that is pulsed at 4 to 10 kHz (Kehayias et al. 1987). Carbon is detected by counting the 4.44 MeV γ-rays from the inelastic scattering of fast neutrons from ^{12}C nuclei. Sodium iodide detectors detect carbon's 4.44 MeV γ-rays from carbon.

Total body nitrogen is measured using prompt-γ neutron activation analysis and is based on the reaction $^{14}N + n \rightarrow {}^{15}N^{*} \rightarrow {}^{15}N + \gamma$ (10.83 MeV). The system has a ^{238}Pu-Be neutron source (85 Ci), and the high energy prompt 10.83 MeV γ-rays from nitrogen are simultaneously detected by NaI crystals with radiation (Dilmanian et al. 1990).

Delayed-γ neutron activation is used to measure total body calcium and chlorine based on the reactions $^{48}Ca + n \rightarrow {}^{49}Ca^{\bullet} \rightarrow {}^{49}Ca + \gamma$ (3.10 MeV) and $^{37}Cl + n \rightarrow {}^{38}Cl^{\bullet} \rightarrow {}^{38}Cl + \gamma$ (2.17 MeV). Total body sodium is measured along with TBCa and TBCl using delayed-γ neutron activation (Dilmanian et al. 1990) with the reaction $^{23}Na + n \rightarrow {}^{24}Na^{\bullet} \rightarrow {}^{24}Na + \gamma$ (2.75 MeV). There are two parts to this system: an irradiation facility and a shielded whole-body γ-ray counter (Dilmanian et al. 1990). The irradiation system uses an array of ^{238}Pu-Be moderated neutron sources. After neutron exposure, the subjects are rapidly transported to the whole-body counter where the induced γ-rays emitted from neutron capture activation of calcium, chlorine, and sodium are counted.

The between-measurement coefficients of variation (CV) for elemental measurements are shown in table 12.3. The propagated error in the six-component TBC method of estimating total body fat is 3.4% to 4% (Heymsfield et al. 1991; Kehayias et al. 1991). The table also gives the radiation exposure for representative neutron activation measurements.

The importance of the six-component TBC method for quantifying total body fat is that it provides values that are independent of the classical multicomponent hydrodensitometry methods described earlier. The method is primarily based on highly stable chemical models that are not known to be influenced to an appreciable degree by age, sex, or ethnicity. Because only a few centers in the world have the necessary in vivo neutron activation systems, the six-component TBC method has limited applicability. However, this method provides an excellent example of the various assumptions that are incorporated into most multicomponent model methods.

Multicomponent Total Body Protein Method

Protein is a functionally important component at the molecular level of body composition. Protein mass in healthy adults is also relatively large, representing 10.6 kg or 15.1% of body mass in Reference Man (Snyder et al. 1975).

Nitrogen, which is found in all amino acids and proteins, consists of linked amino acids. The prompt-γ in vivo neutron activation method provides an estimate of TBN, and TBPro is then calculated as TBPro = TBN/0.16 or 6.25 × TBN. The neutron activation method is now considered the criterion for TBPro measurement, and cadaver validations were reported by Knight and colleagues (1986).

Some multicomponent models are recognized for evaluating both TBPro mass and distribution. Two of these models are based on TBN and TBK. Burkinshaw and colleagues (1979) and Cohn and colleagues (1980) reported what later became known as the Burkinshaw–Cohn model for estimating TBPro mass and distribution from measured TBN and TBK. Protein was assumed in the model to be distributed into two components: skeletal muscle and nonskeletal muscle. James and colleagues (1984) also reported a TBN–TBK model for estimating cellular and collagen proteins.

Burkinshaw and Cohn's model was applied to predict skeletal muscle mass, the largest body component at the tissue–organ level (Burkinshaw et al. 1979; Cohn et al. 1980). Potassium is distributed mainly in skeletal muscle (SM, in kg) with overlap into the nonskeletal muscle lean component (non-SM, in kg). Nitrogen is also distributed in the non-SM component with overlap into the SM component. Two simultaneous equations can be written:

$$TBK (g) = 3.56 \times SM + 1.88 \times non\text{-}SM$$
$$TBN (g) = 30 \times \qquad (12.14)$$
$$SM + 36 \times non\text{-}SM,$$

where the coefficients suggested by Burkinshaw and colleagues represent the assumed constant (g/kg) potassium and nitrogen contents of SM and the non-SM lean component. Accordingly, an SM prediction equation was derived as follows:

$$SM (kg) = 0.503 \times TBK (g) -$$
$$0.0263 \times TBN (g). \qquad (12.15)$$

Although innovative advances at the time, these models were later shown to have a number of theoretical limitations. One of these is that the assumed potassium and nitrogen concentrations in SM and non-SM lean components are not stable across subjects. This may be one explanation why SM estimates by the Burkinshaw model were inaccurate when compared with neutron activation and computed tomography methods (Forbes 1987; Wang et al. 1996). The Burkinshaw model is thus mainly of historical importance.

There are additional reported approaches for estimating TBPro mass. Fuller and his colleagues (2001) suggested a TBPro prediction model by using a combination of DXA and TBW measurements. Total body protein can be calculated by a four-component model,

$$TBPro = BM - (FM + TBW +$$
$$total\ body\ mineral), \qquad (12.16)$$

where total body mineral (i.e., bone mineral plus nonbone mineral) was assumed to be equal to $1.2741 \times$ bone mineral content (BMC; Fuller et al. 1992). Thus, total body protein mass can be predicted as

$$TBPro = BM - FM - TBW -$$
$$1.2741 \times BMC, \qquad (12.17)$$

where FM and BMC can be measured by DXA, and TBW can be measured by the dilution method.

Drawing on equation 12.17 and assumed constant FFM hydration (i.e., the ratio of TBW to FFM) at 0.732 for healthy adults, Fuller and colleagues (2001) further suggested a simplified DXA model for predicting total body protein mass:

$$TBPro = 0.268 \times BM -$$
$$0.268 \times FM - 1.2305 \times BMC. \qquad (12.18)$$

Recently, a TBPro mass and distribution model was developed by Wang and colleagues (2003). Three protein-containing compartments are distinguished at the cellular body composition level: body cell mass (BCM), extracellular fluid (ECF), and extracellular solids (ECS). Both BCM and ECF contain potassium, water, and protein. Potassium and water can be used as predictors of BCM protein and ECF protein: BCM protein = $0.00259 \times$ TBK $- 0.0104 \times$ TBW, and ECF protein = $0.0112 \times$ TBW $- 0.000073 \times$ TBK, where protein and TBW are in kg, and TBK is in mmol. The model also considers ECS protein. Both bone mineral and protein are the major constituents of ECS, and thus bone mineral measured by DXA can be used as a predictor of ECS protein: ECS protein = $0.732 \times$ Mo. If the three protein compartments are combined, total body protein mass can be predicted as follows:

$$TBPro = 0.00252 \times$$
$$TBK + 0.732 \times Mo. \qquad (12.19)$$

Although theoretically reasonable, the protein distribution estimates cannot be evaluated at present and there are no published independent estimates of protein in the BCM, ECF, and ECS components. Further studies are thus needed to validate protein distribution prediction models.

Summary

The aim of this overview of multicomponent molecular level models was to examine in depth the theoretical basis of multicomponent model methods and to then demonstrate these concepts with a few examples. The concepts presented should allow for the development of methods suited to the research needs and resources of each investigator. The possibility exists to prepare different multicomponent models with a firm understanding of the process by which such methods are developed.

Many research possibilities are related to the development of multicomponent model methods. As is clear from this review, innumerable assumptions are required in developing multicomponent model methods. Some of these assumptions are tenuous and would benefit from probing studies. Our review only touched on the area of multicomponent method errors, and more work is therefore needed in this area.

Multicomponent model method research includes other body composition levels that are not reviewed in this report but share identical underlying concepts. The study of multicomponent body composition methods thus offers a vast potential area for future research.

Chapter 13

Measuring Adiposity and Fat Distribution in Relation to Health

Luís B. Sardinha and Pedro J. Teixeira

The relationships among excess body weight, obesity, and health have been extensively studied. These relationships are mediated by the metabolic characteristics of the adipose tissue (AT). Although the primary role of adipocytes is to store triglycerides, they have a more complex role, producing a large number of hormones, prohormones, cytokines, and enzymes with autocrine, paracrine, and endocrine actions. In the obese state, the production of adipose tissue–derived proteins is increased, with implications for several health problems.

There are regional differences in protein production in AT. Compared with subcutaneous AT, visceral AT associated with upper-body obesity is a more important contributor to cardiovascular disease (CVD; Bjorntorp 1997). Enlargement of AT in the visceral compartment has pleiotropic effects on several hormonal and metabolic functions that can contribute to the development of CVD and other chronic health problems. The American Heart Association has declared obesity as a major modifiable risk factor for coronary heart disease (Eckel and Krauss 1998). The World Health Organization (WHO) and the U.S. National Institutes of Health (NIH) have declared obesity as a chronic disease and a major risk factor for noncommunicable diseases such as diabetes type 2, CVD, and cancer and several psychosocial problems in industrialized countries, with a special concern regarding abdominal obesity (NHLBI 1998; WHO 1998).

The adverse effects of obesity and abdominal adiposity likely begin in childhood and adolescence. Recent surveys confirm that the prevalence of overweight and obesity has increased in the United States (Mokdad et al. 2003; Ogden et al. 2002a), whereas developed and developing countries worldwide are also affected by this alarming trend (WHO 1998). The obesity epidemic may soon offset the current efforts to prevent CVD and may increase health care costs. There is a need for management strategies for at-risk individuals and groups, including the identification of those with excess adipose tissue that may impair health and quality of life.

Excess adiposity is a lifelong health problem with specific screening issues related to age maturation and race or ethnicity. In this chapter we first review measures of total body fatness, with special reference to their relationship with chronic health problems, and the usefulness of related cutoffs for screening for overweight and obesity. Considering the physiological and health implications of AT distribution, we also review the most common measures available to estimate fat distribution adiposity through the life span.

Total Body Fatness

The whole-body level of body composition entails simple measurements such as body weight, stature, segment lengths, circumferences, and skinfold thickness. Many studies are limited to the use of these anthropometric measures, including BMI, and their relationship with clinical outcomes. Therefore, the term *total body fatness* was adopted in this section to include the concepts of excess adiposity and total body fat, which, although related, reside on different levels of the five-level model and have unique properties.

Measures of Total Body Fatness

Many laboratory methods are available to estimate total body fat or adipose tissue, with different levels of accuracy, practicality, and cost. Imaging methods such as computed axial tomography (CT) and magnetic resonance imaging (MRI) are used to assess adipose tissue, a component at the tissue–organ level of body composition (Wang et al. 1992), with high accuracy and also high cost. Other methods, such as dual-energy X-ray absorptiometry (DXA) and air displacement plethysmography, estimate total body fat, a component at the molecular level consisting mostly of triglycerides, also with considerable accuracy. Because adipose tissue contains approximately 80% fat, and because fat is also found in other tissues, total body fat and adipose tissue mass are not identical.

Often the concept of percent body fat (%BF) has been used to assess the relationship between body composition and health. The utility of %BF has been recently challenged, because normalization of fatness for body weight ignores the between-subject variation in fat-free mass (Wells 2001). Two individuals may differ in %BF when they have equivalent fat-free mass but different values of fat mass or when they have similar fat mass but different fat-free mass. In children, after height is controlled for, variation in fat-free mass represents two thirds of the variation in %BF, illustrating that the use of %BF to define obesity may be dependent on fat-free mass (Wells et al. 2002). This issue is especially critical during growth (Maynard et al. 2001; Wells 2001), although it also has relevance in adults (Van Itallie et al. 1990). It is possible that normalizing both fat and fat-free mass to height squared in children (Wells and Cole 2002) and in adults (Van Itallie et al. 1990) to derive both fat mass index (FMI) and fat-free mass index (FFMI) would improve sensitivity to detect changes in fat and fat-free body stores. Annual changes in body composition of growing children and adolescents boys are primarily dependent on FFMI rather than on FMI (Maynard et al. 2001). In girls, longitudinal

changes are similar for FMI and FFMI until age 16 years, after which body composition changes are attributable primarily to FMI. These observations support the need to analyze the extent to which each component contributes to overall changes in composition in boys and girls during specific periods of development along with the associated health-related outcomes. Although these two indexes seem more appropriate to express fatness and leanness, they have not yet been used in relation to morbidity. Recent attempts have been made to develop FFM and FFMI reference data for a wide age range (Schutz et al. 2002). Further analyses are needed to assess the association of FMI with obesity-related diseases and to judge its usefulness as a more specific total body adiposity index at various ages.

Most available evidence relating obesity to disease risk is based on indexes of weight for height. The principal limitation of any weight-based index is that it cannot distinguish excess adiposity from greater-than-average muscularity or skeletal tissue. Thus, a high body mass index (BMI), for example, may indicate either obesity or an athletic, muscular build with low body fat. Similarly, in elderly people, a constant BMI may mask changes despite stable weight, such as increased adiposity, decreased skeletal muscle, and bone loss. The relation of BMI to adiposity varies with age, sex, race, and FFM, and the comparisons across subgroups must be interpreted carefully, because compositional differences may be obscured. For example, in children a BMI of 20 kg/m^2 may correspond to a %BF range between 5% and 40%; conversely, a 20 %BF includes a range of 15 to 30 kg/m^2 (Ellis 2001). Although the limits of agreements are somewhat smaller, low accuracy at the individual level is also found in adults and elderly people. Nevertheless, BMI is commonly used to classify underweight, overweight, and obesity based on its relationships with body fatness and health outcomes (NHLBI 1998). In children and youth, different indexes of weight divided by height raised to a power (*p*) have been developed to improve the estimate of adiposity (Freedman and Perry 2000). Ideally, the exponent is chosen to minimize the correlation with height and to maximize the relationship with referent measures of adiposity. In youth, the age-specific optimal exponent increases from 2.5 (5 years old) to 3.5 (9-11 years old) and then decreases to 2.0 (17 years old) by young adulthood (Freedman and Perry 2000).

Other anthropometric measurements have also been used as surrogates of total body adiposity. Skinfold thickness is often used to provide estimates of total body fatness. However, there is a considerable between-subject variability in subcutaneous thickness at any given site, in the ratio of various adipose tissue depots, and in the compressibility of adipose tissue at a given site; thus, comparisons across individuals can be strongly influenced by age, sex, race, and site and degree of obesity. Optimally, descriptive models to predict %BF based on skinfolds need to be specific to the characteristics of the population used for their development, although generalized models have also been developed to be applied to a wide range of ages. The most appropriate set of skinfold sites may differ by sex, age, degree of obesity, clinical outcome, and body fat patterning, supporting the use of a variety of skinfolds in different age groups and sexes. Single skinfolds or combinations of several skinfolds have been used as surrogates for total body adiposity and adipose distribution in studies of their relationships with health outcomes (Freedman and Perry 2000; Sardinha et al. 2000; Teixeira et al. 2001). Analyses of sensitivity and specificity to identify obese and nonobese subjects have been performed using several skinfold and health-related %BF standards (Sardinha et al. 1999; Wellens et al. 1996) in order to identify cutoff values that can be easily used for obesity screening in the field.

Risk Classification Based on Total Body Fatness

Measures of body composition and total body fatness change with growing, developing, and aging. Specific cutoff values for children and adolescents, adults, and the elderly have been suggested based on specific health-related criteria.

Children and Adolescents

Defining obesity standards remains controversial. In both children and adults, widely accepted %BF criteria have yet to be developed. Because of its simplicity and correlation with adiposity, BMI has become the preferred method (Maynard et al. 2001).

Two approaches have been used to classify adiposity. One method, the norm-referenced approach, identifies a certain segment of the population as overweight or obese, dependent on the distribution of the index in the population. The specific cutoff may also depend on the time period in which the data were collected. Because the prevalence of obesity is increasing (Ogden et al. 2002a), the absolute value corresponding to norm-reference standard has likely increased. This sex- and age-

specific reference may not necessarily be related to the level of adiposity that best reflects elevated risk for unfavorable health outcomes. In contrast, the criterion-referenced approach is based on identifying critical levels of adiposity associated with increased risk for several health outcomes. The selected cutoff points are independent of reference population data and the respective distribution of the variables of interest.

Four studies have related %BF with measured CVD risk factors in young subjects (table 13.1). In a large sample of 9- and 10-year-old Japanese boys and girls, the ratio of total cholesterol (total-Chol) to high-density lipoprotein cholesterol (HDL-Chol) was significantly higher with an estimated 23% to 25% BF by bioimpedance (Washino et al. 1999). In prepubertal children, a cutoff point of 33% BF estimated by DXA had a good sensitivity to identify an adverse risk factor profile (Higgins et al. 2001). Two studies with large samples of children and adolescents showed that greater than 30% BF lipid profiles worsen in girls (Dwyer and Blizzard 1996; Williams et al. 1992). In boys, a 20% BF (Dwyer and Blizzard 1996) and a 25% BF (Williams et al. 1992) were suggested as cutoffs for obesity based on CVD risk. Both studies supported the concept of a single cutoff value for a wide range of ages including prepubertal and postpubertal girls and boys. There are several difficulties in defining health-related cutoffs and comparing results of different studies. First, %BF and CVD risk factors are continuous variables, and no clearcut point exists at which the correlation between them changes drastically. Of particular relevance is the fact that these relationships may be influenced by different types of interactions between age, race, sex, and the distribution of CVD risk factors in any specific cohort, which may explain the different cutoff points suggested by these four studies. Because the relationship between adiposity and selected risk factors changes in different ways with aging (e.g., the correlation between adiposity and low-density lipoprotein cholesterol [LDL-Chol] increases between ages 5 and 17 years whereas the relationship with systolic blood pressure decreases with age), it has been argued that studying the longitudinal relationship between adiposity indices and CVD risk factors is more important than defining cutoff points from cross-sectional data (Freedman and Perry 2000). Nevertheless, a criterion-referenced approach, with its limitations, is clinically relevant and a useful tool for screening purposes relating critical levels of total body fatness to increased risk for disease.

Table 13.1 Review of Cutoff Points in Percent Body Fat (%BF) Defining Obesity for Children Based on CVD

References	N	Age	Methods	Variables	Cutoff (%)
Williams et al. 1992	3,320	5-18	Sum of skinfolds[a]	Blood pressure	
Boys	1,667		Tricipital	(Systolic and diastolic)	25
Caucasian	1,062		Subscapular	Lipid profile	
Black	605			Total-Chol	
Girls	1,653			HDL-Chol	35
Caucasian	1,028			LDL-Chol	
Black	625			VLDL-Chol	
				AI:	
				(VLDL-Chol + LDL-Chol)/HDL-Chol	
				LDL-Chol/HDL-Chol	
Dwyer and Blizzard 1996	1,834[b]	9 and 15	Skinfolds[c]	Blood pressure	
Boys	NA		Tricipital	(Systolic and diastolic)	20
Caucasian	NA		Bicipital	Lipid profile	
Black	NA		Subscapular	Total-Chol	
Girls	NA		Suprailiac	Triglycerides	30
Caucasian	NA			HDL-Chol	
Black	NA				
Washino et al. 1999	1,289	9 and 10	Bioelectrical impedance (Tanita TBF-102)	Lipid profile	
Boys (Asian)	651			Total-Chol	23
Girls (Asian)	638			Triglycerides	23
				HDL- Chol	
				AI: (Total-Chol–HDL-Chol)/HDL-Chol	
Higgins et al. 2001	87	4-11[e]	DXA[f]	Blood pressure	
Boys	NA			(Systolic and diastolic)	33
Caucasian	NA			Fasting insulin	
Black	NA			Lipid profile	
Girls	NA			Total-Chol	33
Caucasian	NA			Triglycerides	
Black	NA			HDL-Chol	
				LDL-Chol	
				Total-Chol–HDL-Chol	

Note. DXA =dual-energy X-ray absorptiometry; Chol = cholesterol; HDL = high-density lipoprotein; LDL = low-density lipoprotein; VLDL = very low density lipoprotein; AI = atherosclerogenic index.

[a]Body density (BD) was converted to percent body fat (%BF) using age-adjusted, sex-specific equations (Lohman 1989). [b]Skinfolds and blood lipid measurements on 1,144 children and skinfolds and blood pressure measurements on 1,757 children. [c]BD estimation for 9-year-old by Brook formula (1971). BD estimation for 15-year-old by Durnin and Rahaman formula (1967). %BF conversion by the Siri equation (1961). [d]BD estimation for boys and girls by Sakamoto and colleagues' (1993) formulas (predictor variables: body weight, body height, and Z from Tanita-ohms). %BF was calculated using the formula of Brozek et al. (1963). [e]Prepubescent children. [f]DXA-BF (kg) was adjusted by Pintauro and colleagues' (1996) formula.

Because of the difficulties in developing well-recognized cutoffs using biological end points to classify and identify excess adiposity, two additional norm-based approaches have been used. Kuczmarski and Flegal (2000) developed a statistical definition based on representative data from several countries, whereas another approach used an international combined database to develop sex- and age-specific BMI cutoffs among children and adolescents by extrapolating sex-specific BMI–age curves to 25 and 30 kg/m^2 at age 18 (Cole et al. 2000).

In the United States, the most recent definition of overweight for children and adults is based on the year 2000 Centers for Disease Control and Prevention growth charts (Ogden et al. 2002b). *Overweight* in children is defined as 95th percentile or greater of BMI, whereas increased risk for overweight is defined as BMI between the 85th and 95th percentiles. Other countries, including the United Kingdom (Wright et al. 2002), Italy (Cacciari et al. 2002), and Sweden (Karlberg et al. 2001), have developed national BMI-for-age charts using the same method, adjusting BMI distribution for skewness, and allowing individual BMI values to be expressed as an exact centile or standard deviation score (Cole et al. 1990). The World Health Organization Expert Committee (WHO 1995) defined "at risk for overweight" for adolescents (10-19 years) as those with a sex- and age-specific BMI in the 85th percentile, and it defined *obesity* using the combination of sex- and age-specific BMI in the 85th percentile and high subcutaneous fat ($\geq$90th percentile), assessed with U.S. subscapular and triceps skinfold norms from the National Center for Health Statistics (Must et al. 1991). Regardless of the qualitative definitions, it is recognized that youths with a BMI at the 95th percentile or greater should be referred for further assessment and follow-up, and those with a BMI at the 85th or greater percentile and less than the 95th percentile should be examined for various CVD risk factors, including family history, blood pressure, total-Chol, considerable prior increase in BMI, and concern about weight increase (Himes and Dietz 1994). These guidelines are based on the evidence that higher adiposity is associated with current and subsequent health problems, even though there is no empirical biological evidence for a clear-cut two-level screening approach.

Databases from Brazil, the United States, the United Kingdom, Hong Kong, the Netherlands, and Singapore, representing 97,876 males and 94,851 females from birth to 25 years of age, were combined to develop the International Obesity Task Force reference standard definition for childhood and adolescent overweight and obesity (Cole et al. 2000). BMI cutoffs were derived by extrapolating adult (at age 18) percentiles related to adult cutoffs for overweight (25 kg/m^2) and obesity (30 kg/m^2) to children and adolescents.

These standard definitions were intended to overcome arbitrary definitions of overweight and obesity and to provide international standards for use in between-country comparison studies of overweight and obesity prevalence across countries. Although the notion of a single worldwide standard linked to accepted adult definitions is attractive, it remains controversial (Deurenberg 2001; Jebb and Prentice 2001; Katzmarzyk et al. 2003; Reilly 2002). Application of a percentile-based definition arbitrarily sets the prevalence of overweight and obesity and could obscure secular trends. Moreover, population-based differences in the BMI–percent fat relationships suggest that BMI percentiles may relate differently to morbidity risk across countries. Other concerns include (a) doubt about the methodology used to develop the age- and sex-specific BMI cutoff values, (b) lack of biological validity linking these values to morbidity, and (c) conceptual weakness of linking pediatric overweight and obesity to adult BMI values of 25 and 30 kg/m^2. This approach also does not address the need to assess underweight boys and girls (Reilly 2002). Despite the limitations, using the international guidelines, Katzmarzyk and colleagues (2003) showed that overweight girls had between 1.6 and 9.1 times the risk of having elevated total-Chol, LDL-Chol, total-Chol–HDL-Chol ratio, triglycerides, and glucose and reduced HDL-Chol and physical work capacity compared with normal-weight girls. Overweight boys had between 1.6 and 5.7 times the risk for the same outcome variables compared with normal-weight boys. Also, the proportion of overweight girls and boys increased as the number of the risk factors increased, indicating that overweight as defined by the new international guidelines increases the probability of CVD risk factors clustering.

During and after sexual maturation, anthropometric indexes of obesity are associated with elevated blood pressure and adverse changes in serum lipids and lipoproteins (Freedman and Perry 2000; Wattigney et al. 1991). The clustering among changes in lipid–carbohydrate metabolism, hyperinsulinemia, and high blood pressure is also evident in childhood and increases with age and obesity (Bao et al. 1996). Insulin resistance is an important subclinical marker of this clustering, associated with most of the other clinical abnormalities found in obesity. Two recent studies showed that children with a higher BMI had impaired glucose tolerance (Sinaiko et al. 2002; Sinha et al. 2002). Plasma leptin levels and BMI in combination (Chu et al. 2000) and tumor necrosis factor-α receptor (Chu et al. 2003) also seem to be significant predictive markers of the resistance syndrome among children. Together, these results suggest that BMI as an index of total body fatness is related to insulin resistance and is useful for screening children and adolescents for diabetes risk.

Recently it has been suggested that the cluster of these risk factors may be attributable to a low-grade systemic inflammation and that obesity could be an inflammatory disorder (Das 2002). Thus, inflammation could be the link between obesity and CVD. Interestingly, there is some recent evidence that childhood obesity estimated by anthropometric methods is related to serum concentrations of acute-phase proteins such as C-reactive protein (CRP), which is recognized as an important endogenous activator of atheromatous lesions (Blake and Ridker 2001). In 3,512 children aged 8 to 16 years from the National Health and Nutrition Examination Survey III, overweight (defined as a BMI or sum of triceps, subscapular, and suprailiac skinfolds above the sex-specific 85th percentile) boys and girls were 3.7 to 5.1 and 2.9 to 3.2 times more likely to have elevated CRP, respectively, than their normal-weight counterparts (Visser 2001). In a smaller sample of 364 males and 335 females aged 10 to 11 years, CRP was 270% higher in the highest quintile compared with the lowest quintile of ponderal index (weight/height^3) and was associated with several cardiovascular risk factors such as elevated systolic and diastolic blood pressure, LDL-Chol, triglycerides, fasting insulin, and glucose, and decreased HDL-Chol (Cook et al. 2000). Recent studies with adolescents found similar relationships between cardiovascular risk factors, plasma CRP, and anthropometric variables (Cook et al. 2000; Wu et al. 2003). BMI was more highly related to CRP than other metabolic variables. These data support that total body fatness assessed with a weight-for-height index is related to higher levels of inflammation and CVD risk factors. This is clinically relevant because individuals can develop coronary heart disease in the absence of lipoprotein abnormalities (Libby 2002).

Considering that BMI tends to overestimate adiposity in rapidly growing children, a weight/height^p

index that is independent of height theoretically should be better correlated with other indicators of total body fatness such as triceps and subscapular skinfold thickness and with CVD risk factors. This issue is especially critical in boys, because annual increases in BMI are generally attributed to the lean rather than to the fat component of BMI (Maynard et al. 2001). However, recent studies comparing BMI with weight/height^p have found similar correlations with CVD risk factors, despite an age-related trend in (p) for the allometric weight-for-height index (Frontini et al. 2001). Thus, BMI remains the recommended index.

More precise estimates of adiposity estimated with laboratory methods have shown that %BF and total body fat are also related to CVD risk factors in children and adolescents (Daniels et al. 1999; Lindsay et al. 2001; Teixeira et al. 2001). Overall, BMI and DXA-derived adiposity estimates were highly correlated, but not always equivalent at different development stages, with linear and curvilinear relationships between BMI and fat mass and %BF, respectively, in a wide range of adiposity values (Pietrobelli et al. 1998) and in obese individuals (Lindsay et al. 2001). An important finding was that in a large group of overweight and obese children, BMI and DXA-derived estimates yielded similar relationships with several CVD and metabolic risks, such as fasting and 2-hr glucose after an oral glucose tolerance test, insulin, triglycerides, and HDL-Chol (Lindsay et al. 2001). These data further support the usefulness of BMI as a surrogate method to estimate total body fatness and to study the associated health consequences, although in certain pathological conditions and in diverse ethnic groups, more precise adiposity estimates may be necessary.

Most of the data that describe obesity comorbidities are derived from observational studies and short-term interventions. Further evidence is needed from long-term studies to better specify categories of risk associated with childhood obesity (Power et al. 1997). However, the majority of recent empirical evidence suggests that increased adiposity is associated with selective inflammatory CVD risk markers and coronary artery lesions in late adolescence. Attributable largely to the association of BMI with several CVD risk factors, and despite BMI's limitations for discriminating among the different body composition compartments, most studies suggest that BMI and skinfolds can be used as a surrogate of total body adiposity in studies of adiposity and disease in children and adolescents.

Adults

Current definitions of overweight and obesity in adults are well established at BMIs of 25 to 29.9 kg/m² and 30 kg/m² or greater, respectively (NHLBI 1998). For the NIH-specified overweight, the WHO adopted the term *preobese*, with the rationale that some health risks can exist at these BMIs and to increase awareness for strategies to prevent further weight gain beyond this level of adiposity (WHO 1997). Both organizations suggest that a desirable range spans 18.5 to 25 kg/m², which corresponds to the BMI range with minimal morbidity. To improve the risk stratification for CVD, diabetes, and hypertension, the NIH suggested the measurement of waist circumference as a surrogate measure of visceral adiposity (NHLBI 1998). Table 13.2 gives the proposed classification by BMI and the designations of risks associated with waist circumference according to the WHO and NIH. A waist circumference of at least 88 cm in women and 102 cm in men has been associated with increased health risks. The use of a single WC threshold within each BMI category has been questioned (Ardern et al. 2004). In a large cohort of 18,254 men and women, Ardern and colleagues (2004) found that the optimal WC thresholds to predict high risk of coronary events increased across BMI categories from 87 to 124 cm in men and from 79 to 115 cm in women. Because the relationship between BMI and %BF varies across

populations, new criteria have been suggested to define overweight and obesity in Asians (Gill 2001). In a study including Hong Kong Chinese, the risk of CVD risk factors increased at a BMI of about 23 kg/m² (Ko et al. 1999).

The concept of ethnic-specific cutoffs was recently challenged (Stevens 2003). Assuming that the cutoff points for defining overweight and obesity should be based on CVD risk factors, and considering that a recent review of BMI–mortality association did not indicate a difference in Asians compared with Caucasians (Stevens and Nowicki 2003), these authors suggested that the BMI cutoffs to define different excess adiposity categories should not be population-specific. Thus, despite evidence that the BMI–%BF relationship is ethnic-specific, it is unclear whether CVD risk factors tend to increase at similar levels of BMI or %BF in Caucasians and Asians.

Several %BF ranges have been linked to the current definitions of underweight, overweight, and obesity (by BMI) in African American, Asian, and white men and women. Using four-component models to estimate %BF, which accounted for molecular-level differences in composition, Gallagher and colleagues (2000) found that age and ethnicity were significant terms in prediction models relating BMI and %BF. As indicated in table 13.3, for any given BMI up to the age of 59 years, Asians had slightly higher values of %BF whereas African Americans had slightly lower values. These

Table 13.2 Classification of Overweight and Obesity by BMI, Waist Circumference, and Associated Disease Risk

	BMI (kg/m²)	Obesity class	DISEASE RISK[a] (RELATIVE TO NORMAL WEIGHT AND WAIST CIRCUMFERENCE)	
			Men ≤40 in. (≤102 cm) Women ≤35 in. (≤88 cm)	Men >40 in. (>102 cm) Women >35 in. (>88 cm)
Underweight	<18.5		—	—
Normal	18.5-24.9		—	—
Overweight	25.0-29.9		Increased	High
Obese	30.0-34.9	1	High	Very high
	35.0-39.9	2	Very high	Very high
Extremely obese	≥40	3	Extremely high	Extremely high

Note. BMI = body mass index.

[a]Disease risk for type 2 diabetes, hypertension, and cardiovascular disease.

From USDHHS, 2000.

findings imply that, for any given level of adiposity, Asians have a lower BMI value, supporting the new definitions for overweight and obesity in this population. This approach of linking BMI to accurately measured fatness levels in different groups is very important because most epidemiological studies have used BMI as an easy method to estimate adiposity levels and related morbidity. Before these new BMI–%BF relationships were published, no reference values for %BF were known in adults.

In the National Health and Nutrition Examination Survey III data, the disease burden associated with excess adiposity increased with BMI categories across all racial and ethnic subgroups (Must et al. 1999; table 13.4). Moreover, except for high blood levels of total-Chol, the prevalence of all health outcomes increased with increasing severity of overweight and obesity, with the strongest effect in men and women less than 55 years of age. Other databases confirm the utility of BMI for defining health risk in adults. For example, with BMI based on self-reported weight and height, overweight and obese adults in the Behavioral Risk Factor Surveillance System had higher risks for diabetes, high blood pressure, high total-Chol, asthma, arthritis, and fair or poor health status (Mokdad et al. 2003).

The most recent report from the Framingham Heart Study also demonstrates that overweight and obesity are associated with an increased relative risk for CVD risk factors, new vascular disease, and mortality (Wilson et al. 2002; table 13.5). The association was strongest for obesity, with lower relative risks in women. For men and women, the end points most highly associated with overweight and obesity were angina pectoris and total coronary heart disease. Table 13.5 indicates that even though the relative risks are higher in the obese, the population attributable risks are higher for overweight. These data clearly show that at the individual level, obese subjects are at higher risk for the several health-related outcomes, whereas from a population standpoint, considering the relative prevalence of the overweight and obesity categories, the disease burden is higher in the overweight category. The overall higher burden in overweight subjects compared with subjects who have normal or desirable BMI highlights the need to target overweight subjects for primary prevention.

Table 13.3 Predicted Percentage Body Fat by Sex and Ethnicity

Age and BMI	WOMEN			MEN		
	African American	Asian	White	African American	Asian	White
20-39 YEARS						
BMI <18.5	20	25	21	8	13	8
BMI ≥25	32	35	33	20	23	21
BMI ≥30	38	40	39	26	28	26
40-59 YEARS						
BMI <18.5	21	25	23	9	13	11
BMI ≥25	34	36	35	22	24	23
BMI ≥30	39	41	41	27	29	29
60-79 YEARS						
BMI <18.5	23	26	25	11	14	13
BMI ≥25	35	36	38	23	24	25
BMI ≥30	41	41	43	29	29	31

Note. BMI = body mass index. Calculated from the following equation: Percent body fat = $63.7 - 864 \times (1/BMI) - 1.21 - 12.1 \times Sex + 0.12 \times Age + 129 \times Asian \times (1/BMI) - 0.091 \times Asian \times Age - 0.030 \times African\ American \times Age$, centering on the ages of 30, 50, and 70 years.

From D. Gallagher et al., 2000, "Healthy percentage body fat ranges: an approach for developing guidelines based on body mass index," *American Journal of Clinical Nutrition* 72:694-701. Reproduced with permission by the *American Journal of Clinical Nutrition*. © Am J Clin Nutri. American Society for Clinical Nutrition.

Table 13.4 Estimated Odds Ratios for Selected Overweight- and Obesity-Related Morbidity[a]

	Overweight	Obesity Class 1	Obesity Class 2	Obesity Class 3
MEN				
TYPE 2 DIABETES MELLITUS				
<55 years	3.27 (1.17-9.05)	10.14 (4.03-25.08)	7.95 (2.44-25.23)	18.08[b](6.71-46.84)
≥55 years	1.77 (1.26-2.47)	2.56 (1.71-3.74)	4.23 (2.09-7.59)	3.44[b] (1.11-8.32)
GALLBLADDER DISEASE				
<55 years	1.43 (0.50-4.02)	4.08 (1.33-12.21)	6.84 (0.98-41.83)	21.11[b] (4.12-84.15)
≥55 years	1.45 (0.95-2.16)	1.82 (1.15-2.79)	1.70 (0.70-3.67)	2.55[b] (0.56-7.11)
CORONARY HEART DISEASE				
	0.97 (0.76-1.24)	1.59 (1.17-2.11)	1.14 (0.73-1.75)	2.22[b] (0.92-4.22)
HIGH BLOOD CHOLESTEROL LEVEL				
<55 years	1.28 (1.02-1.57)	1.34 (0.98-1.76)	1.37 (0.94-1.90)	1.45[b] (0.93-2.09)
≥55 years	1.18 (1.01-1.35)	1.17 (0.92-1.44)	0.85 (0.53-1.25)	0.88 (0.38-1.59)
HIGH BLOOD PRESSURE				
<55 years	1.62 (1.25-2.05)	2.52 (2.02-3.08)	4.50 (3.34-5.60)	4.60[b] (3.00-6.07)
≥55 years	1.11 (0.96-1.25)	1.35 (1.15-1.51)	1.47 (1.18-1.66)	1.66[b] (1.21-1.80)
WOMEN				
TYPE 2 DIABETES MELLITUS				
<55 years	3.82 (1.75-8.21)	2.49 (1.01-6.12)	10.67 (4.02-27.11)	12.87[b] (5.69-28.05)
≥55 years	1.81 (1.41-2.31)	2.19 (1.56-3.01)	3.24 (2.13-4.67)	5.76[b] (4.17-7.42)
GALLBLADDER DISEASE				
<55 years	1.94 (1.25-3.00)	2.56 (1.62-4.02)	4.33 (2.20-8.12)	5.20[b] (2.92-8.92)
≥55 years	1.34 (1.04-1.70)	2.02 (1.58-2.53)	2.29 (1.69-3.00)	3.04[b] (2.10-4.07)
CORONARY HEART DISEASE				
	1.30 (0.97-1.71)	1.58 (1.19-2.10)	1.74 (1.24-2.42)	2.98[b] (2.07-4.20)
HIGH BLOOD CHOLESTEROL LEVEL				
<55 years	1.90 (1.58-2.25)	1.67 (1.34-2.04)	1.71 (1.29-2.20)	1.68[b] (1.11-2.40)
≥55 years	1.23 (1.11-1.35)	1.10 (0.95-1.25)	1.19 (0.95-1.43)	0.91 (0.57-1.31)
HIGH BLOOD PRESSURE				
<55 years	1.65 (1.23-2.18)	3.22 (2.56-3.98)	3.90 (2.94-4.99)	5.45[b] (4.16-6.78)
≥55 years	1.16 (1.06-1.25)	1.24 (1.15-1.32)	1.42 (1.34-1.48)	1.41[b] (1.26-1.50)

[a]Reference category is individuals with a body mass index (BMI) of 18.5 – 24.9 kg/m^2; 95% confidence interval in parentheses. Weight categories are based on National Heart, Lung, and Blood Institute classification (NHLBI Obesity Task Force). [b]Trend in logistic model, with BMI as a continuous variable, $p < .05$.

Adapted, by permission, from A. Must et al., 1999, "The disease burden associated with overweight and obesity," *Journal of the American Medical Association* 282: 1523-1529. Copyright © 1999 American Medical Association. All rights reserved.

Table 13.5 Relative Risk and Population Attributable Risk Percentage[a]

Outcome	BMI 25-29.9 kg/m²			BMI ≥30 kg/m²			COMPOSITE (BMI ≥25 kg/m²)
	Age-adjusted RR (95% CI)	Multivariable adjusted RR (95% CI)[b]	Population attributable risk (%)	Age-adjusted RR (95% CI)	Multivariable adjusted RR (95% CI)[b]	Population attributable risk (%)	Population attributable risk (%)
MEN							
Angina pectoris	1.42 (1.10-1.84)	1.47 (1.12-1.92)	18	1.85 (1.33-2.57)	1.81 (1.28-2.55)	8	26
Myocardial infarction	1.22 (0.96-1.56)	1.26 (0.98-1.61)	12	1.19 (0.84-1.68)	1.17 (0.82-1.67)	2	14
Hard CHD	1.33 (1.05-1.68)	1.37 (1.08-1.74)	15	1.45 (1.05-1.99)	1.45 (1.04-2.01)	5	20
Total CHD	1.40 (1.17-1.68)	1.43 (1.19-1.73)	17	1.65 (1.30-2.09)	1.58 (1.24-2.03)	6	23
Cerebrovascular disease	1.31 (0.89-1.93)	1.28 (0.86-1.91)	12	1.86 (1.14-3.02)	1.61 (0.98-2.67)	8	20
Total CVD	1.21 (1.05-1.40)	1.24 (1.07-1.44)	10	1.46 (1.20-1.77)	1.38 (1.12-1.69)	5	15
CVD death	1.00 (0.71-1.41)	1.05 (0.74-1.48)	2	0.98 (0.60-1.61)	0.98 (0.59-1.63)	0	2
WOMEN							
Angina pectoris	1.58 (1.21-2.05)	1.42 (1.08-1.86)	13	2.07 (1.53-2.80)	1.63 (1.18-2.25)	9	22
Myocardial infarction	0.96 (0.65-1.42)	0.91 (0.61-1.36)	-3	1.68 (1.10-2.57)	1.46 (0.94-2.28)	8	5
Hard CHD	1.03 (0.72-1.46)	0.98 (0.69-1.41)	-1	1.43 (0.95-2.15)	1.30 (0.85-1.98)	5	4
Total CHD	1.32 (1.07-1.62)	1.22 (0.99-1.52)	7	1.83 (1.43-2.32)	1.54 (1.19-1.98)	8	15
Cerebrovascular disease	1.18 (0.83-1.66)	1.10 (0.77-1.56)	4	1.21 (0.79-1.88)	1.02 (0.65-1.59)	0	4
Total CVD	1.20 (1.03-1.41)	1.13 (0.96-1.33)	4	1.64 (1.37-1.98)	1.38 (1.14-1.68)	6	10
CVD death	0.77 (0.50-1.17)	0.77 (0.50-1.18)	-9	1.67 (1.09-2.56)	1.56 (1.00-2.43)	10	1

BMI = body mass index; RR = relative risk; CI = confidence interval; CHD = coronary heart disease; CVD = cardiovascular disease.

[a]Reference group is normal weight (BMI, 18.5-24.9 kg/m²). [b]Multivariate adjusted for age, smoking, hypertension, hypercholesterolemia, and diabetes.

Adapted from P.W.F. Wilson et al. 2002, "Overweight and obesity as determinants of cardiovascular risk," *Archives of Internal Medicine* 162: 1867-1872.

Another way to analyze the sensitivity of BMI as a marker of excess adiposity is to compare indexes of early coronary atherosclerosis across BMI categories. As indicated in figure 13.1, after adjusting for age, sex, diabetes, hypertension, smoking, and hypercholesterolemia in 111 subjects, Suwaidi and colleagues (2001) found differences ($p < .05$) between the two fatness categories in the plaque plus media area and percent area stenosis of the left anterior descending coronary artery. Different categories of excess adiposity also yielded different levels of coronary atherosclerosis. Because several CVD risk factors were controlled, an important finding of this study was the independent effect of excess adiposity on atherosclerosis, which may imply that the mechanism by which adiposity influences the progression of atherosclerosis is independent of these traditional CVD risk factors (Libby 2002).

Together, the available data support the use of the new international reference guidelines, given that the different categories correspond to added disease risk. However, there is no obvious cutoff point at which the relationship between BMI and disease risks changes in direction. Also, the interaction between BMI and percent BF must be considered. Tanaka and colleagues (2002) recently showed that normal-weight Caucasian men with a BMI between 18.5 and 24.9 kg/m², with increased %BF, had a higher prevalence of CVD risk factors than men with low %BF and had a CVD risk similar to that of overweight subjects. In contrast, normal-weight females with increased %BF did not have a higher CVD risk. Even though adiposity distri-

bution was not considered in this study, a major confounder of the interactions between BMI, %BF, and CVD risk factors, these data underscore that in men, at least, within each BMI category there may be a wide range of %BF values that correspond to different disease risks.

In summary, the most commonly used index in studies of the relationship between total body fat and CVD risk factors is BMI. The current definitions of overweight and obesity using BMI yield different CVD risk, suggesting that in the absence of other more specific methods to estimate total body fat, BMI can be used in adults for total body fatness screening and risk stratification.

Elderly Subjects

Body composition changes attributable to aging are characterized by a progressive reduction in fat-free mass and an increase in the ratio between fat and fat-free mass. Thus, the relation between BMI and %BF may be more closely related within narrow age ranges of older people, and clearer definitions of obesity should be established for this group. Currently, there are no international cutoffs to define overweight and obesity in older women and men. As shown in table 13.3, in subjects aged 60 to 79 years, the overweight and obesity BMI categories developed for adults correspond to 38% BF and 43% BF for Caucasian women and 25% BF and 31% BF for Caucasian men. Slightly lower values were found for African American and Asian men and women (Gallagher et al. 2000). The values for the Caucasian women were close to the 38.7% BF and 45.1% BF values

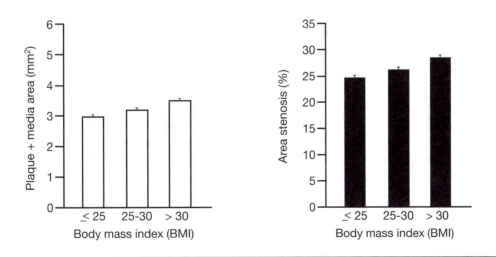

Figure 13.1 Mean ± SE plaque plus media area and percent area stenosis for different body mass indexes (BMIs).

Adapted from *American Journal of Cardiology*, 88, J.A. Suwaidi et al., Association between obesity and coronary atherosclerosis and vascular remodeling, 1300-1303, Copyright 2001, with permission from Excerpta Medica Inc.

found in another recent study (Blew et al. 2002). Table 13.3 shows that for the same BMI category, %BF increases with increasing age. This shift to the right in the BMI–%BF relationship with aging needs further analysis to reveal the implications of using, in elderly subjects, standards developed for young adults. Indeed, considering that obesity definitions should rely on excess accumulation of adiposity, using the same cutoffs in elderly people would implicate a definition based on a higher and more liberal %BF value.

The usefulness of the NIH BMI cutoffs for diagnosing obesity in older women was recently challenged (Blew et al. 2002; Sardinha and Teixeira 2000). When the authors considered several %BF values, sensitivity of the 30 kg/m^2 cutoff point was very low, indicating a low probability of BMI to classify a woman as obese when she is truly obese. On the other hand, specificity was high, indicating that BMI will most likely not classify a nonobese woman as obese. The best tradeoff between sensitivity and specificity to screen for obesity in these older women was found to be 24.9 kg/m^2 (Blew et al. 2002). Using DXA, Sardinha and Teixeira (2000) reported that a similar BMI value (25.5 kg/m^2) was the best cutoff to screen

for obesity in older women. According to these results, a BMI of 30 kg/m^2 may be too conservative, and an alternative value of 25 kg/m^2 may be a better criterion to diagnose obesity in older women. For men, there are no available data to support a specific value different from the one suggested for adults. However, considering that the age-related increase in the ratio between fat and fat-free mass also occurs in men, it is likely that the sensitivity to screen for obese men would be higher if a value closer to 25 kg/m^2 were adopted.

In older persons, little attention has been devoted to the relationship of excess adiposity and morbidity, and there is a paucity of data to assess the prognostic importance of increased adiposity, based on recommendations for healthy weight. The actual clinical guidelines defining risk stratification based on BMI and waist circumference were assessed in the prospective cohort of the Iowa Women's Health Study (Folsom et al. 2000). Overall, the results showed for any given waist circumference, the risk for diabetes, hypertension, and CVD increased across BMI categories for overweight and obesity (table 13.6). When waist circumference was 88 cm or less, BMI alone

Table 13.6   Age-Adjusted Odds Ratios of Any Designated End Point During Follow-Up in Relation to Categories of BMI and Waist Circumference for the IWHS Compared With the Obesity Panel Consensus Risk Category[a]

| | | EMPIRICAL RISK IN THE IWHS | | | |
| | | FOR DIABETES, HYPERTENSION, OR CVD-RELATED DEATH | | FOR DEATH, CANCER, DIABETES, HYPERTENSION, OR HIP FRACTURE | |
Weight status	BMI	Waist circumference ≤88 cm OR[b] (95% CI)	Waist circumference >88 cm OR[b] (95% CI)	Waist circumference ≤88 cm OR[b] (95% CI)	Waist circumference >88 cm OR[b] (95% CI)
Underweight	< 18.5	0.92 (0.7-1.3)	—	1.5 (1.2-2.0)	—
Normal	18.5-24.9	1.0	1.2 (1.1-1.3)	1.0	1.2 (1.1-1.3)
Overweight	25.0-29.9	1.3 (1.2-1.4)	1.9 (1.7-2.1)	1.1 (1.1-1.2)	1.6 (1.5-1.7)
Obesity, Class					
1	30.0-34.9	1.5 (1.1-2.0)	2.4 (2.2-2.7)	1.3 (1.0-1.7)	1.9 (1.8-2.1)
2	35.0-39.9	—	3.1 (2.6-3.7)	—	2.4 (2.1-2.9)
3	≥ 40	—	4.2 (3.1-5.7)	—	3.7 (2.7-5.0)

Note. OR = odds ratio; BMI = body mass index; IWHS = Iowa Women's Health Study; CVD = cardiovascular disease; CI = confidence interval.

[a]Obesity panel consensus risk category; [b]Age-adjusted OR of any end point relative to normal weight and waist circumference.

Adapted, by permission, from A.R. Folson et al., 2000, "Associations of general and abdominal obesity with multiple health outcomes in older women: the Iowa Women's Health Study," *Archives of Internal Medicine* 160(14): 2117-2128. Copyright © 2000 American Medical Association. All rights reserved.

was not a good predictor of the pooled risk for cancer, hip fracture, and death. However, cross-classifying BMI with waist circumference identified increased risk for several diseases and death. The age-adjusted odds ratios in this cohort of older women indicate that BMI, as a marker of general obesity, is specific to identify increased risk in older women with higher abdominal obesity.

A recent study analyzed the application of the guidelines for overweight and obesity; Heiat and colleagues (2001) conducted a comprehensive review of the available data that reported the relationship between BMI, cardiovascular mortality, and all-cause mortality in older adults. Most studies did not show a significant association between an elevated BMI and increased mortality rate. This relationship was even less conclusive when CVD was considered. In the few studies that showed a U-shaped or positive relationship between BMI and mortality, only BMI values higher than those suggested to identify the overweight category were associated with increased mortality rate. In the elderly, the U-shaped curve specifying the BMI–mortality relationship tends to have a large flat bottom and a right curve that starts to increase for BMIs greater than 31 to 32 kg/m^2 (Heiat et al. 2001). Given this general trend, the current guidelines for adults are very restrictive, especially the overweight category. Older people in the overweight category with BMIs between 25 and 29.9 kg/m^2 seem not to have the same overall risks as younger adults in the same category. An article not included in this review confirmed that a wide range of BMIs are associated with similar risk for mortality and stroke, especially in women (Wassertheil-Smoller et al. 2000). In the Systolic Hypertension in the Elderly Program study, the lowest 5-year adjusted mortality and stroke rates corresponded to a BMI of 25.8 kg/m^2 in men and 29.6 kg/m^2 in women. Especially among men with hypertension, overweight conferred increased risk for stroke and for mortality. In addition to this sex difference for mortality risk at any given BMI, the relative risk with a greater BMI tends to decline with age for both men and women (Stevens et al. 1998).

In conclusion, because of changes in body composition with aging, current BMI cutoff values to define overweight and obesity in adults should be used with caution in older persons. For any given BMI, increased loss of fat-free mass may mask increased total adiposity in the elderly. In the elderly, compared with younger adults, the association between indexes of excess adiposity

with several clinical outcomes and mortality is much weaker. Because most studies used BMI, which is a less specific index of total body fatness in older compared with younger adults, there is a need for more studies with more valid estimates of total body fatness and its relationship with clinical outcomes and mortality in older subjects. No studies are available that have simultaneously assessed %BF, CVD risk factors, and mortality in older people. There is a clear need for this type of data to overcome the limitations of BMI in elderly subjects.

Fat Distribution

In this section, we describe the most common regional measures, with an emphasis on established health relationships, followed by a discussion of cutoff values for those variables with evidence suggesting critical thresholds of increased risk.

Measures of Fat Distribution

The term *fat distribution* refers to the relative amounts of fat in the primary compartments where adipose tissue and fat are stored in the body. Improvements in body composition assessment methods have made it possible to measure fat in nonadipose tissue locations, such as the liver and the muscle. *Ectopic* fat has been used to designate these nonadipose fat deposits (Goodpaster 2002). Assessing fat distribution involves measuring one site or variable in relation to another so that a dichotomous fat distribution "type" can be identified (e.g., a contrast or a ratio), as in a gynoid versus android fat distribution. Alternatively, *regional fat* typically represents a single variable as in total abdominal fat or subcutaneous adipose tissue. Frequently, regional anthropometric measures (e.g., the waist and the hip circumferences) are used to establish a fat distribution category (e.g., upper or lower fat distribution pattern). *Fat patterning* is often used interchangeably with fat distribution and regional fatness. Although the distinction between fat distribution, regional fat, and fat patterning may be useful in some cases, it is presented here for clarification only. In this section, the terms are used interchangeably, similarly to what is common in the field.

The use of any of the previous terms reflects an important distinction in the study of body composition, parting from analyses that do not consider where the fat is stored. Given that central, abdominal fat is more closely related to ill health than any other fat deposit, the study of fat distribution has been more frequently related to this feature of body morphology and composition (i.e., the degree of fat *centralization*) than with other depots. In the future, new and similarly important relationships may be established between other fat distribution measures and health, such as studies involving liver fat or lipid contained around and within muscle (Kelley et al. 2002).

Fat distribution and regional fat can be described in several ways, depending on the objective and also on the techniques used for measurement (table 13.7). Recently, it has been recommended that the expression *adipose tissue* be used whenever tissue-based measurements are made, as when using imaging techniques, as opposed to *fat,* which implies a measurement at the chemical level, such as that performed by DXA (Shen et al. 2003).

Anthropometric Methods

Anthropometric measures of fat distribution, particularly of abdominal adiposity, are readily made and widely used. Molarious and Seidell (1998) systematically reviewed several of these measures, giving a comprehensive analysis of their advantages and disadvantages.

Waist–Hip Ratio (WHR)

The ratio between the waist and hip circumferences was the first measure used to indicate fat distribution and has been among the most widely employed. The rationale for WHR was to adjust the waist circumference for body build, creating an index that would identify people with upper- versus lower-body fat preponderance. In epidemiological studies, WHR has been found to be associated with disease risk and with mortality in both men and women (Folsom et al. 1989, 2000; Lapidus et al. 1989; Larsson et al. 1984). One salient feature of the WHR as a health-related measure is that it is partially independent of total adiposity; thus, for a given level of WHR, there may be considerable variability in total body adiposity (Pouliot et al. 1994). Importantly, increases in adiposity, including abdominal adiposity, may occur that will not be detected by WHR, if increases in hip circumference accompany increases in waist circumference (Després et al. 2001). Thus, to the extent to which total adiposity per se is related to health, the WHR may not capture that association. The fact that a WHR-based fat distribution pattern is a predictor of health risk, partially independent of total fatness, can also be considered as an advantage. For instance, in the absence of large total fat stores, the WHR may be particularly selective in detecting hormonal disturbances linked to disease (e.g., involving cortisol and growth hormone) or other metabolic abnormalities (e.g., high serum triglycerides and insulin resistance) associated with a relative increase in abdominal size but not reflected by total fat measures (Seidell and Bouchard 1997).

Findings that the WHR was not as predictive of visceral fat content as other simple measures, such as the waist circumference and the abdominal sagittal diameter (Després et al. 1991; Ferland et al. 1989; Pouliot et al. 1994), have led to a preference toward these latter measures, especially waist circumference, to assess abdominal adiposity in clinical and field settings (NHLBI 1998). Overall, waist circumference and the sagittal diameter are better predictors of risk than WHR, in part because they are simultaneously correlated with whole-body and visceral adiposity. Because waist

Table 13.7   Common Variables Used to Indicate Regional Fatness and Fat Distribution

Regional fatness	Fat distribution and patterning
Total abdominal fat	Upper-body vs. lower-body fat
Subcutaneous adipose tissue	Visceral or internal vs. subcutaneous adipose tissue
Intra-abdominal or visceral adipose tissue	Abdominal vs. nonabdominal fat
Legs, arms, and trunk fat	Central vs. peripheral fat
Waist circumference	Gynoid vs. android fat pattern
Abdominal sagittal diameter	Trunk vs. limbs (arms, legs) fat
Single/regional (e.g., trunk) skinfolds	Waist–hip, waist–height, waist–thigh ratios[a]
Ectopic fat (e.g., intermuscular)	Skinfold ratios (e.g., triceps–subscapular)[a]

[a]Although they are ratios and provide an index of fat distribution (e.g., upper vs. lower), these single variables have face validity and are expressed in a continuous manner and thus can also be considered as a direct indicator of regional fat.

circumference is the numerator in the WHR equation, the two measures are necessarily intercorrelated. A recent study attempted to separate the waist and hip associations with cardiovascular risk in 8,400 men and women, concluding that the two variables have independent and opposite effects on health (Snijder et al. 2004). Individual variations in the WHR may be more complex than once assumed, because they likely represent variations in lower-body morphological factors (fat, bone, and muscle) as well as differences in visceral and abdominal fat, which are not neutral in their associations with health outcomes (Seidell et al. 2001). Although the mechanisms through which a larger hip circumference is protective are unclear, other researchers have noted that lower-body fat may carry a lower risk than upper-body fat (Hunter et al. 1997; Terry et al. 1991). The issue of whether leg or lower-body adiposity is protective to health remains an interesting and unresolved question.

Waist Circumference

The use of imaging techniques to measure abdominal and intra-abdominal adipose tissue provided the means to analyze relationships between anthropometric and more direct measures of fatness and distribution. The majority of studies have shown that waist circumference has a closer association with cardiovascular risk factors and with other health indicators than does the WHR (Wajchenberg 2000). Waist circumference also proved to be a better marker of visceral fat than WHR and other measures (Hill et al. 1999; Rankinen et al. 1999). The shared variance of the waist circumference with total fat mass (or BMI) is typically above 80%, whereas the correlation coefficient of the waist circumference with visceral fat is about .80 (Pouliot et al. 1994; Seidell et al. 2001). Notably, waist circumference remains statistically associated with visceral, subcutaneous, and total fatness, even after adjustment for BMI and WHR (Seidell and Bouchard 1999).

Given the strength of findings, both the WHO (1997) and the NHLBI (1998) now recommend measurement of waist circumference for obesity-related risk assessment. For example, an adult woman with a BMI between 30 and 35 kg/m² (Class 1 obesity) is expected to have a high or very high risk for diabetes, hypertension, and cardiovascular disease depending on whether her waist circumference is either below or above 88 cm (35 in.). The waist circumference cutoff of 102 cm (40 in.) was adopted for adult men (see table 13.2). Some data have shown that waist circumference

corrected for height (waist–height ratio, WHtR) is more closely related to the risk for coronary disease than waist circumference (Ashwell et al. 1996) or WHR (Hsieh and Yoshinaga 1995) and that WHtR prospectively predicts mortality better than the waist circumference, at least in men (Cox and Whichelow 1996). However, the use of WHtR has been criticized because the two variables (waist and height) are virtually uncorrelated (Han et al. 1996), whereas the slight statistical superiority of the WHtR to predict morbidity and mortality (over waist circumference) is attributable to a negative association of height with morbidity (Molarius and Seidell 1998). In fact, the use of ratios (e.g., WHR, WHtR) in statistical analyses has been criticized because they are more difficult to interpret biologically, are less sensitive to change, and may also be more prone to spurious results than the use of single variables (Allison et al. 1995; Molarius and Seidell 1998).

Abdominal Sagittal Diameter

Sagittal diameter is defined as the shortest distance between the anterior and posterior trunk at the level of the iliac crest or lowest trunk site. Sagittal diameter can be measured with a Harpenden caliper or a CT or MRI scan. Kvist and colleagues (1988) showed that sagittal diameter is highly correlated with the visceral fat volume (r = ~.93) in subjects with varying BMIs. Correlation coefficients between sagittal diameter and visceral fat area have varied between approximately .50 and .90 (Kvist et al. 1988; Pouliot et al. 1994; Van der Kooy et al. 1993). The sagittal diameter may be more predictive of visceral and subcutaneous adiposity in leaner individuals than obese individuals (Busetto et al. 1992; Ferland et al. 1989). Some studies have shown sagittal diameter to be superior to the waist circumference and WHR in predicting metabolic risk factors (Ohrvall et al. 2000; Richelson and Pederson 1995) and visceral fat content (Kvist et al. 1988; Sjostrom 1991). Sagittal diameter was a strong prospective predictor of coronary and all-cause mortality in middle-aged men followed for about 20 years in Baltimore (Seidell et al. 1994) in a study where BMI, skinfolds, and WHR did not predict outcomes. In a study of Swedish men and women aged 19 to 66 years (average, 41.0 years), with mean BMIs of 24.9 and 23.8 kg/m², respectively, sagittal diameter was slightly more predictive of a "metabolic risk" index, which included triglycerides, HDL-Chol, glucose, insulin, plasminogen activator inhibitor-1, and blood pressure, and also "total risk" (including several metabolic and other risk factors) than waist circumference or WHR (Ohrvall et al. 2000). Despite these results, the utility of this measure as a diagnostic tool for abdominal obesity, compared with the simpler waist circumference, is questionable because the correlation between the two measures (sagittal diameter and waist circumference) is generally high (Richelson and Pederson 1995) and waist circumference is easier to assess and is at least a comparable predictor of health outcomes.

A number of other indicators of fat distribution have been proposed, including skinfolds (Freedman et al. 1995; Mueller and Stallones 1981; Sardinha et al. 2000), the index of conicity (Valdez et al. 1993), and the abdominal sagittal diameter, adjusted for thigh circumference (Kahn 1993) or transverse abdominal diameter (Rissanen et al. 1997). Because there are few data suggesting superiority of any of these indexes, the WHR, waist circumference, and sagittal diameter are generally supported for their diagnostic ability related to health outcomes.

Imaging Methods

Imaging techniques, such as CT, MRI, and DXA, can be used to assess fat distribution in humans. An advantage of imaging techniques is that they directly measure fat tissue in selected regions, whereas other techniques can only estimate fat content by relying on known relationships or estimation equations. Imaging methods are generally more expensive and cumbersome, are more dependent on advanced technician training, and in some cases (CT and DXA) expose subjects to significant doses of radiation. However, these methods provide reliable and valid measures of in vivo fat content, in addition to other body composition measures (e.g., bone density, regional and total muscle mass, organ composition), which may warrant the use of these methods when available. Recent technological advances allow quantification of ectopic fat (Goodpaster 2002) and permit the subdivision of subcutaneous fat into deep and superficial regions (Kelley et al. 2000). In fact, because of imaging technology, a new taxonomy of whole-body and regional adipose tissue based on anatomical location has been proposed (Shen et al. 2003).

CT and MRI

These methods are often considered the gold standard procedures to measure adipose tissue compartments and sometimes total body adiposity (see chapter 7). These methods have been critical to efforts to quantify fat in the abdominal region (Van Loan 1996). Although they are based on different physical properties (proton movement in the case of MRI and defined attenuation or Hounsfield values for CT), the two procedures give comparable outcomes (Seidell et al. 1990; Van der Kooy and Seidell 1993), and both are reproducible and reliable (Ross et al. 1993; Thaete et al. 1995). Unlike MRI, CT scans use ionizing radiation that is absorbed by the body, thus limiting the widespread application of CT, particularly in children and women of childbearing age.

A single slice or several slices can be obtained. One scan, if strictly defined, is sufficient to determine fat tissue area, whereas several scan slices are needed to estimate fat tissue volume. A single scan at the L4-L5 or umbilicus level is typically used to measure visceral and subcutaneous abdominal fat areas. The ratio of visceral to subcutaneous fat has been used as an index of abdominal fat distribution; it distinguishes between a more visceral profile (ratio $\geqslant 0.4$) or a more subcutaneous pattern (ratio <0.4), the former being more closely linked to metabolic abnormalities, independent of sex, age, and BMI (Fujioka et al. 1987). Advances in measurement of internal as well as subcutaneous adipose tissue with CT and MRI allow investigators to study several compartments that until recently have been indistinguishable, for example, different aspects of visceral adipose tissue, such as intrathoracic and intra-abdominopelvic areas, the latter being composed of intra- and extraperitoneal compartments and possibly relating differently to metabolic factors (Marin et al. 1992). Deep and superficial layers of subcutaneous adiposity have also been investigated with CT and MRI, as discussed later in this text.

DXA

Because of its increased accessibility, DXA has been used in numerous studies of different issues, such as exercise, obesity and weight loss, bone health, AIDS, and sarcopenia. Because a whole-body DXA scan can be analyzed in predetermined (default) or user-determined regions of interest, it is possible to analyze the composition of selected areas. Legs, arms, and trunk are the most common regions to undergo DXA, but the operator can further define regions of interest as desired. Svendsen and colleagues (1993b) proposed guidelines to evaluate different fat regions within the trunk and abdomen. Kelley and colleagues (2000) found that total abdominal fat measured by DXA is highly correlated ($r = .97$) with total abdominal tissue area measured by CT, confirming previous reports in men and women (Jensen et al. 1995; Svendsen et al. 1993a). Treuth and colleagues (1995) found DXA total trunk fat to explain about 70% of the variance in intra-abdominal fat in a larger female sample, a coefficient of determination similar to what was observed for total, arm, and several trunk regions assessed by DXA in relation to the intra-abdominal CT measure. DXA measures total fat (not adipose tissue per se) within a certain region of interest, which is different from the adipose tissue areas (or volume) obtained by CT and MRI. Also, region delimiters set by an operator may not follow true anatomical boundaries (e.g., fascial tissue) and are thus more prone to technical error and may include, within a particular region (trunk), adipocytes with distinct histological and metabolic characteristics. Thus, comparisons between variables using these different methodologies (DXA vs. CT or MRI) are not completely appropriate. Nevertheless, DXA has great potential for regional fat measurements because of its high precision, relatively low radiation exposure, and capacity to quantify regional fat masses. Several studies have reported associations between DXA-based regional fat measures and health indicators (Rissanen et al. 1997; Sardinha et al. 2000; Walton et al. 1995), generally showing that trunk fat was the best predictor of metabolic end points, in some cases independently of total fatness.

Ultrasound

Ultrasound has been used to estimate adipose tissue depth in both the subcutaneous and internal abdominal layers. Although B-mode ultrasound and anthropometry are both valid methods for estimating subcutaneous adiposity, ultrasound is more reliable and overcomes the difficulty of measuring skinfolds with a manual caliper in obese individuals, where the procedure is less accurate and often not practical (see chapter 8). Ultrasound has also been used to assess deep abdominal tissue. Correlation coefficients between ultrasound and CT measures of intra-abdominal adiposity are highly significant, ranging from .70 to .85 in most cases (Wajchenberg 2000). In obese women, Armellini and colleagues (1993) showed ultrasound to be superior to both sagittal diameter and WHR in distinguishing three groups split by tertiles of intra-abdominal abdominal adiposity (by CT).

Although ultrasound techniques have been used for more than 2 decades, few studies have evaluated the associations between ultrasound-based measures and health outcome. One study with 191 Brazilian men showed intra-abdominal thickness by ultrasound to be a better predictor of several cardiovascular risk factors than waist circumference and sagittal diameter (Wajchenberg 2000). Cutoffs for ultrasound intra-abdominal measures were associated with moderate and high risk for elevated total-Chol and low HDL-Chol, high triglycerides, and elevated blood pressure. Replication of these data in larger samples would support wider use of ultrasound to assess risk from deep abdominal obesity.

Magnetic Resonance Spectroscopy (MRS)

Volume localized magnetic resonance spectroscopy has recently been used to quantify the lipid content of nonadipose tissues (Goodpaster 2002). MRS is able to distinguish between intramyocellular and extramyocellular lipid in vivo, using conventional whole-body magnets under a predetermined field strength. Despite limitations and a standard measurement error of about 6% (Boesch et al. 1997), associations between intramyocellular lipid and metabolic variables such as insulin resistance have been analyzed (Jacob et al. 1999). This is a very recent and noninvasive methodology that may provide important information about content, distribution, and metabolism of ectopic fat in tissue and organs of the human body.

Risk Classification Based on Fat Distribution

This section reviews proposed cutoff values for measures of fat distribution and evaluates the criteria on which they were based for children, adults, and elderly people. Despite the relatively large number of variables reflecting fat distribution, cutoff points have been empirically tested for only a few measures, namely WHR, waist circumference, and visceral adipose tissue area, primarily in adults. All three variables are indicative of abdominal obesity, highlighting the fact that this characteristic of body composition has been the most extensively studied in relation to health outcomes, because it has an established rationale for increased risk (Bjorntorp 1990; Kissebah and Krakower 1994).

Whether different cutoffs should be adopted for different groups remains an open issue. Theoretical shortcomings of using generalized cutoff points for risk assessment must be recognized. One limitation is the absence of an obvious point at which health risk increases dramatically for a given change in body habitus (Ho et al. 2001). Thus, the cutoff must be statistically defined based on the sensitivity and specificity of the measure related to the outcome. A second difficulty relates to the need for cutoff values that are applicable to large groups, in order to be practical and usable in a generalized way. However, age, sex, ethnicity, and the health outcome of interest, among other factors, may cause the best cutoff values to vary substantially (Després et al. 2001; Dobbelsteyn et al. 2001; Molarius et al. 1999). Thus, recommendations may need to be group-specific.

Children and Adolescents

Although chronic diseases are infrequent in youth, risk factors are present and track from childhood to adult life (Goran and Malina 1999; Rolland-Cachera et al. 1990). Specific criteria to define risk are more difficult to establish in children and adolescents than in adults. In some cases, risk categories have been defined statistically, based on a given percentile of the distribution for a specific outcome variable (Williams et al. 1992), either replicating known adult criteria (Higgins et al. 2001) or establishing criteria based on comparisons between different measures of fat distribution (Taylor et al. 2000).

As in adults, increased abdominal adiposity in children and adolescents is associated with a worsening of several metabolic and cardiovascular health variables, whether fat distribution is measured with anthropometry (Asayama et al. 1995; Bellu et al. 1993; Flodmark et al. 1994; Higgins et al. 2001; Maffeis et al. 2001; Sangi et al. 1992; Savva et al. 2000; Suter and Hawes 1993; Teixeira et al. 2001) or with more advanced imaging techniques (Brambilla et al. 1994; Caprio 1999; Caprio et al. 1996; Goran et al. 2001; Gower et al. 1998, 1999; Herd et al. 2001; Huang et al. 2002). Waist circumference has been consistently associated with adult metabolic risk factors in children and adolescents. Higgins and colleagues (2001) looked at the best waist circumference cutoff points to detect the presence of one or more of five adult CVD risk factors (fasting insulin, HDL-Chol, LDL-Chol, triglycerides, and total–HDL-Chol) in 87 prepubertal African American and white children aged 4 to 11 years. Children with a waist circumference of 71 cm or more were 14 times more likely to have a negative risk profile than a normal one (Higgins et al. 2001). In obese girls 9 to 15 years old, Asayama and colleagues (2000) found that a value of 77 cm for waist circumference showed the best combination of sensitivity and specificity to

detect metabolic abnormalities. To our knowledge, these are the only published critical values for fat distribution in children based on measured biological end points. Cross-validation of these results is necessary with other pediatric samples and other end points such as lipoproteins and insulin resistance.

Freedman and colleagues have clearly shown the negative impact of fat centralization in children and adolescents (Freedman et al. 1987, 1989; Kikuchi et al. 1992). They have consistently found central fat, assessed by trunk skinfolds and trunk–peripheral skinfold ratios, to predict lipids, lipoproteins, and insulin sensitivity. A number of reports have indicated that trunk skinfolds may be a particularly sensitive tool to detect metabolic abnormalities in children and adolescents (Asayama et al. 1995; Bellu et al. 1993; Maffeis et al. 2001; Sangi and Mueller 1991; Sangi et al. 1992; Teixeira et al. 2001). In these studies, individual trunk skinfolds (or the sum of several skinfolds) were stronger predictors of metabolic variables than was waist circumference. For example, in a sample of Portuguese boys and girls (10-16 years old), the sum of the abdominal, subscapular, and suprailiac skinfolds and DXA trunk fat were the two best correlates of CVD risk factors, whereas waist circumference and WHR were not as good predictors of metabolic outcomes (Teixeira et al. 2001).

The association of central adiposity with risk factors is clearly stronger for obese children and adolescents than for normal-weight individuals (Caprio et al. 1996; Freedman et al. 1989; Teixeira et al. 2001). Figure 13.2 shows the distinct association of trunk skinfolds with triglycerides in lean and obese children and adolescents. From the same study, age-adjusted associations between anthropometry- and DXA-derived fat distribution variables and serum outcomes showed an interesting heterogeneity in the relationships, dependent on total fatness and cholesterol-containing blood lipids; apolipoprotein-B was only predicted by fat distribution in the intermediate fatness levels, whereas triglycerides were significantly correlated with most fat distribution measures only in the fatter boys and girls (Teixeira et al. 2001).

Triglycerides may play a particularly selective role as a marker for metabolic disturbances (Després et al. 2001), although this has not been established in younger individuals. In obese children and adolescents, triglycerides have consistently been predicted by visceral adipos-

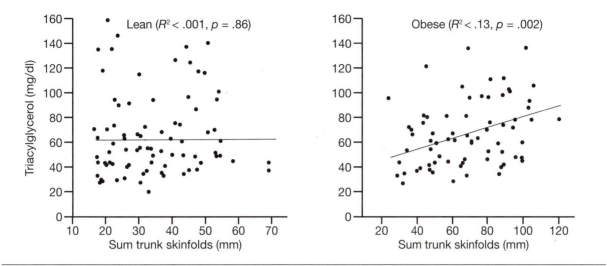

Figure 13.2 Distinct associations for lean and obese children of trunk skinfolds with serum triglycerides.

Reprinted, by permission, from P.J. Teixeira et al., 2001, "Total and regional fat and serum cardiovascular disease risk factors in lean and obese children and adolescents, *Obesity Research* 9(8): 432-442.

ity (Brambilla et al. 1994; Caprio 1999; Caprio et al. 1996) and also by anthropometric and DXA measures of abdominal fatness in obese (Bellu et al. 1993; Flodmark et al. 1994) but not in lean groups (Maffeis et al. 2001; Suter and Hawes 1993; Teixeira et al. 2001). Studies measuring visceral adipose tissue (VAT) in children and adolescents have shown intra-abdominal AT to be variable across BMI levels (Fox et al. 1993; Goran et al. 1997) representing only 5% to 20% of the total abdominal AT, the remainder being subcutaneous fat (Brambilla et al. 1994; Caprio et al. 1995, 1996; Fox et al. 2000; Owens et al. 1998). The ratio of visceral to subcutaneous adipose tissue is also lower in obese youth compared with the nonobese (Brambilla et al. 1994; Fox et al. 1993), indicating that abdominal obesity in younger individuals is primarily a function of increases in the subcutaneous fat layer.

Studies with obese children and adolescents have shown that VAT is generally more strongly associated with serum cardiovascular risk factors (Brambilla et al. 1994; Caprio et al. 1996; Owens et al. 1998) and glucose–insulin homeostasis (Caprio et al. 1995; Gower et al. 1998) than is subcutaneous abdominal fat. Interestingly, more recent evidence indicates that in children, abdominal adiposity is related primarily to fasting insulin and less to insulin sensitivity, which may be more determined by total fatness (Goran et al. 2001; Gower 1999). In the first longitudinal trial investigating fat distribution assessed by CT and changes in insulin levels and insulin response in children, Huang and colleagues (2002) showed that VAT changes and

initial abdominal subcutaneous adipose tissue both independently predicted insulin changes over time. It was postulated that a time-dependent causal effect of subcutaneous, but not visceral, fat could be present. Contrary to the prepubertal years, in adolescence, VAT is also a significant predictor of insulin sensitivity (Gower 1999). Nevertheless, the collective results suggest that the subcutaneous abdominal fat depot contributes to the relationship between fat distribution and negative health risk.

In summary, in obese children and adolescents, an enlarged abdomen and a higher amount of trunk fat are consistently related to a deterioration of lipid and metabolic profiles. Thus, the primary aim should be to identify obese children and adolescents. Fat distribution can then be measured directly by trunk skinfolds, by the ratio of trunk to peripheral fat (e.g., by skinfolds), or by waist circumference. In the pediatric population, trunk skinfold thickness may in fact be the best correlate of visceral fat content, because it is slightly more predictive than waist circumference (Goran 1998). The waist circumference is a good correlate of subcutaneous and total trunk fat across a large spectrum of BMIs and may also serve as a surrogate of markedly enlarged visceral adipose depots in obese children. Although the risk prediction ability of waist circumference, compared with that of skinfolds, may be dependent on individual variation in VAT content, cutoffs are available for the waist circumference, and they can be a useful tool in large-scale screening protocols, once cross-validated. In leaner as well as in obese children and adolescents, subcutaneous trunk skinfolds can be used to detect centralized body fat distribution, with less regard for whether it is located internally or subcutaneously.

Adults

Comprehensive reviews covering the association between fat distribution and health in adults are available (Kissebah and Krakower 1994; Wajchenberg 2000). Despite numerous studies, there is little evidence for the establishment of cutoff points for regional fat or fat distribution. Bjorntorp (1985) was the first to present specific threshold values for increased risk, based on the WHR, for middle-aged men (1.00) and women (0.80). Two 12-year longitudinal studies of cardiovascular disease and mortality supported that larger WHRs were deleterious in men (Larsson et al. 1984) and women (Lapidus et al. 1984). These cutoffs were apparently determined based on a visual inspection of association between the WHR and health outcomes in the two cited trials. Also, hip circumference was measured at the level of the iliac crest, an important fact considering that this is usually not the anatomical site of choice for this variable. In 1987, Bray recommended the use of 1.00 and 0.90 cutoffs for the WHR, for men and women, respectively, citing the reference data that were used by Bjorntorp. Bray's thresholds have received widespread attention and have often been adopted as the "official" cutoff points for the WHR. Other cutoffs have since been proposed, namely by the USDA (0.95 for men and 0.90 for women) and in other samples from different countries (Dobbelsteyn et al. 2001; Ito et al. 2003; Lin et al. 2002; USDA 1990).

Responding to the need for a sound biological rationale for the association between a larger WHR and ill health, multiple studies in the 1990s compared the waist circumference and related indexes with CT-derived measures of intra-abdominal fat. A mechanistic link between an enlarged visceral fat depot and metabolic abnormalities was formally proposed (Bjorntorp 1990) and has since been widely regarded as plausible. Després' group in Canada has extensively examined the associations of intra-abdominal adiposity and health, and his team was responsible for the first report indicating cutoff values for WHR and waist circumference (and also for the sagittal diameter), based on critical levels of visceral fat (Lemieux et al. 1996; Pouliot et al. 1994). In 403 men and women of varying age and obesity level, the WHR cutoffs corresponding to the selected threshold of vis-

ceral fat area (130 cm^2) were 0.94 for men and 0.88 for women (Lemieux et al. 1996). Although these values agree well with the ranges for men (0.90-1.00) and women (0.80-0.90) proposed by others, the authors expressed concerns that the association of the WHR with visceral adipose tissue area was age-, sex-, and obesity-dependent. The authors noted that such was not the case for waist circumference, described in this database as influenced by age but not by sex or obesity (Lemieux et al. 1996).

Recommended cutoffs for abdominal obesity, based on WC alone, were published by Lean and colleagues (1995) using a random sample of Scottish men and women. They identified the best WC values that simultaneously corresponded to both a high BMI and a high WHR, concluding that the action levels of 102 cm (for men) and 88 cm (for women) should be used to advise patients to reduce weight. These represented the upper action level, with a lower one (94 cm for men and 80 cm for women) proposed as a warning threshold above which individuals, particularly young men, should gain no further weight (Lean et al. 1995). These cutoffs were cross-validated in a larger sample of Dutch men and women, confirming their high sensitivity and specificity values for BMI and WHR thresholds (Han et al. 1995). In addition, this study also showed a strong association of WC action levels with several cardiovascular risk factors, providing support for the use of the proposed WC critical values. Table 13.8 summarizes the criteria and samples used to define cutoff points based on the WHR and WC. Generally, the use of the WHR has not been advocated, in addition or alternatively to the WC or BMI.

Evidence is available for the use of WC cutoffs, showing that a greater health risk is associated with higher WCs for normal-weight, overweight, and obese men and women (Ardern et al. 2004; Janssen et al. 2002). Others have also tested the WC cutoffs in different samples, generally confirming their capacity to identify patients at increased cardiovascular risk (Carroll et al. 2000; Folsom et al. 2000; Okosun et al. 2000). However, the usefulness of the WC as a clinical diagnostic tool, in addition to the BMI and assessment of CVD risk factors, as currently advocated (NHLBI 1998), has been challenged by observations that relatively few people have a low BMI and a high WC (and vice versa) and that most individuals are correctly classified within categories of a treatment algorithm using BMI and information on CVD risk factors (Folsom et al. 2000; Kiernan and Winkleby

Table 13.8   Authors Who Have Proposed Cutoffs for the Waist–Hip Ratio and Waist Circumference

			CUTOFF	
Author	Sample size	Criteria	Men	Women
WHR				
Bjorntorp 1985	792 (M), 1,462 (F)	CVD risk and death	1.00	0.80
Bray 1987	792 (M), 1,462 (F)	CVD risk and death	1.00	0.90
Pouliot et al. 1994	81 (M), 70 (F)	Metabolic risk factors	1.00	0.80
Lemieux et al. 1996	213 (M), 190 (F)	Absolute VAT level	0.94	0.88
Dobbelsteyn et al. 2001	4,951 (M), 4,962 (F)	CVD risk factors[c]	0.90	0.80
Lin et al. 2002	26,359 (M), 29,204 (F)	Metabolic risk factors	0.85	0.76
Ito et al. 2003	768 (M), 1,960 (F)	Metabolic risk factors	0.90	0.80
WAIST CIRCUMFERENCE				
Lean et al. 1995	990 (M), 1,216 (F)	BMI and WHR cutoffs[b]	102 cm	88 cm
Pouliot et al. 1994	81 (M), 70 (F)	Metabolic risk factors	100 cm	100 cm
Lemieux et al. 1996	213 (M), 190 (F)	Absolute VAT level	90/100[a] cm	90/100[a] cm
Dobbelsteyn et al. 2001	4,951 (M), 4,962 (F)	CVD risk factors[c]	90	80
Lin et al. 2002	26,359 (M), 29,204 (F)	Metabolic risk factors	80.5	71.5
Ito et al. 2003	768 (M), 1,960 (F)	Metabolic risk factors	84	72

Note. WHR = waist–hip ratio; M = males; F = females; VAT = visceral adipose tissue; CVD = cardiovascular disease.

[a]Cutoff is 100 for ages ≤40 years and 90 cm for ages >40 years; [b]later verified against CVD risk factors; [c]including lifestyle factors such as physical inactivity and smoking.

2000). The advantage of using the BMI, WC, or self-reported CVD risk factors, together or in different combinations, is still not clear, especially for recommending a particular treatment protocol. For example, Kiernan and Winkleby (2000) have shown that the identification of individuals in need of treatment is not altered by the inclusion or exclusion of the WC in the diagnosis procedure, once the BMI and preexisting cardiovascular risk factors are known. This was confirmed in relation to three traditional risk factors (total cholesterol, LDL cholesterol, and hypertension) with an evaluation of WC action levels with ROC analysis in a large, random Dutch sample (Han et al. 1996). Risk prediction by anthropometric variables was found to be low and BMI and WC had a similar screening utility. The addition of WC to BMI did not improve its predictive ability, leading the authors to recommend WC as an alternative but not a superior or additive index to the BMI, as assumed in the recommendations by the NIH and the WHO.

Findings in larger samples confirm the previous recommendations, collectively showing a slight advantage of using the WC in relation to BMI (Dobbelsteyn et al. 2001; Ho et al. 2001; Ito et al. 2003; Lin et al. 2002). More recently, in a U.S. representative sample, Janssen and colleagues (2004) observed that the WC was superior to the BMI in predicting CVD risk factors when the two variables were entered together in prediction models and that after adjusting for WC (as a continuous, but not as a categorical variable), BMI levels were no longer predictive of risk. These findings suggest that the present single-value cutoff for the WC may lack sensitivity as a diagnostic tool beyond what BMI already provides and thus may be in need of further refinement (Bray 2004). More importantly, Janssen and colleagues (2004) also highlighted the difficulty in the process of translating research findings into simple guidelines that practitioners and the population at large will find helpful and usable. While cutoffs are easy to understand and widely applicable, they are not as precise in predicting risk as more developed equations, which in turn may be too complex or impractical for most day-to-day and clinical applications. Recently, combined measures of BMI and WC were found to provide higher sensitivity to predict CVD risk (Zhu et al. 2004). Predictive models including BMI and WC generated a score that better estimated the odds of having CVD risk factors than either alone. New BMI-specific WC thresholds are available that improve the identification of health risk (Ardern et al. 2004). In this large cohort involving normal-weight, overweight, obese I, and obese II+ patients, WC cutoffs of 90, 100, 110, and 125 cm in men and 80, 90, 105, and 115 cm in women, respectively, had the best tradeoff between sensitivity and specificity to predict coronary events.

The choice of the best diagnostic tool to represent increased risk is related to the risk biomarker. In a critical review of the clinical repercussions of abdominal obesity, Després and colleagues (2001) strongly argued that the use of the WC together with the measurement of fasting triglycerides offers the best diagnostic and practical tool to identify those individuals more likely to carry a pernicious triad of metabolic abnormalities, namely hyperinsulinemia, increased serum apolipoprotein B, and an increased number of small, dense LDL particles. Even in the absence of more classic lipid risk factors, such as type 2 diabetes, high total cholesterol, and hypertension, this hypertriglyceridemic waist phenotype was identified as a vital sign all clinicians should evaluate.

Increased visceral adiposity is undoubtedly associated with a deterioration of several indicators of metabolic and cardiovascular health (Aronne and Segal 2002). Of all studies that have analyzed this issue with direct measures of VAT (CT and MRI), a few have suggested threshold values for increased risk for men or women (table 13.9). In some studies, the same cutoff values for males and females were recommended. However, other studies suggest that the critical levels of VAT may be higher in men (around 130 cm²) than in women (more variable), although the different samples and measurement sites used preclude a definitive generalization of the single best VAT-based threshold. Though limited in clinical application, valid VAT-based thresholds may be useful in research settings to identify correlates (e.g., genetic, behavioral) of intra-abdominal obesity and also to more thoroughly test the efficacy of interventions targeting weight and waist reduction. In addition, identification of critical levels of VAT may help researchers explain the biological mechanisms behind visceral obesity (Seidell and Bouchard 1997).

An alternative way to express abdominal fat distribution is by using the ratio of visceral (V)

Table 13.9 Proposed Cutoffs for Visceral Adipose Tissue Measured Directly by Imaging Techniques

	SAMPLE CHARACTERISTICS			**VAT AREA**
Study	**Sex**	**Obese included?**	**Scan site**	**Cutoff (cm²)**
Després and Lamarche 1993	M/F	Yes	L4-L5	130
Hunter et al. 1994	M	Yes	L4-L5	131
Williams et al. 1996	F	Yes	L4-L5	110
Anderson et al. 1997[a]	M/F	Yes	L4-L5	132
Matsuzawa et al. 1996	M	No	Umbilicus	133
Saito et al. 1998	M	Yes	Umbilicus	100
	F	Yes	Umbilicus	90
Lottenberg et al. 1998	M/F	Yes	Umbilicus	107

Note. VAT = visceral adipose tissue; M = male; F = female.

[a]VAT assessed by magnetic resonance imaging; all others used computed tomography.

Adapted from B.L. Wajchenberg, 2000, "Subcutaneous and visceral adipose tissue: their relation to the metabolic syndrome," *Endocrine Reviews* 21(6): 697-738.

to subcutaneous (S) AT area (V/S ratio). Fujioka and colleagues (1987) found that a V/S ratio of 0.4 was a marker for elevated risk factors for disturbances in glucose and lipid metabolism and hypertension in men and women. Subjects with a V/S ratio greater than 0.4 had an increased risk, which was independent of sex, age, and BMI. Measurements were performed by CT scan at the umbilicus level.

Available evidence suggests that the waist circumference is the best overall index of fat distribution in adults in relation to health. The relationship of this anthropometric index with whole-body, abdominal, and visceral adiposity justifies its use. Whether waist circumference should be used along with BMI or as a single predictor of risk is still debated. For the general population, cutoffs proposed by the WHO and the NIH are supported; however, age-, and population-specific standards will likely emerge. Enlarged visceral AT is a consistent marker of higher CVD risk, but its measurement will probably remain limited to a restricted number of facilities, despite the fact that large-scale trials using CT are under way (Harris et al. 2000).

In addition to waist circumference, WHR, and VAT, other indicators of fat distribution may help elucidate the complex relationships among whole-body composition, fat distribution, and metabolic or cardiovascular health. For example, subcutaneous fat in the trunk may be a good marker of some metabolic disturbances such as glucose and insulin homeostasis (Goodpaster et al. 1997) or lipid and lipoprotein abnormalities (Sardinha et al. 2000). Sardinha and colleagues (2000) showed that the sum of four trunk skinfolds (subscapular, chest, abdominal, and suprailiac) was strongly related to several CVD risk factors, independently of total fatness and total abdominal fat, as assessed by DXA (Sardinha et al. 2000). Other investigators have also shown that subcutaneous fat plays an important role as an indicator of peripheral insulin sensitivity (Abate and Garg 1995; Abate et al. 1996; Goodpaster et al. 1997; Misra et al. 1997).

The debate over the health impact of peripheral fatness (arms, presumably deleterious) and lower-body fat (postulated as positive) continues. Using the sum of upper-arm and leg skinfolds, Hunter and colleagues (1997) showed that total peripheral adipose tissue was independently related to CVD risk factors. Pouliot and colleagues (1992) used the ratio of abdominal to femoral adipose tissue, measured by CT, to indicate a protective effect of thigh fat. Previously, it had been shown that lower-

body fat assessed by DXA (Carey et al. 1996) or CT (Sparrow et al. 1986) was not associated with insulin–glucose homeostasis, whereas Terry and colleagues (1991) observed that thigh fat could indeed have a health-protective role. In contrast, other studies show that thigh fat measured by CT (Goodpaster et al. 1997) and the sum of triceps, biceps, thigh, and calf skinfolds (Abate et al. 1996, 1995) are inversely related to insulin sensitivity in men. We have shown that appendicular fat, mainly in the arms, has detrimental effects similar to that of subcutaneous abdominal fat, suggesting that the role of appendicular fat on CVD risk factors should be analyzed independently (Sardinha et al. 2000).

Recent studies differentiating anatomical regions within subcutaneous adipose tissue are contributing to a clearer understanding of its metabolic role. Subcutaneous abdominal fat can be divided into superficial and deep regions, separated by a clear fascial plane (Kelley et al. 2000). Table 13.10 shows the correlations of these two components (and other abdominal and trunk measures) with insulin resistance measured with a euglycemic clamp (Kelley et al. 2000). In this study of glucose disposal rate, deep subcutaneous and visceral adipose tissue were better markers for a poor metabolic profile than was superficial abdominal fat. Because approximately three fourths of all deep subcutaneous AT in the trunk is located in the posterior region, this study confirmed the earlier observations (cited previously) that posterior trunk fat predicts glucose insensitivity better than does anterior subcutaneous AT, both measured by MRI (Misra et al. 1997).

Thigh fat, often considered a subcutaneous depot, is actually composed of superficial, intermuscular, and intramuscular fat. Whereas superficial and intermuscular fat are relatively easy to measure with CT or MRI, measurement of intramuscular fat (a component of ectopic or nonadipose tissue fat) with noninvasive methods is only now becoming possible. Relationships between these different components and risk factors are variable, with fat in and around muscle showing a stronger correlation with insulin resistance than superficial fat (Goodpaster et al. 2000). The application of newer methodologies, namely MRS, should clarify the relationship of intramyocellular lipid to metabolic health.

Elderly Subjects

Several studies have analyzed the effect of older age as a mediating factor in the relationships

Table 13.10   Correlation of Regional Adiposity (Thigh and Abdominal) Variables With Insulin Resistance

	Insulin resistance[c]
THIGH FAT VARIABLES[a]	
Total AT	–.17
Subcutaneous AT	–.12
Subfascial AT	–.36*
Intermuscular AT	–.45*
Total thigh fat (DXA)	–.23
ABDOMINAL FAT VARIABLES[b]	
Total subcutaneous AT	–.53**
Superficial subcutaneous AT	–.29
Deep subcutaneous AT	–.64**
Visceral AT	–.61**
Deep subcutaneous + visceral AT	–.68**
Trunk AT volume	–.55

Note. AT = adipose tissue; DXA = dual-energy X-ray absorptiometry.

[a]From Goodpaster et al. (2000). All variables assessed by computed tomography, except total thigh fat measured by DXA. Sample was composed of 68 lean and obese middle-aged men and women, 11 of whom were obese and had diabetes, the remaining glucose tolerant. [b]From Kelley et al. (2000). All variables assessed by computed tomography. Sample was composed of 47 glucose-tolerant middle-aged men and women. [c]Glucose disposal rate determined by hyperinsulinemic euglycemic clamp.

*$p < .01$. **$p < .001$.

among fat distribution, health, and mortality (Chang et al. 2000; Iwao et al. 2000; Zamboni et al. 1997). Central fat increases with age (Poehlman et al. 1995), and central adiposity and an android fat pattern are associated with worse health outcomes in older persons (Chumlea et al. 1992; Mykkanen et al. 1992; Turcato et al. 2000; Wu et al. 2001). In the old and very old, compared with younger samples, predictors of mortality assume a greater importance. Establishing appropriate risk cutoffs for older adults based on simple anthropometric measures is currently under way, and within the next few years we can expect elderly-specific critical values for fat distribution.

Waist circumference, WHR, and sagittal diameter have been the most studied fat distribution variables in the older population. Several reports have shown that WHR is a good correlate of metabolic abnormalities and mortality in older men and women (Folsom et al. 1989, 1990, 1993; Mykkanen et al. 1992; Rimm et al. 1995; Ward et al. 1994). Others have used alternative central obesity indexes to show the impact of this characteristic of body fat patterning on health in older individuals (Cassano et al. 1992; Chumlea et al. 1992; Haarbo et al. 1989; Wu et al. 2001). Several of these studies, especially the early ones, did not measure or did not report data on waist circumference or sagittal diameter. With data from the Baltimore Longitudinal Study on Aging, Seidell and colleagues (1994) showed that the sagittal diameter predicted mortality in younger but not older men. More recently, other studies have indicated that WHR categories (quintiles) are not as useful when compared with waist circumference cutoffs (Visscher et al. 2001; Woo et al. 2002) and that waist circumference and sagittal diameter are the best correlates of CVD risk in the elderly (Turcato et al. 2000). The associations among waist circumference, WHR, sagittal diameter, and several metabolic variables were studied in 229 Italian men and women aged 67 to 78 years (Turcato et al. 2000). Before and after adjustment for age and BMI, waist circumference, and sagittal diameter were the best predictors of CVD risk. After the authors controlled for BMI, associations were stronger for women than men, indicating that the relationship between fat distribution and risk factors may be more dependent on total fatness in men than in women.

Folsom and colleagues (2000) recently compared the ability of the BMI, waist circumference, and WHR to predict multiple health outcomes in more than 31,000 women between the ages of 55 and 69. The authors' purpose was to test the NIH–WHO recommendations by analyzing whether the WHR would provide information in addition to the current BMI and waist circumference screening protocol. The authors showed that waist circumference provides useful information regarding the incidence of diabetes, hypertension, and CVD, whereas WHR adds additional predictive ability regarding other end points, such as cancer and hip fractures (Folsom et al. 2000). Overall, women in the 5th quintile of both waist circumference and WHR had significantly higher relative risk for total mortality, compared with the lower quintile. In men, data from the Rotterdam Study showed that waist circumference was associated with mortality to a greater extent than was BMI or WHR, although the results were not significant

(Visscher et al. 2001). At least three other studies have attempted to evaluate the performance of already established cutoff values used in the adult population in older cohorts (Iwao et al. 2000; Molarius et al. 2000; Woo et al. 2002). Collectively, these studies offer some support for the use of current waist circumference cutoffs, in addition to BMI, to detect higher prevalence of risk factors in elderly people.

Older individuals accumulate higher relative amounts of abdominal fat, especially visceral fat, without corresponding changes in weight (Zamboni et al. 1997). Thus, waist circumference and sagittal diameter are possibly less sensitive in identifying health risks in older individuals attributable to overweight or abdominal fat distribution, compared with younger individuals. In fact, data from two large trials that contrasted visceral adipose tissue assessed by CT with anthropometric indexes (waist circumference, sagittal diameter, waist–thigh ratio, skinfolds) showed that the anthropometric indices are poor predictors of VAT in older men and women and instead are more related to total body adiposity (Harris et al. 2000; Schreiner et al. 1996). In summary, although abdominal adiposity is related to negative health outcomes in elderly people in a similar, albeit statistically weaker manner compared with younger samples, the best criteria for diagnosis are still to be determined.

Summary

The majority of empirical evidence shows that increased adiposity is associated with CVD risk markers and coronary artery lesions in children and adolescents. BMI and skinfolds can be used as a surrogate of total body adiposity in studies of children and adolescents.

Skinfolds can be used to estimate %BF and identify children and adolescents at risk based on %BF cutoffs developed against health-related criteria. As an alternative, there is evidence to support the use of the new BMI international cutoff guidelines to define overweight and obesity in children and adolescents, because these are useful in predicting CVD risk factors. Waist circumference also correlates well with subcutaneous and total trunk fat across a large spectrum of BMIs and may also serve as a surrogate of markedly enlarged visceral adipose depots in obese children and adolescents.

In adults, the current definitions of overweight and obesity using BMI indicate "discriminative" CVD risk, suggesting that in the absence of other more specific methods to estimate total body fat, BMI can be used in adults for screening total body fatness and for risk stratification. As is true with children and adolescents, waist circumference is the best overall index of fat distribution in adults in relation to health. Data from several cohorts support the BMI and waist circumference cutoffs proposed by the WHO and the NIH.

Current BMI cutoff values to define overweight and obesity in adults should be used with caution in older persons, because the association between different indexes of excess adiposity with several clinical outcomes and mortality is much weaker in elderly people. For any given BMI, older subjects have a higher %BF. Because BMI is recognized as a less specific index of total body fatness in older adults, there is a need for more valid estimates of total body fatness and its relationship with clinical outcomes and mortality in older subjects.

Despite the usefulness of anthropometry, there is a need to implement the widespread use of more accurate and reproducible methods to estimate total and regional adipose tissue accumulation. These methods include DXA and ultrasound to estimate fat at the molecular level and CT and MRI to estimate adipose tissue at the tissue level. Abdominal fat and adipose tissue compartmentalization tends to be more associated with obesity-related comorbidities than are whole-body indexes. Of special interest is the information provided by CT and MRI concerning the anatomic and morphologic location of different depots, including subdivision of the subcutaneous adipose tissue compartments, visceral adipose tissue, and intermuscular adipose tissue. Visceral and intermuscular adipose tissue appears to have stronger associations with physiological and pathological processes than does total body fatness.

Chapter 14

Assessing Muscle Mass

Henry C. Lukaski

Methods for assessing human body composition emphasize the estimation of body fat, and techniques to assess muscle mass are limited (Forbes 1987; Lukaski 1987). This emphasis reflects, in part, the demands of the scientific community to estimate a compositional variable (i.e., percent body fat) that might be a useful predictor of risk development of chronic diseases, particularly ischemic heart disease and non-insulin-dependent diabetes. More recently, however, an increased awareness of the significance of developing and maintaining skeletal muscle mass has stimulated a reappraisal of available approaches for the in vivo assessment of skeletal muscle mass.

In the body, muscle is present in three distinct forms: skeletal, smooth, and cardiac. Skeletal, also known as voluntary or striated, muscle represents approximately 30% to 40% of the body weight of a healthy 58-kg woman or a 70-kg man (ICRP 1975). In an adult, the majority of skeletal muscle is found in the legs, with lesser amounts in the head, trunk, and arms.

The need for methods to estimate either whole-body or regional muscle mass is compelling and reflects multidisciplinary interests. One application of such methods is to monitor changes in muscle mass in relation to growth and development of infants and children, and another is to establish normative data for use in medicine and clinical investigations.

There are additional applications in physiology, nutrition, and medicine. Because skeletal muscle is required for movement, exercise scientists are interested in relating estimates of muscle mass to various types of aerobic and anaerobic athletic performance and to the effects of physical training on work capacity and physical performance. Clinicians require estimates of muscle mass to evaluate the progression of catabolic disease and to evaluate the efficacy of therapeutic interventions on prognosis. Geriatricians seek longitudinal assessments of muscle mass to monitor the effects of aging on muscular development and function, to evaluate the efficacy of strength-building exercise programs in maintaining and improving functional capacity and ambulation of elderly people, and thus to improve the quality of life for elders.

The general availability of a useful technique for assessing skeletal muscle mass is limited by the relative lack of direct data on anatomical tissue masses. Skeletal muscle masses determined by dissection and body masses are known for only 25 adult men (Martin et al. 1990). Therefore, development and validation of most current methods of assessing muscle mass are limited by the use of indirect reference methods that rely on some biochemical or physical characteristic of skeletal muscle.

This chapter describes available methods for estimating human skeletal muscle mass in vivo. Emphasis is placed on the biological and physical basis of each approach, its application, and its limitations.

Anthropometric Indicators of Muscle Mass

The use of anthropometry to estimate muscle mass requires the selection of some body measurements

U.S. Department of Agriculture, Agricultural Research Service, Northern Plains Area, is an equal opportunity–affirmative action employer, and all agency services are available without discrimination. Mention of a trademark or proprietary product does not constitute a guarantee or warranty of the product by the United States Department of Agriculture and does not imply its approval to the exclusion of other products that may also be suitable.

that are predictive of muscle mass. In general, a muscle group is selected with the assumptions that site-specific physical measurements reflect the mass of that muscle and that the mass of the estimated muscle group is directly proportional to the whole-body skeletal muscle mass.

Estimation of Regional Muscle Mass

Historically, the use of anthropometric measurements of the upper arm to estimate muscle mass has dominated the literature (Jelliffe and Jelliffe 1969). However, measurements of the lower leg also have been used, but only infrequently (Heymsfield et al. 1979). Estimates of upper-arm muscle circumference have served as a functional index of protein-energy malnutrition (Jelliffe 1966).

Simple anthropometric measurements of the upper arm have been derived (figure 14.1). The common variables include arm circumference (C_A), corrected for subcutaneous adipose tissue thickness, and muscle cross-sectional area estimated from the corrected circumference. The tissue boundaries in the arm cross-section are assumed to be circular and concentric. If the skin plus adipose tissue thickness is defined as d, then the arm muscle circumference (C_M) is given as $C_M = C_A - 2\pi d$. Because the skinfold measurement includes a double thickness of adipose tissue, and 2d is represented by D, then $C_M = C_A - \pi D$.

Because scientists have realized that estimates of arm circumference represent a unidimensional approximation and muscle mass is a three-dimensional quantity, the estimation of muscle area has taken prominence as an index of nutritional assessment (Heymsfield et al. 1982b). The basic assumptions of this index include the following:

the midarm is circular; the triceps skinfold thickness is twice the average adipose rim diameter at the middle of the upper arm; the midarm muscle compartment is circular; and bone responds in a manner similar to muscle and fat during growth and caloric deprivation (Gurney and Jelliffe 1973; Jelliffe 1966).

The validity of the anthropometric estimation of upper-arm muscle area has been examined. Heymsfield and colleagues (1979) reported that each of the assumptions was in error to some degree. In a sample of adults with body weights ranging from 60% to 120% of ideal body weight, anthropometric midarm muscle area was overestimated 15% to 25% compared with reference values determined by computed axial tomography. Subsequently, sex-specific equations were derived to account for errors in each of the four assumptions (Heymsfield et al. 1982a). The revised equations, containing corrections for bone contributions to anthropometric measurements, resulted in an average intraindividual error of 7% to 8% in the calculated midarm muscle area. However, in the subjects whose ideal body weight was greater than 150%, the midarm muscle area was in error by more than 50% even after the revised equations were used.

Estimation of Whole-Body Muscle Mass

A few attempts have been made to use anthropometric measurements to estimate whole-body, compared with regional, muscle mass. All of the proposed models are similar because they rely on measurements of regional body circumferences and skinfold thicknesses, but they differ because of the reference methods used for derivation and validation of the prediction models.

In 1921, Mateiga proposed a model based on measurements of circumferences of the forearm, upper arm, thigh, and calf, corrected for the corresponding skinfold thicknesses, to derive an average value for muscle limb radius. This value was squared, multiplied by stature, and then multiplied by a "constant" of 6.5. This hypothetical model was neither validated by Mateiga (1921) nor critically examined by subsequent investigators.

Using anthropometric estimates of arm muscle area, Heymsfield and colleagues (1982b) attempted to predict whole-body muscle mass from estimates derived from urinary creatinine excretion. It was assumed that each gram of creatinine excreted in the urine per day was equivalent to 20 kg of

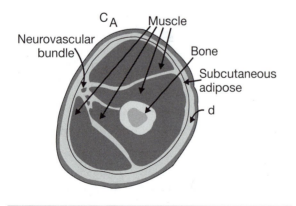

Figure 14.1 Diagram of the anthropometric measurement sites for the determination of midarm muscle circumference. Measurements include arm circumference (C_A) and skinfold plus skin thickness (d).

muscle mass. These authors proposed an equation to predict total muscle mass from corrected arm muscle area and stature. The error of the predicted values ranged from 5% to 9%, with an average error of about 8%.

These approaches have some general limitations. The use of a single, regional measurement promotes errors in estimation of whole-body muscle mass because of interindividual variability in the distribution of muscle mass and adipose tissue and because of bone thickness. The validity of this approach in individuals with excess adipose tissue in the upper extremities is poor. Thus, this approach only provides a qualitative index of whole-body muscle mass.

Brussels Cadaver Study

In contrast to the previous studies, the Brussels Cadaver Study used anthropometric measurements and tissue mass determinations in the same subjects (Clarys et al. 1984). These data have been used to develop regression models for the estimation of whole-body muscle mass (Martin et al. 1990).

Twenty-five cadavers were studied; 12 were embalmed (6 men, 6 women) and 13 were unembalmed (6 men, 7 women). Anthropometric measurements included skinfold thicknesses (triceps, subscapular, biceps, anterior thigh, and medial calf) and circumferences (forearm, mid–upper arm, mid–anterior thigh, and calf). Limb muscle girths were estimated by correcting limb girths for skinfold thicknesses. Data from the six unembalmed cadavers were used to develop a regression model, and the data from five embalmed cadavers were used for validation. Data from one embalmed cadaver were not used for validation study because of atrophy of muscle and bone in one leg.

Regional anthropometric measurements were strong predictors of total dissected muscle mass (table 14.1). Because skinfold thickness of the forearm is infrequently measured and the correlation coefficient of the uncorrected forearm circumference with total dissected muscle mass was very strong, Martin and colleagues (1990) derived the following prediction equation for men:

$$MM = S \, (0.0546 \, CTG^2 + 0.119 \, FG^2 + 0.0256 \, CCG^2) - 2980 \qquad (14.1)$$

where MM is total skeletal muscle mass (g), S is stature or standing height (cm), CTG is thigh circumference corrected for anterior thigh skinfold thickness (cm), FG is uncorrected forearm circumference (cm), and CCG is calf circumference corrected for skinfold thickness (cm).

Comparison of predicted and dissected total skeletal muscle mass in five unembalmed male cadavers indicated a relationship similar to the line of identity with $R^2 = .93$ and standard error of estimate (SEE) = 1.58 kg. The authors proposed that because the SEE values between the derived model (1.56 kg) and the validation comparison (1.58 kg) were similar, a combined model for men should be generated. The model is as follows:

$$MM = S \, (0.0553 \, CTG^2 + 0.0987 \, FG^2 + 0.0331 \, CCG^2) - 2445 \qquad (14.2)$$

$$SEE = 1.53 \text{ kg}, R^2 = .97$$

The validity of the derived model also was examined in relation to values predicted from the models of Heymsfield and colleagues (1982a) and Matiega (1921). Estimated error (estimated minus the dissected value) for five embalmed cadavers was evaluated as a function of dissected muscle

Table 14.1   Relationships of Limb Girths and Girths Corrected for Skinfold Thicknesses With Total Dissected Skeletal Muscle Masses

Site	CORRELATION COEFFICIENTS	
	Basic girth*	Corrected girth
Forearm	.824	.896
Arm	.963	.998
Thigh	.942	.990
Calf	.836	.911

* Based on six unembalmed male cadavers.

Adapted, by permission, from A.D. Martin et al., 1990, "Anthropometric estimation of muscle mass, *Medicine and Science in Sports & Exercise* 22: 729-733.

mass. The model derived by Martin and colleagues (1990) had a random and substantially reduced error (approximately ±2 kg) compared with the other prediction models, which consistently underestimated muscle mass, ranging from 5 to more than 10 kg.

Updated Model

Lee and colleagues (2000) developed and evaluated the accuracy of a new anthropometric model compared with magnetic resonance imaging (MRI). A multiracial sample of 244 nonobese adults was randomized into two subgroups. By using limb anthropometric measurements and physical and demographic characteristics, the authors generated a model to predict skeletal muscle mass (SMM) in one sample ($n = 122$) and cross-validated the model in an independent group ($n = 122$). They found no significant difference (-0.3 ± 2.5 kg, $p = .23$) between MRI-measured and predicted SMM in the validation group. Measured and predicted values were significantly correlated ($R^2 = .89$); there was no significant relationship between differences in measured and predicted SMM and measured SMM ($R^2 = .005$, $p = .45$). Data from both groups were combined to yield a common prediction model:

$$SMM = S\ (0.00744\ CAG^2 + 0.00088\ CTG^2 + 0.00441\ CCG^2) + 2.4\ Sex - 0.048\ Age \quad (14.3)$$

$$SEE = 2.2\ kg,\ R^2 = .91$$

where SMM is in kg, S is stature or standing height (m), weight is in kilograms, CAG is corrected upper-arm girth adjusted for skinfold thickness (cm), CTG is thigh circumference corrected for anterior thigh skinfold thickness (cm), CCG is calf circumference corrected for skinfold thickness (cm), sex is 1 for male and 0 for female, and race is –2 for Asian, 1.1 for African American, and 0 for white or Hispanic. This model was evaluated in 80 obese adults. There was no significant difference (0.4 ± 3.0 kg, $p = .28$) between MRI-measured and predicted SMM.

Lee and colleagues (2000) also developed and cross-validated another model based on standing height, weight, age, sex, and race in the nonobese sample and reported a significant relationship ($R^2 = .86$, $p < .0001$) between MRI-measured and predicted SMM values with no significant difference between the values (-0.3 ± 2.7 kg, $p = .17$). They combined the data and derived a common prediction equation:

$$SMM = 0.244\ BW + 7.80\ S - 0.098\ Age + 6.6\ Sex + Race - 3.3 \quad (14.4)$$

$$SEE = 2.8,\ R^2 = .86$$

where BW is weight in kg, S is stature or standing height in m, age is in years, and sex and race are as previously defined. When the equation was cross-validated in the obese adults, measured and predicted values were significantly correlated ($R^2 = .79$, SEE = 3.0). However, there was a significant difference (-2.3 ± 3.3 kg, $p < .0001$) between measured and predicted SMM values. Examination of bias revealed a significant correlation between measured and predicted SMM difference and measured SMM ($R^2 = .18$, $p < .001$). These findings show the accuracy of the anthropometric (circumferences and skinfold thicknesses) compared with demographic (height, weight, age, sex, and race) data to predict SMM in adults of varying obesity status.

Limitations to the Use of Anthropometric Estimates of Muscle Mass

In general, the use of anthropometric measurements, including circumference and skinfold thickness measurements, to predict both regional and total muscle mass yields quantitative assessments. Although one study provided encouraging results (Lee et al. 2000), it is not known if anthropometric estimations are either sufficiently accurate or sufficiently sensitive to monitor small changes in muscle mass associated with weight loss or gain in an individual.

Muscle Metabolites

The hypothesis that endogenous components or metabolites of skeletal muscle metabolism may be used to estimate skeletal muscle mass is established on some basic premises. It is assumed that the chemical marker is found only in skeletal muscle, the size of the pool of the marker is constant, the rate of turnover is relatively unchanged over long periods of time, and the compound is not further metabolized after release into the circulation. Two skeletal muscle-specific metabolites have been used as indexes of skeletal muscle mass.

Creatinine

Creatinine is formed by the nonenzymatic hydrolysis of creatine liberated during the dephosphorylation of creatine phosphate in the liver and kidney (figure 14.2). Although many tissues take up creatine, the predominant source (98%) is skeletal muscle, principally in the form of creatine phosphate (Borsook and Dubnoff 1947).

Urinary Creatinine Excretion

Since Folin (1905) hypothesized that urinary creatinine was an indicator of body composition and Hoberman and colleagues (1948) demonstrated that body creatine was directly related to daily urinary creatinine output, endogenous urinary creatinine excretion has been accepted as a marker of fat-free mass (FFM) and skeletal muscle mass (Cheek 1968; Forbes and Bruining 1976; Talbot 1938).

Estimation of skeletal muscle mass from urinary creatinine output presumes a constant relationship between these variables. Data from different studies showed that 1 g of creatinine excreted during a 24-hr period was derived from approximately 18 to 20 kg of muscle mass (Talbot 1938; Cheek 1968). This difference probably reflects discrepancies in muscle sampling and methodological variation between the studies.

The broad use of urinary creatinine excretion to predict muscle mass in clinical investigations is limited by some factors that affect its validity (Forbes 1987). Large intraindividual variability in daily urinary creatinine excretion ranges from 11% to 30% for individuals consuming self-selected diets (Greenblatt et al. 1976; Lykken et al. 1980; Ransil et al. 1977). It can be reduced to less than 5% if meat-free diets are consumed (Forbes and Bruining 1976; Lykken et al. 1980). Significant reductions (10-20%) in creatinine excretion occur in healthy men consuming meat-free diets for several weeks (Calloway and Margen 1971). Thus, creatine intake affects urinary creatinine output. Among healthy young men, urinary creatinine excretion increased with creatine feeding and then decreased with a creatine-free diet (Crim et al. 1975). Lykken and colleagues (1980) derived a mathematical model with feedback controls that describes the creatinine pool size and creatinine excretion as a function of time and changes in creatine and protein intake. Thus, the body creatinine pool size is not under strict metabolic control, and urinary creatinine excretion may be, to some degree, independent of body composition.

Practical concerns may affect the validity of the urinary creatinine approach to estimate muscle. Creatinine is filtered and secreted in the kidney.

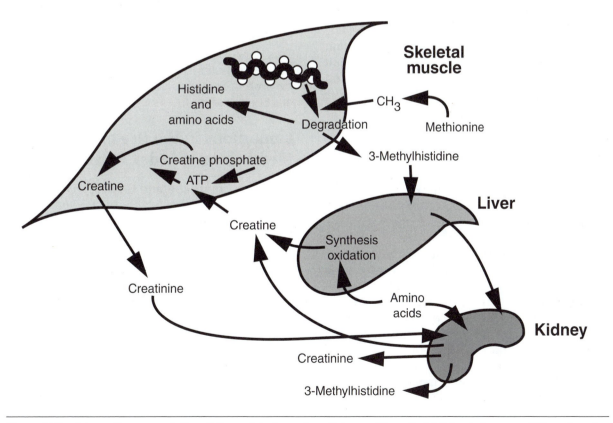

Figure 14.2 Schematic representation of the metabolism of creatinine and 3-methylhistidine in humans. ATP = adenosine triphosphate.

Accurately timed urine collections are critical because an error as small as 15 min in a collection period represents an error of 1% in the 24-hr urinary creatinine excretion (Forbes and Bruining 1976). Thus, three consecutive 24-hr urine collections are advised to ensure representative creatinine excretion for an individual (Forbes and Bruining 1976).

The critical question concerning the constancy of the ratio of 1 g of urinary endogenous creatinine excretion to a unit of muscle mass or even FFM has been addressed. Forbes and Bruining (1976) proposed that urinary creatinine excretion does not represent a constant fraction of either muscle or FFM in children or adults. They concluded that it is inappropriate to use a constant ratio of creatinine to muscle unless factors such as age, sex, maturity, physical training, and metabolic state are controlled.

3-Methylhistidine (3-MH)

The endogenous excretion of the amino acid 3-MH has been proposed to index muscle protein breakdown (Young and Munro 1978). This unique amino acid is located principally in skeletal muscle (figure 14.2). Skeletal muscles had the greatest 3-MH concentration (3-4 μmol/g fat-free, dry weight) with intermediate concentrations in cardiac and some smooth-muscle tissues (1-2 μmol/g) and low values (<1 μmol/g) in spleen, liver, and kidney taken postmortem from five adults (Elia et al. 1979).

The 3-MH concentration of human muscle was constant (3-4 μmol/g) between the ages of 4 to 65 years (Tomas et al. 1979). Urinary excretion of 3-MH decreased with age, which was interpreted to indicate that 3-MH turnover decreased with age. Alternatively, the reduction in urinary 3-MH output might reflect a decreased muscle mass with age (Cohn et al. 1980).

Experimental evidence shows that specific histidine residues are methylated after the formation of the peptide chains of actin of all muscle fibers and that, during catabolism of the myofibrillar proteins, the released 3-MH is neither reused for protein synthesis nor metabolized oxidatively but rather excreted quantitatively in the urine (Young and Munro 1978). Thus, endogenous urinary 3-MH output may be a useful index of muscle mass.

Early investigations established the relationship between 3-MH excretion and human body composition. Endogenous urinary 3-MH output was well correlated ($r = .90, p < .001$) with FFM in 16 healthy men after consuming a meat-free diet for 3 days (Lukaski and Mendez 1980). Urinary creatinine excretion also was correlated ($r = .67, p < .01$) with FFM. Subsequently, a strong relationship ($r = .91, p < .001$) between muscle mass, assessed from determinations of total body potassium (TBK) and nitrogen (TBN; Burkinshaw et al. 1978) and endogenous 3-MH, was shown in 14 athletic men fed a meat-free diet (Lukaski et al. 1981b). Urinary 3-MH excretion was modestly correlated ($r = .33$) with nonmuscle protein mass. Urinary creatinine also was correlated ($r = .79, p < .01$) with muscle mass. A significant relationship between endogenous 3-MH and creatinine ($r = .87$) was noted and is consistent with the results of a study of men and women (Tomas et al. 1979). Thus, endogenous urinary 3-MH reflects muscle mass assessed directly with nuclear methods or indirectly with a surrogate indicator of muscle mass.

The use of 3-MH as a marker of muscle mass has received some criticism. One concern is the influence of nonskeletal muscle protein turnover on the excretion rate of 3-MH (Rennie and Millward 1983). Data from studies of rats revealed a significant contribution of smooth muscle protein turnover from the skin and gastrointestinal tract to the urinary output of 3-MH (Wassner and Li 1982). This concern has not been examined in humans because simultaneous measurements of myofibrillar protein synthesis and degradation rates have not been assessed.

Limitations in the Use of Muscle Metabolites to Index Muscle Mass

Although creatinine and 3-MH derive primarily from muscle, their value to predict skeletal muscle mass needs further study to determine factors that affect their pool sizes and turnover rates. Additional research is required in humans to discern the significance of contributions of nonskeletal muscle sources of these metabolites in daily endogenous excretion.

Two additional concerns hamper the general use of urinary excretion of endogenous metabolites to predict muscle mass in vivo. These limitations include the need to consume a meat-free diet that is adequate in protein content to eliminate exogenous sources of creatine, creatinine, and 3-MH and the requirement for accurately timed urine collections.

Radiographic Methods

In contrast to other methods that indirectly assess muscle mass, the radiographic techniques offer unique opportunities for direct visualization and measurement of compositional variables, including adipose tissue, bone, and muscle. These methods rely on the differing responses among tissues, based on their chemical composition, as the tissue interacts with electromagnetic energy. Thus, these techniques facilitate the regional measurement and, in some cases, whole-body assessment of body composition.

Three radiographic techniques currently used in body composition research are computed tomography (CT), magnetic resonance imaging (MRI), and dual-energy X-ray absorptiometry (DXA). Although CT and MRI have been used in body composition research principally to determine regional subcutaneous and visceral adipose tissue (Fuller et al. 1994), these methods have the capability to measure muscle mass. Essentially the same compositional information that is available from CT also is available from MRI. The DXA technique, which was developed from photon absorptiometry, also has the capability to assess bone, fat, and fat-free tissues. Whereas CT and MRI provide three-dimensional images of anatomy, DXA provides only two-dimensional representations of body structures.

Computed Tomography

This technique requires the placement of a patient on a bed between a collimated X-ray source (0.1-0.2 A, 60-120 kVp) and detectors aligned at opposite poles of a gantry. As X rays pass through tissues of the body, the X rays are attenuated. The induced attenuations in X-ray intensity are related to differences in the physical density of the tissues examined. This physical effect is expressed quantitatively as the linear attenuation coefficient or CT number. The CT number, expressed in Hounsfield units, is a measure of the tissue attenuation relative to that of water. For example, air, adipose tissue, and muscle have average CT numbers of –1,000, –70, and 20, respectively.

The attenuation of the X-ray beam, and thus the CT number, depends on some physical effects of each tissue: coherent scattering, photoelectric absorption, and Compton interactions (Haus 1979). The principal determinants of these effects are the physical density of the tissue and the atomic numbers of the chemical components.

In general, there is a linear relationship between tissue density and CT number.

Each CT image is reconstructed from picture elements or pixels, which are usually 1 mm × 1 mm (length × width) in dimension. A third dimension is slice thickness. Each pixel or voxel, a volume element derived from pixel dimensions and slice thickness, has a corresponding CT number. Thus, the reconstructed picture represents not the image at the surface of a tissue slice but rather an average representing the full thickness of the slice.

The CT method offers high image contrast and clear separation of fat from other soft tissues (figure 14.3). The differing attenuation of adipose tissue and skeletal muscle permits visual and mathematical separation of image components.

Different approaches have been used with CT to assess body composition (Heymsfield 1987). The structure of interest is traced directly on the viewing console with a cursor. The cross-sectional area of the tissue of interest (adipose, bone, muscle) then can be determined. Because the slice thickness is known, one can calculate the relative surface area or volume occupied by the tissue or organ in the reconstructed picture. This technique has been used to quantitate reference

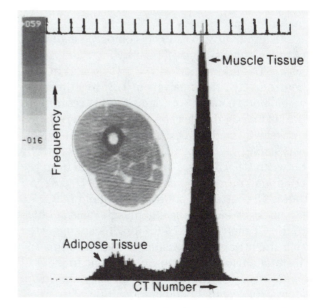

Figure 14.3 Cross-sectional computed tomography (CT) image of the midthigh area of a healthy adult and a histogram of the image pixels.

Reprinted, by permission, from S.B. Heymsfield, 1987, Human body composition: Analysis by computerized axial tomography and nuclear magnetic resonance. In *AIN symposium proceedings: Nutrition '87*, edited by O.E. Levander (Bethesda, MD: American Institute of Nutrition), 92-96.

values for skeletal muscle area for use in developing anthropometric models to estimate midarm muscle area and for assessment of arm muscle area in malnutrition and obesity.

Another CT technique is used when no sharp boundaries between structures are apparent, but the tissues differ markedly in radiographic density. The pixels in successive slices are plotted as a function of physical density in a histogram separating the pixels into adipose and fat-free tissues (figure 14.3). Because the volume of each pixel can be calculated, the volume of skeletal muscle and adipose tissue in each slice can be determined from the number of pixels forming each slice and then added for all slices performed.

Although CT offers considerable promise as a method to assess muscle mass, data are limited regarding comparisons with other methods. Wang and colleagues (1993) compared estimates of skeletal muscle mass derived from 22 CT images with other predictions from indirect methods, including anthropometry, endogenous creatinine and 3-MH excretions, and DXA. These authors found no difference between estimates derived by CT and anthropometry, whereas the other techniques resulted in discrepancies relative to CT. Because CT was a more direct method than the others, the authors concluded that the indirect methods require further refinement to yield accurate estimates of muscle mass.

Ma and colleagues (1994) compared estimates of muscle mass from 16 healthy men and 8 men with AIDS determined with CT, DXA, and in vivo neutron activation analysis of TBN together with whole-body counting of potassium-40 according to Burkinshaw and colleagues (1978). The correlation between CT and DXA estimates of skeletal muscle was strong ($r = .86$, $p < .001$), but it was less strong between CT estimates and assessments from the combined nitrogen–potassium model ($r = .82$, $p < .001$). The nitrogen–potassium model (24.9 ± 7.6 kg) underestimated skeletal muscle mass relative to the CT method (31.7 ± 6.4 kg). It was suggested that the nitrogen–potassium model of estimating skeletal muscle mass requires modification.

CT has provided unique measurements of change in regional body composition of humans. In a strength training study of nonagenarians, Fiatarone and colleagues (1990) used regional CT scans to show significant changes in midthigh muscle area with marked increases in quadriceps (9%) and hamstring and adductor areas (8.4%) in response to strength training. No significant changes were observed in subcutaneous or intra-muscular adipose tissue. These impressive gains in muscle area were not paralleled with a measurable increase in FFM, indicating the necessity to conduct regional, compared with whole-body, assessment of body composition to document structural adaptation in response to physical training of a limb.

Some caution is needed with regard to the use of CT to assess muscle in patients. There exists wide variation in attenuation of normal muscle, from 30 to 80 Hounsfield units, depending on which muscle group is examined (Bulcke et al. 1979; Montegrano et al. 1977). Although the cross-sectional area of a muscle can be measured, differences in the size or volume of muscles on contralateral sides of the body may be present. Because of the wide normal variation in muscle size associated with physical activity (i.e., increased size) and poor nutritional status (i.e., decreased size), generalized reductions in muscle size are difficult to diagnose in the early stages of disease without baseline reference data. The most common type of muscle disorder identified by CT is atrophy, in which the muscle diminishes in size and, because of infiltration by fat, shows reduced attenuation. In gross atrophy, for example, the muscle may be replaced for the most part with material characterized by fatty attenuation and may yield negative Hounsfield unit values (Dixon 1991).

Magnetic Resonance Imaging

Nuclear magnetic resonance (NMR) is a powerful technique that can present both images (MRI) and chemical composition of tissues (NMR spectroscopy); an NMR instrument can perform either imaging or spectroscopy, but not both functions. As with CT, MRI can be used to assess regional and, by calculation, whole-body composition.

The basis of NMR is the fact that atomic nuclei can behave like magnets. The application of an external magnetic field across a segment of the body causes each nucleus to attempt to align with the field. When a radio-frequency electromagnetic wave is directed into body tissues, some nuclei absorb energy from the magnetic field. When the radio wave is turned off, the activated nuclei emit the radio signal that they absorbed. The emitted signal is used to develop an image of the chemical composition of the tissue with a computer.

Elements with dipole nuclei, such as hydrogen-1, phosphorus-31, carbon-13, and sodium-23, have been examined with NMR. Each of the nuclei has an angular momentum or spin, with dipole momentum arising from their inherent nuclear characteristics. Because these nuclei have electrical charges, the spin generates a dipole moment. In response to an external magnetic field, these nuclei align themselves either parallel or anti-parallel to the lines of induction from the field. A resonant radio-frequency pulse is used to rotate the nuclei 90° relative to the magnetic field. After the radio-frequency signal is ceased, the nuclei realign themselves by a process termed *relaxation*. The absorbed energy is dissipated into the environment.

Instruments used for clinical investigation surround the patient with a magnetic coil with a 5- to 30-mm wavelength (60- to 110-MHz) radio-frequency signal. The signal produced when the nuclei relax is collected by the NMR receiver and stored for analysis.

In contrast to conventional X-ray radiographic and CT images, which depend on electron density, MRI depends on the density of hydrogen nuclei and the physical state of the tissue as reflected in the relaxation times to evaluate body compositional variables, particularly adipose tissue and muscle. Anatomical information has been obtained by comparing MRI images and corresponding frozen sections of animals (Hansen et al. 1980). In addition, proton MRI has been used to estimate total body water of baboons (Lewis et al. 1986). The hydrogen associated with body water was measured as the amplitude of the free-induction decay voltage. Body water was calculated as the product of peak amplitude by the experimentally determined constant for a water standard. Values for total body water were similar whether calculated from MRI or by the gravimetric method.

MRI has been used to assess FFM in humans (Ross et al. 1994). Transverse slices of 10-mm thickness were acquired every 50 mm from head to foot. To calculate the fat-free and adipose area, the areas of the respective tissue regions in each slice were computed automatically by summing the given tissue's pixels and multiplying by the pixel surface area. The volume of the respective fat-free and adipose regions in each slice was calculated by multiplying the tissue area by the slice thickness. The fat-free and adipose volumes were calculated by adding the volumes of truncated pyramids defined by pairs of consecutive slices (Ross et al. 1992). Comparison with anthropometric predictions of FFM showed a variability of 3.6% and 6.5% for men and women, respectively.

Dual-Energy X-Ray Absorptiometry

Another radiographic technique that is used for body composition assessment is DXA. Although originally developed for regional assessment of bone mineral content (BMC) and bone area density, this technique has been refined for the assessment of soft tissue composition (Lukaski 1993).

As with CT, DXA exposes the patient to X rays, although the amount of radiation exposure is substantially less with DXA (Kellie 1992). The beam of X rays is attenuated in proportion to the composition of the region of the body through which it is passed and the thickness of the body region. The attenuation also is affected by the energy of the X rays. Specifically, soft tissues, which contain principally water and organic compounds, restrict the flux (number of X rays per unit area) of X rays much less than does bone.

Dual X-ray scanning systems include a source that emits X rays, which are collimated into a beam that can be turned on and off by a shutter mechanism. The beam passes in a posterior-to-anterior direction through bone and soft tissue, continues upward, and enters the detector. The system's components are mechanically connected so as to scan the beam in a rectilinear pattern across the patient's body.

Although a variety of software algorithms are available, a general pattern for analysis of X-ray attenuation is used (Lukaski 1993). First, the attenuation attributable to bone is determined, and then attenuation attributable to soft tissue composition is assessed. Soft tissue composition (fat and fat-free) is calculated from the ratio of beam attenuation at the lower energy relative to that at the higher energy. In this manner, regional and whole-body estimates of BMC and bone area density, together with assessment of fat and lean soft tissue (LST), are performed with appropriate computer software.

The DXA scan provides a means by which to distinguish compositional differences between individuals with the same stature and body weight (figure 14.4). As shown in table 14.2, marked differences in BMC, body fatness, and FFM are evident despite similar body masses and stature.

This method additionally permits regional body composition assessment, with a particular emphasis on appendicular muscle mass. Heymsfield and colleagues (1990) related estimates of limb muscle mass assessed with dual-photon absorptiometry, a method similar to DXA, in healthy men and women with measurements of

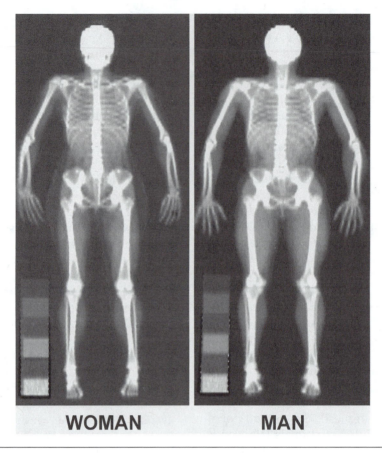

WOMAN   **MAN**

Figure 14.4 Whole-body dual X-ray absorptiometric scan images of a woman and a man of similar age, body weight, and stature.

TBK, TBN, and anthropometric assessments of regional muscle mass. The variability of repeated measurements in a subsample of four volunteers was 7%, 2.4%, and 3% for arm, leg, and combined appendicular muscle mass, respectively. Significant correlations were found between appendicular muscle mass and TBK ($r = .94$), TBN ($r = .78$), and upper arm ($r = .82$) and thigh ($r = .88$) muscle-plus-bone areas. These findings indicate the potential use of DXA to assess regional muscle mass, at least in healthy individuals.

One concern about the use of DXA in clinical assessment of muscle mass is the potential interference from fluid accumulation. Because it is well known that DXA cannot differentiate between body water and fat-free mass (Nord and Payne 1990), Horber and colleagues (1992) examined the effect of food consumption and hemodialysis on DXA estimates of body composition. The ingestion of fluid volumes ranging from 0.5 to 2.4 L 1 hr before a DXA scan did not affect DXA estimates of BMC or fat mass but did result in significant increases in LST, as expected. Ingestion of small fluid volumes (<500 ml) did not bias estimates of body composition.

Studies of the effects of hemodialysis on DXA estimates of body composition revealed some important findings (Horber et al. 1992). After dialytic treatment, body weight decreased 0.9 to 4.4 kg in six patients. The DXA estimates of whole-body BMC and fat did not change despite removal of water and salts. In contrast, a significant decrease in LST mass was observed, which accounted for about 95% of the change in body weight as a consequence of the dialytic therapy. The amount of weight lost and the change in LST observed during therapy were correlated ($r = .94$, $p < .006$). When the reduction in body weight as a result of dialysis was examined in relationship to body regions (i.e., trunk, arms, legs), there were no changes in BMC or fat content. However, the postdialytic decrease in LST was disproportionately attributed to the trunk (61%), legs (30%), arms (5.5%), and remainder of the body (3.5%). These findings suggest that areas of the body that accumulate excess water and salts in patients with end-stage renal disease are the regions in which decreases in LST occur after dialytic therapy.

Table 14.2 Whole-Body and Regional Dual X-Ray Absorptiometric Assessment of Bone Mineral Content and Soft Tissue Composition of a Man and Woman of Equal Age, Body Weight, and Stature

Characteristics	Man	Woman
Weight (kg)	70.5	70.0
Stature (cm)	174.0	172.6
BMI (kg/m²)	23.3	23.5
BMC (g)*	2761	2230
Fat (kg)	7.1	24.6
Fat (%)	9.9	35.5
LST (kg)	61.1	42.5

Note. BMI = body mass index; BMC = bone mineral content; LST = lean soft tissue.

*Hologic 2000W, Enhanced Whole-Body v5.54

The effects of ingestion of fluid and regional accumulation of water and salts are monitored by DXA as estimates of LST. These maneuvers alter the extracellular fluid volume. Thus, the DXA technique is unable to distinguish between intra- and extracellular fluids in the assessment of LST.

Limitations of Radiographic Methods

The optimism concerning the use of CT and MRI for muscle mass assessment is tempered by practical considerations. The high cost of the instrumentation and the cost per study limit the availability of CT and MRI to major medical centers and thus restrict the routine use of these instruments in determining muscle mass in healthy individuals. Because of the exposure to ionizing radiation, CT cannot be used for routine scans or multiple scans of the same individual and for scans of pregnant women and children. In contrast, MRI does not use ionizing radiation and hence may be presumed a safe alternative to CT for determining muscle mass in various segments of the population. The limited use of MRI in body composition studies may be attributed to limited access of instrumentation, cost, and time required for a scan (about 45 min). In addition, studies determining the validity of CT and MRI to assess body composition, including muscle mass, have yet to be performed.

The DXA method offers a reasonable alternative to CT and MRI for assessment of muscle mass as well as bone and fat. Cost, however, is a limiting factor for routine assessment of body composition. Although ionizing radiation is used with DXA, the patient is exposed to far less than with CT. The sensitivity of DXA to changes in extracellular fluid accumulation does not affect estimates of bone or fat but predictably does influence estimates of whole-body and appendicular LST and muscle mass.

Nuclear Techniques

Knowledge of the physical characteristics and biological distribution of the elements potassium and nitrogen has promoted the development of techniques for measuring these elements and the derivation of models for estimating muscle in the body. This approach has proven useful in the study of body composition in health and disease (Cohn et al. 1980a; Cohn et al. 1980b).

Total Body Potassium

Chemical analyses have shown that potassium is essentially an intracellular cation that is not present in stored triglyceride. In addition, potassium-40, which emits a characteristic gamma ray at 1.46 MeV, exists in the body at a known natural abundance (0.012%) with a very long physical half-life (1.3 billion years). These facts have promoted the estimation of body cell mass and FFM in animals and humans by external counting of potassium-40.

Estimation of TBK by external counting of potassium-40 has basic requirements. A heavily shielded room for counting is needed to minimize natural background radiation from cosmic and terrestrial sources. Detection of gamma rays requires a sensitive detector system positioned in direct proximity to the patient and a data acquisition system capable of absolutely distinguishing the 1.46-MeV gamma ray from other body burdens of contaminant gamma-emitting radioisotopes.

The potassium content of the fat-free body of humans is very consistent (Forbes 1987). Chemical analyses of a limited number of human cadavers yielded values of 2.66 and 2.50 g potassium/kg FFM in men and women, respectively (Forbes 1987). Less direct approaches using whole-body counting and determinations of total body water (2.5 and 2.31 g/kg for men and women, respectively) and densitometry (2.46 and 2.28 g/kg for men and women, respectively; 2.46 and 2.50 g/kg for men) have yielded remarkably similar estimates (Boling et al. 1962).

Estimation of muscle from TBK measurements is limited because other components of the fat-free body contain appreciable amounts of potassium (Forbes 1987). Although the potassium content of skeletal muscle is great (3.1-3.5 g/kg) compared with bone (0.2 g/kg), other organs and tissues have variable amounts of potassium (liver, 3.0 g/kg; adipose tissue, 1.9 g/kg; skin, 0.8 g/kg). Thus, use of TBK measurements to estimate muscle mass is prone to significant error because of the lack of sensitivity of potassium to differentiate muscle *per se* from the other components of the fat-free body.

Total Body Nitrogen

Because of the unique role of body nitrogen as an essential component of body protein, assessment of body nitrogen has been sought as an index of whole-body protein mass. The development of neutron activation techniques permitted the in vivo measurement of TBN.

Whole-body neutron activation systems designed for in vivo studies deliver a moderated beam of fast neutrons to the patient. Capture of those neutrons by atoms of the target elements in the body creates unstable isotopes. The induced unstable isotopes revert to a stable condition by the emission of one or more gamma rays of characteristic energy. Radiation from the irradiated patient is determined from a recording of the radio-spectrum of the emissions. The energy levels

of the emissions identify the activated elements, and the levels of activity indicate their abundance in the body.

The first nuclear methods for direct in vivo determination of TBN in humans used the $^{14}N(n, 2n)^{13}N$ reaction (Cohn and Dombrowski 1971; Boddy et al. 1973), which suffered from interferences from positron emissions from other elements in the body. The later development of the prompt-gamma techniques (Vartsky et al. 1984), using the reaction $^{14}N(n\gamma)^{15}N$, led to the measurement of TBN measurements in body composition assessments in health and disease.

The prompt-gamma technique uses a plutonium-238 beryllium source to provide fast neutrons that are slowed by passage through a deuterium oxide moderator before contacting the patient, who is irradiated bilaterally. This method quantitates TBN absolutely by using total body hydrogen as an internal standard (Vartsky et al. 1984). Because thermal neutron capture is used, the nitrogen and hydrogen interact with the moderated or slow neutrons to produce transiently (10^{-15} s) induced nuclides, ^{15}N and ^{2}H, emitting gamma rays of characteristic energy levels (10.83 and 2.23 MeV, respectively) that are quantitated simultaneously during the neutron irradiation. This procedure requires a 20-min neutron irradiation. This technique reduces errors in counting resulting from differences in irradiation and detection conditions and differences in size and shape of patients, which are critical in longitudinal clinical studies of patients with varying degrees of body habitus.

Nitrogen–Potassium Models

Clinical interest in discriminating body protein mass into muscle and nonmuscle protein components led to development of two conceptual models, which require TBN and TBK measurements and depend on differences in the relative concentrations of nitrogen and potassium in different tissues of the body.

Leeds Model

Burkinshaw and colleagues (1978) assumed concentrations of potassium and nitrogen in muscle and nonmuscle protein to estimate muscle (M_m) and nonmuscle (NM_m) masses of the fat-free body:

$$M_m = [19.25 \text{ (TBK)} - \text{TBN}]/38.34 \quad (14.5)$$

$$NM_m = [\text{TBN} - 8.45 \text{ (TBK)}]/20.20$$

where M_m and NM_m are in kilograms, and TBK and TBN are in grams.

Various attempts to validate this model with indirect methods of body composition assessment led Burkinshaw (1987) to conclude that the Leeds model is "at best only approximately correct."

Birmingham Model

James and colleagues (1984) suggested the separation of body protein into intra- and extracellular components assuming that all TBK is intracellular. The validity of this model cannot be evaluated because there are no published independent estimates of intra- and extracellular protein.

The criticism of the accuracy of the Leeds model has been related to the relative lack of consistency of the potassium–nitrogen ratio in the muscle and nonmuscle protein. Review of the basic data used by Burkinshaw and colleagues (1978) indicates considerable variation of the assumed chemical composition of protein components of the fat-free body (Lukaski 1992).

Burkinshaw and colleagues (1990) acknowledged these limitations and proposed a revision of the original model that required measurement of body water and the absolute elemental composition of the body. This model awaits evaluation in patients with wasting disease and obesity.

Limitations of the Nitrogen– Potassium Models

Nuclear techniques for determining TBK and TBN provide the only noninvasive approaches for estimating body cell mass and protein. The potassium–nitrogen ratio as an index of protein distribution in the body, however, should be used with caution because of the variation of this ratio in the muscle component of the fat-free body. The high cost and restricted availability of whole-body counters and neutron activation analysis systems to a few highly specialized research centers limit the general use of this technology.

Bioelectrical Impedance Analysis

Another approach for assessing muscle mass, regional and whole-body, is bioelectrical impedance analysis (BIA). This method relies on the conduction of an applied electrical current to index conductor volume (Lukaski 1991).

Because skeletal muscle in the upper arm is the dominant conductor in that region, and because the upper-arm musculature has been used as an index of nutritional status, an adaptation of the BIA method has been proposed for the noninvasive assessment of upper-arm composition. Brown and colleagues (1988) used a tetrapolar electrode arrangement of band electrodes and a 1-mA current at 50 kHz to measure the longitudinal resistance of the upper arms of 20 healthy volunteers. A parallel resistance model was used as an equivalent circuit assuming that the resistivity of muscle, fat, and bone was 1.18, 16, and greater than 100 ohm · m, respectively. Reference compositional data was derived from anthropometric assessment (Heymsfield et al. 1982a) and CT measurements of muscle and fat areas.

The upper-arm areas of fat and muscle determined by CT were related to those estimated by using anthropometry and BIA (table 14.3). The areas of muscle and fat estimated by BIA were significantly correlated with CT measurements. The mean differences between bioelectrical impedance and CT estimates of muscle and fat were not significantly different from zero. Anthropometric estimates of muscle and fat areas also were significantly related to CT measurements. The BIA method accounted for more of the variance in predicting muscle and fat areas (96% and 89%) than did anthropometry (89% and 77%). The mean differences between the CT and anthropometric measurements of muscle and fat areas were greater ($p < .01$) than zero. These findings indicate the potential of using bioelectrical impedance measurements to assess upper-arm muscle and fat areas.

A similar approach was used to estimate upper- and lower-leg muscle volumes in healthy men and women. Fuller and colleagues (1999) used a parallel-equivalent model at 50 kHz with a resistivity of 1.49 ohm · m for skeletal muscle. Compared with MRI estimates of muscle volume, BIA explained more variance than anthropometry in estimating thigh (46 vs. 35%) and calf (81 vs. 69%) muscle volume. Furthermore, anthropometry overestimated thigh muscle volume by 40% compared with BIA (10%); it also overestimated calf muscle volume by 20%, whereas BIA underestimated it by 4%. Similarly, Miyatani and colleagues (2000) used bioelectrical impedance and brightness-mode ultrasonography (B-mode ultrasound) to estimate upper-arm muscle volume in 26 young men. The authors found with regression analysis that BIA and B-mode ultrasound from the shoulder

Table 14.3 Correlation Coefficients (*r*) and Differences (cm²) Between Upper-Arm Skeletal Muscle and Fat Areas Determined by Computed Tomography (CT) and Estimated by Anthropometric (ANT) and Bioelectrical Impedance (BIA) Measurements

	MUSCLE AREA		FAT AREA	
	r	Difference	*r*	Difference
CT vs. BIA	.981	0.22	.945	0.73
CT vs. ANT	.943	–2.62	.876	2.61

Adapted, by permission, from B.H. Brown et al., 1988, "Determination of upper arm muscle and fat areas using electrical impedance measurements," *Clinical Physics and Physiological Measurement* 9: 47-55.

to the elbow explained similar variance compared with MRI muscle volumes (94% and 92%). However, 11% less error was associated with BIA (36.2 ± 4.8 cm³) estimates of muscle volume compared with ultrasound (40.3 ± 5.8 cm³). Thus, BIA predicts MRI estimates of appendicular muscle volumes better than anthropometry and ultrasound.

Recently, BIA methods were used to assess dynamic changes and functions of skeletal muscle. Among 12 obese women participating in a controlled weight loss program, serial tetrapolar BIA measurements (50 kHz, 800 μA) of the thigh were highly correlated with changes in muscle mass determined with DXA ($R^2 = .80$, $p < .001$; Lukaski 2000). Miyatani and colleagues (2001) used specially designed BIA acquisition systems and found that impedance at 50 kHz, corrected for limb length, of upper and lower limbs was significantly correlated ($R^2 = .82-.95$) with MRI-determined muscle volumes in 22 men. Also, peak isometric torque developed in elbow flexion or extension and knee flexion or extension was significantly correlated with muscle volume measured by MRI of elbow and knee flexors and extensor muscles. These findings provide key evidence of the practical value of BIA measurements at 50 kHz to assess muscle volumes and to relate them to muscle function.

A recent modification of the BIA method is the prediction of whole-body composition by measuring impedance in a limb-to-limb electrode arrangement. Similar to Miyatani and colleagues (2000), Nunez and colleagues (1997) developed a 50-kHz pressure-contact foot-pad electrode system. This system introduces the alternating current into the plantar surfaces of each foot and determines weight and impedance. In 231 adults, impedance was systematically greater with pressure contact electrodes compared with the standard hand-to-foot electrode arrangement (526 vs. 511 ohm, $p < .0001$). Fat-free mass determined by using DXA was better corre-

lated with a hand-to-foot impedance corrected for height ($R^2 = 91\%$, SEE = 4.3) than foot-to-foot impedance corrected for standing height ($R^2 = 79\%$, SEE = 6.1). Subsequently, Nunez and colleagues (1999) confirmed the importance of height²/impedance in men ($n = 34$, $R^2 = 79\%$) and women ($n = 60$, $R^2 = 72\%$) and found that age also was a significant independent predictor of muscle mass.

The leg-to-leg bioelectrical impedance approach has been used in nonobese and obese adults. Utter and colleagues (1999) reported no difference in fat-free mass estimated with underwater weighing and a commercial leg-to-leg impedance device in 127 women ($-0.01 ± 3.7$ kg, $p > .05$). Leg-to-leg impedance yielded values of FFM that were similar to those measured with hydrodensitometry after 12 weeks of weight loss. These findings suggest the validity of leg-to-leg bioimpedance to determine FFM in the general population as well as during weight loss.

Despite the emergence of regional bioimpedance for body composition assessment, whole-body bioelectrical impedance continues to be used to predict muscle mass. Janssen and colleagues (2000) developed a model to estimate skeletal muscle mass (SMM) in a multiethnic sample of 388 men and women aged 18 to 86 years:

$$SMM = 0.401 \ (Ht^2/R) + 3.825 \ Sex - 0.071 \ Age + 5.102 \qquad (14.6)$$

$$SEE = 2.7, R^2 = .86$$

where R is resistance (ohm) determined at 50 kHz, Ht is height (cm), sex is male = 1 and female = 0, and age is in years.

Cross-validation showed no significant differences between predicted and MRI-measured skeletal muscle mass values. However, further examination showed a significant bias of impedance in prediction of skeletal muscle mass ($r = .20$, $p < .01$). This error was small, 2% to 3% of the MRI-measured value.

Limitations of the Bioelectrical Impedance Technique

The ease of bioelectrical impedance measurements and the availability of instrumentation indicate the strong potential for routine use of this technique to assess regional composition. Some caution is needed because additional factors require further investigation.

The type of electrodes, spot or circumferential, and their placement have not been defined to minimize the error of this technique. In addition, the potential advantage of using a multifrequency, compared with a single-frequency, analyzer to assess regional composition has not been examined. Finally, the effect of edema on the BIA measurements, and hence the estimation of muscle volume and mass, requires investigation.

High-Frequency Energy Absorption (HFEA)

This new technique relies on the absorption of electromagnetic energy by a conductor in a biological specimen to estimate its chemical composition. Two basic assumptions are required: The distribution of water and electrolytes between the intra- and extracellular spaces is relatively constant, and electrolytes are not present in bone or fat. If these premises are correct, then one can estimate the volume of the biological conductor of a limb, and perhaps the whole body.

Michaelsen and colleagues (1993) developed a portable induction coil powered by a 9-V battery for determination of regional HFEA. The HFEA was measured as the difference in voltage from when the induction coil is empty to when it encircles an object. In a calibration study, the HFEA readings of beakers increased in relation to circumference and salt (NaCl) concentrations of the beakers.

The HFEA technique also was evaluated in six volunteers who underwent two determinations of HFEA of the calf and thigh. The coefficients of variation for repeated HFEA measurements ranged from 4% to 8%.

In a second trial, MRI sections of the calf and thigh and HFEA measurements were made in 12 males and 12 females aged 8 to 61 years. Muscle values were expressed as a fraction of the limb volume. HFEA was measured at the same site as the MRI and expressed relative to HFEA of normal saline at the same circumference of the limb. There were significant sex effects in the prediction of calf

HFEA measurements; 80% of the variation of calf HFEA was explained by MRI and sex. In the thigh, 74% of the variance in HFEA was attributed to MRI and coil circumference. The standard error was 6% to 7% for calf and thigh HFEA measurements.

These preliminary data suggest the potential value of HFEA measurements to assess regional muscle volume as an index of nutritional status. The advantage of this approach is its portability for use at the bedside and outside of the laboratory. Because of the reliance on the assumption of a normal distribution of electrolytes and fluids in the intra- and extracellular compartments, additional investigation is need to ascertain the validity of the technique in patients with altered fluid status and acid-base balance.

Summary

Clinical investigators and practitioners seek methods for reliably and accurately assessing regional and whole-body muscle mass. These methods need to be available at costs, in terms of both time and money, that facilitate their routine use in the clinic, the laboratory, and the field. The availability of such a method, or methods, remains to be established.

Table 14.4 compares the characteristics of the methods described in this chapter for in vivo assessment of muscle mass. In general, the techniques with the greatest precision tend to be the most expensive, although acceptable precision is available from the less costly techniques, such as BIA and HFEA. The accuracy of all of these methods has not been established. However, indirect comparisons among these techniques indicate that the radiographic methods, along with BIA and HFEA, have acceptable levels of accuracy (±5% of reference value).

The ease with which a technique is used is an important practical issue. Generally, all of the methods, with the exception of the urinary metabolites of muscle metabolism and the measurement of TBK and TBN, can be performed with minimal technical expertise. The limitation of the use of urinary creatinine and 3-MH is the need for consumption of a meat-free diet and timed urine collections, in addition to the required analytical instrumentation. Although many of the techniques have the potential for whole-body assessments, the routine use is to measure selected regions or limbs of the body.

Selection of a method may depend on the resources available and the experimental question.

Table 14.4   Comparison of Methods for Assessing Skeletal Muscle Mass in Humans

Method	Precision	Accuracy	Utility	Site	Cost
Anthropometry	3[a]	?	4	R, WB	1
Creatinine	2	2	1	WB	3
3-MH	2	2	1	WB	4
CT	5	4	4	R, WB	5
MRI	5	4	4	R, WB	5
DXA	5	4	4	R, WB	4
TBK–TBN	4	2	1	WB	5
BIA	4	?	4	R	1
HFEA	4	3	3	R	3

Note. R = regional and WB = whole-body assessment; 3-MH = 3-methylhistidine; CT = computed tomography; MRI = magnetic resonance imaging; DXA = dual-energy X-ray absorptiometry; TBK–TBN = total body potassium–total body nitrogen; BIA = bioelectrical impedance analysis; HFEA = high-frequency energy absorption.

[a]Ranking system: ascending scale, 1 = least and 5 = greatest.

For example, if radiographic instrumentation is available, then it could be used for either region or whole-body assessments of muscle. In addition, if an experimental intervention is hypothesized to affect a specific region of the body, then any method, with the exception of the urinary metabolites and potassium–nitrogen model, would appear to be adequate.

Evaluation of the available methods for assessment suggests the need to develop portable, inexpensive, safe, and convenient methods for general use. Although sophisticated radiographic tech-niques are available, factors such as cost, access, and, in some cases, exposure to radiation deem it necessary to investigate further the development of BIA and HFEA.

Within the limitations described in this chapter, methods are available for assessment of human muscle mass. The selection of a technique depends on an understanding of the practical considerations and limitations of each method in relation to the experimental hypotheses being evaluated.

Chapter 15

Body Composition, Organ Mass, and Resting Energy Expenditure

Dympna Gallagher and Marinos Elia

Resting energy expenditure (REE) is the minimum amount of energy required to sustain vital bodily functioning in the postabsorptive awakened state. REE is therefore a fundamental biological parameter of living organisms with important relationships to total energy requirement, development, aging, and longevity. In humans it is influenced by sex, race, age, body composition, and illness, among other factors. In healthy humans, REE accounts for more than 60% of total energy expenditure and an even greater proportion in persons with disease.

Relating heat production to body mass has presented a challenge to investigators, who have been limited in their ability to quantify heat-producing organs in living humans. Expressing heat production relative to body mass is essential when comparing thermal or energy flux rates between individuals who differ in size or, more specifically, in heat-producing tissue. The surface law first formulated by Rubner in 1883 (cited by Krogh 1916) and Richet (1889) suggests that animals produce heat in proportion to body surface area. In the early 1900s, age- and sex-specific REE norms based on body weight and stature (Boothby et al. 1936) were developed, and Kleiber (1961) showed that adult mammals differing widely in body size had similar metabolic rates relative to body weight raised to

the 0.75th power. Rubner (1902) proposed expressing REE relative to heat-producing "active tissue mass." Two components are usually considered as representative of whole-body metabolically active tissue: body cell mass (BCM) and fat-free body mass (FFM). BCM is typically estimated as the exchangeable potassium space that can be measured by total body potassium (Moore et al. 1963). The FFM component can be measured using two-component body composition methods (Forbes 1987). Almost all human research over the past 3 decades has explored thermal processes including REE using either BCM (Kinney et al. 1963) or FFM as measures of metabolically active tissue mass.

REE prediction equations based on body weight (Kleiber 1932, 1947) have therefore been followed by models based on the energy requirements of two distinct body composition compartments: fat or adipose tissue and FFM or adipose tissue–free mass, which have markedly different specific energy requirements. FFM is the principal contributor to energy requirements and is commonly used as a surrogate for metabolically active tissue. However, this practice is inherently flawed because it pools together numerous organs and tissues that differ significantly in metabolic rate. The brain, liver, heart, and kidneys alone account for approximately 60% of REE in adults, although their combined weight is less than 6% of total body weight or 7% of FFM (Elia 1992a; Holliday 1971; Gallagher et al. 1998; Grande 1980). The skeletal

This work was supported by Grants DK42618 (Project 4) and HL70298 from the National Institutes of Health.

muscle component of FFM comprises 40% to 50% of total body weight (or 51% of FFM) and accounts for only 18% to 25% of REE (Elia 1992a; Gallagher et al. 1998; Holliday 1971; WHO 1985). REE varies in relation to body size across mammalian species (Kleiber 1932, 1947). Within humans, REE/kg of body weight or FFM is highest in newborns (~56 kcal/kg weight/day; Holliday et al. 1967), declines sharply until 4 years, and declines slowly thereafter reaching adult values (~25 kcal/kg weight/day; Holliday et al. 1967). Among adults, REE is lower in the later adult years, to an extent beyond that explained by changes in body composition (Fukagawa et al. 1990; Klausen et al. 1997; Piers et al. 1998; Poehlman et al. 1993; Vaughan et al. 1991; Visser et al. 1995). That is, the loss of FFM cannot fully explain the decrease (5-25%) in REE in healthy elderly people.

This chapter focuses on understanding REE at the organ tissue body composition level while bringing attention to the inadequacies of using FFM as a metabolically active denominator. We focus primarily on the role of specific high metabolic rate organs within FFM and therefore do not dissect FFM in its entirety.

Measurement of REE

Current techniques used to assess whole-body REE include primarily indirect calorimetry. The measurement is made in the morning after an overnight fast, using either a metabolic cart or a respiratory chamber indirect calorimeter (Heymsfield et al. 1994). The environment should be dimly lit, noise-free, and thermoneutral, while the subject rests comfortably on a bed. After 40 min of rest, a plastic transparent ventilated hood is placed over the head for 20 min. Oxygen and carbon dioxide analyzers contained within the metabolic cart sample the expired air from the subject and analyze the rates of oxygen consumption and carbon dioxide production. The data are displayed and stored in a computer at regular intervals of 25 to 60 s (depending on cart system). Gas exchange results are evaluated during the stable measurement phase and converted to REE (in kJ/day) with a formula such as that of Weir (1949). For a standard alcohol phantom, gas concentration measurements should be reproducible to within 5%.

Measurements made using a respiratory chamber allow the subject to spend several hours to multiple days inside a small room while sleeping, resting, eating, and exercising. Total energy expenditure, including its components resting, sleeping, activity, and thermic effect of feeding, are measured. REE is generally measured in the morning during the period after waking but before movement occurs after an overnight fast.

Quantifying Specific Organ and Tissue Masses

Much of the available information on the mass of specific organs and tissues in health and disease has been acquired from autopsy specimens. Data sets for individual or multiple organs have been compiled from persons of Indian (Jain et al. 1995), Korean (Seo et al. 2000), Japanese (Inoue and Otsu 1987; Ogiu et al. 1997), Danish (Garby et al. 1993), Western European, and North American (International Commission on Radiological Protection 1975) origins (tables 15.1 and 15.2). Although such data sets provide insight into the influences of age, sex, race, and geographical location or cohort effects on organ and tissue mass, the lack of corresponding REE data before death prevents us from estimating the respective contributions of specific organs and tissues to REE.

The relatively recent availability of imaging techniques for the *in vivo* estimation of organ and tissue masses presents a major breakthrough, making possible both cross-sectional and longitudinal evaluations of whole-body and regional body composition. Magnetic resonance imaging (MRI) and computed tomography (CT) are described in chapter 7. The *in vivo* measurement in humans of skeletal muscle (SM) mass, adipose tissue (AT) mass and its distribution, and masses of several organs is now possible. SM and AT, including subcutaneous adipose tissue (SAT), visceral adipose tissue (VAT), and intermuscular adipose tissue (IMAT), are routinely measured using whole-body multislice MRI protocols in some laboratories.

At the New York Obesity Research Center, subjects are placed on a 1.5-T scanner (General Electric, 6X Horizon, Milwaukee, WI) platform with their arms extended above their heads. The adult protocol involves the acquisition of approximately 40 axial images of 10 mm thickness, and at 40-mm intervals across the whole body (Gallagher et al. 2000; Ross 1996). The pediatric protocol involves the acquisition of approximately 35 axial images of 10-mm thickness, and at 35-mm intervals across the whole body (Hsu et al. 2003). Liver, kidney, and spleen images (with subjects in a postabsorptive state) are produced using an axial T1-weighted spin echo sequence with 5-mm

Table 15.1 Weights of Organs (g) as a Function of Ethnicity and Age

	20-39		40-59		60+	
Ethnicity	M	F	M	F	M	F
BRAIN						
American[a]						
Brain weight	1,441.6	1,294.8	1,406.8	1,276.4	1,349.5	1,215.2
Body weight	69,908	55,953	72,541	57,885	69,809	57,594
Height (cm)	176.2	164.4	174.7	162.2	172.8	160.4
Indian[b]						
Brain weight	1,273.2	1,170.9	1,252.8	1,165.6	1,211.0	1,116.0
Body weight	54,329	48,385	55,295	48,559	51,500	50,900
Height (cm)	164	153.2	163.9	153.8	162.2	153.3
Japanese 1988[c]						
Brain weight	1,456.1	1,315.5	1,424.1	1,295.2	1,370.0	1,265.3
Japanese 1997[d]						
Brain weight	1,436.5	1,326.1	1,383.6	1,274.5	1,351.8	1,208.0
Body weight	62,245	50,370	58,641	50,959	53,034	44,540
Height (cm)	167.3	155.4	163.1	152.7	161.5	146.9
Australian (Caucasian)[e]						
Brain weight	1,433.9	1,281	1,433.6	1,260.9	1,388.5	1,234.5
Australian (Aboriginal)[e]						
Brain weight	1,281.3	1,119.4	1,261.5	1,116.1	1,181	—
Venezuelan[f]						
Brain weight	1,275.7	1,170.8	1,239.9	1,139.1	1,172.7	1,098.9
HEART						
Indian						
Heart weight	255.1	220.4	274.8	243.9	266.0	260.0
Body weight	54,329	48,385	55,295	48,559	51,500	50,900
Height (cm)	164	153.2	163.9	153.8	162.2	153.3
Japanese 1988						
Heart weight	364.9	291.4	399.8	344.4	413.4	356.5

(continued)

Table 15.1 *(continued)*

Ethnicity	20-39		40-59		60+	
	M	F	M	F	M	F
Japanese 1997						
Heart weight	319.9	251.0	339.1	294.6	338.7	314.9
Body weight	62,245	50,370	58,641	50,959	53,034	44,540
Height (cm)	167.3	155.4	163.1	152.7	161.5	146.9
Caucasian[g]						
Heart weight	281.8	245.6	301.1	258.5	319.5	291
African American[g]						
Heart weight	297	243.7	309.2	271.6	322.7	285.5
American[h]						
Heart weight	236.7	212.6	340.9	236.1	347.6	254
Body weight	—	52,700	62,831	58,033	61,241	50,241
American[i]						
Heart weight	319.2	258.6	318.7	278.7	323.9	248.8
LIVER						
Indian						
Liver weight	1,132.0	1,200.4	1,266.3	1,236.8	1,141.0	1,312.0
Body weight	54,329	48,385	55,295	48,559	51,500	50,900
Height (cm)	164	153.2	163.9	153.8	162.2	153.3
Japanese 1988						
Liver weight	1,590.7	1,326.8	1,566.8	1,332.2	1,283.6	1,058.8
Body weight						
Japanese 1997						
Liver weight	1,465.2	1,222.8	1,448.9	1,259.8	1,184.2	1,010.8
Body weight	62,245	50,370	58,641	50,959	53,034	44,540
Height (cm)	167.3	155.4	163.1	152.7	161.5	146.9
Caucasian						
Liver weight	1,723.7	1,473.0	1,661.2	1,411.6	1,498.9	1,295.0
African American						
Liver weight	1,678.8	1,439.6	1,574.1	1,359.1	1,402.2	1,179.4

Ethnicity	20-39		40-59		60+	
	M	F	M	F	M	F
KIDNEY						
Indian 1995						
Kidney weight	215.3	203.6	208.4	399.4	187.0	200.0
Body weight	54,329	48,385	55,295	48,559	51,500	50,900
Height (cm)	164	153.2	163.9	153.8	162.2	153.3
Japanese 1988						
Kidney weight	321.0	278.6	324.1	269.7	281.2	231.5
Body weight	—	—	—	—	—	—
Japanese 1997						
Kidney weight	295.5	259.6	310.4	272.1	271.4	224.3
Body weight	62,245	50,370	58,641	50,959	53,034	44,540
Height (cm)	167.3	155.4	163.1	152.7	161.5	146.9
Caucasian						
Kidney weight	331.0	316.9	351.8	293.6	327.8	274.3
African American						
Kidney weight	353.6	341.2	348.4	311.9	329.5	293.7
SPLEEN						
Indian						
Spleen weight	141.0	136.7	134.4	126.1	121.0	116.0
Body weight	54,329	48,385	55,295	48,559	51,500	50,900
Height (cm)	164	153.2	163.9	153.8	162.2	153.3
Japanese 1988						
Spleen weight	137.4	126.9	119.5	108.5	98.1	78.9
Body weight	—	—	—	—	—	—
Japanese 1997						
Spleen weight	102.5	95.4	93.1	88.9	75.1	67.1
Body weight	62,245	50,370	58,641	50,959	53,034	44,540
Height (cm)	167.3	155.4	163.1	152.7	161.5	146.9
Caucasian						
Spleen weight	142.0	125.6	139.6	101.1	129.6	96.4

(continued)

Table 15.1 *(continued)*

Ethnicity	20-39 M	20-39 F	40-59 M	40-59 F	60+ M	60+ F
African American						
Spleen weight	128.7	100.4	117.0	96.4	105.3	74.5

Note. Data from all studies were collected via autopsies. Inclusion and exclusion criteria were as follows:

Ogiu et al. 1997	Cadavers with severe postmortem changes, extensive burn injuries, damage in many organs, and pathological lesions in many organs were excluded, as were those receiving therapeutic blood transfusion or infusion of physiological fluids.
Jain et al. 1995	Only cases in which death occurred within 72 hr of arrival at the hospital and when the autopsy was performed within 24 hr after death were considered. Subjects had to be physiologically and nutritionally normal and free of pathological changes.
Tanaka et al. 1989	Only cases of sudden death were included.
Bean 1926	All organs with gross pathological lesions were excluded. Spleens in typhoid fever cases were discarded as were all livers in chronic passive congestion cases.

Although the Tanaka et al. and Bean studies did not include tables containing the height and weight of the subjects, general ranges of height and weight were made available as follows:

	BODY WEIGHT RANGE (G) M	BODY WEIGHT RANGE (G) F	HEIGHT RANGE (M) M	HEIGHT RANGE (M) F
Tanaka et al. 1989	60,000-63,000	47,000-52,000	1.6-1.65	1.50-1.55
Bean 1926	—	—	1.35-1.75 (majority 1.5-1.7)	1.45-1.9 (majority 1.6-1.8)

[a]Dekaban and Sadowsky 1978; [b]Jain et al. 1995; [c]Tanaka et al. 1989; [d]Ogiu et al. 1997; [e]Harper and Mina 1981; [f]Sanchez et al. 1997; [g]Bean 1926; [h]Womack 1983; [i]Zeek 1942.

slice thickness, no interslice gap, and a 40 × 40 cm² (256 × 192/2 number of excitations) field of view. Approximately 40 slices are acquired from the diaphragm to the base of the kidneys. Brain images (~29) are produced using a body coil with a fast spin echo T2-weighted sequence with 5-mm contiguous axial images and a 40 × 40 cm² (256 × 256/1 number of excitations) field of view. Image analysis software (Tomovision, Montreal, CA) is used to analyze images on a PC workstation (Gateway, Madison, WI). MRI-volume estimates are converted to mass using the assumed density for each tissue and organ (table 15.3). In our laboratory, the technical errors for four repeated readings of the same four scans by the same observer of MRI-derived SM, SAT, VAT, and IMAT volumes are 1.9%, 0.96%, 1.97%, and 0.65%, respectively.

Eight MRI scans of the liver in normal, healthy, 26- to 70-year-old men and women were analyzed by two different operators in order to estimate reading error. The standard deviation (SD) of the mass differences was 0.14 kg with a mean weight of 1.58 kg. Left ventricular mass is used as a surrogate for heart mass and is measured either by echocardiography (Devereux and Reichek 1977; Jahn et al. 1978) or cardiac gated MRI (Boxt et al. 1992; Katz et al. 1993) with standard techniques. The technical error for repeated echo measurements of the same scan by the same observer for left ventricular mass is 1.1%. The coefficient of variability for cardiac MRI based on two readings of recordings from eight subjects by a single reader is 5%.

Table 15.2 Brain, Body, Liver, Heart, Spleen, and Kidney Weights of Children

Age (years)	CAUCASIAN (AMERICAN)[a] M Brain weight	M Body weight	F Brain weight	F Body weight	DANISH[b] M Brain weight	F Brain weight	JAPANESE[c] M Brain weight	M Body weight	F Brain weight	F Body weight	INDIAN[d] M Brain weight	M Body weight	F Brain weight	F Body weight
Birth	353	3.4	347	3.4	644	560	565	4.8	502	4.6	—	—	—	—
0.5	—	5.7	—	7.3	913	819	926	8.4	907	7.7	—	—	—	—
1	971	10.1	894	9.8	1,126	1,050	1,117	10.0	1,036	9.5	783	7	833	7.5
2	1,076	12.6	1,012	12.3	1,249	1,176	1,146	11.5	1,206	11.7	967	8.8	911	7.9
3	1,179	14.6	1,076	14.4	1,317	1,213	1,380	14.3	1,264	14.6	1,072	10.3	970	10.7
4	1,290	16.5	1,156	16.4	1,419	1,243	1,389	15.6	1,250	16.3	1,101	12.3	1,055	12.2
5	1,275	19.4	1,206	18.8	1,480	1,284	1,373	17.8	1,300	16.8	1,236	14.5	1,030	14.9
6	1,313	21.9	1,225	21.1	1,437	1,286	1,381	19.9	1,269	19.9	1,183	16.9	1,066	14.7
7	1,338	24.5	1,265	23.7	1,424	1,328	1,454	21.7	1,290	21.2	1,143	15.4	1,154	17.6
8	1,294	27.3	1,208	26.4	1,457	1,400	1,383	22.8	1,357	26.9	1,309	19.6	1,150	18.8
9	1,360	29.9	1,226	28.9	1,489	1,360	1,453	29.4	1,310	26.7	1,273	22.2	1,187	24.0
10	1,378	32.6	1,247	31.9	1,501	1,550	1,453	29.3	1,371	33.6	1,278	22.7	1,085	21.4
11	1,348	35.2	1,259	35.7	1,397	1,380	1,505	34.5	1,398	39.4	1,243	24.0	1,078	24.3
12	1,383	38.3	1,256	39.7	1,483	1,356	1,517	40.7	1,485	37.4	1,259	26.2	1,193	30.7
13	1,382	42.2	1,243	45.0	1,564	1,453	1,513	41.6	1,353	45.1	1,316	32.9	1,260	37.5
14	1,356	48.8	1,318	49.2	1,484	1,322	1,459	50.8	1,352	52.7	1,274	35.3	1,168	37.4
15	1,407	54.5	1,271	51.5	1,483	1,378	1,478	58.7	1,361	48.5	1,274	37.7	1,154	39.2
16	1,419	58.8	1,300	53.1	1,547	1,383	1,496	58.6	1,383	50.5	1,269	44.2	1,205	43.5
17	1,409	61.8	1,254	54.0	1,582	1,380	1,468	57.3	1,294	49.3	1,307	45.7	1,158	43.4
18	1,486	63.1	1,312	54.4	1,491	1,359	1,444	58.2	1,370	50.2	1,292	47.6	1,156	46.0

LIVER (G) AND HEART (KG) WEIGHTS

(continued)

Table 15.2 (continued)

LIVER (G) AND HEART (KG) WEIGHTS

Age (years)	CAUCASIAN (AMERICAN)[d] Liver weight M	F	JAPANESE[e] Liver weight M	F	INDIAN[f] Liver weight M	F	CAUCASIAN (AMERICAN)[d] Heart weight M	F	JAPANESE[e] Heart weight M	F	INDIAN[f] Heart weight M	F
Birth	124	125	192	181	—	—	19	20	26.9	23.6	—	—
0.5	300	240	321	248	—	—	—	—	41.3	40.2	—	—
1	400	390	406	388	290	292	54	48	50.8	49.1	33	50
2	460	450	483	447	365	351	63	62	57.1	61.8	56	40
3	510	500	501	475	407	360	73	71	70.2	71.6	69	57
4	555	550	609	571	522	448	83	80	83.6	84.8	65	60
5	595	590	643	644	559	507	95	90	97.3	96.4	81	91
6	630	635	709	644	513	653	103	100	112.6	101.9	90	95
7	665	685	675	648	610	596	110	113	118.0	112.1	98	84
8	715	745	824	834	607	633	122	126	131.9	135.6	99	110
9	770	810	902	861	610	693	132	140	152.4	148.5	105	120
10	850	880	948	1,008	777	768	144	154	163.7	169.8	125	115
11	950	960	1,015	1,106	837	870	157	168	173.6	188.8	121	138
12	1,050	1,080	1,111	1,072	888	907	180	188	213.8	195.0	133	156
13	1,150	1,180	1,043	1,095	1,003	910	202	207	209.1	195.3	162	165
14	1,240	1,270	1,302	1,285	965	1,000	238	226	265.6	236.6	188	168
15	1,315	1,330	1,322	956	1,040	1,176	258	238	272.4	208.3	194	176
16	1,380	1,360	1,330	1,134	1,214	1,006	282	243	276.9	223.0	217	185
17	1,450	1,380	1,247	1,129	1,347	1,101	300	247	283.8	211.1	225	197
18	1,510	1,395	1,204	1,296	1,292	1,125	310	250	292.8	221.1	241	214

SPLEEN WEIGHT AND KIDNEY WEIGHT (G)

Age (years)	SPLEEN WEIGHT CAUCASIAN (AMERICAN)[g] M Spleen weight	CAUCASIAN (AMERICAN)[g] F Spleen weight	JAPANESE[h] M Spleen weight	JAPANESE[h] F Spleen weight	INDIAN[i] M Spleen weight	INDIAN[i] F Spleen weight	KIDNEY WEIGHT CAUCASIAN (AMERICAN)[g] M Kidney weight	CAUCASIAN (AMERICAN)[g] F Kidney weight	JAPANESE[h] M Kidney weight	JAPANESE[h] F Kidney weight	INDIAN[i] M Kidney weight	INDIAN[i] F Kidney weight
Birth	8	6	16.9	13.9	—	—	24	24	36.6	32.7	—	—
0.5	—	—	36.4	30.6	—	—	—	—	56.2	54.2	—	—
1	35	34	40.6	37.3	28	26	72	65	68.3	64.2	46	49
2	42	41	41.2	41.5	34	27	85	75	78.5	77.2	59	53
3	48	47	42.4	44.2	50	38	93	84	85.2	83.7	75	74
4	53	52	48.3	44.9	47	42	100	93	98.3	95.2	84	63
5	58	57	63.2	60.8	56	44	106	102	111.1	122.3	95	90
6	62	62	77.6	64.3	52	70	112	112	141.9	127.2	96	96
7	64	67	56.0	54.4	100	64	120	123	126.6	125.6	96	86
8	68	71	66.2	67.5	64	79	128	135	134.6	153.9	96	100
9	73	77	82.5	83.6	83	73	138	148	152.3	165.4	119	117
10	82	85	89.5	85	81	78	150	163	174.7	173.1	132	120
11	91	93	102	115	86	83	164	180	177.8	184.8	128	160
12	101	103	116.8	101.7	83	77	178	195	219.6	219.6	142	163
13	111	112	93.4	79.7	99	85	196	210	194.9	196.8	175	175
14	127	120	135.6	127.2	114	99	212	222	271.3	231.5	163	172
15	135	127	128.6	90	133	130	229	230	270.3	215.4	180	194
16	145	134	131	116	138	110	224	236	272	215.4	201	171
17	152	140	104.8	108.9	160	129	260	240	256.7	257.6	215	201
18	157	146	99.7	96.6	169	136	270	244	266.4	249.3	225	197

[a]Haddad et al. 2001; [b]Voigt and Pakkenberg 1983; [c]Ogiu et al. 1997; [d]Jain et al. 1995; [e]Haddad et al. 2001; [f]Ogiu et al. 1997; [g]Ogiu et al. 1997; [h]Haddad et al. 2001; [i]Jain et al. 1995.

Haddad et al. (2001) used data compiled by Altman and Dittmer (1962). In the Japanese, Indian, and Danish studies, data were gathered from autopsy reports from accidental deaths. Inclusion and exclusion criteria were as follows:

Table 15.2 *(continued)*

Ogiu et al. 1997	Cadavers with severe postmortem changes, extensive burn injuries, damage in many organs, and pathological lesions in many organs were excluded, as were cases receiving therapeutic blood transfusion or infusion of physiological fluids.
Jain et al. 1995	Only cases in which death occurred within 72 hr of arrival at the hospital and when the autopsy was performed within 24 hr after death were included. Subjects had to be physiologically and nutritionally normal and free of pathological changes.
Voigt and Pakkenberg 1983	All cases with severe brain lesions as well as cases missing vital information were excluded.

Table 15.3 Organ and Tissue Resting Metabolic Coefficients

	Weight (kg)[a]	Density (kg/L)[a]	Resting metabolic rate (kJ · kg^{-1} · day^{-1})[b]
Skeletal muscle	28.0	1.04	55.0
Adipose tissue	15.0	0.92	19
Liver	1.8	1.05	840
Brain	1.4	1.03	1,008
Heart	0.33	1.03	1,848
Kidneys	0.31	1.05	1,848
Residual	23.16	*	50

Note. Residual mass was not assigned a density but was calculated as body mass minus sum of other measured mass components.
[a]Derived from 70-kg Reference Man (Snyder 1975) and [b]Elia (1992a, 1997).

Specific Metabolic Rates of Organs and Tissues

Tissue-specific metabolic rate can be assessed by measuring the arteriovenous concentration difference of oxygen across a tissue coupled with blood flow. Much of the available information on organ-specific metabolic rates has been acquired using this approach. These coefficients were primarily derived from studies of oxygen consumption of tissue using arteriovenous catheterization techniques. The arteriovenous concentration difference of oxygen is multiplied by the blood flow to establish a rate of oxygen uptake by that tissue. Because the energy equivalent of oxygen varies little with the type of fuel being metabolized (Elia and Livesey 1992), and because the measurement of blood flow often relates to 100 g of tissue (or flow per estimated 100 g or ml of tissue can be estimated from an assumed mass of tissue), an approximate value for tissue-specific coefficient of energy expenditure can be readily calculated. The values were established for healthy individuals, typically aged 20 to 50 years, after an overnight fast in a rested steady state. These values do not apply during physical activity or after meal ingestion. Table 15.3 shows the resting metabolic coefficients considered representative for several organs and tissues based on a summary of earlier literature reported by Elia (1992a). Residual mass was assigned an energy expenditure of 50 kJ · kg^{-1} · day^{-1} as reported by Elia (1997) and consists of all tissues not listed in table 15.2. For example, included in the "residual mass" is the gut (sum of small intestine, large intestine, and stomach), which accounts for approximately 5% of REE (Elia 1996).

Modeling REE

Recent attention has been given to modeling REE based on available information on organ- and tissue-specific metabolic rates combined with the mass of these tissues. Wang and colleagues (2001) reported on a new viewpoint of Kleiber's law demonstrated through an organ–tissue-level REE model, with five components (liver, brain, kidneys, heart, and remaining tissues). Across species, the authors showed that with increasing body size, specific organ–tissue metabolic rate for all five components had negative exponents that were directly proportional to body mass (–0.06 to –0.27). Moreover, the mass of each component was directly proportional to body mass (0.76-1.01).

Whole-body REE can be calculated from organ–tissue mass (REE_c) and then compared with REE measured using indirect calorimetry (REE_m) for individuals or groups. REE (in kJ/day) of each organ–tissue component (subscript i) can be calculated using the following equation (Gallagher et al. 1998)

$$REE_i = OMR_i \times M_i \qquad (15.1)$$

where OMR (organ metabolic rate) is the metabolic rate constant (in $kJ \cdot kg^{-1} \cdot day^{-1}$) for each organ–tissue component (table 15.3) and M is the mass of the corresponding organ–tissue (in kg). Whole-body REE (in kJ/day) is calculated as the sum of the seven individual organ–tissue REEs (Gallagher et al. 1998)

The whole-body REE equation is as follows:

$$REE_c \sum_{i=1}^{7} (REE_i) \qquad (15.2)$$

$$REE_c = 1008 \times M_{brain} + 840 \times M_{liver} + \\ 1848 \times M_{heart} + 1848 \times M_{kidneys} + \\ 55 \times M_{SM} + 19 \times M_{AT} + 50 \times M_{residual} \qquad (15.3)$$

In the following sections, data are discussed using this approach.

REE and Organ Metabolic Rates During the Life Span

Organs and tissues vary in their energy expenditure across the life span. The OMR values therefore may differ during development stages, growth, and senescence.

Growth and Childhood

The course of metabolic rate during growth and in aging remains understudied. REE is known to be relatively high in children, compared with adults, but the reason for this difference is not well understood. It is hypothesized that the explanation may lie in differences in body composition or in oxidative requirements of tissues. The inability to make precise and highly differentiated measurements of body composition in children and the nonlinear growth and development that occur

during puberty have challenged investigators and hindered understanding of children's energy expenditure. The manner in which REE is usually estimated is based on either whole-body weight or two-compartment models of body composition, which use FFM as a surrogate for metabolically active tissue. This practice inappropriately pools together diverse organs and tissues with different oxidative activities and assigns a single metabolic coefficient to this nonhomogeneous compartment. A more refined model would factor in differences and changes in the proportion of these tissues during growth, development, and aging and might thereby account for the observed differences in REE presently attributed to race, sex, and age.

Compared with adults, children have higher REE per kilogram of body weight or per kilogram of FFM (Elia 1992a, 1997; Holliday 1971, 1986; Weinsier et al. 1992), which declines steadily during the growth years. Weinsier and colleagues (1992) reviewed 31 data sets from 1,111 subjects and found a decrease in REE with increasing age. REE was the highest in preschool children, followed by adolescents and then adults. Holliday (1971, 1986) hypothesized that the decrease in REE during growth and development is secondary to changes in body composition. In the first year of life, organs grow in proportion to body weight; thereafter, organ growth rates decelerate (Holliday et al. 1967) and are variable across organs (figure 15.1). Accordingly, among children from birth to 2 years, REE expressed per kJ/kg body weight or per kJ/kg organ weight per day is relatively stable (Elia 1997) because the increases in body and organ weights are proportional. By the age of 5 years, total brain volume has reached approximately 95% of adult size (Giedd 1999), and by 6 years, heart diameter is 80% of adult values (Simon and Tanner 1972). Skeletal muscle mass grows at a faster rate than body weight after the first year of life (Holliday 1986).

Normal growth of mammalian tissues is associated with numerous metabolic and morphological changes. Brain has one of the highest specific metabolic rates (table 15.3) and is thus a good representation of high metabolic rate organs. Overall brain size reaches its maximum value by the end of the first decade (Caviness et al. 1996; figure 15.1). Before age 7, the relative rate of volume increase of gray matter structures is significantly greater than that of white matter, although white matter volume may continue to increase through the end of the second decade (Pfefferbaum et al. 1994; Reiss et al. 1996). Reported changes in the ratio of cerebral gray to white matter volume from the end of the

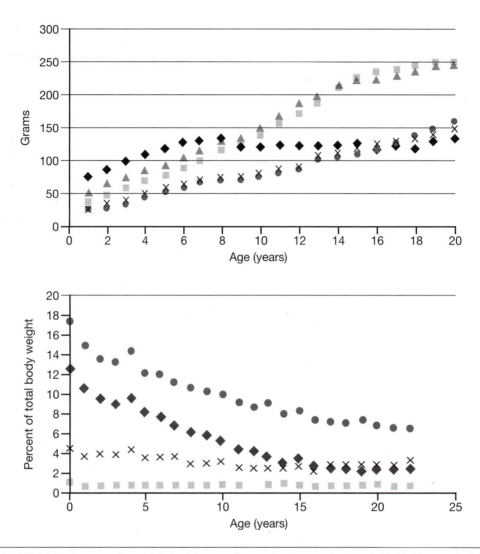

Figure 15.1 (a) Weights of organs from birth to 20 years from autopsy reports (Boyd 1962) (diamonds, brain; squares, heart; ×, liver; circles, spleen; triangles, kidneys). (b) Weight of individual organs (diamonds, brain; squares, heart; ×, liver) and combined organs (circles, total) as a percentage of body weight from birth to 23 years (Abt 1923).

first decade to the end of the second decade range from 1.8 to 1.3 (Jernigan and Tallal 1990).

The metabolic rate per gram of individual organs has been claimed to change little during growth and development (Holliday 1971); however, this assumption has not been scientifically tested. Changes in oxygen consumption of brain (expressed per 100 g of brain per minute) were found not to change in 3- to 11-year-old children (Kennedy and Sokoloff 1957). However, when expressed per unit body weight or FFM, REE decreases from years 1 through 20 (Elia 1992a). With regard to brain metabolic rate, Van Bogaert and colleagues (1998) using positron emission tomography (PET) combined with fluoro-deoxyglucose investigated age-related changes in regional distribution of glucose metabolism adjusted for global activity in 6- to 38-year-old subjects. Their

findings were that (a) adjusted regional glucose metabolism was highly variable from childhood through adulthood in numerous brain structures, with changes most evident in the thalamus, selected cortical regions (insular, posterior and anterior cingulated, frontal and postcentral cortexes), basal ganglia, and brain stem; and (b) relative metabolic activity in these regions increased primarily before age 20 years, reaching maximal values at 25 to 30 years, and decreased thereafter. These data highlight the error associated with assuming that brain metabolic rate (oxygen uptake rather than glucose uptake) is uniform and constant throughout the growth and development years.

The use of glucose uptake as a measure of the overall energy expenditure or oxygen consumption of the brain has two main limitations, however.

First, it may not be the only fuel that is used by the brain. After an overnight fast, ketone bodies account for a small proportion (<5%) of oxidative metabolism, but they become the dominant fuel during prolonged starvation (i.e., they displace glucose as the major fuel). Second, some of the glucose carbon is not oxidized but released as 3-carbon fragments (mainly lactate and to a smaller extent pyruvate; i.e., there is glycolysis without oxidation of the glucose carbon in the Krebs cycle). Dietze and colleagues (1980) provided a graph showing the proportion of lactate and pyruvate output as a percentage of glucose uptake: up to about 10% after an overnight fast, rising to about 25% after a 60-hr fast, and 30% or more after a 120-hr fast. In prolonged fasting (38-41 days in the obese), the percentage rises to more than 50% (Owen 1967). What is also interesting is that visual stimulation (which excites the occipital cortex) and somatosensory stimulation of the finger pads (which excites the sensorimotor cortex) also stimulate local glycolysis (Fox and Raichle 1986; Fox et al. 1988). Thus, there may be a 30% to 50% increase in glucose utilization but less than a 10% increase in oxygen consumption.

Other high metabolic rate organs including heart, kidneys, liver, and spleen increase in size through the end of the second decade (figure 15.1a) and are highly correlated with height and weight (Ogiu et al. 1997). Because REE/kg in humans declines during growth from an assumed approximately 13 kJ to 6 kJ (~50% reduction), this decline may be attributable to a decrease in the organ metabolic rate (OMR) relative to organ size, a decrease in the relative organ weight in relation to body weight, or both.

Limited data are available on OMR in humans because of the invasive nature of traditional techniques, involving arterial cannulation. Data in rats (Bertalanfey and Pirozynski 1951) suggest that OMR remains constant during growth. Holliday and colleagues (1967) suggested that upward from 10 to 12 kg of body weight, REE/kg declines during growth because the internal organs become a smaller proportion of body weight. However, no data are available on the metabolic rate per gram of specific organs during growth. Skeletal muscle and adipose tissues, on the other hand, represent low metabolic rate tissues in the rested state (table 15.3). Despite significant increases in the mass of these tissues relative to body weight during growth, REE per kilogram of body weight decreases.

A reduction in organ growth coupled with an increase in skeletal muscle growth could account for a decrease in whole-body REE adjusted for FFM with increasing age in children. This has been the basis for the hypothesis that the decline in REE during growth is a result of a decrease in the proportion of the more metabolically active FFM components (Holliday 1986). One study reported on the relationship between REE and *in vivo* body composition in healthy, growing children, for whom select organ and tissue measures were available, thereby allowing for an evaluation of the relative importance of two hypotheses in accounting for changes in REE during growth and development. Hsu and colleagues (2003) hypothesized that if REE in children can be accurately estimated from measured organ–tissue mass using adult metabolic rate coefficients (table 15.3), then one could conclude that the metabolic rate per unit organ–tissue mass remains relatively constant from childhood to early adulthood and that changes in body composition are responsible for the decline in REE. On the other hand, if REE in children continued to be underestimated after organ mass was taken into account, then both changes in body composition and in metabolic rate coefficients are likely to play a role in changing REE during the growth years.

The associations between REE measured by indirect calorimetry (REE_m) and several sets of independent variables (adipose tissue–free mass [ATFM], AT, age group, brain, liver, kidneys, and heart) were investigated with multiple regression analyses (table 15.4). AT and ATFM combined explained 72% of the variance in REE_m (Model 1), and the addition of age group increased the explained variance to 84%, while confirming the significant role of age in predicting REE_m (Model 2). Because the contribution of brain and liver masses to ATFM was greater in children than adults, the sum of brain and liver mass was added to Model 2, thereby increasing the explained variance to 88% (Model 3). In Model 3, age group continued to make a highly significant contribution ($p < .001$) despite the fact that the intercept ceased to be statistically significant. As discussed by Hsu and colleagues (2003), the variables in Model 3 may have identified the important compartments that contribute to oxygen consumption so that an arbitrary constant in the formula is no longer necessary and the regression line passes through zero. One interpretation is that the results do not support the hypothesis that differences in body composition alone do not adequately explain differences in REE between children and adults. Furthermore, figure 15.2 shows that after organ–tissue-specific metabolic constants derived from healthy adults are accounted for (Elia 1992a), they appear to be inadequate for predicting REE in children.

Table 15.4   Resting Energy Expenditure (REE)–Body Composition Regression Models in Children

Model	Variable	a	SEE	p	R^2
1. $REE_m = ATFM + AT$	Intercept	740.2	93.8	.001	
	ATFM	16.0	2.4	.001	.72
	AT	4.6	10.5	.66	
2. $REE_m = ATFM + AT + Age\ Group$	Intercept	580.7	81.1	.001	
	ATFM	23.7	2.6	.001	.84
	AT	11.1	8.2	.18	
	Group	−382.9	89.3	.001	
3. $REE_m = ATFM + AT + Age\ Group + B\&L$	Intercept	89.5	211.5	.67	
	ATFM	14.2	4.5	.005	.88
	AT	12.0	7.4	.12	
	Group	−361.6	81.5	.001	
	B&L	312.9	126.3	.021	

REE_m = resting energy expenditure measured in kJ/day; AT = adipose tissue in kg; ATFM = adipose tissue–free mass in kg by magnetic resonance imaging; Age group: 0 = children, 1 = adults; B = brain; L = liver; SEE = standard error of estimate; R^2 = explained variance.

[a]Estimate of regression coefficient.

From A. Hsu et al., 2003, "Larger mass of high metabolic rate organs does not explain higher REE in children," *American Journal of Clinical Nutrition*, 77: 1056-11. Reproduced with permission by the *American Journal of Clinical Nutrition*, © Am. J. Clin. Nutr. American Society for Clinical Nutrition.

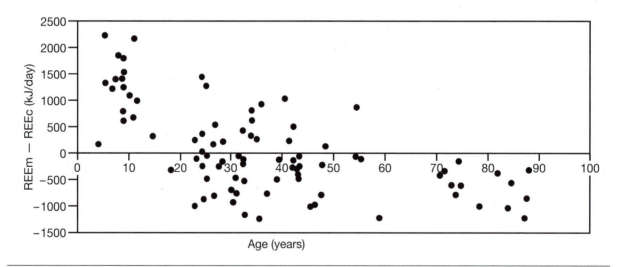

Figure 15.2   The difference between measured resting energy expenditure (REE_m) and calculated REE (REE_c) as a function of age in healthy females and males.

In summary, these data support the hypothesis that the proportion of FFM as certain high metabolic rate organs, specifically liver and brain, is greater in children compared with young adults. However, after this disproportion is accounted for, the specific organ–tissue metabolic constants available in healthy adults (Elia 1992a) appear to be inadequate to account for REE in children (figure 15.2) and elderly adults. These results therefore imply that the decline in REE per kilogram of body weight (or per kilogram of FFM) during growth is likely attributable to both changes in body composition and changes in the metabolic rate of individual organs or tissues. An alternative or additional explanation is that other tissues that are prominent in early childhood have not been adequately accounted for in our model, for example, thymus, lymphoid tissue, and red bone marrow. A definitive answer requires the direct measurement of tissue-specific metabolic rates during growth and development, and such information is currently lacking.

Several important questions remain to be answered with regard to REE in children: When in growth and maturation does the metabolic rate of children reach adult levels? Does this occur earlier in obese children? (Amador et al. 1992; DeSimone et al. 1995; Forbes 1977). A significant contribution to understanding the basis of pediatric energy requirements would be made if it could be established how the most metabolically active fractions of body mass contribute to measured REE. In addition, defining energy requirements for normal growth and development might be more feasible.

Young and Middle Adulthood

During the adult years (after the second decade), REE is hypothesized to decline by 1% to 2% per decade, assuming weight stability during this same period (Keys et al. 1973). In the absence of REE data in humans across the life span, the measurement of REE in Fischer 344 rats across the life span (McCarter and Palmer 1992) supports what has been pieced together from various studies in humans. In rats, REE per unit of lean body mass was highest at youngest ages, declined rapidly (~50%) from 6 weeks to 3 months, declined more slowly (~1.6% per month) from 3 to 18 months, and increased from 18 to 24 months (8.9% per month). These data are consistent with findings of Rubner (1908), who suggested that a constant metabolic potential for different species related to metabolic events occurs only after the attainment of sexual maturity.

Gallagher and colleagues (1998) compared REE_c and REE_m in 13 healthy young adults (31.2 ± 7.2 years) where REE_c was based on individual or combined organ masses (measured by MRI) as described in equation 15.3. REE_c and REE_m were 6,962 ± 1,455 kJ/day and 7,045 ± 1,450 kJ/day, respectively (p nonsignificant). REE_c was highly correlated with REE_m (REE_c = 352.4 + 0.94 × REE_m; r = .94, standard error of estimate = 540 kJ/day, p = .0001). A Bland–Altman plot showed no significant trend (r = –.01, p nonsignificant) between calculated and measured REE difference (i.e., REE_m – REE_c; 83 ± 525 kJ/day) versus the average of calculated and measured REE. Skeletal muscle mass and brain mass were the significant contributors to REE_c. Illner and colleagues (2000), using a similar approach in 26 healthy young adults (mean age 25 years), found good agreement between REE_c and REE_m (difference ranged from 155 to –485 kJ/day).

In this study, skeletal muscle and liver were the significant contributors to the REE_c model.

Sparti and colleagues (1997) used a multicomponent approach to measure FFM (liver and kidney volumes by CT, heart mass by echocardiography, and muscle mass by DXA) in 40 subjects combined with REE by indirect calorimetry. Using multiple regression analysis, the authors found that fat and FFM explained 83% of the variability in REE (standard error of estimate = 420 kJ/day). The authors did not directly test the effect of adding organ mass (liver, kidneys, heart) to fat and FFM in the regression equation to ascertain whether additional variance would be explained by these high metabolic rate organs. Instead, the authors used a stepwise regression procedure and found that the inclusion of subcomponents of FFM in the list of variables did not improve the prediction of REE over and above the traditional FFM and fat mass model. They interpreted this to mean that that the weight of internal organs is not a main determinant of residual unexplained REE.

Of course, the magnitude of the additional contribution of organ weights to variance accounted for in a regression model depends on the degree to which their mass varies independently of FFM in the sample under study. Indeed, organ mass may not be a main determinant of REE once FFM is taken into account because of their high degree of collinearity. In the sample of Sparti and colleagues (1997), fat and FFM combined explained 83% of the variance. Granted some amount of error in the measurement of organ masses and REE, and some within-individual day-to-day variation, there was relatively little variance remaining for individual differences in organ mass to explain. Despite the failure of organ mass measures to improve R^2 in the stepwise regressions, REE computed as the sum of the tissue masses multiplied by the published values for their specific energy expenditure correlated well with measured REE (r = .95), considerably better than the correlation of REE with FFM alone (r = .90). It is difficult to imagine why this would be the case were it not for the individual differences in the proportion of these tissues and their respective contribution to REE.

Late Adulthood

REE decreases during the adult years and has been attributed largely to age-related changes in the fat-free body mass compartment (Keys et al. 1973). More recent understanding is that the loss of FFM cannot explain in full the decrease in REE in elderly (Fukagawa et al. 1990; Piers et al. 1998;

Poehlman et al. 1992; Vaughan 1991; Visser et al. 1995). Because the FFM compartment consists of numerous tissues and organs, each with a different oxidative metabolic activity, the possibility exists that changes in the proportion of these tissues with age may explain the lower REE in the elderly.

A number of studies in humans have investigated REE at the organ–tissue level (Gallagher et al. 2000; Garby et al. 1993; Puggaard et al. 2002) for the purpose of determining in adults whether age-related changes in the organ–tissue composition of FFM can account for the lower REE reported in elderly. Gallagher and colleagues (2000) developed REE$_c$ models (equations 15.1-15.3) from measured tissues and organs, and energy-flux rates were assigned for each of the seven tissue–organ components as reported by Elia (1992a). Older men (n = 6, 80.3 ± 7.5 years) and women (n = 7, 76.5 ± 5.5 years) had significantly lower REE$_m$ compared with REE$_c$ (p = .001). The magnitude of the difference was 13% and 9.5%, respectively, for men and women. These findings suggest that even after adjustment for age-related organ and tissue atrophy in the elderly, whole-body REE by indirect calorimetry continues to be lower than expected (figure 15.2). Puggaard and colleagues (2002) used the weights of organs (brain, heart, liver, kidneys, and spleen) registered during autopsies (n = 238) in women from similar birth cohorts to those of elderly women (n = 77, 65-85 years) on whom REE, fat, and FFM were measured in the laboratory. Additional laboratory measures were conducted on younger women (n = 104, 14-60 years) for whom no organ weights were available. Among the elderly women, the authors found that REE (measured by indirect calorimetry) was progressively lower with increasing age, and this apparent decrease in REE paralleled the observed reduction in weight of organs with age. REE was found to be 6% and 7% lower between 65 to 75 and 75 to 85 years, respectively, suggesting that the decline in REE is accelerated in the later years of life compared with earlier adult years (Keys et al. 1973). No difference was found between measured and predicted REE in this age group.

Collectively, from the REE- and MRI-derived organ and tissue mass studies conducted in our laboratory thus far, we have observed that the decrease in REE from 5 years through adulthood and elderly years appears to parallel the decrease in the fraction of FFM as organ mass. A theoretical representation presented by Elia (1997) has been confirmed using in vivo MRI measures of organs. Figure 15.3 shows that the percentage of FFM of four organ masses (brain + liver + kidneys + spleen) decreases in a linear manner during the growth years, possibly leveling off to adult levels in the late teenage years and decreasing further during the elderly years. The remarkable similarity between figures 15.2 and 15.3 might lead one to conclude that the changing proportions of FFM during the growth years (figure 15.3) are accountable for the decrease in REE adjusted for FFM during these same years (figure 15.4). However, as seen in figure 15.2, this is not entirely true. It is also noteworthy that the proportion of FFM as high metabolic rate organs is larger in smaller persons (figure 15.5).

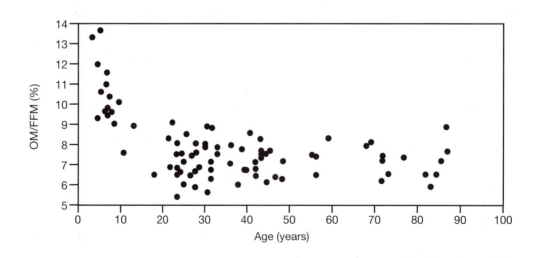

Figure 15.3 Four organ masses (brain, liver, kidneys, spleen) expressed as a fraction of fat-free mass (FFM) as a function of age in healthy females and males.

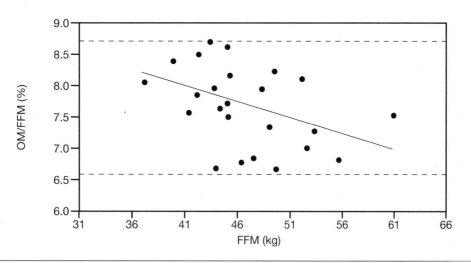

Figure 15.4 Five organ mass (OM) (brain + liver + kidneys + spleen + heart) expressed as a fraction of fat-free mass (FFM by dual-energy X-ray absorptiometry) as a function of FFM in women ($n = 24$, 20-45 years). This figure shows that the proportion of FFM as high metabolic rate organs is larger in smaller persons.

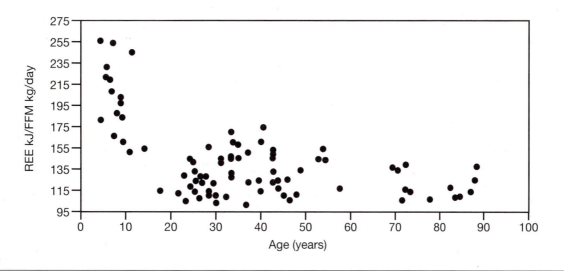

Figure 15.5 Resting energy expenditure (REE; kJ/FFM kg/day) expressed as a fraction of fat-free mass (FFM) as a function of age in healthy individuals.

REE: Race and Ethnicity

Race differences in REE are known to exist at early ages in prepubertal children (Kaplan et al. 1996; Morrison et al. 1996; Yanovski et al. 1997). Several published studies (Kaplan et al. 1996; Morrison et al. 1996; Sun et al. 2001; Wong et al. 1999; Yanovski et al. 1997) indicate that there are differences in REE between African American and Caucasian children. Such findings are consistent with reports of lower REE in African American compared with Caucasian women (Albu et al. 1997; Foster et al. 1997; Hunter et al. 2000; Kushner et al. 1995). After adjusting for differences in body weight or FFM, Kushner and colleagues (1995), Albu and col-

leagues (1997), and Foster and colleagues (1997) found REE to be 6% (38 kJ per day), 3% (29 kJ per day), and 6% (22 kJ per day), respectively, lower in African American compared with Caucasian obese women. Similarly, REE was reported 7.5% lower in normal-weight African American compared with Caucasian women. A 5% and 7% lower sleeping metabolic rate was reported in African American men and women compared with Caucasians (Weyer et al. 1999). Total daily energy expenditure was found to be 10% lower in older (>55 years) African American men and women compared with Caucasians (Carpenter et al. 1998), which was considered attributable to a 6% lower REE and a 19% lower physical activity energy expenditure. It is noteworthy that similar ethnic patterns in REE have emerged in girls at all stages of pubertal maturation (Kaplan et al. 1996).

The basis of these racial and ethnic differences has not been established. Also, the extent to which puberty affects the growth of different tissues and body compartments that contribute to REE and whether this growth might be influenced by sex and race are unknown. Sun and colleagues (2001) published the results of a longitudinal investigation of how REE changes with pubertal maturation in relation to changes in body composition, across race and sex. They reported that REE was significantly lower in African American compared with Caucasian children (by ~250 kJ/day) and that REE declined with Tanner stage after adjustment for race, sex, fat mass, and FFM. Body composition was measured using dual-energy X-ray absorptiometry. These findings support earlier reports of lower daily REE in African American girls compared with Caucasian girls, after adjustment for FFM (by 19 kJ, Wong et al. 1999; 22 kJ, Yanovski et al. 1997; 34 kJ, Morrison et al. 1996; and 51 kJ, Kaplan et al. 1996). Of interest is that the authors also found that the African American children had relatively higher limb lean mass and lower trunk lean mass than the Caucasian children. Lower trunk lean mass may be suggestive of lower abdominal–trunk organ mass in African American children.

Significant race differences have been reported in the distribution of lean tissue between overweight (Weinsier et al. 2000) and lean (Hunter et al. 2000) African American and Caucasian women. Despite having similar amounts of FFM, African American women in both of these studies had lower amounts of trunk FFM, the body compartment that contains organ tissue, thereby suggesting that African American women have a lower proportion of metabolically active organ tissue, which may contribute to their lower REE and sleeping energy expenditure.

We measured body composition (including organs) using multislice MRI in 19 African American and 19 Caucasian, healthy, medication-free, premenopausal women (Gallagher et al. 2000). The sum of liver, brain, kidneys, and spleen mass (organ mass) was used as the dependent variable. After we adjusted for differences in height, weight, and age, organ mass was significantly smaller (p = .009) in African American compared with Caucasian women. These variables explained 56% and race explained an additional 10% of the variance in highly metabolically active organ mass. The calculated combined contribution of kidneys, brain,

liver, and spleen to REE is approximately 54%. We concluded that a smaller mass of these highly metabolically active organs in African Americans may contribute to their lower observed REE. Studies are ongoing that include the measurement of REE with the intent of determining if differences in organ–tissue mass can explain in part the race differences in REE between African American and Caucasian children and adults.

REE in Disease

Medications and disease processes have independent effects on metabolic rate. In disease conditions where abnormalities of body composition are present, prediction equations based on healthy populations are likely to be inaccurate. To adequately understand whether REE is altered in a diseased state compared with normal, we need to understand the composition of the body.

Acute Diseases

There is little information available on organ mass during acute disease. However, REE typically increases according to the severity and phase of the illness (Elia 2001, 2003). In severe catabolic illness, such as severe burns, there may be more than 50% increase in REE and transient elevations that may be as high as 100%. During this time there is a loss of protein from the body and no obvious organomegale that can be detected clinically. Presumably, the increase in metabolic rate is predominantly attributable to an increase in tissue-specific metabolic rates rather than an increase in the mass of metabolically active tissues. Such an increase in tissue-specific metabolic rates is mediated by a combination of hormones, such as catecholamines, glucagons, and cytokines. Therefore, REE per gram of organ mass would be increased in this hypermetabolic state. An additional consideration is that a disease state can be accompanied by anorexia and malnutrition (Elia 1992a).

Chronic Diseases

From among the many disease conditions that exist, consideration will be given to just one, anorexia nervosa. REE (adjusted for FFM) is reported lower (Krahn et al. 1993; Melchior et al. 1989; Schebendach et al. 1997) or not different (Obarzanek et al. 1994) in patients with anorexia nervosa (AN), given that REE increases with increased energy intake and body weight (Krahn et al. 1993; Obarzanek et al. 1994; Rigaud et al. 2000). In studies where REE normalized for FFM at low weight or with weight gain was no longer different from control subjects, the conclusion drawn is that the apparent low REE at low weight is solely a result of lower FFM. The assumption made here is that the composition of FFM is similar between low weight and normal weight. Alternatively, in those studies that found a difference in REE normalized for FFM between participants with AN and control participants after refeeding, the inference is that REE cannot be explained solely by the reduction in FFM or that the composition of FFM is different between participants with AN and control participants. In some studies, REE has been found to increase in a disproportionate manner to weight gain or FFM gain (Obarzanek et al. 1994; Scalfi et al. 1993).

Careful attention must be given to the approach used in individual studies when normalizing for weight, BCM, or FFM. The issue at hand is that the ratio of REE to BCM or REE to FFM is not constant among individuals but varies systematically with body weight (Bogardus et al. 1986; Kinney et al. 1963; Leibel et al. 1995; Poehlman et al. 1993; Ravussin and Bogardus 1989; Ravussin et al. 1988). Individuals with lower body weights tend to have higher REEs per kilogram of BCM and FFM. That is, heat production rates per unit of BCM and FFM at rest are not the same between individuals who differ in body size (Heshka et al. 1990). Studies that have used a ratio (REE/FFM) approach inappropriately assume that changes in the numerator and denominator are proportional (Allison et al. 1995). Alternatively, the use of regression analysis avoids the numerous problems associated with the use of ratios in statistical analysis (Allison et al. 1995) and is well accepted in the modeling of REE (Heshka et al. 1990; Leibel et al. 1995; Poehlman et al. 1993; Ravussin et al. 1988; Ravussin and Bogardus 1989).

Generally, brain volume is reduced in AN (Dolan et al. 1988; Kingston et al. 1996; Kohn et al. 1997; Krieg et al. 1988; Swayze et al. 1996). Two longitudinal studies of adolescents after weight regain showed persistent deficits in gray matter but recovery of white matter (Golden et al. 1996; Katzman et al. 1996). These findings support results that show persistent deficits in patients who have made full recovery from their eating disorder (Lambe et al. 1997). A more recent study (Swayze et al. 2003) showed that brain volume increases again on refeeding.

A single published study assessed the mass of specific organs and tissues (by MRI) in female patients with AN at low weight, after refeeding to within 90% of ideal body weight, and compared these with healthy control women (Mayer et al. 2002). The participants with AN (n = 6) and controls (n = 10) were matched for age, sex, and percent ideal body weight. The participants with AN were assessed within 2 weeks of admission to an inpatient unit (during a period of weight stability) and again after having normalized weight to 90% ideal body weight, and controls were studied once. All participants were medication-free for a minimum of 2 weeks before testing. At low weight, patients with AN weighed significantly less than healthy controls (40.36 ± 7.01 kg vs. 55.45 ± 3.93 kg). Skeletal muscle, total fat, and spleen and kidney mass were significantly reduced in low-weight patients compared with normal-weight controls. Brain (1.46 ± 0.10 kg), liver (1.44 ± 0.17 kg), and heart (0.11 ± 0.02 kg) mass, however, were not significantly different from controls (1.40 ± 0.13 kg, 1.36 ± 0.20 kg, 0.12 ± 0.02 kg, p nonsignificant). This finding persisted when the sample size increased to 18 participants with AN compared with 13 controls (personal communication, Laurel S. Mayer, August 28, 2003). With normalization of weight, there were significant increases in skeletal muscle, total body fat, and kidney and spleen mass to values not significantly different from controls. There were no significant changes in brain, liver, or heart mass with weight gain. When the sample sizes increased to 18 participants with AN and 13 controls, all organs except brain increased with weight gain (personal communication, Laurel S. Mayer, August 28, 2003). These data show in a small sample that the composition of FFM is disproportionately altered in participants with AN at low weights but is not different at ideal weights.

REE in Malnutrition or Weight Change

Unintentional starvation is frequently coupled with disease, making it difficult to separate the effects of starvation from the effects of disease. There are a limited number of studies of intentional starvation in humans. In early total starvation (typically 1-2 days), there is a consistent transient increase in REE (Elia 1992b) that occurs despite loss of protein from the body, presumably including some loss from some of the organs. Metabolic factors responsible for this increase have been discussed (Elia 1992b). Reductions in REE can be detected after some days of starvation in humans (Dulloo et al. 1996; Keys et al. 1950), and REE remains low during long-term caloric restriction (Blanc et al. 2003; DeLany et al. 1999; Gonzales-Pacheco et al. 1993). This may be attributable to the decrease in FFM associated with weight loss and to the better metabolic efficiency associated with low energy intakes. The contribution of FFM to weight loss and energy loss during a period of semistarvation differs depending on whether the person is lean or obese, with greater FFM loss in lean individuals. A reduction in REE by 33% was observed in one study (Keys et al. 1950) between the baseline 3-week period before starvation and during the last 3 weeks of semistarvation. The corresponding reduction in physical activity energy expenditure was 57%.

Body temperature is lower in caloric-restricted nonhuman primates (Lane et al. 1996), which suggests a reduction in oxygen consumption. With refeeding, both REE and diet-induced thermogenesis increase.

During prolonged starvation, ketone bodies dominate as the energy source for brain. After 3 weeks of total starvation of four obese subjects, Redies and colleagues (1989) showed that oxygen utilization, blood flow, and blood volume did not change whereas glucose utilization decreased by 54%. Similarly, after 5 weeks of starvation, oxygen consumption of the forearm did not change significantly (Owen and Reichard 1971).

There has been much debate surrounding the effects of weight loss during a hypocaloric diet on REE in humans and animals. In the reduced weight state or after weight loss (when weight is stable) and independent of changes in body composition, REE has been reported lower (Leibel et al. 1995) or not different (Rissanen et al. 2001) from pre–weight loss levels. Data from a nonhuman primate colony (rhesus monkeys) showed that after 11 years of dietary restriction (30%), REE adjusted for FFM was 250 kJ/day lower in the dietary restriction group compared with an ad libitum–fed control group (Blanc et al. 2003).

A partial explanation for this lower REE in a reduced weight state is that body composition is altered in dietary restriction and as such, the metabolic rates of subjects under dietary restriction and controls, based on FFM, fail to reflect variations in organ and tissue mass. Studies in calorically restricted rats have shown that heart, liver, kidneys, prostate, spleen, and skeletal muscle were disproportionately lower compared with controls, whereas brain and testes were not different (Weindruch and Sohal 1997; Weindruch and Walford 1988). Because many of these organs present as

high metabolic rate organs, a lower mass of high metabolic rate organs could partially account for the lower observed REE.

REE in children with malnutrition is generally reduced mainly because of a low body mass relative to healthy control subjects of the same age. However, REE/kg body weight or REE/kg FFM (percent deviation from normal) has been found to be variable, sometimes higher and sometimes lower than normal. This may be caused by a combination of factors, such as the presence of edema; changes in body weight, which may be increasing or decreasing; changes in body composition; and organ–tissue-specific metabolic rates. Muscle and adipose tissue tend to be lost to a greater extent than other tissues, whereas the brain and skeleton are preferentially preserved. Indeed, the brain may continue to grow while body weight remains stable or even decreases. Special consideration needs to be given to the brain in the young child. In the newborn, the brain accounts for more than 40% of REE compared with about 20% in the adult. The typical appearance of a marasmic infant is one of a wasted body and a large head (relatively large brain). Because the brain has a high metabolic rate relative to muscle and adipose tissue, it is not surprising that some studies in marasmic infants have reported an increase in REE/kg body weight or REE/kg FFM. Such issues and the associated confounding variables are discussed in more detail elsewhere (Elia 1997).

Ongoing Efforts in the Measurement of Tissue-Specific Metabolic Rates

There is an obvious need to measure organ-specific metabolic rate in vivo to establish metabolic rate constants, specific to age, sex, and race. Kinetic studies of energy metabolism at the organ–tissue level using noninvasive techniques are required. Take the brain as an example. Brain has one of the highest specific metabolic rates (table 15.1) and is thus a good representation of a high metabolic rate organ. Oxidative metabolism in the healthy adult brain uses primarily glucose. In the normal, nondiseased brain, cerebral blood flow (CBF) is proportional to the cerebral metabolic rate for oxygen ($CMRO_2$) and glucose (CMR_{glc}) and is connected to brain oxidative metabolism (Sokoloff 1976). Reductions in brain blood flow, oxygen, and glucose consumption have been documented beyond the 7th decade of life (Frackowiak et al. 1984). Some reports indicate a significant association between overall brain glucose use and age (Kuhl et al. 1982; Moeller and Eidelberg 1997) and between age-related declines in cerebral perfusion and oxygen use (Leenders et al. 1990; Martin et al. 1991; Moeller and Eidelberg 1997). Moeller and colleagues (1996) reported a 12% decline in brain glucose metabolic rate from ages 20 to 80 years. In contrast, CMR_{glc} at 31 identified midline and bilateral structures was uncorrelated with age in 21 healthy men (21-83 years; Duara et al. 1983). $CMRO_2$ in cerebral cortex gyri was found to decrease linearly and by approximately 6% in all four lobes and on both sides in healthy 20- to 68-year-old subjects (n = 25), even after the authors adjusted for cortical atrophy, sex, and head size (Marchal et al. 1992). No effect was found in the white matter, deep gray nuclei, thalamus, and cerebellum. Gray matter CBF, but not mean gray matter oxygen consumption ($CMRO_2$), was found to decrease linearly with age (Pantano et al. 1984). However, when younger subjects (<50 years) were compared with older subjects (>50 years), an age-related decrease of 18% was observed in both CBF and $CMRO_2$. White matter CBF and $CMRO_2$ remained remarkably stable with age. However, it is unclear whether the results noted were attributable to methodological issues, loss of cortical neurons, diminished activity of cortical neurons, or age-related changes in cognitive decreases. Collectively, these studies demonstrate a lack of consistency with regard to the effect of age on brain metabolism.

Brain weight decreases by 7% to 8% (Dekaban and Sadowsky 1978; Yamaura et al. 1980) of peak adult weight. Despite the many anatomical, morphological, and neurochemical differences between brains of young and elderly persons, substantial variability exists such that not all elderly brains are dissimilar to young brains (Creasey and Rapoport 1985). Of central importance therefore, when one is attempting to describe the independent effect of age on brain metabolism, is the need to study persons who are free of cardiovascular diseases, cerebrovascular diseases, primary brain diseases, dementia, or mental disorders. After age 50 years, approximately 50% of brains have atherosclerosis (Moossy 1971). Oxygen and glucose delivery to the brain may be restricted in these disease conditions such that reduced CBF could be misinterpreted as an age rather than a disease effect (Duara et al. 1983).

An ongoing study is attempting to address whether metabolic rate per unit of brain and heart tissue is different at the extremes of the adult age range. Positron emission tomography (PET) for the in vivo quantification of labeled ^{15}O uptake in brain and ^{11}C-acetate in myocardial tissues allows for the quantification of both brain and myocardial oxygen consumption in healthy individuals, without cardiac catheterization. PET is an intrinsically quantitative tool that allows for the noninvasive assessment of the distribution of substances of physiological interest within the body. When mathematically and physiologically appropriate mathematical models are used, quantitative assessment of important variables such as blood flow, blood volume, and metabolic utilization can be made. We are currently conducting studies using this approach to assess blood flow and oxygen extraction and utilization in the brain and heart in healthy volunteers, and we are evaluating the effect of aging with correction for tissue volume as well as for myocardial work. However, radiation exposure in PET studies precludes such investigations in children at this time, and therefore we do not propose these measures in children.

Summary

There is a need to measure heat production in individual cells within organs using noninvasive and in vivo techniques. Within the FFM compartment, there are many organs not mentioned in this chapter that would promote our understanding of REE if they were measured during growth, aging, and altered states of nutrition. Therefore, more imaging protocols need to be developed.

Part IV

Body Composition and Biological Influences

Genes, environment, age, and many other factors influence body composition. These are the topics examined in part IV. A determinant of the body composition phenotype is inheritance, and current knowledge in this area is explored in chapter 16. As humans and animals grow and age, both the mass and distribution of components change, leading to important functional consequences. Genetic and quantitative genetic studies have revealed a significant within-population and across-population variation in body composition and physique. This chapter provides an overview of the findings pertaining to the role of inherited variation and specific gene polymorphisms on body components, including body fat, adipose tissue distribution, skeletal muscle, and skeletal characteristics.

Body composition varies across racial and ethnic groups, contributing to phenotypic heterogeneity. These overlapping topics are examined in chapter 17, "Age," and chapter 18, "Variation in Body Composition Associated With Sex and Ethnicity." The focus is on the major body components such as fat-free mass, fat, total body water, bone mineral content, fluid compartments, and skeletal muscle and interrelationships among the components over the life span.

Pregnancy, as discussed in chapter 19, produces dynamic changes in body composition that are increasingly documented using new methods of body composition analysis. The relationship of these changes to maternal and infant outcomes is discussed.

Moderators of the five levels of body composition are examined in chapters 20 and 21, respectively. Chapter 20 characterizes the changes that occur in body composition following various types of exercise training. This chapter also discusses links between body composition and physical performance across the life span. Chapter 21 discusses recent advances in selected areas with an emphasis on the effects of steroid hormones and their interactions with insulin and human growth hormone.

Chapter 16

Genetic Influences on Human Body Composition

Peter Katzmarzyk and Claude Bouchard

Given the wide range of variation in human body composition, there has historically been great interest in determining the genetic and environmental influences on morphological traits. Early studies focused on phenotypes that were easily measured, such as body mass, stature, and other simple anthropometric dimensions. As techniques and indexes that quantified more specific aspects of body composition were developed, studies of the genetic and environmental determinants of these quantities soon followed. Furthermore, the realization that many aspects of body composition are related to health and disease has sparked renewed interest in identifying both genetic and nongenetic determinants of human morphology as potential risk factors for disease.

Among the issues that have been addressed by scientists, the level of inheritance of body dimensions and body composition phenotypes has been studied most frequently. Several reports have dealt with segregation patterns and the hypothesis of single gene effects on these traits. However, relatively few body composition phenotypes have been investigated with the tools of molecular biology, such as association studies, crossbreeding, transgenic or knockout animal experiments, gene expression studies, or the quantitative trait locus (QTL) linkage mapping approach. The exceptions to this are phenotypes related to obesity and body fat topography, which have attracted a great deal of interest from a growing number of research groups.

This chapter provides an overview of the role of inherited variation and specific gene polymorphisms on several aspects of body composition, including body fat content, adipose tissue topography, and skeletal muscle characteristics. It begins with a brief examination of key concepts and methods most commonly used to study the genetic basis of complex multifactorial traits such as those considered here. Although there have been few new studies on the familial aggregation of body composition, the number of molecular epidemiological studies that have been published has increased dramatically since the printing of the first edition of this book. Much of this increase can be attributed to the development and refinement of new imaging techniques for the assessment of body composition and the rapid development of molecular biology techniques along with new analytical strategies.

Basic Concepts

Human body composition phenotypes are complex multifactorial traits whose mode of inheritance cannot be readily reduced to simple Mendelian patterns. These traits have evolved under the interactive influences of dozens of affectors from the social, behavioral, physiological, metabolic, cellular, and molecular domains. Segregation of genes is not easily detected in familial or pedigree studies, and whatever the influence of the genotype, it is generally attenuated or exacerbated by environmental factors (Bouchard 1991a). The total observed variance in a trait can be broken down into genetic, environmental, and interaction effects. The proportion of the total phenotypic variance that is attributable to genetic factors is termed the *heritability* and is generally expressed as a percentage from 0% to 100%.

Efforts to understand the genetic causes of complex phenotypes can be successful only if they are based on an appropriate conceptual framework, accurate phenotypic measurements, proper sampling strategies, and extensive typing of candidate genes and other molecular markers. In this context, the distinction between "necessary" genes and "susceptibility" genes proposed by Greenberg (1993) seems particularly relevant. A necessary gene is one that is sufficient to cause the disease if the deficient allele or alleles have been inherited. For a quantitative phenotype, a necessary gene would be one with a large effect on the phenotype. A susceptibility gene is defined as one that increases susceptibility or risk for a disease but is not necessary for disease expression (Greenberg 1993). It merely lowers the threshold for a person to develop the disease. The concept is also relevant to complex quantitative phenotypes such as those considered in this chapter. The concept implies that the true causes of variation in a phenotype may be nongenetic or genetic, including genetic variation at a gene other than the susceptibility gene or additive or interactive effects at several susceptibility loci.

Another important issue is whether there are genotype–environment effects or gene–gene effects that influence the phenotype. These effects remain very difficult to investigate even with current technology. One way to test for the presence of a genotype–environment effect in humans is to challenge several genotypes in a similar manner by submitting both members of monozygotic (MZ) twin pairs to a standardized treatment and comparing the within- and between-pair variances of the response (Bouchard et al. 1990). The finding of a significantly higher variance in the response between pairs than that within pairs suggests that the changes induced by the treatment are more heterogeneous in genetically dissimilar individuals.

One could also use the method proposed by Berg (1981, 1990), which can be carried out on cross-sectional observations in pairs of MZ twins. Because MZ twins have the same genes, any difference between two members of a given pair in a multifactorial phenotype is caused by nongenetic factors. One can compare the mean within-pair difference in the phenotype between MZ pairs who have a given allele at a particular locus with those who have other alleles of the same gene. Berg argued that if an allele has a permissive effect, the within-pair variance would be greater in those pairs with the allele than in those lacking the particular allele. The opposite would be true for an allele with a restrictive effect. This essentially amounts to a method allowing a test for genotype–environment interaction effect considering one gene at a time.

Studies of gene–gene interaction effects on body composition phenotypes have begun to appear since the first edition of this book. It seems that gene–gene interactions will be a common topic in the field of body composition, and there is no doubt that the topic will attract considerable attention in the coming years. It will likely be a complicated area of research, however, in the sense that large sample sizes will be needed if several genes are to be considered in the same analysis. Innovative experimental designs will have to be developed, and the panel of the most important genes to investigate remains to be defined. A more detailed review of the gene–environment and gene–gene interaction concepts and methods is provided by Bouchard and colleagues (2004).

A schematic representation of these various genetic effects is depicted in figure 16.1. The figure incorporates major and minor gene effects, gene–environment interaction effects, and gene–gene interaction effects on phenotypes. It also allows for the contribution of nongenetic factors.

Methods

There are obvious individual differences in body composition among human beings. Even though most of the relevant phenotypes tend to aggregate in families, they do not seem to behave as simple Mendelian traits when segregation patterns across generations are considered. It should eventually become possible to reduce the genetic component of body composition phenotypes to a series of major and minor single gene effects, but this stage has not been reached. Most of the evidence accumulated so far on these continuously distributed, multifactorial phenotypes has been obtained by the methods of classic genetic epidemiology, but relevant genes and mutations are being identified at an increasing rate.

Two strategies have been traditionally used by geneticists to study the role of genes in continuously distributed phenotypes in humans. As shown in figure 16.2, they are referred to as the

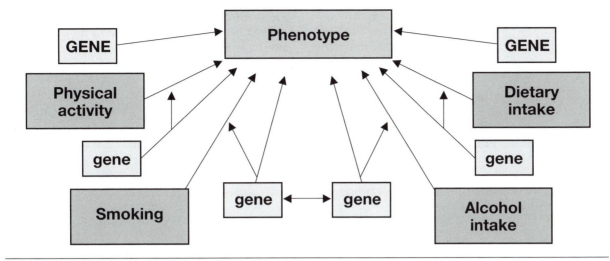

Figure 16.1 An overview of the genetic and selected environmental influences on body composition. Genes with large effects are shown in capital letters. See text for explanation.

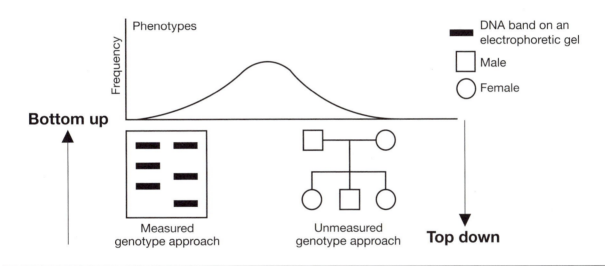

Figure 16.2 The two main strategies to study the role of genes in continuously distributed phenotypes.
Adapted from Sing and Boerwinkle, 1987.

unmeasured genotype (top-down) and the measured genotype (bottom-up) approaches (Sing and Boerwinkle, 1987; Sing et al. 1988). These top-down and bottom-up approaches incorporate a large network of designs and technologies that are used by geneticists in the effort to understand the genetic basis of complex phenotypes.

The unmeasured genotype approach attempts to estimate the contribution of genetic variation to the phenotypic variance and to find quantitative evidence for single genes with detectable (major) effects on the phenotype. Because inferences about the contribution of genes are made from the phenotype, this approach is also referred as to the top-down strategy. Here one uses various sampling designs (twins, nuclear families, families with adopt-

ees, extended pedigrees) in combination with statistical tools such as path analysis, variance component estimation, and complex segregation analysis.

The measured genotype approach is based on direct measurements of genetic variation at the DNA levels or protein in an effort to assess the impact of allelic variation on the phenotypic variation. Because inference about the role of genes is made from the gene to the phenotype, this approach is at times referred to as the bottom-up strategy (Sing et al. 1988).

Genetic Epidemiology

The field of genetic epidemiology is concerned with multifactorial phenotypes and until about

10 years ago used almost exclusively the unmeasured genotype approach. For body composition phenotypes, five strategies have been traditionally used. First, inferences about the role of inherited variation are often made on the basis of population differences taking into account the within- and between-population variances and making assumptions about the similarity of the environmental conditions for each population. Second, familial aggregation is investigated in nuclear families or extended pedigrees. Total transmission (genetic and nongenetic) across generations is assessed, and maternal versus paternal effects are contrasted. Third, pairs of MZ and dizygous (DZ) twins are often used to estimate the heritability of complex phenotypes. The method is based on the principle that MZ twins are genetically identical whereas DZ twins share only 50% of their genome. The differences between pairs of MZ and DZ twins can then be ascribed to the genetic heterogeneity in DZ pairs provided that both types of twins have experienced and are exposed to similar environments. MZ twins reared apart have also been used to study the role of the genotype. Fourth, adoption studies have been used in either a full adoption or a partial adoption design. In brief, variation among adoptees (foster parents with adopted children or siblings by adoption) is compared with the variation among the adoptees and their biologically related parents or siblings. Several designs can be derived from the basic adoption model to infer genetic and nongenetic effects. Fifth, segregation analyses are performed to test whether single gene effects can be inferred for a given phenotype. Here, nuclear families or extended pedigrees are used to verify if a major effect compatible with Mendelian transmission or a major effect combined with a multifactorial effect (mixed model) best fits the data. The role of all these genetic epidemiology procedures, and of segregation analysis in particular, is to dissect the phenotypic variance into components that can be studied subsequently at the level of the gene (MacCluer 1992).

Association and Linkage Studies

The contribution of molecular biology to the study of human health and disease has increased rapidly over the last decade. Even the most complex of human traits can now be investigated with molecular methods, and it has been proposed that even the whole field of epidemiology should move into the era of "molecular epidemiology" (Schulte and Perera 1993). Association and linkage studies with genetic polymorphisms have been a central focus of molecular genetic epidemiology for some time, and these approaches are important to delineate the genetic basis of body composition. However, the terms *association* and *linkage* cannot be used interchangeably.

The concept of association refers to a situation in which the relationship between a genetic polymorphism and a phenotype is investigated. Such studies are most often carried out on samples of unrelated individuals. These studies may take several forms, including comparison of cases versus controls (e.g., obese vs. lean subjects), analysis of variance across genotypes for the locus under consideration, or comparison of carriers versus noncarriers of a given allele. If a significant association is observed with a polymorphism at a candidate gene locus, there are two likely explanations: Either the locus is causally related to the phenotype, or the locus is in linkage disequilibrium with another polymorphism of relevance. Association studies may provide important information on genes with a major or a minor contribution to a phenotype; the method is particularly useful for the identification of genes that make only a minor contribution (Greenberg 1993).

When a gene has a large influence on a given phenotype, then both will be transmitted together across generations. This concept is referred to as linkage. Linkage analysis can be performed with candidate gene markers or with other polymorphic markers such as microsatellites. Evidence for linkage becomes more apparent as the marker loci get closer to the true locus that cosegregates with the phenotype. The procedure can be undertaken with large pedigrees or with panels of nuclear families. It is commonly used for complex multifactorial phenotypes, which are characterized by the presence of a segregating major gene.

A practical method is the so-called sib-pair linkage analysis, which allows screening for potential linkage relationships between quantitative phenotypes and genetic markers (Haseman and Elston 1972). The method is based on the notion that sibs who share a greater number of alleles (at a given locus) identical by descent at a linked locus will also share more alleles at the phenotype locus. Thus, these sibs will have more similar phenotypes than pairs of sibs who share fewer marker alleles. Under these conditions, the slope of the regression of squared sib-pair phenotype differences on the proportion of genes identical by descent is expected to be negative when linkage is present. An important advantage of the method is that it is not necessary to specify the mode of inheritance for the phenotype being considered.

A more general exposé of the strengths and limitations of each research strategy was published by Greenberg (1993). The recent developments in these areas have been reviewed by Rao (2001), Borecki and Suarez (2001), and Rice and Borecki (2001).

Genomic Scans and Quantitative Trait Locus Mapping

Whereas traditional linkage studies in humans have used the candidate gene approach, genomic scans represent a special case of linkage analysis wherein a large number of molecular markers spaced across the entire genome, including both candidate genes and anonymous markers, are examined in relation to a given phenotype. When this approach is used, there are no a priori assumptions regarding the effects of a particular gene on a phenotype. A genomic scan has the ability to identify a quantitative trait locus or a chromosomal region related to body composition. The identified regions can then be examined more closely for traditional and positional candidate genes that are related to the phenotype of interest. The method was fully described by Warden and colleagues (1992) and has been used to identify a total of 168 animal QTLs related to body weight or body fat to date (Chagnon et al. 2003). Regions of homology between mouse or rat and human chromosomes have been extensively defined, and this allows for the identification of the approximate location of a putative gene linked to the phenotype of interest on the human gene map.

When a QTL on a given chromosome has been identified with an acceptable linkage score in an animal model or a human genomic scan, considerable work is needed before the true nature of the gene involved is uncovered. Positional cloning, DNA sequencing, linkage disequilibrium studies among single nucleotide polymorphisms, and cell system–based expression studies, as well as development of congenic strains, transgenics, and knockout mouse models and other tools, are then needed to define the gene implicated.

Genetics and Body Fat Content

More is known about the genetics of body weight and obesity traits than any other body composi-

tion phenotype. This section reviews the evidence concerning heritability levels, segregation of major genes, and contributions of molecular markers.

Heritability of Body Fat Content

There is still some disagreement among researchers regarding the importance of genetic factors in the familial resemblance observed for body fat (Bouchard and Pérusse 1988; Garn et al. 1989). Given the expense of direct measurements of body fat, most studies have used the body mass index (BMI) or skinfold thicknesses at a few sites as estimates of body fat content. Heritability values ranging from almost zero to as high as 90% have been reported for BMI (Bouchard and Pérusse 1988; Stunkard et al. 1986a). With the use of different designs (twin, family, and adoption studies), large variation in the age of subjects, only a few types of relatives, and, very often, small sample sizes, such wide variation in the reported heritabilities within and between populations is not unexpected. With few exceptions, these studies could not separate the effects of genes from effects of the environment shared by relatives living together in the same household. Moreover, only a handful of studies have included the full range of BMI values to ensure that the phenotype was adequately represented. Space limitations preclude a complete review of the genetics of body fat here; the reader is referred to more comprehensive reviews of the area (Bouchard et al. 2004; Maes et al. 1997).

Twin and Adoption Data

The comparison of MZ twins reared apart with MZ twins reared together is an interesting design to assess the role of heredity with some control over some of the confounding influences of shared environment. The correlations of MZ twins reared apart are generally similar to those of MZ twins reared together, as shown by three studies (MacDonald and Stunkard 1990; Price and Gottesman 1991; Stunkard et al. 1990) suggesting that the shared familial environment did not contribute much to the variation in BMI. The correlations of MZ twins reared apart provide a direct estimate of the genetic effect if we assume that members of the same pair were not placed in a similar environment, that twins were not, for some undefined reasons, behaving similarly despite the fact that they were living apart, and that intrauterine factors did not influence long-term variation in BMI. According to these studies, the heritability of BMI would be in the range of 40% to 70%. Other twin study designs also tend to generate similar heritability estimates and often even higher values.

Five adoption studies (Price et al. 1987; Sørensen et al. 1989, 1992a, 1992b; Stunkard et al. 1986b), in which BMI data were available in both the biological and the adoptive relatives of the adoptees, reported that the effect of shared family environment on BMI was negligible. In a review on this topic, Grilo and Pogue-Geile (1991) also concluded that experiences shared among family members appeared largely irrelevant in determining individual differences in body weight and obesity. These findings are somewhat at odds with the strong familial resemblance of the major affectors of energy balance, that is, energy intake (Pérusse and Bouchard 1994) and energy expenditure (Bouchard et al. 1993); the whole issue deserves further investigation by genetic epidemiologists. The heritability estimates for the BMI derived from the adoption studies tend to cluster around 30% or less.

Family Studies

During the last 60 years or so, many authors have reported that obese parents had a higher risk of having obese children than lean parents (Bray, 1981). These investigations were complemented by studies comparing the resemblance in spousal, parent–child, and sibling pairs for body weight, BMI, and skinfold measurements. These studies have been reviewed by Mueller (1983) and Bouchard and Pérusse (1988, 1993).

Table 16.1 summarizes some of the correlations reported in four large family-based studies of BMI. These results were obtained from the Framingham Heart Study (Heller et al. 1984), the Canada Fitness Survey (Pérusse et al. 1988), the Québec Family Study (QFS; Bouchard et al. Pérusse 1988b), and the Nord-Trøndelag Norwegian National Health Screening Service Family Study (Tambs et al. 1991).

In the Framingham Heart Study, adult levels of BMI were used in both the parental and the offspring generations (Heller et al. 1984). The authors reported that their results provided little support for a genetic effect on BMI. In the study of the BMI obtained on 74,994 persons of both sexes, 20 years of age and older, from the population of Nord-Trøndelag, Norway, correlations were available for many types of relatives, some of which are summarized in table 16.1. Fitting a path model for genetic and environmental transmission to their data, the authors obtained a broad heritability of about 40% (Tambs et al. 1991). They concluded

Table 16.1   Familial Correlations for Body Mass Index Derived From Four Large Family Studies

Relationship (*n* pairs in parentheses)	Framingham Heart Study	Canada Fitness Survey	Quebec Family Study	Norway
Spouses	.19 (1,163)	.12 (3,183)	.10 (248)	.12 (23,936)
Parent–offspring	.23 (4,027)	.20 (7,194)	.23 (1,239)	.20 (43,586)
Siblings	.28 (992)	.34 (3,924)	.26 (370)	.24 (19,157)
Uncle or aunt–nephew or niece	.08 (1,970)	–.11 (34)	.14 (88)	0 (1,146)
Grandparent–grandchild	NA	.05 (32)	NA	.07 (1,251)
Dizygotic twins	NA	NA	.34 (69)	.20 (90)
Monozygotic twins	NA	NA	.88 (87)	.58 (79)

Note. NA = no data available.

Data derived from Heller et al. (1984) for the Framingham Heart Study, Pérusse et al. (1988) for the Canada Fitness Survey, Bouchard et al. (1988) for the Québec Family Study, and Tambs et al. (1991) for Norway. Table reprinted, by permission, from C. Bouchard and Y. Perusse, 1994, Genetics of obesity: family studies. In *The genetics of obesity,* edited by C. Bouchard (Boca Raton, FL: CRC Press), 81.

that a simple model with only an additive genetic effect and an individual environmental effect could be rejected.

Two other large studies considered the transmission effects and heritability of the BMI and subcutaneous fat as assessed by the sum of several skinfolds. The first was based on a stratified sample of the Canadian population and included the BMI and the sum of five skinfolds for 18,073 subjects living in thousands of households, yielding 4,825 pairs of spouses, 8,881 parent–child pairs, 3,929 pairs of siblings, 43 uncle or aunt–nephew or niece pairs, and 85 grandparent–grandchild pairs (Pérusse et al. 1988). The total transmission effect across generations for the age- and sex-adjusted phenotypes reached about 35%. The second study relied on 1,698 members of 409 families, which included nine types of relatives by descent or adoption (Bouchard et al. 1988b). Under these conditions, there was a total transmissible variance across generations for the BMI and the sum of six skinfolds of about 35% but a genetic effect of only 5%.

Fewer reports have dealt with total fat mass or percent body fat measured with one of the commonly accepted direct methods for measuring body composition. Data from Utah demonstrated significant familial correlations for percent body fat, assessed by bioelectrical impedance analysis (BIA; Ramirez 1993). In the QFS, underwater weighing was performed in a relatively large number of individuals representing nine different kinds of relatives (Bouchard et al. 1988b). About one half of the variance, after adjustment for age and sex, in fat mass or percent body fat was associated with a transmissible effect, and 25% of the variance was compatible with an additive genetic effect. In the HERITAGE Family Study, the maximal heritability for fat mass estimated from densitometry was 62%, which was higher than that obtained for the sum of skinfolds (34%) in the same sample (Rice

et al. 1997a). The heritability of percent body fat from BIA analysis in the San Antonio Heart Study was reported to be 37% (Mahaney et al. 1995), whereas a recent study of Nigerians, Jamaicans, and African Americans produced heritability estimates of approximately 50% for fat mass and percent body fat estimated by BIA across the three ethnic groups (Luke et al. 2001). Thus, it appears as though the heritability of fat mass, assessed by more direct means, is similar to the estimates obtained for BMI.

Trends in Heritability Estimates

Table 16.2 describes the trends in heritability estimates in terms of the various designs used to generate the results. Thus, the heritability level is highest with twin studies, intermediate with nuclear family data, and lowest with adoption data. When several types of relatives are used jointly in the same design, the heritability estimates cluster around 25% to 40% of the age- and sex-adjusted phenotype variance, although higher values have been reported (Maes et al. 1997). There is no clear evidence for a specific maternal or paternal effect, and the common familial environmental effect is marginal. The presence of a nonadditive genetic effect is often suggested from this research.

Heritability of Changes in Body Fat Content

Although the inheritance of many body fat phenotypes has been well characterized using cross-sectional designs, there have been few studies on natural changes in body weight or body fatness. This is in part attributable to the difficulty in collecting longitudinal data on a sufficiently large sample. Apparently only three twin studies and two family studies have determined the heritability of natural changes in body weight and fatness using a long-term follow-up. Evidence from the

Table 16.2 Overview of the Genetic Epidemiology of Human Body Fat Content

	Heritability–transmission (%)	Maternal–paternal	Familial environment
Nuclear families	30-50	No	Minor
Adoption studies	10-30	Mixed results	Minor
Twin studies	50-80	No	No
Combined strategies	25-40	No	Minor

Reprinted, by permission, from C. Bouchard, 1994, Genetics of obesity: overview and research directions. In *The genetics of obesity*, edited by C. Bouchard (Boca Raton, FL: CRC Press), 225.

Kaiser Permanente Women's Twin Study (Austin et al. 1997) and the National Heart, Lung, and Blood Institute Twin Study (Fabsitz et al. 1980, 1994) indicates that genetic factors explain between 57% and 86% of the variance in changes in BMI over 10 to 43 years. In contrast, 6-year changes in body weight and BMI among Finnish twins were mainly attributable to environmental rather than genetic factors (Korkeila et al. 1995).

Two family studies have demonstrated significant familial resemblance for changes in body weight and adiposity. Rice and colleagues (1999) estimated that the heritabilities of changes in the BMI, sum of skinfolds, and trunk–extremity skinfold ratio (TER) were 37%, 0%, and 59% over 12 years in the longitudinal QFS. Similarly, the heritabilities for changes in BMI, the sum of five skinfolds, and waist circumference were 14%, 12%, and 45%, respectively, in the Canada Fitness Survey 7-year follow-up (Hunt et al. 2002). Clearly more research is necessary to determine the influence of genes on changes in body fatness over time.

Evidence for a Major Gene

Many studies have used complex segregation analyses to identify major gene effects on BMI and other adiposity phenotypes. These studies were recently reviewed by Bouchard and colleagues (2004). The results of these studies are quite divergent, because some studies found no evidence of major gene effects whereas many reported a recessive locus accounting for 35% to 45% of the variance. Although most of the segregation analyses used BMI, at least four studies reported results for fat mass, one conducted in the QFS using underwater weighing (Rice et al. 1993) and the other on Mexican American families in San Antonio conducted using BIA (Comuzzie et al. 1995). The results of these two studies were similar, identifying a major recessive gene accounting for up to 45% of the phenotypic variance; however, in the San Antonio Study, there was a significant genotype–sex interaction, with 37% of the variance explained for in males and 43% in females. On the other hand, results from the Stanislas Family Study indicated a major genetic effect on fat mass; however, the major effect was not consistent with the transmission of a single major gene (Lecomte et al. 1997). A recent study by Borecki and colleagues (1998) found support for the existence of two pleiotropic major recessive genes, together accounting for 64% and 47% of the variance in BMI and fat mass, respectively.

Molecular Markers

The investigation of the association between DNA sequence variation at specific genes and body fat phenotypes has been the focus of much research since the first version of this book. Since 1996, an annual review of the human obesity gene map has been published and updated (Bouchard and Pérusse 1996; Chagnon et al. 1998, 2000b, 2003; Pérusse et al. 1997, 1999, 2001a; Rankinen et al. 2002). As of the 2002 update, 222 studies have reported positive associations with 71 candidate genes, and an additional 68 loci have been identified using human genomic scans (Chagnon et al. 2003). Table 16.3 summarizes the incredible changes in the status of the human obesity gene map that occurred between 1994 and 2002. Efforts must now be aimed at better characterizing the QTLs that have been identified by using denser maps of the regions where significant linkages occurred. Given that significant linkages have been found on all chromosomes except for Y, a comprehensive review of this area is beyond the scope of this chapter; however, the reader is referred to the published human obesity gene map updates cited previously for a complete summary of studies. An expanded and frequently updated version of the human obesity gene map can be viewed at http://obesitygene.pbrc.edu. It is becoming clear that in the vast majority of cases, obesity is a polygenic disorder that cannot be characterized by a single mutation.

Genetic–Environmental Interactions

A series of experiments were used to test whether differences in body fat content could be explained by genetic factors by comparing the intrapair (within-genotype) and interpair (between-genotype) resemblances in the response of MZ twins to positive and negative energy balance.

Response to Positive Energy Balance

Two experiments were undertaken to study individual differences when subjects were exposed to a positive energy balance protocol. In both experiments, subjects ate a 4.2-MJ (1,000-kcal) per day caloric surplus during periods of 22 days (Bouchard et al. 1988c; Poehlman et al. 1986) and 100 days (6 days a week; Bouchard et al. 1990). Both experiments resulted in significant changes in body fat phenotypes, but considerable interindividual differences in the adaptation to the extra calories were observed. In the long-term experi-

Table 16.3 Changes in the Status of the Human Obesity Gene Map Between 1994 and 2002

	1994	1995	1996	1997	1998	1999	2000	2001	2002
Single-gene mutations[a]				2	6	6	6	6	6
Mendelian disorders with map location	8	12	13	16	16	20	24	25	33
Candidate genes with positive findings	9	10	13	21	29	40	48	58	71
Animal QTLs	7	9	24	55	67	98	115	165	168
Human QTLs from genome scans				3	8	14	21	33	68

Note. QTL = quantitative trait locus.

[a]Number of genes, not number of mutations. Numbers indicate the cumulative total going from left to right across the table.

Reprinted, by permission, from Y.C. Chagnon et al., 2003, "The human obesity gene map: The 2002 update," *Obesity Research* 11(3): 313-367.

ment (Bouchard et al. 1990), the mean body mass gain of the 24 subjects (12 pairs of MZ twins) was 8.1 kg, but the range of weight gain was 4 to 13 kg. However, the variation observed was not randomly distributed; the variance in response to long-term overfeeding was about three times larger between pairs than within pairs for gains in body weight and fat mass (Bouchard et al. 1990). Figure 16.3 illustrates the within-pair resemblance in weight gain under the influence of the 100-day overfeeding protocol (left panel). The same pattern was observed for the changes in fat mass and other indicators of body fat. These results suggest that the amount of weight gained or fat stored in response to a caloric surplus is significantly influenced by the genotype of the individual.

Response to Negative Energy Balance

In two other experiments, exercise was used to induce an energy deficit in MZ twins to test for the contribution of the genotype in the response to negative energy balance sustained for 22 days (Poehlman et al. 1987) or about 100 days (Bouchard et al. 1994). In both experiments, the energy deficit was obtained by exercising twins on cycle ergometers twice a day for about 50 min per session. The exercise prescription was designed to induce an extra energy expenditure of about 4.2 MJ (1,000 kcal) over resting metabolic rate while maintaining energy intake at the baseline level throughout the study. Results from the long-term experiment revealed a significant within-pair resemblance for the reduction in body weight (right panel of figure 16.3) and fat mass, whereas results from the short-term study revealed that only fat free mass changes were characterized by a significant MZ twin resemblance.

The results from both the positive and negative energy balance experiments generally suggest that undetermined genetic characteristics specific to each individual are associated with the response of body composition phenotypes to changes in energy balance.

Genetics and Fat Topography

Regional fat distribution has been shown to be an important determinant of the relationship between obesity and health and an independent risk factor for various morbid conditions, such as cardiovascular diseases or type 2 diabetes.

Truncal–Abdominal Subcutaneous Fat

Upper-body obesity is more prevalent in males than in females and increases in frequency with age in males and after menopause in females. It is moderately correlated with total body fat and appears to be more prevalent in individuals habitually exposed to stress. It is also associated in females with the levels of plasma androgens and cortisol. In addition, the activity of abdominal adipose tissue lipoprotein lipase is elevated with higher levels of truncal–abdominal fat (Bouchard et al. 1991).

Evidence for familial resemblance in body fat distribution has been reported (Donahue et al. 1992). Using skinfold measurements obtained in 173 MZ and 178 dizygotic pairs of male twins, Selby and colleagues (1989) concluded that there was a significant genetic influence on central deposition of body fat. Using data from the Canada Fitness Survey and the strategy of path analysis, Pérusse

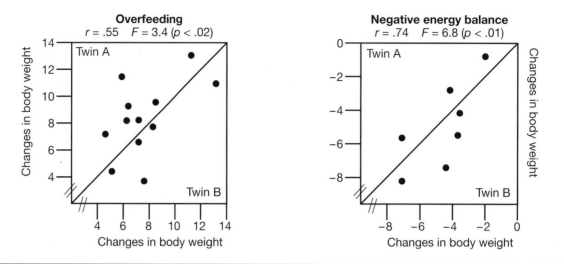

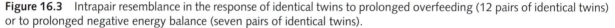

Figure 16.3 Intrapair resemblance in the response of identical twins to prolonged overfeeding (12 pairs of identical twins) or to prolonged negative energy balance (seven pairs of identical twins).

and colleagues (1988) showed that the transmissible effect across generation reached about 40% for trunk skinfolds (sum of subscapular and suprailiac skinfolds), extremity skinfolds (sum of biceps, triceps, and medial calf skinfolds), and TER. The transmissible effect across generation also reached 28% for the waist–hip ratio (WHR). If all transmissible effects are genetic, these results suggest that heredity accounts for a maximum of 40% of the phenotypic variance for various indicators of fat topography.

The biological and cultural components of transmission in regional fat distribution were further assessed with data from the QFS (Bouchard et al. 1988b). Two indicators of regional fat distribution were considered: the TER and the subcutaneous fat–fat mass ratio obtained by dividing the sum of six skinfolds by fat mass derived from body density measurements. Genetic effects of 25% to 30% were obtained. When the influence of total body fat was taken into account, the profile of subcutaneous fat deposition was found to be characterized by higher heritability estimates reaching about 40% to 50% of the residual variance (Bouchard 1988, 1990). These results imply that for a given level of fatness, some individuals store more fat on the trunk or abdominal area than others, and these effects have a significant genetic component.

In the HERITAGE Family Study (Pérusse et al. 2000), there was evidence for significant familial aggregation for trunk, extremity, and total skinfolds (30-35%) and for the TER and waist circum-

ference (50%). The familiality for the response of these phenotypes to 20 weeks of exercise training was generally lower, indicating that familial factors are more important in determining the amount and distribution of subcutaneous fat than their response to training.

Results from a few studies suggest the influence of major genes on regional fat distribution phenotypes. In one study, Hasstedt and colleagues (1989) reported a major gene effect explaining 42% of the variance in a relative fat pattern index defined as the ratio of the subscapular skinfold to the sum of the subscapular, suprailiac, and skinfold thicknesses. Recent results from the QFS suggest major gene effects for TER, adjusted for total fat mass, accounting for about 35% of the phenotypic variance (Borecki et al. 1995). A study of families from Andhra Pradesh, India, revealed a major recessive gene for the TER, but after statistical adjustment for overall subcutaneous adiposity, the effect disappeared, suggesting genetic pleiotropy between total and regional fatness (Feitosa et al. 1999). A recent segregation analysis of the WHR in the Minnesota Breast Cancer Family Study also revealed a putative major gene with a frequency of 27% that accounted for 42% of the total phenotypic variance in postmenopausal women, but no effects were found when the total sample of women was used (Olson et al. 2001).

Abdominal Visceral Fat

Levels of abdominal visceral fat (AVF) increase with age, in both sexes, in lean as well as obese individuals (Enzi et al. 1986). Males have on the average more AVF than females, and obese people have more than lean persons. However, the level of AVF is only moderately correlated with total body fat, with a common variance ranging from about 30% to 50% and perhaps lower (Bouchard 1994a; Ferland et al. 1989).

The genetics of AVF levels were recently reviewed (Katzmarzyk et al. 1999); however, two genomic scans were published subsequently. Table 16.4 outlines the evidence for genetic influences on abdominal fat levels (total, subcutaneous, and visceral) from two large studies that used computed tomography to assess abdominal fat: the QFS and the HERITAGE Family Study. For all measures of abdominal fat, both before and after statistical control over total fat mass, there are consistent heritabilities of 42% to 70% (Pérusse et al. 1996; Rice et al. 1997b). A major gene effect

was found in both studies for AVF (Bouchard et al. 1996; Rice et al. 1997c), but the effect disappeared after adjustment for fat mass, which suggests that a gene that affects fat mass may also affect AVF either directly (pleiotropy) or indirectly. Indeed, a cross-trait analysis from the QFS indicates that there are shared familial determinants for AVF and total fat mass (Rice et al. 1996). Several markers have shown promising linkages with abdominal fat, particularly abdominal subcutaneous fat, in genomic scans (Pérusse et al. 2001b; Rice et al. 2002; table 16.4). These regions need to be examined in greater detail using denser maps in an effort to identify potential candidate genes.

Genetic-Environmental Interactions

Further indication that the genotype may be an important determinant of AVF and subcutaneous

Table 16.4 Evidence for a Genetic Effect on Abdominal Fat Levels Assessed by Computed Tomography

| | Phenotype | Maximal heritability (%) | MAJOR GENE | | Genomic scan results[a] |
			Yes–no	(%)[a]	
Québec Family Study	ATF	70	—	—	Scan not performed
	ASF	68	—	—	Scan not performed
	AVF	68	Yes	51	Scan not performed
	ATF$_{FM}$	57-76	—	—	None
	ASF$_{FM}$	42	—	—	2p15.1, 2q32.1, 7q31.3, 9q22.1-22.2, 12q22-23, 12q24.33, 13q14.11, 13q34, 17q21.1-21.3
	AVF$_{FM}$	56	No	—	None
HERITAGE Family Study	AVF	47	Yes	54	Scan not performed.
	ATF$_{FM}$	—	—	—	Whites: 5q31.2-31.3; blacks: none
	ASF$_{FM}$	—	—	—	Whites: 2p14, 5q31.2, 22q11.23; blacks: 3p26.3, 3q29, 4q31.22 7q36.3, 11p15.2, 11p14.1, 14q24.1
	AVF$_{FM}$	48	No	—	Whites: 2q22.1, 2q36.1-36.3; blacks: none

Note. ATF = abdominal total fat; AVF = abdominal visceral fat; ASF = abdominal subcutaneous fat; subscript$_{FM}$ = adjusted for total fat mass.

[a]Regions containing markers linked with the phenotype at $p < .0023$.

Data for Québec Family Study derived from Pérusse et al. (1996; 2001b) and Bouchard et al. (1996). Data for HERITAGE Family Study derived from Rice et al. (1997b, 1997c, 2002).

fat distribution phenotypes comes from the positive and negative energy balance experiments with identical twins (Bouchard et al. 1990, 1994). There was about six times more variance between pairs than within pairs for the increases with overfeeding in truncal subcutaneous fat and in computed tomography–assessed AVF after adjustment for the gains in total fat in both cases. Significant intrapair resemblance was also obtained in the negative energy balance experiment for the loss of AVF.

These results are consistent with the significant identical twin intrapair resemblance seen in the response of adipose tissue lipolysis and lipoprotein lipase activity when the twins are challenged by overfeeding, acute exercise, or exercise training (Després et al. 1984; Mauriège et al. 1992; Poehlman et al. 1986; Savard and Bouchard 1990). The results imply that for a given level of fatness, fat mass gain, or fat mass loss, some individuals store or mobilize fat on the trunk–abdominal area or the visceral depot specifically whereas others store or mobilize primarily in other depots (Bouchard 1990; Bouchard et al. 1990, 1994).

Genetic Pleiotropy and Body Fat Phenotypes

Pleiotropy refers to the tendency of genes to affect more than one trait simultaneously. There is great interest in identifying genes that affect several aspects of body composition or indicators of body composition and disease risk factors.

Body Fat Content and Fat Topography

The issue of genetic pleiotropy for body fat phenotypes has been addressed in a number of studies. The genetic variation in BMI was shared with both percent body fat (10%) and the TER (18%) in the QFS (Rice et al. 1995), whereas fat mass and abdominal visceral fat shared approximately 29% to 50% of the genetic variation in the same sample (Rice et al. 1996b). Similarly, there was significant pleiotropy between the BMI and trunk skinfolds in the Leuven Longitudinal Twin Study (Beunen et al. 1998). On the other hand, Cardon and colleagues (1994) found no shared genetic variance between the BMI and skinfold ratios but found significant pleiotropy between BMI and WHR. At least three other studies have demonstrated significant genetic pleiotropy among measures of body fatness (Choh et al. 2001; Comuzzie et al. 1994;

Faith et al. 1999); however, these analyses did not address the issue of pleiotropy between fatness and fat topography per se. Rather, they studied associations among eight individual skinfolds (Comuzzie et al. 1994), between BMI and percent body fat (Faith et al. 1999), and among various skinfolds and BMI (Choh et al. 2001).

Body Fat Content, Fat Topography, and Chronic Disease Risk Factors

Cross-trait studies are valuable in determining the degree to which two risk factors share genes in common. For example, a study of male MZ and DZ twins indicated that the concordance rates of having diabetes, hypertension, and obesity within the same individual were 31.6% in MZ pairs and 6.3% in DZ pairs, suggesting the presence of a common underlying genetic factor for the risk factor cluster (Carmelli et al. 1994). Another study of elderly Swedish twins revealed that component risk factors of the metabolic syndrome share a latent genetic factor, whereas three of the components (triglycerides, insulin resistance, and high-density lipoprotein cholesterol) are also influenced by a latent individual-specific environmental factor (Hong et al. 1997).

Five studies have addressed the issue of genetic pleiotropy between body fatness and blood pressure. Rice and colleagues (1994) found that upper-body fat (TER) had the strongest genetic correlation with blood pressure, particularly diastolic blood pressure (DBP), in the QFS. On the other hand, 11% of the genetic variation in systolic blood pressure (SBP) was shared with body weight in a sample of twins from Virginia (Schieken et al. 1992). Likewise, Schork and colleagues (1994) found significant pleiotropy for both the BMI and body weight with mean arterial blood pressure in the Gubbio Study. Similarly, there were significant genetic correlations between body weight, BMI, skinfolds, with both SBP and DBP in a sample of Samoans (Choh et al. 2001). An analysis of the Victorian Family Heart Study in Australia also demonstrated a significant genetic correlation between BMI and SBP (Cui et al. 2002). Because risk factors for chronic disease tend to cluster within individuals, a major thrust in genetic epidemiology should be on determining the genetic and environmental relationships among risk factors.

Two important features of the metabolic syndrome are insulin resistance and abdominal obesity. A few studies have addressed the issue of pleiotropy among body fat and insulin–glucose metabolism phenotypes. In the San Antonio Heart Study, there were high genetic correlations for both fasting and 2-hr postchallenge insulin levels with the BMI, WHR, and subscapular–triceps skinfold ratio, indicating that the same genes or sets of genes were influencing the body fat phenotypes and insulin levels (Mitchell et al. 1996). In the QFS, the sum of skinfolds was related to the ratio of fasting insulin to glucose, with a bivariate familiality estimate of 8% (Rice et al. 1996a), whereas in the HERITAGE Family Study, 48% of the heritability of fasting insulin was shared with AVF, and 20% of the heritability of AVF was shared by insulin, after adjustment for total fat mass (Hong et al. 1998). A more recent segregation analysis that used the same sample revealed a putative locus that accounted for 54% of the variance in fasting insulin; however, after adjustment for fat mass and AVF, the major gene effect disappeared (Hong et al. 2000). These results suggest that there are shared genetic influences among body fat and insulin phenotypes and that there is a putative locus with pleiotropic effects on both insulin and body fat.

Given that there is a significant association between body fat phenotypes and blood lipid levels, a question of interest is the degree to which common genes influence body fat and blood lipid phenotypes. Results from the San Antonio Heart Study indicated significant pleiotropy between fat mass and BMI and between high-density lipoprotein cholesterol and triglycerides; however, there was no pleiotropy between body fat and blood lipid measures (Mahaney et al. 1995). On the other hand, Schork and colleagues (1994) reported significant genetic pleiotropic effects between BMI and levels of total cholesterol in the Gubbio Study. Finally, an analysis from the QFS provided evidence that environmental factors shared within sibships are important determinants of the covariance between body fat and blood lipids; however, a genetic pleiotropy hypothesis could not be ruled out based on the analyses conducted (Pérusse et al. 1997). Clearly more research is required to better delineate the shared genetic factors underlying the covariation between body fat and blood lipid phenotypes.

Genetics and Skeletal Muscle Phenotypes

Sarcopenia, skeletal muscle atrophy, and skeletal muscle hypertrophy are all examples of variation in muscle mass. Differences in skeletal muscle mass among individuals are also found outside of these extreme conditions. Skeletal muscle mass is an important component of body composition at all ages. Its direct health implications have not been considered extensively so far in relation to morbidity and mortality end points (Roche 1994). Skeletal muscle mass is commonly approximated by measurements of fat-free mass obtained with a variety of procedures or is estimated from imaging techniques applied to specific sites or regions of the body. Few genetic studies have been performed on skeletal muscle mass phenotypes.

Fat-Free Mass

In the QFS, the heritability of fat-free mass derived from underwater weighing estimates of body density reached about 30% of the age- and sex-adjusted phenotype variance (Bouchard et al. 1988b). However, no major gene effect could be identified by segregation analysis (Rice et al. 1993). The latter finding was confirmed in a sample of Mexican American subjects (Comuzzie et al. 1993) and in the Stanislas Family Study (Lecomte et al. 1997). A recent genome-wide linkage scan for fat-free mass in the QFS revealed several related genomic regions (Chagnon et al. 2000a). Significant linkages were observed on chromosomes 7p, 15q, and 18q. Candidate genes in these regions include the neuropeptide Y (*NPY*) and growth hormone-releasing hormone (*GHRH*) receptor on 7p and insulin-like growth factor 1 receptor (*IGF1R*) gene on 15q.

Estimated Muscle Mass

Quite often the nutritional assessment of muscle mass is attempted from anthropometric indicators, such as limb circumferences corrected for skinfold thicknesses. Such data were available in the sets of midparent–natural child and foster midparent–adopted child of the QFS (Bouchard 1991b), where the upper-arm circumference corrected for the triceps skinfold and the calf circumference corrected for the medial calf skinfold as described earlier by Brožek (1961) were used. The correlations for both estimated muscle diameters are presented in table 16.5. They are not significantly different from zero for both variables in the relatives by adoption but reached 0.3 in the biological relatives. These results are quite compatible with our observations on the heritability of fat-free mass, which indicate that the genotype contributes significantly to variations in densitometrically estimated fat-free mass and most likely muscle size or mass.

Table 16.5 Midparent–Child Interclass Correlations for Estimated Muscle Size

Variable[a]	Foster midparent–adopted child (N = 154 sets)	Midparent–natural child (N = 622 sets)
Upper-arm muscle diameter	.09	.30[b]
Calf muscle diameter	.10	.29[b]

[a]Scores were adjusted for age and sex by generation and normalized; [b]$p < .001$.

Reprinted, by permission, from C. Bouchard, 1991, Genetic aspects of anthropometric dimensions relevant to assessment of nutritional status. In *Anthropometric assessment of nutritional status*, edited by J. Himes (New York, NY: Alan R. Liss), 225. This material is used by permission of Wiley-Liss, Inc., a subsidiary of John Wiley & Sons, Inc.

Skeletal Muscle Characteristics

In humans, as well as in most mammals, skeletal muscle contains varying proportions of two major categories of fibers exhibiting specific contractile properties. These skeletal muscle fiber categories have been commonly named slow-twitch (ST) and fast-twitch (FT) fibers because of the time they require to reach peak isometric tension in response to a single twitch. In humans, ST and FT muscle fibers have an approximate mean time to peak tension of 90 and 45 ms, respectively (Gollnick 1982; Saltin and Gollnick 1983). The vastus lateralis muscle of sedentary black subjects from western and central African countries was shown to have a slightly greater percentage of type 2 fibers and higher enzyme activities of the anaerobic energy processes than did sedentary white subjects, whereas enzyme markers of aerobic–oxidative metabolism were comparable between both groups (Ama et al. 1986). Sex differences were also noted between mean values for histochemical, morphological, and biochemical characteristics of human skeletal muscle. On average, women exhibit slightly higher fiber type 1 proportion, smaller fiber areas, and lower glycolytic potential than men (Simoneau and Bouchard 1989).

An important question is to what extent human skeletal muscle characteristics are under the control of genetic factors. The data available are few, and the results are widely divergent. Repeated measurements within the same muscle, such as the vastus lateralis, revealed that both sampling and technical error combined (standard deviation for repeated measurements) represented from 5% to 10% of the mean value of each fiber type category (Blomstrand and Ekblom 1982; Gollnick et al. 1973; Simoneau et al. 1986).

Nimmo and colleagues (1985), using inbred strains of mice, showed that genetic factors accounted for about 75% of the variation in the

proportion of type 1 fiber of the soleus muscle with 95% confidence intervals ranging from 55% to 89%. Our own study based on a sample of brothers (n = 32 pairs), DZ twins (n = 26 pairs), and MZ twins (n = 35 pairs) indicated that the heritability of muscle fiber type proportion was much lower than previous estimates (Bouchard et al. 1986). Intraclass correlations for the percentage of type 1 fibers were significant and reached .33 in brothers, .52 in DZ twins, and .55 in MZ twins. Broad heritability estimates can be obtained from twice the biological sib correlation (66%), from twice the difference between MZ and DZ correlations (6%), or directly from the MZ twin sibship correlation (55%; Falconer 1960). Although brothers and DZ twins share about one half of their genome by descent, comparison of their correlations suggests that increased environmental similarity (i.e., DZ twins experience more similar environmental circumstances than regular brothers) appears to translate into increased phenotypic resemblance for the proportion of skeletal muscle type I fibers (i.e., intraclass coefficient of .52 vs. .33).

Although the heritability level of fiber type proportions has not been clearly delineated, Simoneau and Bouchard (1995) synthesized the results from available studies. The results suggested that approximately 15% of the observed variance in type 1 fiber type proportion is attributable to sampling and technical variance, 40% to environmental variance, and 45% to genetic variance. Although a difference of about 30% in type 1 fibers could theoretically be explained by environmental factors, a genetic hypothesis supports the finding that 25% of people have either less than 35% or more than 65% of type 1 fibers (Simoneau and Bouchard 1995).

The large interindividual variability in the enzyme activity profile of human skeletal muscle confirms that the catabolism of different substrates in the skeletal muscle of healthy sedentary and moderately active individuals of both sexes will vary (Simoneau and Bouchard 1989). Numerous factors are undoubtedly involved in accounting for the large interindividual variations observed. Only one study has dealt with the heritability of different enzyme markers of the human skeletal muscle energy metabolism in the last decade. Maximal enzyme activity of creatine kinase, hexokinase, phosphofructokinase (PFK), lactate dehydrogenase, malate dehydrogenase, 3-hydroxyacyl CoA dehydrogenase, and oxoglutarate dehydrogenase (OGDH) was determined in brother, DZ twin, and MZ twin sibships (Bouchard et al. 1986). Genetic factors appeared to be responsible for about 25% to 50% of the total phenotypic variation in the activities of the regulatory enzymes of the glycolytic (PFK) and citric acid cycle (OGDH) pathways and in the variation of the glycolytic to oxidative activity ratio (PFK–OGDH ratio) when the data were adjusted for age and gender differences (Bouchard et al. 1986). These results indicate that variation in the key enzyme activity of human skeletal muscle appears to be inherited to a significant extent. Such a genetic effect could not be accounted for by charge variation in the enzyme molecules (Bouchard et al. 1988a; Marcotte et al. 1987).

Summary

Understanding the causes of human variation in body composition has a long tradition in human biology disciplines. Population genetic and quantitative genetic studies have revealed that a significant portion of within-population variation in body composition, and perhaps some of the variability across populations, can be accounted for by inherited differences. In the last decade or so, genetic epidemiological research has shown that body fat and fat topography phenotypes are generally characterized by the contribution of a multifactorially transmitted component as well as a major gene effect. A major gene effect is apparently not detected for fat-free mass.

The number of association and linkage studies with molecular markers for phenotypes related to level of body fat and body fat topography has increased dramatically since the first edition of this book. There is growing evidence that the genetic component of body composition phenotypes will eventually be defined in terms of a series of contributing and interacting genes. Although much remains to be done, the technologies and study designs that have the potential to generate such information are now available and are routinely used in many laboratories.

Chapter 17

Age

Richard N. Baumgartner

Changes with age in body composition begin, literally, at the moment of conception and end only with the death and subsequent decomposition of an organism. In some species, the changes between different phases of life can be dramatic, as in the metamorphosis of a caterpillar into a butterfly. In mammals, particularly long-lived species such as humans, age-related changes are much more subtle and gradual. In considering age changes in human body composition, we can divide these into three phases: growth and development, maturity, and senescence. There is considerable interest in defining normal trajectories for changes within each of these phases, because abnormalities are associated with disease.

The purpose of this chapter is to describe human body composition changes during aging in relation to the three phases of growth and development, maturity, and senescence. This presentation emphasizes changes in the major components of body composition on the cellular and tissue–organ levels (see chapter 1), specifically body fat (or adipose tissue) and its anatomical distribution, body cell mass, and fat-free mass and its main constituents: water, bone, skeletal muscle, and organ. Age-related changes in these aspects of body composition that have direct health and functional implications are discussed. Changes in components at other levels, such as the molecular, are considered where these result in compositional changes that affect methods of in vivo measurement of the cellular and tissue–organ components of interest. Measurement methods are reviewed elsewhere in this book and are only cursorily reviewed in this chapter with regard to their merits and limitations for estimating changes.

The review considers the "normal age trajectories" of these components and their constituents as well as factors influencing variability within and between individuals. When we define normal trajectories for changes with age in body composition, it is important to consider three things: (a) There is considerable variation within as well as between individuals; (b) total variation is a function of a complex interaction between genes, environment, and behavior; and (c) the separation between age-related ("normal") and disease-related ("abnormal") changes is often unclear. Where appropriate, variations in these age trajectories are also considered in relationship to physiological processes associated with aging and more broadly with regard to adaptation to environment and evolution.

Data are presented from some studies to illustrate age trajectories. These are not presented as "reference data," however. Several sets of reference data for normal body composition have been developed that include age changes as well as sex and racial differences. The National Health and Nutrition Surveys (NHANES I-III), for example, include data for anthropometric measurements, including body mass index and skinfold thicknesses, for large, nationally representative samples (www.cdc.gov/nchs/nhanes.htm). These data, however, provide only indirect evidence for body composition differences with age, sex, and race. Recently, Chumlea and colleagues (2002) published reference data from NHANES III for body composition predicted using bioelectric impedance analysis. Janssen and colleagues (2002) published data for predicted skeletal muscle mass from bioelectric impedance analysis. Overall, data for body composition on the molecular, cellular, or tissue–organ levels remain limited. Data for total body fat, fat-free mass (FFM), and FFM components including total body water, potassium, and intra- and extracellular fluid have been published for children and adolescents (Ellis et al. 2000; Fomon et al. 1982; Forbes 1986; Haschke 1989; Guo et al. 1997) and adults (Cohn et al. 1976; Ellis 1990; Guo et al. 1999; He et al. 2003; Mott et al. 1999). Reference data are considerably more sparse for people older than 65 years and especially for the oldest old, those greater than 85 years (Baumgartner et al. 1995; Steen et al. 1977; Visser et al. 2003).

There are several caveats with regard to the use of these data sets. First, they mostly come from small selected samples that are not population based; data for large, heterogeneous, population-based samples are difficult to obtain because of the cost of application of the most accurate body composition methods. Second, the data are mostly cross-sectional: There are few longitudinal studies for age changes within individuals. Third, the accuracy and precision of the in vivo methods used vary between studies and are time-dependent as new methods are developed and applied. Last, some data sets are old and do not reflect contemporary changes in the environment that affect nutrition, physical activity, health status, and body size (Chumlea and Baumgartner 1989).

Fat Mass

Fat mass is the most variable component of body composition. Between-individual variability ranges from about 6% to more than 60% of total body weight. Within-individual variability can also be considerable, but changes with age in individuals tend to "track," showing distinct trajectories over time.

Infants average about 10% to 15% fat at birth (Forbes 1987). This increases to about 30% by 6 months of age and then begins to gradually decline during early childhood (Butte et al. 2000). An approximate 2% difference in percent body fat between boys and girls is evident by about 5 years of age. During midchildhood, between about 5 and 8 years of age, a preadolescent "fat wave" or "adiposity rebound" occurs (Rolland-Cachera et al. 1984). Total body fat (TBF) continues to increase during adolescence at an estimated rate of about 1.4 kg/year in girls and 0.6 kg/year in boys (Guo et al. 1997). Percent body fat increases from an average of 20% to 26% in girls between 9 and 20 years but decreases in boys from about 17% to 13% after age 13 years as FFM rapidly increases (Guo et al. 1997). Dietz (1994) suggested that gestation, early infancy, midchildhood adiposity rebound, and ado-

lescence were critical periods for the development of obesity. Evidence for the association of body composition during the first three periods with obesity in adulthood is inconsistent and controversial (Cameron and Demerath 2001). Data for tracking of body fatness during adolescence, however, support the hypothesis that obese adolescents are at increased risk for persistent obesity in adulthood. Guo and colleagues (1997) analyzed serial data from the Fels Longitudinal Study for changes in TBF and FFM during adolescence. Eighty-six percent of males and 77% of females in the upper tertile of TBF at age 13 years remained in the upper tertile at 18 years.

Total body fat mass increases slowly with age during adulthood (Guo et al. 1999; Mott et al. 1999; Siervogel et al. 1998). The rate of increase differs by sex and, possibly, race. Estimates vary between studies and may depend on the method used to measure fat mass. Guo and colleagues (1999) estimated the rate of increase in TBF to be approximately 0.37 kg/year in men and 0.41 kg/year in women, using longitudinal data from the Fels Study. The changes in fat mass were associated with decreasing physical activity and menopausal status in women. In a separate analysis of Fels data, Siervogel and colleagues (1998) estimated the rate of increase in TBF to be 0.57 kg/year between ages 18 to 45 years and 0.37 kg/year between ages 45 to 65 years in men, suggesting deceleration in the rate of increase with age. In women, however, the rates of increase in TBF were 0.44 kg/year and 0.52 kg/year between ages 18 to 45 years and 45 to 66 years, respectively, indicating no deceleration over the same age intervals. The deceleration with age in the rate of increase in fat mass in men agrees with data from a large cross-sectional study by Mott and colleagues (1999). This data set suggests nonlinear trends for fat mass by age in both sexes and Asian, Black, Puerto Rican, and white ethnic groups. All groups reached a maximum fat mass between 50 and 60 years, during which there was little or no change on average. Fat mass decreased after age 60 years in all groups, with the exception of Puerto Rican women (figure 17.1). There are few data for changes with age in fat mass during senescence (i.e., age >65 years), particularly longitudinal data for changes within individuals. Longitudinal data from the New Mexico Aging Process Study suggest that total body fat mass remains relatively stable over time in elderly men and women over a 3-year period but decreases significantly over long periods of follow-up (table 17.1).

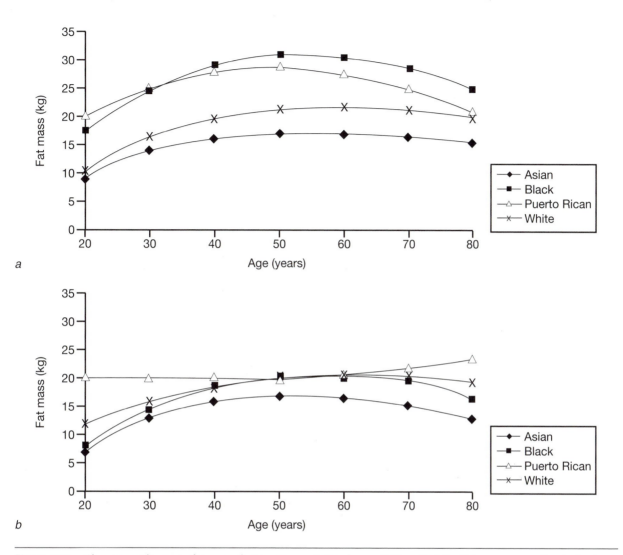

Figure 17.1 Changes with age in fat mass (kg) in women *(a)* and men *(b)* by ethnicity. Fat mass was estimated using a four-component model based on body volume, total body water, total body bone mineral, and body weight.

From J.W. Mott et al., 1999, "Relation between body fat and age in 4 ethnic groups," *American Journal of Clinical Nutrition* 69:1007-1013. Reproduced with permission by the *American Journal of Clinical Nutrition.* © Am J Clin Nutr. American Society for Clinical Nutrition.

From the perspective of evolutionary theory, the great range of variability in fat mass in humans can be attributed to the selection for genes that confer high efficiency of fat storage within adipocytes in the context of acute, possibly seasonal, changes in food supply: the "feast or famine" hypothesis (Lev-Ran 2001). The ability to store excess energy during times of plenty would have conferred a significant advantage for survival and reproduction during subsequent periods of scarcity. There are plenty of data that support this hypothesis, ranging from resistance to high leptin levels in obesity (Hofbauer 2002) to delayed menarche and amenorrhea at very low levels of body fatness (Frisch 1985). We have evolved to be insensitive to hormonal signals that we are too fat.

On the other hand, very low body fat shuts down our ability to reproduce. The current epidemic of obesity is undoubtedly attributable to effects of these "obesity genes" within an evolutionarily new environment of chronic energy surplus. This epidemic of obesity is having clear consequences for secular changes in age-related trajectories in body fatness and associated disease patterns.

Fat Distribution

The anatomical distribution of adipose tissue shows distinct patterns of change with age and marked sexual dimorphism. Subcutaneous adipose tissue distribution is conventionally referred

Table 17.1   Changes in Body Composition Over 3- and 7-Year Follow-Up Periods in Elderly Men and Women: New Mexico Aging Process Study

	WOMEN			MEN			COMBINED		
	n	Mean ± SD	Range	n	Mean ± SD	Range	n	Mean ± SD	Range
3-YEAR CHANGE									
Weight (kg)	250	−0.45 ± 3.64	−17.2 to 11.8	133	−1.10[a] ± 4.15	−13.0 to 9.7	383	−0.67 ± 3.83[a]	−17.2 to 11.8
Fat mass (kg)	225	−0.25 ±3.19	−14.5 to 11.3	128	−0.12 ± 3.26	−9.2 to 9.3	353	−0.20 ± 3.21	−14.5 to 11.3
Muscle mass (kg)	225	−0.29 ± 0.80[a]	−5.5 to 1.7	128	−0.97 ± 1.29[a]	−6.2 to 2.8	353	−0.54 ± 1.05[a]	−6.2 to 2.8
7-YEAR CHANGE									
Weight (kg)	114	−1.85 ± 4.59[a]	−14.7 to 11.2	72	−1.71 ± 4.49[a]	−16.5 to 6.6	186	−1.79 ± 4.54[a]	−16.5 to 11.2
Fat mass (kg)	85	−1.41 ± 3.79[a]	−14.4 to 6.7	57	0.16 ± 3.96	−11.5 to 8.8	142	−0.84 ± 3.91[a]	−14.4 to 8.8
Muscle mass (kg)	85	−0.51 ± 0.92[a]	−2.8 to 1.9	57	−1.09 ± 1.31[a]	−4.6 to 2.1	142	−0.75 ± 1.13[a]	−4.6 to 2.1

[a]Statistically significant ($p < .05$) changes.

to as "fat patterning" to distinguish it from the accumulation of internal adipose tissue, particularly visceral adipose tissue (Bouchard and Johnston 1988). Age, ethnic, and racial differences in subcutaneous fat patterning have been extensively described from radiographs, skinfold thicknesses, and body circumferences using a variety of ratios or indexes (Bouchard and Johnston 1988; Forbes 1987; Johnston and Foster 2001; Roche et al. 1986). Mueller first developed principal components analysis as a multivariate method of describing fat patterning from several skinfold thickness measurements and described changes with age (Mueller 1982; Mueller and Wohlleb 1981). Baumgartner and colleagues (1986) subsequently described changes during adolescence in relation to stages of sexual maturity. In general, subcutaneous adipose tissue thicknesses increase on the trunk in boys during adolescence, and gluteal–femoral fat increases in girls, leading to distinct phenotypes in adulthood that have been described as "android" versus "gynoid" fat patterns (Bouchard and Johnston 1988). These changes during adolescence are associated with stage of sexual maturity, sex hormone levels, and changes in plasma lipid or lipoprotein cholesterol concentrations (Baumgart-

ner et al. 1989; Bouchard and Johnston 1988). An android fat pattern in adults of either sex is associated with a spectrum of metabolic risk factors for chronic disease, including hypercortisolism, hypercholesterolemia, hypertension, and insulin resistance (Seidell et al. 1989), and with behavioral and psychosocial risk factors, such as low physical activity, smoking, alcohol intake, and depression and anger (Mueller et al. 2001).

During the 1980s, it became increasingly clear that the risk for metabolic disorders was more strongly associated with intra-abdominal adipose tissue than subcutaneous fat patterning (see chapter 13). Interest shifted toward the use of body circumferences and various ratios to describe fat distribution, specifically the relative amounts of intra-abdominal, or visceral, adipose tissue. An enormous volume of work has subsequently been published on the epidemiology of fat distribution using waist, waist–hip, or other circumference ratios, including changes with age over the life span in diverse populations; this body of literature is too large to review in this chapter (Casey et al. 1994; Heitmann 1991; Molarius et al. 1999; Poehlman et al. 1995; Shimokata et al. 1987). Casey and colleagues (1994) used longitudinal data on waist and hip diameters to describe changes from approximately age 5 to 30 years. Waist–hip diameter ratios decreased from age 5 to 18 years in both sexes and then increased from age 18 to 30 years. Tracking was strong in both sexes: Waist–hip diameter ratio at age of peak height velocity predicted more than 50% of the variance in the ratio at 30 years of age. Poehlman and colleagues (1995) described changes with age in waist circumference in 720 men and women aged 18 to 88 years. The rate of increase with age in waist circumference was 0.28 cm/year in women and 0.18 cm/year in men. Heitman (1991) described changes with age in waist–hip ratio from 35 to 65 years of age in a Danish population sample. Men had consistently greater ratios relative to TBF than women at all ages. Waist–hip ratios increased in men up to age 55 years; however, most of the increase with age in women occurred after 55 years. A variety of studies have indicated that menopause is a "critical period" for increases in visceral adiposity in women (Toth et al. 2000).

Some investigators have noted that waist circumference and waist–hip ratio are fairly insensitive measures of visceral adiposity, especially for detecting changes and in elderly populations (Baumgartner et al. 1993; Ross et al. 1996). Imaging methods, such as computed tomography (CT) and magnetic resonance imaging (MRI), are the only methods available for accurate measurement of visceral adipose tissue (VAT; see chapter 7). Few studies using imaging methods have explicitly considered age-related changes in VAT; the majority of studies have used anthropometric surrogates because CT and MRI methods are expensive to use in large population samples. The various published data indicate age-related patterns and sex differences in CT- or MRI-measured VAT that generally parallel those established for their anthropometric surrogates. VAT increases with age, especially during middle age. Women have a marked increase in VAT during menopause. During old age, absolute amounts of VAT may remain relatively stable, although VAT may increase in proportion to total body fat, because the latter declines during senescence (Baumgartner et al. 1993). Goran and associates are among the few to have described age changes in VAT in children from CT images (Figueroa-Colon et al. 1998; Gower et al. 1999; Huang et al. 2001). Their data indicate that boys and girls aged 5 to 10 years have similar amounts of VAT; however, white children tend to have greater amounts than African American children. In one longitudinal study, these authors

estimated the rate of increase in VAT area to be approximately 5.2 cm^2/year over a 3- to 5-year follow-up period in children approximately 8 years old at baseline (Huang et al. 2001). I am aware of no similar published data for rates of change in VAT in other age groups.

Fat-Free Mass

Changes with age in FFM during growth show strong "tracking," or age-to-age correlations, within individuals, suggesting a considerable amount of genetic control (Guo et al. 1997). FFM increases during growth, is relatively stable throughout maturity, and declines during senescence (Gallagher et al. 1997; Guo et al. 1997, 1999). Changes with age during growth show marked sexual dimorphism beginning at about age 13 years, as boys develop greater muscle and bone mass than girls (Guo et al. 1997). Changes with age during senescence may also show sexual dimorphism: Rates of loss are reported to be greater in males than in females (Baumgartner et al. 1995; Gallagher et al. 1997). FFM includes "body cell mass" (BCM), on the cellular level of body composition organization, which represents the body's mass of metabolically active, or living, cells (Moore et al. 1963). Changes in FFM are highly correlated with changes in BCM; although some studies suggest that the fraction BCM/FFM decreases with age as the population of metabolically active cells shrinks and is replaced by inactive components, such as extracellular fluid and solids (Baumgartner et al. 1995; Gallagher et al. 1996). Decreases in FFM and BCM occur in relation to starvation (wasting), disease (cachexia), and age (sarcopenia; Roubenoff et al. 1997). Death occurs when FFM decreases to about 40% of an individual's normal status (Roubenoff and Kehayias 1991).

One of the first methods applied to estimate FFM was total body potassium (TBK; chapter 4). This method assumes that the potassium content of the FFM is relatively constant at 64.2 mmol/kg, and that about 60% of TBK is contained in skeletal muscle, with the remainder distributed in nonskeletal muscle, organs, and other tissue components. Several studies have described changes with age in TBK (Flynn et al. 1989; Forbes and Reina 1970; Novak 1972; Pierson et al. 1974). He and colleagues (2003) recently published the most extensive data to date for changes from 20 to 90 years of age in a large, multiracial, multiethnic cohort of black, white, and Hispanic participants. Cross-sectional analyses suggested that TBK begins to decline after about 30 to 31 years of age regardless of sex, race, or ethnicity. The estimated rate of decrease in this study was higher in men (176 mmol/decade) than in women (87 mmol/decade). Previous studies in other populations using different statistical methods also indicate decreases with age in TBK, with greater rates of loss in men than women; however, these studies differ as to initial age, some reporting earlier (Pierson et al. 1974) or later (Flynn et al. 1989) ages.

Several studies indicate that the concentration of potassium in FFM decreases systematically with age (Cohn et al. 1976; Heymsfield et al. 1993, 2000; Keyahias et al. 1997; Mazariegos et al. 1994). If true, TBK would tend to overestimate the rate of decrease in FFM with age. Overall, data for TBK generally agree with those based on newer methods indicating that soft tissue components of FFM decline more rapidly in older men than women (Baumgartner et al. 1995; Gallagher et al. 1997).

There are important age-related changes in the composition of FFM. The main molecular components of the FFM are water, protein, osseous and nonosseous mineral, and glycogen (see chapter 12). The proportions of water, protein, and osseous mineral in the FFM are known to vary systematically with age. This variation is important because it affects assumptions underlying some in vivo measurement methods, such as hydrodensitometry, but also because it has health and functional significance. On the tissue–system level, the FFM is composed of skeletal and nonskeletal muscle, organ, connective tissue, and bone. The relative proportions of skeletal muscle, organ, and bone in

FFM also vary systematically with age. Age-related changes in the proportions of minor components, including soft tissue mineral, carbohydrates, nonskeletal muscle, and connective tissue may occur but are not well understood.

FFM Hydration

This refers to the proportion of the FFM composed of water, both intra- and extracellular fluids. This proportion is conventionally taken as approximately 73.2% of FFM. The hydration of the FFM has been shown to be remarkably stable in healthy individuals and across a wide variety of mammalian species (Wang et al. 1999). This suggests the evolution of tight physiological control mechanisms for cellular hydration that are necessary for health.

Hydration of the FFM is clearly higher in infants and young children, as much as 80% on average, and decreases to about 73% by about 10 to 15 years of age (figure 17.2; Ellis 1990; Fomon et al. 1982; Friis-Hansen 1961; Wang et al. 1999). This change is accompanied by corresponding increases in the proportions of protein and mineral in the FFM and a consequent increase in its density. There has been controversy as to the relative stability of FFM hydration during the remainder of the life span, particularly in senescence, as well as the presence of systematic differences with sex, ethnicity, and level of obesity. Some studies suggest that hydration of the FFM increases slightly in old age, resulting in a slight, systematic decrease in FFM density (Heymsfield et al. 1993, 2000), whereas

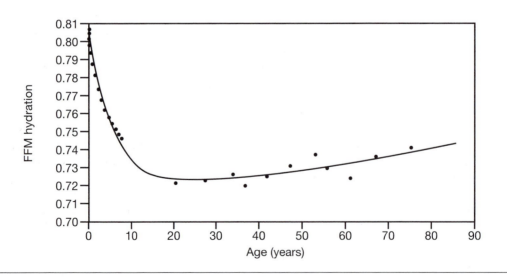

Figure 17.2 Changes with age in hydration of fat-free mass (FFM).

From S.B. Heymsfield et al., 2000, "Anthropometry and methods of body composition measurement for research and field applications in the elderly," *European Journal of Clinical Nutrition* 54(Suppl. 3): S26-S32. Raw data compiled from previous studies.

others found no significant increase (Chumlea et al. 1999; Lesser and Markofsky 1979; Visser et al. 1997). These conflicting findings are likely the result of different methodologies and study sample selection criteria. It is difficult to separate effects of age from those of morbidity in old age, and some studies may have included elders with malnutrition, those with subclinical disease, or participants who were taking medications that could affect FFM hydration. Baumgartner and colleagues (1991) hypothesized that control mechanisms governing FFM hydration homeostasis might deteriorate in old age, resulting in increased within- and between-person variability in FFM hydration, and for this reason these authors endorsed the use of multicomponent models to improve individual estimates. There are few published data, however, to support or refute this hypothesis (Schoeller 1989).

Changes in the hydration of FFM may be accompanied by changes in the distribution of body fluids between intra- and extracellular spaces. Cheek (1961) reported a rapid decrease in the fraction of total body water in the extracellular space during the first year of life, from about 46% at birth to 41% by 1 year of age. The ratio of extracellular fluid (ECF) to intracellular fluid (ICF) is under tight homeostatic physiological control and is relatively stable throughout most of life. Several studies suggest that the ratio may increase slightly during senescence (Baumgartner et al. 1995; Fulop et al. 1985; Lesser and Markofsky 1979; Mazariegos et al. 1994; Steen 1988). It isn't clear, however, whether this change reflects loss of BCM with a relative preservation of ECF or whether it reflects an increase in ECF with age. Body fatness increases with aging, and a study by Waki and colleagues (1991) suggested that this may increase the ECF–ICF ratio. As for FFM hydration, an increased ECF–ICF ratio in elderly individuals could again be a marker of mild malnutrition or undiagnosed disease rather than the result of aging. Large increases in the ratio of ECF to ICF are clearly associated with malnutrition and disease (Moore et al. 1963).

Bone Mineral Density

Total body bone mass increases with age during growth and development to maturity, reaching peak bone mass between 20 and 30 years of age in most individuals (Mora and Gilsanz 2003). The density of mineral (mainly calcium and phosphorus) in bone, which is a significant determinant of bone strength, also increases with age from birth to maturity. Bone mineral density decreases with age after reaching its peak in young adulthood. The rate of decrease accelerates in women during menopause, in relation to declining levels of circulating estrogens (Blunt et al. 1994). Because of the serious impact of osteoporotic fractures, there are extensive reference data for changes with age, sex, racial, and geographic differences in bone mineral density, mainly for the hip and spine, and numerous studies of the causes and consequences of osteoporosis that cannot be reviewed in this brief overview. Changes with age in bone mineral density have been reviewed extensively with regard to their effects on in vivo methods of body composition analysis, particularly hydrodensitometry (see chapter 2).

Skeletal Muscle

Skeletal muscle is the next most variable component, within and between individuals, after fat mass. Data for changes with age in skeletal muscle mass are relatively sparse compared with those for body fat, fat distribution, or bone. In general, growth and development are a period for rapid accretion of skeletal muscle, with marked sexual dimorphism developing during adolescence. Skeletal muscle mass is relatively stable within individuals during adulthood up to about age 30 to 40 years, after which mass begins to decrease. The rate of decrease is greater in men than women and appears to accelerate in old age.

Urinary creatinine excretion has been used most extensively to estimate changes with age in skeletal muscle mass. Creatinine is a product of muscle metabolism and has been shown to be excreted in urine in direct proportion to muscle mass (Heymsfield et al. 1983). Metter and colleagues (1999) analyzed data from the Baltimore Longitudinal Study for changes with age in muscle mass and strength. Age trajectories for creatinine excretion and muscle strength were essentially parallel to each other, showing accelerating rates of decline after age 40. Creatinine levels at age 80 were about 60% of the value at age 20 in both sexes. Muscle cross-sectional area, as estimated by anthropometry, and FFM from dual-energy X-ray absorptiometry, however, showed slower rates of decline with different trajectories compared with creatinine or muscle strength. This is not surprising, because anthropometric estimates of cross-sectional muscle area have been shown to be inaccurate in older adults (Baumgartner et al. 1992), and FFM includes changes in nonmuscle components, such as bone and organ tissue. Longitudinal analyses, conducted for men only, also showed accelerating rates of decline in creatinine and muscle strength. Metter and colleagues (1999) also examined age trajectories in muscle quality, in terms of muscle strength per unit creatinine, cross-sectional area, and FFM. The results were inconsistent and depended on the measure used in the denominator. The relationship between changes with age in muscle strength and size continues to be a controversial issue.

Janssen and colleagues (2000) published cross-sectional data for changes with age in skeletal muscle mass, as measured using MRI, in men and women aged 18 to 88 years. These data indicate that muscle mass is relatively stable, on average, up to 45 years, after which there are accelerating rates of loss in both sexes. Rates of loss are greater for lower-body and leg muscle than for upper-body and arm muscle (figure 17.3). Muscle mass as a percent of body weight decreases more rapidly in men than in women with age.

The acceleration of the rate of decrease in skeletal muscle mass in old age has garnered increased interest in the last decade, because it was provided with a name, *sarcopenia,* signifying deficiency in skeletal muscle mass and strength. Sarcopenia is associated with impaired thermogenesis and immunocompetency, functional limitation and disability, and increased risk for falls and bone fractures (Evans 1997). Low muscle mass is associated with physical inactivity and declining levels of testosterone in elderly men and, possibly, growth hormone in both sexes (Baumgartner et al. 1998, 1999; Morley et al. 2001). It is associated with anatomical changes, specifically the selective loss of type 2 muscle fibers, loss of alpha motor neurons, and decreased capillary density (Roubenoff 2000). The prevalence of sarcopenia increases rapidly with age greater than 60 years, as shown in studies of several different data sets, including the large, nationally representative NHANES III (Baumgartner et al. 1998; Janssen et al. 2002; Melton et al. 2000; Tanko et al. 2002). Studies of the epidemiology of sarcopenia conducted to date have used varying definitions, and standardized methods remain to be developed. The direct health care costs of sarcopenia-associated disability have been estimated to be at least $12 billion (Janssen et al. 2004). Given that sarcopenia contributes indirectly to several other conditions, such as osteoporotic fractures, the total health care costs may be considerably greater.

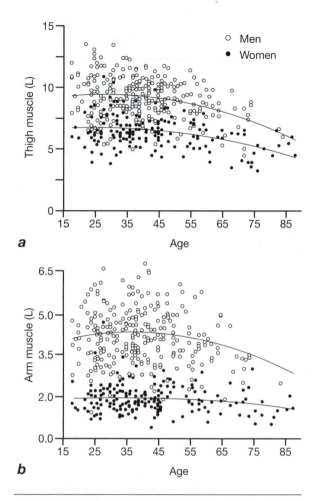

Figure 17.3 Changes with age in *(a)* thigh muscle volume and *(b)* arm muscle volume from magnetic resonance imaging in men and women.

Reprinted, by permission, from R. Ross et al., 2000, "Skeletal muscle mass and distribution in 468 men and women aged 18-88 yr.," *Journal of Applied Physiology* 89: 81-88.

Age-related changes also occur in the composition of muscle. Several anatomists in the early 20th century noted from dissection and chemical analyses of cadavers that the amount of fat in muscle appeared to increase with age (Frantzell and Ingelmark 1951). Frantzell and Ingelmark were the first to document this increase in vivo using radiographic methods. Their data, however, are semiquantitative, expressed in terms of the percentage of thigh and leg radiographs with visible interstitial and subfascial fat. They also noted the positive correlation of muscle fat with the thickness of subcutaneous adipose tissue. Interest in this aspect of body composition was renewed in this last decade with the development of accurate CT and MRI methods for quantifying muscle fat, which show the replacement of muscle with adipose and connective tissues in CT or MRI images (Baumgartner et al. 1992; Goodpaster et al. 2000). Increased muscle fat, as measured using CT attenuation coefficients, increases with age and is associated with muscle insulin resistance (Goodpaster et al. 1997). This method, however, does not differentiate extramyocellular from intramyocellular lipid. Recently, magnetic resonance proton spectroscopy methods have been developed that specifically measure intramyocellular lipid (Boesch and Kreis 2000). The extent to which intramyocellular lipid is increased in aged muscle, independent of obesity, has not yet been established.

It has been recognized for some time that age-related changes in skeletal muscle may interact with changes in other body composition components, particularly body fatness. Forbes (1991) noted that changes fat and fat-free body mass components are correlated and generally occur in a constant proportional relationship to weight change: 70% fat to 30% FFM. This relationship, however, may become dysregulated during senescence, allowing the development of discordant changes in lean and soft tissue components leading to a body composition characterized by low muscle mass in the presence of high levels of body fatness. This disordered form of body composition has been called *sarcopenic obesity* (Baumgartner 2000; Roubenoff 2000).

Sarcopenic obesity develops in increasing numbers of elderly people with advancing age. I developed a method for defining sarcopenic obesity from estimates of appendicular skeletal muscle and percent body fat (Baumgartner 2000) and estimated that the prevalence of this body composition type may increase from about 2% to 10% from age 65 to 85 years (figure 17.4; Morley et al. 2001). In analyses of cross-sectional data from the New Mexico Aging Process Study (NMAPS), as well as the population-based New Mexico Elder Health Survey, sarcopenic obesity was more strongly associated with functional limitations, such as balance and gait disorders, and disability than either obesity or sarcopenia. The odds ratio for three or more physical disabilities was 4.12 for sarcopenic obesity, compared with 2.33 for nonsarcopenic obesity and 2.07 for sarcopenia without obesity in the NMAPS, after adjustment for sex, age, smoking, and morbidity. New analyses of longitudinal data from the NMAPS suggest that sarcopenic obesity predicts the onset of disability over a 5-year period. In a Cox proportional hazards analysis, the relative risk of incident disability was 2.54 (90% confidence interval, 1.23-4.86), after adjustment for sex, age, physical activity, and morbidity.

The etiology of sarcopenic obesity is presently unknown. Roubenoff (2000) hypothesized that muscle was the primary process contributing to the age-related gain in fat mass, which would in turn result in accelerated loss of muscle. However, age-related increases in body fatness that lead to obesity generally precede the onset of sarcopenia, which appears to occur regardless of level of body fatness. Thus, I propose the alternative hypothesis that sarcopenic obesity is a late-life consequence of long-standing obesity, which may accelerate muscle loss in old age. Adipose tissue is now recognized as an endocrine, as well as lipid storage, organ that produces a variety of hormones and cytokines that regulate metabolism and influence body composition, including

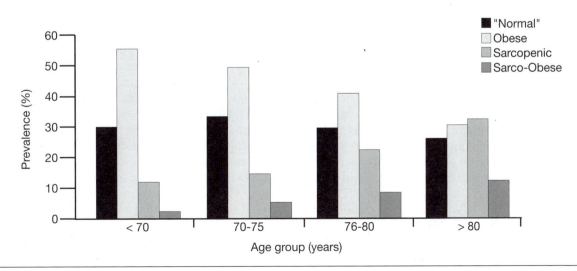

Figure 17.4 Increasing prevalences of sarcopenia and sarcopenic obesity with age: New Mexico Aging Process Study.

estrone, leptin, adiponectin, tumor necrosis factor-α, and interleukin-6 (Das 2001). Thus, obesity is believed to be associated with a low-level, chronic inflammatory state. Sarcopenia has been shown to be associated with suppressed protein synthesis in skeletal muscle (Morley et al. 2001), but there are few data suggesting increased protein degradation. Inflammatory cytokines, such as tumor necrosis factor-α and interleukin-6, that are secreted by adipose tissue and increased in obesity may increase degradation in skeletal muscle via their effects on the ubiquitin–proteasome pathway (Biolo et al. 2003). Thus, persistent obesity and its associated systemic inflammatory effects might result in accelerated skeletal muscle loss over time, leading to sarcopenic obesity in old age. Long-term longitudinal data are needed to test this hypothesis.

Other Organ Tissues

It is well established that the sizes of various internal organs, other than adipose tissue, muscle, or bone, change with aging, nutritional status, and disease. The extant reference data have been derived mainly from cadaver autopsies, raising the question of potential effects of disease and postmortem changes (Coppoletta and Wolbach 1933; Dekaban and Sadowsky 1978; Eurich and Linder 1984; Snyder et al. 1975). Imaging methods such as CT and MRI have considerable promise for providing in vivo information for healthy individuals, but data from these approaches are still sparse because of their high cost (Gallagher et al. 1998).

The brain and liver are the largest organs in the body, making up respectively about 12.2% and 4.5% of total body weight at birth. These percentages decrease to about 2.5% each at maturity around 20 years of age and remain stable over most of the life span. Kidneys and heart are the next largest organs, comprising respectively about 0.45% and 0.54% of body weight during adulthood. The available data suggest slight decreases with age in the relative sizes of these organs beginning in middle age and progressing into senescence. Together, these organs comprise about 18% of body weight at birth, compared with 21% for muscle, and about 6% in adulthood versus 40% for muscle (Coppoletta and Wolbach 1933; Dekaban and Sadowsky 1978; Elia 1992; Henry 2000).

Research on variation in the sizes of internal organs in body composition and nutrition has primarily focused on estimating their contribution to total basal energy expenditure (Elia 1992; Gallagher et al. 1998, 2000; Henry 2000; Holliday 1971). It is unclear to what extent decreases in organ size contribute to decreases in basal metabolic rate during senescence (Gallagher et al. 2000).

Summary

Hippocrates described aging in terms of the loss of body heat and water: "The elderly are cold and moist, because the fire burns out in their bodies and the watery element flows in" (Gaylord and Williams 1994, p. 335). Later physicians (e.g., Aristotle, Galen) disagreed with regard to water, considering aging to be more a process of drying out. "Cold and dry" might be the prosaic, contemporary description of old age, except that in the modern era of excess caloric intake we might now add "fat."

There has been a great resurgence of interest in changes with age in body composition during the last 2 decades attributable to the advent of new in vivo technologies and studies demonstrating the associations of changes with a spectrum of health outcomes. Some of the documented changes with age agree, eerily, with ancient impressions. The hallmark of senescence is increasingly recognized to be the gradual, accelerating loss of the metabolically active, heat-generating part of the body, however it may be measured. Body cell mass, skeletal muscle mass, organ mass, and FFM all decline with age. This decrease is associated with declining resting energy expenditure and heat production. Ironically, there is increasing evidence that this loss is associated with the processes that generate body heat. The oxidative stress produced by metabolism causes cumulative damage to mitochondria resulting in muscle cell apoptosis and sarcopenia (Wanagat et al. 2001). The "fire" burns out!

Although there is still some debate as to whether the body becomes systematically "wetter" or "drier" with age, there is little doubt in the minds of most geriatric professionals that elderly patients are more susceptible to problems with body fluid regulation and more vulnerable to dehydration in particular. In any case, the contemporary observation that anhydrous fat increasingly replaces lean body mass with age suggests that we do, in another sense, get "drier" in old age. So, cold, dry, and fat?

The future promises the continued extension of current and new in vivo methods to increasingly larger, population-based samples from which truly representative reference data can be derived for multiple racial and ethnic groups across the entire age range. This will allow not only more powerful studies of the causes and consequences of changes with age in body composition but also trials testing the efficacy of interventions to prevent or reverse deleterious changes. The most immediate challenge is the current epidemic of obesity, which has already been observed to be associated with an emerging epidemic of type 2 diabetes. The obesity-associated increase in the prevalence of type 2 diabetes among children and adolescents is most alarming. The emergence of trailing epidemics of other obesity-associated chronic diseases, such as cardiovascular disease, hypertension, and endocrine-dependent cancers, may be expected. As the population is aging, it will also become increasingly important to study the late-life consequences of obesity, such as sarcopenic obesity and disability.

Chapter 18

Variation in Body Composition Associated With Sex and Ethnicity

Robert M. Malina

Two major sources of variation in body composition are the biological sex of the individual and population affinity. Biological differences between the sexes influence body composition per se and processes that affect body composition, for example, rate of growth and maturation, the timing and tempo of the adolescent growth spurt and sexual maturation, body proportions, and physique, among others (Malina et al. 2004). Sex differences in body composition are apparent early in life, are magnified during the adolescent growth spurt and sexual maturation, and persist through adulthood.

Population variability is often viewed in the context of race and ethnicity, terms that have different but related meanings. Race implies a biologically distinct group that shares a relatively large percentage of its genes in common by descent. Ethnicity implies a culturally distinct group. Biological and cultural homogeneity often overlap, as in minorities of color and linguistic and religious groups who share a common ancestry.

Racial background on a global basis is, with few exceptions, viewed in terms of area of geographic origin, that is, European, African, Asian, Amerindian, Pacific Islander, and so on. Within the American culture complex, racial–ethnic background is

commonly viewed on a color and surname basis, that is, European Americans (Whites, non-Hispanic Whites, Caucasians), African Americans (Blacks, non-Hispanic Blacks), Native Americans (American Indians), and Hispanic Americans (Mexican Americans, Puerto Ricans, Cubans, Latinos), and more recently in terms of geographic origin, that is, Asian Americans and Pacific Islanders, and others. In Canada, native Canadians are indicated as First Nation (Amerindians), Inuit (Eskimo), and Metis (descendants of French Canadians and Cree Indians).

Genetic and cultural heterogeneity of racial–ethnic groups should be recognized. American Whites are derived from all countries of Europe. American Blacks are descended from African slaves, most of whom were imported from West Africa in the 18th century (although origins within Africa are variable). There is a significant degree of admixture between American Blacks and Whites. Surname is used to classify Hispanic Americans, a heterogeneous group that includes Mexican Americans, Puerto Ricans, Cubans, immigrants from Central and South America, and Europeans of Spanish ancestry. Mexican Americans, the largest Hispanic group, are largely descendants of admixture between American Indians and Spaniards, which began in the 16th century.

Although individuals are labeled as belonging to a particular racial or ethnic group, variation within each of the categories is considerable.

I greatly appreciate the kindness of Claude Bouchard, Jack Wilmore, and Tuomo Rankinen in providing the unpublished descriptive data for the HERITAGE Study.

Moreover, variation in culturally determined habits, attitudes, and behavior patterns specifically related to diet, physical activity, and other aspects of lifestyle (e.g., perceptions of ideal body size, alcohol consumption) has implications for variation in body composition.

For the sake of convenience, I use the term *ethnic* in this report, recognizing the complexity of issues related to the concept. A good deal of body composition data are based on American samples of European, African, and Mexican ancestry, and ethnicity is generally self-reported. The terms *American White, American Black,* and *Mexican American* are used throughout. Data for Asian samples are commonly reported by country of origin, and these designations are used when reported.

Variation in body composition associated with sex and ethnicity is the focus of this chapter. Reference values and elemental composition of the body are initially discussed, followed by detailed consideration of total body composition. Attention is then devoted to adipose tissue distribution, skeletal (bone) tissue, and skeletal muscle tissue. Comparisons by sex and ethnicity are made in the context of the specific body composition methodology used, recognizing that there may be variation among methods. Age is a factor that affects sex differences and is considered where appropriate. Variation in body composition associated with the adolescent growth spurt and sexual maturation is discussed in more detail elsewhere (Malina et al. 2004).

Comparisons are based on trends in central tendencies (means, medians). This is a limitation, and comparisons need to be interpreted with care. Interindividual variation in body composition is considerable within and between ethnic groups. Data for samples defined as obese are not considered. Note, however, that some comparisons may be influenced by ethnic variation in the prevalence of overweight and obesity.

Reference Values

In vivo estimates of body composition are based on measurements of specific elements and components, for example, water, protein, mineral, potassium, calcium, and phosphorus. These in turn are converted to dimensions of body composition within the context of specific models (see chapter 1). If the principles, methods, and models for estimating body composition are to be accurately applied to children and adolescents,

it is important to understand changes in the composition of fat-free mass (FFM) with growth and maturation. Attainment of adult composition of FFM is commonly labeled chemical maturity, a concept that was defined more than 75 years ago by Moulton (1923, p. 80): "The point at which the concentration of water, proteins, and salts [minerals] becomes comparatively constant in the fat-free cell is named the point of chemical maturity of the cell."

Historically, focus has been on the chemical composition of the young adult reference male (Brozek et al. 1963) and on the reference newborn (Fomon 1966). Subsequently, direct and indirect estimates of body composition have been used to derive reference values for the composition of FFM in infants, children, and adolescents (Fomon et al. 1982; Haschke 1989; Lohman 1986; see table 18.1). Sex differences in the relative composition of FFM are negligible during infancy but are apparent in early childhood. After about 3 years of age, the estimated relative composition of FFM indicates less water and more protein and mineral in boys; water comprises a slightly greater percentage of FFM in girls. The sex difference is also reflected in the potassium content and density of FFM. From about 3 years of age, the estimated potassium content and density of the FFM are greater in boys than in girls. The sex difference is stable during childhood but is magnified during adolescence attributable, to a large extent, to a marked gain in muscle mass and bone mineral in males. The gain in skeletal mineral between 10 years and young adulthood reflects growth and maturation of the skeleton during the adolescent spurt. The relative mineral content of FFM in boys increases from 5.4% at about 10 years of age to 6.6% at 17 to 20 years. The gain in relative mineral content of FFM from early through late adolescence (1.2%) is about 22% of the initial value at age 10. The corresponding increase in mineral content of FFM in girls is less, 5.2% to 6.1%, a relative increase of about 16%.

The data summarized in table 18.1 are estimates derived in part from biochemical analyses of several cadavers and specific tissue samples and from in vivo estimates of total body water (TBW), potassium, nitrogen, calcium, and bone mineral pooled from several samples available at the time of the compilations. The estimates vary somewhat from laboratory to laboratory and are based on samples of American and European Whites. Recent estimates for a sample of American White children and adolescents 5 to 18 years of age, based on measures of TBW (by deuterium oxide), total

Table 18.1   Estimated Composition of the Fat-Free Mass (FFM) From Infancy to Young Adulthood

| Age (years) | COMPARTMENTS OF THE FFM | | | | |
	Water (%)	Protein (%)	Mineral (%)	Potassium (g/kg)	Density (g/cm³)
MALES					
Birth	80.6	15.0	3.7	1.92	1.063
1	79.0	16.6	3.7	2.21	1.068
3	77.5	17.8	4.0	2.39	1.074
5	76.6	18.5	4.3	2.49	1.078
7-9	76.8	18.1	5.1	2.40	1.081
9-11	76.2	18.4	5.4	2.45	1.084
11-13	75.4	18.9	5.7	2.52	1.087
13-15	74.7	19.1	6.2	2.56	1.094
15-17	74.2	19.3	6.5	2.61	1.096
17-20	74.0	19.4	6.6	2.63	1.099
FEMALES					
Birth	80.6	15.0	3.7	1.92	1.064
1	78.8	16.9	3.7	2.24	1.069
3	77.9	17.7	3.7	2.38	1.071
5	77.6	18.0	3.7	2.42	1.073
7-9	77.6	17.5	4.9	2.32	1.079
9-11	77.0	17.8	5.2	2.34	1.082
11-13	76.6	17.9	5.5	2.36	1.086
13-15	75.5	18.6	5.9	2.38	1.092
15-17	75.0	18.9	6.1	2.40	1.094
17-20	74.8	19.2	6.0	2.41	1.095

Note. Data from birth to 5 years are adapted from Fomon et al. (1982), whereas those for the other ages are adapted from Lohman (1986). The estimated relative composition of FFM in the data of Fomon et al. (1982) does not add to 100 because the small, constant percentage of carbohydrate (0.6%) is not included in the table. The protein content of the FFM in the estimates of Lohman (1986) is derived by subtraction: 100 – water – mineral = protein.

body potassium (TBK; by whole body counting) and bone mineral content (BMC; by dual-energy X-ray absorptiometry [DXA]) compare favorably with reference values presented by Fomon and colleagues (1982) and Haschke (1989), although protein and bone mineral show some differences (Ellis et al. 2000). The more recent sample of children and adolescents are heavier, and fat mass (FM) accounts for most of the heaviness.

Corresponding data for TBW, TBK, and BMC of American Black and Mexican American children and adolescents 5 to 18 years may have implications for ethnic-specific reference values (Ellis et al. 2000). Ethnic comparisons of TBW, TBK, and BMC within sex are illustrated in figure 18.1. The samples of American White, American Black, and Mexican American youth are, on average, similar in height and weight, except for Black males, who are

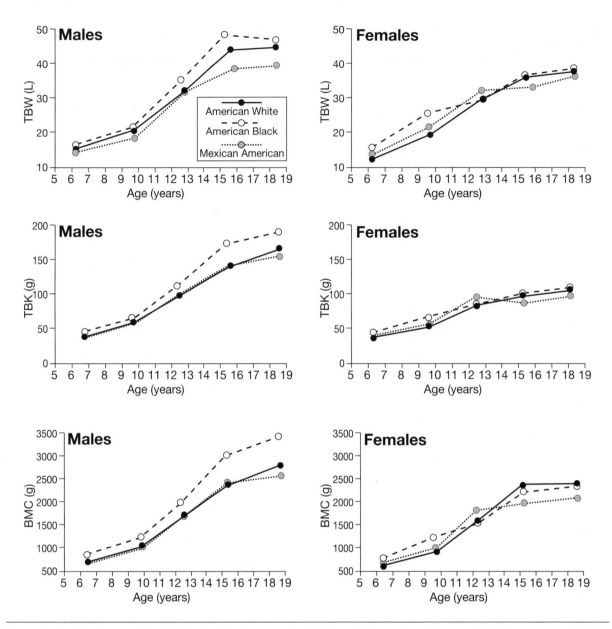

Figure 18.1 Total body water (TBW), total body potassium (TBK), and bone mineral content (BMC) in American White, American Black, and Mexican American children and adolescents 5 to 19 years of age.

Data from K.J. Ellis et al. (2000).

heavier in adolescence. Trends across childhood and adolescence suggest greater TBW, TBK, and BMC in American Black males, especially in early adolescence (11-13 years). Mean values of TBW, TBK, and BMC do not consistently differ among females by ethnicity. Nevertheless, after adjustment for height and weight, TBW, TBK, and BMC differ significantly among ethnic groups in both sexes (Ellis et al. 2000).

Sex differences in TBW, TBK, and BMC within the samples of American White, American Black, and Mexican American youth are small during childhood (5-7, 8-10 years) and are established during adolescence (figure 18.2). The comparisons suggest that sex differences in the three components are greater in American Black adolescents than among American White and Mexican American adolescents.

Reference values for the body composition of young adult males and females 20 to 30 years of age are summarized in table 18.2. With the exception of FM, all components of body composition are greater in males (Snyder et al. 1984). Estimates for young adult American White males and females

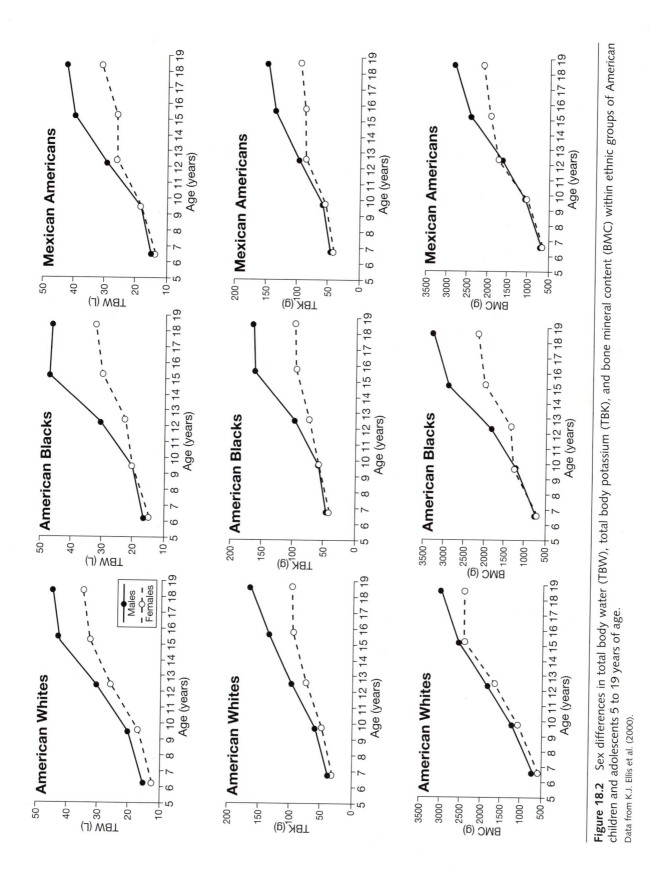

Figure 18.2 Sex differences in total body water (TBW), total body potassium (TBK), and bone mineral content (BMC) within ethnic groups of American children and adolescents 5 to 19 years of age.

Data from K.J. Ellis et al. (2000).

based on total body neutron activation analysis, whole body counting, and tritiated water dilution are included for comparison. The more recent sample of females differs slightly from the reference sample in height and weight, but estimates of body composition do not significantly differ. The more recent sample of males are taller and heavier, and the larger size is associated with larger skeletal and lean tissue masses and greater water content (Ellis 1990).

Mean trends in the elemental composition of American White adults are illustrated in figure 18.3. The data are based on total body neutron activation analysis, whole-body counting, and tritiated water dilution in a cross-sectional sample of 1,134 females and 167 males 20 to 74 years of age. Variation about the means (not shown) is considerable, 8% to 14% (Ellis 1990). Total body calcium (TBCa), total body phosphorus (TBP), total body nitrogen (TBN), TBK, and TBW are less in females than males, and the components decline with age across adulthood. TBCa and TBP are primary components of bone. The decline in TBCa is generally linear across age in males; in contrast, TBCa appears to be rather constant in females to about 50 years and then declines more

sharply to the mid-70s. If we allow for variation in sample size and the cross-sectional nature of the sample, relative decline in elemental composition between 20 to 24 and 70 to 74 years appears to differ between males and females. Corresponding elemental data indicate higher levels of TBCa, TBP, TBK, and TBN in small samples of American Blacks, 35 females and 30 males (Ellis 1990). The issue of ethnic variation in the composition of the FFM needs further systematic analysis (Visser et al. 1997).

Total Body Composition

The study of body composition historically has been driven by the availability of methods to measure, or more correctly, estimate it. The research was to some extent directed by what could be measured rather than what researchers wanted to measure. Over the past 10 to 20 years, significant progress in the development and refinement of techniques now permits measurement of virtually all components of the body. In turn, models that provide the framework for studying body composition have been modified (chapter 1).

Table 18.2 Comparison of Estimated Body Composition of Young Adult White Males and Females

Characteristics	MALES		FEMALES	
	ICRP[a]	IVNAA[b]	ICRP	IVNAA
Age (years)	20-30	23.9	20-30	24.4
Weight (kg)	70	77.4	58	59.1
Height (cm)	170	177.7	160	163.3
BODY COMPOSITION				
TBW (L)	42	45.5	29	31.3
ICW (L)	23.8	24.9	17.4	15.5
ECW (L)	18.2	20.2	11.6	15.6
FFM (kg)	57.4	64.4	39.6	44.0
Fat (kg)	13.5	19.9	16.0	15.9
BCM (kg)	41.2	36.3	26.9	21.0

Note. TBW = total body water; ICW = intracellular water; ECW = extracellular water; FFM = fat-free mass; BCM = body cell mass.

[a]Reference values of the International Commission on Radiological Production (ICRP); [b]Data for American Whites based on in vivo neutron activation analysis (IVNAA).

Data from K.J. Ellis, 1990.

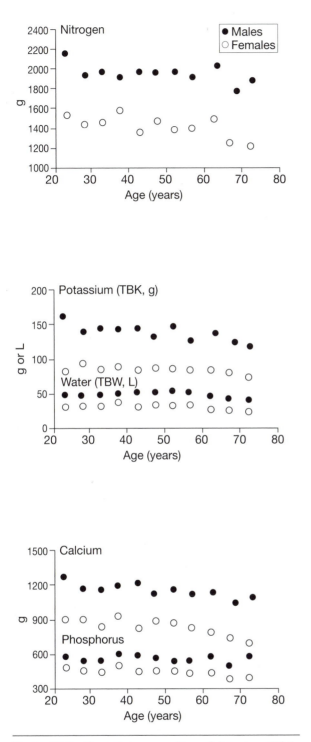

Figure 18.3 Elemental composition of American White adults.

Data from K.J. Ellis (1990).

Two-Component Model

A good deal of the available data that span childhood, adolescence, and adulthood are derived from body density (D_b) and TBW. Data for D_b

and TBW were compiled from 32 and 11 studies, respectively, to illustrate sex differences in these basic parameters (Malina 1989; Malina et al. 1988). The total sample for D_b was 3,667 (2,110 males, 1,557 females) from 8 to 20 years old, and that for TBW was 1,152 (675 males, 477 females) from infancy to 22 years. Subjects were primarily European and American Whites, although several samples of American Blacks were included. Reported means for D_b and TBW were adjusted for varying sample sizes to derive age- and sex-specific means. TBW and D_b were converted to estimates of FFM and %BF, respectively, using age- and sex-specific estimates of the water content and density of FFM (Lohman 1986). FFM and %BF respectively were then used in conjunction with the body weights of the composite samples to estimate FM and %BF from TBW and to estimate FM and FFM from D_b.

FFM follows a growth pattern like that of height and weight, and sex differences become clearly established during the adolescent growth spurt (figures 18.4 and 18.5). Young adult values of FFM are reached earlier in females, about 15 to 16 years compared with 19 to 20 years in males. In late adolescence and young adulthood, males have, on the average, a FFM that is about 1.4 times larger than the FFM of females. The average FFM of young adult females is only about 70% of the mean value for young adult males. The difference reflects the male adolescent spurt in muscle mass and the sex difference in stature in young adulthood. When FFM is expressed per unit stature, sex differences are small in childhood and early adolescence, but after 14 years of age, males have considerably more FFM for the same stature as females. The sex difference increases with age so that young adult males have about 0.36 kg of FFM for each centimeter of stature compared with about 0.26 kg of FFM for each centimeter of stature in females (Malina et al. 2004).

The sex difference in FM is negligible before 5 or 6 years of age. Subsequently, FM increases more rapidly in girls than in boys. FM increases through adolescence in girls, but it appears to reach a plateau or to change only slightly near the time of the adolescent spurt in boys (about 13-15 years). In contrast to FFM, females have, on average, about 1.5 times the FM of males in late adolescence and young adulthood.

Percent body fat is only slightly greater in girls than in boys during infancy and early childhood, but from 5 to 6 years through adolescence, girls consistently have a greater %BF than boys. The relative fatness of females increases gradually

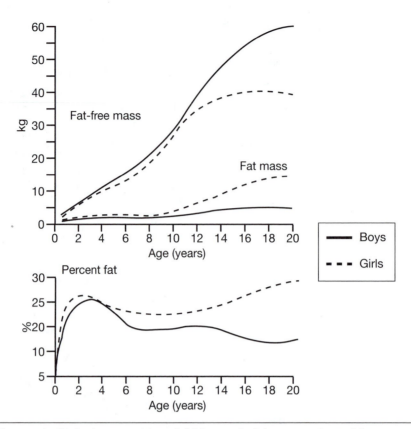

Figure 18.4 Sex differences in the body composition of children and adolescents estimated from measures of total body water.

Data from Malina et al. (1988).

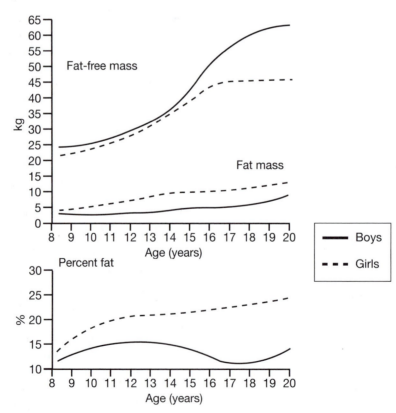

Figure 18.5 Sex differences in the body composition of children and adolescents estimated from measures of body density.

Data from Malina et al. (1988).

through adolescence in the same manner as FM. Percent body fat also increases gradually in males until just before the adolescent spurt (about 11-12 years) and then gradually declines. It reaches its lowest point at about 16 to 17 years in males and then gradually increases into young adulthood. In contrast to estimates of FM, %BF tends to decline during male adolescence because of the rapid growth of FFM and slower accumulation of FM at this time.

The relative accuracy of FFM, FM, and %BF derived from D_b for diverse samples with the two-component model can be evaluated by comparison with corresponding estimates based on a multicomponent model in the mixed-longitudinal sample from the Fels Longitudinal Study (Guo et al. 1997). A multicomponent model incorporating age- and sex-specific estimates of the density and major components of FFM was used to derive %BF, FM, and FFM in the Fels sample. Estimates of FFM in the two independent samples of females compare closely, whereas estimates of FFM in Fels males are slightly but consistently lower than estimates for the composite sample. Estimates of FM are slightly but consistently higher in Fels females than in the composite sample of females but are quite similar in Fels males and the composite sample of males.

Percent body fat is consistently higher in Fels males and females compared with the composite samples, although differences in late adolescent males are small (Malina et al. 2004).

Estimates of body composition based on D_b with the two-component model for other ethnic groups are limited. Estimates of %BF, FFM, and FM for Japanese adolescents and young adults 11 to 25 years (Tahara et al. 2002a, 2002b) were derived with procedures similar to those used in the samples of Europeans and Americans of primarily White ancestry (Malina 1989; Malina et al. 1988). Subjects in the Japanese sample were shorter and lighter from 14 years in females and 15 years in males. Estimates of body composition in the two samples are compared in figure 18.6. FFM is identical in Japanese and American males until 15 years and then is less in the Japanese, reflecting the ethnic difference in height. The same trend is apparent in Japanese and American females, but the differences in later adolescence and young adulthood are less than in males. Estimates of FM are quite similar in Japanese and American adolescents and young adults of both sexes. Estimates of %BF are greater in American females in early adolescence but do not differ in late adolescence and young adulthood.

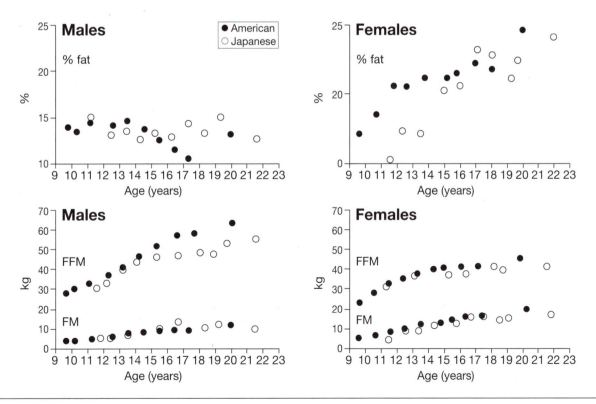

Figure 18.6 Comparison of the body composition of American and Japanese adolescents and young adults estimated from measures of body density.

Data from Malina et al. (1988) and Tahara et al. (2002a, 2002b).

Estimates of %BF in Japanese and American males overlap through 16 years of age and differ in late adolescence.

Densitometric estimates of body composition extend from 11 to 60 years of age in the Japanese sample (Tahara et al. 2002a, 2002b). D_b is greater and %BF is less in males than in females at all ages except 11 years. The sex difference is greatest in late adolescence and the early 20s. Subsequently, it is smaller but consistent through the late 40s and increases in the 50s. Estimated FFM is similar in males and females at 11 to 13 years but then is greater in males; the sex difference in FFM increases with age during adolescence, reflecting the sex difference in height, and persists through adulthood. Estimated FM, on the other hand, is slightly greater in females during adolescence and young adulthood but is similar in males and females from the late 20s through the 50s.

Multicomponent Model

Trends in estimated body composition for a cross-sectional sample of American White men and women 20 to 79 years of age are illustrated in figure 18.7. The estimates are based on total body neutron activation analysis, whole-body counting, and tritiated water dilution (Ellis 1990). Body mass was partitioned in the context of the four-component model: water, protein, mineral (ash) and fat; body cell mass (BCM) was derived from TBK; FFM was estimated from protein, TBW, and ash; and FM was estimated as body mass minus FFM. Protein, mineral, water compartments, and BCM are larger in males throughout adulthood. Protein, mineral, intercellular water, and BCM decrease with advancing age, whereas extracellular water is rather constant across adulthood. The age-related decline in BCM parallels intercellular water and is steeper in males than in females from the 50s through the 70s. Estimated FM overlaps considerably between the sexes and is the most variable parameter of body composition. The cross-sectional trends suggest that FM declines after the 50s in females but continues to increase with age more or less linearly in males from the 30s through the 70s. After adjustment for sexual dimorphism in body mass, females have a greater %BF.

Dual-Energy X-Ray Absorptiometry

Variation associated with sex and ethnicity is apparent in estimates of lean tissue mass (LTM), BMC, and FM derived from DXA measurements of

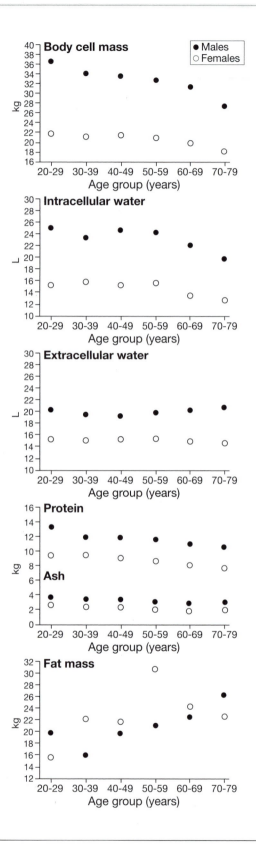

Figure 18.7 Sex differences in the body composition of American White adults estimated with a multicomponent model.

Data from K.J. Ellis (1990).

cross-sectional samples of American White (145 males, 141 females), American Black (78 males, 104 females), and Mexican American (74 males, 68 females) children and adolescents 3 to 18 years of age (Ellis 1997; Ellis et al. 1997). Height does not consistently differ among ethnic groups, but American Blacks and Mexican Americans of both sexes tend to be heavier than American Whites in adolescence. Ethnic comparisons by 3- to 5-year age groups are shown in figure 18.8. Ethnic differences in LTM and BMC among males are negligible before adolescence. LTM and BCM are larger in

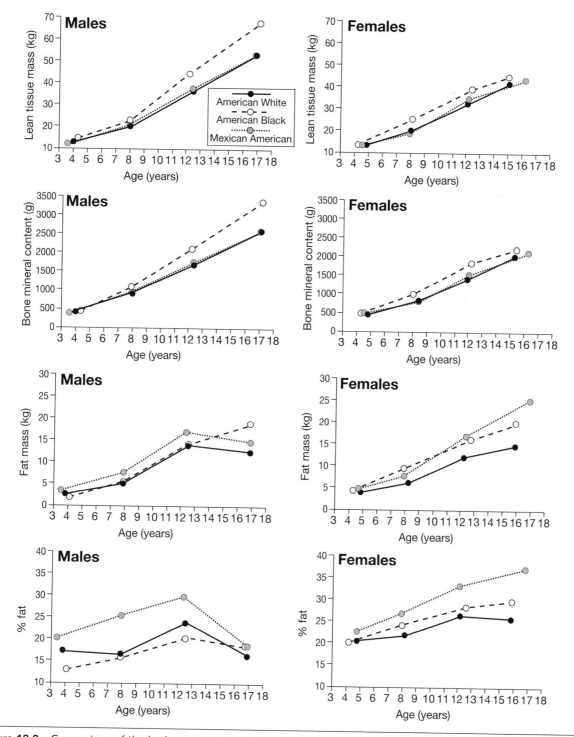

Figure 18.8 Comparison of the body composition of American White, American Black, and Mexican American children and adolescents 3 to 18 years estimated from dual-energy X-ray absorptiometry.

Data from K.J. Ellis (1997) and K.J. Ellis et al. (1997).

American Black males during adolescence but do not differ between American White and Mexican American males. Ethnic variation in estimated FM is less and inconsistent across the age groups. However, %BF is greater in Mexican American males than in American White and American Black males except in the oldest age group. Percent body fat and FM are not consistently different between Black and White males, but %BF is slightly lower in the former. The ethnic differences in body composition among males persist after statistically controlling for body size (Ellis 1997).

Ethnic comparisons of LTM and BMC in females indicate similar trends, although the magnitude of the differences is smaller. LTM and BMC are, on average, larger in American Black females but are similar in American White and Mexican American females. In contrast to males, American Black and Mexican American females are similar in FM, and both have a greater FM than American White females. The ethnic difference in FM increases with age. Relative fatness shows a gradient across the age groups, Mexican American > American Black > American White, and the ethnic differences increase with age among females (Ellis et al. 1997).

The magnitude of the sex difference within each ethnic group appears to vary with component of body composition (figure 18.9). Sex differences within each ethnic group are negligible in the two youngest age groups for LTM and BMC and increase with age so that during adolescence males have larger LTM and BCM than females. FM is only slightly larger in American White females than males across the four age groups. FM is also greater in American Black females than males, but the sex difference increases with age. Mexican American males and females do not differ in FM in the two youngest groups, but older females have an especially larger FM. Percent body fat is greater in females than in males within each ethnic group, and the sex difference appears to be greater in Mexican Americans and American Blacks compared with American Whites.

Bioimpedance Analysis

FFM, FM, and %BF were estimated from bioimpedance analysis in American White, American Black, and Mexican American participants, 12 to 80 years of age, in the nationally representative sample of the National Health and Nutrition Examination Survey (NHANES) III (1988-1994). Resistance values were transformed to TBW and in turn FFM using prediction formulas derived from a sex-specific, multi-component model (Chumlea et al. 2002; see also Sun et al. 2003). FM and %BF were estimated. The resistance index, stature squared divided by resistance (S^2/Res), shows ethnic variation. It is higher in American Black than in American White females in all age groups except one (18-19 years), but is higher in American White than in American Black males in all but one age group (18-19 years). On the other hand, the resistance index is higher in American Whites and Blacks than in Mexican Americans of both sexes across the age range (Chumlea et al. 2002). Other data suggest a similar relationship between resistance and body mass in European White, Melanesian, and Polynesian adults of both sexes after adjustment for height and age. In contrast, the slope between mass and resistance differs in Australian Aborigine adults (Heitmann et al. 1997).

Ethnic comparisons of FFM, FM, and %BF in the NHANES III sample are summarized in figure 18.10. Estimated FFM is, on average, generally larger in American White and Black than in Mexican American males across all ages. The differences are negligible in the youngest age group (12-13 years) but increase with age through adolescence into young adulthood. Estimated FFM decreases in males of each ethnic group in the 60s and 70s. Estimated FM does not differ consistently by ethnicity among males across the age range. Percent body fat is greater in Mexican American males in most age groups but does not differ between American White and Black males.

Corresponding comparisons of females indicate several trends. Estimated FFM is, on average, consistently less in Mexican American compared with American White and Black females. The difference is apparent in late adolescence and persists across the age range. Estimated FFM is similar in American Black and White adolescent females but is consistently larger in Black than in White females through adulthood. The gradient for FFM in adult women is American Black > American White > Mexican American. Ethnic differences in estimated FM are apparent during adolescence. American Black and Mexican American females have a larger FM than American White females. Across the adult age range, FM is greatest in American Black females. FM is greater in Mexican American than American White females through the 40s but does not differ in the 50s, 60s, and 70s. In contrast, %BF does not differ between American Black and Mexican American females throughout the age range, but their %BF is consistently greater than that of American White females.

The magnitude of sex differences in body composition varies, on average, with ethnicity in the

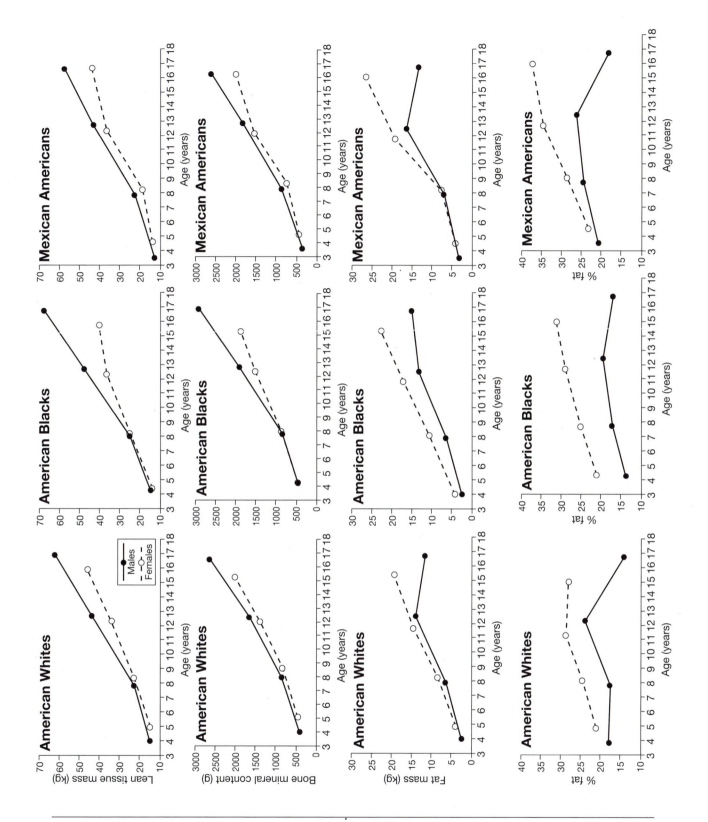

Figure 18.9 Sex differences in the body composition of children and adolescents within ethnic groups of American children and adolescents 3 to 18 years old.

Data from K.J. Ellis (1997) and K.J. Ellis et al. (1997).

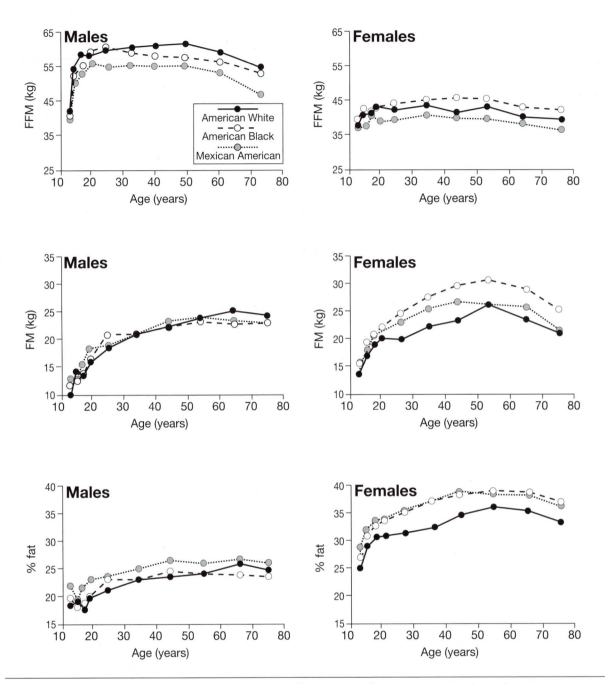

Figure 18.10 Comparisons of the body composition of American White, American Black, and Mexican American adolescents and adults estimated from bioimpedance analysis.

Data from Chumlea et al. (2002).

NHANES III sample. The sex difference in estimated FFM increases from early through late adolescence in the three ethnic groups. During adulthood, the sex difference in FFM tends to be larger in American Whites, whereas the magnitude of the sex difference in FFM in American Blacks and Mexican Americans is not consistently different. Although there is variation, especially during adolescence and young adulthood, the sex difference in FM and

%BF suggests the following gradient: greatest in American Blacks, followed by Mexican Americans and then American Whites.

Adipose Tissue Distribution

The anatomical distribution of the major components of the body mass—adipose, skeletal muscle, and skeletal tissues—varies between males and females. The term *distribution* refers to the absolute or relative amount of a tissue in different regions or compartments of the body. Of the three tissues, attention has traditionally been given to the relative distribution of subcutaneous adipose tissue (SAT). Currently, emphasis is on the distribution of abdominal visceral and subcutaneous adipose tissue.

The relative distribution of SAT in different regions of the body is commonly addressed with ratios of skinfold thicknesses taken at several trunk (T) and extremity (E) sites (see Malina 1996). Computed tomography (CT) is the primary tool in the study of abdominal visceral adipose tissues (VAT) and SAT. Limited magnetic resonance imaging (MRI) data also provide information on abdominal VAT and SAT.

Trunk–Extremity Contrast in SAT Distribution

Studies vary in skinfold thicknesses used to calculate T–E ratios as estimates of relative SAT, for example, the sums of the subscapular and suprailiac (T) and triceps and biceps (E) skinfolds for a mixed-longitudinal sample of French infants through young adults (Rolland-Cachera et al. 1990) or the sums of the subscapular, suprailiac, and abdominal skinfolds (T) and triceps, biceps, and medial calf skinfolds (E) for a cross-sectional sample of French Canadian ancestry, 9 to 60 years (Malina 1996; Malina and Bouchard 1988). Data from the two studies suggest the following trends in relative SAT distribution. Sex differences are negligible from infancy through childhood into early adolescence. Subsequently, males accumulate proportionally more SAT on the trunk compared with the extremities. The disproportionate accumulation of trunk SAT in males is marked during adolescence and then continues more slowly through the fifth decade. The adolescent increase in the T–E ratio in males is, in part, a function of a reduction in the absolute thicknesses of extremity skinfolds and an increase in the thickness of trunk skinfolds during the growth spurt (Malina et al. 2004). Females, in contrast, gain proportionally similar amounts of SAT at trunk and extremity sites so that the T–E ratio is reasonably stable through the fourth decade. Subsequently, trunk skinfolds increase proportionally more in thickness than extremity skinfolds.

The trends based on skinfolds do not take into account changes that occur in limb size and area. DXA measurements may provide a better estimate of changes in SAT distribution during childhood and adolescence. Data for a sample spanning 4 to 35 years of age show similar trends for DXA measurements of T and lower E adipose tissue (Cowell et al. 1997), that is, a slight increase in T–E ratios in late childhood, a negligible change in the ratio in adolescence and young adulthood in females, and a continued increase in the ratio in males from adolescence through young adulthood. Corresponding data for females 8 to 27 years old indicate absolutely greater accumulation of trunk compared with extremity SAT (leg and arm) with sexual maturation (Goulding et al. 1996), but estimated T–E ratios are reasonably stable in the later stages of puberty and young adulthood.

Principal components analysis of skinfold thicknesses measured at a variety of sites is commonly used to address sex and ethnic variation in relative SAT distribution (Malina 1996). The first principal component differentiates trunk from extremity skinfolds and varies with sex and ethnicity. Results of a principal components analysis of nine skinfold thicknesses in American Blacks and Whites 6 to 30 years of age, after adjustment for stage of sexual maturity, suggest that males of both ethnic groups, but not females, accumulate proportionally more SAT on the trunk with maturation (Baumgartner et al. 1986). Black males and females, respectively, also have a more central distribution of SAT than White males and females. A principal components analysis of four skinfolds (triceps, subscapular, suprailiac, medial calf) in American adolescent girls, 12 to 17 years old, from four ethnic groups (Mexican, White, Asian-Filipino, Black) indicates ethnic differences in the first component (Malina et al. 1995). Asian and Mexican American adolescent girls have proportionally more trunk SAT compared with White and Black girls. The second component, which differentiates between upper and lower extremity skinfolds in this sample of adolescent girls, does not differ among the ethnic groups, but the third component, which differentiates between skinfolds on the upper (triceps, subscapular) and lower (suprailiac, medial calf) parts of the body, suggests proportionately more SAT on the upper body in Asian, Black, and Mexican compared with White adolescent girls. Similar results are apparent in principal components analyses of Black, White, and Mexican adolescent females (Mueller et al. 1982) and Japanese young adult women (Hattori 1987). Ethnic variation in the distribution of SAT is also evident in bivariate plots of skinfold thicknesses for several sites (Mueller 1988). The plots indicate a more central distribution of SAT in American Black, Mexican, and Japanese children and adolescents compared with American Whites, who have a more peripheral pattern of SAT.

Similar trends are apparent in T–E ratios of Mexican American and American White adults from San Antonio (table 18.3). Ratios are consistently higher in Mexican Americans of both sexes, except for the oldest age group of males, which shows variation by ratio. The ethnic dif-ference is greater in females than males and is attributable to especially larger trunk skinfolds in Mexican American women. The ratios are also higher in males than females across age groups within each ethnic group, indicating proportionally more trunk SAT in males. Standard deviations of the two ratios are greater in males than females in each ethnic group but do not consistently differ between ethnic groups within each sex.

Table 18.3 Trunk–Extremity Skinfold Ratios in American Whites and Mexican Americans in San Antonio, Texas

Age group	MEXICAN AMERICANS			AMERICAN WHITES		
	N	Mean	SD	N	Mean	SD
SUBSCAPULAR + SUPRAILIAC/TRICEPS + BICEPS SKINFOLD THICKNESSES						
MALES						
25-34	236	2.16	0.54	90	2.00	0.64
35-44	208	2.37	0.64	115	2.11	0.56
45-54	144	2.27	0.62	88	2.22	0.61
55-69	123	2.20	0.62	79	2.24	0.57
FEMALES						
25-34	288	1.54	0.34	115	1.39	0.39
35-44	292	1.55	0.35	158	1.38	0.35
45-54	215	1.55	0.35	89	1.44	0.37
55-69	157	1.52	0.39	86	1.43	0.42
SUBSCAPULAR + SUPRAILIAC/TRICEPS + MEDIAL CALF SKINFOLD THICKNESSES						
MALES						
25-34		1.88	0.50		1.69	0.55
34-44		2.09	0.54		1.80	0.49
45-54		2.05	0.57		1.98	0.63
55-69		2.10	0.65		2.00	0.60
FEMALES						
25-34		1.31	0.29		1.12	0.35
35-44		1.35	0.34		1.13	0.35
45-54		1.38	0.34		1.19	0.32
55-69		1.34	0.33		1.20	0.40

Data from R.M. Malina and M.P. Stern (unpublished). See Malina et al. 1983 for a more complete description of the study.

Abdominal Adipose Tissue

Estimates of abdominal VAT and SAT in European and North American Whites and American Blacks are summarized in table 18.4. Sex differences in abdominal VAT and SAT during childhood and adolescence are negligible, but abdominal SAT shows more variation than VAT. Data for a small sample of younger children (n = 16, 4-9 years) indicate no sex differences in VAT and SAT (Goran et al. 1995). Early and late pubertal girls also do not differ in VAT area, but the former have a significantly larger SAT area (de Ridder et al. 1992). The data of Huang and colleagues (2001) included American Black and White children who were followed for 3 to 5 years. Mean age at baseline was 8.1 ± 1.6 years with a range of 4.6 to 12.1 years. The estimated rate of growth of abdominal VAT was greater in White than Black children by about 1.9 ± 0.8 cm^2/year, but that for abdominal SAT did not differ between ethnic groups. Estimated growth rates for VAT, SAT, and total body fat did not differ between boys and girls. The data of Fox and colleagues (2000), based on MRI, suggest more VAT and SAT in girls than in boys at 11 and 14 years of age; however, the sex difference in VAT was quite small. Over an interval of 2 years, boys gained proportionally more abdominal VAT than girls, 69% versus 48%, whereas girls gained proportionally more SAT than boys, 78% versus 19%.

The limited data for adolescent boys and girls indicate less abdominal VAT compared with adults, which suggests that abdominal VAT accumulates late in adolescence or young adulthood and continues into adulthood (figure 18.11). If we allow for variation in the measurement of abdominal adipose tissue, the mean values suggest that VAT increases with age in both sexes during adulthood, more so in males. SAT appears to increase with age to about 60 years and then declines in both sexes. SAT overlaps between adult males and females, although means tend to be higher in females.

Data for American Blacks are limited (table 18.4, figure 18.11). Abdominal VAT and SAT for Black children and adults are well within the range of means for White samples, suggesting negligible ethnic variation.

Ratios of abdominal VAT to SAT should be interpreted with caution because the majority are estimated from reported means. The ratios indicate considerable overlap among samples of White children and adolescents, suggesting a negligible sex difference in the proportional distribution of VAT and SAT. Estimated ratios are higher in the two samples of American Black males compared with females, suggesting proportionally more VAT (figure 18.11). The ratios for the two samples of Black males are similar to those of White males, but those of Black females are lower, suggesting an ethnic difference in females. On the other hand, MRI measures of VAT and SAT at each lumbar level and integrated across L1-L5 are less in American Black than American White girls 7 to 10 years old matched for weight, body mass index (BMI), skeletal age, stage of puberty, and socioeconomic status, but the ratio of VAT to SAT does not differ (Yanovski et al. 1996).

In contrast to children and adolescents, adult males have proportionally more VAT (higher ratio) than adult females. The sex difference increases with age, but women gain relatively more abdominal VAT after menopause (Enzi et al. 1986). The ratios for the two samples of American Black adults tend to be slightly lower than White adults within each sex, suggesting proportionally less abdominal VAT in Blacks. Within Blacks, males have proportionally more VAT (higher ratio) than females.

The data summarized in table 18.4 and figure 18.11 need to be viewed with some degree of caution. Abdominal VAT area is significantly correlated with FM in children, adolescents, and adults, and higher levels of VAT are generally observed in overweight or obese individuals (Bouchard 1994; Goran et al. 1997). Thus, individuals with a higher FM tend to have more VAT. There are, nevertheless, considerable individual differences in level of VAT at any level of overall adiposity (Bouchard 1994).

Anthropometric procedures have been used to partition FM into SAT and VAT in adults. Fifteen skinfolds, segmental surface areas, and the density and proportion of lipid in adipose tissue were used to estimate SAT in Japanese subjects 18 to 23 years old (Hattori et al. 1991), whereas 14 skinfolds, body surface area, skin weight, and density of adipose tissue were used to estimate SAT in Japanese subjects 40 to 77 years old (Komiya et al. 1992). FM was estimated from D_b and TBW, respectively; VAT was derived by subtraction. Although results of the two studies are not directly comparable, they suggest sex differences in SAT and VAT (table 18.5). Estimated SAT comprises a greater percentage of total FM in females than males at all ages; by inference, males have proportionally more VAT. The estimated proportion of total AT as VAT increases with age in both sexes.

Table 18.4 Visceral and Subcutaneous Abdominal Adipose Tissue Areas in Samples of European and North American Whites and American Blacks

Reference	Sex	N	AGE (YEARS) Mean	SD	VAT (CM^2) Mean	SD	SAT (CM^2) Mean	SD	VAT–SAT RATIO[a] Mean	SD
EUROPEAN AND NORTH AMERICAN WHITES										
Goran et al. (1997) CT umb	M	16	8.2	1.6	2.7	16	65	67	0.41	
	F	20	8.2	1.2	54	27	172	102	0.31	
Huang et al. (2001) CT umb	M	23	8.6	1.9	33	20	103	92	0.32	
	F	60	8.1	1.4	27	15	78	61	0.35	
Fox et al. (2000) MRI L4	M	25	11.5	0.3	18	10	78	49	0.31	0.28
	M	25	13.7	0.3	30	11	93	55	0.39	0.09
	F	17	11.5	0.3	26	10	75	42	0.39	0.16
	F	17	13.7	0.2	38	10	134	73	0.35	0.17
de Ridder et al. (1992) MRI minimal waist	F	13[b]	11.5	0.9	24	4	44	6	0.54	
	F	11[b]	14.0	0.3	26	4	63	14	0.41	
Lemieux et al. (1993) CT L4-5	M	89	36	3	123	49	254	101	0.48	
	F	75	35	5	104	54	428	210	0.24	
Bouchard and Wilmore (unpublished)[c] CT L4-5	M	163	25	6	77	43	203	145	0.38	
	F	172	25	6	53	30	252	145	0.21	
	M	99	54	5	159	62	269	108	0.59	
	F	95	52	5	120	59	363	121	0.33	
Seidell et al. (1988) CT L4-5	M	14	31	5	73	44	130	80	0.66	0.27
	F	7	34	4	58	46	210	132	0.32	0.28
	M	35	52	7	96	47	147	76	0.74	0.36
	F	19	53	8	93	49	244	105	0.40	0.19
	M	17	74	6	109	63	105	57	1.06	0.39
	F	8	68	2	115	59	199	148	0.40	0.14
Weits et al. (1988) CT L4	M	14	30	7	52	40	97	59	0.50	0.20
	F	12	34	6	55	23	187	78	0.35	0.20
	M	25	52	6	126	70	157	63	0.77	0.33
	F	27	52	5	87	53	243	109	0.35	0.13
	M	29	69	5	121	71	144	68	0.84	0.39
	F	23	67	5	89	60	222	94	0.37	0.18
AMERICAN BLACKS										
Goran et al. (1997) CT umb	M	27	7.3	1.6	22	17	61	86	0.36	
	F	38	7.4	1.8	28	17	106	82	0.26	
Huang et al. (2001) CT umb	M	24	7.9	1.5	36	32	104	125	0.35	
	F	31	8.1	2.0	25	14	94	83	0.26	
Bouchard and Wilmore (unpublished)[c] CT L4-5	M	88	27	7	68	50	231	187	0.29	
	F	149	28	8	60	36	338	192	0.18	
	M	29	50	7	106	69	244	139	0.43	
	F	60	46	7	96	44	389	144	0.25	

Note. VAT = visceral adipose tissue; SAT = subcutaneous adipose tissue; CT = computed tomography; umb = level of the umbilicus; M = male; F = female. Values are means and standard deviations (SD). Except for age among children and adolescents, decimals have been rounded to nearest whole value.

[a]When no standard deviation is reported, the ratio is based on the group means; [b]early pubertal (breast stage 2) and late pubertal (breast stage 4); [c]see Wilmore et al. (1999) for a more complete description of the study.
Data from Malina, Bouchard, and Beunen, 1988.

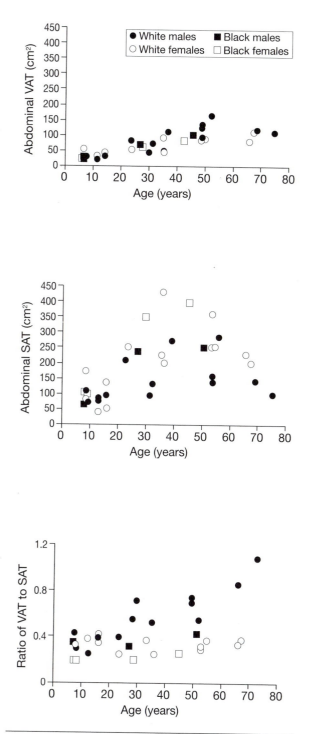

Figure 18.11 Abdominal visceral adipose tissue (VAT), subcutaneous adipose tissue (SAT), and the VAT–SAT ratio in children, adolescents, and adults.

Drawn from data in table 18.4.

The anthropometric estimates of VAT and SAT in Japanese adults differ from estimates based on whole-body multislice MRI in Asian American adults of both sexes. MRI gives considerably lower estimates of VAT (table 18.5), which indicates a

limitation of anthropometric approaches to partitioning FM. VAT as percentage of total AT is reasonably similar in Asian American and American White women, 5.1% and 3.4%, respectively, and men, 9.0% and 10.1%, respectively. Asian American males tend to have less total AT, VAT, and SAT than American White males, whereas Asian American females have slightly less total AT and SAT but more VAT than American White females (Park et al. 2001). Consistent with abdominal measures of VAT and SAT, adult males have a greater proportion of VAT than females.

Bone As a Component of Body Composition

Historically, one of the major obstacles to estimates of body composition was the lack of a verified in vivo method for quantifying bone tissue. Earlier studies were based on analyses of the whole skeleton, bone sections, and standard radiographs. These studies suggest variability in bone mineralization and mass associated with age, sex, and ethnicity. The new technologies, specifically dual-photon absorptiometry (DPA) and DXA, have provided new insights into bone as a component of body composition and skeletal mineral status.

Skeletal Proportions

Individual bones and in turn body segments contribute differentially to total length of the skeleton or stature. Proportional relationships among skeletal segments vary with age, sex, and ethnicity (Malina et al. 2004) and probably influence estimates of body composition (Gerace et al. 1994). Data are most available for the ratio of sitting height to stature, which is an index of the relative contribution of the trunk, neck, and head as a unit (sitting height) to stature, and by subtraction, the relative contribution of the lower extremities, specifically subischial length (stature minus sitting height), to stature. The ratio is highest in infancy and declines through childhood into adolescence. The ratio reaches a nadir early in the adolescent growth spurt because the legs experience their spurt first. The spurt in the trunk then follows and the ratio increases in late adolescence and young adulthood. During the fourth decade of life (the 30s), sitting height begins to decline, attributable largely to the compression and eventual loss of elasticity of the intervertebral discs

Table 18.5　Estimates of Total Fat Mass (FM), Subcutaneous Adipose Tissue Mass (SAT), and Visceral Adipose Tissue Mass (VAT) in Japanese Adults

Age	Sex	N	FM (KG)		SAT (KG)		VAT (KG)		% SAT	
			Mean	SD	Mean	SD	Mean	SD	Mean	SD
ANTHROPOMETRIC ESTIMATES IN JAPANESE ADULTS										
18-23[a]	M	121	7.7	3.6	4.1	1.9	3.6	2.1	53.7	11.2
18-23	F	93	11.4	3.5	6.9	1.6	4.5	2.4	62.6	9.8
47 ± 5[b]	M	42	19.7	5.2	8.8	4.5	10.9	3.1	42.8	12.9
49 ± 6	F	25	22.2	4.9	12.0	3.8	10.1	2.2	53.5	7.9
67 ± 5	M	12	14.9	4.4	5.1	2.0	9.8	3.2	33.7	8.
65 ± 4	F	14	20.0	5.6	9.4	11.1	10.6	2.8	45.2	10.0
MRI ESTIMATES IN ASIAN AND WHITE AMERICAN ADULTS[c]										
Asian, 29 ± 7	M	19	12.1	3.0	10.8	2.7	1.1	0.6		
Asian, 29 ± 5	F	18	14.2	4.6	13.4	4.3	0.7	0.5		
White, 32 ± 7	M	34	15.5	5.8	13.7	5.0	1.7	1.1		
White, 33 ± 7	F	36	16.5	5.2	15.9	5.0	0.6	0.3		

Note. Data based on whole-body multislice magnetic resonance imaging (MRI) in Asian American and American White adults are included for comparison. M = males; F = females.

[a]Adapted from Hattori and colleagues (1991), anthropometry; [b]adapted from Komiya and colleagues (1992), anthropometry; [c]adapted from Park and colleagues (2001), MRI.

with advancing age. Thus, the lower extremities contribute proportionally more to stature with advancing age; accumulation of SAT over the buttocks in adulthood may offset to some extent the decline in measured sitting height.

There are no sex differences in the sitting height–stature ratio during infancy and childhood. By about 10 to 12 years of age, the ratio is slightly higher in girls and remains so through adolescence into adulthood. Thus, during adolescence and adulthood, females have, on average, relatively shorter lower extremities than males for the same stature. There is also a sex difference in the proportional relationship between the shoulders (biacromial breadth) and hips (bicristal breadth), which arises during adolescence. The difference is attributable to differential growth of biacromial breadth in males because bicristal breadth is, on average, similar in the sexes (Malina et al. 2004). Males thus have proportionally broader shoulders compared with the hips, whereas females have proportionally broader hips compared with the shoulders.

There is ethnic variation in the sitting height–stature ratio. The ratio is, on average, lower in American Blacks from infancy through adulthood, whereas ratios of American Whites and Mexican Americans differ slightly though consistently. The ratio should also be viewed in the context of stature. American Black and White children, adolescents and adults, on average, differ only slightly in stature; in contrast, Mexican Americans are shorter. American Blacks have shorter trunks and longer lower extremities than American Whites, whereas Mexican Americans have absolutely shorter trunks and lower extremities. Thus, for the same stature, American Blacks have relatively shorter trunks, or conversely, relatively longer lower extremities than American Whites and Mexicans. On the other hand, relative proportions do not differ markedly between American Whites and Mexicans, although the latter are absolutely smaller in both segments (Bouchard et al. 1997; Malina et al. 2004). Data also suggest variation between Blacks and Whites in proportional differences of specific segments of the lower (thigh, lower leg) and upper (upper arm, forearm) extremities and in pelvic breadth dimensions (Malina 1973).

Data for specific segment lengths are not extensive for groups other than American Blacks and Whites. Individuals of Asian ancestry tend to be shorter than American Whites and Blacks. Shorter lower extremities account for the difference. Sitting height does not differ between American White and Japanese children and adolescents, and it is greater than in American Blacks. The absolute differences in stature and estimated leg length translate into proportional differences (Malina et al. 2004).

Skeletal Weight and Density

The most comprehensive analysis of the skeleton and its parts has come from the work of Trotter and colleagues (Trotter and Hixon 1974; Trotter and Peterson 1962, 1970; Trotter et al. 1959, 1960). Although based on traditional methods of physical anthropology, the results are relevant to understanding ethnic variation and in vivo studies of the skeleton as a component of body composition.

Weight of the Skeleton

The dry, defatted skeleton weighs, on average, about 95 g in infant boys and slightly less in girls and about 4.0 kg and 2.8 kg in young adult males and females, respectively. Corresponding data for childhood and adolescence are lacking. As a percentage of body mass, the dry, fat-free skeleton comprises about 3% of body weight in the fetus and newborn and about 6% to 7% of body mass in the adult. Bone mineral, estimated from ash weight, comprises about 2% of body mass in infants and 4% to 5% of body mass in adults. The estimates are reasonably similar to measures of total body bone mineral derived from DXA. The skeleton is consistently heavier and has more mineral in American Blacks than in American Whites from infancy through adulthood (Trotter and Hixon 1974; Trotter and Peterson 1970; see also Malina 1969). Mean age-adjusted (to an age of 63.0 years with analysis of covariance) weights of the dry, fat-free skeleton in American adults (*n* = 30 per group) are as follows: Black males, 3,899 g; White males, 3,446 g; Black females, 2,846 g; and White females, 2,335 g (Trotter and Hixon 1974).

Densities of Specific Bones

Densities of specific bones vary with sex and ethnicity in American Black and White adults (Trotter

and Hixon 1974; Trotter et al. 1960). Several trends are apparent: (a) Bones of males are more dense than those of females within each ethnic group; (b) bones of American Blacks are more dense than those of American Whites; (c) the densities of individual bones in males and females of both ethnic groups decrease with age; (d) long bones (limbs and ribs) are more dense than vertebrae; and (e) cervical vertebrae are more dense than other vertebrae in males but not significantly so in females.

The fat-free skeleton is a composite of bone mineral and organic bone matrix, and the ratio of ash weight to the weight of the dry, fat-free bone (percent ash weight) provides an estimate of mineral content. Although bone densities decrease with age, percent ash weight varies among bones (Trotter and Hixon 1974; Trotter and Peterson 1970). Within each sex and ethnic group, the mandible has the highest and the sternum the lowest percent ash weight. The cranium and long bones of the limbs rank closest to the mandible, whereas the vertebrae, ribs, and bones of the pelvis rank closest to the sternum. The remaining skeletal components range between the two extremes. Males generally show higher percent ash weights than females, which suggests that sex differences in density are related to less bone mineral in females. Significant differences in percent ash weight between American Blacks and Whites are limited to the clavicles, scapulae, bones of the hand, ribs, thoracic vertebrae, lumbar vertebrae, sacrum, and sternum; on the other hand, the major long bones do not show significantly greater percent ash weights in Blacks.

In Vivo Studies of Bone Mineral

Total body bone mineral content (TBMC) and bone mineral content (BMC) for several regions increase during childhood and adolescence (Faulkner et al. 1993; Maynard et al. 1998), although there are differences in delineating anatomical regions on DXA scans. There is no sex difference in the BMC of the head and trunk from childhood through adolescence and also no sex difference in TBMC and BMC of the limbs during childhood. Girls appear to have slightly greater TBMC and limb BMC at the time of the female growth spurt (about 12 years). During the male growth spurt, bone mineral in the limbs and entire body accumulates at a greater rate than in females and contributes to the sex difference in TBMC and limb BMC (Malina et al. 2004). TBMC and limb BMC increase from late adolescence into the 20s in males but show little change in females (Rico et al. 1992). The sex differences persist through adulthood (Horber et al. 1997).

Variation in estimated bone mineral density (BMD) associated with sex from childhood through adolescence is generally similar to TBMC. Estimated TBMD and trunk and limb BMD are somewhat greater in females about the time of the adolescent growth spurt. TBMD and BMD of the trunk and upper and lower extremities are greater in males later in adolescence, which reflects the sex difference in the timing of the growth spurt and late adolescent growth of the trunk (Faulkner et al. 1993; Maynard et al. 1998). The sex difference continues into and through adulthood (Horber et al. 1997; Rico et al. 1992).

Studies with DXA or DPA that focus on specific sections of a long bone or specific vertebrae indicate generally similar trends for sex differences in childhood, adolescence, and young adulthood (Bonjour et al. 1991; Geusens et al. 1991; Glastre et al. 1990; Gordon et al. 1991; Theintz et al. 1992). DXA measurements of BMC and BMD at the lumbar spine and femoral neck over a 1-year interval indicate substantial increments between 11 and 14 years in females and 13 and 17 years in males. These changes coincide with the adolescent growth spurt and sexual maturation. Accretion of bone mineral and increases in bone density continue after 16 years in girls and 17 years in boys, but the magnitudes of the increments are only a fraction of those during the adolescent spurt (Theintz et al. 1992).

DPA measures indicate greater TBMC, greater TBMD, and especially greater mineral content and density of the appendicular skeleton in adult males. Sex differences in other areas of the skeleton are small (Mazess et al. 1984). BMD at mid- and distal sections of the radius (single-photon absorptiometry [SPA]) and L1-L4 (DPA) is greater in clinically normal men than women 20 to 89 years of age, but the estimated decline in BMD is greater in females than males at each site (Riggs et al. 1981). Similar trends are apparent in Japanese residents in Hawaii. The decline in BMC (SPA) between 60 to 64 and 75 to 80 years is about two to four times greater in women than in men but varies by site: distal radius, males 12%, females 29%; proximal radius, males 7%, females 22%; distal ulna, males 12%, females 31%; proximal ulna, males 6%, females 22%; os calcis, males 8%, females 32% (Yano et al. 1984).

Ethnic variation in TBMC among American White, American Black, and Mexican American children, adolescents, and adults was discussed earlier. TBMC is, on average, greater in American Blacks of both sexes, whereas it does not consistently differ between American Whites and Mexican Americans. Ethnic variation in TBMC and TBMD in other samples of adults is summarized in table 18.6. The samples of American Blacks and Whites were matched for age, mass, and stature (Gerace et al. 1994; Ortiz et al. 1992) and for age, mass, height, and menstrual status (Aloia et al. 1997; Gasperino et al. 1995; Ortiz et al. 1992). Subjects in one study were also similar in SAT (Adams et al. 1992). Black and White adults differ significantly in TBMC and TBMD. The ethnic difference in BMC is more marked in the upper extremities, about 19% greater in Blacks, than in the lower extremities, about 12% greater in Black adults (Ortiz et al. 1992). The available data thus indicate a significantly larger appendicular skeletal mass in American Blacks.

Estimates of TBMD of U.S.-born Japanese American women overlap those for American Black and White women and are generally greater than Japanese-born immigrant Japanese American women (table 18.6). After adjustment for age, stature, and mass, TBMD does not differ between U.S-born and Japanese-born Japanese American women (Kin et al. 1993). Although some evidence suggests that TBMD and TBMC do not differ between men of Asian ancestry (mostly Chinese, primarily born outside of the United States) and American Whites after adjustment for age, stature, and mass (Russell-Aulet et al. 1991), other data suggest that TBMC of Japanese men and women is less than that of American Blacks and Whites, even allowing for smaller body size (Tsunenari et al. 1993).

Data for specific bone sites using SPA or DPA yield generally similar results in comparisons of American Black and White children and adolescents (Li et al. 1989; McCormick et al. 1991; Slaughter et al. 1990) and adults (Liel et al. 1988; Luckey et al. 1989; Nelson et al. 1988, 1991). BMD at L2-L4 in Hispanic children and adolescents (most likely Mexican American) is less than in American Blacks but similar to American Whites (McCormick et al. 1991).

Data for Asian American children and adolescents are limited. A comparison of small samples, 9 to 25 years old, indicates generally similar TBMC in American Whites, Mexicans, and Asians across three maturational stages (early puberty, mid-puberty, maturity) but greater TBMC in Blacks (Wang et al. 1997). About one half of the Asian youth were born in the United States; the others

Table 18.6 Ethnic Variation in Total Body Bone Mineral Content (TBMC, g) and Total Body Bone Mineral Density (TBMD, g/cm^2)

Reference, methods, subjects	N	AGE (YEARS)		TBMC (G)		TBMD (G/CM^2)	
		Mean	SD	Mean	SD	Mean	SD
Adams et al. (1992), DXA							
Black females	26	22.5	3.6	3,021	305	1.25	0.05
White females	26	23.6	2.8	2,718	321	1.15	0.07
Ortiz et al. (1992), DPA							
Black females	28	44.2	15.2	2,640	490	1.18	0.14
White females	28	43.6	15.3	2,320	330	1.09	0.09
Gerace et al. (1994), DPA							
Black males	24	39.8	10.3	3,430	476	1.28	0.11
White males	24	39.7	10.3	3,156	513	1.23	0.11
Gasperino et al. (1995), DPA, DXA							
Black females	13	30	5	2,852	411	1.24	0.06
White females	13	31	5	2,453	252	1.12	0.06
Black females	21	52	10	2,531	346	1.14	0.11
White females	21	54	9	2,260	339	1.04	0.08
Aloia et al. (1997), DXA[a]							
Black females	23	39.9	1.6	3,100	288		
White females	23	41.7	1.6	2,800	240		
Russell-Aulet et al. (1991), DPA							
White males	154	50.0	16.0	3,040	532	1.19	0.12
Asian males[b]	84	51.0	17.0	2,697	421	1.15	0.10
Tsunenari et al. (1993), DXA							
Japanese males	8	19.5	1.7	2,540	410		
Japanese females	15	42.8	2.5	2,450	250		
	11	65.5	5.5	2,370	950		
	10	21.4	0.5	2,310	190		
	10	43.4	3.2	2,320	310		
	14	66.0	3.7	1,610	220		
Kin et al. (1993), DXA							
Japanese-American females, U.S.-born	19	20-29				1.12	0.10
	30	30-39				1.12	0.07
	26	40-49				1.10	0.07
	11	50-59				1.06	0.10
	39	60-69				1.00	0.09
	20	70-79				0.93	0.05
	151	18-84				1.05	0.11
	151	18-84				1.04[c]	—
Japanese-American females, Japanese-born	10	30-39				1.09	0.06
	31	40-49				1.10	0.06
	48	50-59				1.03	0.09
	42	60-69				0.95	0.07
	137	19-89				1.03	0.10
	137	19-89				1.04[c]	—

Note. DXA = dual-energy X-ray absorptiometry; DPA = dual-photon absorptiometry.

[a]TBMC reported in kg; standard deviations were estimated from standard errors; [b]77 of the 84 subjects were of Chinese ancestry; [c]adjusted for age, stature, and weight.

were children of Chinese, Korean, and Vietnamese immigrants (Bhudhikanok et al. 1996). Ethnic differences in TBMC are somewhat variable within each pubertal stage, which may reflect lack of statistical control for chronological age within pubertal stage and perhaps ethnic variation in the timing of sexual maturation. For example, 15-year-old boys in genital stage 3 tend to be shorter and lighter than 13-year-old boys in the same stage, whereas 15-year-old mature boys tend to be taller and heavier than 13-year-old mature boys. American Black youth also mature earlier than American White and Mexican American youth (Malina et al. 2004).

In an insured population of adults 20 to 69 years old, American Blacks have higher BMC of the distal radius (SPA) and higher BMC per unit of radial width than American Whites and Asians, whereas differences between the latter are not consistent (Goldsmith et al. 1973). Japanese residents in Hawaii (males 61-81 years, females 43-80 years) have generally less BMC of the distal and proximal radius (SPA) than American Whites (Yano et al. 1984). Japanese American men and women have greater BMC (SPA) at the proximal radius (6-10%) and BMC per unit of radial width (16-17%) than Japanese men and women, but adjustment for weight and height reduces the differences by about one half (Ross et al. 1989). Comparisons of BMD by body region in American White and Asian (largely Chinese) men indicate significantly higher values in Whites for all regions (4.0-6.5%) except the arms (2.0%). However, after adjustment for age, stature, and mass, none of the differences are significant (Russell-Aulet et al. 1991). The preceding studies suggest that observed differences in BMC between Asians and Whites may be attributed in part to ethnic differences in body and bone size and perhaps proportions.

Within Asian samples, U.S.-born Japanese American women have slightly greater BMD at L2-L4, femoral neck, Ward's triangle, and greater trochanter than Japanese-born Japanese American women after adjustment for age, stature, and mass (Kin et al. 1993). However, TBMD does not differ between the two groups of Japanese American women after the authors controlled for age, stature, and mass. Other data show relatively small differences in BMD at L2-L4 and the femoral neck among cross-sectional samples of Japanese (n = 259), Korean (n = 62), and Taiwanese (n = 77) women and between Japanese (n = 81) and Korean (n = 48) men about 20 to 90 years of age, but these data suggest ethnic variation among Asian populations in the loss of bone mineral with age (Sugimoto et al. 1992).

Skeletal Muscle

The contribution of skeletal muscle mass to total body composition is most often derived from the concentration of potassium and creatinine excretion (chapter 14). CT, MRI, DPA, and DXA permit more precise measurement of appendicular muscle mass.

Muscle tissue is rich in potassium and accounts for a major proportion of potassium in the body. Potassium concentrations are similar in males and females during childhood. The sex difference becomes apparent during the adolescent growth spurt, reflecting primarily the male spurt in muscle mass, and persists through the life span. Potassium concentration also shows ethnic variation, being greater in American Black than in American White and Mexican American children and adolescents (Ellis et al. 2000) and in American Black compared with White adults (Ellis 1990).

Estimates of muscle mass based on creatinine excretion show a similar pattern. Sex differences are small in childhood. Males gain considerably more muscle mass than females during the adolescent spurt, and the sex difference persists through the life span (Bouchard et al. 1997; Malina et al. 2004). Creatinine excretion is influenced by diet, particularly protein intake, physical activity, and other factors, in addition to normal day-to-day variation (chapter 14).

Radiography

Historically, soft tissue radiographs have been used to document changes in limb musculature during childhood and adolescence. Muscle widths of the arm and calf have growth curves that resemble those for estimates of total muscle mass and body mass. Estimated velocities of growth in muscle widths of the arm and calf are similar in boys and girls before adolescence. Males show well-defined adolescent growth spurts in arm and calf musculature, whereas females show less defined spurts. Estimated rate of growth of arm muscle during adolescence is twice as great in males as in females, whereas the sex difference in rate of growth of calf muscle is small. The male adolescent spurt is thus more apparent for the upper-extremity musculature (Malina et al. 2004).

Data applying the newer technology to large samples of children and adolescents are quite limited. DXA measures of lean tissue on the arms and legs are available only for females 8 to 27 years of age grouped by stage of sexual maturity (Goulding et al. 1996). Corresponding data for males and for children and adolescents of different ethnic backgrounds are apparently not available. Variations in DXA estimates of total body LTM associated with sex and ethnicity were described earlier.

Cross-Sectional Areas

Sex differences in estimated cross-sectional areas (CSA) of lean tissue of the extremities in several samples of European and American White adults are summarized in table 18.7. The data are derived from CT, MRI, and ultrasound, and results with the different methods are not directly comparable. Quadriceps CSA is about 30% larger with CT than ultrasound (Sipila and Suominen 1993). Arm and thigh CSA include muscle and bone, although some refer to "lean CSA" (Borkan et al. 1983). Sex differences are greater in the arm than in the thigh, but the magnitude of differences varies. The sex difference in arm lean tissue persists after adjustment for stature, whereas that for the thigh is not significant (Schantz et al. 1983). The age-associated decline in CSA varies between sexes and with age and muscle groups compared. Differences between younger and older men are about 25% for quadriceps CSA (25-75 years, Young et al. 1985), about 24% for arm CSA and 13% for calf CSA (31-75 years, Rice et al. 1990), and 12% for the "lean" CSA of the arm and thigh (46-69 years, Borkan et al. 1983). A corresponding difference between younger and older women in quadriceps CSA is about 33% (24-74 years, Young et al. 1984). Longitudinal observations on 14 subjects over 8 years (mid-70s to early 80s) indicated a median decline in quadriceps CSA of 0.8% per year (Greig et al. 1993). In a sample of 10 frail elderly men and women (90 years, 6 females, 4 males), muscle CSA of the midthigh was only 31% of total CSA (Fiatarone et al. 1990), which is less than estimates from younger ages, for example, 70% in "lean" thigh CSA in men 69 years (Borkan et al. 1983) or 70% and 46% in midthigh CSA, respectively, in men 77 and women 80 years (Baumgartner et al. 1992). The trends suggest that atrophy of muscle tissue accelerates with advancing age into the ninth decade.

Asymmetry in quadriceps CSA is greater in older (mid-70s) than in younger (mid-20s) subjects. The relative difference between dominant and nondominant limbs is about 4% to 5% in young men and women and about 9% in older men and women (Young et al. 1984, 1985). Muscle volumes of dominant and nondominant forearms show similar differences (about 4%) in young adults of both sexes (Maughan et al. 1984).

Appendicular Skeletal Muscle Mass

DXA estimates of appendicular skeletal muscle mass (table 18.8) show similar sex differences as estimates based on CT, MRI, and ultrasound (Table 18.7). The difference is more apparent in the upper than in the lower extremity in several samples of European and American White and American Black adults. Among samples of American Black and White adult women matched for age, body size, and menstrual status, Black women have a larger appendicular skeletal muscle mass (Gasperino et al. 1995; Ortiz et al. 1992). The ethnic difference is more marked in the upper than in the lower extremity (Ortiz et al. 1992). Limited data for American Black and White males show less ethnic variation in limb musculature. Nevertheless, after adjustment for age, height, and mass, skeletal muscle mass of the legs is significantly larger in American Blacks of both sexes. The ethnic difference also persists after adjustment for variation in relative leg length. Arm skeletal muscle mass is also significantly larger in American Blacks of both sexes after adjustment for age and mass and also for relative arm length (Gallagher et al. 1997).

In Japanese ($N = 36$) and American White ($N = 42$) females 20 to 30 years of age, ultrasound measurements of muscle thicknesses at the upper extremity (forearm, biceps, triceps), trunk (subscapular, abdomen), and lower extremity (quadriceps, hamstrings, posterior calf) sites are greater in Whites (Ishida et al. 1992, 1994). The ethnic difference persists when measures are adjusted for FFM/stature^2 and when subjects are matched for the BMI. Ratios of upper extremity–trunk (2.5 vs. 2.2) and lower extremity–trunk (6.7 vs. 5.7) muscle thicknesses are greater in Japanese women, but ratios of upper to lower extremity muscle thicknesses do not differ (0.4 vs. 0.4, Ishida et al. 1992).

Table 18.7 Cross-Sectional Areas (cm²) of Lean Tissue in the Upper and Lower Extremities in Samples of European and American White Adults

Reference, method, area	MALES				FEMALES			
	N	Age	Mean	SD	N	Age	Mean	SD
Schantz et al. (1983), CT								
Arm Triceps brachii Biceps brachii	8	26	31.1 23.5	1.6 1.2	6	27	19.0 14.4	0.7 0.6
Thigh Medial extensors Lateral extensors Nonextensors	11	26	32.9 55.4 95.1	1.0 2.2 3.8	10	27	24.1 42.4 75.3	1.4 1.8 2.4
Rice et al. (1990), CT								
Arm MBA Calf MBA	7 13 7	31.4 (4.3) 74.8 (7.1)	59.7 45.5 97.6 84.8	6.9 4.9 8.6 10.3				
de Koning et al. (1986), CT								
Arm MBA	10	20-50	49.6	9.1	8	20-30	26.0	3.2
Borkan et al. (1983), CT								
Arm "lean" Midthigh "lean"	21 20	46.3 (2.6) 69.4 (4.1)	54.6 48.2 147.3 129.1	8.7 8.2 14.6 15.3				
Baumgartner et al. (1992), MRI								
Arm MBA Midthigh MBA	8	77.0 (3.8)	47.8 118.2		17	80.5 (6.2)	26.4 16.8	
Sipila and Suominen (1991, 1993), ultrasound, CT								
Quadriceps—ultrasound Quadriceps—CT	11	73.4 (2.4)	48.4	11.1	15	73.6 (2.9)	29.4 43.2	6.8 7.5
Young et al. (1984, 1985), ultrasound								
Quadriceps	12 12	21-28 70-79	70.8 53.2	5.6 6.3	25 25	20-29 71-81	57.1 38.7	6.6 8.5

Note. MBA = muscle–bone area; CT = computed tomography; MRI = magnetic resonance imaging.

Table 18.8   Sex and Ethnic Variation in Appendicular Skeletal Muscle Mass in Adults

| | | | | APPENDICULAR MUSCLE MASS (KG) | | | | | |
| | | AGE (YEARS) | | ARMS | | LEGS | | TOTAL | |
Reference, method, subjects	N	Mean	SD	Mean	SD	Mean	SD	Mean	SD
Horber et al. (1997), DXA[a]									
White females	34	32	6	4.0	0.6	13.6	1.7		
White males	32	33	6	6.6	1.1	19.2	2.3		
White females	25	58	5	3.8	0.5	12.3	1.5		
White males	28	63	5	5.6	1.0	16.3	1.6		
Oritz et al. (1992), DPA[b]									
Black females	28	44	15	6.1	1.3	11.9	2.1	18.0	3.0
White females	28	44	15	4.8	1.2	10.9	1.6	15.7	2.2
Gasperino et al. (1995), DPA, DXA[c]									
Black females	13	30	5					18.4	3.1
White females	13	31	5					13.5	2.3
Black females	21	52	10					17.6	3.0
White females	21	54	9					13.4	2.4
Gallagher et al. (1997), DXA									
Black females	80	51	16	4.9	1.1	15.6	2.0	20.5	2.8
White females	68	46	19	4.1	1.0	14.5	1.9	18.6	2.6
Black males	72	40	15	7.7	1.4	21.7	3.4	29.4	4.4
White males	64	42	20	7.2	1.5	21.1	2.8	28.3	3.9

Note. DXA = dual-energy X-ray absorptiometry; DPA = dual-photon absorptiometry.

[a]Swiss; the younger and older women were pre- and postmenopausal, respectively. Subjects were matched for height, weight, and body mass index. [b]Subjects were matched for age, height, weight, and menstrual status. [c]Subjects were matched for age, height, weight, and menstrual status; younger women were premenopausal; older women were postmenopausal.

Summary

Sex differences in body composition are negligible in infancy and childhood and are established during the adolescent spurt and sexual maturation. Differences reflect a larger FFM, TBMC (especially of the limb skeleton), and skeletal muscle mass in males. Estimated FM is similar in male and female adolescents, but females have greater %BF. Sex differences in relative SAT distribution also emerge in adolescence; the same is suggested for the proportional contributions of SAT and VAT to abdominal AT. The sex differences established during adolescence persist through adulthood, and there is a sex difference in changes in specific components of body composition with advancing age.

The pattern of sex differences is the same in all ethnic groups, but the magnitude of the sex difference appears to vary. Blacks have greater TBMC during childhood, adolescence, and adulthood. Data for other components are less consistent and vary with methods. Comparisons between Mexican Americans and American Whites show small differences in FFM and TBMC, although Mexican Americans tend to have greater adiposity.

Data for skinfolds indicate proportionally more SAT on the trunk in American Blacks, Mexican Americans, and Asian Americans compared with American Whites. Limited data suggest no difference between American Black and White children and adolescents in the proportional contributions of VAT and SAT to abdominal AT but suggest proportionally less abdominal VAT in American Blacks than Whites. Asian Americans, although smaller in body size, have proportionally more VAT (whole body) than American Whites. Data on the proportional distribution of VAT and SAT for Mexican Americans are not available at present. DXA data suggest proportionally more appendicular skeletal muscle mass in American Blacks compared with Whites. Data for other ethnic groups are not available. Limited data based on ultrasound suggest greater extremity musculature in American compared with Japanese young adult females. Although data from different methodologies are not directly comparable, the evidence suggests ethnic variation in components of body composition and in distribution of adipose, skeletal, and skeletal muscle tissues. Data contrasting more ethnic groups are needed.

The BMI is a universally used weight–height index, largely used in the context of the increasing prevalence of overweight and obesity throughout the world. Ethnic variation in body composition has implications for interpretations of the BMI. Evidence indicates ethnic variation in the relationship between %BF and the BMI in adults (Deurenberg et al. 1998; Gallagher et al. 1996; Norgan 1994; Wang et al. 1996). Presently available data are consistent in showing a lower BMI but a higher %BF in Asian compared with White samples of adults (Chang et al. 2003; Wang et al. 1994, 1996). Corresponding studies need to be extended to children and adolescents and to other ethnic groups.

Chapter 19

Pregnancy

Sally Ann Lederman

Understanding body composition changes during pregnancy begins with the recognition that weight gain is one of the hallmarks of this state. The tissues gained include those of the conceptus—fetus, placenta, and amniotic fluid—as well as the uterine tissues of the mother directly involved in the maintenance of the conceptus. Other maternal tissue increments, such as increased red cell mass, plasma volume, blood vessels, extracellular water, and breast tissue, may also somewhat less directly determine the safety and development of the fetus and newborn.

As might be inferred from this list of augmented tissues, water gain is a big component of pregnancy weight gain for most women. In late pregnancy it is not uncommon for healthy women to have detectable edema, which resolves rapidly after delivery. Maternal fat gain also contributes significantly to weight gain during pregnancy. In addition to fat and water gains, 60-120 g of bone mineral is added to the normal fetus by the time of delivery (Forbes 1987; Pedersen et al. 1989).

Why Measure Pregnancy Body Composition?

The changes that occur in body composition during pregnancy have been recognized for more than 40 years (Hytten and Leitch 1964). Nevertheless, interest in body composition changes during pregnancy persists today in part because of continuing uncertainty about the energy needs of pregnancy and their determinants. Estimated body fat increase during pregnancy (figure 19.1) is a major contributor to the increase in energy intake recommended for pregnancy. Research

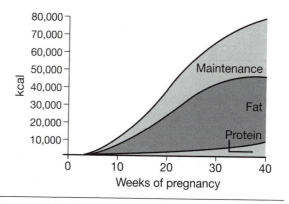

Figure 19.1 Energy cost of pregnancy and its components.

Reprinted, by permission, from F.E. Hytten and G. Chamberlain, 1980, Nutrition. In *Clinical Physiology and Obstetrics,* (Oxford: Blackwell Scientific Publications), 166.

is needed to establish the fat gain that is ideal during pregnancy and to determine how individual women can achieve appropriate increments.

Infant birth weight is one of the best predictors of infant health and survival. Among the factors long recognized as positively associated with birth weight are maternal body weight before conception and pregnancy weight gain. Research indicates that an adequate expansion of plasma volume, which increases by 1.2 to 1.5 L on average (Hytten and Chamberlain 1980), may be key to the unrestricted growth of the fetus (Duffus et al. 1971; Rosso et al. 1993). More recent work, discussed subsequently, has clarified the important association of water increments and birth weight. In addition, a large body of work has demonstrated that fetal characteristics related to maternal nutritional status may program fetal development and determine the onset of chronic diseases late in life.

Some diseases that develop during pregnancy affect body composition, and extremes of body composition changes during pregnancy may lead to or reflect disease. Decades ago it was recognized that preeclampsia and gestational hypertension were associated with alterations in body water, and recent work suggests that a decrease in total body water (TBW) in the second trimester can be demonstrated in pregnant women who develop hypertension (Valensise et al. 2000). Although gestational fat gain is considered normal, excessive fat gain, which predisposes women to obesity, is ever more commonplace. Factors that encourage these seemingly aberrant body composition changes during pregnancy are almost completely unexplored.

Measuring Body Composition Changes During Pregnancy

During pregnancy, the composition of lean tissue is changed, invalidating the standard two-compartment model conversion factors and making it difficult to properly assess the amount of fat-free mass (FFM). Even if the error is small relative to the large amount of FFM, the error in the fat estimate is proportionally greater, because fat mass is generally much less. The problem is exacerbated when the information desired is the *change* in body fat. Even a relatively small error (~1 kg) in FFM can result in errors of 20% to 50% in the fat change. Furthermore, common methods that use radioactive tracers or involve radiation exposure, such as in vivo neutron activation, dual-energy X-ray absorptiometry (DXA), and dual-photon absorptiometry (DPA), cannot be used during pregnancy because of the risk to the fetus. The tracers can be replaced with nonradioactive tracers such as deuterated or oxygen-18 water for TBW, bromide for extracellular water, and Evans blue for plasma volume measurement. DXA and dual-photon absorptiometry have been used shortly before conception and shortly after delivery to assess bone changes (exclusive of the baby) during pregnancy (Sowers et al. 1991).

Total Body Water by Stable Isotope Dilution

When TBW is determined during pregnancy using oxygen-18 or deuterated water, difficulty arises because the hydration of the FFM increases during pregnancy, differs in different women, and varies as pregnancy progresses. FFM cannot be accurately computed from measured TBW using the standard FFM hydration constant of 0.73. Various workers have attempted to estimate mean lean tissue hydration at different stages of pregnancy based on theoretical principles, discussed later. Although these theoretical approaches may improve mean estimates of body fat and its changes, values for fat and lean content of individual women can still be inaccurate, making it impossible to determine the factors affecting individual outcomes.

Body Density

Body density can be measured by hydrodensitometry and air displacement plethysmography, but interpreting the values obtained is more complex during pregnancy. In nonpregnant women, body fat is estimated with equations that use body weight and the assessed densities of fat (0.900 g/cm^3) and FFM (1.10 g/cm^3). Fat-free mass increments during pregnancy are particularly rich in water and contain little bone mineral and thus have a lower density than the preexisting lean tissue. In addition, the increments in FFM in early pregnancy are predominantly maternal tissues, whereas in later pregnancy FFM increments are lower density, predominantly in the fetus. As a result, no single figure for the density of the FFM is suitable across pregnancy. Any unestimated decline in the FFM density results in an overestimation of fat content.

Theoretical considerations have been used to correct mean density of the FFM during pregnancy using average values for fetal, placental, and amniotic fluid weight and compositions (Fidanza 1987; van Raaij et al. 1988). Even if such approaches could reveal some truth about mean body composition changes, the more variable individual differences that are the particular interest of pregnancy researchers may be obscured.

Total Body Potassium

Measurement of total body potassium (TBK) during pregnancy seems to be an ideal method for body composition determination because of its speed and simplicity of use. If we ignore technical problems related to the changing diameter of the body and its effect on signal detection, there is still a problem in translating the TBK value measured during pregnancy into an estimate of FFM, based on an assumed average concentration of potassium in the FFM. Potassium is almost exclusively intracellular, so even when it accurately assesses the intracellular space, we must extrapolate to FFM with some assumptions about the ratio of extra- to intracellular water. The fetus has a higher proportion of extracellular water, which changes as pregnancy advances, changing the ratio of potassium to lean tissue (Ziegler et al. 1976), so potassium concentration in the lean tissue declines. The change differs from one woman to another, because the tissues added differ. If the standard TBK–FFM value is used, FFM is underestimated and fat overestimated. With progressive dilution, fat values later in pregnancy tend to be more overestimated, resulting in a general overestimation of fat gain.

Skinfold Thickness

Because of its simplicity of measurement, skinfold thickness (SFT) has been used extensively to estimate changes in body composition during pregnancy. Some studies have used several skinfold measurements to estimate body density using regression equations developed by others in nonpregnant women (Durnin and Womersley 1974). To predict body fat, the resulting body density values are used in equations such as those of Siri (1961) based on nonpregnant women. However, specific fat depots are preferentially increased during pregnancy, and some changes may reflect redistribution of fat between internal and subcutaneous sites, depending on the hormonal milieu of the period of gestation.

In addition, skinfolds include a lean tissue component that may change during pregnancy because the ground substance, or the mucopolysaccharides that are part of the skin, increase in bound water content under the influence of estrogen (Hytten et al. 1966). The effect of increased body water is further magnified by the tendency of most pregnant women to swelling or even visible edema in the legs late in pregnancy, indicative of an increase in extracellular water. The skinfolds that increase most are those that decline shortly after delivery (Villar et al. 1992); "about one-third of the increased skinfold thickness in pregnant women is rapidly lost after delivery" (Hytten and Chamberlain 1980, p. 224). Although these SFT decrements are usually interpreted as indicating a loss of fat, it is more likely that the changes reflect a redistribution of body water and a return toward nonpregnant fat distribution.

Bioimpedance Analysis

Measurement of body water using bioimpedance analysis (BIA) during pregnancy is very attractive for clinical purposes. Bioimpedance can be measured reasonably rapidly with relatively inexpensive equipment. Currently, BIA studies are based on empirical equations relating the resistance and reactance values to extracellular or total body water or on equations provided by the manufacturer of the equipment. The unspecified assumptions that are implicit in the equations used are likely to be very different during pregnancy. For example, the resistivity of the blood varies with hematocrit and plasma sodium concentration (Lukaski et al. 1994). If BIA is to be successfully applied to clinical use, prediction equations based on studies of pregnant women will need to be further validated. In the meantime, there may be an opportunity to develop empirical relationships between BIA values and particular pregnancy outcomes that do not depend on detailed theoretical underpinnings or the precise assessment of TBW for their clinical utility.

Magnetic Resonance Imaging, Ultrasound, and Air Displacement Plethysmography

Few studies have used any of these methods during pregnancy, and therefore they are not covered in this chapter.

Advances of the Field

In 1990, the Institute of Medicine (IOM) presented data from seven reports on TBW changes determined by isotopic dilution in 5 to 91 subjects from four developed countries. Interpretation was limited by the differences in timing of the measurements, the varying measurement intervals covered, the small sample sizes, and wide differences in prepregnancy weight and pregnancy weight gain, which are thought to affect the subsequent composition changes. The effects on body composition changes of differences in prepregnancy weight, pregnancy weight gain, parity, infant birth weight, and other such factors were not examined.

The IOM publication (1990) also cited two studies using body density measurement during pregnancy (van Raaij et al. 1988; Seitchik et al. 1963). The later study made a correction for estimated decreases in FFM density in late pregnancy. When a similar correction was applied to the earlier data, the two studies gave consistent body fat estimates.

Of course, both estimates depended on the accuracy of the correction for the density of the FFM in late pregnancy, which was not experimentally verified.

Three reports of TBK changes during pregnancy were also summarized (Emerson et al. 1975; Forsum et al. 1988; Pipe et al. 1979), with notable differences in their findings. Two of the studies combined measurements of TBK and TBW to account for changes in FFM hydration, significantly reducing the estimate of body fat increments during pregnancy (by ≥2 kg) compared with estimates based on TBK changes only. The IOM body composition chapter (1990, p. 134) concluded the following:

> "In the future, multicompartment models of body composition need to be used in studies of larger numbers of pregnant women. Attention must also be given to differences in the gestational period studied, weight gain, initial weight, maternal age, ethnic background, and parity."

Studies of Correction Factors for Pregnancy

Studies completed since the publication of the IOM report (1990) have focused on developing and using correction factors to account for differences in FFM density and FFM hydration, taking repeated measurements, and using more complex models to improve estimates of body composition changes. The main purpose here is to illustrate the completeness and consistency of the data and identify remaining gaps.

Hydration of the FFM

Two early studies by Fidanza (1987) and Van Raaij and colleagues (1988) used Hytten's estimates (Hytten and Chamberlain 1980) of the amount and composition of tissues gained to calculate changes in FFM hydration in pregnant women. Van Raaij and colleagues (1988), however, assumed that nonpregnant hydration was 0.724 and calculated that it rose to 0.725 by week 10 and to 0.750 by week 40; they also computed values for women with generalized edema. Fidanza based his estimations on the hydration of the FFM of the nonpregnant woman being 0.738, unchanged at week 10 of gestation, rising to 0.761 at 40 weeks. Prentice and colleagues (1989) developed a model based on literature values for the water content of the amniotic fluid, plasma volume, and extra-

cellular fluid and the "excess" hydration (>73%) of the remaining products of conception. Their approach yielded a somewhat higher estimate of the hydration of the FFM, 0.768 at 36 weeks.

Catalano and colleagues (1995) computed hydration of the FFM as 0.762 at 30 weeks gestation from measured body density and TBW values in a split sample of 40 women, using values from 20 women to validate the results obtained in the other 20. Hopkinson and colleagues (1997), in a very thorough study of the effect of the model used on the body composition values obtained, used a four-compartment model as the standard, as did Paxton and colleagues (1998). Both groups measured TBW, density (by underwater weighing [UWW]), and bone mineral, computed fat mass (FM) and FFM, and estimated the hydration and density of the FFM. Hopkinson and colleagues (1997) reported a hydration of 0.76 at 36 weeks gestation. Paxton and colleagues (1998) reported values of 0.7384 at 14 weeks and 0.7566 at 37 weeks gestation.

The results of these studies span similar values but not for the same gestational ages. The data show broad agreement among the different theoretical hydration estimates, with van Raaij and colleagues' (1988) values for women without edema being lowest. The van Raaij estimates show that the diversity of values obtained in different studies may be determined in part by the degree or frequency of edema in the women studied. These comparisons illustrate the difficulties that might be encountered in applying general models to a physiological state characterized by large individual differences in adaptations. Furthermore, the results show that, although theoretical estimates are available for the whole of pregnancy, studies of measured composition tend to be limited to one or two measurement times, making comparison across studies difficult.

Several authors have built on earlier work and used the hydration constant corrections described here to assess body composition changes during pregnancy using TBW measurements (table 19.1). Forsum and colleagues (1989) assumed FFM hydration of 0.732 at weeks 16 to 18 and at 2 or 6 months postpartum and 0.748 hydration at 30 and 36 weeks gestation, values within the range proposed by van Raaij and colleagues (1988) for this period.

An earlier article by Forsum and colleagues (1988) used the model of Pipe and colleagues (1979) and reported the absolute changes in fat shown in table 19.1. In other works shown in table

Table 19.1 Body Composition Changes During Pregnancy From Studies of TBW Using Corrections for Hydration of the FFM

	TIMING OF ESTIMATE (WEEKS GESTATION UNLESS OTHERWISE INDICATED)										
	Prepreg	6 weeks	12-14 weeks	16-18 weeks	24 weeks	30 weeks	36-37 weeks	Computed mean change from prepreg	5-10 days postp	15 days postp	6 months postp
Forsum et al. (1988)	*n* = 22										
BW (kg)	61.0			63.7		70.2	72.7	*11.7*	67.6		61.9
TBW (kg)	33.0			32.5		36.7	38.7	*5.7*	34.2		31.4
Maternal FM (kg)[a]	17.2			20.7		22.6	22.3	*5.1*	22.9		20.4
Lederman et al. (1993)	*n* = 58		*n* = 65				*n* = 65			*n* = 65	
BW (kg)	62.6		66.0				77.4	*14.8*		69.3	
FM (kg)			22.0				23.7				
Goldberg et al. (1993)	*n* = 12										
BW (kg)	61.7	62.2	63.3	65.4	68.7	71.7	73.6	*11.9*			
FFM (kg)	43.1	42.7	43.3	44.1	45.2	47.0	49.5	*6.4*			
FM (kg)	18.6	19.4	19.7	20.5	21.8	22.5	22.2	*3.6*			
Jaque-Fortunato et al. (1996)	*n* = 9						39 weeks		postp		
BW (kg)						74.8	79.5		70.8		
TBW (kg)						37.7	41.0		33.7[b]		
FFM (kg)						52.2	56.8		46.7[b]		
% fat						29.6	28.1		33.3		
Calculated kg fat						22.1	22.3		23.6		
Hopkinson et al. (1997)	*n* = 56										
BW (kg)	60.7						74.6	*13.9*		66.3	
TBW (kg)							39.3			33.2	
FM (kg)							22.0				

Note. BW = body weight; FM = fat mass; FFM = fat-free mass; preg = pregnancy; prepreg = prepregnancy; postp = postpartum; TBW = total body water. Values in italics were computed from authors' mean values.
[a]Body fat by this method excludes the contribution of the conceptus; [b]Siri method.

19.1, Lederman and colleagues (1993) interpolated from Fidanza's (1987) data and estimated the FFM hydration to be 0.7404 at 14 weeks and 0.7583 at 37 weeks gestation in 65 women. Goldberg and colleagues (1993) used the corrected hydration figures that their group had previously developed (Prentice et al. 1989) and reported FFM and FM for 12 women. Jaque-Fortunato and colleagues (1996) measured TBW and computed FFM and percent fat using van Raaij and colleagues' (1988) hydration-corrected equations for pregnancy and the standard Siri (1961) model postpartum in 9 women. Hopkinson and colleagues (1997) reported results on 56 women at 36 weeks gestation using TBW with the hydration correction of van Raaij (table 19.1). Despite similar weight changes, the women studied by Forsum and colleagues (1988) and Goldberg and colleagues (1993) showed fat changes that were quite different, attributable largely to the lower prepregnancy fat value reported by Forsum and colleagues. The consistency of FM near the end of pregnancy is noteworthy.

Van Raaij and colleagues (1988) also presented data showing that the same FFM gain during pregnancy would affect the hydration of the FFM differently in women with different prepregnant FFM. Large and small women with a given FFM increment do not have the same change in hydration. The authors compared prepregnant women with FFM density of 1.100 g/cm^3 and FFM hydration of 0.724 who had the same increase in FFM, where that increase represented 16.6% to 25% of their initial FFM, attributable to differences in initial values. In these women, the hydration would range from 0.746 to 0.755 at term (0.761-0.771 if the women developed generalized edema). Thus, depending on the woman's initial FFM and whether she developed edema, FFM hydration at term could range from 0.746 to 0.771, even though it started out the same. Clearly, use of a single average figure, even one designed for pregnancy, could give significantly spurious results for the changes in a given woman or group of women, although mean values would not be expected to be as aberrant as the values of individuals.

Density of the FFM

The density of lean tissue during pregnancy is used to determine fat and FFM by UWW, and small differences in density have large effects on the results. Van Raaij and colleagues (1988) presented data showing how the same FFM gain during pregnancy would affect the density of the total FFM in women with different prepregnant FFM. The authors deter-

mined that the density of the total FFM at term would range from 1.0895 to 1.0850 g/cm^3 (1.0830-1.0785 g/cm^3 if the women developed generalized edema) in women whose FFM gain represented 16.6% to 25% of their initial FFM. The large range of these values shows that the degree of edema and the woman's initial FFM can be more significant determinants of gestational FFM density than is the stage of pregnancy.

Several authors have used pregnancy-specific FFM density constants to assess body composition changes during pregnancy (table 19.2). Lederman and colleagues (1993) reported that, when the model of van Raaij and colleagues (1988) was used, the mean density of the FFM was 1.099g/cm^3 at week 14 and 1.089g/cm^3 at week 37. De Groot and colleagues (1994) and Hopkinson and colleagues (1997) also presented results calculated with the density formulas of van Raaij and colleagues but covered different gestational periods. Note that the accuracy of the FM and FFM values shown in the tables is not known. Mostly these results only enable us to compare data sets that used similar models.

Hopkinson and colleagues (1997) and Highman and colleagues (1998) both used body weight, body density (by UWW), and TBW to determine FFM and FM (table 19.3). The subjects in Highman and colleagues' study were significantly heavier than in most other studies, but fat gain was somewhat lower.

Studies Comparing Different Body Composition Models in the Same Pregnant Women

The development of the technology and theoretical underpinnings of multicomponent approaches to the understanding of body composition changes during pregnancy is crucial to increasing our knowledge of pregnancy physiology. Water is important to measure directly during pregnancy because water increases substantially and because the magnitude of the increase varies widely among women. Although bone mineral changes during pregnancy are now known to be minor (Ritchie et al. 1998; Sowers et al. 1991), differences in bone mineral mass among women are large. Because bone mineral has a density (2.982 g/cm^3) almost triple that of the remainder of FFM (1.000 g/cm^3), relatively small variations in bone mineral can have a large effect on FFM density.

Four-component models, which include measurement of body weight, TBW, body density,

Table 19.2 Body Composition Determined From Body Density Measurements Corrected for Changes in the Density of the FFM During Pregnancy

			TIMING OF ESTIMATE (WEEKS GESTATION UNLESS OTHERWISE INDICATED)						
	Prepreg	12 weeks	14 weeks	23 weeks	34 weeks	36 weeks	37 weeks	15 days postp (19.8 days postp)	Mean pregnancy changes
Lederman et al. (1993)									
BW (kg)	62.6		66.0				77.4	(69.3)	
FM (kg)			22.1				25.6		3.5 (weeks 14-37)
de Groot et al. (1994)									
BW (kg)	60.6	62.1		66.4	72.3				11.7
FM (kg)	16.3	17.4		18.9	19.7				3.4
FFM (kg)	44.3	44.7		47.5	52.6				8.3
Hopkinson et al. (1997)									
BW (kg)						74.6		66.3	
FM (kg)						23.3		23.2	

Note. BW = body weight; FM = fat mass; FFM = fat-free mass; postp = postpartum. Values in italics were computed from authors' mean values.

Table 19.3   Body Fat During Pregnancy Determined From Measurement of Body Density (UWW) and TBW

	TIMING OF ESTIMATE (WEEKS GESTATION UNLESS OTHERWISE INDICATED)				
	Prepreg	12-14 weeks	34-36 weeks	15 days postp	Computed mean pregnancy change
Hopkinson et al. (1997)					
BW (kg)			74.6	66.3	
FM (kg)			23.0	21.9	
Highman et al. (1998)					
BW (kg)	75.9	77.2	87.5		*11.6*
FM (kg)	29.4	28.7	31.4		*2.0*

Note. BW = body weight; FM = fat mass; prepreg = prepregnancy; postp = postpartum; UWW = underwater weighing. Values in italics were computed from authors' mean values.

and bone mineral content, account for all of the major variable components that contribute to important changes during pregnancy. Therefore the four-component model is considered a "gold standard" for determining body composition during pregnancy. A more detailed discussion of multicomponent models can be found in chapter 12 of this book.

In the early 1990s, my colleagues and I reported the first study designed to use a four-component model to determine longitudinal changes in body composition occurring during pregnancy (Lederman et al. 1993). Body weight, TBW, and body density were measured at about 14 and 37 weeks gestation, and bone mineral was measured at about 3 weeks postpartum. Fat was also computed from two-component models (TBW, UWW, TBK) using the traditional estimates for FFM density and hydration and with the correction factors proposed by Fidanza (1987; FFM hydration = 0.7404 at week 14 and 0.7583 at week 37) and by Van Raaij and colleagues (1988; FFM density = 1.099g/cm³ at week 14 and 1.089g/cm³ at week 37). The results from all methods are given in table 19.4.

This comparison illustrated several crucial points. First, the differences in fat estimates by different methods increased as pregnancy advanced. Second, whereas intermethod differences in the fat estimates were relatively small in early pregnancy (except FM based on measurement of TBK), differences were greater in late pregnancy and the resulting estimates of mean fat change were extremely different. Even the results based on calculations using corrected FFM hydration or density in late

pregnancy, which showed a reduced difference in the total body fat estimates, gave fat change values that were each about 35% different from the four-component model. Certainly, values for individual women would be even more aberrant.

Hopkinson and colleagues (1997) also compared the results obtained with two-, three-, and four-component models. They measured TBW, TBK, and body density at 36 weeks gestation and postpartum. Bone mineral was measured at 15 days postpartum (table 19.4).

These two studies allow some firm conclusions regarding body composition changes during pregnancy. Different methods give very different results. Even when correction for changes in the composition of the FFM is used, the values obtained for fat changes during pregnancy are very different from those obtained with the more complete and accurate four-component models. Unless the goal of a study can clearly be achieved despite these problems, there is little merit in repeating the kinds of studies that have been done numerous times in the past. Future investigations into the factors that determine body composition during pregnancy or those that explore the effects of these changes should use the more precise three- and four-component models.

Four-Component Models for Studies of Body Composition Changes During Pregnancy

The four-component model is a challenge to use in pregnant women for several reasons. Under-

Table 19.4 Comparison of Fat Mass and Fat Change in Pregnant Women From Traditional Two-Component Models and From Three- and Four-Component Models

Time of measurement	TOTAL BODY FAT (KG)				
	Before pregnancy	14 weeks pregnant	37 weeks pregnant	Observed change	3 weeks postpartum
Lederman et al. (1993)					
BW (kg)	62.6	66.0	77.4		69.3
TBW (L) (40 women, paired comparison)		34.0	41.3	7.3	
Fat estimation (kg)					
TBW 2-C model		21.5	21.8	0.3	
UWW 2-C model		22.3	28.0	5.6	
TBK 2-C model		29.1	37.2	8.1	
Fidanza hydration correction		22.0	23.7	1.7	
Van Raaij density correction		22.1	25.6	3.5	
Selinger 4-C model		22.5	25.0	2.6	
Time of measurement	**Before pregnancy**		**36 weeks**		**15 days postpartum**
Hopkinson et al. (1997)					
BW (kg)	60.7		74.6		66.3
Fat estimation (kg)					
TBW 2-C model			20.8		20.8
UWW 2-C model			25.6		23.2
TBK 2-C model			27.9		23.5
3-C models					
Pipe et al. (1979)			22.9		21.5
TBW and density			23.0		21.9
Van Raaij density correction			23.3		—
4-C model, Fuller et al. (1992)			22.8		22.0

Note. BW = body weight; TBW = total body water; TBK = total body potassium; UWW = underwater weighing; 2-C = two-component; 3-C = three-component; 4-C = four-component; l = liters.

water weighing for density determination is difficult for many women near term; validation of air displacement plethysmography may be a useful alternative in future studies. TBW measurement with deuterated or [18]O water is expensive, and a DXA scan done before or after pregnancy does not include the small contribution of fetal bone mineral. On the other hand, multicomponent models are particularly useful because they can improve the accuracy of each subject's measurements and decrease the number of subjects needed.

My colleagues and I began a longitudinal study of body composition changes during pregnancy in the early 1990s. In this project, TBW and UWW were measured in 200 women at about 14 and 37 weeks gestation, and bone mineral was determined at about 3 weeks postpartum by DXA. The four-component model of Selinger (1977; see chapter 12) was used to calculate FM at each pregnancy visit.

In the early publication from this study (Lederman et al. 1993), findings from 65 subjects were reported. Subsequently, results from 196 women who completed every essential measurement were presented (Lederman et al. 1999; table 19.5). Butte and colleagues (1997) used a similar four-component model (Fuller et al. 1992) to determine FM and FFM 36 at weeks pregnancy and twice postpartum. In a later report (Kopp-Hoolihan et al. 1999), nine women were measured before pregnancy, at weeks 10, 26, and 36 of pregnancy, and at 4 to 6 weeks postpartum using a four-component model developed by the authors (table 19.5). This model estimated FFM before and after pregnancy from TBW, assuming the hydration to be 72.4%; the model estimated total body mineral as 1.235 times the bone mineral content, and it assumed that fetal bone mineral at birth was 2.32% of birth weight (i.e., 81 g for a 3,500-g newborn). Fetal bone mineral at other measurement times was estimated as a percentage of this value to correct for the effect of the unmeasured mineral content of the fetus on FFM density.

Total body protein was estimated before and after pregnancy as the difference between the calculated FFM and the sum of measured TBW and computed total body mineral. Intermediate values for body protein at weeks 10, 26, and 36 were estimated based on estimates of the protein content of the weight gained in each interval and the estimate of protein content before pregnancy. FFM was then estimated for weeks 10, 26, and 36 as the sum of measured water and estimated protein and mineral.

The resulting values for FFM were used with the measured water and estimated protein and mineral to calculate the percent composition of the FFM. Using known values for the density of water, protein, and body mineral, Kopp-Hoolihan and colleagues (1999) determined the mean density of the FFM. When this value was combined with measures of total body weight and density, fat mass could be determined.

These studies provide the most complete information available on pregnancy body composition. Because of the similar, precise methods used, the studies can be reasonably interpreted together. Nevertheless, gaps remain.

Research Applications of Body Composition Changes

Few researchers studying pregnancy are interested in body composition per se. Most interest concerns the relationship of body composition to maternal and infant health with particular focus on maternal fat gain. Interest also concerns the relationship of maternal body composition changes to fetal growth.

Weight Gain Versus Fat Gain

Concerns about obesity development in women suggested that we examine how pregnancy weight gain and fat gain were related in women of different starting weights. At the time my colleagues and I were planning our study of pregnancy composition changes, new pregnancy weight gain recommendations that differed for women of different prepregnancy weights had been promulgated (IOM 1990). These recommendations needed experimental testing and validation. Therefore, we examined the relationship between maternal weight gain and fat gain for women with different prepregnancy BMIs, grouping women into BMI categories defined by the IOM (1990). Key changes are shown in table 19.6.

Regarding the question of the relationship of weight gain to fat gain, we (Lederman et al. 1997) found that fat gain between weeks 14 and 37 was highly and significantly correlated with weight gain (r = .81) and negatively correlated with prepregnancy weight (r = –.25). In contrast, the increase in body water was not related to prepregnant weight grouping.

We also demonstrated that obese women gaining within the IOM guidelines did not increase body fat, providing strong support for the recommended weight gain for obese women (a subject

Table 19.5 Body Composition in Pregnant Women From Longitudinal Studies Using Four-Component Models

	TIMING OF ESTIMATE (WEEKS GESTATION UNLESS OTHERWISE INDICATED)								
	Prepreg	10 weeks	14 weeks	26 weeks	37 weeks	Change from prepreg	Change from week 14		
Lederman et al. (1999)									
BW (kg)	63.4[a]		65.4		77.0	13.6[a]	11.6		
FM (kg)			21.4		24.8		3.3		
TBW (kg)			32.4		39.4		7.0		
Birth weight (kg)									3.45

	Prepreg			36 weeks	Change from prepreg		3 months postp	6 months postp
Butte et al. (1997)								
BW (kg)	61.3			75.4	15.9		65.0	64.0
FM (kg)				23.2			21.7	20.6
FFM (kg)				52.2			43.3	43.4
Percent fat				30.2			32.6	31.1
Birth weight (kg)								3.46

	Prepreg	10 weeks		26 weeks	36 weeks	Change from prepreg		4-6 weeks postp	
Kopp-Hoolihan et al. (1999)									
BW (kg)	64.7	64.9		72.1	75.9	11.2		68.0	
FM (kg)	20.2	20.3		24.4	24.3	4.1		22.0	
TBW	33.5	33.9		36.5	39.1	5.6		33.8	
FFM (kg)	46.3	46.7		49.7	52.8	6.5		46.7	
Birth weight (kg)									3.56

Note. BW = body weight; prepreg = prepregnancy; postp = postpartum; TBW = total body water.
[a]Based on reported prepregnancy weight (Lederman and Paxton 1998).

Table 19.6   Changes in Body Composition During Pregnancy by a Four-Component Model in Women With Different Prepregnancy BMIs

Prepregnancy weight group	Underweight (BMI <19.8) $n = 21$	Normal weight (BMI =19.8-26) $n = 118$	Overweight (BMI>26-29) $n = 29$	Obese (BMI >29) $n = 28$
Recommended weight gain[a] (kg)	12.5-18	11.5-16	7-11.5	≥7[b]
Weight gain weeks 14-37 (kg)	12.6	12.2	11.0	8.7
TBW gain weeks 14-37 (L)	6.1	7.0	7.8	7.3
Fat gain weeks 14-37 (kg)	4.8	3.9	2.8	0.2
For women gaining as recommended	($n = 7$)	($n = 46$)	($n = 9$)	($n = 6$)
Fat gain weeks 14-37 (kg)	6.0	3.8	2.8	–0.6
Total body fat week 14 (kg)	12.2	17.9	25.1	33.1
Total body fat week 37 (kg)	18.2	21.7	28.0	32.5

Note. BMI = body mass index; TBW = total body water.

[a]Recommendations of the Institute of Medicine (1990), endorsed by the American College of Obstetricians and Gynecologists and other medical groups; [b]we used 7-9.2 kg as the acceptable range for obese women.

Based on Lederman et al., 1997.

of controversy). Body fat gain was highest in the underweight women. However, despite their seemingly high fat gain, underweight women gaining as recommended had a body fat content at term (18.2 kg) similar to normal-weight women early in pregnancy (17.9 kg), supporting the weight gain recommendation for low-weight women.

Birth Weight and Maternal Body Composition

The interest in body composition was partly based on the long-known association between maternal weight gain and infant birth weight. A study by Brown (1988) illustrates some of the relationships that drove this interest. Using data from more than 130,000 women from Kansas, she showed that term birth weight increased with increasing pregnancy weight gain and that higher birth weight was associated with higher prepregnancy weight at the same weight gain. Weight gain had a much attenuated role in determining birth weight in very obese women.

Because improved infant health and survival are known to be related to higher birth weight (Rees et al. 1996) even when gestational age is controlled (Seeds and Peng 2000), the relationship of birth weight to maternal weight gain has been the object of much research attention. Our data set permitted us to determine the components of maternal weight gain that were responsible for

its association with birth weight (Lederman et al. 1999). We used multivariate regression techniques and controlled for gestation duration, baby's sex, and maternal weight, height, parity, race, and age, factors that also affect birth weight. Use of initial and final weights (weeks 14 and 37) allowed the effects of starting weight and weight gain to be studied independently. We found that weight gain, as reflected in weight at week 37 (with weight at week 14 held constant), contributed significantly to birth weight. Each kilogram of weight gain increased mean birth weight by 20.2 g, in agreement with prior studies. In contrast, with final weight held constant, week 14 weight was negatively associated with birth weight (–16.4 g).

To study the components of weight gain that determined the relationship of weight gain to birth weight, we removed initial and final weights from the regression and included our measurements of TBW and fat mass at week 14 and week 37. Although the regression coefficients of other variables were hardly changed, the second model clearly showed that the component of weight gain that was the main predictor of birth weight was TBW. A 1-L increase in TBW was associated with a 34.9-g increase in birth weight. Greater fat gain does not seem to be strongly related to birth weight.

These findings were consistent with an earlier report that showed a positive relationship between birth weight and lean mass estimated from SFT at week 37 in 56 Swedish women (Langhoff-Roos et al. 1987). Other workers (Hediger et al. 1994) using multivariate analyses observed a decrease in birth weight associated with an increase in arm fat area from 28 weeks pregnancy to 4 to 6 weeks postpartum based on measurements of SFT and their changes, consistent with the negative relation with birth weight reported in a study using SFT shortly postpartum (Briend 1985). These findings were not easily interpreted, however, because of the difficulty in unambiguously translating SFT changes from pregnancy to postpartum. Furthermore, in our cohort, we found that previous anthropometric models resulted in much higher estimates of fat changes during pregnancy than were indicated by the four-component model (Paxton et al. 1998).

Clinical Implications

Women differ greatly in their fat gain during pregnancy, even at a given weight gain. Data on fat and water gain during pregnancy would be a beneficial adjunct to regular prenatal care. Because high gestational fat gain can contribute to obesity development in the mother and is not particularly beneficial to fetal growth, reducing maternal fat gain, but not water gain, would seem desirable. To identify beneficial interventions, we need to better understand the determinants of fat gain versus water gain during pregnancy. The necessary longitudinal studies, with repeated measurements on a large sample of pregnant women of different initial weights, heights, parity, and ethnicity, would be very costly and are not likely to be replicated often. It is therefore imperative that implemented studies be carefully designed to provide as much information as possible.

Future Research Questions

Measurements of pregnancy body composition have helped in the exploration of several research questions. Much of this past work has been limited in scope both methodologically and conceptually. With the accurate body composition methods now available, the growing ability to consider genetic determinants, and the ability of multivariate analyses to control for other factors influencing outcomes of interest, researchers should be implementing mechanistic studies of pregnancy body composition changes and their relationship to pregnancy outcomes. For example, Gutersohn and colleagues (2000) reported that primiparous women carrying the 825T allele of the gene coding for the β-3 subunit of heterotrimeric G proteins had a threefold higher risk of being overweight (BMI >24.9 g/m^2) within the year following the birth of a child compared with similar women who were not carriers of this allele. The genotype did not affect weight in nulliparous women, seeming to specifically affect retention of pregnancy weight gain. Moreover, Hocher and colleagues (2000) found that this maternal genotype was also associated with a lower birth weight in the offspring. A more recent Japanese study (Masuda et al. 2002) showed an association of this maternal genotype with reduced head circumference (but not reduced birth weight) in the offspring.

Our studies have shown that maternal fat gain is not an important determinant of fetal growth. The genetic studies suggest that genetic factors may predispose some women to greater pregnancy fat gain and reduced infant growth. Future work is needed to help identify prospectively women at risk for excessive fat gain and retained pregnancy weight and to help identify dietary or behavioral factors that can mitigate that process and be part of preventive interventions.

Chapter 20

Exercise

Daniel P. Williams, Pedro J. Teixeira, and Scott B. Going

The effects of exercise on body composition are diverse, in part because different assessment techniques of varying accuracy and precision are used to quantify exercise-related changes in body composition. In addition, many exercise interventions are blended with other treatments, especially dietary modification, which further complicates the ability to determine the independent effects of exercise. This chapter attempts to minimize study assessment and design limitations by emphasizing more recent studies and reviews that relied on more accurate body composition assessment techniques such as dual-energy X-ray absorptiometry, magnetic resonance imaging, and computed tomography and by giving preference to those studies and reviews that attempt to isolate the independent effects of exercise on body composition. However, because diet and exercise are recommended for inducing weight loss in the obese (Jakicic et al. 2001; National Institutes of Health [NIH] 1998), the section on body weight examines the effects of diet plus exercise.

Another emphasis of the chapter is on better understanding the health benefits of exercise throughout the life span. The health benefits of regular physical activity and improved physical fitness are well documented (Blair et al. 2001), and many of the known health benefits of exercise result, either directly or indirectly, from the beneficial effects of exercise on body composition (Williams 2001). However, despite the introduction of revised physical activity recommendations in 1996 (U.S. Department of Health and Human Services 1996), a number of uncertainties remain regarding the optimal dose and combination of exercise mode, intensity, frequency, and duration to recommend for male and female children, adults,

and the elderly. Thus, the chapter reviews the dose–response effects of exercise on body composition throughout the life span. A life span approach is vital because obesity prevalence has steadily risen over the last 2 decades among children (Ogden et al. 2002) and adults (Flegal et al. 2002). Furthermore, sarcopenia (low skeletal muscle mass) and osteopenia (low bone mass) are prevalent among older adults (Baumgartner et al. 1998; World Health Organization 1994). Therefore, obesity, sarcopenia, and osteopenia are major public health problems. Moreover, an understanding of the effects of regular exercise on body composition throughout the life span must underlie efforts to develop intervention strategies designed to curb many of the long-term health consequences associated with lifelong obesity (Dietz 1998) and to prevent many of the long-term disabilities associated with age-related skeletal muscle (Baumgartner et al. 1998) and bone (Marshall et al. 1996) losses in the elderly.

A contemporary principle of adaptation to regular exercise is that individuals respond to regular exercise in highly different ways (Bouchard and Rankinen 2001; Haskell 2001). For instance, the body composition response to a given dose of regular exercise varies widely from one person to the next (Wood et al. 1988). To better understand the individual response, we must clarify how individual differences in genotype affect the phenotypic body composition response to exercise. Therefore, the chapter reviews emerging evidence that individual differences in specific genotypes affect the magnitude of the response of fat, lean, and bone tissue to exercise. Ultimately, identifying the specific biological mechanisms whereby exercise alters fat, skeletal muscle, and bone tissue is paramount for refining individualized health-related exercise prescriptions.

Body Weight

The ability of exercise to influence body weight is governed by the first two laws of thermodynamics. The first law of thermodynamics states that energy is neither created nor destroyed. Instead, energy is converted from one form to another. During exercise, the chemical energy from food intake is converted, in part, to the mechanical energy of human movement. As a result, the energy expended during and immediately after physical activity is vital for counterbalancing the energy stored from food intake. The second law of thermodynamics states that biological energy conversions are inefficient. Thus, not all of the energy expended during and immediately after exercise contributes to movement or to the short-term recovery from exercise. In addition, regular exercise may help to preserve (Forbes 2000) or to increase (Teixeira et al. 2003) the body's fat-free mass (FFM). Because the FFM has a relatively high metabolic activity, it is an important determinant of energy expenditure at rest (Halliday et al. 1979).

One of the main challenges in determining the effects of exercise on body weight management lies in assessing exercise-related changes in energy balance (energy intake – energy expenditure). Even when changes in energy intake are adequately controlled, it is difficult to assess how exercise affects total energy expenditure, especially when many of the ways in which exercise may affect energy expenditure are rarely assessed. For instance, the effects of exercise on resting energy expenditure, spontaneous physical activity outside of the exercise regimen, and postexercise oxygen consumption are infrequently assessed and rarely considered when exercise is prescribed for body weight management. In addition, studies examining the effects of exercise on body weight management have historically relied on endurance training (Miller et al. 1997), and, as a result, relatively less is known about the effects of strength training on body weight management. Nevertheless, strength training may have more potential than endurance training for improving body composition in the absence of weight loss, which may be especially important in growing children and aging older adults for whom weight loss may be contraindicated (NIH 1998). In older adults, strength training reduces body fat and increases the body's FFM (Toth et al. 1999), whereas endurance training reduces body fat but does not increase FFM (Toth et al. 1999). Clearly, more research is needed to better understand how repeated bouts of endurance and strength training exercise affect body weight management. Furthermore, it is not known how energy expenditure between each exercise bout affects body weight management throughout the life span.

Although difficult to prove, declines in physical activity likely account, at least in part, for the steady increases in obesity prevalence among children (Ogden et al. 2002) and adults (Flegal et al. 2002) in the United States over the past 2 decades. In 1998, the National Institutes of Health (NIH) published guidelines for treating overweight and obese adults (NIH 1998). Three years later, the American College of Sports Medicine (ACSM) published a position stand on weight loss (Jakicic et al. 2001). In brief, the exercise guidelines recommend that overweight and obese adults engage in a minimum of 150 min of moderate-intensity exercise per week and gradually increase to 200 to 300 min of exercise per week (Jakicic et al. 2001). The guidelines also caution that the rate of exercise-related weight loss tends to plateau after 6 months, in part because less energy is expended for a given activity at a lower body weight (NIH 1998). This phenomenon of "weight-settling" (Ballor 1996) has led to the idea that regular physical activity may be more effective for preventing age-related weight gain (DiPietro 1999) and weight regain (NIH 1998) than it is for promoting weight loss. Other problems associated with using exercise alone as a weight loss strategy are that humans may compensate for exercise-related increases in energy expenditure with reductions in physical activity outside of the prescribed exercise (Goran and Poehlman 1992) and with increases in energy intake (Blundell and King 1999).

Therefore, the combination of regular physical activity, dietary restriction, and behavioral therapy is currently recommended for treating obesity (NIH 1998). A large meta-analysis of weight loss research of the past 25 years concluded that diet plus exercise was more effective than exercise alone for promoting weight loss (Miller at al. 1997). Even after statistical adjustment for differences in experimental design, diet plus exercise studies (n = 134) resulted in greater average weight loss than studies (n = 90) of exercise alone (–11 ± 0.6 kg vs. –2.9 ± 0.4 kg). Diet plus exercise likely induces a more negative energy balance than exercise alone because energy intake is reduced while energy expenditure is increased. However, a major limitation of the reviewed weight loss trials was their relatively short duration, because most trials lasted for only 13 to 21 weeks (Miller et al. 1997). Thus, a recent 2-year randomized clinical trial not only confirmed but also extended the effectiveness of diet plus exercise programs for promoting weight loss over a longer duration (Esposito et al. 2003). Esposito and colleagues (2003) reported an 11-kg greater weight loss, on average, in the 60 obese premenopausal women who received detailed diet plus exercise advice than in the 60 obese women who received general information about healthy food choices and exercise (Esposito et al. 2003). The clinical weight loss program followed the NIH guidelines and included detailed advice about reducing caloric intake on a Mediterranean-style diet; advice about increasing physical activity through walking, swimming, and aerobic ball games; and education on such behavioral strategies as personal goal setting and self-monitoring (Esposito et al. 2003).

Furthermore, the 2-year diet and exercise program supported the health-related significance of weight loss for obese women (Esposito et al. 2003). The biological link between obesity and type 2 diabetes and heart disease is attributable, in large part, to the proinflammatory consequences of excess body fat (Pradhan and Ridker 2002). Fat tissue is an important endocrine organ (Mora and Pessin 2002). It acts as a dynamic, and partially self-regulating, metabolic depot that stores fat as triglyceride and mobilizes fat as free fatty acids (FFAs). Moreover, fat tissue produces proinflammatory cytokines like interleukin-6 (IL-6) that increase systemic inflammation by stimulating the expression of C-reactive protein (CRP) in the liver. Fat tissue also produces a hormone, adiponectin, which has anti-inflammatory and insulin-sensitizing actions. In addition, excess and enlarged fat cells release more FFAs, secrete more IL-6, and secrete less adiponectin into the circulation. Therefore, the improvements in average circulating FFA (–27.9%), IL-6 (–32.6%), adiponectin (+48.2%), and CRP (–34.4%) levels that accompanied the 11-kg weight loss are meaningful reductions in adipocyte-derived and adipocyte-stimulated risk factors for type 2 diabetes and heart disease (Esposito et al. 2003). Overall, the findings are significant because they support the effectiveness of diet plus exercise programs for promoting long-term weight loss in obese people, which helps to fill a recognized knowledge gap in weight loss research (Miller et al. 1997; NIH 1998). The findings are also significant because they demonstrate that weight loss in obese people is accompanied by improvements in the metabolic signaling that putatively links enlarged fat cells to atherogenesis and diabetogenesis.

The combination of diet and exercise may also affect body weight regulating mechanisms. Research interest in the neuroendocrine control of body weight regulation heightened with the identification of the gene that expresses the adipocyte hormone, leptin. Leptin modulates appetite and energy expenditure. Despite the recent proliferation of research on leptin, its exact role in body weight regulation in humans remains unclear. Hickey and Calsbeek (2001) explained that exercise-induced weight loss is typically accompanied by reductions in circulating leptin levels. However, they also explained that circulating leptin levels may occasionally be reduced after exercise training in the absence of weight or body fat loss. Hickey and Calsbeek (2001) identified two possible reasons for the conflicting findings on how exercise may affect circulating leptin levels in humans. First, circulating leptin levels show a strong diurnal rhythm with a nadir around noon and a zenith around midnight. However, most exercise studies assess circulating leptin levels only once per day. Second, the 24-hr leptin rhythm may be more responsive to changes in energy balance (energy intake – energy expenditure) than to changes in either energy intake or energy expenditure alone. Despite this helpful insight, the authors cautioned that the effect of energy balance-related changes in the 24-hr leptin rhythm on body weight regulation is unknown. Because the hypothalamus is such a vital target tissue for leptin action, many challenges remain in understanding the exact biological role of leptin in body weight regulation. Nevertheless, diet plus exercise programs may be more effective than exercise alone for inducing a negative energy balance (Miller et al. 1997). Thus, future studies should determine whether the greater weight loss resulting from diet plus exercise programs is accompanied by greater changes in the 24-hr leptin rhythm.

Whole-Body Fat

One of the prerequisites for developing physical activity recommendations is quantifying the dose–response relationship between exercise and specific health-related outcomes (Haskell 1994, 2001). Ross and Janssen (2001) recently reviewed the dose–response effects of exercise on body fat. In brief, they reported that exercise, without caloric restriction, effectively reduces fat mass in overweight or obese subjects. However, the effectiveness of exercise was dependent, in large part, on the dose of the exercise intervention. Perhaps because of the difficulty of long-term exercise adherence, higher doses of exercise were used in shorter-duration studies that lasted 16 weeks or less, whereas lower doses of exercise were used in longer-duration studies that lasted 26 weeks or more. For example, the average dose in the 20 shorter-duration studies was twice as high as the average exercise doses in the 11 longer-duration studies (2,200 kcal/week vs. 1,100 kcal/week). Accordingly, the higher-dose exercise studies resulted in larger fat mass (FM) reductions than the lower-dose exercise studies (–0.21 kg/week vs. –0.06 kg/week). Nearly all of the exercise-induced weight loss consisted of fat tissue, but all of the reviewed studies used subjects who were initially overweight or obese. In the face of a modest, exercise-induced negative energy balance, overweight and obese subjects are better able than leaner subjects to lose body fat and to preserve lean tissue (Forbes 2000).

We reviewed recent exercise trials that used reliable and accurate body composition assessment techniques (table 20.1 and figure 20.1). In general, our review of these recent trials supports the review by Ross and Janssen (2001). For instance, the higher-dose trials resulted in more body fat loss than the lower-dose trials (–0.09 kg/week vs. –0.04 kg/week). Although we did not have enough information to estimate exact weekly caloric expenditures, the shorter trials generally included higher doses of exercise than the longer trials (table 20.1 and figure 20.1). With one notable exception of an extremely high dose of exercise (Ross et al. 2000), our review of recent exercise trials (table 20.1 and figure 20.1) included weekly exercise doses that are similar to those recommended in the weight loss guidelines (Jakicic et al. 2001; NIH 1998). Therefore, higher doses of exercise are generally associated with greater amounts of body fat loss. However, because of the difficulty of maintaining long-term exercise adherence, there is a paucity of information on the effects of higher doses of exercise (>1,500 kcal/week) on body fat changes over 6 months or longer.

We caution that our selection of more recent exercise trials (table 20.1 and figure 20.1) was primarily based on body composition assessment quality and may not be representative of all exercise trials. Furthermore, the selected studies are quite diverse with respect to the age and sex of the study participants and with respect to the types of exercise interventions. Nevertheless, some of the trends within and across the selected studies

illustrate some of the unresolved issues regarding the effects of exercise on body fat in males and females throughout the life span. For instance, men lost more body fat than women in the largest of the reviewed exercise trials (Wilmore et al. 1999). However, that study did not determine whether the smaller body fat reductions in women were confined to the younger, and presumably leaner, premenopausal women. Nevertheless, it is not clear why the women, who had more initial body fat, on average, than the men (21.4 kg vs. 18.8 kg), lost almost half as much body fat as the men in response to endurance training (–0.5 kg vs. –0.9 kg). Typically, exercise results in greater fat loss in those with higher amounts of initial body fat (Forbes 2000). However, in the large exercise trial reporting smaller body fat losses in women than in men (Wilmore et al. 1999), there was no assessment of changes in energy intake. Thus, the smaller body fat losses in women may have been attributable to increased energy intake in women.

Another unresolved issue is how exercise affects body fat at different ages. For instance, in growing obese children, relatively short-term endurance and strength training may help to prevent further age-related gains in body fat (LeMura and Maziekas 2002; Owens et al. 1999). Long-term randomized controlled trials are needed to assess the efficacy of exercise for reducing body fat gains in growing obese children. In older adult men and women, relatively short-term endurance and strength training may reduce body fat (Toth et al. 1999). In addition, two recent randomized controlled trials assessed the efficacy of 12 months of endurance and strength training on body fat in relatively large samples of older postmenopausal women (table 20.1 and figure 20.1). Irwin and colleagues (2003) reported that 45 min/day of moderate-intensity endurance activity for 5 days per week resulted in an average body fat loss of –1.4 kg over 12 months in 87 obese postmenopausal women who were not using hormone replacement therapy (HRT; table 20.1). By contrast, Teixeira and colleagues (2003) reported that 60 to 75 min/day of high-intensity strength training and weight-bearing activity for 3 days/week reduced body fat over 12 months in 58 postmenopausal women who were using HRT but not in 59 postmenopausal women who were not using HRT (–0.7 kg vs. –0.4 kg; figure 20.1). However, unlike the sample of exclusively obese postmenopausal women in the endurance training trial (Irwin et al. 2003), the postmenopausal women in the strength training trial were initially inactive but not necessarily obese (Teixeira et al. 2003).

Table 20.1 Effect of Exercise on Total and Regional Soft Tissue Composition

Reference	Design	Population	Duration	Interventions	Initial value	%Δ
Irwin et al. (2003)	Randomized, clinical trial with a stretching control group	N = 86 F 50-75 years old, PM, inactive, BMI ≥24.0 kg/m² at baseline, and not using HRT	12 months	Controls attended weekly 45-min stretching sessions and were asked to maintain usual diet and exercise.	Wt = 81.7 kg FM = 38.4 kg FFM = 43.3 kg WC = 93.5 cm VAT = 147.6 g/cm²	0.1 -0.3 -0.9 -0.1 -0.1
		N = 87 F 50-75 years old, PM, inactive, BMI ≥24.0 kg/m² at baseline, and not using HRT		Exercisers trained ≥45 min/day for 5 days/week at an intensity that reached 60-75% HR$_{max}$ by week 8 (N = 15 different activities).	Wt = 81.6 kg FM = 38.5 kg FFM = 43.1 kg WC = 93.1 cm VAT = 147.6 g/cm²	**-1.6** **-3.6** 0.2 **-1.1** **-5.8**
Evans et al. (2001)	Nonrandomized, clinical trial with a nonexercising control group	N = 18 F 60-84 years old, PM, and inactive	11 months	Controls did not undergo any formal exercise training and no HRT use.	Wt = 66.3 kg FM = 27.7 kg FFM = 38.6 kg WC = 80.1 cm	0.4 0.7 0.2 3.9
		N = 16 F 60-84 years old, PM, and inactive		Weight-bearing endurance exercise progressing to 45 min/day for 3 days/week at 80-85% HR$_{max}$ and no HRT use	Wt = 62.7 kg FM = 24.6 kg FFM = 38.1 kg WC = 79.6 cm	**-2.7** **-7.7** **2.4** **-7.7**
		N = 19 F 60-84 years old, PM, and inactive		Same exercise as above + 0.625 mg per day of oral estrogens + 5 mg per day of progestins for 13 days every third month	Wt = 69.9 kg FM = 31.6 kg FFM = 38.8 kg WC = 84.7 cm	**-5.2** **-15.8** **3.6** **-7.5**
Poehlman et al. (2000)	Randomized, clinical trial with a nonexercising control group	N = 20 F 18-35 years old, inactive, and with a BMI <26 kg/m² at baseline	6 months	Controls did not undergo any formal exercise training.	Wt = 60 kg FM = 17 kg FFM = 39 kg VAT = 36 cm²	1.7 0.0 2.6 13.9
		N = 14 F 18-35 years old, inactive, and with a BMI <26 kg/m² at baseline		Endurance training increased from 25 min of slow jog to 60 min of interval training at 85% HR$_{max}$.	Wt = 58 kg FM = 16 kg FFM = 40 kg VAT = 40 cm²	0.0 -6.2 0.0 2.5
		N = 17 F 18-35 years old, inactive, and with a BMI <26 kg/m² at baseline		Resistance training was 3 days per week for three sets of 10 reps at 80% of 1RM for nine different exercises.	Wt = 58 kg FM = 16 kg FFM = 39 kg VAT = 36 cm²	**3.4** 6.2 **5.1** 0.0

317

Reference	Design	Population	Duration	Interventions	Initial value	%Δ
Hunter et al. (2002)	Nonrandomized, clinical trial with no control group	N = 12 F 61-77 years old, PM, HRT status not reported	6 months	Resistance training was 45 min/day for 3 days/week for two sets of 10 reps at 65-80% of 1RM for eight different exercises.	Wt = 65.9 kg FM = 27.0 kg FFM = 38.9 kg VAT = 131.2 cm^2	−1.1 −6.3 2.6 −11.7
		N = 14 M 61-77 years old			Wt = 78.5 kg FM = 19.6 kg FFM = 58.8 kg VAT = 143.4 cm^2	1.0 −9.2 4.8 6.2
Wilmore et al. (1999)	Nonrandomized, clinical trial with no control group	N = 299 F 17-64 years old, inactive, and with a BMI <40 kg/m^2 at baseline	5 months	Computer-controlled cycling for three sessions per week at 55% $\dot{V}O_2$max for 30 min per session and gradually increasing to 75% $\dot{V}O_2$max for 50 min per session during the last 6 weeks	Wt = 66.9 kg FM = 21.4 kg FFM = 45.5 kg WC = 85.2 cm VAT = 66.9 cm^2	−0.1 −2.3 0.9 −1.0 −4.6
		N = 258 M 17-65 years old, inactive, and with a BMI <40 kg/m^2 at baseline			Wt = 82.1 kg FM = 18.8 kg FFM = 63.3 kg WC = 91.8 cm VAT = 90.7 cm^2	−0.5 −4.8 0.8 −1.1 −7.0
Owens et al. (1999)	Randomized, clinical trial with a control group	N = 26 F; N = 13 M Obese, 7-11 years old, with a triceps SKF >85th percentile for age, sex, and race	4 months	Controls did not undergo any formal exercise training.	Wt = 56.9 kg FM = 26.0 kg FFM = 30.9 kg VAT = 258 cm^3	3.5 3.5 3.6 8.1
		N = 23 F; N = 12 M Obese, 7-11 years old, with a triceps SKF >85th percentile for age, sex, and race		40 min of activity at 70-75% HR$_{max}$ with 20 min of aerobic machine exercise and 20 min of playing games for 5 days/week	Wt = 57.5 kg FM = 26.2 kg FFM = 31.3 kg VAT = 253 cm^3	1.9 −3.1 6.1 0.5

Ross et al. (2000)	Randomized, clinical trial with a control group	N = 8 M, Age 46 ± 11 years, BMI > 27 kg/m² at baseline	3 months	Controls were asked to maintain body weight on isocaloric diet.	Wt = 96.7 kg AT = 30.5 kg SM = 34.1 WC = 108.87 VAT = 198 cm²	0.1 −2.0 −1.5 −0.1 0.0
		N = 16 M, Age 45 ± 8 years, BMI >27 kg/m² at baseline		Exercise-induced weight loss; asked to maintain isocaloric diet to expend 700 kcal/day by brisk walking/light jogging at ≤70% V̇O₂max	Wt = 101.5 kg AT = 33.1 kg SM = 35.2 kg WC = 112.0 cm VAT = 186 cm²	**−7.4** **−18.4** **−3.7** **−5.8** **−28.0**
		N = 14 M, Age 45 ± 8 years, BMI >27 kg/m² at baseline		Exercise without weight loss; asked to increase energy intake to balance expenditure of 700 kcal/day by brisk walking/light jogging at ≤70% V̇O₂max	Wt = 97.9 kg AT = 30.6 kg SM = 34.5 kg WC = 110.0 cm VAT = 186 cm²	−0.5 −4.9 1.2 −1.6 **−16.8**

Note. F = females; M = males; PM = postmenopausal; BMI = body mass index; HRT = hormone replacement therapy; Wt = weight; FM = fat mass; FFM = fat-free mass; WC = waist circumference; VAT = visceral adipose tissue; HR_{max} = maximum heart rate. Boldface stands for statistically significant changes, $p < .05$.

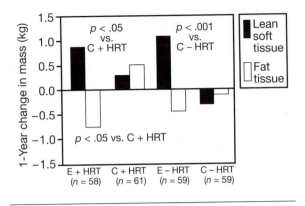

Figure 20.1 Changes in fat and lean soft tissue mass in response to 12 months of resistance exercise (E) versus control (C) in postmenopausal women who use hormone replacement therapy (+HRT) or do not use HRT (–HRT).
Based on data from Teixeira et al. (2003).

Because the average initial body fat was more than 12 kg higher in the postmenopausal women in the endurance training trial than in the strength training trial (38.5 kg vs. 26 kg), it is difficult to ascribe the greater body fat loss to endurance training per se. Furthermore, because the endurance training trial did not include postmenopausal women undergoing HRT (Irwin et al. 2003), it is not known if HRT may potentiate exercise-related body fat losses in postmenopausal women.

In summary, greater doses of energy expenditure during exercise are associated with greater amounts of body fat loss, especially in well-controlled trials of overweight and obese subjects (Ross and Janssen 2001; table 20.1 and figure 20.1). Future exercise trials should focus on interventions that last 12 months or longer, and they should include intermediate assessments of body fat and gradually increasing doses of exercise. In longer-duration interventions, the inclusion of intermediate assessments of body composition would help in determining how exercise affects the rate of body fat loss, and the use of gradually increasing doses of exercise may help to facilitate the long-term maintenance of body fat loss (Jakicic et al. 2001). Furthermore, when exercise is viewed from the public health perspective of treating and preventing obesity, it will also be important to continue testing the efficacy and effectiveness of exercise throughout the life span (Haskell 2001). For instance, higher doses of exercise may be more efficacious for achieving larger amounts of short-term body fat loss in select samples of highly motivated obese people (Ross et al. 2000). However, lower doses of exercise may be more effective for maintaining body fat loss (albeit, a lesser degree

of loss) over the long term in larger and less select samples of obese people (Irwin et al. 2003). This is because a larger percentage of obese people may be able to sustain the lower doses of exercise for a longer period of time.

Exercise Intensity

The intensity of exercise affects the mixture of foods used to meet the energy demands of the exercise bout. In particular, lower-intensity exercise relies more on fat utilization, whereas higher-intensity exercise relies more on carbohydrate utilization. Moreover, low- to moderate-intensity endurance training increases fat utilization and fat oxidation during exercise. For instance, in a study of 24 obese men, 12 weeks of cycling at 40% of maximal aerobic capacity ($\dot{V}O_2$max) increased fat oxidation during exercise by 40%, whereas 12 weeks of cycling at 70% of $\dot{V}O_2$max did not change fat oxidation during exercise (Van Aggel-Leijssen et al. 2002). The study was a true test of the metabolic effects of exercise intensity because the lower- and higher-intensity bouts of exercise were matched for energy expenditure (~350 kcal/bout). Neither the lower-intensity training nor the higher-intensity training changed fat oxidation at rest (Van Aggel-Leijssen et al. 2002), and there were no endurance training–related changes in body weight (Van Aggel-Leijssen et al. 2002).

Findings such as those just described suggest that low- to moderate-intensity exercise (the so called "fat-burning zone") may not result in greater body weight or body fat losses. In his recent review, Shephard (2001) found no advantage of lower-intensity exercise training for reducing body fat, but he cautioned that few studies comparing the effects of different exercise intensities adequately matched the training programs for energy expenditure. By contrast, two reviews (Hardman 2001; Shephard 2001) concluded that studies comparing higher- versus lower-intensity bouts of exercise, especially those that were well matched for energy expenditure, generally supported the benefits of higher-intensity exercise for increasing the magnitude and duration of the excess postexercise oxygen consumption after each bout of acute exercise. Shephard (2001) also suggested that higher-intensity exercise may be more effective for preserving FFM, which is one mechanism whereby exercise may help to maintain the resting energy expenditure. However, despite the possible advantages of higher-intensity exercise for increasing postexercise oxygen consumption and for preserving FFM, there is no conclusive advantage of higher-intensity exercise for body fat or weight loss (Hardman 2001; Shephard 2001). Therefore, at present, it appears that exercise dose (i.e., caloric expenditure) is a more important determinant than exercise intensity for inducing body fat losses.

Exercise Mode

Recent reviews have concluded that both endurance and strength training are efficacious for reducing body fat in obese children and adolescents (LeMura and Mazeikas 2002) and in older adults (Toth et al. 1999). Strength training may be an important adjunct to endurance training for treating obesity in pediatric populations wherein negative energy balance could, in theory, attenuate the growth and development of skeletal muscle and peak bone mass. Similarly, strength training may be important for countering the age-related losses in skeletal muscle, resting energy expenditure, and bone in older adults. In our review of recent exercise trials, resistance training reduced body fat in two of three trials (table 20.1 and figure 20.1). The two trials reporting strength training–related body fat losses (Hunter et al. 2002; Teixeira et al. 2003) included older adult subjects with higher initial body fat values than the one trial reporting no effect of strength training on body fat (Poehlman et al. 2000). Similarly, endurance training either prevented age-related body fat gains or reduced body fat in five of six trials (table 20.1 and figure 20.1). The five endurance training trials with positive results included subjects with higher initial body fat values (Evans et al. 2001; Irwin et al. 2003; Owens et al. 1999; Ross et al. 2000; Wilmore et al. 1999) than the one endurance training trial that did not change body fat (Poehlman et al. 2000). Nevertheless, more research is needed to better understand the separate and combined effects of repeated bouts of strength- and endurance-type exercise on energy expenditure between each exercise bout. At present, it is not entirely clear why strength training may reduce body fat. Compared with endurance training, strength training typically includes more rest intervals and less energy expenditure during exercise.

Genetic Differences

A major limitation of most reviews of the effects of exercise on body fat is that they focus on the average losses of body fat rather than consider the individual responses, which typically range from small gains in body fat to extremely large reductions in body fat. Because of recent technical advances in molecular biology, studies such as the HERITAGE Family Study (Wilmore et al. 1999) have begun to examine how specific genetic differences account for some of the individual differences in the body fat response to exercise. The HERITAGE Family Study is a large multicenter clinical trial investigating the genetic basis for the variability in endurance training–related changes in aerobic capacity, body composition, and risk factors for type 2 diabetes and cardiovascular disease in black and white families in the United States and Canada. Selected exercise trials reporting genotypic differences in the body composition response to exercise training are summarized in table 20.2. With respect to exercise-related body fat changes, four different genes were identified in the HERITAGE Family Study that modified the body fat response to 20 weeks of moderate-intensity cycling for 30 to 50 min/day on 3 days of the week (table 20.2). These genes include those that produce two separate subtypes of the β-*adrenergic receptor* (Ukkola et al. 2003), one that produces an *uncoupling protein* (Lanouette et al. 2002), and another that produces the *lipoprotein lipase* enzyme (Garenc et al. 2001). Furthermore, between-subject differences in individual genotypes accounted for 2.5% to 3% of the variance in the body fat response to exercise (Garenc et al. 2001; Ukkola et al. 2003), which is significant when we consider how many different genes may modify the body fat response to exercise.

To better understand why genetic differences in the previously mentioned genes may modify the body fat response to exercise, it is important to appreciate the metabolic interplay between fat and skeletal muscle tissue during exercise and in response to endurance training. In simple terms, fat is stored as triglyceride in many tissues but primarily in adipose tissue. During exercise, sympathetic nerve activity increases, and circulating epinephrine and norepinephrine levels rise (Powers et al. 1982). Both epinephrine and norepinephrine bind to β-adrenergic receptors on adipocytes, thereby initiating a series of intracellular signals to activate hormone-sensitive lipase. In turn, activated hormone-sensitive lipase hydrolyzes and releases stored triglycerides as circulating free fatty acids. Coupled with the exercise-related increase in skeletal muscle blood flow, the mobilized fatty acids in the circulation may be directed to actively contracting skeletal muscle where they may be oxidized to help meet

Table 20.2 Genotypic Differences in the Body Composition Response to Exercise

Gene	Reference	Exercise	Sample	Significant gene × exercise effects
Angiotensin-converting enzyme (ACE)	Montgomery et al. (1999)	10 weeks of intensive physical training	N = 30 white M British Army recruits	Larger MRI-assessed gains in thigh muscle volume in an ACE genotype that comprised 53% of the sample
β_2-adrenergic receptor (β_2AR) and β_3-adrenergic receptor (β_3AR)	Ukkola et al. (2003)	20 weeks of moderate-intensity cycling for 30 to 50 min/day on 3 days/week	N = 205 black participants in the HERITAGE Family Study	Larger HW-assessed losses in fat mass in carriers of a specific combination of β_2AR and β_3AR genotypes that comprised 17% of the sample
Estrogen receptor-α (ERα)	Remes et al. (2003)	4 years of walking, jogging, cross-country skiing, swimming, and cycling for 45-60 min 5 × per week	N = 68 white Finnish M	DXA-assessed lumbar spine bone gains were observed in two separate ERα genotypes that comprised 69% of the sample. In contrast, no change in lumbar spine bone was observed in the remaining ERα genotype that comprised 31% of the sample
Insulin-like growth factor-1 (IGF-1)	Sun et al. (1999)	20 weeks of moderate-intensity cycling for 30-50 min/day on 3 days per week	N = 502 white participants in the HERITAGE Family Study	Smaller HW-assessed gains in fat-free mass in an IGF-1 genotype that comprised 52% of the sample
Lipoprotein lipase (LPL)	Garenc et al. (2001)	20 weeks of moderate-intensity cycling for 30-50 min/day on 3 days per week	N = 237 white F in the HERITAGE Family Study	Larger HW-assessed losses in fat mass in an LPL genotype that comprised 17% of the sample
			N = 128 black F in the HERITAGE Family Study	Larger CT-assessed losses in visceral adipose tissue area in an LPL genotype that comprised 12% of the sample
Uncoupling protein 3 (UCP-3)	Lanouette et al. (2002)	20 weeks of moderate-intensity cycling for 30-50 min/day on 3 days/week	N = 393 white participants in the HERITAGE Family Study	Larger SKF-assessed losses in subcutaneous fat in a UCP-3 genotype that comprised 3% of the sample

Note. M = males; MRI = magnetic resonance imaging; HW = hydrostatic weighing; DXA = dual-energy X-ray absorptiometry; F = females; CT = computed tomography; SKF = skinfold.

the increased energy demands of exercise (Brooks et al. 2000). However, not all of the fatty acids that are mobilized from adipose tissue and directed to skeletal muscle can be oxidized following each acute bout of exercise. Thus, it has been speculated that the up-regulation of the specific mitochondrial uncoupling protein expressed in human skeletal muscle, uncoupling protein-3, following each acute bout of exercise may help in protecting the mitochondria against fatty acid accumulation and in maintaining skeletal muscle fat oxidative capacity (Hesselink et al. 2003; Schrauwen and Hesselink 2003). In brief, uncoupling protein-3 may disrupt the linkage between the oxidation of reduced electron carriers and the phosphorylation of adenosine diphosphate to adenosine triphosphate by exporting a fatty acid anion in exchange for importing a proton across the inner mitochondrial membrane (Hesselink et al. 2003). Therefore, the acute exercise-related increase in uncoupling protein-3 expression in skeletal muscle may primarily defend the mitochondria against fatty acid damage, while secondarily reducing the efficiency of oxidative phosphorylation (Schrauwen and Hesselink 2003). In turn, a lowered efficiency of oxidative phosphorylation during the recovery from exercise may increase the postexercise oxygen consumption, which contributes to the overall energy expenditure of each exercise bout. Finally, lipoprotein lipase is an enzyme that hydrolyzes the triglyceride core of circulating lipoproteins to free fatty acids. In response to endurance training, lipoprotein lipase activity is increased in skeletal muscle (Kiens and Lithell 1989) and decreased in adipose tissue (Lamarche et al. 1993). These training-related adaptations may help in directing circulating fats toward skeletal muscle oxidation and away from fat tissue storage (McCarty 2001).

In reviewing the genotypic differences in the body fat response to endurance exercise (table 20.2), Ukkola and colleagues (2003) reported that black persons with a specific combination of β_2-adrenergic receptor and β_3-adrenergic receptor genotypes experienced a 2.5-fold greater loss of body fat after endurance training than black participants with other β-adrenergic receptor genotypes (Ukkola et al. 2003). The findings suggest that genetic differences in the lipolytic response to exercise may account for some of the individual differences in the body fat response to exercise (Ukkola et al. 2003). By contrast, in response to the same endurance training program, white participants with a specific uncoupling protein-3 genotype experienced

a 3.4-fold greater loss of the sum of eight skinfold thicknesses than white participants with other uncoupling protein-3 genotypes (Lanouette et al. 2002). However, it is not yet known whether the individuals with the specific uncoupling protein-3 genotype who lost the most subcutaneous fat after endurance training also experienced a greater expression of uncoupling protein-3 after each acute bout of exercise (Lanouette et al. 2002). In another report, white females with a specific lipoprotein lipase genotype experienced a 2.5-fold greater loss of body fat in response to endurance training than white females with other lipoprotein lipase genotypes (Garenc et al. 2001). Moreover, black females with the same lipoprotein lipase genotype experienced greater endurance training–related losses of visceral adipose tissue than black females with other lipoprotein lipase genotypes (Garenc et al. 2001). However, the genotypic difference in endurance training–related losses of visceral adipose tissue in black females was of borderline statistical significance ($p = .05$). Nevertheless, the findings suggest that genetic differences in the lipoprotein lipase response to exercise may account for some of the individual differences in the body fat response to exercise. The findings are biologically plausible because exercise-related changes in the activity of the lipoprotein lipase enzyme in different tissues may help direct the flow of circulating fats into more metabolically active tissues (McCarty 2001).

Visceral Adipose Tissue

Abdominal obesity and excess intra-abdominal or visceral adipose tissue (VAT) are important determinants of increased risk for type 2 diabetes and cardiovascular disease (Matsuzawa et al. 1995). Thus, it is important to determine whether exercise may prevent age-related gains in VAT and whether exercise may reduce VAT. Ross and Janssen (1999) reviewed studies that examined the effects of exercise on intra-abdominal fat. At the time of their review, there were no randomized, controlled trials that used magnetic resonance imaging (MRI) or computed tomography (CT) to quantify exercise-related changes in VAT. Nevertheless, their review of nonrandomized trials cautiously concluded that exercise may reduce VAT in either the presence or the absence of weight loss (Ross and Janssen 1999).

In our review of more recent exercise studies (table 20.1), four of the six trials that used imaging techniques to measure VAT used randomized designs (Irwin et al. 2003; Owens et al. 1999; Poehlman et al. 2000; Ross et al. 2000). Three of the four studies using a randomized design reported that exercise either prevented the age-related gain in VAT (Owens et al. 1999) or reduced VAT (Irwin et al. 2003; Ross et al. 2000). By contrast, the one trial using a randomized design that reported no effect of exercise on VAT had the fewest subjects with the smallest VAT areas at baseline (Poehlman et al. 2000). Both of the studies using a nonrandomized design reported that exercise reduced VAT (Hunter et al. 2002; Wilmore et al. 1999). In the study of obese children (Owens et al. 1999), exercise prevented further age-related gains in whole-body fat and VAT. In the other four studies reporting exercise-related reductions in VAT in middle-aged and older adults (Hunter et al. 2002; Irwin et al. 2003; Ross et al. 2000; Wilmore et al. 1999), two of the studies reported no exercise-related weight loss (Hunter et al. 2002; Ross et al. 2000), and one of the studies reported no exercise-related whole-body fat loss (Ross et al. 2000).

In our view, there is insufficient evidence to support the possibility that exercise may selectively reduce body fat in the intra-abdominal depot. Instead, most of the studies suggest that exercise-induced intra-abdominal fat loss is one component of exercise-induced fat loss throughout the body. However, we do not exclude the possibility that exercise may selectively reduce body fat in the intra-abdominal depot, because most of the studies, to date, have used DXA or hydrostatic weighing to assess whole-body fat, and there is no way to subtract the contribution of VAT from those assessments of whole-body fat. Furthermore, intra-abdominal adipocytes are more sensitive than subcutaneous adipocytes to lipolytic stimulation (Fried et al. 1993). Therefore, a selective, exercise-induced reduction in intra-abdominal fat is biologically plausible.

Fat-Free Mass, Lean Soft Tissue, and Skeletal Muscle

The fat-free mass (FFM) consists of skeletal muscle, bone, connective, and other tissues. In the remaining sections, we discuss the effects of exercise on skeletal muscle and bone. Because of limitations in body composition assessment techniques, it is not always possible to separate the lean soft tissue (LST) from bone. For that reason, preference is given to those studies reporting exercise-related changes in LST and to those studies reporting exercise-related changes in CT- or MRI-assessed images

of skeletal muscle. Strength training is more effective than endurance training for increasing FFM and skeletal muscle (Ballor 1996; Toth et al. 1999; Kraemer et al. 2002). Similar strength training–related gains in FFM and skeletal muscle have been documented in both younger (20-30 years of age) and older (65-75 years of age) adult men and women (Roth et al. 2001). As a result, this section does not extensively review the effects of strength versus endurance exercise, nor does it extensively review age- or sex-related differences in the skeletal muscle response to exercise. Because of the health-related concerns of obesity in children (Dietz 1998) and of sarcopenia in older adults (Baumgartner et al. 1998), this section focuses on the role of exercise in preserving lean soft tissue growth and development in obese children and adolescents and in increasing lean soft tissue and skeletal muscle in older adults. We also briefly describe emerging evidence about how individual differences in specific genotypes affect the FFM and skeletal muscle responses to exercise training.

Obese Children and Adolescents

There are two important reasons for examining the effect of exercise on FFM or LST in obese children and adolescents. First, exercise may not preserve FFM above some as-yet-undefined threshold of a negative energy balance (Forbes 2000). Thus, it is important to ensure that weight management programs do not impair normal growth and development in obese children and adolescents. Second, reductions in FFM during weight loss may contribute to greater weight regain after program cessation (Schwingshandl and Borkenstein 1995). Because of the relatively high metabolic activity of the FFM (Halliday et al. 1979), preservation of normal growth-related gains in FFM may improve long-term energy balance and prevent excessive age-related body fat gains in obese children.

LeMura and Maziekas (2002) analyzed 30 studies reporting the effects of exercise on body composition in obese children and adolescents. However, only 12 of the studies reported the effects of exercise on FFM, and only three of the studies included nonexercising control groups. Among the three controlled trials reporting the effects of only 10 to 16 weeks of exercise on FFM, the average effect size was 0.57, which suggests that FFM increases by approximately one half of a standard deviation more in exercising than in nonexercising obese children and adolescents. However, there were no randomized controlled trials using DXA to assess LST changes or using MRI to assess skeletal muscle changes after strength training in obese children and adolescents.

One nonrandomized trial included a nonexercising control group, in which DXA-assessed LST changes were reported after 5 months of strength training in obese girls (Treuth et al. 1998). Twelve prepubertal obese girls aged 7 to 10 years elected to participate in a supervised strength training program for 5 months, and 11 prepubertal obese girls aged 7 to 10 years elected to participate as controls for 5 months (Treuth et al. 1998). Growth-related gains in height ($p < .05$), weight ($p < .001$), and LST mass ($p < .001$) were observed in both the strength-trained and the untrained girls. However, the assessment of any difference in LST gains between the resistance-trained and untrained girls was biased because the girls who elected to participate in the strength training program were taller and heavier at baseline than the girls who elected to participate as controls. The study underscores the need for a randomized, controlled strength training trial in obese children. Moreover, training-related gains in strength may result more from neural than from hypertrophic factors in prepubescent children (Kraemer et al. 1989). Therefore, greater age-related gains in LST after strength training may not be biologically plausible in preadolescent children. Nevertheless, there is no compelling evidence of growth impairment after strength training that would warrant its contraindication in obese children participating in weight management programs.

In fact, strength training is currently advocated for children and adolescents as part of an overall health and fitness program (Hass et al. 2001). In addition, a twofold increase in mean nitrogen balance (nitrogen intake – nitrogen excretion) after 6 weeks of strength training was reported in 11 nonobese children (Pikosky et al. 2002). The increase in nitrogen balance was accompanied by normal growth-related gains in height and weight (Pikosky et al. 2002). Although the findings were limited by the absence of a nonexercising control group, the gains in nitrogen balance offer some metabolic support for normal growth and development in children participating in a strength-training program. Furthermore, with respect to endurance training, older children had the largest gains in DXA-assessed LST after 4 months of endurance training in 70 obese children aged 7 to 11 years (Barbeau et al. 1999). The data are consistent with the interpretation of normal growth and development in obese children participating in a weight

management–exercise program because greater rates of growth and development are expected in older children.

Overall, the data suggest that participation in a regular exercise program preserves normal growth-related gains in FFM and LST in obese children and adolescents. In addition, both endurance- and strength-training programs can be safely implemented in obese children and adolescents. However, additional randomized controlled trials of longer duration are needed to determine the optimal combination of exercise modes and doses for preserving lean tissue growth and development in obese children and adolescents participating in weight management programs. Because of the steady increase in the prevalence of childhood obesity over the last 2 decades (Ogden et al. 2002), continued research efforts aimed at improving the efficacy, effectiveness, and safety of exercise programs for obese children are vital for preventing such metabolic consequences as type 2 diabetes and cardiovascular disease in later life.

Older Adults

Sarcopenia is the nearly universal loss of skeletal muscle with advancing age that involves a selective atrophy of type 2 muscle fibers (ACSM 1998). Sarcopenia should be minimized in older adults because it is associated with age-related increases in disability (Baumgartner et al. 1998) and with age-related reductions in functional independence, resting energy expenditure, bone mineral density, insulin sensitivity, aerobic capacity, and spontaneous physical activity (ACSM 1998). Therefore, strength training is currently recommended as one component of an active lifestyle aimed at countering sarcopenia, reducing health risks, and improving quality of life in older adults (ACSM 1998). In particular, strength training–related increases in skeletal muscle mass (hypertrophy) for the novice may be maximized when the participant trains two to four times a week, using loads of 60% to 80% of 1 repetition maximum (1RM), with 6 to 12 repetitions per set and 1 to 3 sets per exercise (Kraemer et al. 2002). Some investigators recommend that older adults begin a strength-training program using lower-intensity loads (30-40% of 1RM) that are lifted more frequently (10-15 repetitions per set) to ensure safety (Hass et al. 2001). However, the ACSM position stand on physical activity for older adults emphasizes that strength training exercises using loads between 60% and 80% of 1RM increase strength and skeletal muscle size (ACSM 1998).

A series of landmark studies by Fiatarone and colleagues documented that high-intensity strength training beginning at 50% to 60% of 1RM and gradually progressing to 80% of 1RM is an effective tool for increasing muscle mass in 50- to 70-year-old women (Nelson et al. 1994) and for increasing CT-assessed thigh muscle areas in 86- to 96-year-old men and women (Fiatarone et al. 1990, 1994). Moreover, across a wide variety of exercise doses, strength training increased FFM in 15 of 28 studies of men and women aged 55 years or older (Toth et al. 1999). The average strength training–related gains in FFM ranged from 1.1 to 2.2 kg (Toth et al. 1999). However, 20 of the 28 reviewed studies were 4 months or less in duration. Thus, because of the difficulty of long-term exercise adherence, the magnitude of strength training–related gains in FFM, LST, or skeletal muscle resulting from longer-duration strength training programs may be smaller.

In the largest randomized study of 12 months of strength training, to date, 117 postmenopausal women were assigned to 12 months of strength training for 3 days per week while using loads of 70% to 80% of 1RM, and 116 postmenopausal women were assigned to 12 months of a nonexercising control condition (Teixeira et al. 2003). In the 119 women undergoing hormone replacement therapy (HRT), strength training plus HRT increased LST more than HRT alone (0.9 ± 0.1 kg vs. 0.3 ± 1.1 kg, $p < .05$; figure 20.1). In the 114 women who were not undergoing HRT, strength training effectively reversed the age-related loss of LST (1.0 ± 0.9 kg vs. -0.3 ± 1.1, $p < .001$; figure 20.1). Although the findings require confirmation, they suggest that 12 months of strength training, regardless of HRT use, increases LST by approximately 1 kg in postmenopausal women.

In summary, no substantial effects of age, sex, or postmenopausal HRT use appear to mitigate or to potentiate strength training–related gains in FFM, LST, and skeletal muscle in older adults. However, long-term studies are needed to define the exact dose of strength training needed to prevent and to reverse sarcopenia in older adults.

Genetic Differences

In reviewing the genotypic differences in the FFM and skeletal muscle responses to exercise (table 20.2), Montgomery and colleagues (1999) found that white males with a specific angiotensin-converting enzyme (ACE) genotype experienced a 2.6-fold greater gain in MRI-assessed skeletal muscle volume in response to military training than white males with other ACE genotypes. The findings are somewhat difficult to explain. The ACE genotype associated with greater training-related gains in skeletal muscle is known to express low ACE activity throughout the body. It is not entirely clear how the genotypic lowering of the ACE-catalyzed reaction product, angiotensin II, could increase training-related gains in skeletal muscle. Nevertheless, the same ACE genotype is also associated with greater endurance training–related gains in insulin sensitivity (Dengel et al. 2002). Because insulin stimulates protein synthesis, future studies should examine whether ACE genotype–related improvements in insulin sensitivity help to explain the ACE genotype-related gains in skeletal muscle mass. In addition, white males and females with a specific insulin-like growth factor-1 (IGF-1) genotype experienced only one half as much gain in FFM in response to endurance training as white males and females with other IGF-1 genotypes (Sun et al. 1999). The findings are biologically plausible because exercise-related increases in growth hormone and bioactive IGF-1 stimulate protein synthesis and promote skeletal muscle hypertrophy.

Bone

Bone responds to the loads placed upon it. Adaptations occur in mass, external geometry, and internal microarchitecture that optimize bone strength and minimize bone strain for a given load, without unduly increasing its weight. The importance of mechanical stimuli for optimal bone growth (Heaney et al. 2000) and maintenance of skeletal mass is not disputed (Marcus 2002). However, whether physical exercise promotes alterations beyond those encouraged by the typical activities of daily living remains controversial, and the optimal type and dose of exercise to achieve the benefits are uncertain (Block 1997; Forwood 2001; Forwood and Burr 1993; Marcus 2002). Therefore, this section examines the extent to which regular skeletal loading–type exercise may increase peak bone mass in children or counter age-related bone losses in older adults. Furthermore, this section briefly describes emerging evidence about how individual differences in specific genotypes affect the bone response to exercise training.

Effective Load Characteristics

Mechanical loads may be characterized by several parameters, including load magnitude, number of cycles, and the rate at which strain is induced. Animal models, wherein these parameters can be precisely manipulated and bone strain measured directly, have been essential for describing the characteristics of effective mechanical loading. In a classic study, Rubin and Lanyon (1985) showed that the loss of bone cross-sectional area after immobilization may be prevented by loading the bone at low strain magnitudes and high frequencies of strain cycles in turkey ulnae. The results of other animal studies (Lanyon and Rubin 1984; Newhall et al. 1991; Van der Wiel et al. 1995) as well as mathematical models based on human studies (Whalen et al. 1988) generally agree with these results, showing that the number of loading cycles necessary to maintain bone mass increases as the load magnitude decreases. In addition to peak load magnitude, the *rate* of strain (i.e., the time over which strain develops after load application), which is roughly equivalent to *impact*, may be critical to the skeletal response (O'Conner et al. 1982; Turner et al. 1995). For example, Turner and colleagues (1995) observed linear increases in bone formation and mineral apposition as the rate of strain increased in rat tibias.

Descriptive Studies

Numerous studies have shown that elite athletes and other chronic exercisers have higher bone mineral density (BMD) than their age-matched sedentary peers, particularly at the sites that undergo loading during exercise (Drinkwater 1994; Lohman 1995; Marcus 2002). The results of these animal studies suggest that BMD would be higher in individuals engaged in high-load or high-impact activities. Comparisons among athletes in different sports support this notion, because gymnasts and weightlifters have higher BMD, on average, than swimmers (Drinkwater 1994).

However, the difference between athletes and sedentary controls may be attributable to genetic factors and selection bias. For example, individuals with higher BMD may excel in certain sports (e.g., gymnastics), whereas individuals with lower BMD may excel in other sports (e.g., swimming). Studies in which dominant versus nondominant limbs are compared in athletes whose sport involves unilateral loading represent a special type of cross-sectional study, because each subject serves as his or her own control. Several studies have demonstrated greater skeletal dimensions in the playing arm compared with the nonplaying arm of tennis players (Dalean et al. 1985; Haapsalo

et al. 1996; Heinonen et al. 1995; Huddelston et al. 1980). Thus, it appears that at least some of the differences between athletes and nonathletes are attributable to training.

Exercise Training Studies

Prospective, randomized controlled exercise trials are required to assess whether physical exercise is truly beneficial to bone. Unfortunately, many of the training studies reported to date have been limited by some combination of short duration, small sample size, nonrandomization, and inadequate load or assessment methods (Block 1997). Studies have generally reported a slowing of bone loss or gains (~1-3%) in BMD over 12 to 18 months that are considerably less than what might be predicted from the cross-sectional comparisons between athletes and sedentary controls described previously. In a particularly well-designed study, Kerr and colleagues (1996) compared higher-load, lower-repetition weightlifting against lower-load, higher-repetition weightlifting in postmenopausal women. In this study, only one side of the body was trained, and each participant served as her own control. The BMD of the trained side was significantly increased relative to the untrained side, and the BMD increase was greater with higher-load, lower-repetition weightlifting than it was with lower-load, higher-repetition weightlifting. Other studies of strength training have demonstrated similar responses in BMD in premenopausal (Lohman et al. 1995) and postmenopausal women (Going et al. 2003; Kohrt et al. 1995; Nelson et al. 1994).

Very few training studies have compared bone responses to different types of exercise. In one of the few studies to do so, Kohrt and colleagues (1997) showed site-specific responses, such that ground-reaction exercise (walking, jogging, and stair climbing) was most effective at the femoral neck, whereas joint-reaction exercise (weightlifting and rowing) was most effective at the femur trochanter and lumbar spine. Bassey and colleagues (1998) also studied the effects of high-impact loading on BMD in healthy premenopausal and postmenopausal women. The ground-reaction exercise consisted of 50 8.5-cm vertical jumps each day, with loads equivalent to three to four times body weight per jump. After 5 months, premenopausal women experienced 2% to 3% increases, on average, in femoral neck and trochanteric BMD. By contrast, postmenopausal women, irrespective of HRT status, experienced no change in BMD after jumping exercise.

The lasting effects of training-related increases in BMD remain to be established. However, in studies of female collegiate gymnasts, substantial BMD plasticity has been observed over a single year (Taaffe et al. 1997) and over 2 consecutive years (Snow et al. 2001). When gymnastics training diminished during the summer off-season, bone was lost, and with the reinitiation of seasonal gymnastics training, BMD was regained (Snow et al. 2001). In a study of retired elite gymnasts, BMD at multiple sites was 0.5 to 1.5 standard deviations higher in retired gymnasts than in controls. Thus, despite the lower frequency and intensity of exercise after retirement from competitive gymnastics, no detectable reduction in the participants' BMD advantage was observed (Bass et al. 1998). Similar retention of a BMD advantage into retirement and old age has been reported for other types of athletes, although the results have not been uniformly positive (Karlsson et al. 2000). By contrast, Dalsky and colleagues (1988) and Winters and Snow (2000) clearly showed that short-term exercise training–related gains in BMD degrade within months of training cessation.

Age

A number of factors may modify the skeletal response to mechanical loading, including age, the hormonal milieu, and the nature and amount of loading. There is considerable interindividual variation in the bone response to exercise (Going et al. 2003). The effect of age per se on the bone response to exercise is not certain, because most human studies are limited to defined groups with rather narrow ages (i.e., adolescents, young adults, pre- or postmenopausal women). Exercise trials in women ranging in age from the third to the eighth decade have shown increases in bone mass of a few percent at most ages (Friedlander et al. 1995; Gleeson et al. 1990; Kerr et al. 1996; Kohrt et al. 1995, 1997; Lohman et al. 1995; Nelson et al. 1988; Prince et al. 1995; Pruitt et al. 1992; Snow-Harter et al. 1992). Although two studies showed larger increases in spine BMD (~5-8%) in older postmenopausal women (Dalsky et al. 1988; Notelovitz et al. 1991), the sample sizes were small, the study designs were nonrandomized, and the magnitude of the bone responses have not been replicated in other studies. Overall, the reported BMD response to exercise is quite similar across studies of women of different ages. However, as noted by Marcus (2002), the exercise training protocols may impose greater incremental loading in older individuals, given their overall lower activity, which complicates interpretation of the age-related response to a given dose of exercise.

The adaptive response of growing bones includes expansion of bone size as well as increases in bone mass and density, and all contribute to increased bone strength. Once linear growth ceases, the plasticity of bone geometry is limited. Thus, some authors suggest that the greatest opportunity to increase bone dimensions as well as mass through increased physical activity may be during times of growth (Bailey 2000; Bailey et al. 1996, 1999; Khan et al. 2000; Marcus 2002; McKay et al. 2000). Very few exercise training studies have been done in growing children and adolescents. However, 7 to 9 months of jump training exercise has been reported to increase hip and spine bone mass in children and adolescents (Fuchs et al. 2001; Witzke and Snow 2000), which suggests that ground-reaction exercise, like jumping, may increase bone mass in growing children and adolescents. In a recent randomized trial, the effects of a 10-min, three times per week circuit of jumping exercise were assessed in prepubertal and early pubertal girls (MacKelvie et al. 2001; Petit et al. 2002). After 7 months, early pubertal girls (and not prepubertal girls) participating in the jumping program experienced greater gains (1.5-3.1%) in BMD at the femoral neck and lumbar spine than early pubertal girls in the non-jumping control program. Underpinning the jumping-related increases in hip BMD were increases in bone cross-sectional area, reduced endosteal expansion, and improved section modules (bending strength) at the femoral neck (Petit et al. 2002). In two other recent studies, positive structural adaptations, in addition to augmented BMD, were found in students enrolled in physical education classes (vs. students not taking physical education; Bradney et al. 1998; Morris et al. 1997). However, these studies were nonrandomized, were short in duration, and included rather small sample sizes. Thus, the results must be viewed with some caution.

Hormone Status

Whether endogenous hormone adequacy limits the bone response to exercise is uncertain. Frost (1987) proposed that circulating reproductive hormones may alter the skeletal perception of loading, thereby modulating the sensitivity and the magnitude of the adaptive response. However, the evidence supporting Frost's hypothesis is mixed. Experimental studies suggest that estrogen-deficient animals retain the ability to respond to mechanical loading and exercise (Barengolts et al. 1994; Lin et al. 1994; Yeh et al. 1994). Young amenorrheic athletes have higher BMD than amenorrheic sedentary controls, particularly those in sports involving high loads or impact (e.g., weightlifting and gymnastics; Heinrich et al. 1990; Marcus et al. 1985; Robinson et al. 1995). However, the degree to which these findings are indicative of the ability of estrogen-depleted women to respond to exercise or of other factors that confer a performance advantage in these sports is uncertain.

Some studies have tested whether there is an advantage of hormone replacement therapy (HRT) in combination with exercise in postmenopausal women. Kohrt and colleagues (1995) demonstrated additive effects of HRT and vigorous weight-bearing exercise in older women. Similarly, Notelovitz and colleagues (1991) reported that strength training led to dramatic increases in vertebral BMD in postmenopausal women on HRT. By contrast, Heikkinen and colleagues (1991) reported that exercise did not enhance the effects of HRT on BMD. The results of these studies must be viewed with some caution because the sample sizes were quite small, subject assignment to treatment groups was either nonrandom (Kohrt et al. 1995) or not explained (Heikkinen et al. 1991), and the duration of HRT use was variable (Notelovitz et al. 1991). Nevertheless, Notelovitz and colleagues (1991) and Kohrt and colleagues (1995) reported remarkably similar exercise-related increases in total body (~2%) and lumbar spine (7-8%) BMD in postmenopausal women on HRT.

A question of particular interest is whether exercise alone is sufficient to maintain bone in women who are recently menopausal and not taking HRT. In a recent study, Going and colleagues (2003) assessed the effects of 12 months of moderate- to high-intensity strength training on BMD in women 1 to 4.9 years past menopause. In that study, women who had been using HRT or not using HRT for at least 1 year were randomized to strength training or to their usual activity (low activity, no strength training). Femoral trochanteric BMD was significantly increased in the women who participated in strength training and did not use HRT compared with the women who did not participate in strength training and did not use HRT. Femoral trochanteric BMD also increased more in the women who participated in strength training and used HRT compared with the women who did not participate in strength training and used HRT. Although the increases in BMD after 12 months of strength training were

modest (Going et al. 2003), they were comparable to exercise-related changes reported in other studies of postmenopausal women and with exercise effect size estimates from recent meta-analyses (Kelley 1998a, 1998b; Wolff et al. 1999). The finding of approximately two times greater change in trochanteric BMD in women who used HRT and exercise versus those who used HRT alone (Going et al. 2003) was comparable to the effect observed in the Kohrt study (Kohrt et al. 1995). However, Going and colleagues observed beneficial effects of exercise at the hip, whereas Kohrt and colleagues observed beneficial effects of exercise at the spine. Furthermore, Going and colleagues observed no interaction between exercise and HRT. Therefore, in contrast to the findings of Kohrt and colleagues, the findings of Going and colleagues did not support a synergistic effect of HRT and exercise on BMD in postmenopausal women.

Dose Response

Although osteogenic load characteristics have been described (Turner 1998), the optimal type and amount of exercise needed to stimulate bone formation remain unclear. However, Turner and Robling (2003) derived an "osteogenic index" (OI) for predicting the outcome of an exercise protocol on bone mass. Developed from animal data, the OI is based on the following considerations: (a) Bone tissue responds more favorably to dynamic rather than static loading, (b) high-impact exercises that produce large rates of deformation and are applied at a high frequency are effective for stimulating osteogenesis, (c) the required mechanical load necessary to initiate new bone formation decreases as the loading frequency increases, and (d) bone cells exhibit a desensitization phenomenon in the presence of extended mechanical-loading sessions. Calculations of the OI for weekly exercise regimens suggest that OI is best improved by adding more exercise bouts per week rather than lengthening the duration of individual sessions, as long as adequate recovery time is allowed.

Human studies exploring the dose–response relationship between exercise and BMD are lacking and are needed to better define the amount, intensity, and location of exercise needed for increasing or at least maintaining bone mass (Skerry 1997; Swezey 1996). In a recent secondary analysis of data from the Bone, Estrogen and Strength Training (BEST) Study, Cussler and colleagues (2003) reported that femoral trochanteric

BMD increased by 0.001 g/cm^2 for every one standard deviation increase in the amount of weight lifted in a yearlong strength training program in 140 postmenopausal women. Furthermore, the association between larger gains in trochanteric BMD and greater amounts of weight lifted during strength training ($p < .01$) was independent of age, baseline BMD, HRT use, and body weight changes (Cussler et al. 2003). In support of these findings, Kerr and colleagues (1996) also reported larger increases in trochanteric BMD when comparing higher versus lower loading during strength training in postmenopausal women. However, neither study (Cussler et al. 2003; Kerr et al. 1996) was designed to test the bone response to increasing doses of exercise. Additional experiments are needed to define the dose–response relationship between exercise and bone.

Genetic Differences

In reviewing the genotypic differences in the bone response to exercise (table 20.2), Remes and colleagues (2003) reported that white males with two different estrogen receptor-α (ERα) genotypes experienced 5% to 6% gains in lumbar spine BMD in response to 4 years of endurance activity, whereas white males with another ERα genotype experienced no change in lumbar spine BMD in response to 4 years of endurance activity (Remes et al. 2003). The findings are plausible because deficiencies in ERα expression in bone cells are associated with osteoporosis in both men and women. However, it is not yet clear how exercise training may affect the complex interactions between ERα expression and the aromatization of androgens to estrogen in men. Moreover, the magnitude and variability in the bone response to endurance training in men (Remes et al. 2003) may not match the bone response to strength training in women (Going et al. 2003).

Summary

Mechanical stimuli are necessary for optimal bone growth and maintenance of bone mass and density throughout the life span. Animal studies have shown that only a few repetitions of loads producing high strain magnitudes and high loading rates are needed to stimulate osteogenesis. At lower magnitudes of loading, more repetitions are needed. Observations of higher bone mineral density in chronically active persons, particularly in athletes engaged in strength training or in sports with high ground-reaction forces, are congruent with the animal data. Numerous factors, such as age, hormones, nutrient levels, and genetics, undoubtedly affect the individual bone response to a particular exercise regimen. Thus, the results of prospective exercise training studies have been variable. Nevertheless, the training studies suggest that exercise can increase bone mineral mass and density if continued on a regular basis. More work is needed to better define the optimal type and amount of exercise to promote osteogenesis at various ages. Additional studies in children and adolescents are particularly important, because some data suggest that lasting benefits in bone structure and mass parameters may occur when bone-loading exercise is initiated early in life.

Chapter 21

Hormonal Influences on Human Body Composition

Marie-Pierre St-Onge and Per Bjorntorp

Hormones exert major, determining effects on body composition. Hormonal changes during growth dictate the sex differences observed when puberty is achieved. Alterations in steroid hormone levels with aging are associated with significant changes in body composition that may also lead to the emergence of disease risk factors in late adulthood. In addition, hormones from adipocytes and the gastrointestinal tract that circulate in proportion to the degree of body fat stores can also influence energy balance on a day-to-day and long-term basis, contributing to body weight regulation. Hormones also mediate alterations in body composition in times of physical stress. Both peptide and steroid hormones, alone or in combination, exert profound changes on both the lean and fat compartments of the body. Extreme examples may be found in clinical entities such as acromegaly, dwarfism, eunuchism, Cushing's disease, Addison's disease, and hyperthyroidism.

This chapter focuses on the impact of major hormones on body composition with an emphasis on the effects of endogenous steroid hormones and their interactions with insulin and growth hormone (GH). The effects of adipocyte-derived hormone levels either in response to or as a means of mediating body composition changes are also discussed briefly. Given the importance of the regional distribution of adipose tissue (AT) in relation to serious and prevalent diseases, the effects of cortisol, testosterone, and 17-β-estradiol on AT and its distribution are emphasized.

Hormonal Effects Through the Life Cycle

Throughout the life cycle, several hormones act to impact development and body composition. In early life, growth hormones play a crucial role in physical development, and deficiencies and excesses in this hormone have profound effects on body composition. In puberty, sex hormones help orchestrate changes in body composition. Changing levels of sex hormones also influence changes in body composition later in life, such as in menopause.

Growth Hormone

Pituitary GH is an important regulator of body composition. GH is a major anabolic hormone that triggers the growth of all body cells with such capacity. At puberty, its secretion is stimulated by sex hormones, leading to increased growth and size of fat-free mass (FFM), with a corresponding decrease in percent body fat in both boys and girls. Recent data also suggest that this hormone may be important in the regulation of AT distribution. GH also has marked effects on protein, carbohydrate, lipid, mineral, and water metabolism, mostly stimulating fatty acid release from AT while minimizing or sparing the use of protein and carbohydrates for energy. The net result is therefore mobilization of fat mass (FM) and maintenance or accretion of FFM.

GH excess, as in acromegaly or in transgenic animals, results in an increased body cell mass, reflecting increases in muscle mass and in visceral organs other than the spleen and brain (Ebert et al. 1988). GH excess also results in increases in extracellular water volume and a marked decrease in absolute body fat (Bengtsson et al. 1989). GH deficiency in children (Parra et al. 1979; Tanner and Whitehouse 1967) and in adults (Rosen et al. 1993) is associated with increased body fat, which is reversed by GH treatment (Bengtsson et al. 1993; Jorgensen et al. 1989b; Nass and Thorner 2002; Parra et al. 1979; Salomon et al. 1989; Tanner and Whitehouse 1967).

GH affects AT growth and metabolism in different ways. An important observation was that GH seems to promote the conversion of preadipocytes to fully differentiated adipocytes, at least in clonal preadipocyte cell lines (Hauner 1992; Morikawa et al. 1982). Children with GH deficiency have fewer adipocytes than normal children, and GH treatment increases the number of adipocytes. The changes in body fat observed in GH-deficient children are, therefore, attributable to increases in adipocyte size, and GH reduces body fat mainly by reducing the size of the adipocytes (Bonnet et al. 1974). However, it is unclear if the growth-promoting effect of GH on preadipocytes is important after puberty, because effects of GH on preadipocyte conversion to adipocytes have not been observed in adults (Hauner 1992).

The long-term effects of GH on AT metabolism are undoubtedly a stimulation of lipolysis and inhibition of lipogenesis, reflected in the marked changes in body composition with GH excess or deficiency. There is, however, still some controversy as to whether GH alone has any direct lipolytic effects on AT. GH receptors are found in human adipocytes (DiGirolamo et al. 1986), and GH receptor messenger RNA (mRNA) is expressed in AT (Vikman et al. 1991). In long-term cell culture, GH has an inhibitory effect on lipogenesis and a moderate lipolytic effect on both the 3T3 adipocyte cell line (Schwartz et al. 1985) and human AT (Ottosson et al. 2000), indicating that GH indeed exerts its effects directly on the adipocyte. It has been proposed that this effect is partly attributable to inhibition of glucose transport via a decrease in glucose carriers (Silverman et al. 1989), in line with the findings that GH must exert its insulin antagonistic effect distal to the insulin receptor (Dietz and Schwartz 1991). Reductions in GH levels have also been associated with obesity (Iranmanesh et al. 1991), and they may be more pronounced in visceral than peripheral obesity (Marin et al. 1993b). GH levels are, however, restored with weight loss, possibly indicating that decreased

GH levels may be secondary to weight gain as opposed to causative of obesity when decreased (Nass and Thorner 2002). However, GH administration to obese women resulted in decreased FM in an earlier trial (Richelsen et al. 1994).

GH levels are reduced in middle-aged and older men and women. In men, GH secretion depends on age and gonadal function (Iranmanesh et al. 1991). Changes in GH secretion are also observed in physiological and pathophysiological states such as fasting, uncontrolled diabetes, hypothyroidism, and Cushing's syndrome (Asplin et al. 1989; Hartman et al. 1992; Jorgensen et al. 1989a; Wiedemann 1981). Lower concentrations of circulating GH are thought to be partially responsible for the age-dependent changes observed in the elderly: increased total body fat, especially visceral AT (VAT), decreased FFM, and decreased bone mineral density (Nass and Thorner 2002). However, it has yet to be determined whether changes in GH levels with aging result in changes in body composition or whether changes in body composition with age lead to lower GH concentrations (Nass and Thorner 2002). Furthermore, changes in the regulatory set point of the hypothalamic–pituitary–adrenal (HPA) axis with aging bring about increases in cortisol levels that may act in combination with the reduced lipolytic action of lower GH levels to contribute to the increase in VAT with age (Nass and Thorner 2002). Treatment of elderly men with GH for 1 month has been shown to increase FFM (Brill et al. 2002). Similarly, in a study of GH and sex steroid hormone supplementation in elderly subjects, GH administration increased FFM and decreased FM in both men and women compared with baseline and placebo (Blackman et al. 2002). However, because of the side effects associated with GH administration, not the least being enhanced insulin resistance, its use is not recommended outside controlled clinical experiments (Blackman et al. 2002).

The metabolic effects of GH are also dependent on insulin and glucocorticoids (Vernon and Flint 1989). GH has no growth-promoting effect in the absence of insulin (Cheek and Hill 1974), and the metabolic effects of GH on AT also seem to depend on the presence of insulin. In fact, insulin increases GH binding to adipocytes (Gause and Eden 1985), which may be important for the ability of GH to interact with its target tissues, and insulin is important in maintaining the intracellular metabolic machinery intact. Moreover, GH seems to directly stimulate insulin secretion from the pancreatic β-cells (Nielsen 1982), further complicating the hormonal interactions.

Sex Steroid Hormones: Males

The increase in muscle mass with puberty in boys is well known and is associated with testosterone levels. In boys with delayed puberty, administration of testosterone is followed by an increased velocity of growth of FFM and stature (Gregory et al. 1992). In normal men, large doses of testosterone seem to be required to obtain similar effects, probably because of the inhibition of endogenous testosterone secretion (Friedl et al. 1990). Large doses of androgens are frequently used by power athletes with the hope of improving performance, but this is followed by a number of unwanted effects (Friedl 1990; Wade 1972), such as insulin resistance (Cohen and Hickman 1987; Holmang and Bjorntorp 1992). The effects of androgen administration on muscle seem to include protein synthesis in muscle (Alen et al. 1994; Hervey et al. 1981), but the precise mechanism is not known.

The effects of testosterone on total AT mass are apparently not very marked in normal adults (Friedl et al. 1990) or in aging men (Tenover 1992). AT distribution is, however, probably regulated by testosterone. Testosterone deficiency has been suggested to be associated with centralization of body FM (Bjorntorp 1993). Testosterone production is decreased by approximately one third by age 70 and approximately one half by age 80 and may be responsible for the hypogonadal-like features that emerge concurrently with age, such as loss of bone and muscle mass (Brill et al. 2002) and increased VAT (Haffner et al. 1993; Khaw et al. 1992; Seidell et al. 1990; Simon et al. 1992). The accumulation of VAT can be reversed, at least partly, by testosterone administration (Marin et al. 1993a).

Testosterone administration in healthy elderly men, either alone or in combination with GH, can increase circulating testosterone levels and result in increased FFM (Tenover 1992). Trials of several months' duration with higher doses of testosterone have shown increased FFM as well as reduced FM in older men (Lam et al. 2001; Snyder et al. 1999). In a recent 26-week trial, testosterone alone tended to increase FFM but only did so significantly in combination with GH. It also failed to reduce FM when given alone but was effective in decreasing FM when provided with GH (Blackman et al. 2002). It is believed that when GH and testosterone are given together, the lipolytic responsiveness of adipocytes is improved, an effect that seemed to be at least partly explained by an additive effect of the two hormones on β-adrenergic receptors

(Yang et al. 1995). These results suggest a synergy between GH and testosterone in mediating body composition changes in elderly men.

The cellular effects of testosterone on adipocytes have received surprisingly little study regarding lipid accumulation; the activity of lipoprotein lipase (LPL) and other enzymes of importance for triglyceride synthesis is apparently inhibited by testosterone (Xu and Bjorntorp 1994). LPL is responsible for the flux of free fatty acids (FFAs) into AT stores, and triglyceride uptake has been shown to be inhibited by testosterone administration, an effect that is more pronounced in abdominal than femoral AT (Marin et al. 1995). Furthermore, the cortisol-induced stimulation of LPL is prevented by testosterone (Ottosson and Bjorntorp 1994).

An additional feature of these observations is the consistent finding of differences in the expression of effects in different AT compartments. Studies have shown that the inhibitory effects of testosterone on lipid uptake, measured in the integrated system with the method of oral labeled triglyceride administration, are clearly more pronounced in visceral than subcutaneous depots (Marin et al. 1996). These observations are consistent with the finding of an apparently specific decrease in VAT, measured with computed tomography, after testosterone treatment (Marin et al. 1993a). It is therefore not surprising that low plasma levels of testosterone have been observed in obesity, especially in abdominal obesity, and may be associated with type 2 diabetes (Abate et al. 2002). In nondiabetic men, testosterone levels are inversely correlated with total FM and truncal and peripheral skinfold thickness; however, these data do not imply a causal relationship between elevated testosterone concentrations and regional adiposity (Abate et al. 2002).

There is some indirect evidence suggesting that the density of androgen receptors (AR) may be higher in VAT than in other AT compartments, possibly explaining the more pronounced effects of testosterone on triglyceride flux in VAT. These effects may be attributable to enhanced β-adrenergic sensitivity, which is regulated by androgens via the AR. β-adrenergic sensitivity is higher in men than women and higher in younger than older men, in parallel with testosterone concentrations (Rebuffe-Scrive et al. 1989). Direct measurements in human AT are required to resolve this important issue, because the more marked effects on VAT than other AT masses might be explained by differences in AR density. The marked effects of androgens on VAT depots may also be attributable to the higher cellular density of these depots or to greater innervation or blood flow to this region.

Sex Steroid Hormones: Females

Androgens are also produced in women, albeit in much smaller amounts than in men. In females, the effects of testosterone on AT might be weaker, because of the lower concentrations of testosterone, or different than observed in males, as suggested by testosterone's effects on AT metabolism in oophorectomized female rats (De Pergola et al. 1990). Furthermore, estrogens seem to down-regulate the AR density in AT of female rats and may thus protect against the effects of androgen on AT distribution (Li and Bjorntorp 1995), favoring the gynoid over the android body shape in women. This is an important issue for further research because of the deposition of fat to central VAT in hyperandrogenic conditions in women, such as those with polycystic ovarian syndrome (Rebuffe-Scrive et al. 1989), visceral obesity (Corbould et al. 2002), and non-insulin-dependent diabetes mellitus (Andersson et al. 1994).

Female sex steroid hormones also regulate FM, although it is not clear whether these effects are direct or mediated via energy intake or expenditure (Wade and Gray 1979). It is clear that sex hormones affect body fat distribution in women. The observation that normal premenopausal women have less VAT than men supports this view (Kvist et al. 1988).

With menopause, VAT increases, but this is preventable to some extent by hormone replacement therapy (HRT; Haarbo et al. 1991), suggesting that estrogen in women may prevent upper-body adiposity before the menopause transition. However, a recent randomized, placebo-controlled study of the effects of GH and HRT, alone and in combination, showed that GH and GH + HRT increased FFM and decreased FM, with no effect of HRT alone on either FFM or FM (Blackman et al. 2002). Therefore, the effects of HRT on body composition remain controversial. Studies have suggested that progestins may be responsible for the lack of effect of HRT on body composition after menopause (O'Sullivan et al. 1998).

Also, with menopause, the specific female enlargement of femoral subcutaneous adipocytes disappears. This enlargement is paralleled by an elevation in LPL activity and a low sensitivity to lipolytic agents, which may be responsible for the adipocyte enlargement in this AT depot prior to menopause. This metabolic pattern is reversed during late pregnancy and lactation, suggesting that femoral subcutaneous adipose tissue (SAT) might be important during these life stages and perhaps also for the survival of the fetus and newborn child (Rebuffe-Scrive et al. 1985a). The typical functional characteristics of adipocytes in the femoral subcutaneous region suggest that lipid retention decreases with menopause but this can be restored, at least partially, by HRT in postmenopausal women (Rebuffe-Scrive et al. 1986). Female sex steroid hormones may thus preserve the female body composition phenotype of greater lower-body adiposity and smaller upper-body adiposity relative to males.

Effects of Glucocorticoids

Excess cortisol secretion and its consequences are clearly seen in Cushing's syndrome as a wasting of FFM, particularly muscle. The excess glucocorticoids in Cushing's syndrome also lead to insulin resistance, further exacerbating the effects of glucocorticoids on body composition. These effects of glucocorticoids on body composition in turn are determined by the insulin resistance commonly seen in Cushing's syndrome, created by the excess of glucocorticoids. With insufficient compensatory secretion of insulin, the condition develops into diabetes, with its well-known lack of anabolic effects both on protein sparing in muscles and on maintaining storage fat in AT. Therefore, the actions of insulin are important for the net effect of glucocorticoids on muscle mass as well as on FM and distribution. The opposing disorder, Addison's disease, is associated with glucocorticoid deficiency and characterized by extreme leanness and fat redistribution (Udden et al. 2003).

Studies examining the mechanism of action of glucocorticoids on AT have given conflicting results. The main enzymatic regulator of triglyceride uptake, LPL, is influenced by glucocorticoids. The reports in the literature on this subject are not congruent, with some showing inhibitory and others stimulatory effects (De Gasquet et al. 1975; Krotkiewski et al. 1976). Some of these controversies may be attributable to the well-known species differences in the regulation of AT metabolism. Nevertheless, the effects of glucocorticoids on both triglyceride uptake and release have been reported, and there is strong evidence that these effects vary in intensity in different regions of AT compartments, with more potent effects in omental compared with subcutaneous adipocytes (Tomlinson et al. 2003).

Studies on human AT, under fully controlled tissue-culture conditions, seem to have resolved at least some of these controversies. Cortisol alone seems to exert limited effects on LPL expression in human AT, with the presence of insulin being critically important (Appel and Fried 1992; Cigolini and Smith 1979; Ottosson et al. 1994). In a recent in vitro study of AT metabolism, cortisol, in the presence of insulin, favored lipid accumulation by stimulating LPL activity and inhibiting basal and catecholamine-stimulated lipolysis (Ottosson et al. 2000). GH obliterates the expression of LPL activity by cortisol and insulin (Ottosson et al. 1994). In the presence of insulin, cortisol and GH have opposing lipolytic activity, with cortisol inhibiting and GH stimulating lipolysis in vitro (Ottosson et al. 2000). Thus, the attenuation of cortisol-induced LPL activity by GH, in the presence of insulin, leads to enhanced lipolysis and possibly diminished triglyceride storage into AT. Therefore, it would be expected that under conditions of cortisol excess in vivo, insulin would promote a balance between a slight inhibition of lipolysis and a blunted antilipolytic effect.

The net effects of glucocorticoids on lipid mobilization in vivo have been found to result in marked increases of circulating FFA both in humans (Divertie et al. 1991) and rats (Guillaume-Gentil et al. 1993). In vitro studies of AT from patients with Cushing's syndrome with hyperinsulinemia have shown that the lipolytic sensitivity to catecholamine is unchanged or reduced (Rebuffe-Scrive et al. 1988). Thus, the elevated FFA concentrations after cortisol excess may be attributable to a weak antilipolytic effect of insulin, an inhibited glucose transport (which diminishes fatty acid re-esterification in AT), or a combination of both. Glucocorticoids may inhibit FFA re-esterification through interactions with glucose transport into adipocytes (Carter-Su and Okamoto 1987) because glucose is necessary for re-esterification.

The results of these studies illustrate the interactions between steroid and peptide hormones in the regulation of AT metabolism and its subsequent effects on body FM. The situation is, however, probably more complex. Sex steroid hormones interact with cortisol in specific ways. In the condition of excess cortisol secretion of central origin, sex steroid hormone secretion is probably inhibited through interactions between corticotropin-releasing factor and gonadotropin-releasing hormone (Olsen and Ferin 1987). There-

fore, the net effects will mainly be those resulting from cortisol and insulin, with the effects of GH and sex steroid hormones being blunted. This situation would thus contribute to the total effects of cortisol excess on body fat attributable to central inhibitory mechanisms.

Steroid hormones usually exert their effects via specific receptors for subsequent interactions of the hormone-receptor complex at the level of the appropriate genes. A glucocorticoid receptor (GR) has been found in AT of rats (Feldman and Loose 1977) and humans (Rebuffe-Scrive et al. 1985b). The density of this receptor seems to vary in different AT compartments, with higher density being observed in VAT compared with SAT in studies using ligand binding in cytosol preparations in humans (Rebuffe-Scrive et al. 1985b). Furthermore, steady-state GR mRNA levels have been reported to be higher in VAT than SAT in humans (Rebuffe-Scrive et al. 1990).

Obviously, the effects of glucocorticoids on the metabolic regulation of AT metabolism, mass, and distribution depend on the density of the GR. The GR density is regulated by feedback inhibition from glucocorticoids (McDonald and Goldfine 1988). In conditions of hypercortisolemia, such reductions in GR density will therefore be expected. Interestingly, however, the rank order of GR density in different AT compartments seems to remain after down-regulation (Peeke et al. 1993).

The effects of glucocorticoids on AT are clearly region specific. The most dramatic evidence for this statement is found in Cushing's syndrome, with its marked redistribution of storage fat from the periphery to central, mainly intra-abdominal, depots. The reason for this shift is most likely a combination of several factors. There is a high density of GR in the intra-abdominal depot (Rebuffe-Scrive 1991; Rebuffe-Scrive et al. 1985b; Sjogren et al. 1995). In the normal condition, the cellular density of VAT depots (Salans et al. 1973), as well as blood flow (West et al. 1989) and innervation (Rebuffe-Scrive 1991), is probably higher than in other AT depots, further exaggerating the effects of hypercortisolemia on these AT compartments. In the integrated system in vivo, hypercortisolemia will be followed by elevated insulin concentrations and inhibited GH secretion. The combined metabolic effects of these hormonal changes in the adipocytes are expected to be an augmented LPL expression by cortisol plus insulin with lack of effects of GH and perhaps a diminished lipolytic potential. Such aberrations have also been observed, with a regional difference, in patients with Cushing's syndrome (Rebuffe-Scrive et al. 1988), suggesting that the interpretation is correct. Nevertheless, FFA concentrations are elevated, perhaps attributable to a combination of a re-esterification defect, a blunted antilipolytic effect of insulin, or both. Increased concentration of FFA would tend to diminish AT mass. It might be that these effects are also regionally different, sparing central AT compartments, or that the effects on the balance between lipid accumulation and mobilization are disturbed, the former effects being more pronounced.

Evidence strongly suggests that obesity, with a disproportional increase of VAT depots, is a condition of increased cortisol secretion caused by a high sensitivity of the HPA axis to different forms of stress (Marin et al. 1992). The accumulation of excess fat in the visceral area may be a consequence of periodic hypercortisolemia through the mechanisms suggested previously. Furthermore, there is evidence of age-dependent increases in evening plasma cortisol levels, possibly attributable to the decline in the resilience of the HPA axis

seen with aging, leading to progressively greater exposure of tissues to glucocorticoids (Nass and Thorner 2002). However, there are no data specifically linking cortisol with measures of total FM or VAT in older adults, although these measures seem to increase in parallel in elderly people (Nass and Thorner 2002).

Glucocorticoids can therefore affect major components of the metabolic syndrome: insulin resistance and abdominal obesity (Bjorntorp and Rosmond 1999; figure 21.1). Increases in glucocorticoid levels in periods of stress can also partly explain the dyslipidemia of metabolic syndrome. These increases may be further exacerbated by deficiencies of GH and sex steroid hormones as seen in aging (Bjorntorp and Rosmond 1999). Furthermore, abnormalities in GR have also been associated with abdominal obesity, insulin resistance, and elevated blood pressure (Bjorntorp 1999) and therefore may play a role in the etiology of the metabolic syndrome and the impact of the HPA axis on these perturbations.

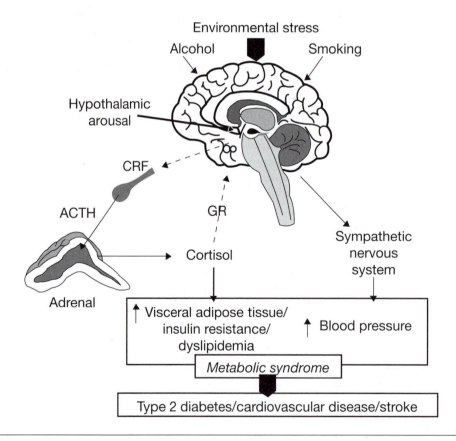

Figure 21.1 The hypothalamic origin of the metabolic syndrome. Environmental factors activate stress centers followed by a hypothalamic arousal, with activation of the sympathetic nervous system and elevated cortisol secretion, the latter amplified by a molecular genetic abnormality of the glucocorticoid receptor (GR) with inefficient feedback control. This will result in hypertension, insulin resistance with dyslipidemia, and visceral obesity with risk predictors for prevalent disease. CRF = corticotropin-releasing factor; ACTH = adrenocorticotropic hormone.

Interactions Between Hormones

As mentioned earlier, there are clear interactions among the effects of the steroid hormones, the peptide hormones, insulin, GH, and the catecholamines. In addition, there are most likely interactions between the corticosteroid and sex steroid hormones. Normally, cortisol is always present and is necessary for survival. The sex steroid hormones, however, vary in concentration with age, menopausal status, and menstrual cycle. The effects of sex steroid hormones would therefore be expected to modify the basal effects of the corticosteroids. However, very little is known about such interactions, although the homology of the promoter areas of several genes where the steroid hormone–receptor complex interacts (Carson-Jurica et al. 1990) suggests possibilities for interactions.

From a physiological aspect, it is interesting to compare the mechanisms of action of cortisol and testosterone and the effects of insulin and GH on AT. As reviewed earlier, the net effect of cortisol in combination with insulin seems to be lipid accumulation both by induction of LPL activity and by inhibition of lipolysis. It is clear that this is efficiently reversed by GH. Testosterone, however, decreases lipid accumulation by inhibition of LPL and stimulation of lipid mobilization, with this effect also being critically dependent on GH. Together, these observations seem to indicate that cortisol and insulin are mainly promoting lipid accumulation and retention of triglycerides in the adipocytes, whereas testosterone and GH oppose this effect. The dual interaction with GH in these processes is particularly interesting, but the mechanism is largely unknown.

It is interesting to compare the experimental results just summarized with body fat distribution in clinical situations characterized by changes in the secretion of these hormones. In Cushing's syndrome, cortisol and insulin concentrations are elevated whereas testosterone and GH secretions are blunted. This might explain the dramatic accumulation of VAT, with cortisol and insulin shifting the balance toward VAT accumulation with little counterregulatory action of testosterone and GH. Aging and menopause are associated with essentially normal cortisol and insulin values, whereas decreases in testosterone and GH levels in men and in estrogens and GH in women are observed with aging. This would make the hormonal balance tip toward VAT accumulation because of the relative lack of the presumed counterbalancing effects of the sex steroid hormones in combination with GH. In GH-deficient humans after hypophysectomy, normalization of cortisol and sex steroid hormone concentrations after treatment results in VAT accumulation attributable to the lack of GH effects. This VAT accumulation is reversed with GH substitution and exaggerated in acromegaly. Finally, elevated cortisol secretion, hyperinsulinemia, and diminished secretion of sex steroid hormones and GH result in the marked accumulation of VAT that is observed in visceral obesity. Observations of clinically well-defined endocrine disturbances of cortisol, insulin, sex steroid hormones, and GH strongly suggest that the balance between cortisol and insulin on the one hand and sex steroid hormones and GH on the other is an important factor in the regulation of body fat distribution, particularly VAT accumulation.

Effects of Adipocyte-Derived Hormones

There are some major regulators of AT metabolism other than those exerted by direct hormonal actions on adipocytes. The density of fat cells (or the number of adipocytes per unit AT mass), blood flow, and innervation are such factors. The number of adipocytes will determine the active cellular mass of an AT depot and therefore the tissue density of regulatory enzymes for lipid accumulation and mobilization. Blood flow determines the delivery of substrates for the lipid uptake process in adipocytes and is essential for the efficient removal of products of lipolysis, whereas FFAs have been shown to cause a feedback inhibition on the lipolytic process (Rosell and Belfrage 1979). In human AT, the acute lipolytic effects of catecholamines are probably regulated by direct effects of the sympathetic nervous system (Bjorntorp and Ostman 1971).

Leptin

Leptin, a hormone that is secreted by adipocytes, has been shown to circulate in the blood in concentrations related to body FM (Altman 2002). Leptin's actions, mediated through receptors in the arcuate nucleus, include regulation of appetite and energy expenditure such that, with an increase in FM, the corresponding elevation in plasma leptin concentrations would lead to an energy equilibrium favoring restoration of FM to lower levels. Likewise, only loss in FM would result in a decrease in circulating plasma leptin levels, which would lead to increased appetite and decreased energy expenditure, with the net effect being recovery of initial FM. Leptin's actions operate in concordance with insulin and a cascade of centrally active neurohormones (Niswender and Schwartz 2003).

Some individuals have mutations in the gene encoding leptin. These individuals are obese and become obese at a very young age. Leptin administration results in reductions in body weight in such people. However, the prevalence of genetic leptin deficiency is very low, suggesting that obesity may be a leptin-resistant state where an individual does not respond appropriately to increases in circulating leptin concentrations. Nevertheless, administration of exogenous leptin to lean and obese subjects can result in weight loss when leptin is administered at high doses (0.30 mg/kg body weight; Heymsfield et al. 1999).

In a clinical trial examining the effects of weight loss and weight gain on energy expenditure and plasma leptin concentrations, it was found that maintenance of a reduced body weight lowered the plasma leptin–FM ratio in women, but not in men, and that maintenance of elevated body weight increased the plasma leptin–FM ratio in men but not in women (Rosenbaum et al. 1997). In this trial, plasma leptin concentrations were significantly correlated with resting and total energy expenditure at all weight plateaus, baseline, elevated, and reduced.

Plasma leptin concentrations, corrected for FM, are generally found to be greater in females than in males (Rosenbaum et al. 1996, 1997) and to be higher in pre- than postmenopausal women (Rosenbaum et al. 1996). It has been hypothesized that women may be more resistant to leptin than men and that leptin concentrations in women may be affected by estrogen or progesterone levels (Rosenbaum et al. 1996). Furthermore, leptin levels increase more rapidly as a function of body mass index and decline more strongly as a function of age in women than in men (Isidori et al. 2000), possibly further contributing to the increase in FM in women after menopause. Sexual differences in leptin concentrations seem to be evident starting at puberty only and have not been consistently observed in prepubertal children (Horlick et al. 2000). In a study examining the effect of puberty on leptin concentrations Horlick and colleagues (2000) found that sex differences in leptin levels

in late puberty remain after adjustment for differences in testosterone, estrogen, and body composition, suggesting that other factors, such as X or Y chromosome genes, may be involved.

Adiponectin

Adiponectin is another adipokine, or adipocytokine, which differs from the other adipocyte-derived hormones in that its circulating concentrations are inversely associated with adiposity. In fact, studies have shown that adiponectin levels are negatively correlated with percent body FM in children (Stefan et al. 2002), adolescents (Weiss et al. 2003), and adults (Weyer et al. 2001). In a mouse model it was shown that adiponectin acts primarily on muscle tissue to increase the influx and combustion of FFA, decreasing skeletal muscle triglyceride content (Yamauchi et al. 2001). This effect of adiponectin on muscle triglycerides is reflected in the observation that plasma adiponectin levels are strongly and negatively correlated with intramyocellular lipid content even after adjustment for percent body FM and central adiposity in obese adolescents (Weiss et al. 2003). In this cross-sectional study, plasma triglycerides and intramyocellular lipid content were the most significant independent predictors of adiponectin levels.

Consistent with the inverse association with obesity, adiponectin levels have been found to be elevated in women with anorexia nervosa compared with normal-weight women (Pannacciulli et al. 2003). These observations suggest that adiposity may exert negative feedback inhibition of further adiponectin production and that body weight reduction results in disinhibition of this regulatory system, leading to an augmentation in adiponectin levels (Panacciulli et al. 2003). In fact, changes in plasma adiponectin levels over a 5-year longitudinal study of Pima Indian children were negatively correlated with changes in percent body FM and body mass index (Stefan et al. 2002). However, in this study, plasma adiponectin levels at age 5 were not predictive of changes in percent body FM adjusted for sex.

Adiponectin concentrations have also been found to be inversely correlated with leptin in both men and women examined separately but not together (Cnop et al. 2003). Further examination of sex differences in adiponectin levels showed that, for similar values of adiposity, women had greater adiponectin levels than men. Also, when examining whether adiponectin levels varied with age, Cnop and colleagues (2003) found that older men and women had higher concentrations when matched to younger individuals of similar insulin sensitivity, body mass index, waist–hip ratio, SAT, and leptin levels. The authors suggested that sex steroid hormones may be responsible for those age-related differences in adiponectin levels because both testosterone and estrogen lower adiponectin production and these sex hormones decrease with age.

Regulation of Energy Exchange

Energy balance remains relatively stable in adults with close regulation of energy intake and energy expenditure (Gale et al. 2004). Both central and peripheral signals contribute to the complex interplay between system regulatory controls maintaining energy balance, food intake, and energy expenditure over time. Some of the currently recognized body composition and energy-store regulatory factors are summarized in figure 21.2.

Summary

Hormonal regulation of body weight involves the control of energy expenditure and energy intake. Several brain centers are involved in the control of energy balance, and these centers receive input from both adipocyte- and gastrointestinal-derived hormones. Brain centers, mostly located in the hypothalamus, must respond to these changes via production of various neurotransmitters that will ultimately signal for an elevation or diminution of energy expenditure and a corresponding decrease or increase in energy intake. These adjustments in energy output and intake should restore body weight.

Body composition is additionally modified through oxidation of the various energy substrates in the body, whether lipid, protein, or glycogen. A preferential oxidation of protein amino acids, for example, would lead to muscle catabolism, whereas oxidation of lipids would reduce AT stores. These modifications in substrate oxidation

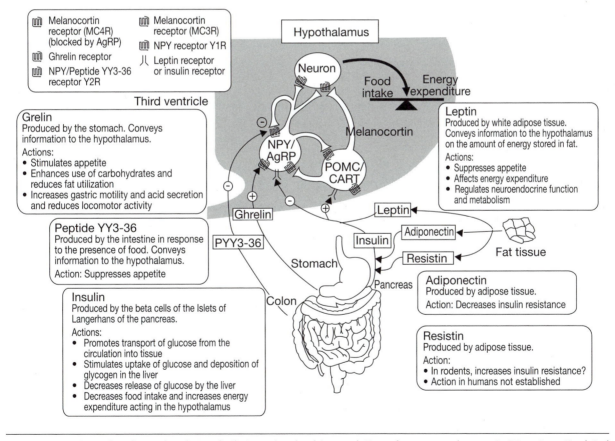

Figure 21.2 Central and peripheral signals that are involved in regulation of energy exchange. AgRP = Agouti-related peptide; NPY = neuropeptide Y; POMC = pro-opiomelanocortin.

From S.M. Gale, V.D. Castracane, and C.S. Mantzoros, 2004, "Energy homeostasis, obesity, and eating disorders: Recent advances in endocrinology," *Journal of Nutrition* 134: 295-298. Reprinted with permission from American Society for Nutritional Sciences.

or synthesis are mostly mediated by endocrine hormones such as the thyroid hormones, GH, and glucocorticoids.

Regulation of body weight and composition is thus a complex venture integrating hormonal signals from various body tissues and organs. Enhanced production of one hormone over another can lead to major body composition variations.

Furthermore, interactions—synergistic, permissive, or antagonistic—between the different effector hormones dictate body composition, and those interactions seem to be different between sexes and with age. Body composition maintenance is clearly a dynamic process integrating inputs from many regions and tissues of the body.

Part V

Body Composition and Pathological States

Disease often brings about changes in body composition. This section of the book examines the effects of pathological states on body composition. The associations of morbidity and mortality and body composition are reviewed in chapter 22. Body weight and body composition are markers of physiological function, and in turn, organ and tissue function are linked with morbidity and mortality.

Cancer is the second leading cause of death in the United States and obesity is a well-known risk factor for many types of cancer. The presence of a malignancy is often heralded by weight loss, and weight loss is secondarily associated with altered body function. Chapter 23 reviews the associations of body composition components, including fat tissue and lean soft tissue mass as well as bone mineral density with cancer occurrence, recurrence, and mortality. The effects of cancer treatments on body weight and body composition are also discussed.

Type 2 diabetes (sometimes referred to as adult-onset diabetes) is closely linked with adiposity and excessive intracellular lipid accumulation. Chapter 24 examines the common pathways linking insulin resistance and body composition. Chapter 24 also reviews effects of weight loss treatments on body composition.

Few conditions are more striking in their effects on all body systems than acquired immunodeficiency syndrome (AIDS) and related HIV infection. The clinical and body composition effects of this chronic condition are reviewed in chapter 25. Early studies of AIDS revealed severe wasting with profound loss of skeletal muscle and body cell mass, ultimately leading to morbidity and mortality. More recently, with lengthening life span, patients infected with HIV are developing new patterns of body composition change, notably lipodystrophy.

People of all ages, male and female, are susceptible to inflammatory diseases. These disorders, such as rheumatoid arthritis, are catabolic conditions that lead to loss of lean tissue and related functional abnormalities that alter energy and protein metabolism and that lead to changes in body composition. These conditions are extensively reviewed in chapter 26.

Chapter 22

Morbidity and Mortality Associations

Jacob C. Seidell

This chapter discusses the impact of body composition on diseases and death. There are many factors that potentially confound this relation and the direction of causality is not always clear. In addition many indirect measures of body composition are used (such as indicators based on weight, height, and body circumferences) and these are not always easy to interpret in terms of body composition. Methodological issues that play a role in the study of body composition and health are discussed as well as risk assessment. Effects of body composition on health in different age and ethnicity groups are described.

Anthropometric Measures in Relation to Morbidity and Mortality: Methodological Issues

The relationships between body composition and the risks of disease and death are complex because they can be modulated by many factors. Such modifying factors may change over time and complicate general statements about the relationship between body composition and disease. Demographic and disease pattern characteristics of populations may be important. For instance, in a population in which the majority of deaths can be attributed to infectious diseases at a relatively early age and in which relatively high levels of fatness are associated with afflu-

ence, the relationship between overweight and mortality may be inverse. On the other hand, in populations where most people live beyond their 50s, in which the predominant cause of death is cardiovascular disease, and in which relatively high levels of fatness are associated with low socioeconomic status, the relationship between overweight and mortality will definitely be positive.

In a single society, such very different relationships between disease and body composition may exist simultaneously. For instance, even in racially homogeneous populations in Europe, the older cohorts will have been born in the 1930s or earlier, when relationships between socioeconomic factors and body composition as well as disease patterns were markedly different from those in the 1950s. Such complexities become even more pronounced when there is a regular inflow of migrants from cultures at different stages of socioeconomic transition and with different genetic predispositions for certain disorders.

In addition, in most studies only simple anthropometric variables, such as height and weight, are available to study the relationships between body composition and disease.

The relationships between such anthropometric measurements and body composition may also differ depending on genetic (ethnic or racial) factors and long-term socioeconomic conditions. For instance, in undernourished populations where stunting is a common problem, body proportions (such as the relationship of sitting height to leg length) may differ and identical body mass indexes may have different implications compared with those in nonstunted populations. Furthermore, shrinking with age, sarcopenia, and other age-related changes in body composition may affect body composition and body mass index (BMI) even though body weight may remain constant. Finally, indicators of body composition such as the BMI do not allow for differences in fat and muscle distribution, which may be particularly important in determining the health risks associated with different levels of overweight.

Some of the factors that may affect the relationship between body composition and disease are shown in the next column. In the remainder of this chapter, findings of studies performed in primarily European or U.S. populations are discussed, often with the BMI as the sole indicator of body composition. It is clear that conclusions based on these studies may not necessarily apply to other populations and circumstances. Especially in populations undergoing rapid cultural and lifestyle changes, there is some evidence that individuals may be particularly prone to develop high levels of fatness and abdominal fat distribution and associated complications.

Anthropometry, Body Composition, and Mortality

To evaluate the associations between body composition and mortality, we have to exclusively rely on data from observational studies. It is unlikely that the issue of body weight and mortality will ever be resolved from randomized controlled clinical trials. The relationship between indicators of overweight and mortality has been the subject of many studies over the last 40 years. Many of the important studies were reviewed by Seidell and colleagues (1999). The relationship between BMI and mortality is still subject to debate in the literature partly because most studies suffer from methodological drawbacks. In the following section, some of these factors affecting mortality associations are discussed.

Cigarette Smoking

Cigarette smoking is an established confounder or an effect modifier in the relationship between overweight and death. The role of smoking as a confounder can be explained by the observation that smokers have lower BMIs compared with nonsmokers as well as higher mortality rates. This

Factors Affecting Body Composition and Disease

The following factors may affect the relationships between anthropometric indicators and risk of disease and death:

Biological factors

Age
Sex
Genetic susceptibility
Ethnicity or race
Menopausal status

Milieu

Sociocultural factors
Physical environment
Economic factors

Lifestyle factors

Smoking habits (including past smoking habits)
Dietary intake (quality and quantity)
Alcohol consumption
Physical activity

Health-related factors

Background prevalence of disease
Genetic predisposition to diseases
Presence of diseases
Presence of other risk factors

Biometric factors

Height (including history of stunting and wasting)
Fat and muscle distribution
Body proportions (leg length, sitting height)
History of large weight fluctuation

could attenuate the relationship between BMI and mortality. There have been indications that relationships between BMI and mortality differ by smoking status (e.g., curvilinear in smokers; linear in nonsmokers) although these observations have not been confirmed by others. In the context of disease prevention and health promotion, it would be wise to rely on data concerning the association between BMI and mortality in nonsmokers. Public health advice to smokers should be first to stop smoking despite an increased risk for weight gain after cessation of smoking.

Clinical or Subclinical Disease

The issue of preexisting clinical and subclinical disease is controversial. Although not many dispute the possibility that, especially at the lower ranges of BMI, some diseases may lead to dramatic weight loss as well as an increased risk of dying, the methods to overcome these have not been clearly established. Because information about subclinical disease is usually unavailable, most investigators disregard mortality within the first few years of follow-up, based on the assumption that such deaths are largely attributable to disease present at entry. Allison and colleagues (1999) published a meta-analysis on this issue and concluded that "the effect of eliminating early deaths was statistically significant but minuscule in magnitude" and that "either pre-existing disease does not confound the BMI-mortality association or that eliminating early deaths is inefficient for reducing that confounding" (p. 418).

Intermediate Risk Factors in the Association Between BMI and Mortality

In early studies, relative weight based on data of the Metropolitan Life Insurance Company was evaluated using statistical models that incorporated other risk factors such as blood pressure, serum cholesterol, and blood glucose. From such analyses it was determined that obesity was not independently related to mortality, which was interpreted to mean that obesity per se was not a risk factor and that obesity is benign when it exists without other major risk factors for cardiovascular disease. It is now generally accepted that adjustment for intermediate risk factors leads to underestimation of the mortality risk associated with high BMI and that such adjustments are inappropriate (Seidell et al. 1999). In addition, coexistence of other risk factors for mortality such as diabetes mellitus, hypertension, and hyperlipidemia may potentiate the risk of dying in obese subjects. For some time it was proposed that lean hypertensive people have higher mortality compared with obese hypertensive people, but this was not confirmed in later studies. Stratification by risk factor status is not done often but is recommended when possible.

Limitations of BMI As an Estimate of Body Fatness or Fat Distribution

It is well accepted that there are limitations in the use of BMI as a measure of overall fatness. Especially in the middle range of BMI values, contributions of lean body mass and body fat mass are both relatively large. It is likely that body fat and lean body mass have different associations with mortality. One interpretation for the curvilinear association between BMI and mortality is that it is the result of the combination of two linear functions, namely that increasing fat mass is directly associated with increasing risk and that increasing lean mass is associated with decreasing risk (Allison et al. 1997).

This hypothesis was confirmed in a prospective study of Swedish men aged 60 at baseline in whom it was shown that high mortality was a linear function of both high fat mass and low fat-free mass. The issue of the limitation of body mass index as a measure of fatness becomes increasingly important at older ages (Heitmann et al. 2000). In Swedish women aged 45 to 75 years, body fat mass assessed by bioelectrical impedance was more strongly related to all-cause mortality than body mass index (Lahmann et al. 2002).

Weight Change Before and After Measurement of BMI

Several reports suggest that weight change and weight fluctuations are independent determinants of mortality. Weight changes and fluctuations may be more common toward extremes of the distribution of BMI in populations and may affect the risk estimates at these levels.

Age, Period, and Cohort Effects

Age, period, and cohort effects pertain to issues that are not unique to the BMI–mortality relationship but may be relevant. Age, period (calendar time), and cohort (year of birth) need to be accounted for. When the effect of BMI on mortality in different age groups is compared, usually cohorts of different age groups are followed for an identical period, and mortality rates and ratios are calculated in categories of BMI. Members of these different age groups do not only differ by age; they also have been born in different times and may not share time periods that affect both mortality and body weight. The older cohorts of today (e.g., ≥60 years) living in Europe have experienced World War II and possibly also periods of severe economic recession. They were young in periods when obesity was rare, and they are likely to have had different patterns of physical activity and diet compared with the older groups of tomorrow.

Underlying Behavioral Factors and Fitness

Obese men who are physically active may not have the same risk of dying compared with those who are equally obese but who are inactive. Physical inactivity may lead to increased obesity but, alternatively, obesity may lead to physical inactivity because of poorer physical functioning, for example, because of impaired respiratory function and musculoskeletal problems. Blair and Brodney (1999) proposed that high levels of obesity may be relatively benign when accompanied by high levels of fitness. Although the coexistence of high levels of fitness and obesity may be rather low in most sedentary societies, it is possible that the health benefits of leanness may be limited to fit men and that being fit may reduce the hazards of obesity (Lee et al. 1999).

With regard to diet, it is poorly understood whether risks of obesity differ by dietary pattern. In a study comparing measures of obesity and other risk factors among European women aged 38 years, we observed that Dutch women had an average BMI of 23 kg/m^2 and women in a village outside Naples, Italy, had a mean BMI of 28 kg/m^2 (Seidell et al. 1989). Nevertheless, cholesterol levels were higher in Dutch women (mean 5.6 mmol/L vs. 4.9 mmol/L in Italian women). This may reflect underlying differences in diet or genetic factors. Alcohol consumption may also be important in this respect. In one study it was observed that moderate and high alcohol consumption was associated with increased mortality, especially at low levels of body mass index.

Effects of Socioeconomic Status and Ethnicity

Low socioeconomic status is related to increased prevalence of obesity as well as increased mortality, and this may confound the relationship between obesity and mortality. Much of this confounding effect may be attributable to underlying differences in lifestyle, but it is also likely that high educational levels reflect other reasons for low mortality and low BMI (like better access to adequate medical interventions and prevention programs). Few studies have adequately dealt with the influence of socioeconomic status (education, income, profession) on the BMI–mortality relationship. Tayback and colleagues (1990) for instance, observed in elderly women that low BMI was associated with increased mortality only in women with poor socioeconomic status. Ethnicity is in many societies strongly associated with socioeconomic status. Although information is incomplete, it is clear that the relationships between BMI and mortality vary across ethnic groups. This partly reflects limited appropriateness of BMI as an indicator of body fatness. It has now been firmly established that the same level of BMI may indicate different levels of body fatness across ethnic groups. Particularly, people living in Asia have higher levels of body fatness compared with Caucasian people, which may be one of the reasons why an increased risk of type 2 diabetes occurs at lower levels of BMI in Asians (Deurenberg and Deurenberg-Yap 2001; Deurenberg et al. 1998).

Sample Size and Duration of Follow-Up

Sjöström (1992) showed in a review of 51 cohorts (40 studies) that studies in which no relationship between BMI and mortality was found were clustered among small or short-term studies. All studies with more than 20,000 subjects and 20 of 21 studies with more than 7,000 subjects showed a positive association between BMI and mortality. He estimated that follow-up needs to be longer than 5 years (but preferably much longer) to demonstrate effects of obesity on mortality rate. It seems obvious that duration of follow-up is important (in young women, a short follow-up will not be sufficient to see the long-term impact of obesity on chronic diseases and mortality), but very little systematic work is done in this area. For instance, in Seventh-Day Adventist women followed for 26 years, in those aged 30 to 54 a weak linear association between BMI and mortality was observed during the first 8 years of follow-up, a significant linear relation during 9 to 14 years of follow-up, and a U-shaped association during 15 to 16 years of follow-up (Lindstedt and Singh 1997).

Self-Report of Weight and Height

Many of the largest studies use self-reported height and weight. Although the validity is reasonably adequate for ranking individuals, there is evidence that with increasing degrees of obesity subjects tend to underestimate their weight (Rowland 1990).

Assessment of Risk

Usually the relationship between BMI and mortality is expressed in terms of relative risk in categories of body mass index compared with an arbitrary reference category. The choice of reference categories and cutoff points for other categories will greatly affect the estimates of relative risk. Absolute risks or population attributable risks are alternatives less often used but actually more meaningful in terms of public health importance as well as guidance for clinical management.

Even more compelling are estimates of reductions in life expectancy or a reduction in healthy life-years or disability-free years of life in obese groups compared with those with acceptable weights. Two recent studies (Fontaine et al. 2003; Peeters et al. 2003) showed that obese subjects have considerably shorter life spans compared with those with low BMI. Fontaine and colleagues (2003) observed marked race and sex differences in estimated years of life lost (YLL). Among white participants, a J- or U-shaped association was found between overweight or obesity and YLL. The optimal BMI (associated with the least YLL or greatest longevity) was approximately 23 to 25 for white participants and 23 to 30 for black participants. For any given degree of overweight, younger adults generally had greater YLL than did older adults. The maximum YLL for white men aged 20 to 30 years with a severe level of obesity (BMI >45) was 13 and for white women is 8. For men, this could represent a 22% reduction in expected remaining life span. Among black men and black women older than 60 years, overweight and moderate obesity were generally not associated with an increased YLL and only severe obesity resulted in YLL. However, black participants at younger ages with severe levels of obesity had a maximum YLL of 20 for men and 5 for women.

In the study by Peeters and colleagues (2003), 40-year-old female nonsmokers lost 3.3 years and 40-year-old male nonsmokers lost 3.1 years of life expectancy because of overweight. Forty-year-old female nonsmokers lost 7.1 years and 40-year-old

male nonsmokers lost 5.8 years because of obesity. Obese female smokers lost 7.2 years and obese male smokers lost 6.7 years of life expectancy compared with normal-weight smokers. Obese female smokers lost 13.3 years and obese male smokers lost 13.7 years compared with normal-weight nonsmokers. Body mass index at ages 30 to 49 years predicted mortality after ages 50 to 69 years, even after adjustment for body mass index at age 50 to 69 years. Thus, obesity and overweight in adulthood are associated with large decreases in life expectancy and increases in early mortality. These decreases are similar to those seen with smoking. Obesity in adulthood is a powerful predictor of death at older ages. Because of the increasing prevalence of obesity, more efficient prevention and treatment should become high priorities in public health.

These methodological issues and other factors become especially relevant in older people. In the following section, several essential issues are examined in more detail.

Body Composition and Aging: Implications for Morbidity and Mortality

Aging is related to changes in body composition as well as to changes in risk for disease and death. Associations between body composition and disease may thus be modified by age. In addition, there are some methodological considerations that may be of importance in the interpretation of age-related associations. These will be discussed in the following sections.

Epidemiology

In cross-sectional studies, the prevalence of high body weight or obesity (BMI >30 kg/m^2) increases with age up until about age 60 and then declines (Seidell and Visscher 2000). For this there can be several explanations:

- Selective survival (obese young and middle-aged persons have died prematurely)
- Cohort effect (old people come from cohorts in which obesity was less common)
- Weight loss after age 60

Probably all three possibilities are simultaneously acting together, and their exact quantitative contribution has not been studied systematically. There are some effects associated with aging that work the other way, such as a decline in stature. Even when weight is maintained in all people, the stature decline would lead to an increase in the prevalence of obesity with aging (Launer et al. 1995).

Effect of Age on Body Composition and Body Fat Distribution

One fundamental problem in the interpretation of such epidemiological data is that anthropometric measurements have different implications for body composition at different ages. For instance, using densitometry, Deurenberg and colleagues (1991) derived the following equation:

$$\% \text{ body fat} = 1.20 \cdot \text{BMI} + 0.23 \cdot \text{age} - 10.8 \cdot \text{sex}$$
$$(\text{men} = 1; \text{women} = 0) - 5.4 \ (r^2 = .79) \ (22.1)$$

This implies, for instance, that at BMI = 30 kg/m², a man aged 20 years has 24.4% body fat and at age 80 years 38.8% body fat and that women have much larger relative fat mass than men at every age and level of BMI. However, there is redistribution of body fat with age in the sense that more of the fat becomes located in the abdominal cavity (visceral fat; Seidell and Visscher 2000). The reason for this redistribution is not exactly known, but the declining testosterone and growth hormone levels in combination with declining rates of lipolysis of visceral fat with aging may play a role in men. Low testosterone levels have been shown to be associated with increased visceral fat mass in men. In women, declining estrogen levels after menopause may be a critical factor. In addition, weight change at older age does not reflect the same changes as at younger age. In the elderly, weight loss is more closely associated with loss of lean body mass than in young adults. This accelerated loss of fat-free mass in older people is often called *sarcopenia* (Poehlman et al. 1995; Roubenoff 2000) and may not be reflected in BMI-derived percent fat. Involuntary weight loss is frequently reported in elderly patients and usually caused by acute or chronic diseases.

Effect of Age on the Association Between BMI and Health

Many studies have documented that the well-known U-shaped relation between BMI and all-cause mortality becomes less pronounced with aging (Seidell and Visscher 2000). There has been some debate whether the BMI–mortality curve flattens with aging and the nadir of the curve increases with age (the point of minimum mortality). If these points are true, this would imply that cutoff points for obesity should be higher in older compared with younger people (Andres et al. 1985). Such conclusions are very difficult to derive from cross-sectional analyses. Particularly the implication that minimum mortality can only be maintained with continued weight gain during adult life has been criticized, especially because adult weight gain is an independent predictor of mortality and morbidity (Willett et al. 1991).

There are several potential explanations for the different associations of BMI and mortality by age:

- Body mass index is not an optimal indicator of body composition in the elderly, and we may be looking at opposite effects of fat mass and fat-free mass (as described earlier in this chapter).

- Selective survival occurs, where obese persons at high risk have died prematurely and there remains a selection of relatively healthy obese subjects.

- There is a "ceiling effect," where absolute mortality rates increase with age and, for instance, at age 90 years 15-year mortality will be close to 100% regardless of any risk factor status.

- There is a cohort effect, where obese subjects now at old age have been exposed to different lifestyles and environments compared with future obese elderly persons. Visscher and colleagues (2000) attempted to separate age and cohort effects in aging men and showed that when a cohort effect is excluded, age has little effect on the relative risks associated with high BMI.

- Excess fat is less detrimental in older people compared with younger people. This issue is explored further in the section of body fat distribution.

Although the relative risks of a high body mass index may be less pronounced with aging, the relatively high prevalence of obesity in the "young old" coupled with the high mortality rates (absolute risks) makes a high BMI an important public health issue.

Body Composition and Morbidity

Much of the available data on body mass index and health are devoted to mortality. There is, however, increasing evidence that an increased body mass index in elderly women and men is associated with impairment of health, disability, and reduced quality of life in terms of functional limitations (Launer et al. 1994, Visser et al. 1998a, 1998b). In several of these studies, the authors explored the possibility that disability is caused by low lean body mass or high fat mass. In all instances, they concluded that particularly high body fat mass is associated with mobility disability and general disability even in the very old.

Comparative data of these risks in young and elderly people are not available, but given the large number of functional limitations and disabilities in elderly people, this is certainly a more important public health issue with advancing age.

Effect of Age on the Association Between Weight Change and Health

On the average, people older than about 60 years of age lose weight when aging. Many observational studies have suggested that weight change (weight increase, weight loss, and weight fluctuations) is a predictor of mortality (Peters et al. 1995). In the elderly, most weight loss may be involuntary and may be caused by underlying disease processes. Psychosocial factors may also play an important role. Therefore, the increased mortality and morbidity rates assumed to occur in elderly subjects who lose weight may be spurious. There is no documented information on the effects of voluntary weight loss in the elderly.

Weight loss in elderly people is likely to reflect in particular a loss of lean body mass (particularly muscle), and this may contribute to the increased mortality with weight loss especially if starting body weight is relatively low. Weight loss has been found to be associated with increased mortality, whereas in the same studies body fat loss was associated with reduced mortality (Allison et al. 1999).

Weight loss may obscure true associations between body mass index and morbidity in old age. Harris and colleagues (1997) showed that a high body mass index in late middle age was associated with increased coronary heart disease,

whereas the same was not true in old age. This difference in risk was attributable to weight loss between middle and old age. High body mass index remained a risk factor for coronary heart disease once those with substantial weight loss were excluded. In elderly subjects with a body mass index in the acceptable range, weight stability is of great importance. Because there is a lack of information from intervention studies, it is hazardous to recommend voluntary weight loss in the elderly even when they are overweight, although in theory many of the comorbidities may respond favorably to weight loss.

Effect of Age on the Association Between Fat Distribution and Health

Because of the redistribution of body fat with aging, body mass index becomes a poor indicator of both overall fatness and abdominal fatness (Borkan et al. 1985, Seidell et al. 1988). Waist–hip circumference ratio and, more recently, waist circumference alone have been recommended as better indicators of abdominal fatness than BMI. The waist circumference may actually be a better measure than BMI of overall fatness as well. Waist–hip ratio, particularly in the elderly, may be difficult to interpret because the waist measures abdominal fatness and the variation in hip circumference may reflect variation in pelvic width and gluteal muscle. A recent cross-sectional study comparing anthropometric type 2 diabetes mellitus with health population controls suggested that the association between high waist–hip ratio and diabetes may be attributable to both relatively large waist and narrow hip circumferences, especially in men (Lissner et al. 2001; Seidell et al. 1997; Snijder et al. 2003). One hypothesis is that narrow hips reflect peripheral muscle wasting and that this is an important correlate of diabetes mellitus. More detailed studies comparing Indians living in Sweden compared with Swedish men support this hypothesis (Chowdhury et al. 1996). Indian men with the same height and weight have elevated risk factors (blood pressure, insulin, and triglycerides) and waist–hip ratio compared with Swedish men. Whole-body computed tomography revealed that there were no differences in body fat mass and fat distribution between the two groups, but the only significant difference was the relatively low leg muscle mass. This aspect of peripheral muscle wasting is likely to be much more pronounced in elderly subjects, and the interpretation of waist–hip ratio may be increasingly difficult with advancing age. Other recent studies, however, have shown that it is actually the increased peripheral fat stores that may be protective against type 2 diabetes mellitus and cardiovascular disease (Snijder et al. 2004; Tankó et al. 2003).

There have been few systematic comparisons between the relationships of BMI, waist circumference, and waist–hip ratio to health in elderly subjects. Visscher and colleagues (2001) showed that in men aged 55 or older, a large waist circumference was a better predictor of 5-year mortality than waist–hip ratio or BMI. A positive association between waist circumference and mortality was even seen in men aged 70 years or older. Low BMI was a better predictor of mortality than narrow waist circumference. It could be that low lean body mass is better reflected by low BMI, whereas increased (abdominal) fatness is better reflected by increased waist circumference. Similarly, Woo and colleagues (2002) proposed that waist size is a better predictor of mortality than BMI in the elderly.

Comparative data on the impact of abdominal fatness on mortality in younger and older subjects are scarce. Especially rare are studies where total body fatness and abdominal fatness are considered together in the same study. The Baltimore Longitudinal Study on Aging showed that increased abdominal depth (sagittal diameter) was related to increased all-cause mortality and coronary heart disease mortality in men aged 55 years or younger, whereas in older men this diameter was not related to either end point (Seidell et al. 1994). Again, explanations include selective survival and cohort effects, but it may also be that the increased abdominal accumulation of fat is less hazardous in older compared with younger men. One well-accepted theory of the importance of intra-abdominal or visceral fat to health is the release of free fatty acids into the portal vein, thereby exposing the liver to high concentrations of free fatty acids (Bjorntorp 1990). Although there are many potential confounding factors (Seidell and Bouchard 1997), and the presence of exceptionally high concentrations of free fatty acids in the portal vein has not been demonstrated in humans in vivo (Frayn et al. 1997), this is still an attractive hypothesis. Ostman and colleagues (1969) showed that stimulated lipolysis in omental fat is progressively reduced with aging,

and this may mean that the increased accumulation of abdominal fat is a less important source of free fatty acids in elderly compared with younger adults

Summary

Anthropometric measurements have been used for decades to study the health consequences of different body compositions. Although such measures are simple and useful tools in epidemiological studies, they have limited validity to assess body composition and distribution of fat and muscle. Conventional anthropometric indicators such as body mass index and waist–hip ratio may not be adequate measures of overall and abdominal fatness, particularly in elderly subjects. More accurate and precise body composition reference methods such as dual-energy X-ray absorptiometry are becoming available for large-scale population studies. Dual-energy X-ray absorptiometry offers the advantage of both total body and regional fat assessment (see chapter 5). In the absence of such measurements, waist circumference (compared to body mass index) may be a much better indicator of body composition, fat distribution, and impaired health, particularly in older people.

Even crude anthropometric measures of overweight and regional fatness are important determinants of a number of severe and, in some cases, potentially disabling conditions as well as a reduction in life expectancy. However, a high body mass index and a high waist–hip ratio may reflect heterogeneous underlying genetic, socioeconomic, and lifestyle factors that may substantially affect the health implications of these measures.

Weight stability, with weight in the acceptable or even slightly overweight category, seems best for optimal (lowest) mortality in the elderly. However, weight loss with aging, particularly at low body mass index, may be partly responsible for the seemingly high optimal body mass index. In addition, high body mass index (especially high body fat mass) is associated with increased risk of disability. It is likely inappropriate to recommend high levels of fatness in old people because of the relationship between BMI and mortality. Rather, researchers need to have access to body composition measurements, including total body and regional composition, rather than body weight both in clinical situations and in epidemiological studies.

Chapter 23

Body Composition and Cancer

Zhao Chen

Cancer affects people of all ages around the world and is the second leading cause of death in the United States before 1999. Recently, the American Cancer Society reported that cancer had surpassed heart disease as the top killer of Americans under 85 (Jemal et al. 2005). For many years, body composition methods have been considered important tools in cancer risk assessments, tools that not only contribute to our understanding of cancer etiology but also enhance our ability to identify high-risk populations for prevention and early detection. There is a rapidly growing body of knowledge about the associations of various body composition measurements, particularly obesity measures, with risk and prognosis of certain cancers. The scientific and clinical values of body composition assessments in foreseeing therapy responses, evaluating treatment effects, and predicting recurrence and survivorship in patients with cancer are increasingly recognized by scientists and health care providers. These developments have contributed to better understanding of the link between body composition and cancer biology as well as to improved precision and feasibility of various body composition assessment methods in various populations of people with cancer.

Body Composition and Cancer Risk

The high prevalence of obesity in developed countries may contribute to the elevated incidence rates of cancers in these populations (Bergstrom et al. 2001a; Calle et al. 2003). Obesity and inactivity often coexist, and both contribute to cancer risk. It has been estimated that overweight and inactivity together account for one fourth to one third of worldwide cases of breast, colon, endometrial, kidney, and esophageal cancer (Anonymous 2002a). In addition to increased general obesity, other body composition components, such as increased bone density (Buist et al. 2001b; Cauley et al. 1996; Zhang et al. 1997) and reduced muscle mass (Oppert and Charles 2002), are also related to increased cancer risk. In the past decade, anthropometric measurements, bioelectric impedance analysis (BIA), dual-energy X-ray absorptiometry (DXA), and computed axial tomography (CT) have been used in various experimental and clinical settings to investigate the relationship between body composition and cancer risk as well as the mechanisms linking cancer risk and body composition measurements.

Breast Cancer

Breast cancer is the most common cancer among women, other than skin cancer. It is the second leading cause of death in women. Many studies have been done to understand the relationship between breast cancer and body size and composition. Cumulative evidences support a link between breast cancer risk and obesity, although the association may vary by a woman's age at her breast cancer diagnosis. Some studies have also suggested that the risk of breast cancer is related to body height.

Height

Dietary intake in early life may affect the risk for breast cancer in later life. Because adult stature partially reflects nutritional status in childhood and young age, research on breast cancer often examines height as a predictor for breast cancer risk. It has been hypothesized that height reflects mammary gland mass, which could be related to breast cancer risk (Trichopoulos and Lipman 1992). A recently pooled analysis of major cohort studies has shown that the positive association between height and breast cancer risk is weaker and more variable across studies in premenopausal women than in postmenopausal women (van den Brandt et al. 2000). Genetic determinants, energy restriction, and physical activity levels during growth and the entire lifetime may modify the relationship between height and breast cancer risk. In addition, loss of height with aging may cause underestimation of the true adult height in older adults, and questionnaire data on height are subject to measurement errors. All these factors may significantly contribute to the variations in study results on height and breast cancer risk.

Weight, Body Mass Index, and Breast Cancer Risk

The observed high incidence of breast cancer in developed Western countries may be linked to the high prevalence of obesity (Stoll 2000). An interesting phenomenon is that obesity appears to be associated with postmenopausal breast cancer but is protective in premenopausal women (Ballard-Barbash and Swanson 1996). In a pooled analysis of seven major cohort studies, van den Brandt and colleagues (2000) found that for every 10-kg increase in weight or for every 4-unit (kg/m^2) increment in body mass index (BMI), there was approximately a 10% to 11% reduction in breast cancer risk among premenopausal women but a 6% to 7% increase in risk for breast cancer among postmenopausal women. Similar findings were also reported from case–control studies (Bruning et al. 1992; Sonnenschein et al. 1999; Swanson et al. 1997; Trentham-Dietz et al. 1997). Despite the heterogeneity in the results from other cohorts or case–control studies (Ballard-Barbash et al. 1990; den Tonkelaar et al. 1994; Kaaks et al. 1998), the overall evidence supports a link between post-menopausal breast cancer and obesity.

Breast cancer risk may be more closely related to changes in weight rather than current body weight (Ballard-Barbash et al. 1990; Trentham-Dietz et al. 2000). Results from a case–control study by Kumar and colleagues (1995) suggested that women who progressively gain weight from puberty to adulthood, and especially in the third decade of life, have a higher risk for developing breast cancer. Weight at birth may also affect breast cancer risk. A weak J-shaped relationship between birth weight and postmenopausal breast cancer risk was recently suggested by Titus-Ernstoff and colleagues (2002).

Very few data exist regarding the relationship between premenopausal breast cancer risk and body weight at different ages. Results from a population-based case–control study in the United States showed that being much heavier or lighter before age 21 is associated with reduced risk for premenopausal breast cancer. Weight gain after age 20 also protected women from breast cancer risk, but the effect was confined to early-stage and lower-grade premenopausal breast cancer (Coates et al. 1999).

Huang and colleagues (1997) reported that the association between obesity and breast cancer risk was only significant in postmenopausal women who had never used postmenopausal estrogen therapy. This finding is supported by the results from the Women's Health Initiative cohort, in which weight, BMI, change in BMI since age 18, and maximum BMI were associated with breast cancer risk only among women who had never used postmenopausal estrogen therapy (Morimoto et al. 2002). These findings have provided supportive evidence for the estrogen connection in the relationship between obesity and postmenopausal breast cancer.

Fat Mass, Fat Distribution, and Breast Cancer Risk

Abdominal fat deposition appears to contribute to breast cancer risk. After 28 years of follow-up, the Framingham Study reported that central subcutaneous fat instead of general adiposity was associated with subsequent breast cancer among women aged 30 to 62 (Ballard-Barbash et al. 1990). In that study, general adiposity was assessed by the sum of five skinfolds or BMI at the beginning of the study. The central fat patterns were assessed using an index that was calculated from the sum of the truncal skinfolds (chest, subscapular, and abdominal) divided by the sum of the extremity skinfolds (triceps and thigh). Similar findings were reported from a case–control study among newly diagnosed women with invasive carcinoma of the breast and age-matched controls (Schapira et al. 1990), in which waist circumference to hip circumference ratio (WHR) was significantly higher in the women with cancer. When researchers used skinfolds to measure body fat distribution, they found that breast cancer patients had greater upper-body obesity. These results suggest that although obese women have a slightly higher risk for developing breast cancer, women with android obesity are a segment of obese women who have the highest risk for developing breast cancer. In subsequent studies, Schapira and colleagues found further evidence supporting a link between fat distribution and breast cancer risk: Visceral obesity, as assessed by CT, was a significant risk factor for breast cancer in women (Schapira et al. 1994), and first-degree relatives of patients with breast cancer had greater WHR compared with controls (no family history of breast cancer) matched by age and BMI (Schapira et al. 1993).

The association between WHR and breast cancer risk is modified by family history of breast cancer. In the Iowa Women's Health Study, WHR was more strongly related to breast cancer risk among women with a family history of breast cancer compared with women without a family history of breast cancer (Sellers et al. 1992, 1993, 2002). The family history–positive women in the upper quintile of the WHR had a 2.2-fold risk of progesterone receptor–negative tumors compared with those in the lowest quintile (95% confidence interval: 0.9, 5.8; Sellers et al. 2002). The biological basis of the interaction of abdominal fat distribution and a family history of breast cancer is unknown. A nested study in sisters of the participants from the Iowa Women's Health Study suggested that neither insulin, sex hormone binding globulin (SHBG), nor testosterone could explain this interaction (Olso et al. 2000)

Furthermore, the Iowa Women's Health Study found that the association of WHR and breast cancer risk differs depending on menopause status and overall weight. A two-standard deviation increase in WHR was associated with no increase in the relative risk of breast cancer in younger and lighter postmenopausal women. However, in older, heavier postmenopausal women, the same increase in WHR carried greater than a twofold excess relative risk (Folsom et al. 1990). In the prospective cohort of the New York University Women's Health Study, WHR was a significant predictor of premenopausal breast cancer risk but BMI was not. However, the effect of WHR was especially pronounced among subjects who were overweight (Sonnenschein et al. 1999). In a recent review by Harvie and colleagues (2003), it was concluded that an increased risk of breast cancer in postmenopausal women is associated with general obesity; however, in premenopausal women, the increased risk of breast cancer is associated with central obesity.

Mechanisms

Several hypotheses have been proposed to explain the opposite directions in the associations of obesity with pre- and postmenopausal breast cancer risk. It is well established that high estrogen levels are associated with breast cancer risk.

Obesity in premenopausal women may protect against breast cancer by causing more frequent anovulatory menstrual cycles (Sherman and Korenman 1974; Stoll 1994). In postmenopausal women, ovarian estrogen production is diminished, so the aromatization of androstenedione in adipose tissue becomes the major source of estrogen. The levels of estrone and estradiol are higher among obese postmenopausal women compared with women of normal weight (Cauley et al. 1989; Hankinson et al. 1995). In addition to being connected to increased estrogen production, excess weight is also associated with decreased circulating levels of SHBG and therefore increased levels of biologically active estrogen (Moore et al. 1987).

Insulin resistance and high insulin-like growth factor (IGF) levels among obese women are other possible links to breast cancer risk. In a review, Stoll (1998) concluded that only obesity in teenage girls is related to a reduced breast cancer risk, whereas obesity after the teenage years only postpones breast cancer to an older age. In middle-aged women, the concomitants of hyperinsulinemia may activate invasive and proliferative activity in preneoplastic mammary lesions, such as in situ carcinoma. This enhances the risk of progression to invasive breast cancer, which is likely to manifest clinically, mainly after menopause. Furthermore, in obese subjects, enlargement of adipose deposits is known to produce an excess of free fatty acids and tumor necrosis factor-α, both of which may be involved in the pathogenesis of insulin resistance (Stoll 2000).

Although overall adiposity in women adversely affects breast cancer risk mainly by greater exposure of mammary epithelia tissue to endogenous estrogen, central obesity is likely a more specific marker of premalignant hormonal patterns than the degree of general adiposity. For postmenopausal women, hyperinsulinemia is more related to overall obesity, whereas in premenopausal women, it is more related to abdominal localization of fat (Stoll 1995). This may explain why an increased BMI is a risk marker for breast cancer in postmenopausal but not in premenopausal women. The greater abdominal visceral obesity may be related to aberrant insulin signaling through the insulin receptor substrate-1 pathway, leading to insulin resistance, hyperinsulinemia, and increased concentrations of endogenous estrogen and androgen (Stoll 2002). More direct evidence is needed to further support the link between insulin–IGF and breast cancer risk (Jernstrom and Barrett-Connor 1999).

Endometrial Cancer

Body weight, BMI (Jain et al. 2000), weight gain, and various measures of central adiposity (Schapira et al. 1991a; Swanson et al. 1993) are related to the incidence of endometrial cancer and endometrial cancer mortality (Ballard-Barbash and Swanson 1996). Obesity may account for as much as 40% of the incidence of endometrial cancer in affluent societies (Bergstrom et al. 2001a). Unlike breast cancer, obesity is associated with a two- to fivefold increase in endometrial cancer risk in both pre- and postmenopausal women. It has been suggested that only current obesity (in the top quartile), not the history of obesity, is associated with an increased risk of endometrial cancer (Swanson et al. 1993).

It is unclear whether fat distribution is a risk factor for endometrial cancer. In a nested case–control study in the Iowa Women's Health Study, both BMI and fat distribution were significantly associated with endometrial cancer in separate statistical models. However, after adjustment for BMI, fat distribution as measured by WHR and trunk-to-limb ratio was no longer associated with endometrial cancer risk (Folsom et al. 1989). In agreement with these findings, results from a cohort of postmenopausal women in Sweden also showed that general obesity as measured by BMI was significantly related to adenomatous and atypical hyperplasia of the endometrium, but no significant association between fat distribution and these endometrial precancerous lesions was observed (Gredmark et al. 1999). In contrast, a positive association between fat distribution and endometrial cancer risk has been reported in other studies. Trunk fat deposits (measured by subscapular skinfold thickness) and central obesity (subscapular to triceps skinfold thickness ratio) were significantly related to endometrial cancer risk in a case–control study in China (Shu et al. 1992) and a case–control study in Japan (Iemura et al. 2000). Schapira and colleagues (1991a) found that patients with endometrial cancer had significantly greater WHR, abdomen-to-thigh skinfold ratios,

and suprailiac-to-thigh skinfold ratios compared with control subjects matched for age and BMI. In a multicenter case–control study, both weight and upper-body obesity (higher WHR) were positively associated with endometrial cancer risk, whereas upper-body obesity was a significant risk factor for endometrial cancer independent of body weight (Swanson et al. 1993). Upper-body fat distribution may be also correlated with the grade of the endometrial adenocarcinoma as suggested by results from a study among postmenopausal women who were diagnosed with endometrial cancer (Douchi et al. 1997a).

Studies in Mexican women (Salazar-Martinez et al. 2000), postmenopausal women in Sweden (Weiderpass et al. 2000), and women in the United States (Shoff and Newcomb 1998) found that being both obese and diabetic is associated with the highest risk for endometrial cancer compared with being either obese or diabetic alone, suggesting a significant interaction between diabetes and obesity on endometrial cancer risk.

Kaaks and Lukanova (2002) reviewed the mechanisms for the link between obesity and endometrial cancer risk. In both pre- and postmenopausal women, excess weight decreased plasma SHBG and increased levels of estrone and bioavailable testosterone unbound to SHBG. However, weight increases the levels of total and bioavailable estradiol only in postmenopausal women. Chronically elevated plasma insulin levels and insulin resistance may affect the IGF-I/IGFBP (insulin-like growth factor binding protein) system, reflecting in a net increase in bioactive IGF-1, and directly affect endometrial tissue proliferation.

Ovarian Cancer

It has been hypothesized that endogenous hormones are involved in the etiology of ovarian cancer. However, despite the well-known relationship between obesity and endogenous hormone levels, results regarding the association between obesity and risk of ovarian cancer are very inconsistent. There was no increased risk of ovarian cancer among a Danish cohort of women discharged with diagnoses of obesity (Moller et al. 1994). Two prospective cohort studies also found no association between recent obesity and ovarian cancer risk (Fairfield et al. 2002; Tornberg and Carstensen 1994), but higher BMI in young adulthood seems to be associated with an increased risk of premenopausal ovarian cancer (Fairfield et al. 2002). In contrast to these studies, some early studies reported an increased risk of ovarian cancer mortality among women with an increased BMI (Garfinkel 1985; Lew and Garfinkel 1979). A recent report from CPS-II further supported the association between obesity and ovarian cancer (Rodriguez et al. 2002). CPS-II is a prospective mortality study of 1.2 million women and men begun by the American Cancer Society in 1982. During the 16 years of follow-up, 1,511 deaths occurred attributable to ovarian cancer. Ovarian cancer mortality rates were significantly higher among overweight (rate ratio [RR] = 1.16) and obese (RR = 1.26) women, compared with women with BMI <25. In the CAP-II study, the increased risk with obesity (BMI >30) was limited to women who had never used postmenopausal estrogen therapy (RR = 1.36, 95% confidence interval [CI] = 1.12-1.66). Furthermore, ovarian cancer rates were lowest for the shortest women and highest for the tallest. Height and obesity appear to be independently associated with ovarian cancer mortality in this population.

A possible relationship between fat distribution and ovarian cancer was observed in the Iowa Women's Health Study cohort despite lack of an association between high BMI and an increased risk of ovarian cancer in this study (Mink et al. 1996). The association between WHR and ovarian cancer may be modified by family history (Sellers et al. 1993). In contrast to the case in the association between obesity and endometrial cancer risk, diabetes does not seem to modify the association between obesity and ovarian cancer risk (Rodriguez et al. 2002).

The association between obesity and ovarian cancer is likely through serum androgens, which stimulate ovarian epithelial proliferation and contribute to ovarian cancer risk (Helzlsouer et al. 1995). The IGF system is also possibly involved with the link between obesity and ovarian cancer (Rodriguez et al. 2002). More studies are needed to provide direct evidence to support these hypotheses.

Colon Cancer

Consistent results support an association between obesity and an increased risk of colon cancer (Caan et al. 1998; Murphy et al. 2000). Two large cohort studies indicated that a high baseline level of skinfold thickness was correlated with an increased rate of colorectal cancer occurrence after a follow-up of approximately 20 years (Chyou et al. 1994; Ford 1999). In another large

nationwide mortality study of U.S. adults, death rates from colon cancer increased with increased BMI in men. The rate ratio was highest for men with BMI $\geqslant$32.5 (RR = 1.90, 95% CI = 1.46-2.47); a similar but weaker association was observed in women (Murphy et al. 2000). Results from a second national sample of adult U.S. men and women indicated that the increased hazard ratios of obesity for colon cancer were the same in both men and women (Ford 1999). Obesity was found to be a risk factor for colon cancer in multiethnic samples of men and women (Le Marchand et al. 1997), a finding that helps to explain the increases in colorectal cancer risk experienced by Asian immigrants to the United States occurring in the first generation.

Obesity, weight gain, and unstable adult weight may be independently associated with colorectal carcinogenesis (Bird et al. 1998). Increased waist circumference and WHR were found to be strong risk factors for colon cancer in a cohort study (Giovannucci et al. 1995). In a Japanese study, weight gain in combination with high WHR was associated with a significant increased risk for colon adenoma (Kono et al. 1999).

Interestingly, colon cancer risk was not increased among overweight and physically active men (Lee and Paffenbarger 1992). This result may be explained by several factors, including the fact that BMI is not a good estimate of body fat among muscular men, and physical activity can counter the adverse health effects of obesity.

In a case–control study among patients with colorectal polyps and healthy controls, no significant differences in body composition between patients and controls were found. However, among the patients, the increase in adenoma diameter after 3-year follow-up was highly associated with increased triceps skinfold thickness (p = .004), total body fat percentage (p = .02), and BMI (p = .006). The authors proposed that high body fatness is a promoter of adenoma growth (Almendingen et al. 2001). The mechanisms involved in the link between colon cancer and obesity are still under investigation. An increased plasma level of leptin, which is associated with obesity, may be one of the mechanisms linking colon cancer and high BMI. Leptin is a growth factor in colon epithelial cells. The leptin receptor is expressed in human colon cancer cell lines and human colonic tissue. Stimulation with leptin leads to phosphorylation of p42/44 mitogen-activated protein kinase and increases proliferation in vitro and in vivo (Hardwick et al. 2001).

Prostate Cancer

Reports of the association between prostate cancer and BMI are inconsistent (Rodriguez et al. 2001), showing increased risk of prostate cancer with obesity (Cerhan et al. 1997; Putnam et al. 2000) or no association (Furuya et al. 1998; Schuurman et al. 2000). In two large cohorts of adult men in the United States, obesity was associated with an increased risk of prostate cancer mortality. However, as suggested by the authors, decreased survival rather than increased risk of prostate cancer among obese men may explain this association (Rodriguez et al. 2001).

In a recent large prospective cohort in the United States, 439 men of the 8,922 participants developed prostate cancer during follow-up from 1988 to 1993. There was a slightly increased risk of prostate cancer for men with increased BMI or waist circumference, but the association was not statistically significant (Lee et al. 2001). Similarly, in the Netherlands Cohort Study, neither height nor BMI at baseline was significantly associated with prostate cancer risk (Schuurman et al. 2000). However, Putnam and colleagues (2000) found that obesity may be only weakly associated with well-differentiated, localized tumors, but it is a stronger risk factor for more clinically significant prostate cancer. This finding is supported by another prospective cohort study of men 27-75 years of age (MacInnis et al. 2003).

There are very limited data on whether changes in weight and weight at different ages are related to prostate cancer risk. Results from a cohort study in Iowa men indicated that overweight and weight gain in later life were both significant risk factors for prostate cancer (Cerhan et al. 1997). In the Netherlands Cohort Study, higher BMI at age 20 appeared to be associated with an increased risk of prostate cancer, suggesting that body composition in young adulthood may exert an effect on later risk of prostate cancer (Schuurman et al. 2000). However, in another study, Lee and Paffenbarger (1992) found no significant associations between prostate cancer risk and body weight at baseline or age 18 in a U.S. prospective cohort.

Body fat distribution rather than overall obesity may play a role in prostate cancer etiology, even in a very lean population such as the Chinese population. In a case–control study of patients with newly diagnosed prostate cancer in Shanghai, China, men in the highest quartile of WHR had almost a threefold increased risk compared with men in the lowest quartile (Hsing et al. 2000).

The association between prostate cancer and obesity may be attributable to similar mechanisms as in other cancers. In a case–control study, Chang and colleagues (2001) found that leptin is related to tumor volume of prostate cancer. Abdominal fat deposit is related to an increased androgen level, which is a risk factor for many cancers, including prostate cancer.

Obesity and Other Cancers

Obesity may affect the occurrence of other types of cancers. A Danish record-linkage study indicated that cancer risk in a cohort of 43,965 obese people was increased by 16% compared with the Danish population as a whole. The increased risk for cancer included carcinoma of the esophagus, pancreas, and liver in addition to ovaries, colon, and prostate (Moller et al. 1994). A dramatic increase in the incidence of esophageal adenocarcinoma has occurred among men in the United States over the last 2 decades. Although the underlying reasons remain largely unknown, the increasing prevalence of obesity may have played a role. Recently, the association between obesity and esophageal cancer was discussed in two reviews (Li and Mobarhan 2000; Mayne and Navarro 2002). There is limited but consistent evidence supporting an association between obesity and kidney cancer risk (Moyad 2001). In a quantitative review, Bergstrom and colleagues (2001b) suggested that high BMI is equally associated with an increased risk of renal cell cancer in both men and women. A possible association between obesity and gallbladder cancer was reported by two studies (Strom et al. 1996; Zatonski et al. 1997). Mechanisms linking obesity with these cancers are less understood.

Squamous cell esophageal cancer is one of the few neoplasms that are inversely related to BMI. This is likely attributable to cancer-related weight loss or to other correlates of leanness (Gallus et al. 2001). Cross-sectional studies have suggested a modest inverse association between lung cancer and BMI. However, one prospective study among women aged 55 to 69 at baseline found that the inverse association of BMI with lung cancer could be explained by smoking status (Drinkard et al. 1995).

Bone Density and Cancer Risk

High bone density appears to be a significant indicator of increased cancer risk of the breast (Buist et al. 2001b; Cauley et al. 1996; Zhang et al. 1997), endometrium (Douchi et al. 1999, 2000; Newcomb et al. 2001), colon (Zhang et al. 2001), and possibly prostate (Demark-Wahnefried et al. 1997a).

Large prospective studies indicated that postmenopausal women with increased bone mass or bone mineral density (BMD) are more likely to develop breast cancer compared with women with low BMD (Buist et al. 2001b; Cauley et al. 1996; Zhang et al. 1997). Results from the Study of Osteoporotic Fractures (Cauley et al. 1996) showed that among postmenopausal women in the United States, those with BMD above the 25th percentile had a 2- to 2.5-fold increased risk of developing breast cancer compared with women with BMD below the 25th percentile. In the Framingham Study, the relative risk of developing breast cancer ranged from 1.3 to 3.5 among postmenopausal women whose bone mass was in the second to fourth quartiles compared with women whose BMD was in the first quartile (Zhang et al. 1997).

The association between BMD and breast cancer was investigated in a small nested case–control study involving 30 patients with breast cancer and 120 controls, aged 68 ± 6 (mean ± SD) years, as part of the Dubbo Osteoporosis Epidemiology Study (Australia). After adjustment for the effects of duration of lifetime ovulation and BMI, each 0.1 g/cm^2 increase in lumbar spine and femoral neck BMD was associated with a 2.1-fold (95% CI = 1.3-3.4) and 1.5-fold (95% CI = 1.0-2.4), respectively, higher risk of breast cancer (Nguyen et al. 2000).

The association between breast cancer and BMD is modified by a family history of breast cancer. Lucas and colleagues (1998) studied 7,250 elderly white women enrolled in the Study of Osteoporotic Fractures. These researchers found that among women without a family history of breast cancer, those with a proximal radius BMD in the highest tertile were at a 1.5-fold increased risk compared with women in the lowest tertile, whereas among women with a positive family history of breast cancer, those in the highest tertile of BMD were at a 3.4-fold increased risk compared with women in the lowest tertile. In addition, women more than 65 years of age who had high BMD were also more likely to have advanced stages of breast cancer than women with low BMD (Zmuda et al. 2001).

High bone density is related to endometrial cancer risk (Douchi et al. 1999, 2000; Newcomb et al. 2001) and estrogen receptor (ER) expression in endometrial cancer specimens (Douchi et al. 2000). Fractures in postmenopausal women may serve as a surrogate measure of bone density, reflecting long-term lower estrogen levels, which appear to be inversely associated with breast and endometrial cancer risk. Results from a U.S. case–control study on breast cancer and endometrial cancer analysis showed that the odds ratio of fracture was 0.8 for breast cancer and 0.6 for endometrial cancer, suggesting that the endogenous hormonal factors associated with the increased fracture risk are also related to decreased breast cancer risk and more strongly to reduced endometrial cancer risk (Newcomb et al. 2001).

An association between BMD and colon cancer risk was suggested by the Framingham Study. Higher bone density was found to be associated with a lower colon cancer risk in the postmenopausal women. After adjustment for age and other potential confounding factors, the rate ratios of colon cancer were 1.0, 0.7 (95% CI = 0.3, 1.3) and 0.4 (95% CI = 0.2, 0.9) from the lowest to the highest tertile (p for trend = .033). However, in the same study, no association was found between bone mass and rectal cancer (Zhang et al. 2001).

A relationship between increased bone mass and prostate cancer risk has also been suggested. Demark-Wahnefried and colleagues (1997a) found a higher bone mass among men who were newly diagnosed with prostate cancer, although the increased BMD was not statistically significant.

In contrast to results regarding breast cancer, endometrial cancer, and colon cancer, some preliminary data suggested that BMD was lower among women with cervical cancer who had no bone metastases, were premenopausal, and had not had chemotherapy (Hung et al. 2002).

The associations between BMD and cancer risk are likely related to hormonal factors. A cumulative exposure to estrogen has been proposed as a link between bone density and breast cancer (O'Brien and Caballero 1997). BMD may serve as a surrogate of cumulative estrogen exposure. However, BMD may not influence breast cancer risk independent of its relationship with endogenous hormones and measured covariates, such as serum hormone levels, BMI, and number of years since menopause. Results from a nested case–control study showed that relative to women in the lower fourth of the BMD distribution, the risk of breast cancer was 2.6 (95% CI = 1.1-5.8) for women in the upper 25% of BMD. After adjustment for serum hormone levels, the corresponding relative risk reduced slightly to 2.5 and became statistically nonsignificant (95% CI = 0.9-5.2). With BMI and number of years since menopause added to the multivariate analysis, the relative risk further decreased to 1.4 (95% CI = 0.5-4.0; Buist et al. 2001a).

The finding of an association between BMD and cancer risk has been limited primarily to white postmenopausal women. Research needs to be expanded to other populations with different BMD levels and cancer risks so that interactions of genetic and environmental factors in the association of BMD and cancer risk can be investigated (table 23.1).

In summary, cancer risk is associated with body composition, including fat, fat distribution, and BMD. However, more research is needed to further understand the link between body composition measures and each specific type of cancer.

Body Composition and Cancer Treatment

Current research indicates that body composition is related to cancer prognosis and survival. The magnitude and direction of this association vary depending on the type of cancer. For example, increased body weight and body fat negatively affect the prognosis of breast cancer, whereas survivorship in lung cancer patients is improved among those with stable weight or even weight gain. Body composition may change as a result of cancer prognosis and treatment. Previous studies have found that certain cancer treatment regiments cause an increase in fat mass and a reduction in BMD. Osteoporosis and low BMD are commonly known to occur in elderly people or postmenopausal women because of hormonal imbalances. Low BMD, however, can result from many other factors, such as poor nutrition, prolonged pharmacological intervention, disease, or decreased mobility. Because patients with cancer may experience

many of these factors, they are often predisposed to low BMD and increased risk for osteoporotic fractures in later life. Given the growing number of long-term cancer survivors and expectations of leading more fulfilling lives after cancer treatments, it is imperative to understand the risk factors for weight changes and bone loss, so effective prevention and treatment methods can be implemented.

Changes in Body Composition Associated With Chemotherapy for Breast Cancer

Weight gain is a common problem among women with breast cancer who receive adjuvant chemotherapy (ACT; Demark-Wahnefried et al. 1993, 1997b; Dixon et al. 1978; Ganz et al. 1987; Goodwin et al. 1999; Heasman et al. 1985). This change in weight associated with breast cancer treatment not only generates psychosocial stress in women but also affects cancer prognosis and survivorship.

Data from a prospective study (Demark-Wahnefried et al. 2001) among postmenopausal women with breast cancer treated with ACT and those who received only localized treatment indicated that both groups gained body weight in the first year after cancer diagnosis, but weight increase was greater in the group receiving ACT than in the localized treatment group (2.1 kg vs. 1.0 kg, p = .02). Despite the increased weight, a reduction in lean soft tissue mass among women receiving ACT was observed, suggesting that ACT-induced weight gain led patients to a high risk for sarcopenic obesity. Differential effects of ACT on various components of body composition also were reported by Aslani and the coauthors (1999) of a study among women 26 to 70 years of age with

Table 23.1 Current Research Evidence for Body Composition and Cancer Risk

	General obesity	Central fat distribution	High BMD

Breast cancer
  Premenopausal
  Postmenopausal
	−/?
	+
		+/?
		+/?
			?
			+

Prostate cancer	+/?	+/?	+/?
Endometrial cancer	+	+	+
Ovarian cancer	+/−/?	+/−/?	?
Colon cancer	+	+	+/?

Note. BMD = bone mineral density. Increased (+), reduced (−), or unclear (?) risk of cancer.

breast cancer. Using the in vivo neutron activation analysis technique, the authors found no change in total body nitrogen but a mean 0.79-kg increase (p = .003) in total body water and a 1.49-kg increase (p = .008) in fat mass among women during ACT (CMF-based: Cyclophosphamide, Methotrexate, and 5-Fluorouracil) for breast carcinoma, resulting in a net increase in body weight among these women (Aslani et al. 1999).

Weight gain may be avoided in certain ACT regimens as suggested by Kutynec and colleagues (1999). They reported that a chemotherapy (Adriamycin and cyclophosphamide) using fewer antineoplastic agents and less treatments than most chemotherapy protocols for breast cancer did not result in weight gain during the treatment (Kutynec et al. 1999).

In addition to experiencing changes in weight and soft tissue composition, breast cancer survivors may also have an increased risk for osteoporosis and osteopenia (Twiss et al. 2001). Studies indicated that young women who had premature menopause as a result of chemotherapy for breast cancer experienced bone loss and a higher risk for early development of osteoporosis, whereas women who maintained menses did not appear to be at risk for accelerated trabecular bone loss (Headley et al. 1998; Shapiro et al. 2001; Vehmanen et al. 2001).

Chemopreventions for breast cancer have significant impact on bone mass (Gluck and Maricic 2002) and body composition. Tamoxifen is an antiestrogenic drug that is widely used in the treatment of patients with breast cancer. Recently, there has been an increasing interest in using this drug both for benign breast disease and as a chemopreventive agent among women at high risk of breast cancer. In a retrospective study, it was found that tamoxifen use significantly increased percent body fat in women. This increase in percent body fat was probably related to the agonistic estrogenic effect of tamoxifen on body fat (Ali et al. 1998). In another study, results of CT scans for 32 women (with no history of breast cancer) on tamoxifen and 39 controls showed that tamoxifen users had more visceral adipose tissue and more liver fat than the control women (Nguyen et al. 2001). It is known that tamoxifen increases BMD (Resch et al. 1998; Yoneda et al. 2002) or stabilizes BMD in postmenopausal women with breast cancer (Barni et al. 1996; Kristensen et al. 1994; Saarto et al. 1997). The effects of tamoxifen on BMD may vary by genetic polymorphisms of the ER receptor. Results from an intervention study among postmenopausal women with breast cancer suggested that tamoxifen increased BMD

by reducing the bone turnover, and this bone-restoring effect of tamoxifen was more marked in ER-β 21 CA repeats allele carriers than noncarriers (Yoneda et al. 2002).

Antiresorptive agents may reduce the bone loss associated with chemotherapy-induced ovarian failure in premenopausal women (Saarto et al. 1997). Adjuvant clodronate treatment significantly reduced chemotherapy-induced bone loss (Vehmanen et al. 2001). However, clodronate may have limited effects in preventing bone loss among postmenopausal women with breast cancer after the termination of postmenopausal estrogen therapy. In a randomized control trial among postmenopausal women with breast cancer, women who had recently discontinued postmenopausal estrogen therapy experienced more rapid bone loss than women who had not undergone postmenopausal estrogen therapy. Neither 3-year antiestrogen therapy alone nor antiestrogen together with clodronate could totally prevent the bone loss related to postmenopausal estrogen therapy withdrawal, even though clodronate seemed to retard it (Saarto et al. 1997).

Changes in Body Composition Associated With Androgen Deprivation Therapy for Prostate Cancer

Prostate cancer is the most common malignancy among American men. The prevalence proportions based on 0 to 22 cumulative years from diagnosis were 1,084/10,000 for white men and 886/10,000 for black men in 1997 (Merrill 2001). With advances in early detection and treatment for prostate cancer, the 5-year relative survival rate (adjusted for other causes of death) has increased from a low of 64.0% for men newly diagnosed in 1973 to 92.9% for men newly diagnosed in 1990 (Anonymous 2002b). Even within the advanced carcinoma category, men with prostate cancer have a better absolute cumulative 5-year survival rate (20.8%) than people with breast cancer (17.0%), colorectal cancer (5.3%), or lung cancer (1.5%; Kato et al. 2001). Hence, more attention has been given to studying the impact of prostate cancer diagnosis and treatment on overall health and quality of life. Because prostate cancer is testosterone dependent, androgen deprivation therapy (ADT) with gonadotropin-releasing hormone agonists is often used in treating prostate cancer, which renders these men hypogonadal. Smith (2002) reviewed studies on bone loss and osteoporosis associated with ADT among men with prostate cancer. The accumulative evidence indicates that patients with prostate cancer without apparent bone metastases are at increased risk of osteoporosis and obesity as well as low lean soft tissue mass after ADT (Basaria et al. 2002; Chen et al. 2002b; Stoch et al. 2001).

In a case–control study by Chen and colleagues (2002b), men with prostate cancer who had been treated with ADT for 1 to 5 years were measured on DXA and compared with same aged and apparently healthy men who had no prostate cancer history. The results showed that men with prostate cancer had a two- to fivefold increased risk of being overweight or obese compared with controls. Percent body fat was 13% higher in men with cancer than in controls. The men with cancer had significantly lower values in lean soft tissue and in total body, lumbar spine, and hip BMD compared with controls.

Higher body fat and low BMD may already be present in men with prostate cancer even before ADT, whereas ADT significantly contributes to the increased risk of obesity and osteoporosis and reduction in lean soft tissue mass in these men (Berruti et al. 2002). It has been suggested that the adverse changes in body composition may be related to the reduced insulin sensitivity in patients with prostate cancer. Interestingly, the increased fatness primarily resulted from abdominal subcutaneous fat rather than intra-abdominal fat in men who were treated with ADT for prostate cancer (Smith et al. 2002).

Childhood Cancers and Body Composition

The increasing incidence of childhood cancers and improved survival have resulted in a growing number of childhood cancer survivors in the United States. This improved survival is in part attributable to the application of intensive, multimodality therapies that include radiotherapy, surgery, glucocorticoids, and cytotoxic agents. However, such interventions have the potential to induce complex hormonal, metabolic, and nutritional effects that may interfere with longitudinal bone growth and skeletal mass acquisition during the critical growth phases of childhood and adolescence. In a review, van Leeuwen and colleagues (2000) concluded that multiagent chemotherapy, corticosteroids, malignancy itself, cranial radiotherapy, and physical inactivity not only may reduce long bone growth, resulting in short body

height, but also may cause low peak bone mass, which leads to long-term consequences including increased risk for bone fractures in later life (van Leeuwen et al. 2000).

Low BMD is common among young adult survivors of childhood cancer (Nysom et al. 2001; Vassilopoulou-Sellin et al. 1999; Warner et al. 1999), but the reduction in BMD may be quite small and partially attributable to smaller bone sizes in the childhood cancer survivors (Nysom et al. 2001). Arikoski and colleagues (1999) studied longitudinal changes in BMD, bone turnover, and bone hormonal metabolism in newly diagnosed children (3-16 years) with cancer. The authors found significantly increased bone resorption and an impaired development of femoral bone density in children with cancer during chemotherapy (Arikoski et al. 1999). The results from another longitudinal study indicated that some children with cancer already had lower BMD at the time of diagnosis compared with normal children of the same age, and BMD may further decrease during cancer treatments (Henderson et al. 1998).

In addition to shorter adult status and low BMD, a reduction in muscle strength and increased prevalence of obesity have also been seen in survivors of childhood cancer (Nysom et al. 1998). Among the risk factors, cranial radiotherapy may have the most important impact on bone density and body composition (Henderson et al. 1998; Murray et al. 1999; Nysom et al. 1998; Vassilopoulou-Sellin, et al. 1999; Warner et al. 1997). Hesseling and colleagues (1998) assessed BMD of the lumbar spine using DXA among 97 long-term survivors of childhood cancer 5 to 23 years after cancer diagnosis. Their results indicated that increasing doses of cranial irradiation (18-54 Gy) were significantly associated with lower BMD in this population ($p = .001$).

Treatment Effects on Body Composition in Other Types of Cancers

Few studies have investigated the impact of cancer treatments on body composition in other types of cancers, and these studies are often limited by small sample sizes. Total or near-total thyroidectomy plus or minus radioiodine ablative therapy is the mainstay of current treatment for well-differentiated thyroid cancer. Supraphysiological amounts of oral thyroxine (T4) are also administered to suppress the serum thyroid-stimulating hormone to below detectable levels. It is well known that overt clinical hyperthyroidism may increase bone turnover and decrease BMD, but the effect of suppressive thyroxine treatment on BMD is less known. After reviewing current evidence on bone density among people with thyroid cancer who were treated with suppressive thyroxine, Quan and colleagues (2002) concluded that there is no significant change in BMD for premenopausal women or men after diagnosis and treatments for thyroid cancer, but the findings for postmenopausal women are controversial and remain unclear.

Gastric cancer is one of the most common causes of cancer death. The only treatment that leads to a cure in some patients is surgical operation. Liedman (1999) reviewed the impact of gastrectomy on food intake, body composition, and bone metabolism. Overall, the previous studies suggested that a substantial weight loss, amounting to about 10% of the preoperative weight, occurred during the early postoperative period, including a decrease in lean soft tissue. Liedman and colleagues (2000) conducted a small longitudinal study among 22 people with gastric cancer who were long-term survivors after total gastrectomy (mean of 8 years). Whole-body DXA scans were performed at a mean of 5 and 8 years after the operation. The results showed that, on average, patients lost 3.2 kg of their body weight ($p < .006$) with a corresponding loss of lean body mass ($p < .0001$). However, there was no difference in bone density from values seen in age- and sex-matched controls. There was a slight elevation of osteocalcin levels and a minor increase in parathyroid hormone levels. The impact of total

gastrectomy on calcium homeostasis and BMD seemed to be marginal in this study. Nevertheless, the authors argued that the observed close relationship between BMD and body weight suggested the pivotal importance of maintaining body weight after gastrectomy (Liedman et al. 2000).

There is very limited information on body composition among women with cervical cancer and ovarian cancer. A small case–control study in 40 women with cervical cancer who underwent radiotherapy and 40 matched controls found no significant difference in the BMD between the two groups and no significant change in BMD 1 to 7 years after therapy in the patient group (Chen et al. 2002a). In another small study, 15 women (mean age 38.2 ± 7.8 years, range 30-46 years) with ovarian cancer who had been treated with chemotherapy for six cycles every 4 weeks after surgical cytoreduction were measured for bone loss. A significant bone loss was observed ($p < .001$), and baseline lean mass predicted bone loss with anticancer chemotherapy (Douchi et al. 1997b).

Besides premature menopause induced by chemotherapy as a cause of bone loss in premenopausal women with breast cancer, very little is known about other mechanisms of chemotherapy-induced bone loss. Preliminary results suggest that chemotherapy without irradiation may have a dose-dependent toxicity to bone marrow stromal osteoprogenitors and can cause osteopenia by direct damage of the osteoblastic compartment (table 23.2; Banfi et al. 2001).

Table 23.2　Effects of Cancer Treatments on Body Composition

	Obesity or overweight	Low BMD	Low lean soft tissue mass	Major possible mechanisms
Breast cancer treatment	+	+	+/?	Premature menopause associated with adjuvant chemotherapy, stopping HT, and direct toxicity of chemotherapy to bone cells
Prostate cancer treatment	+	+	+	Hypogonadism induced by androgen deprivation therapy
Childhood cancer treatment	+/?	+/?	?	Multiagent chemotherapy, corticosteroids, malignancy itself, cranial radiotherapy, and reduced physical inactivity
Thyroid cancer treatment	?	+/?	?	Thyroidectomy plus or minus radioiodine ablative therapy
Gastric cancer treatment	–/?	+/?	+/?	Gastrectomy on food intake and nutrient absorption

Note. HT = hormone therapy; BMD = bone mineral density. Increased (+), reduced (–), or unclear (?) risk as a result of cancer treatment.

Cancer-Related Wasting

Cancer-related cachexia is caused by a diverse combination of accelerated protein breakdown and slowed protein synthesis, leading to a significant loss of body weight and fat-free mass (May et al. 2002). Tisdale (1999) pointed out that progressive weight loss is a common feature of many types of cancer, which may be the reason patients experience poor quality of life, poor response to chemotherapy, and shorter survival time than patients with comparable tumors without weight loss. The mechanisms of this wasting in cancer may be attributable to changes in proinflammatory cytokine levels and activities, such as tumor necrosis factor-α, interleukin-1, interleukin-6, interferon-γ, and ciliary neurotrophic factor. Simons and colleagues (1999a) suggested that weight loss and low body cell mass associated with lung cancer may result from systemic inflammation, decreased levels of IGF-I, or hypermetabolism.

Increasing numbers of nutritional approaches and pharmacological approaches for treating cancer cachexia are being tested or used. Body composition assessments have been used often to evaluate effects of these approaches, such as to assess nutritional status in children with cancer in tube feeding (den Broeder et al. 2000), to measure metabolic response to feeding in patients with pancreatic cancer who were losing weight (Barber et al. 2000), and to evaluate the effect of symptom control on body composition among cancer patients after gastrectomy (Liedman et al. 2001).

There are many challenges in assessing body composition among people with cancer, because changes in hydration levels may significantly affect the precision and accuracy of body composition measurements. Although there is a good agreement between changes in total body water assessed from BIA and deuterium dilution among some patients (Simons et al. 1999b), BIA is not reliable for body composition measurements in cancer patients with ascites (Sarhill et al. 2000). Metabolic measurements are usually adjusted to lean body mass to account for body composition differences, and body water has been used to estimate lean tissue mass in many circumstances. However, current evidence indicates that total body water significantly overestimates metabolically active tissue in patients with cancer who lose weight. Hence, using total body water as the basis for metabolic requirements in this group of patients is challenged (McMillan et al. 2000).

Other Applications of Body Composition Assessments in Cancer Research

The use of body surface area in determining chemotherapy dosing, particularly in obese people, remains controversial. Finding a better clinical tool for assessing risk of chemotherapy-induced toxicity is important when widely distributed drug combinations such as CMF are used. Recently, total body nitrogen measurement among patients with serious illness has been suggested to be an accurate predictor of clinical course. Nitrogen index assessed using the in vivo neutron capture analysis technique could be a very useful tool in identifying women with breast cancer who are at higher risk of chemotherapy-induced toxicity with a standard CMF-based regiment (Aslani et al. 2000).

BMD evaluation by DXA may be a reliable tool in assessing the response of bone metastases to treatment (Berruti et al. 2000). It has been suggested that it is technically feasible to use DXA to prospectively monitor bone changes in skeletal metastases in patients with breast cancer receiving systemic therapy. It was found that BMD of skeletal metastases increased in patients responding to treatment, and the DXA-derived measurement was significantly correlated with the changes imaged on skeletal X rays and CTs (Shapiro et al. 1999).

Body composition may also serve as a valuable indicator for prognosis and survival in certain cancers. Toso and colleagues (2000) found that weight loss was not significantly associated with survival among patients with lung cancer. However, altered tissue electric properties measured by BIA were more predictive than weight loss for prognosis. Patients with lung cancer were characterized by a reduced reactance component. Patients with a phase angle smaller than 4.5° had a significantly shorter survival time (Toso et al. 2000). In another study, Kadar and colleagues (2000) used in vivo neutron activation to measure body protein among patients with lung cancer (17 with non-small-cell lung cancer and 4 with small-cell lung cancer) who had received radiation treatment. The results from this study indicated that increased body protein was associated with better survivorship and longer recurrence-free survival (Kadar et al. 2000).

There are inconsistent results regarding the relationship between weight and fat distribution in women with breast cancer and their cancer prognosis and survivorship. A study in the Netherlands found no significant differences in survival time between heavier (for height; BMI ≥26) and lighter patients (BMI <26). There was also no association between central fat distribution, measured by the ratio of subscapular and triceps skinfold thickness, and survival time in women with breast cancer after stratification by axillary node status, estrogen receptor status, and way of detection (den Tonkelaar et al. 1995). However, in the Iowa Women's Health Study (698 postmenopausal women with unilateral breast cancer), women in the highest tertile of BMI had a 1.9-fold higher risk (95% CI = 1.0-3.7) of dying after breast cancer diagnosis than those in the lowest tertile, after adjustment for other prognostic variables (age, smoking, education level, extent of breast cancer, and tumor size; Zhang et al. 1995). Kumar and colleagues (2000) suggested that among women with breast carcinoma, android body fat distribution as indicated by a higher suprailiac–thigh ratio was a statistically significant ($p < .0001$) predictor of poor survivorship during up to 10 years of follow-up. In contrast, Schapira and colleagues (1991b) reported that premenopausal and postmenopausal women with upper-body fat distribution appeared to be a subset of women with a more favorable prognosis as measured by less lymph node involvement, smaller tumors, and higher levels of ER in their tumors. Clearly, more studies are needed to understand the impact of weight and fat distribution on cancer prognosis among breast cancer survivors.

Summary

There is convincing evidence supporting a relationship between obesity and increased risk of certain cancers, such as postmenopausal breast cancer and colon cancer. However, the evidence for other cancers, such as prostate and ovarian cancers, is less consistent. Some of these inconsistencies are partially attributable to the limited number of studies, discrepancies in study designs, small sample sizes, unadjusted confounding factors, and measurement errors or recall bias for obesity assessments. Furthermore, the interactions of obesity with other risk factors and health conditions, such as diabetes, in the risk of cancer may also contribute to the different study results at certain cancer sites. Obesity may serve as a surrogate marker of inactivity, increased fat intake, and reduced consumption of fruits and vegetables as well as other unhealthy lifestyle behaviors. Avoiding weight gain is one of the most important strategies for cancer prevention.

The association between increased BMD and increased risk of certain cancers is interesting and may reflect shared genetic and hormonal factors in the development of certain cancers and the acquisition and maintenance of bone mass. More research is needed to elucidate these factors. This association between BMD and cancer risk supports the need to understand potential detrimental effects on BMD when certain genetic and hormonal factors are altered to prevent cancer or to treat cancer. Indeed, cancer treatments such as androgen depletion therapy for prostate cancer and cranial radiotherapy for childhood cancer cause low BMD as well as changes in body soft tissue composition. It is essential to monitor body composition during and after cancer treatment so adverse changes in body composition can be identified, prevented, and treated.

Cancer is a complex disease and may be linked with obesity through various biological pathways, including insulin resistance, hyperinsulinemia, increased circulating levels of leptin, increased serum level of growth factors, increased sex hormone secretion, and reduced hormonal binding proteins. Fat cells may behave like endocrine cells that constantly produce and secrete a wide variety of hormonal factors. These factors under certain conditions may initiate cancers and cause them to grow. More studies are needed to delineate the relationship between body composition and cancer prognosis as well as the underlying mechanisms.

The association between obesity and cancer risk may differ by age, because of the differences in the relative contribution of obesity to the total level of endogenous hormones. In certain cases, general obesity and fat distribution may reflect different hormone profiles.

A majority of studies have used BMI or weight and WHR as proxies for obesity and fat distribution. Measurement errors in these methods may limit our ability to detect the association of obesity or fat distribution with cancer. Abdominal obesity can be more reliably demonstrated by imaging techniques such as CT and magnetic resonance imaging than by anthropometric measurements such as WHR. These advanced techniques for assessing body composition are available throughout the world, but high costs and radiation exposures may limit their use. More applications of DXA in cancer research are expected in the near future. DXA can be used to assess body composition in people of all age groups with excellent precision and accuracy, relative low cost, and a very small amount of radiation. DXA is easy to perform and requires minimal cooperation from the subjects during the scan, making it ideal not only for understanding cancer etiology but also for clinical monitoring of body composition changes with cancer treatments and cancer prognosis in different age groups. As cancer becomes a more prevalent and chronic condition, body composition measurement will have an increasing contribution in evaluating the effects of nutritional status and treatment on the overall health and quality of life among millions of cancer survivors.

Chapter 24

Obesity and Diabetes: Body Composition Determinants of Insulin Resistance

Bret H. Goodpaster and David E. Kelley

Obesity and type 2 diabetes mellitus (DM) are metabolic disorders that have strong associations with body composition. Although it is certainly clear that obesity, attributable to increased adiposity, can be considered a disorder of body composition, a similar perspective applies to type 2 DM and not only because the majority of individuals with type 2 DM are overweight or obese. The metabolic perturbations that characterize type 2 DM, those of hyperglycemia, impaired postprandial insulin secretion, elevated fasting insulin, increased fatty acids and triglycerides, and, most fundamentally, insulin resistance (IR), have been found to be associated with adiposity and perhaps, more particularly, with certain patterns of adipose tissue and fat distribution. Thus, during the past 2 decades, there has been renewed interest in examining the functional associations between body composition and the pathophysiology of type 2 DM.

One focus has been upon examining associations between amounts and distributions of adipose tissue and IR, and there is an increasing interest in a compositional analysis of other tissues or organ systems, such as skeletal muscle and liver, for which a higher fat content has been found in type 2 DM and obesity.

Prevalences of type 2 DM and obesity are increasing at alarming rates in the developed and developing areas of the world (Harris et al. 1998; Mokdad et al. 2000). It has been estimated that there are approximately 100 million individuals

with type 2 DM, and within a generation this will more than double to an estimated 250 million people by the year 2020 (O'Rahilly 1997). Strong epidemiological data support a pivotal role for the "obesity epidemic" driving the concomitant diabetes epidemic (Harris et al. 1998). IR is a principal mechanism by which obesity is considered to heighten risk for type 2 DM. In obesity, there is IR in adipose tissue and muscle (Lillioja et al. 1987) as well as in the liver (Felig et al. 1974); these impairments are strongly manifested in type 2 DM. As a consequence, in many overweight individuals, and among nearly all overweight and obese subjects who develop type 2 diabetes, there is impaired insulin inhibition of splanchnic glucose output, impaired insulin suppression of lipolysis (Reaven 1988), and a decrease in total body glucose disposal, attributable primarily to diminished utilization by skeletal muscle (Caro et al. 1989). IR can be present many years before the onset of impaired glucose tolerance and type 2 DM (Lillioja et al. 1988; Warram et al. 1990).

Body Mass Index, Waist Circumference, and IR in Type 2 DM and Obesity

The most fundamental change in body composition associated with obesity is increased weight.

Overweight and obesity, defined by body mass index (BMI), are risk factors for development of type 2 DM. For overweight individuals, BMI (weight adjusted for height by the formula kg/m^2) is an accurate surrogate marker for adiposity (Gallagher et al. 1996; Jackson et al. 2002). Current clinical guidelines strongly recommend measurement of BMI and waist circumference. Measurement of waist circumference is a cost-effective means to assess aspects of fat distribution. The National Institutes of Health and the World Health Organization have classified a waist circumference of greater than 88 cm (35 in.) in women and 102 cm (40 in.) in men as a risk factor for type 2 DM, hypertension, and cardiovascular disease. There is a curvilinear relationship between BMI and IR (Bogardus et al. 1985). Below a BMI of approximately 35 kg/m^2, there is a negative linear association between insulin sensitivity and BMI (Bogardus et al. 1985; Campbell and Carlson 1993; Carey et al. 1996; Evan et al. 1984; Goodpaster et al. 1997; Krotkiewski et al. 1990). At more severe levels of obesity, the relationship between IR and BMI begins to plateau. There have been fewer studies concerning BMI and severity of IR in individuals with type 2 DM, at least in comparison with the number of studies that have been conducted in obesity not complicated by type 2 DM. One of the earlier studies, by Campbell and Carlson (1993), found that a BMI of 27 kg/m^2 was an approximate inflection point that marked more severe IR in type 2 DM. As well, a number of studies found only modest decreases in insulin action among "lean" patients with type 2 DM (Hollenbeck et al. 1984; Kelley et al. 1993a; Ludvik et al. 1995).

In normal weight or slightly overweight individuals, BMI correlates less well with body adiposity. For example, in young men, modest elevations of BMI can represent increased muscle mass rather than increased adiposity. Yet it is also important to consider the opposite—the "metabolically obese, nonobese" individuals who, despite having normal values for BMI, have relative increases in adiposity and perhaps, in particular, increases in visceral adipose tissue (Ruderman et al. 1981). This concept appears to be particularly relevant in the elderly, who can manifest considerable adiposity and adverse patterns of adipose tissue distribution despite no or only modest elevations above normal BMI (Goodpaster et al. 2003). Also, in certain ethnic groups, such as Asian Indians, a metabolically obese phenotype can be detected despite normal or only modestly increased values for BMI (Banerji et al. 1999).

Another consideration in the clinical evaluation of patients is evaluation of patterns of weight gain. A sustained positive energy balance—the process that leads to increased body weight—induces IR (Bogardus et al. 1985; Evan et al. 1984; Mott et al. 1986; Olefsky 1976; Yki-Järvinen and Koivisto 1983). It has been postulated that IR might be a homeostatic adaptation to positive energy balance and that IR may thereby lessen the risk for further weight gain (Eckel 1992; Travers et al. 2002). Weight change (gain vs. loss or stabilization) has been found to influence risk for development of type 2 DM in both men and women (Colditz et al. 1995). An opportune time of intervention to prevent obesity-related disease is early in the sequence of unwanted weight gain. Recent studies point to the fact that obesity in adulthood is strongly shaped by BMI at age 20, and as well by race and ethnicity, with higher rates of weight gain among minority populations during early adulthood (McTigue et al. 2002). In summary, BMI, waist circumference, and weight gain are simple and clinically useful measures for assessing body composition.

Fat-Free Mass and IR in Type 2 DM and Obesity

Because skeletal muscle is the major tissue accounting for insulin-stimulated glucose utilization, it might be considered whether the amount of fat-free mass, more specifically skeletal muscle, is a determinant of IR. Such a consideration is germane to aging, because there are increases in prevalence of both type 2 DM and sarcopenia (loss of muscle mass) with aging (Evans and Campbell 1993; Fink et al. 1986; Harris et al. 1998; Resnick et al. 2000). In several investigations of older adults, including large population studies, neither the amount of fat-free mass nor the amount of skeletal muscle was associated with IR or type 2 DM (Goodpaster et al. 2003; Kohrt and Holloszy 1995). Obese individuals are insulin resistant after adjustment for fat-free mass (Goodpaster et al. 1997). However, as will be developed later in this chapter, the quality rather than the quantity of fat-free mass has been found to be a very important body composition consideration in IR, obesity, and type 2 DM.

Abdominal Adiposity: Visceral and Subcutaneous Adipose Tissue in Type 2 DM

During the past 2 decades, perhaps the single most important advance in linking body composition research to a more complete understanding of type 2 DM and of IR in general has been attained through measurements of visceral adipose tissue (VAT). Technologies such as computed tomography (CT) and magnetic resonance imaging (MRI) for body composition analysis have enabled measurements of subdivisions of abdominal adiposity (into visceral and subcutaneous compartments) and subdivisions of subcutaneous abdominal adiposity (Ferland et al. 1989; Ross et al. 1996). Because these methods are now so widely used, it is impossible to cite in this chapter the enormous wealth of this information, and instead, for the sake of brevity and with apologies to those whose work is not cited, we will only cite a relatively small portion of this large body of relevant literature.

As seen in the CT scans in figure 24.1, VAT refers to adipose tissue located within the abdominal cavity, below abdominal muscles, and is composed of omental and mesenteric adipose tissue as well as retroperitoneal and perinephric adipose tissue. VAT is recognized as a depot strongly associated with insulin resistance of skeletal muscle, dyslipidemia, and increased risks for hypertension and glucose intolerance (Carey et al. 1996; Colberg et al. 1995; Despres 1998). Even in patients with type 2 DM, in whom IR is characteristically quite severe, variations in the amounts of VAT have been found

to account for 25% to 50% of the variability in the severity of IR and to be a more robust correlate than total fat mass (Kelley et al. 2001). Banerji and associates (1995, 1997) observed that variance in VAT accounted for much of the interindividual variation in IR among a cohort of African American individuals with type 2 DM.

The association between VAT and IR is also observed in those without type 2 DM and exists with general consistency across sex, race, age, and a variety of other subgroups of the population (Albu et al. 1997; Banerji et al. 1995, 1997; Colberg et al. 1995; Goodpaster et al. 1997; Kelley et al. 2001; Kissebah and Peiris 1989; Ross et al. 1996). Interestingly, this relationship is also observed among normal-weight men and women. When BMI criteria were used to appraise obesity in a large-scale cohort of more than 3,000 men and women in the Health, Aging and Body Composition (Health ABC) study (Goodpaster et al. 2003), VAT was found to be a strong predictor of IR and type 2 DM (Goodpaster et al. 2003), particularly among nonobese people. This suggests that older adults can be at greater risk for the development of type 2 DM if they have excess VAT.

VAT is a strong predictor of IR and type 2 DM in children and adolescents (Cruz et al. 2002; Huang et al. 2002), a segment of the population also becoming more obese (Anonymous 2000; Arslanian 2000; Johnson et al. 2001). African Americans have been reported to have less VAT than white persons for a given degree of IR, although the prevalence of type 2 DM is generally higher in African Americans compared with whites (Albu et al. 1997; Goodpaster et al. 2003). In adults without type 2 DM, men generally have approximately twofold more VAT

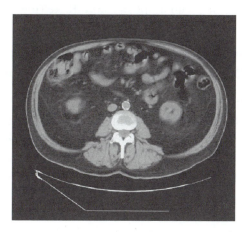 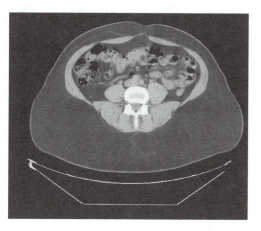

Figure 24.1 Cross-sectional abdominal computed tomography images obtained in two obese individuals. Although both persons have similar total abdominal fat, the individual on the left has substantially more visceral abdominal adipose tissue and thereby much less subcutaneous abdominal adipose tissue.

than women despite having a similar degree of IR (Goodpaster et al. 1997, 2003). These reports illustrate that factors other than VAT are involved in the etiology of IR and type 2 DM.

Typically, VAT comprises a minority of abdominal adipose tissue, generally accounting for approximately 25% to 50% of abdominal adiposity, and, collectively, the volume of VAT is generally considered to account for approximately 10% of the systemic mass of adipose tissue. Therefore, it is puzzling why or how this relatively small depot of adipose tissue might correlate so strongly with the IR syndrome. Certainly one consideration is that VAT (or at least a portion of VAT) releases fatty acids into the portal circulation; a consideration termed the "portal hypothesis." Fatty acid flux to the liver affects insulin sensitivity of the liver, stimulating synthesis and secretion of triglyceride-rich, very low density lipoproteins and reducing hepatic extraction of insulin. It is also recognized that adipocytes within VAT have higher rates of lipolysis than those within subcutaneous depots and are more resistant to insulin suppression of lipolysis (Montague and O'Rahilly 2000). Another possibility is that adipocytes within VAT are especially active in the release of cytokines and other adipocyte-derived hormones (so-called adipokines) that induce IR. There are excellent reviews in this area (Hotamisligil and Spiegelman 1994), and space considerations limit a fuller consideration, but this remains a fertile area of investigation. For example, recent studies also suggest differential effects on cortisol metabolism by VAT compared with peripheral adipose tissue and a potential link of this depot to risk for obesity and IR. Furthermore, it is noteworthy that in animal studies, surgical removal of VAT promptly diminishes IR even in the context of sustained adiposity (Gabriely et al. 2002).

Some have questioned the widely held concept that VAT has a stronger association with IR than does subcutaneous abdominal adipose tissue (SAT; Abate et al. 1995; Goodpaster et al. 1997). Abate and colleagues (Abate et al. 1995, 1996; Goodpaster et al. 1997) found that SAT was more strongly correlated with insulin resistance than VAT. Findings from our laboratory also indicated this (Goodpaster et al. 1997), although with inconsistent findings across various study cohorts (Goodpaster et al. 1999). There are several potential explanations for these differences. One of these pertains to the recognition that within SAT there are subdivisions of adipose tissue that may differ with relation to IR (Kelley et al. 2000; Ross et al. 2002). There is a well-recognized fascial plane that separates SAT

into superficial and deeper layers, and there are known histological differences in the structure of the adipose tissue at these two sites. In general, the deeper layer of SAT has an association with IR that bears resemblance to the pattern observed for VAT, whereas the superficial layer has a weaker association in some (Kelley et al. 2000) but not all (Ross et al. 2002) studies. In this regard, it is worth noting that matched for similar BMI, women have greater amounts of superficial SAT than do men and greater amounts of SAT in the lower extremities despite having less VAT than men. Miyazaki and colleagues (2002) found that the deeper layer of SAT is related to IR in those with type 2 DM. However, our laboratory has not found a robust relationship between the subdivision of deep SAT and IR in type 2 DM (Kelley et al. 2001).

We would like to speculate about a potential explanation for contemporary controversies regarding associations of SAT and VAT with IR in obesity and IR, and yet we also would like to acknowledge that these postulates are indeed speculative at this stage. A principal limitation in clinical investigation of the associations between regional adiposity and IR is that current body composition technologies, including CT and MRI, measure adipose tissue mass or, more precisely, its cross-sectional area or volume. These technologies do not measure adipose tissue metabolism. Hence, for adipose tissue mass to have meaning, there is an implicit assumption that metabolism is essentially the same within different individuals, and certainly this assumption is flawed. With regard to influences upon IR, adipose tissue metabolism (e.g., rates of fatty acid release) as well as endocrine and cytokine secretions likely mediates the link between adiposity and IR. Yet measures of mass or volume alone may not in all instances accurately reflect these qualities, and this may be a limitation particularly with respect to SAT. One of the reasons VAT may, in most instances, correlate well with IR is that its metabolic profile remains fairly consistent across individuals and, therefore, its impact upon systemic metabolism is closely approximated by determinations of the amount of VAT. On the other hand, we postulate that the situation with SAT and its metabolism is quite variable. For example, risk of development for type 2 DM is related to fat cell size within SAT (Weyer et al. 2000); larger, more insulin-resistant adipocytes confer increased risk for type 2 DM. Thus, to the extent that this reasoning is correct, we postulate that two individuals might have similar amounts of SAT yet quite different metabolic profiles for SAT, one profile that leads to IR and yet another, of nearly similar mass, that continues to provide protection against IR. Therefore, SAT mass (or volume) would be an imprecise surrogate for the metabolic potentials of differing depots of SAT and, consequently, in a cross-sectional study, a relatively imprecise correlate of systemic IR.

SAT may protect some individuals and in other individuals increase the risk of IR and type 2 DM. One striking example of the important role that SAT may have in serving as a buffer between periods of positive energy balance and the development of IR is lipodystrophy. The lipodystrophy syndromes entail diminished peripheral adiposity and severe insulin resistance (Reitman et al. 2000). There is an associated increase in VAT and increased fat storage within other organs such as skeletal muscle. An animal model clearly illustrates this phenomenon. Transplantation of subcutaneous adipose tissue to subcutaneous fat-deficient lipodystrophic mice reversed their severe IR (Gavrilova et al. 2000). As previously mentioned, lipolysis in peripheral adipose tissue is more readily suppressed than in VAT and perhaps has a less active role in secreting adipokines that induce insulin resistance (although data in this regard are incomplete). Perhaps we can speculate that peripheral adipocytes provide, from a metabolic perspective, a relatively "safe haven" for the sequestration of excess calories. Therefore, if an individual has a limited capacity for storing energy in SAT or has saturated this depot, then continued accretion of adipose tissue might preferentially expand central depots.

The great majority of patients with type 2 DM do not have lipodystrophy per se, and indeed most obese and overweight patients with type 2 DM have an objectively increased amount of SAT. Yet, perhaps SAT in type 2 DM has a functional compromise in the capacity to buffer positive energy balance. Functional studies of body composition in obesity and type 2 DM would be greatly bolstered by innovative methods that put together both aspects of volume and metabolic function; one example is the innovative use of positron emission tomography (PET) imaging. Recent PET imaging studies indicate that SAT in obese people is itself more insulin resistant than SAT in lean individuals, yet much more work is needed in this area to fully test these hypotheses.

One observation we have recently made is that the ratio of VAT to SAT may differ in type 2 DM and nondiabetic individuals of similar overall obesity. Figure 24.2 illustrates that patients with type 2 DM have significantly more VAT while having

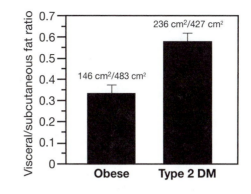

Figure 24.2 Ratio of visceral adipose tissue (VAT) and subcutaneous abdominal adipose tissue (SAT) in obese subjects without diabetes ($n = 54$) and obese subjects with type 2 diabetes mellitus ($n = 83$). Those with type 2 diabetes mellitus (DM) have significantly more ($p < .05$) VAT, less ($p < .05$) SAT, and a higher ($p < .05$) VAT–SAT ratio.

slightly less SAT, resulting in dramatically different VAT–SAT ratios. Indeed, these data support the notion that there is not only more VAT but also a relatively limited storage capacity of SAT, and that together, these alterations of body composition dispose an individual to IR.

Treatment with peroxisome proliferator-activated receptor (PPAR)-γ agonists, agents that increase insulin sensitivity despite inducing weight gain, is associated with adipose tissue "remodeling" in patients with type 2 DM. Treatment with PPAR-γ agonists, while inducing an overall increase in fat mass, appears to preferentially increase SAT while concomitantly reducing VAT, again despite an overall weight gain. Thus, despite substantial progress in examining the relationship between abdominal fat distribution and IR, fascinating areas remain to be more fully explored to understand not only the relationship between the mass of abdominal adiposity and its distributions into VAT and subdivisions within SAT but also the capacity for energy storage and endocrinology of these adipose tissue depots in relationship to the pathogenesis of IR in obesity and type 2 DM.

There is at least one more consideration with regard to the frequent observation that VAT is correlated with reduced insulin-stimulated glucose utilization, a physiological phenotype that denotes skeletal muscle IR. Correlative relationships do not connote causal relationships, yet this is frequently inferred, and with the correlation between VAT and IR, it is commonly inferred that VAT causes muscle IR. Perhaps the opposite occurs. Work from our laboratory (Simoneau et al. 1999) and a number of other laboratories (Kriketos et al. 1996; Tanner et al. 2002) has identified that skeletal muscle in obesity and type 2 DM has lower oxidative enzyme activity and, as a result, a reduced capacity for lipid oxidation (Hulver et al. 2003; Kelley et al. 1999; Kim et al. 2000). Diminished oxidation of fatty acids in skeletal muscle disposes to fat accumulation in this tissue, as will be next discussed, but also disposes to systemic fat accumulation, because of the overall importance of skeletal muscle lipid utilization to systemic fat balance. Thus, VAT may correlate with IR because reduced capacity of skeletal muscle to oxidize fat disposes to accumulation of VAT. Cross-sectional and intervention studies provide some support for this concept, and although we would not propose that this is a complete accounting of the correlation between VAT and IR, it is a relationship that bears consideration, especially with regard to understanding the effects of physical activity upon body composition, because activity enhances capacity for fat oxidation in skeletal muscle and tends to protect against accumulation of VAT.

Lower-Extremity Adipose Tissue

Although abdominal adiposity has received the greatest scrutiny in regard to IR, there is a considerable proportion of fat mass in the legs. Subcutaneous fat mass in the lower extremities is as large as abdominal fat yet is regarded as having only a weak relationship to IR. The original and now classic article by Vague (1956), describing sex-related patterns of adipose tissue distribution in relationship to IR, characterized lower-extremity adiposity as "gynoid adiposity" and noted that it had less impact on manifestations of IR than did accumulation of truncal fat, which was characterized as "android adiposity." This correlative association seems entirely consistent with a role of subcutaneous adiposity serving as a buffer for positive energy balance, a principle discussed previously in relationship to lipodystrophy and animal models of lipodystrophy. Perhaps for these reasons, until recently little attention was given to thigh adiposity, but recent studies have addressed subdivisions of thigh adiposity. Data have begun to rapidly emerge that fat accumulation in and around skeletal muscle may be another aspect of adiposity that links body composition in type 2 DM and obesity with IR.

Exploring this aspect of body composition has provided exciting new clues to the pathogenesis of IR in obesity and type 2 DM.

Goodpaster and colleagues (2001) addressed the role of adipose tissue located beneath the muscle fascia. Subfascial adipose tissue, together with adipose tissue found interspersed between muscle, likely makes up contiguous depots of adipose tissue that we have termed "intermuscular adipose tissue" (IMAT). A representation of this pattern of adipose tissue distribution in the lower extremity is shown in figure 24.3.

Despite being only a small proportion of thigh adiposity, IMAT is significantly correlated with IR in middle-aged (Goodpaster et al. 2000b) and older (Goodpaster et al. 2003) adults. As with VAT, the responsible mechanisms for this association are uncertain. But these mechanisms could be a greater resistance to insulin suppression of lipolysis (Kelley et al. 2001) or greater and differing secretions of adipokines that induce IR, a concept that to our knowledge remains unproven.

Skeletal Muscle Fat Content

In obesity and in type 2 DM, there is an increased amount of lipid contained within skeletal muscle fibers (Goodpaster et al. 2000c; Hulver et al. 2003). A number of studies, including those using in vivo MR and CT imaging and nuclear magnetic

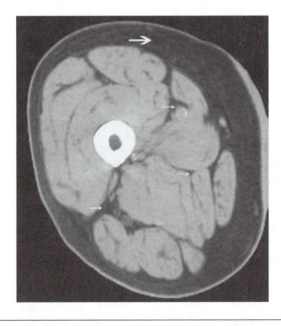

Figure 24.3 Representative CT image of the leg depicting subcutaneous (large arrow) and intermuscular (small arrows) adipose tissue.

resonance spectroscopy (MRS), as well as ex vivo analyses of muscle biopsy samples, have found not only that the content of triglyceride is increased in skeletal muscle in obesity and type 2 DM but that this is also correlated with the severity of IR (Goodpaster et al. 1997; Hulver et al. 2003; Pan et al. 1997; Perseghin et al. 1999; Phillips et al. 1996; Yu et al. 2002). In the case of CT, the Hounsfield attenuation values, mapped pixel by pixel, have been used to calculate the mean values within skeletal muscle and can be used as a surrogate determination of lipid content in skeletal muscle and other tissues, as discussed later in regard to liver. Adipose tissue and fat generate negative Hounsfield attenuation values on CT, and lower Hounsfield attenuation values for skeletal muscle have been shown to reflect higher muscle lipid content (Goodpaster et al. 2000a). The mean values for skeletal muscle Hounsfield attenuation differ between individuals (Brochu et al. 2001; Goodpaster et al. 1997; Kelley et al. 1991) and correlate with IR in obesity and type 2 DM (Goodpaster et al. 1997, 2000b, 2003). Moreover, weight loss is associated with an increase in Hounsfield attenuation values within skeletal muscle, which is consistent with a decrease in muscle fat content as also demonstrated in muscle biopsy samples (Goodpaster et al. 1999, 2000c).

MRS methods have recently been developed to distinguish intra- versus extramyocellular lipid in vivo (Boesch et al. 1997) in association with insulin resistance (Perseghin et al. 1999) and family history of diabetes (Jacob et al. 1999). Hydrogen with a molecular weight of one (^{1}H) occurs with an abundance close to 100%, but the strong signal from water limits the signal-to-noise ratio for detection of organic molecules. Nevertheless, ^{1}H-NMR spectroscopy can be used to determine intracellular triglyceride content, which can occur in millimolar concentrations, and has been widely applied in physiology research and reviewed elsewhere (Boesch et al. 1997; Szcepaniak et al. 1999). A characteristic spectrum of nuclei can be altered by the presence of surrounding nuclei and electrons of other tissues and chemicals, and this process can modify the magnetic spectrum of the probed nucleus producing a chemical shift; this chemical shift permits further distinction of a substance and, in the case of tissue triglycerides, facilitates differentiation of intracellular and extracellular triglyceride in skeletal muscle (Boesch et al. 1997). The signal intensity is proportional to the tissue concentration. Muscle triglyceride content has traditionally been quantified in biopsies and appears to be a determinant of skeletal muscle

insulin resistance (Goodpaster et al. 2000c; Pan et al. 1997; Phillips et al. 1996). ^{1}H-NMR spectroscopy goes one step further and enables multiple noninvasive assessments of intramyocellular triglyceride content in humans and has been applied as an investigative tool in diabetes. For example, in a study of 23 healthy volunteers, intramyocellular lipid content (IMCL) was measured by ^{1}H-NMR spectroscopy and compared with whole-body glucose disposal assessed by hyperinsulinemic–euglycemic clamp (Krssak et al. 1999). The IMCL content showed an inverse correlation to whole-body insulin sensitivity but not to body mass index, age, fasting plasma triglyceride, glucose, or free fatty acid concentrations. This inverse relationship between IMCL and whole-body insulin sensitivity also was confirmed in insulin-resistant individuals by NMR spectroscopy (Forouhi et al. 1999; Perseghin et al. 1999; Sinha et al. 2002).

It is important from a methodological perspective, and hence to interpret these data, to consider both the advantages and limitations of these various imaging methods used to quantify skeletal muscle fat content. Although MRS, MRI, and CT provide advantages over the more invasive muscle biopsy procedure, each has methodological limitations. Use of CT exposes individuals to ionizing radiation, and the data generated represent a marker of fat content of muscle but not a quantitative determination of fat content; these are relative disadvantages. Advantages are the wide availability of CT imaging, the relative simplicity and objectivity of image analysis (based on pixel-by-pixel Hounsfield attenuation coefficients, data that are the basis of image contrast in CT imaging), and the spatial clarity and detail that enable anatomical resolution of not only muscle attenuation characteristics but also adipose tissue distribution in and around skeletal muscle. The advantages of MRS include that it does not use ionizing radiation and that it offers the potential to distinguish IMCL. Disadvantages are that absolute quantification of lipid content within muscle using MRS remains problematic, and the method, at this point, is largely confined to a relatively few centers with the technology and expertise. However, even in the most capable hands, the MRS method for imaging IMCL distinct from extramyocellular lipid content is highly dependent on the orientation of muscle fibers within the magnetic field, and because of this constraint, fat content of only certain muscle groups, specifically tibialis anterior and soleus of the lower leg, can be measured reliably (Boesch et al. 1997). Fiber type orientation also may be altered or at least less consistent in obesity, further compounding separation of IMCL and extramyocellular lipid content peaks. Despite these drawbacks, lipid content can be quantified with MRS in other tissues such as liver (Petersen et al. 1996). Thus, MRS is a novel noninvasive method to quantify the content and distribution as well as potentially the type of fat within nonadipose tissue and organs. MRI methods have been developed very recently to quantify the content of lipid within muscle tissue (Schick et al. 2002). Standard T1-weighted MR images provide visualization of fatty septa in muscle (intermuscular or extramyocellular lipid). However, small concentrations of lipid within muscle tissue cannot be assessed with T1-weighted images. Fat-selective imaging MRI with a high sensitivity to lipid signals described by Schick and colleagues (2002) and Goodpaster and colleagues (2004) can produce high-quality spatial maps of lipid contained within muscle. MRI has a potential advantage over other modalities to quantify fat content within muscle (through the use of lipid standards placed within the field) and is not dependent on the orientation of the muscle fibers, nor is this quantification limited to certain muscle groups of the lower leg.

In the area of assessing skeletal muscle fat content in obesity and type 2 DM, studies of body composition have been carried past organ imaging to cellular imaging with microscopy (Goodpaster et al. 2000c; Gray et al. 2003; Greco et al. 2002; He et al. 2001) and by chemical extraction and analytical procedures. Done under local anesthesia, the percutaneous muscle biopsy procedure is commonly used in clinical investigations of obesity and type 2 DM, and this procedure typically yields between 100 and 150 mg of muscle. Higher amounts of triglycerides within muscle, determined biochemically (Pan et al. 1997) and histologically (Gray et al. 2003; Goodpaster et al. 2000c; Phillips et al. 1996), have been reported to be associated with IR. Perhaps more important, direct examination of tissue has been useful in exploring potential mechanisms that might link tissue fat content with perturbations of metabolic pathways, signal transduction, and gene expression related to IR in obesity and type 2 DM. As one example of this type of inquiry, our laboratory has been exploring the relationship between muscle fat content and enzyme activities in pathways of fatty acid catabolism, and this has yielded information on the association between increased fat content in muscle in obesity and type 2 DM and diminished oxidative enzyme activities and mitochondrial

dysfunction (Goodpaster et al. 1999, 2000c, 2001; He et al. 2001; Simoneau and Kelley 1997).

However, as there are important methodological limitations to the imaging methods of CT and MRS, there are limitations to the use of biopsy material. First, the method is of course invasive. The interspersion of adipose tissue around skeletal muscle complicates interpretation of lipid content because inclusion of even small amounts of residual adipose tissue could greatly influence content of extracted lipid, thus potentially leading to a false impression of muscle fat content. Another consideration, one that is not actually methodological, is that although muscle fat content is increased in obesity and type 2 DM and generally is associated with IR, skeletal muscle fat content can also be elevated in lean, physically active, and highly insulin-sensitive individuals. In the latter cases, muscle fat content provides a substrate for energy production during physical activity, and this is associated with a high level of oxidative enzyme activity and robust pathways for fatty acid catabolism (Coggan et al. 1992; Sial et al. 1996). Quite recent studies from our laboratory suggest that size and other morphological characteristics (and likely biochemical characteristics) of lipid droplets differ in individuals with obesity and type 2 DM compared with those in muscle of trained athletes, and morphology changes in obesity and type 2 DM after activity and weight loss intervention. Thus, these are areas of investigation in which the initial set of answers is generating intriguing and, as yet, unanswered questions concerning the relationship of muscle fat content to IR.

Content of triglyceride in muscle may be a marker of IR in certain physiological contexts, such as those associated with reduced capacity for fatty acid catabolism, but even in these conditions it is not clear that triglyceride content per se induces IR. Perhaps it is more accurate to state that current data indicate that it is unlikely that triglyceride induces IR. More likely culprits include long-chain acyl coenzyme A, diacylglycerol, ceramide, or other lipid moieties, the content of which may also be increased in obesity and type 2 DM and for which a more direct link to generation of IR has been formulated (Schmitz-Peiffer et al. 1999; Yu et al. 2002). These lipid metabolites appear to induce increased serine phosphorylation on insulin receptor substrate-1 (IRS-1) and other components of the insulin signaling cascade, thereby interfering with phosphorylation on tyrosine and consequently interfering with propagation of insulin signaling. Also implicated in lipid-induced insulin resistance is activation of inflammatory pathways (Kim et al. 2001).

The elegance of this body of work concerning skeletal muscle fat content and fat composition is that it begins to reveal at a cellular and molecular physiology level how alterations of body composition in obesity and type 2 DM may have direct causative roles in generating IR. This truly can be interpreted as functional ramification of body composition and can reveal, as do the new data on adipose tissue endocrine functioning, that the study of body composition is undergoing a renaissance that places it near the forefront of physiological research into metabolic disorders such as obesity and type 2 DM.

Hepatic Steatosis and IR in Obesity and Type 2 DM

A majority of overweight and obese individuals with type 2 DM have an elevated content of fat within the liver, which is variously termed hepatic steatosis, fatty liver of diabetes, and, if complicated by inflammation, nonalcoholic steatohepatitis. Ultrasound, CT, and MRI can all detect fat in the liver (figure 24.4); however, liver biopsy remains the only reliable means to identify inflammation or fibrosis (Falck-Ytter et al. 2001).

A recent study from our laboratory illustrates the high prevalence of fatty liver in type 2 DM; among a group of 83 overweight individuals with this disorder, nearly two thirds manifested fatty liver regardless of hepatic function (Kelley et al. 2003). It has been estimated that the prevalence of fatty liver may be 75% or greater in those with a BMI above 30 kg/m^2 and type 2 DM (Chitturi and Farrell 2001). In some individuals with hepatic steatosis, hepatic inflammation (hepatic steatitis) or fibrosis and even end-stage liver disease may develop. An excellent recent review addressed this issue, which is beyond the scope of this chapter (Clark and Diehl 2002). However, from the perspective of IR, the presence of fatty liver even in the absence of clinically evident liver disease appears to have significance. In type 2 DM, the fat content of liver correlates with IR in insulin suppression of endogenous glucose production; the same study also found that fatty liver was a strong predictor of the amount of daily insulin needed to achieve glycemic control, which accentuates the clinical as well as physiological importance of this body

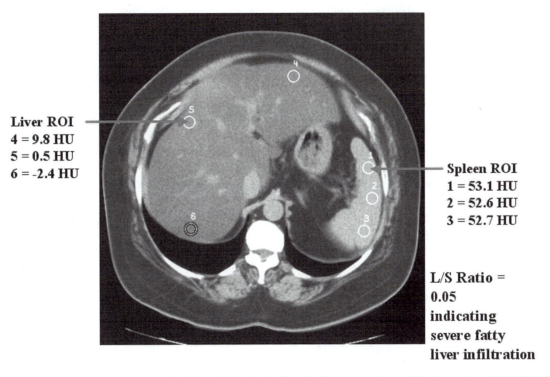

Liver ROI
4 = 9.8 HU
5 = 0.5 HU
6 = -2.4 HU

Spleen ROI
1 = 53.1 HU
2 = 52.6 HU
3 = 52.7 HU

L/S Ratio = 0.05 indicating severe fatty liver infiltration

Figure 24.4 Liver (L) and spleen (S) computed tomography with regions of interest (ROI). L/S ratio = mean Hounsfield unit (HU) of liver ROI ÷ mean HU of spleen ROI. The L/S ratio is 0.05, indicating severe fatty liver infiltration.

Reprinted, by permission, from D. Kelley et al., 2003, "Fatty liver in type 2 diabetes mellitus: relation to regional adiposity, fatty acids, and insulin resistance," *American Journal of Physiology—Endocrinology and Metabolism* 285(4): E906-E916.

composition finding (Ryysy et al. 2000). Fatty liver also has been found to correlate with hepatic IR in those without type 2 DM (Seppala-Lindroos et al. 2002).

Fatty liver in type 2 DM has been found to correlate directly with skeletal muscle IR (Marchesini et al. 1999). The reasons for this remain to be delineated but there are some intriguing potential explanations, chief among which is elevated levels of plasma fatty acids (Day and Saksena 2002). Fatty liver is associated with elevated levels of plasma fatty acids (Kelley et al. 2003). This can be observed during fasting conditions but is even more evident during conditions of insulin infusion, which of course act normally to suppress lipolysis and, hence, plasma fatty acids. Fatty acids are known to induce IR in skeletal muscle (Boden et al. 1994, 2001). Thus, elevated fatty acids may provide excessive substrate for liver, leading to fatty liver, and concomitantly may induce IR in skeletal muscle. Quite possibly, the mechanisms are more complex; perhaps there is reduced clearance of fatty acids by a liver already congested with increased fat content. We have noted that the presence and severity of fatty liver are strongly related to VAT (Kelley et al. 2003). In many respects, one might regard fatty liver as a third element within truncal obesity, adding to its other manifestations of increased VAT and increased plasma FFA. In our studies, the presence of fatty liver in type 2 DM does not necessarily predict the presence or severity of increased fat content within skeletal muscle. Although both phenomena likely share association with plasma fatty acid levels, the lack of concordance between fatty liver and fatty muscle suggests that within each organ, patterns of nutrient partitioning into oxidative and storage pathways may play key roles in the pathogenesis of fat accumulation. It is likely therefore that fuller understanding of the pathogenesis of fatty liver and its link to IR in obesity and type 2 DM will require a line of inquiry similar to that outlined previously in regard to skeletal muscle biochemistry, molecular physiology, and signal transduction.

With respect to type 2 DM, attention also focuses, perhaps unduly, on the relation between IR and alterations of glucose homeostasis; if instead this disorder is considered from the perspective of cardiovascular risk, the greater significance of fatty liver may pertain to its association with the dyslipidemia of IR. Fatty liver in obesity and type 2 DM is associated with increased triglyceride and decreased high-density lipoprotein, the two hallmark lipoprotein abnormalities of dyslipidemia. In our studies, the severity of dyslipidemia in type 2 DM was directly associated with the presence of fatty liver (Kelley et al. 2003).

Effects of Weight Loss on Body Composition and IR

It would not be unreasonable to conclude from the correlations between BMI and risk for type 2 DM, and between adiposity and severity of IR, that substantial amounts of weight loss might be needed to treat or prevent type 2 DM. Fortunately, the data are in fact considerably more encouraging. With regard to those who already have type 2 DM, modest weight loss of 5% to 10% can have a pronounced effect in improving hyperglycemia and reducing other cardiovascular risk factors (Kelley et al. 1993b; Sjostrom et al. 2000; Williams and Kelley 2000). Similar amounts of weight loss substantially reduce the risk for development of type 2 DM (Long et al. 1994). IR and hyperglycemia can begin to improve rapidly once a patient with type 2 DM begins weight loss; indeed, it seems that these improvements almost precede weight loss (Kelley et al. 1993b; Wing et al. 1994).

These observations emphasize that the process of negative energy balance has beneficial effects on IR and hyperglycemia, as well as hypertriglyceridemia, which become evident almost at the outset of negative caloric balance. The fact that negative energy balance can rapidly ameliorate IR and reduce hyperglycemia demonstrates the potential reversibility of IR. Having made these points, we must also acknowledge that substantial weight loss can have a profound effect on type 2 DM, as has been revealed by the impact of bariatric surgery to reverse type 2 DM. Sustained weight reduction over several years as occurs after gastric bypass or gastric banding procedures leads to reversal, or at least near resolution, of type 2 DM in a high percentage of individuals (Pories et al. 1995). Major weight loss as attained with bariatric surgery also has been found to prevent (or at least substantially delay) the onset of type 2 DM among those at high risk (Long et al. 1994). However, even modest weight loss and increased physical activity in those with lesser severity of obesity who have impaired glucose tolerance can prevent or delay the development of type 2 diabetes (Knowler et al. 2002).

Somewhat surprisingly, in patients with type 2 DM, the improvements in hyperglycemia do not seem to be highly correlated with the amount of weight loss or amount of change in regional adiposity. In a recent weight loss intervention study in type 2 DM, the improvements in hyperglycemia (and HbA$_{1c}$, also known as glycosylated hemoglobin, which determines a patient's blood glucose control over the previous 8 to 12 weeks) were significantly correlated with improved insulin sensitivity but not with weight loss per se or specific changes in regional adiposity (Kelley et al. 2004). Perhaps this is an issue of how to best express data on changes in body composition (e.g., absolute changes, percent changes, or changes that take into account baseline values). For example, in a weight loss intervention trial, we observed that among obese nondiabetic participants, the percentage decrease in VAT was the only body composition change that significantly predicted the improvement in IR after weight loss, and this correlation was only of modest strength (Goodpaster et al. 1999).

Summary

Most patients with type 2 DM are overweight or obese, and there is a direct relationship between the body composition alterations of obesity and the severity of IR. This relationship between IR and adiposity can be effectively and efficiently determined in a clinical setting by measurement of BMI and waist circumference. In a research setting, the association between adiposity and IR can be better understood within a paradigm of central versus peripheral adiposity and with regard to fat content in liver, muscle, and potentially other organs. An increased amount of central or visceral adiposity, particularly if manifested as a preponderance relative to subcutaneous adiposity, generally denotes more severe IR. Similarly, increased fat content within skeletal muscle and liver generally denotes a greater IR. Also, a preponderance of large adipocytes and a relative deficiency of small adipocytes denote greater IR within adipose tissue with downstream repercussions for systemic IR. Fatty acids and secreted proteins from adipocytes may mediate the impact of adiposity on IR.

This is an exciting time in the investigation of type 2 DM and obesity, with respect to the use of established methods of determining body compo-

Chapter **25**

Body Composition Studies in People With HIV

Donald P. Kotler and Ellen S. Engelson

The history of HIV infection in the developed world can be divided into two eras, one preceding the development of highly active antiretroviral therapy (HAART, 1981-1995) and the other since that time. Malnutrition characterized the pre-HAART era and continues to be important in resource-poor areas. It was one of the first complications of HIV infection to be recognized and is one of its most common presentations worldwide. Malnutrition contributes independently to morbidity and mortality in this disease. There have been major advances in the understanding and treatment of nutritional alterations associated with HIV infection. Indeed, HIV infection has been considered a model for understanding the effects of chronic disease on nutritional status (Kotler and Heymsfield 1998).

HAART is associated with markedly improved outcomes, including immune reconstitution and resistance to most disease-related morbidities (Palella et al. 1998). Spontaneous weight gain often occurs (Silva et al. 1998), and the incidence of severe, progressive malnutrition decreases markedly. However, other nutritional alterations develop, which include changes in body shape (with areas of fat loss and fat gain) as well as changes in metabolism, including hyperlipidemia and insulin resistance, a development termed *HIV-associated lipodystrophy* (Carr et al. 1998). There is great concern about the development of premature cardiovascular disease in affected patients.

The aim of this chapter is to discuss how studies of body composition have advanced our knowledge of the nutritional alterations in HIV infection, both malnutrition and lipodystrophy.

Early Studies (1981-1983)

The initial study concerning malnutrition was published in 1984 and concerned observations made in 12 subjects who were hospitalized with diarrhea and weight loss (Kotler et al. 1984). The patients were severely ill and had progressive disease with short survival, averaging about 3 months. They represent the natural history of late-stage, untreated disease. They also may represent the proportion of the affected population most susceptible to developing illness. Although they are no longer typical of the HIV-infected individual seen in the United States, they do represent the situation in much of the developing world.

The nutritional tools used in this study were primitive and designed mainly to detect protein depletion. The patients with AIDS were shown to have significantly lower average weight as percent of ideal weight, hemoglobin, iron-binding capacity, and midarm circumference than controls or laboratory normal subjects. Clinical follow-up of these patients clearly demonstrated the limitations inherent in the use of body weight as a measure of nutritional status. Despite severe and progressive illness, with high fevers and decreased food intake, body weight often rose during hospitalization. The obvious reason was progressive fluid overload attributable to intravenous hydration and progressive hypoalbuminemia. Such problems confound the clinical monitoring of nutritional status using body weight.

Body Composition Assessment

Formal body composition studies were performed in people with HIV starting in 1984. Both high-precision and surrogate measurements were made in cross-sectional and longitudinal studies and analyzed in reference to a database of studies in healthy adults at St. Luke's–Roosevelt Hospital Center. Although most of the reported studies used a four-component tissue model including body cell mass, extracellular water, fat, and bone mineral, the results in participants with HIV have been used in the development and application of other compartmental analyses to be used for body composition studies in general. The latter point was an important rationale for performing these studies, because people with HIV often represent those at one extreme of body composition, and it is important that body composition methodology maintain accuracy throughout the range of possible body compositions.

Cross-Sectional Studies

Cross-sectional studies were performed during 1984-1985 in 33 participants with AIDS (27 male, 6 female) (Kotler et al. 1985). Studies included body cell mass by whole-body counting of ^{40}K (TBK), body fat by the anthropometric formulas of Steinkamp, and fluid spaces by probe dilution (^3H_2O, $^{35}SO_4$). The participants with AIDS averaged 82% of ideal body weight. However, body cell mass as TBK was depleted disproportionately and was only 68% of control values. The women studied had equivalent depletion of body cell mass as men but were much more depleted of fat, a finding later confirmed in studies performed using single-frequency bioelectrical impedance analysis (BIA) in the United States and Africa (Kotler et al. 1999a). Body fat content was not significantly different in males with AIDS and concurrently studied homosexual male control subjects, suggesting that the patients had undergone predominant protein depletion.

The results of body water volume analyses corroborated the results of whole-body counting. Average intracellular water volume, calculated as the difference between total body water and extracellular water volume, was significantly decreased in the participants with AIDS, to a similar degree as the depletion of TBK. Furthermore, there was an alteration in the distribution of body water, with relative expansion of the extracellular space, to the extent that the ratios of total body water to body weight and extracellular to intracellular water volumes were significantly elevated.

Other investigators have suggested that potassium loss from the body may not be a true indication of loss of protein mass but rather may be a change in average intracellular potassium content, the latter caused by energy deficiency. Our study specifically ruled out adrenal disease as a cause for potassium wasting. To further define the depletion in body cell mass, studies performed in collaboration with investigators at Brookhaven National Laboratory used in vivo neutron activation (IVNA) to estimate total body nitrogen as a reflection of total body protein content (Kotler et al. 1991a). These studies showed depletion of both potassium and nitrogen in the participants positive for HIV (figure 25.1). Although a linear relationship was detected between the loss of potassium and nitrogen, the potassium–nitrogen ratio was lower in the HIV-positive patients than in the healthy subjects. This observation indicates that the loss of potassium is larger than the loss of nitrogen in participants with HIV. Studies of skeletal muscle mass using computed tomography (CT) scans also were performed, in collaboration with Dr. Steven Heymsfield and colleagues, and estimates of total body protein made using the results of IVNA, lean body mass by dual-energy X-ray absorptiometry (DXA), and CT scans done in the same subjects were compared (table 25.1) (Wang et al. 1996). Both DXA and CT scanning gave similar results, which were much higher than those obtained using IVNA and the Burkinshaw model for estimating muscle mass (Burkinshaw et al. 1978). An important finding in this study came in the comparison of skeletal muscle in subjects positive for HIV and control subjects. The proportion of the

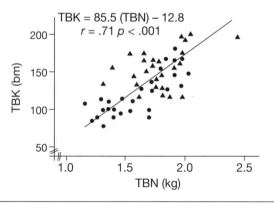

Figure 25.1 Comparison of total body potassium and total body nitrogen in HIV (circles) and control subjects (triangles). A linear relationship between the two elements is noted.

Table 25.1   Effect of AIDS on Skeletal Muscle Mass

	Weight	DXA	CT	IVNA
Controls	77.5 ± 13.8	36.4 ± 6.0	34.4 ± 6.2	27.5 ± 8.4
AIDS	65.5 ± 7.6	28.6 ± 5.2	27.2 ± 4.0	20.9 ± 3.1
Difference	12.0	7.8	7.2	6.6
p	<.01	<.005	<.01	<.05

Note. DXA = dual-energy X-ray absorptiometry; CT = computed tomography; IVNA = in vivo neutron activation. Data are given as kg, mean ± SD, n = 17 controls, 8 participants with AIDS.

weight difference that could be ascribed to skeletal muscle was about 60%, which is much higher than would be expected if the results were attributable simply to semistarvation. Thus, the body composition studies have highlighted and described the depletion of fat-free mass (FFM) that occurs as a consequence of HIV infection and AIDS.

Studies of Pathogenic Mechanisms

Many studies have shown that the pathogenesis of malnutrition in HIV infection is multifactorial (Babameto and Kotler 1997). Body composition studies in HIV infection were performed at the time of clinical evaluation, which allowed correlations to be made with specific diseases and pathogenic mechanisms. Study groups included patients with diminished food intake, usually attributable to esophageal disease such as candidiasis or mucosal ulcer, malabsorption caused by small intestinal infections such as cryptosporidiosis and microsporidiosis, and cachexia caused by systemic infections, such as cytomegalovirus and mycobacterial infections. Studies were performed to compare the relative depletion of nitrogen and potassium and to evaluate hydration status.

Previous studies from several laboratories have suggested that wasting is associated with different losses of potassium and nitrogen and have reported decreases in total body potassium–nitrogen ratios (Cohn et al. 1980; Siwek et al. 1987; Vaswani et al. 1983). The in vivo neutron activation studies described previously (Wang et al. 1996) showed lower potassium per nitrogen in the malnourished participants with AIDS in addition to the depletion of potassium and nitrogen. However, potassium–nitrogen was significantly lower in malabsorbing patients than in those with systemic infections.

We hypothesized that the apparent excess loss of potassium and decreased potassium–nitrogen

is artifactual and related to depletion of only one compartment of nitrogen, that associated with readily metabolizable, intracellular proteins. The pool of nitrogen found in structural proteins, such as collagen, which may represent about one third of the total in a normal subject, is less available for intermediary metabolism and may not contribute in a major way to wasting. We further hypothesized that the depletion of potassium is proportional to the depletion of the readily metabolizable pool of protein (table 25.2). To test this hypothesis, we estimated premorbid TBK from comparisons to matched controls and calculated an estimate of premorbid total body nitrogen (TBN). Assuming that 35% of nitrogen is in structural protein and that all wasting came from intracellular proteins, we calculated an adjusted total body nitrogen and compared the loss of potassium to the loss of the metabolizable pool of nitrogen. When this was done, the potassium–nitrogen ratios of subjects with malnutrition associated with systemic infections or eating disorders did not differ from controls, whereas subjects with malabsorption syndromes continued to have subnormal ratios, suggesting that only malnutrition caused by malabsorption is associated with disproportionate potassium depletion, likely attributable to increased losses in diarrheal stool (table 25.2).

Studies also were performed to examine the effect of pathogenic mechanisms on hydration status (Babameto et al. 1994). Fluid compartments were determined by probe dilution and FFM by DXA. As shown previously, wasting was associated with a decrease in intracellular water volume. The patients with malabsorption had greater depletion of total body water, compared with HIV-positive subjects without malabsorption and to controls. The hydration coefficient (TBW–FFM) was calculated as part of these studies. The hydration coefficient varies over a very narrow range in a variety of mammals. The results of these studies showed that subjects with systemic infections were overhydrated, as

Table 25.2   Comparison of Body Composition Alterations in Different Disease States

	Systemic infection	Intestinal disease	Clinically stable	Control
N	**9**	**7**	**8**	**26**
Weight	64.4 ± 6.5	53.8 ± 5.1	69.5 ± 4.4	78.1 ± 10.2*
Height	176 ± 6	171 ± 5	177 ± 3	176 ± 6
BMI	20.8 ± 1.6	18.5 ± 1.8	22.3 ± 1.8	25.3 ± 2.6*
TBK	111 ± 15	93 ± 5	143 ± 16	148 ± 22*
TBN	1.4 ± 0.1	1.5 ± 0.3	1.9 ± 0.1	1.8 ± 0.2*
K–N	74 ± 9	63 ± 9	77 ± 6	85 ± 11*
K–N$_m$	130 ± 27	103 ± 21*	123 ± 10	131 ± 27

Note. BMI = body mass index; TBK = total body potassium (in g); TBN = total body nitrogen (in kg); K–N = potassium–nitrogen ratio; K–N$_m$ = metabolizable nitrogen. Data are given as mean ± SD.

*$p < .001$.

Table 25.3   Effects of Different Disease States on Hydration Status

	Malabsorption	Systemic infection	Control
TBW	36.0 ± 5.9*	40.5 ± 5.7	42.5 ± 5.3
FFM	51.7 ± 6.1*	56.9 ± 6.5**	60.5 ± 5.6
TBW–FFM	0.68 ± 0.05***	0.72 ± 0.04	0.70 ± 0.05****

Note. TBW = total body water (in L); FFM = fat-free mass (in kg). Data are given as mean ± SD

*$p < .001$; **$p < .01$ vs. control; ***$p < .005$ vs. systemic infection; ****$p < .01$ vs. malabsorption or systemic infection.

shown in our original studies (table 25.3). In contrast, patients with malabsorption had significant decreases in TBW–FFM. The prevalence of dehydration was defined as TBW–FFM below the 90% confidence interval for results in controls. Forty-four percent of the patients with malabsorption were significantly dehydrated, compared with 11% of patients without malabsorption and 4% of controls.

Nutritional Status and Clinical Outcomes

Several studies examined the possible effects of malnutrition on clinical outcomes and found significant associations between weight loss and short-term survival, risk of hospitalization, and development of disease complications, but the causality of these associations is uncertain. Several studies correlated body cell mass depletion with shortened survival. Prospective studies have shown that weight losses of as little as 5% over 4 months were associated with

decreased survival compared with patients without weight loss or with weight gain (Palenicek et al. 1995). We found that normalized body weight and body cell mass, but not body fat, bore a significant relationship to the time of death in patients with wasting illnesses studied in the pre-HAART era (Kotler et al. 1989a). Extrapolated body cell mass at death was 54% of normal, whereas body weight was about one third below ideal. The weight at death was similar to that reported in studies of human starvation, including observations made during the siege of Leningrad, in the Warsaw ghetto, and during lethal hunger strikes by IRA prisoners. Thus, the results suggested that the timing of death from wasting in AIDS is related to the degree of body cell mass depletion rather than the specific cause of the wasting process. In a prospective study performed by Suttmann and colleagues (Suttman et al. 1995), a low ratio of body cell mass to body weight was associated with shortened survival compared with subjects with higher ratios. The effect was independent of, and quantitatively greater than, the level of immune depletion as measured by CD4+ lymphocyte counts.

The interrelationships among nutritional status, immune function, and quality of life were examined in a group of 150 clinically stable HIV-infected subjects, to compare the relative and interacting effects on quality of life of nutritional status, defined as body cell mass determined using bioimpedance analysis, and immune function, defined as peripheral blood CD4+ lymphocyte counts (Turner et al. 1994). We found that body cell mass, but not weight or fat content, was significantly associated with the physical performance index of a standard quality of life instrument (Medical Outcomes Survey, Short Form). The effect was independent of CD4+ lymphocyte counts. Other aspects of quality of life correlated significantly with the degree of immune depletion and not with nutritional status. The results are not surprising, given the evidence for skeletal muscle depletion presented previously. Wilson and Cleary (1997) followed patients infected with HIV and found an association between weight loss and decreased physical function.

Longitudinal Studies

Early clinical experience indicated that progressive wasting in a person with AIDS, once established, was almost universally progressive until death. By the middle of the 1980s it became clear that some patients were experiencing indefinite periods of clinical stability and even improvement. Two studies from that period documented the fact that nutritional status can be maintained in a person with AIDS. We performed follow-up body composition studies in groups of participants with AIDS, homosexual male controls, and heterosexual controls (Kotler et al. 1990a). The participants with AIDS were clinically stable, without ongoing or recent disease complications. The participants with AIDS were within

the normal range for body weight at baseline. TBK values were low, but body cell mass was stable over a 3-month period. Caloric intake, intestinal absorption, and resting energy expenditure were measured as part of these studies. The clinically stable participants with AIDS were found to have normal caloric intakes. Evidence of fat malabsorption was found in the participants with AIDS, and mean resting energy expenditures were lower than controls, which is consistent with the effects of malabsorption but discordant with other studies of resting energy expenditure in people infected with HIV (Hommes et al. 1991; Melchior et al. 1993). Nonetheless, the participants with AIDS were able to maintain their nutritional status, implying that wasting in AIDS is the direct result of disease complications rather than the underlying disease itself.

Longitudinal studies also were used to indirectly demonstrate the effects of systemic infections on nutritional status by documenting changes in body composition in patients with systemic cytomegalovirus infection with colitis, which is a serious disease complication and uniformly fatal, if untreated (Kotler et al. 1989a). Studies of such patients were ongoing at the time the antiviral agent ganciclovir (Roche Laboratories) became available. In addition to its antiviral effect, ganciclovir therapy was associated with repletion of TBK (table 25.4). Untreated patients lost weight, body cell mass, body fat, and serum albumin concentrations, whereas treated patients had increases in all values. We concluded that effective treatment of a disease complication promoting wasting is accompanied by reversal of the wasting.

Body composition studies in these patients also clarified another nutritional issue of the time. We reported a high prevalence of hypertriglyceridemia in people with AIDS (Grunfeld et al. 1989) and an association between serum

Table 25.4 Effect of Treating Cytomegalovirus Infection With Ganciclovir on Body Composition

	Untreated	Treated	p
Weight (g/day)	−132 ± 29	49 ± 19	<.001
TBK (mEq/day)	−6.4 ± 2.7	2.8 ± 1.1	<.005
Fat (g/day)	−36 ± 20	23 ± 14	<.05
Serum albumin concentration (g/L·day)	−0.3 ± 0.1	0.1 ± 0.1	<.02

Note. TBK = total body potassium. Values given as mean ± SE.

triglyceride concentration and serum concentrations of α-interferon (Grunfeld et al. 1991). The cause for the lipid abnormality was linked to a component of the inflammatory response, and it was inferred that serum triglyceride levels might be a correlate of active disease and a harbinger of wasting. Although elevated triglyceride levels are a metabolic epiphenomenon of HIV infection, this abnormality in itself is not a cause for wasting.

Studies of Nutritional Therapies

Nutritional therapy for AIDS wasting falls into two categories that may be used alone or with adjuncts. The first is intended to promote nutritional repletion chiefly by providing a balanced diet with adequate calorie and protein content, whereas the second uses supraphysiological or pharmacological doses of specific micronutrients to affect the underlying disease process. Many of the reported studies of nutritional support in HIV infection have used techniques for body composition analysis in addition to body weight.

Although food supplements often are prescribed for people with HIV, supporting data demonstrating the usefulness of food supplements are limited. Several studies have shown short-term effects on clinical symptoms or body weight, but body composition studies were not reported. Elemental diets were compared with a whole-protein, whole-fat supplement and a multivitamin supplement (Gibert et al. 1999), in weight-stable subjects, with CD4 less than 200. Participants who were able to maintain a high caloric intake (>2,300 kcal/day) gained up to 1% of their prestudy weight, with a convergent change in body cell mass by BIA, thus leading to the conclusion that weight maintenance in people with intact gastrointestinal tract and free of opportunistic infection is related to general caloric intake. Pichard and colleagues (1998) found that an arginine/omega-3 fatty acid supplement increased body weight, fat-free mass, and fat mass, measured by BIA, compared with a standard formula; this treatment was associated with a decreased sense of anorexia but no other clinically meaningful benefits. Both groups exceeded the recommended caloric intake, with most of the increase being attributed to higher protein consumption. Lean mass repletion and possible anticytokine effects of an omega-3 fatty acid supplement were evaluated in an open-label trial by Hellerstein and colleagues (1996). Non-

significant weight gain and no body composition differences were observed in supplemented subjects compared with a nonsupplemented group. Appetite stimulants, such as megestrol acetate (Megace, Bristol Myers Squibb) and dronabinol (Marinol, Roxanne Laboratories) promote weight gain, with an increase in body weight of around 6% (Oster et al. 1994; Von Roenn et al. 1994). Appetite stimulation is most beneficial in the absence of local pathological lesions affecting chewing and swallowing, in malabsorption syndromes, and in cases without active systemic infections. Von Roenn and colleagues (1994) showed that the majority of patients receiving 800 or 400 mg of Megace per day gained about 1.1 kg of fat-free mass and 6 kg of fat mass (figure 25.2). In an analysis of a subgroup of participants from this trial who were studied at one center, the preponderance of weight gain as fat was noted. The additional finding of suppression of serum testosterone concentration could contribute to the proclivity of Megace to promote an increase in body fat content and not fat-free mass, an antianabolic effect (Engelson et al. 1995). The results are consistent with the occurrence of impotence in men receiving the medication. Women may become amenorrheic or develop other menstrual irregularities when taking Megace.

In people with AIDS who have poor food intake and no objective evidence of severe malabsorption or systemic infection, tube feeding might be expected to replete body cell mass. This was shown in a prospective case series using a formula diet administered through a percutaneous endoscopic gastrostomy for 2 months (Kotler et al. 1991b). In addition to finding increases in body cell mass, the authors noted increases in body fat content, serum albumin concentration, and

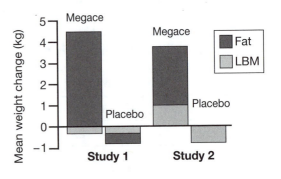

Figure 25.2 Effect of megestrol acetate on body composition. Data adapted from two studies show that the large majority of weight gained during megestrol therapy is fat and not fat-free mass (FFM).

serum iron-binding capacity, the last a reflection of transferrin (table 25.5). Thus, the enteral feeding regimen repleted both somatic and visceral protein compartments. Of note, repletion succeeded despite the persistence of systemic infection in several patients. In this study, total lymphocyte counts in peripheral blood increased significantly during the period of nutritional support, along with an increase in the number of T suppressor cells (CD8+), although no changes were noted in the number of helper CD4+ lymphocytes. Functional improvement, including subjective improvements in cognitive function, was noted in those patients receiving nutritional support. Other studies using percutaneous endoscopic gastrostomy feeding also have documented weight gain.

We prospectively followed a group of people with AIDS who were receiving total parenteral nutrition (TPN; table 25.6) (Kotler et al. 1990b). In this group, TPN led to increases in body weight and fat, but no change in body cell mass, as TBK, was observed. However, the results distinguished between the effects of TPN in the presence of malabsorption and systemic infection. Patients with malabsorption syndromes or eating disorders responded well to TPN, with significant increases in TBK, whereas patients with systemic infections, such as cytomegalovirus infection or *mycobacte-rium avium* complex (MAC), continued to lose TBK and gained fat instead. We concluded that the response to TPN is related more to the clinical status of the patient receiving therapy than to the therapy itself. Melchior and colleagues (1993) found increases in lean body mass (LBM) and in functional status, following 2 months of TPN, but similar survival, compared with placebo, in AIDS-wasting subjects.

In a more recent study, we prospectively compared the effects of 12 weeks of therapy with TPN versus an oral, semielemental diet on body weight, body composition, quality of life, survival, and other outcomes in people with AIDS who had malabsorption (Kotler et al. 1998b). This was a randomized trial performed in outpatients, in which we used home nutritional and dietician support and performed BIA studies in the home. The TPN group consumed significantly more total calories than those assigned to the elemental diet. The TPN group also gained significantly more weight than the elemental diet group, although weight change during therapy was significantly different from pre-treatment in both groups. Changes in body weight correlated with calorie intake rather than mode of feeding. BIA studies showed that the TPN group gained significantly more fat, whereas the changes in body cell mass were similar in the two groups.

Table 25.5 Effect of Enteral Alimentation on Body Composition

	Pretreatment	Posttreatment	p
Weight (kg)	52 ± 4	55 ± 4	<.05
Body cell mass (kg)	22.0 ± 1.2	23.4 ± 1.5	<.02
Body fat (% of normal)	61 ± 14	67 ± 13	<.002

Note. Values are mean ± SD.

From D.P. Kotler et al., 1991, "Effect of enteral feeding upon body cell mass in AIDS," *American Journal of Clinical Nutrition* 53: 149-54. Adapted with permission by the *American Journal of Clinical Nutrition.* © Am J Clin Nutr. American Society for Clinical Nutrition.

Table 25.6 Effect of Total Parenteral Nutrition on Body Composition

	Pretreatment	Posttreatment	p
Weight (kg)	51.5 ± 1.5	54.3 ± 3.0	NS
Body cell mass (kg)	21.8 ± 0.8	22.0 ± 1.1	NS
Body fat (kg)	7.3 ± 1.4	11.2 ± 1.7	<.01

Note. NS = not significant. Values are mean ± SE, *N* = 12.

Reprinted from Kotler DP, Tierney AR, Culpepper-Morgan JA, Wang J, Pierson RN Jr. Effects of home total parenteral nutrition on body composition in patients with Acquired Immundeficiency Syndrome. JPEN J Parenter Enteral Nutr. 1990; 14(5); 454-458 with permission from the American Society for Parenteral and Enteral Nutrition (A.S.P.E.N.). A.S.P.E.N. does not endorse the use of this material in any form other than its entirety.

Quality of life improved in the semielemental diet group and worsened in the TPN group, and those differences were statistically significant. Nutritional therapy was three times more expensive for the TPN group. Survival rates were similar in the two groups. Neither diet affected intestinal function or immune status. We concluded that an oral semielemental diet may mitigate weight loss and wasting in AIDS patients with malabsorption and should be the first option in such patients. The results do not prove equivalence of the two therapies.

A recurring conclusion of these results is that simply providing calories does not lead to body cell mass repletion in patients with AIDS. For this reason, many studies have looked at adjuncts to calories alone. Several trials have tested various anabolic agents for the long-term treatment of wasting. Schambelan and colleagues (1996) showed an average of 3-kg LBM accretion, accompanied by a 1.7-kg decrease in body fat, following a 3-month course of recombinant human growth hormone (rhGH; average dose 6 mg/day, or 0.1 mg · kg^{-1} · day^{-1}), in AIDS-wasting subjects (figure 25.3). The gain in fat-free mass by DXA was confirmed by the finding of a 2.4-kg increase in total body water, measured by D$_2$O dilution, which is consistent with the DXA results, because fat-free mass contains about 73% water. A significant increase in intracellular water volume was noted, about 1.1 L, implying that a portion of the increased fat-free mass was in the body cell mass. However, an increase in extracellular water also was detected. This was associated with an increase in treadmill work output, whereas quality of life and days of disability remained the same, compared with the placebo-controlled group. A more recent study, performed in the HAART era, confirmed the anabolic effects of rhGH, by multifrequency BIA, and showed that the improvement was associated with increases in physical performance and quality of life.

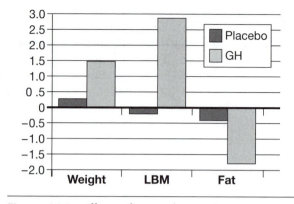

Figure 25.3 Effect of recombinant human growth hormone (GH) on body composition. Compared with placebo, growth hormone promoted an increase in lean-body mass (LBM) and a loss in fat content.

Cytokine inhibitors, such as pentoxfylline (Trental) and thalidomide, also have been proposed to treat malnutrition associated with HIV infection. Thalidomide accelerates the degradation of tumor necrosis factor-α messenger RNA, making thalidomide a potential candidate for immune and nutritional modulation. In a 3-month, placebo-controlled trial, thalidomide use was associated with significant weight and muscle mass increases of an average 4.5 kg and 1 kg, respectively (Reyes-Teran et al. 1996). No effects on viral burden or CD4 counts were noted. Other studies have used testosterone or anabolic steroids (table 25.7) (Engelson et al. 1996; Gold et al. 1996).

Progressive resistance exercise (PRE) has been advocated as a nonpharmacological approach to restore LBM and muscle strength. Anabolic and nutritional therapies were compared with PRE in a number of trials. Agin and colleagues (2001) found that adding supplemental whey protein extract to a PRE regimen did not increase BCM, by TBK, in excess of gains observed with PRE only, whereas whey protein only had an effect on TBK. Bhasin

Table 25.7 Long-Term Effects of Anabolic Agents on Body Composition

	Before therapy	After therapy
Weight*	73.1 ± 11.0	81.5 ± 13.6
FFM**	61.5 ± 8.7	67.6 ± 12.7
Fat	11.6 ± 5.5	13.9 ± 5.0
ASM	26.6 ± 3.7	29.9 ± 4.5

Note. FFM = fat-free mass; ASM = appendicular skeletal muscle, from dual-energy X-ray absorptiometry scans. Six subjects were studied for a mean of 11 months. Data are given as kg, mean ± SD.

*p = .02; **p = .05.

and colleagues (Bhasin et al. 2000) showed that in hypogonadal, AIDS-wasting subjects, the effects of 16 weeks of combining testosterone and PRE were not additive and that PRE and testosterone replacement were both associated with gains in LBM. These findings were confirmed by Grinspoon and colleagues (2000) in a placebo-controlled trial in which PRE and supraphysiological testosterone replacement (200 mg/week) yielded similar LBM gains in eugonadal AIDS-wasting men. In contrast, Strawford and colleagues (1999) found that when they added 20 mg/day of Oxandrolone to PRE and 100 mg/week of testosterone, LBM (determined by DXA scanning) and muscle strength increased significantly, compared with PRE and physiological testosterone replacement, in eugonadal, weight-stable subjects.

In summary, nutritional therapies may be effective in reversing body cell mass depletion. Effectiveness usually requires clinical stability and is best accomplished by a combination of adequate caloric intake and adjunctive therapies.

HIV-Associated Lipodystrophy

As noted, HAART therapy was associated with marked clinical improvement, including decreased prevalence and severity of malnutrition. However, nutritional alterations continue to be seen, consisting of fat redistribution in the absence of a decrease in body cell mass (Carr et al. 1998). Clinical observations have noted at least two forms of fat redistribution. Some subjects lose subcutaneous adipose tissue (SAT) throughout the body (lipoatrophy), which is seen as thinning of the extremities and, especially, the face. Often the appearance is of peripheral lipoatrophy, because of a lack of change

in the size of the torso, or even an increase in girth, leading to a change in pants size. Total body fat may be normal or low, and some patients note a progressive loss of weight with the development of lipoatrophy. Patients often notice prominent veins, especially in the legs, which may be visible as high as the upper thigh. The reason for the prominent veins is not a change in venous diameter but rather a loss of the normal surrounding fat and depends on both the premorbid SAT and the amount lost. For this reason, prominent veins are much more common in men than in women.

Body shape changes also may be accompanied by increases in weight and body fat, which manifest clinically as an increase in waist size, caused by an increase in intra-abdominal fat content (Engelson et al. 1999), or as breast enlargement (Dong et al. 1999), which is not glandular tissue but rather adipose tissue. Increases in the size of other upper-body fat depots, including the dorsocervical region of the neck (buffalo hump), similar to that seen in patients with Cushing's disease or otherwise healthy adults, may be seen (Lo et al. 1998). This change is especially common in women, who may develop generalized upper-body fat that may be accompanied by a substantial increase in bra size.

The changes in fat distribution have been documented on cross-sectional imaging studies, such as MRI or CT scans (figure 25.4). The increase in intra-abdominal fat is noted deep in the abdominal musculature and is found in the omental, mesenteric, and retroperitoneal fat compartments. The relationship between changes in the sizes of the SAT and visceral adipose tissue (VAT) is unclear. Some epidemiological studies suggest that SAT

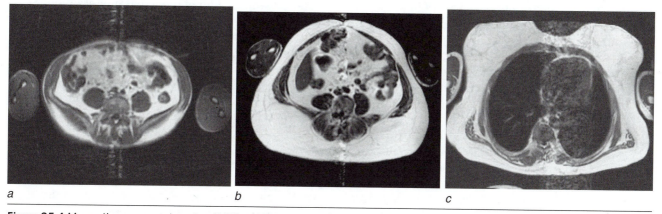

a   *b*   *c*

Figure 25.4 Magnetic resonance imaging (MRI) of HIV-associated lipodystrophy. *(a)* MRI at L4-5 in a male subject shows prominent depletion of subcutaneous adipose tissue (SAT) with a moderate increase in visceral adipose tissue (VAT). *(b)* MRI at L4-5 in a female subject shows increased contents of both SAT and VAT compartments. *(c)* MRI at T6 level in the female subject from *(b)* showing increased upper body and mammary adipose tissue contents.

and VAT are distinct compartments and are affected differently and independently by various influences (Lichtenstein et al. 2001). A substantial proportion of patients, especially men, have evidence of both fat loss and fat gain, but this may be coincidence, because both abnormalities are common. In some men, and especially in women, the increase in VAT is associated with an increase in SAT, and such patients may look indistinguishable from patients without HIV who have the metabolic syndrome X.

There are reports that the manifestations of body fat redistribution may vary by race and sex, but these studies are limited in the sophistication of the body composition measurements. Children are also affected by body composition and metabolic alterations (Arpadi et al. 2001). Most of the studies were done in children between ages 7 and 12 years and not in very young children. Both depletion of SAT and accumulation of VAT have been noted.

Pseudotruncal Obesity

The estimation of waist–hip ratio (WHR), as a reflection of visceral fat accumulation, has repeatedly been shown to predict adverse clinical outcomes in non-HIV situations, although the use of a ratio decreases the specificity of interpreting changes, as discussed elsewhere. For example, an increase in WHR could represent an increase in VAT or a decrease in SAT in the hips. Studies of HIV-infected individuals in the pre-HAART era demonstrated such an effect, that is, an elevated WHR attributable to a greater relative decrease in hip circumference than in waist circumference.

We retrospectively compared the results of body composition studies, including measurements of fat distribution, in 96 HIV-infected subjects studied after January 1996, considered the start of the HAART era, to subjects seen before January 1996 (pre-HAART era) and healthy controls (Kotler et al. 1999a). Subjects were matched by sex, race, age, and height. The measurements included body cell mass by whole-body counting of ^{40}K, plus fat, fat-free mass, and body fat distribution by anthropometry. HAART HIV-positive men weighed more and had more body cell mass than pre-HAART men but less than controls. In women, the between-group differences in fat were greater than the differences in body cell mass. HAART and pre-HAART HIV subjects had lower indexes of SAT and higher indexes of VAT than did controls. WHR values were significantly higher in both HAART and pre-HAART subjects than in controls.

However, the reason for the increased ratio was different in pre-HAART and HAART patients. The predominant reason for an increased WHR in the pre-HAART era was a disproportionate decrease in hip circumference, whereas the reason for an increased WHR in the HAART era was an increase in waist circumference.

Treatment of Fat Redistribution

There have been two responses to the emergence of lipodystrophy as a clinical problem. One response, which is based on the belief that the changes are caused entirely by antiretroviral therapy, involves switching patients to alternative therapies. However, other variations of antiretroviral avoidance include delayed initiation of therapy, intermittent therapy, rotating therapy, and others. These strategies all are experimental and are not recommended for routine clinical application. The other response is to treat the alterations in fat contents separately.

Switch Studies

Many studies have been performed to evaluate the effectiveness of switching therapies. Most of the studies involve a switch from protease inhibitor–based HAART to non-nucleoside reverse transcriptase inhibitor–based HAART, without any change in nucleoside therapy. This design was based on the widespread belief, at the time the studies were designed, that the development of lipodystrophy was a direct consequence of protease inhibitor therapy. Switching to non-nucleoside–based HAART improved metabolic parameters in many studies, with differences related to the specific therapies that are stopped and started (Martinez et al. 1999). However, there is no evidence of reversion to normal fat distribution as defined by DXA in these studies, despite prolonged follow-up. In contrast, Saint-Marc and Tourain (1999) reported an open-label study of discontinuing Zerit therapy, in which SAT contents in the abdomen and midthigh increased, without a change in VAT, and the findings were reproduced by Carr and colleagues (2002) and others. The magnitude of the change has been small, and even clinically unapparent, in some of the studies.

Diet and resistance training exercise are well known to affect total body and regional fat mass in people not infected with HIV. Clinical observations indicate that voluntary weight loss decreases body fat content, although the proportion of visceral fat is uncertain. Weight loss may lead to a decrease in prominence of buffalo humps but worsens the signs of fat depletion. Roubenoff and colleagues (1999) reported a study of exercise in men with central fat accumulation by DXA scanning, in which the intervention seemed to preferentially decrease truncal fat. It is unclear whether the change in truncal fat was in the visceral compartment or in abdominal subcutaneous fat.

Growth hormone has been studied for its ability to successfully treat areas of fat accumulation. Growth hormone, in doses lower than those used for AIDS wasting, has been evaluated for the treatment of VAT accumulation in truncal obesity, and it was shown to promote the depletion of both SAT and VAT (Johannsson et al. 1997). Growth hormone was shown to decrease the size of buffalo humps (Torres et al. 1999) and the visceral fat compartment (figure 25.5) (Engelson et al. 2002; Lo et al. 2001). Although insulin resistance and increased risk of glucose intolerance are well-known effects of growth hormone therapy, the major side effect, clinically, is joint stiffness and pain, a well-known side effect of therapy when given in supraphysiologic doses. The beneficial effects are reversible upon discontinuation of therapy (Engelson et al. 2002).

Summary

The nutritional alterations in HIV infection and AIDS present paradigms of different forms of malnutrition. Untreated HIV infection is a model of a chronic inflammatory disease, in which body cell mass depletion is a characteristic finding. The opportunistic complications of AIDS promote starvation, malabsorption, or cachexia, and the specific pattern of body composition change reflects the specific pathophysiological mechanism. Effective treatment of HIV infection or its disease complications leads to nutritional repletion but may be

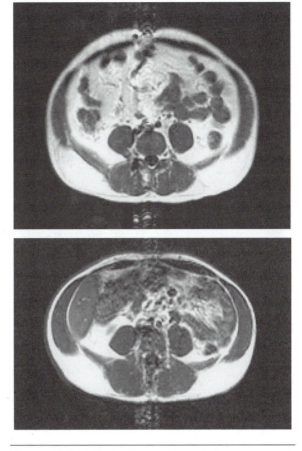

Figure 25.5 Effect of recombinant human growth hormone on visceral adipose tissue (VAT) and subcutaneous adipose tissue (SAT). Segmented MRI images of L4-5 before and after therapy show a predominant decrease in VAT, not SAT.

complicated by the development of fat redistribution, in the absence of change in body cell mass, and metabolic alterations. Investigators and clinicians have applied nutritional measurements in HIV-infected patients both to gain insight into the fundamental biology of the disease process and to guide nutritional and other therapies. Such efforts will continue as the nutritional consequences of HIV infection and its treatment remain a clinical imperative.

Chapter 26

Inflammatory Diseases and Body Composition

Ian Janssen and Ronenn Roubenoff

A plethora of inflammatory diseases are present at a relatively high prevalence in the population. Both sexes and people of all ages are susceptible to inflammatory disease. Many inflammatory diseases are associated with severe changes in mediators of the immune system, which in turn alter energy and protein metabolism and, over time, lead to changes in body composition. Specifically, a substantial decrease in fat-free mass (FFM) is characteristic of most inflammatory diseases. The loss of FFM is distressing given that a reduction in FFM compromises immune function, muscular strength, and functional mobility and is associated with a loss of independence, a reduced quality of life, and poor clinical outcomes.

This chapter examines the changes that occur in energy metabolism and body composition in adults and children with different inflammatory diseases. In addition, we cover the influence of pharmacological, nutritional, and exercise interventions on body composition in these inflammatory diseases. Because the majority of studies examining the effect of inflammatory diseases on body composition have done so using a two-component body composition model, this chapter primarily focuses on fat mass (FM) and FFM. Whenever sufficient data exist, the role of inflammatory diseases on fat distribution and specific

components of FFM, such as skeletal muscle and bone mineral content, is discussed.

Influence of Metabolic Imbalances on Body Composition

A large body of evidence indicates that changes in body composition reflect changes in energy metabolism in the body. For example, in obesity there is a corresponding increase in both FM and fax oxidation (Sidossis et al. 1995). Conversely, during starvation and pathological stress, an increase in protein metabolism and losses in body weight and FFM occur (Hill 1992). Thus, an underlying principle of the mechanisms that explain changes in body composition is that body composition is in equilibrium with metabolic state. However, whereas metabolism changes on a minute-to-minute basis, changes in body composition occur at a slower rate and are a reflection of the metabolic status of an organism over days, weeks, or even years.

Influence of Immune Mediators on Metabolism

The immune system has the capacity to radically modify energy and protein metabolism and thus body composition. When a person is stressed by injury, infection, or a chronic illness, such as an inflammatory disease, the immune system is

Research supported by the U.S. Department of Agriculture, under Agreement 58-1950-9-001 and Contract 53-K06-1. I. Janssen was supported by a Canadian Institutes of Health Research Postdoctoral Fellowship. Any opinions or recommendations expressed in this publication are those of the authors and do not necessarily reflect the view of the U.S. Department of Agriculture.

responsible for recognizing and responding to this disturbance. The T cells are essential for regulating the complex function of the immune system. However, the T cells can only recognize an antigen (foreign body) after it has been processed by an antigen-presenting cell. The antigen-presenting cell contacts an antigen, phagocytoses it, processes an antigen determinant, and brings it to its surface to trigger an immune response. The immune response requires both the presence of a specific signal from the antigen and the elaboration of one or more nonspecific signals, chiefly via secretion of the cytokine interleukin-1β (IL-1β). IL-1β secretion triggers activation of T cells and other portions of the immune response.

Subsequent signals initiated by antigen-presenting cells include the production of tumor necrosis factor-α (TNF-α) and interleukin-6 (IL-6). These three cytokines—IL-1β, TNF-α, and IL-6—play critical roles in the development of the acute-phase metabolic response, which parallels the acute-phase immune response. These cytokines act directly on metabolic target organs such as muscle, liver, gut, and brain, and they also affect metabolism indirectly by regulating the hormones that govern metabolism (Pomposelli et al. 1988). The net result of increased inflammatory cytokine production is an increase in resting energy expenditure (REE), an export of amino acids from muscle to liver, an increase in gluconeogenesis, and a marked shift in liver protein synthesis away from albumin toward the production of acute-phase proteins such as fibrinogen and C-reactive protein (Kushner 1993). In the short term, as is the case for injury or acute illness, these metabolic shifts are favorable and promote survival. However, over a longer duration, as occurs with chronic inflammatory disease, these metabolic shifts can lead to profound changes in body composition.

Usefulness of Body Composition As a Measure of Chronic Disease Status

The slower time scale over which changes in body composition occur offers an opportunity to assess the nutritional and metabolic status of the organism over days, weeks, or years, rather than making the minute-by-minute measurements required for metabolic assessment. For example, people with advanced AIDS average 82% of ideal body weight and only 68% of normal body cell mass (an index of FFM), illustrating the effect of malnutrition on metabolism and body composition in this disease (Kotler et al. 1985). The clinical significance of the reduced FFM in chronic disease is apparent given that the depletion of FFM parallels the reduction in immune competence (Chandra and Chandra 1986), leads to prolonged hospitalization (Reilly et al. 1988), is associated with physical disability and a loss of independence (Janssen et al. 2002), and, when reduced below 40% of baseline, is associated with death (Kotler et al. 1989; Tellado et al. 1989; Winick 1979).

Inflammatory Diseases and Body Composition in Adults

This section reviews the changes that occur in body composition in adults with different inflammatory diseases such as rheumatoid arthritis (RA), lupus, and inflammatory bowel disease. As illustrated in figure 26.1 for FFM, several potential mechanisms explain the altered body composition in inflammatory disease. Changes in mediators of the immune system and energy metabolism play central roles in these changes in body composition.

Rheumatoid Arthritis

RA is a chronic, systemic, autoimmune disease of unknown etiology that causes destruction of joint cartilage and bone. It affects approximately 1% of the population and influences two to three times more women than men (Kelsey and Hochberg 1986). RA generally begins between the fourth and sixth decades of life and is characterized by joint stiffness, pain, and swelling. The synovium of the joint is the crucial site in the onset of joint deterioration and is characterized by an increase in proliferating T-lymphocytes, immunoglobulin production, and inflammatory cytokine production (Kelsey and Hochberg 1986). Increased production of TNF-α and IL-1β is believed to play a central role in the pathogenesis of RA.

Loss of FFM is a hallmark of RA. *Rheumatoid cachexia* is a term used to describe the loss of FFM, which is primarily composed of skeletal muscle, that occurs in most people with RA. Using dual-energy X-ray absorptiometry to assess total and regional FFM, Westhovens and colleagues (1997) observed that FFM was lower in the arms, legs, and trunk of both male and female participants with RA compared with age- and sex-matched controls. Roubenoff and colleagues (1994) reported

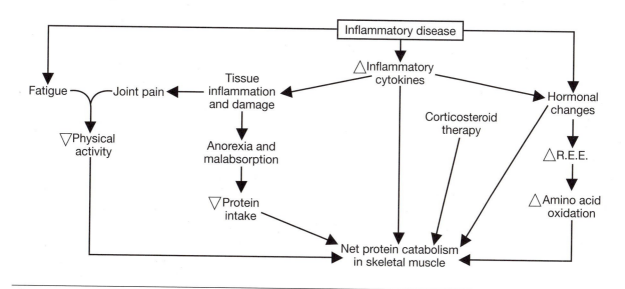

Figure 26.1 Summary of the mechanisms by which inflammatory disease leads to a reduced fat-free mass (FFM). Inflammatory diseases are characterized by excess production of inflammatory cytokines. Excess cytokine production causes protein breakdown, hormonal changes, and tissue damage and inflammation. The end result of these changes is a net protein catabolism in skeletal muscle and a decrease in FFM. Fatigue and corticosteroid therapy may also reduce FFM in inflammatory disease. REE = resting energy expenditure.

that adults with controlled RA have 13% less body cell mass (the metabolizing, oxygen-consuming portion of FFM) than their age-, sex-, race-, and weight-matched healthy controls. In this study, the body weight of participants with RA was matched with healthy controls; thus, the 13% difference in body cell mass indicates that FM was higher in the participants with RA. This condition of a reduced FFM in the presence of an increased FM has been termed *rheumatoid cachectic obesity* (Walsmith and Roubenoff 2002).

A hallmark of rheumatoid cachexia is an elevated REE (Roubenoff et al. 1994), an increase in whole-body protein breakdown (Rall et al. 1996b), a net efflux of amino acids from muscle to liver, and a shift in liver protein production away from albumin toward acute-phase proteins such as fibrinogen and C-reactive protein. Over the long term, these shifts can cause cachexia.

Increased production of the inflammatory cytokines TNF-α and IL-1β plays a central role in the body composition changes that occur in RA. TNF-α is believed to be the central mediator of muscle wasting in RA and is known to act synergistically with IL-1β to exert a powerful influence on muscle protein turnover, resulting in net muscle protein degradation and muscle wasting (Roubenoff et al. 1992, 1994). Reduced peripheral insulin action and low physical activity may also play important roles in the development of rheumatoid cachexia. Insulin inhibits muscle protein degradation (Nair

et al. 1995) and is an important anabolic hormone. Consequently, reduced insulin sensitivity, as observed in RA (Svenson et al. 1987), may contribute to muscle wasting. Reduced physical activity also favors FFM loss and predisposes to fat gain. People with RA have higher FM and lower total energy expenditure than matched healthy controls (Roubenoff et al. 2002).

The reason for low physical activity in RA is multifactorial and involves a combination of joint pain and stiffness that makes physical activity challenging. In addition, cachexia itself may be an important contributor to low physical activity in people with RA because of the muscle weakness that is caused by the low muscle mass. Consequently, RA causes a negative cycle in which low muscle mass and low physical activity reinforce each other to predispose to fat gain. Fat gain in turn causes further functional decline (Visser et al. 1998a, 1998b) and increases circulating TNF-α levels (Kent-Braun and Ng 2000; Kern et al. 1995). Elevated TNF-α levels may promote additional cachexia by accelerating muscle protein catabolism and impairing insulin action. This would in turn cause further losses of functional capacity as well as lower physical activity levels.

In addition to the increase in whole-body FM, there also appears to be a redistribution of fat in people with RA, favoring a shift toward abdominal fat distribution (Westhovens et al. 1997). This is a cause for concern given that abdominal fat has a greater effect on diabetes and cardiovascular disease risk than fat in the periphery (Chan et al. 1994; Rexrode et al. 1998). Not surprisingly, the increase in abdominal adiposity is particularly apparent in people with RA who are on corticosteroid therapy, because steroids have a marked influence on fat distribution (Westhovens et al. 1997). RA is also associated with a deterioration of bone mineral density (BMD) that is particularly reduced in the appendicular regions (Keller et al. 2001; Westhovens et al. 1997). The reduced BMD has been attributed to a number of factors including high inflammatory activity, corticosteroid therapy, and reduced physical activity (Hansen et al. 1996; Laan et al. 1993; Sambrook et al. 1987).

Corticosteroid therapy is the core treatment for many chronic inflammatory diseases, including RA. Corticosteroids have a mild effect on inflammation by acting on kinases or transcription factors necessary for expression of inflammatory cytokines. The symptomatic relief offered by corticosteroids is so great that they are frequently used despite the adverse side effects, including hyperglycemia, myopathy, cataracts, gastric ulcerations, and alterations in body composition (Baxter and Forsham 1972). In relation to FFM, corticosteroids are known to induce negative nitrogen balance in people with RA (Roubenoff et al. 1990). There is also evidence that corticosteroids have deleterious effects on calcium fluxes in the intestine and kidney and osteoblast and osteoclast activity in the bone (Reid 1989).

Anti-TNF-α (e.g., etenercept and infliximab) and IL-1β (e.g., anakinra) therapies provide alternative pharmacological interventions for RA patients. However, no studies have investigated the effect of anticytokine therapy on the body composition abnormalities that are present in RA. Nevertheless, there is some evidence that better control of the inflammation with methotrexate negates the protein catabolism of rheumatoid cachexia. Given the complexity of the cytokine interactions and the number of cytokine targets, the effectiveness and required dose of these therapies for the treatment of rheumatoid cachexia are difficult to determine.

Many studies have examined the effect of exercise on patients with RA. These studies have demonstrated that exercise improves physical function, cardiovascular fitness, and muscular strength in individuals with controlled RA (Rall and Roubenoff 1996; Rall et al. 1996a, 1996b). It was also reported that 12 weeks of high-intensity strength training reversed the accelerated protein

catabolism observed in participants with RA (Rall et al. 1996a, 1996b). These benefits occurred without any adverse effects on immune function or disease status, reinforcing the safety and efficacy of strength training in participants with RA. However, it is noteworthy that the studies observing a reversing of the accelerated protein catabolism did not find a concomitant change in body weight or body cell mass (Rall et al. 1996a, 1996b). The lack of changes in body composition is most likely explained by the short duration and limited dose of the exercise training.

Many people with RA are overweight or obese, and weight loss in this subgroup of RA may be of particular value because the extra weight is a burden on the affected joints. Because overweight people with RA have less FFM than overweight people who are free of disease, the preservation of FFM should be an important goal of an effective weight loss program. However, it was reported that a 4.5-kg diet-induced weight loss, induced over a 12-week period, in obese participants with RA was associated with a 1.7-kg reduction in FFM. This represents 38% of the total weight loss. Similar findings were made in overweight individuals with RA in response to a 4.4-kg weight loss induced by the combination of diet and exercise. That almost 40% of the weight loss was composed of FFM in these two studies is concerning considering the reduced FFM in people with RA. This is a substantial percentage given that in healthy subjects who lose weight at a similar rate (<1 kg/week), only approximately 25% and approximately 15% of weight loss induced by diet alone or the combination of diet and exercise, respectively, consists of FFM (Miller et al. 1997). These findings argue against the efficacy of weight loss as a primary means of treating body composition abnormalities in RA.

Osteoarthritis

Osteoarthritis is the principle form of degenerative arthritis. The condition is primarily observed in individuals older than 40 years, occurs to a greater extent in women than men, and usually occurs in the weight-bearing joints such as the knees and hips. Approximately 68% of individuals older than 55 years of age have radiographic evidence of osteoarthritis in one or more joints (Elders 2000). Osteoarthritis damages and thins the joint cartilage, resulting in joint stiffness, pain, and swelling. By definition, osteoarthritis is not an inflammatory condition. However, in the later stages of osteoarthritis, substantial amounts of cartilage matrix components and wear particles are released, resulting in a substantial and chronic inflammatory reaction in the synovia of the joint. Prolonged inflammation results in the development of bony ridges, spurs, and joint deformity. As with RA, inflammatory cytokines (e.g., IL-1β, TNF-α, IL-17) can be found in increased quantities in the synovia of affected joints in people with osteoarthritis (van den Berg 1999; Westacott and Sharif 1996). However, unlike RA, these cytokines are not increased systemically (van den Berg 1999; Westacott and Sharif 1996), and there is no evidence of accelerated metabolism or cachexia in osteoarthritis.

The association between obesity and osteoarthritis has been demonstrated in both cross-sectional and longitudinal cohort designs (Sowers et al. 1999; Zhang et al. 2000). Men and women with high body mass index (BMI) have an increased risk of developing osteoarthritis. Studies using more specific measures of body fat and fat distribution, such as skinfold thickness and waist–hip circumference ratio, have also shown a relationship between obesity and the development of osteoarthritis (Hochberg et al. 1991, 1995). However, after adjustment for BMI, the associations between skinfold thickness and waist–hip ratio with osteoarthritis are no longer significant. The absence of an independent association between specific body composition measures and osteoarthritis suggests that a biomechanical factor (e.g., increased force from excess body weight), rather than metabolic or systemic factors (e.g., cytokines), is the link between obesity and osteoarthritis, especially of the knees and hips. However, the nature of the link between obesity and osteoarthritis of the hands is less clear.

Because obesity itself is associated with drastic alterations in body composition, and because obesity often precedes osteoarthritis, it is difficult to determine the influence of osteoarthritis per se on body composition. To our knowledge, no longitudinal studies have examined changes in body composition in people with osteoarthritis. However, in a cross-sectional analysis, Toda and colleagues (2000) reported that the percentage of total body weight as FFM in the legs was significantly lower in women with knee osteoarthritis than in controls (19.2% vs. 21.0%). By comparison, the percentage of total body weight as FFM in the arms and trunk was not different in these two groups. Because the reduction in FFM in people with knee osteoarthritis appears to be limited to the legs, it is likely that this atrophy occurs from a decreased muscular activity secondary to joint pain.

Prospective population studies using bone densitometry and radiographs have revealed complex relationships between osteoarthritis and osteoporosis. Independent of body weight, women with osteoarthritis have a higher BMD at baseline. However, during follow-up, women with osteoarthritis have an increased rate of bone loss (Arden et al. 1999; Burger et al. 1996). These observations pose interesting questions about the causal relationship between osteoarthritis and BMD, for which there are no clear answers.

Overweight patients with osteoarthritis experience symptomatic relief with weight loss (McGoey et al. 1990; Toda et al. 2000). Exercise, including both aerobic (Fisher et al. 1991) and strength (O'Reilly et al. 1999) training, also alleviates osteoarthritis symptoms. In addition to being an effective treatment modality, weight loss is associated with a significant reduction in FM in people with osteoarthritis (Toda 2001; Toda et al. 2000). When caloric restriction is combined with exercise, the weight loss consists primarily of FM, with FFM accounting for approximately 20% to 30% of the weight loss (Toda 2001; Toda et al. 2000). However, 8 weeks of aerobic exercise in the absence of weigh loss was associated with no significant changes in body composition in participants with osteoarthritis (Toda 2001). Taken together, these findings suggest that the combination of diet and exercise is likely better than either treatment alone for reducing weight, improving body composition, and decreasing the symptoms of osteoarthritis.

Inflammatory Bowel Disease

Ulcerative colitis and Crohn's disease, collectively termed *inflammatory bowel disease (IBD)*, are characterized by chronic uncontrolled inflammation of the intestinal mucosa. Crohn's disease is characterized by transmural, patchy inflammation of any part of the gastrointestinal tract, although it is most common in the ileocecal area. The inflammation extends deep into the lining of the affected organ, causing pain and diarrhea. Ulcerative colitis is usually confined to the large bowel. The chronic inflammation kills the colic epithelial cells, causing ulcers, and makes the colon empty more frequently, causing diarrhea. Elevated levels of inflammatory mediators, including neuropeptides, oxygen metabolites, and cytokines, are present in the mucosal tissue and circulation of patients with IBD (Rogler and Andus 1998).

A number of studies have examined changes in body composition in patients with Crohn's disease (Capristo et al. 1998a, 1998b; Mingrone et al. 1996; Royall et al. 1995; Tjellesen et al. 1998). The weighted evidence suggests that patients with active Crohn's disease have reduced body weight, which is accompanied by a decrease in FM and FFM. For example, Mingrone and colleagues (1996) observed that FFM and FM were 12% and 45% lower, respectively, in participants with Crohn's disease than in healthy control subjects (Mingrone et al. 1996). Similar findings were made in a study of individuals with Crohn's disease in remission (Tjellesen et al. 1998).

The proposed mechanisms for the changes in body weight and body composition in Crohn's disease are multifactorial. Low food intake secondary to anorexia and abdominal pain is an important factor to consider. So too is malabsorption caused by intestinal inflammation or intestinal resection surgery. Increased production of the inflammatory cytokines may also play a central role. The serum concentrations of IL-6 and TNF-α are increased in patients with Crohn's disease (Rogler and Andus 1998). These cytokines influence muscle protein turnover, resulting in net muscle protein degradation and muscle wasting. As with many other diseases, there is a decreased physical activity in Crohn's disease (Capristo et al. 1998b; Stokes and Hill 1993) that would tend to favor FFM loss. Finally, increased fat oxidation in both resting and postprandial states in people with Crohn's disease may explain the reduced FM (Al-Jaouni et al. 2000; Mingrone et al. 1996).

There is a high prevalence of osteoporosis in patients with Crohn's disease (Abitbol et al. 1995; Staun et al. 1997). Low BMD in these patients could be related to a number of factors, including the influence of corticosteroid treatment (Abitbol et al. 1995) and reduced calcium intake and absorption. It is also possible that the high prevalence of osteoporosis in Crohn's disease occurs secondary to reduced body weight, FFM, or physical activity, because these factors are positively related to BMD.

Unlike Crohn's disease, ulcerative colitis seems to entail very few body composition changes (Ulivieri et al. 2000, 2001). In fact, in a 6-year follow-up study it was reported that the baseline and change scores in BMD, FM, and FFM were not different in men with mild ulcerative colitis compared with controls matched for age and BMI (Ulivieri et al. 2001). Similar observations were found in women, with the exception that baseline FFM was 10% lower in women with ulcerative colitis (Ulivieri et al. 2001). Although these authors sug-

gested that the 10% difference in FFM between the female patients and their controls was not clinically relevant, we argue to the contrary. A 10% difference in FFM is equivalent to the reduction in FFM that would be expected to occur in a healthy adult when aging 10 years (Roubenoff and Hughes 2000) and represents approximately one fourth of the reduction in FFM observed in patients dying of starvation, cancer, and AIDS (Kotler et al. 1989; Tellado et al. 1989; Winick 1979).

Two studies examined the effect of total enteral nutrition on body composition in individuals with Crohn's disease (Royall et al. 1994, 1995). These studies reported modest increases in weight and FFM. For example, weight was increased by 2 kg after 3 weeks of enteral nutrition. The weight gain consisted of 65% water, 18% fat, and 18% protein, thus comprising a normal proportion of body composition (Royall et al. 1995). Little is known about the effects of exercise in patients with IBD, and no studies have examined exercise-induced changes in body composition in patients with Crohn's disease or ulcerative colitis.

Congestive Heart Failure

Congestive heart failure (CHF) is the end stage of many diseases of the heart and is a major cause of morbidity and mortality. The prevalence of CHF exceeds 10% in those aged 65 years and above (Davis et al. 2000). CHF is an imbalance in pump function in which the heart fails to adequately maintain the circulation of blood. This chronic disease is characterized by fatigue, shortness of breath, congestion, and cachexia. CHF can be subdivided into systolic and diastolic dysfunction. Systolic dysfunction is characterized by a dilated left ventricle with impaired contractility, whereas diastolic dysfunction occurs in a normal left ventricle with impaired ability to receive and eject blood. There is a growing body of evidence that immunological responses mediated by cytokines and other growth factors may play an important pathogenic role in the development of CHF. A number of studies have demonstrated that patients with CHF express excessive levels of numerous inflammatory cytokines (e.g., TNF-α, IL-1β, IL-6, interferon-γ) in the circulation (Blum and Miller 2001; Sasayama et al. 1999).

Cardiac cachexia is a term used to describe the loss in body weight and FFM that is characteristic of many CHF patients. Specifically, skeletal muscle atrophy is found in up to 68% of these patients (Mancini et al. 1992). Thomas and colleagues

(1979) observed that body weight was 21% lower in participants with CHF than in healthy controls and that the reduction in body weight consisted almost entirely of FFM. However, others, including Anker and colleagues (1999b) and Toth and colleagues (1997), reported a significant reduction in both FFM and FM in patients with cardiac cachexia. Regardless of the discrepancies in FM, it is clear that muscle wasting is a common occurrence in CHF, particularly in patients who lose body weight. However, many CHF patients maintain a normal body weight and body composition (Anker et al. 1999b; Toth et al. 1997). Muscle weakness and fatigue, common symptoms of CHF, occur primarily in patients with cachexia and muscle wasting (Anker et al. 1997b; Harrington et al. 1997).

Cardiac cachexia is a multifactorial disorder with many neuroendocrine and metabolic origins. Anorexia is often considered to be a main cause of cardiac cachexia. Anorexia can be related to CHF via its main symptoms (fatigue and dyspnea), intestinal edema, or protein-losing gastroenteropathy or as a side effect of sodium-restricted diets and some angiotensin-converting enzyme inhibitors (Anker and Coats 1999; Buchanan et al. 1977). Physical inactivity has also been implicated as an important cause of the muscle atrophy observed in CHF (Mancini et al. 1992). Physical activity–induced energy expenditure is about 65% lower in CHF patients with cachexia compared with healthy controls (Toth et al. 1997). Increased production of inflammatory cytokines such as TNF-α, IL-1β, and IL-6 is also a hallmark of CHF (Anker and Coats 1999; Anker et al. 1999b). As previously alluded to, these cytokines have a powerful influence on muscle protein turnover, resulting in net muscle protein degradation and muscle wasting. Indeed, the levels of catabolic cytokines are significantly correlated with reduced FFM in people with CHF (Anker and Coats 1999; Anker et al. 1999b). Other neuroendocrine changes that appear in cardiac cachexia and that may be related to muscle wasting include markedly increased catecholamine levels (Anker et al. 1997a) that are accompanied by an increase in sympathetic nervous system activity (Anker et al. 1997a) and REE (Toth et al. 1997), a 2.5-fold increase in cortisol levels (Nicholls et al. 1974), and a reduction in the anabolic hormone dehydroepiandrosterone (Anker et al. 1997a). Reduced insulin sensitivity, as observed in CHF patients (Swan et al. 1994), may also contribute to muscle wasting, because insulin is an important anabolic hormone that also inhibits protein degradation within skeletal muscle (Nair et al. 1995).

There is evidence that BMD is reduced in CHF patients, particularly in cardiac cachexia. BMD is similar in people with noncachectic CHF disease and controls but is approximately 5% lower in individuals with cachectic CHF than in controls (Anker et al. 1997a, 1999a, 1999b). The loss in BMD correlates with the loss in FM and FFM (Anker et al. 1999a), suggesting that BMD loss may be secondary to a decreased mechanical stress placed on the skeleton.

The effect of anticytokine therapy on body composition has not been studied in people with CHF. Nutritional support techniques in cardiac cachexia include oral dietary supplementation, tube feeding, and total parenteral nutrition. In addition, exercise may be an essential treatment modality in CHF. CHF is a vicious cycle such that disease symptoms lead to exercise intolerance, which in turn leads to physical inactivity, greater physical deconditioning, and a worsening of the situation (Coats et al. 1994; Piepoli et al. 2001). Thus, exercise training might be able to slow down this cycle and reverse some of the consequences of CHF. Indeed, clinical trials in patients with CHF have demonstrated increases in exercise capacity and muscle mass with exercise (Coats et al. 1990, 1992). Many other physiological gains have also been observed with exercise training including improved autonomic control of the heart and circulation (Radaelli et al. 1996) and enhanced resting stroke volume and central hemodynamics (Coats et al. 1990; Hambrecht et al. 2000). Therefore, an exercise training program may benefit both the clinical symptoms and body composition changes observed in CHF. If performed correctly and under proper supervision, both aerobic and resistance exercises are safe and effective for people with CHF (Meyer 2001).

Chronic Obstructive Pulmonary Disease

Chronic obstructive pulmonary disease (COPD) is a term used to describe airflow obstruction caused by emphysema or chronic bronchitis. Smoking is the leading cause of COPD, accounting for 80% to 90% of all cases. Emphysema weakens and breaks the air sacs within the lungs, which causes airways to collapse and airflow obstruction to occur. Chronic bronchitis is an inflammatory disease that begins in the smaller airways and gradually advances to larger airways. It increases mucus in the airways and bacterial infections in the bronchial tubes, which in turn impedes airflow. COPD is characterized by chronic inflammation in the airways or alveoli, involving increased numbers of neutrophils, macrophages, CD8+ T cells, or mast cells in the airway walls, alveolar compartments, and vascular smooth muscle. Cytokines associated with COPD inflammation include TNF-α, IL-1β, and IL-6 (De Boer 2002).

A comparison of 79 participants with COPD to a group of healthy controls with a similar age and body weight found a moderately lower (3.7 kg) FFM in those with COPD (Engelen et al. 1998). It has also been reported that muscle cross-sectional area in the midthigh is approximately 25% lower in people with COPD than in age-, sex-, and weight-matched controls (Bernard et al. 1998). The decreased muscle mass in COPD is related to muscle weakness, reduced physical performance, and enhanced disease symptoms (Bernard et al. 1998; Hamilton et al. 1995).

Although the two studies mentioned in the proceeding paragraph matched the COPD patients and control subjects for body weight, a negative energy balance is a common occurrence in COPD (Schols et al. 1989). This negative energy balance leads to cachexia and an even greater loss of FFM. For example, Engelen and colleagues (1994) reported that COPD patients with cachexia have a much lower weight (23 kg) and FFM (13 kg) than COPD patients without cachexia. In addition to a reduced muscle mass, BMD is also lower in people with COPD than in healthy controls, with greater than 50% of people with COPD having osteopenia or osteoporosis (Engelen et al. 1998).

As with many of the other inflammatory diseases already discussed, the causes of body composition abnormalities in COPD may be related to malnutrition, decreased physical activity, the use of corticosteroids, and an increase in inflammatory cytokine production. Currently available pharmacological therapies are aimed at improving the clinical symptoms of COPD and reducing the airway inflammation. The commonly used corticosteroids reduce inflammation by acting on kinases or transcription factors necessary for expression of inflammatory cytokines. New anticytokine therapies have also been shown to reduce the inflammation and alleviate the clinical symptoms of COPD. However, the effect of corticosteroids and anticytokine therapies on body composition in COPD is unknown.

Exercise training is now recognized as an essential component of pulmonary rehabilitation (ACCP/AACVPR Pulmonary Rehabilitation Guidelines Panel 1997). However, limited information exists on the effects of exercise on body composition in people with COPD. Bernard and colleagues (1999) found that 12 weeks of aerobic exercise in participants with COPD did not significantly affect muscle size, as assessed by computed tomography imaging in the midthigh. However, in this study, the combination of aerobic and strength training had a significantly greater effect on muscle size and strength than did aerobic exercise alone (figure 26.2). Although the addition of resistance exercise improved skeletal muscle characteristics, the improvements in exercise capacity and quality of life were similar in response to aerobic exercise alone or the combination of aerobic and resistance exercise.

Systemic Lupus Erythematosus

Although there are several forms of lupus, systemic lupus erythematosus (SLE) is the most common. SLE primarily affects premenopausal women, and the prevalence varies from 1 in 1,000 in African American women to 1 in 4,300 in Caucasian women (Morrow et al. 1999). SLE is a systemic autoimmune disease that leads to inflammation and damage to various tissues including the joints, skin, kidneys, heart, lungs, blood vessels, and brain. Most people with SLE demonstrate systemic manifestations such as fatigue, painful or swollen joints, muscle pain, fever, anorexia, and weight loss. Many cytokines have been implicated in regulating disease activity in people with SLE, including TNF-α, IL-1β, IL-6, and IL-10 (Dean et al. 2000).

Few studies have examined the effect of SLE on body composition, and to our knowledge no studies have examined the effects of nutritional and exercise interventions on body composition in people with SLE. Kipen and colleagues (1998) examined the relationship between disease severity and measures of body composition in 82 women with SLE. In this cohort, disease severity was independently and negatively related to BMD and FFM. In a 3-year follow-up study of 55 of these women, SLE disease severity was predictive of an increase in FM (Kipen et al. 1999). Furthermore, there was an independent positive association between changes in exercise participation and FFM, highlighting the potential importance of physical activity as a means of treating body composition abnormalities in SLE.

The altered body composition in people with SLE may also be explained by the high use of corticosteroids. Corticosteroid exposure has a negative relationship with BMD and FFM in people with SLE

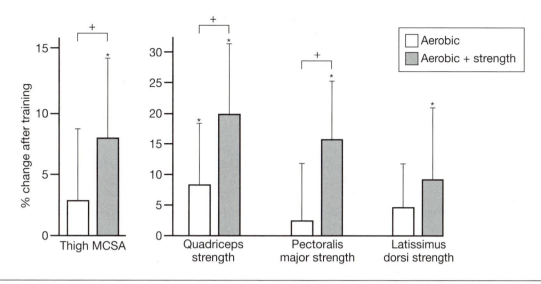

Figure 26.2   Mean ± standard deviation percent change in midthigh muscle cross-sectional area (MCSA) and in strength of the quadriceps, pectoralis major, and latissimus dorsi muscles after 12 weeks of aerobic exercise training or aerobic plus strength training in patients with chronic obstructive pulmonary disease. Midthigh MCSA and strength of all three muscle groups was increased ($p < .05$) in the aerobic + strength group (*), and with the exception of latissimus dorsi strength, these changes were greater ($p < .05$) than those in the aerobic group (†).

Reprinted, by permission, from S. Bernard et al., 1999, "Aerobic and strength training in patients with chronic obstructive pulmonary disease," *American Journal of Respiratory and Critical Care Medicine* 159: 896-901.

(Kipen et al. 1998). In addition, many of the cytokines that have a strong effect on energy and protein metabolism—TNF-α, IL-1β, IL-6—are elevated in SLE (Dean et al. 2000), which may explain the loss in FFM observed in these individuals.

Inflammatory Diseases and Body Composition in Children and Adolescents

As with adults, children and adolescents can experience profound alterations in body composition attributable to changes in energy and protein metabolism. In addition, an altered energy and protein metabolism in children and adolescents can have a profound effect on normal growth and development. This section reviews energy metabolism, body composition, and growth and development patterns in children and adolescents with three common inflammatory diseases: juvenile rheumatoid arthritis, inflammatory bowel disease, and cystic fibrosis.

Juvenile Rheumatoid Arthritis

Juvenile RA is one of the more chronic pediatric illnesses, with an incidence rate equal to that of type 1 diabetes (Kunnamo et al. 1986). Juvenile RA

begins before the age of 16, and girls are affected twice as often as boys (Sullivan et al. 1975). Unlike adult-onset RA, juvenile-onset RA is a heterogeneous disease with at least five clinical subtypes. The etiology and pathogenesis of juvenile RA are unclear; however, abnormal cytokine production may be involved in the pathophysiology of the different forms of juvenile RA. This is demonstrated by the fact that the serum levels of several inflammatory cytokines (e.g., IL-1β, IL-6, TNF-α) increase during the active phase of the disease (Mangge and Schauenstein 1998; Yilmaz et al. 2001).

Abnormalities in growth and development are common in juvenile RA (Bernstein et al. 1977). For a given chronological age, the mean height of juvenile RA patients is approximately four standard deviations below the population mean (Touati et al. 1998). It has been suggested that the abnormal growth in juvenile RA is related to the underlying increase in inflammatory activity and is exacerbated by the prolonged use of corticosteroids (Sturge et al. 1970).

Few studies have examined the effects of juvenile RA on alterations in FFM and FM, and the results of these studies are contradictory. For example, Haugen and colleagues (1992) reported that upper-arm muscle area was similar in children with pauciarticular juvenile RA but lower in children with polyarticular juvenile RA compared with healthy controls. By comparison, Bacon and colleagues (1990) reported that upper-arm muscle area was noticeably smaller in both pauciarticular and polyarticular juvenile RA patients than healthy controls. The results for BMD are clearer than those for the soft tissues, because an increase in osteoporosis and related fractures in children with RA is well documented (Elsasser et al. 1982; Varonos et al. 1987).

Although the effects of nutritional and exercise interventions on growth, development, and body composition in juvenile RA are unknown, three studies have examined the effects of growth hormone on these parameters (Simon et al. 1999, 2000; Touati et al. 1998). In these studies, daily injections of growth hormone to children with juvenile RA over one year or more substantially increased growth velocity and FFM while decreasing FM. Thus, growth hormone treatment may counteract many of the adverse effects of corticosteroids in juvenile RA.

Inflammatory Bowel Disease

IBD in children and adolescents is very similar to that seen in adults, with the two subtypes of Crohn's disease and ulcerative colitis. However, because of the increased importance of proper nutrient intake and absorption during the growing years, the clinical implications of IBD may be more severe in children and adolescents than adults.

Impaired growth is a common complication in children and adolescents with IBD. The impaired growth can precede the diagnosis of the disease (Kanof et al. 1988) and can occur in the absence of malabsorption (Kirschner et al. 1981) or reduced caloric intake (Motil et al. 1993). An example of the relationship between IBD and impaired growth is depicted in a prospective study conducted by Motil and colleagues (1993). Over a 3-year period they measured growth in children with IBD and found that approximately 30% had impaired growth regardless of pubertal development. In a retrospective analysis of young adults with juvenile-onset IBD, Markowitz and colleagues (1993) found that 31% had permanently impaired growth. Individuals with Crohn's disease appear to have a greater impairment in growth than individuals with ulcerative colitis (Motil et al. 1993).

We are aware of two studies that examined the relationship between IBD and body composition in children and adolescents, and both of these studies were conducted in participants with Crohn's disease. Sentongo and colleagues (2000) reported that boys with Crohn's disease had 3.5 kg less FFM

and girls with Crohn's disease had 2.9 kg less FFM than control subjects, after adjustment for age. Boys with Crohn's disease had a similar FM, but girls with Crohn's disease had a 3.7 kg greater FM than control subjects. Azcue and colleagues (1997) reported that body weight was 11.1 kg (23%) lower and FFM was 7.7 kg (22%) lower in children with Crohn's disease than in controls matched for age and sex.

There is also evidence to suggest that, as with FFM, BMD is reduced in children with IBD (Boot et al. 1998; Cowan et al. 1997; Issenman et al. 1993). The mean lumbar and spine BMD values in children with IBD are 0.75 and 0.95 standard deviations below the mean of normal values, respectively, with people with Crohn's disease having a lower BMD than those with ulcerative colitis (Boot et al. 1998). The reduced BMD is concerning because not only do children with osteopenia have a higher fracture risk in childhood, but they also have an increased fracture risk in adulthood because they do not reach their optimal peak BMD.

The causes of the body composition abnormalities in juvenile-onset IBD may be related to anorexia, malabsorption, decreased physical activity, use of corticosteroids, and increased inflammatory cytokine production. The effect of exercise on body composition in children with IBD is unknown. However, parenteral nutrition is associated with an improvement in disease activity and FFM in children with Crohn's disease and ulcerative colitis (Khoshoo et al. 1996; Lin et al. 1989). Moreover, four months of growth hormone treatment increased growth velocity and FFM in children with IBD who were on corticosteroid therapy (Mauras et al. 2002).

Cystic Fibrosis

Cystic fibrosis is the most common heredity disease in Caucasians, occurring in 1 of every 2,500 births (Cystic Fibrosis Foundation 1998). Many people with cystic fibrosis die young, but the prognosis of cystic fibrosis has improved greatly in recent decades such that the mean survival time is now 30 years (FitzSimmons 1993). The genetic defect present in cystic fibrosis leads to an inability to transport chloride ions in response to agents that elevate intracellular cyclic adenosine monophosphate in the epithelial cells of nearly all exocrine glands (Riordan et al. 1989). This results in abnormal glandular secretions that affect the function of the endocrine gland. In the pancreas and intestinal endocrine glands, the secretions are thick and can completely block the gland. In many people with cystic fibrosis, this compromises nutrient digestion and ultimately results in a stunted growth and development.

Also of importance to people with cystic fibrosis are the abnormal secretions of the mucus-producing glands in the airways. These glands produce abnormally thick secretions that clog the lung airways and allow bacteria to multiply. This bacterium colonizes with a high affinity for the airways, and despite antibody therapy, it persists for a long time (Hata and Fick 1988). The early stage of bacterial infection is characterized by the generation of a cascade of endogenous host mediators, including the inflammatory cytokines TNF-α, IL-6, and IL-8 (Bonfield et al. 1995). However, despite high levels of cytokines in the bronchial tree, infection is long lasting in the cystic fibrosis lung. This infection is a major problem because recurrent bronchitis and pneumonia gradually destroy the lungs. Death in people with cystic fibrosis usually results from a combination of respiratory and heart failure caused by the underlying lung disease.

Despite recent advances in the treatment of bacterial infection and emphasis on adequate nutritional intake, many people with cystic fibrosis have a less than ideal body weight and height. The Cystic Fibrosis Foundation reports that 23% of people with cystic fibrosis are below the 5th percentile for weight and 18% are less than the 5th percentile for height (Cystic Fibrosis Foundation 1998). These findings are believed to be attributable, in part, to inadequate nutrition and malabsorption (Shepherd 2002; Stettler et al. 2000). People with cystic fibrosis also manifest a number of hormonal abnormalities that may contribute to reduced growth and development, including decreased insulin release, decreased insulin sensitivity, and low circulating levels of insulin-like growth factor (Holl et al. 1997; Taylor et al. 1997).

The previously mentioned nutritional and hormonal abnormalities may also contribute to the body composition irregularities that are apparent in people with cystic fibrosis. Specifically, cystic fibrosis is characterized by a modestly reduced FFM. For example, Borovnicar and colleagues (2000) reported that the mean body cell mass in people with cystic fibrosis is approximately 5% lower than the age- and height-adjusted population mean. The reduced FFM in cystic fibrosis is related to a reduced inspiratory muscle function, skeletal muscle strength, and cardiorespiratory fitness (Gulmans et al. 1997; Ionescu et al. 1998).

A high prevalence of low BMD in the spine and appendicular regions has been reported in children and adolescents with cystic fibrosis (Bhudhikanok et al. 1996; Henderson and Madsen 1999). There is evidence that the reduced BMD in people with cystic fibrosis is associated with an increased fracture risk during adolescence (Henderson and Specter 1994). Furthermore, low BMD is a recognized problem in patients awaiting lung transplantation and those with end-stage disease (Aris et al. 1998). The reduced FFM and BMD in people with cystic fibrosis may be caused by a number of factors, including protein-energy malnutrition, a reduced weight, a lack of weight-bearing physical activity, and a chronic inflammatory state.

Using a double-blind placebo-controlled crossover design, Bucuvalas and colleagues (2001) investigated the effects of insulin-like growth factor therapy on growth and body composition in seven children with cystic fibrosis. These authors observed no significant difference in linear growth rate, weight gain, or the accretion of FFM during six months of treatment. Hardin and colleagues (2001) investigated the effects of growth hormone therapy on growth and body composition in a 12-month randomized controlled trial in 19 children with cystic fibrosis. The group given daily injections of growth hormone had significantly greater increases in height, weight, and FFM than the control group (figure 26.3). In addition, a decreased number of hospitalizations, decreased use of antibiotics, and improvement in lung function were observed in the growth hormone group alone, indicating that growth hormone therapy improves growth and clinical status in children with cystic fibrosis. Two studies investigated the effects of megastrol acetate, a powerful appetite stimulant, on changes in body composition in children with cystic fibrosis (Eubanks et al. 2002; Marchand et al. 2000). Both studies observed an increase in body weight, FM, FFM, and lung function after approximately 12 weeks of megastrol acetate therapy.

Exercise training aimed at improving aerobic fitness is recognized as an essential component of pulmonary rehabilitation (ACCP/AACVPR Pulmonary Rehabilitation Guidelines Panel 1997). Exercise training improves exercise tolerance in people with cystic fibrosis (de Meer et al. 1999; Selvadurai et al. 2002), and patients with high levels of aerobic fitness have much better long-term survival than those with lower levels of fitness (Nixon et al. 1992). Despite the positive effects on clinical outcomes and survival, aerobic

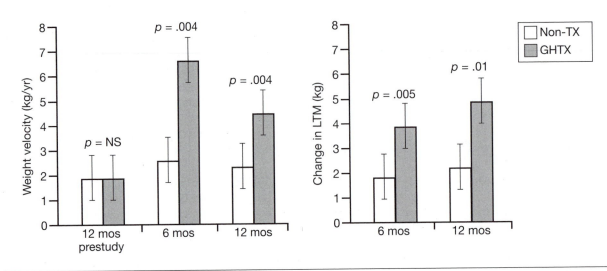

Figure 26.3 *Left panel*: Weight velocity (kg/year) for the 12-month period preceding the study and at 6 and 12 months after beginning the study in the control group (Non-TX, open bars) and growth hormone treatment group (GHTX, shaded bars). Weight velocity was greater ($p = .004$) at 6 and 12 months in the growth hormone treatment group. *Right panel*: Changes in fat-free mass after 6 and 12 months of treatment. At both time points the changes were greater ($p \leq .01$) in the growth hormone treatment group than the control group.

Reprinted from *Journal of Pediatry*, 139, D.S. Hardin et al., Growth hormone improves clinical status in prepubertal children with cystic fibrosis: Results of a randomized controlled trial, 636-42, Copyright 2001, with permission from Elsevier.

exercise has little or no effect on body weight and composition in people with cystic fibrosis (de Meer et al. 1999; Selvadurai et al. 2002). However, a recent study reported that resistance exercise was associated with a significant increase in body weight (2.7 kg) in hospitalized persons with cystic fibrosis, and the increased body weight consisted almost entirely of FFM (2.4 kg; Selvadurai et al. 2002). Nonetheless, in that study, aerobic exercise training had a better effect than resistance training on overall activity levels and quality of life (Selvadurai et al. 2002). Additional studies are required to clarify whether aerobic exercise, resistance exercise, or a combination of the two exercise modalities is the most effective treatment for people with cystic fibrosis.

Summary

Body composition changes are a hallmark of many inflammatory diseases. The alterations in body composition in inflammatory disease appear to be explained in part by changes in mediators of the immune system, including the inflammatory cytokines. The changes in body composition in people with inflammatory disease are related to numerous clinical outcomes. Specifically, the reduction in FFM that is observed in many inflammatory diseases is associated with a compromised immune function, muscle weakness, functional impairment, a reduced quality of life, and poor clinical outcomes. In addition, the reduced BMD that is characteristic of inflammatory disease is associated with an increased prevalence of osteoporosis and fractures. Nutritional and exercise interventions have proven to be safe and effective treatments for minimizing or reversing the body composition changes that occur in many inflammatory diseases. Future studies are needed to investigate the safety and effectiveness of new pharmacological treatments, such as anticytokine therapy, in preventing or normalizing changes in body composition in different inflammatory diseases. Future studies are also needed to better describe the optimal mode (e.g., aerobic exercise, resistance exercise, or both) and dose of exercise required to improve body composition in various inflammatory diseases.

Appendix: Reference Body Composition Tables

Richard N. Pierson, Jr.

Table A.1 Reference Fetus

The Reference Fetus database contains body composition information at the atomic and molecular levels from 24 to 40 gestational weeks. Chemical analysis data include water, nitrogen, fat, calcium, phosphorus, magnesium, sodium, potassium, and chlorine. The development of the Reference Fetus involved the following steps: selection of data from the literature, establishment of regression equations for each component in relation to age, and use of these regression equations for reconstruction of body composition (Ziegler et al. 1976).

Tables A.2-A.5 Reference Boy and Girl

There are three major reports of Reference Boy and Girl (Butte et al. 2000; Foman and Nelson 2002; Fomon et al. 1982). The database of Butte and colleagues only contains body composition information during the first 2 years of life, whereas the database of Fomon and colleagues provides information from birth to age 10 years. The values from Butte and colleagues are the group means, and thus the sum of fat and fat-free mass may not exactly equal body mass. Tables A.2 and A.3 show

the molecular level body composition on the basis of whole-body weight, whereas tables A.4 and A.5 provide body composition on the basis of fat-free mass.

Tables A.6-A.8 Reference Man and Woman

The Reference Man is a compilation of many research reports and is periodically updated by the International Commission on Radiological Protection (Snyder et al. 1975). The current Reference Man is between 20 and 30 years of age, weighs 70 kg, is 170 cm in height, and lives in a climate with an average temperature of 10° to 20°C. Reference Man is a Caucasian and Western European or North American in habitat and custom. Table A.6 provides body composition information on the atomic and tissue–organ levels, whereas table A.7 contains data from the molecular level. Although less complete, some body composition and functional information is available on a Reference Woman and is presented in table A.8 (Elia 1992; Snyder et al. 1975).

Table A.1 Body Composition of Reference Fetus

Gestational age (weeks)	PER 100 G OF BODY WEIGHT					PER 100 G OF FAT-FREE MASS							
	BW (g)	Water (g)	Protein (g)	Lipid (g)	Other (g)	Water (g)	Protein (g)	Ca (mg)	P (mg)	Mg (mg)	Na (mmol)	K (mmol)	Cl (mmol)
24	690	88.6	8.8	0.1	2.5	88.6	8.8	621	387	17.8	9.9	4.0	7.0
25	770	87.8	9.0	0.7	2.5	88.4	9.1	615	385	17.6	9.8	4.0	7.0
26	880	86.8	9.2	1.5	2.5	88.1	9.4	611	384	17.5	9.7	4.1	7.0
27	1,010	85.7	9.4	2.4	2.5	87.8	9.7	609	383	17.4	9.5	4.1	6.9
28	1,160	84.6	9.6	3.3	2.4	87.5	10.0	610	385	17.4	9.4	4.2	6.9
29	1,318	83.6	9.9	4.1	2.4	87.2	10.3	613	387	17.4	9.3	4.2	6.8
30	1,480	82.6	10.1	4.9	2.4	86.8	10.6	619	392	17.4	9.2	4.3	6.8
31	1,650	81.7	10.3	5.6	2.4	86.5	10.9	628	398	17.6	9.1	4.3	6.7
32	1,830	80.7	10.6	6.3	2.4	86.1	11.3	640	406	17.8	9.1	4.3	6.6
33	2,020	79.8	10.8	6.9	2.5	85.8	11.6	656	416	18.0	9.0	4.4	6.5
34	2,230	79.0	11.0	7.5	2.5	85.4	11.9	675	428	18.3	8.9	4.4	6.4
35	2,450	78.1	11.2	8.1	2.6	85.0	12.2	699	443	18.6	8.9	4.5	6.3
36	2,690	77.3	11.4	8.7	2.6	84.6	12.5	726	460	19.0	8.8	4.5	6.1
37	2,940	76.4	11.6	9.3	2.7	84.3	12.8	758	479	19.5	8.8	4.5	6.0
38	3,160	75.6	11.8	9.9	2.7	83.9	13.1	795	501	20.0	8.8	4.5	5.9
39	3,330	74.8	11.9	10.5	2.8	83.6	13.3	836	525	20.5	8.7	4.6	5.8
40	3,450	74.0	12.0	11.2	2.8	83.3	13.5	882	551	21.1	8.7	4.6	5.7

Note. BW = body weight.

Table A.2 Body Composition of Reference Boy

Age	LENGTH (cm)		BM (g)		FAT (g)		FAT (%)		FFM (g)		COMPONENT OF FFM (% OF BODY MASS)						
											PROTEIN	TBW	ECW	ICW	MO	MS	CHO
	F	B	F	B	F	B	F	B	F	B	F	F	F	F	F	F	F
Birth	51.6		3,545		486		13.7		3,059		12.9	69.6	42.5	27.0	2.6	0.6	0.5
0.5 mo		52.5		3,760		440		11.4		3,350							
1 mo	54.8		4,452		671		15.1		3,781		12.9	68.4	41.1	27.3	2.6	0.6	0.5
2 mo	58.2		5,509		1,095		19.9		4,414		12.3	64.3	38.0	26.3	2.4	0.6	0.5
3 mo	61.5	61.2	6,435	6,330	1,495	1,910	23.2	30.2	4,940	4,370	12.0	61.4	35.7	25.8	2.3	0.6	0.5
4 mo	63.9		7,060		1,743		24.7		5,317		11.9	60.1	34.5	25.7	2.3	0.5	0.4
5 mo	65.9		7,575		1,913		25.3		5,662		11.9	59.6	33.8	25.8	2.3	0.5	0.4
6 mo	67.6	67.9	8,030	8,040	2,037	2,320	25.4	29.1	5,993	5,630	12.0	59.4	33.4	26.0	2.3	0.5	0.4
9 mo	72.3	72.2	9,180	9,130	2,199	2,340	24.0	25.7	6,981	6,710	12.4	60.3	33.0	27.2	2.3	0.6	0.5
12 mo	76.1	76.1	10,150	10,030	2,287	2,560	22.5	25.6	7,863	7,400	12.9	61.2	32.9	28.3	2.3	0.6	0.5
18 mo	82.4	82.6	11,470	11,430	2,382	2,830	20.8	24.5	9,088	8,550	13.5	62.2	32.3	29.9	2.5	0.6	0.5
24 mo	87.2	87.6	12,590	12,460	2,456	3,100	19.5	25.4	10,134	9,130	14.0	62.9	31.9	31.0	2.6	0.6	0.5
3 yr	95.3		14,675		2,576		17.5		12,099		14.7	63.9	31.1	32.8	2.8	0.6	0.5
4 yr	102.9		16,690		2,656		15.9		14,034		15.3	64.8	30.5	34.2	2.9	0.6	0.5
5 yr	109.9		18,670		2,720		14.6		15,950		15.8	65.4	30.0	35.4	3.1	0.6	0.5
6 yr	116.1		20,690		2,795		13.5		17,895		16.2	66.0	29.6	36.4	3.2	0.6	0.5
7 yr	121.7		22,850		2,931		12.8		19,919		16.5	66.2	29.1	37.1	3.3	0.6	0.5
8 yr	127.0		25,300		3,293		13.0		22,007		16.6	65.8	28.3	37.5	3.4	0.6	0.5
9 yr	132.2		28,130		3,724		13.2		24,406		16.8	65.4	27.6	37.8	3.5	0.6	0.5
10 yr	137.5		31,440		4,318		13.7		27,122		16.8	64.8	26.7	38.0	3.5	0.6	0.5

Note. B = data from Butte et al. 2000; BM = body mass; CHO = carbohydrates; ECW = extracellular water; F = data from Foman and Nelson 2002 and Fomon et al. 1982; ICW = intracellular water; FFM = fat-free mass; mo = month; MO = bone mineral; MS = soft tissue mineral; TBW = total-body water; yr = year.

Table A.3 Body Composition of Reference Girl

Age	LENGTH (cm)		BM (g)		FAT (g)		FAT (%)		FFM (g)		COMPONENT OF FFM (% OF BODY MASS)						
											Protein	TBW	ECW	ICW	MO	MS	CHO
	F	B	F	B	F	B	F	B	F	B	F	F	F	F	F	F	F
Birth	50.5		3,325		495		14.9		2,830		12.8	68.6	42.0	26.7	2.6	0.6	0.5
0.5 mo		52.0		3,640		520		14.2		3,120							
1 mo	53.4		4,131		668		16.2		3,463		12.7	67.5	40.5	26.9	2.5	0.6	0.5
2 mo	56.7		4,989		1,053		21.1		3,936		12.2	63.2	37.1	26.1	2.4	0.6	0.5
3 mo	59.6	60.7	5,743	6,030	1,366	1,900	23.8	31.5	4,377	4,110	12.0	60.9	35.1	25.8	2.3	0.6	0.5
4 mo	61.9		6,300		1,585		25.2		4,715		11.9	59.6	33.8	25.8	2.3	0.5	0.4
5 mo	63.9		6,800		1,769		26.0		5,031		11.9	58.8	33.0	25.9	2.2	0.5	0.4
6 mo	65.8	66.5	7,250	7,600	1,915	2,440	26.4	32.0	5,335	5,210	12.0	58.4	32.4	26.0	2.2	0.5	0.4
9 mo	70.4	71.0	8,270	8,620	2,066	2,470	25.0	28.8	6,204	6,120	12.5	59.3	32.0	27.3	2.3	0.5	0.4
12 mo	74.3	75.3	9,180	9,500	2,175	2,620	23.7	27.6	7,005	6,880	12.9	60.1	31.8	28.3	2.3	0.5	0.5
18 mo	80.2	82.0	10,780	10,940	2,346	2,870	21.8	26.3	8,434	7,990	13.5	61.3	31.5	29.8	2.4	0.6	0.5
24 mo	85.5	87.7	11,910	12,020	2,433	3,050	20.4	25.4	9,477	8,990	13.9	62.2	31.5	30.8	2.4	0.6	0.5
3 yr	94.1		14,100		2,606		18.5		11,494		14.4	63.5	31.3	32.2	2.5	0.6	0.5
4 yr	101.6		15,960		2,757		17.3		13,203		14.8	64.3	31.2	33.1	2.5	0.6	0.5
5 yr	108.4		17,660		2,949		16.7		14,711		15.0	64.6	31.0	33.6	2.5	0.6	0.5
6 yr	114.6		19,520		3,208		16.4		16,312		15.2	64.7	30.8	34.0	2.6	0.6	0.5
7 yr	120.6		21,840		3,662		16.8		18,178		15.2	64.4	30.3	34.1	2.5	0.6	0.5
8 yr	126.4		24,840		4,319		17.4		20,521		15.2	63.8	29.6	34.2	2.5	0.6	0.5
9 yr	132.2		28,460		5,207		18.3		23,253		15.1	63.0	28.9	34.1	2.5	0.6	0.5
10 yr	138.3		32,550		6,318		19.4		26,232		15.0	62.0	28.1	33.9	2.5	0.6	0.5

Note. B = data from Butte et al. 2000; BM = body mass; CHO = carbohydrates; ECW = extracellular water; F = data from Foman and Nelson 2002 and Fomon et al. 1982; ICW = intracellular water; FFM = fat-free mass; mo = month; MO = bone mineral; MS = soft tissue mineral; TBW = total-body water; yr = year.

Table A.4 Body Composition of Fat-Free Mass of Reference Boy

Age	PROTEIN	TBW		ECW	ICW	MO	MS	CHO	TBK		D_FFM	
				COMPOSITION (% OF FFM)					(mmol/kg FFM)		(g/cm³)	
	F	F	B	F	F	F	F	F	F	B	F	B
Birth	15.0	80.6		49.3	31.3	3.0	0.7	0.6	49.0		1.063	
0.5 mo			82.7							48.6		1.054
1 mo	15.1	80.5		48.4	32.1	3.0	0.7	0.6	50.1		1.064	
2 mo	15.4	80.3		47.4	32.9	3.0	0.7	0.6	51.2		1.065	
3 mo	15.6	80.0	81.0	46.4	33.6	3.0	0.7	0.6	52.2	52.5	1.065	1.061
4 mo	15.8	79.9		45.8	34.1	3.0	0.7	0.6	53.0		1.066	
5 mo	15.9	79.7		45.2	34.5	3.0	0.7	0.6	53.6		1.066	
6 mo	16.0	79.6	80.7	44.7	34.9	3.0	0.7	0.6	54.1	53.2	1.066	1.062
9 mo	16.4	79.3	79.7	43.5	35.8	3.0	0.7	0.6	55.5	55.9	1.068	1.066
12 mo	16.6	79.0	79.3	42.5	36.5	3.0	0.7	0.6	56.5	57.0	1.068	1.068
18 mo	17.1	78.5	78.3	40.8	37.7	3.1	0.7	0.6	58.2	59.8	1.070	1.072
24 mo	17.4	78.1	77.0	39.6	38.5	3.2	0.7	0.6	59.3	63.3	1.072	1.077
3 yr	17.8	77.5		37.8	39.7	3.3	0.7	0.6	61.1		1.074	
4 yr	18.2	77.0		36.3	40.7	3.5	0.7	0.6	62.5		1.076	
5 yr	18.5	76.6		35.2	41.4	3.6	0.7	0.6	63.6		1.078	
6 yr	18.7	76.3		34.2	42.0	3.7	0.7	0.6	64.4		1.080	
7 yr	18.9	75.9		33.4	42.6	3.8	0.7	0.6	65.2		1.081	
8 yr	19.1	75.7		32.6	43.1	3.9	0.7	0.6	65.9		1.082	
9 yr	19.3	75.4		31.8	43.6	4.0	0.7	0.6	66.7		1.084	
10 yr	19.5	75.1		31.0	44.1	4.1	0.7	0.6	67.4		1.085	

Note. B = data from Butte et al. 2000; CHO = carbohydrates; D_{FFM} = density of fat-free mass; ECW = extracellular water; F = data from Fomon and Nelson 2002 and Fomon et al. 1982; ICW = intracellular water; FFM = fat-free mass; mo = month; MO = bone mineral; MS = soft tissue mineral; TBK = total body potassium; TBW = total body water; yr = year.

Table A.5 Body Composition of Fat-Free Mass of Reference Girl

| | COMPOSITION (% OF FFM) | | | | | | | | | | TBK | | | D_{FFM} | |
| | PROTEIN | TBW | | ECW | ICW | MO | MS | CHO | | (mmol/kg FFM) | | | (g/cm³) | |
Age	F	F	B	F	F	F	F	F		F	B		F	B
Birth	15.0	80.6		49.3	31.3	3.0	0.7	0.6		49.0			1.064	
0.5 mo			83.1								46.9			1.053
1 mo	15.2	80.5		48.3	32.1	3.0	0.7	0.6		50.2			1.064	
2 mo	15.5	80.2		47.1	33.1	3.0	0.7	0.6		51.5			1.065	
3 mo	15.8	79.9	81.1	46.0	33.9	3.0	0.7	0.6		52.7	52.0		1.066	1.060
4 mo	15.9	79.7		45.2	34.5	3.0	0.7	0.6		53.5			1.066	
5 mo	16.1	79.5		44.6	34.9	3.0	0.7	0.6		54.2			1.067	
6 mo	16.3	79.4	80.7	44.0	35.4	3.0	0.7	0.6		54.8	53.1		1.067	1.062
9 mo	16.6	79.0	79.8	42.7	36.4	3.0	0.7	0.6		56.3	55.5		1.068	1.066
12 mo	16.9	78.8	78.8	41.6	37.1	3.0	0.7	0.6		57.4	58.4		1.069	1.070
18 mo	17.2	78.4	78.2	40.3	38.1	3.0	0.7	0.6		58.8	59.8		1.070	1.072
24 mo	17.4	78.2	77.0	39.5	38.7	3.0	0.7	0.6		59.6	60.4		1.071	1.073
3 yr	17.7	77.9		38.4	39.5	3.0	0.7	0.6		60.8			1.071	
4 yr	17.9	77.7		37.8	40.0	3.0	0.7	0.6		61.5			1.072	
5 yr	18.0	77.6		37.3	40.3	3.0	0.7	0.6		62.0			1.073	
6 yr	18.1	77.5		36.8	40.7	3.0	0.7	0.6		62.5			1.073	
7 yr	18.3	77.3		36.3	41.0	3.1	0.7	0.6		62.9			1.073	
8 yr	18.4	77.2		35.8	41.4	3.1	0.7	0.6		63.4			1.074	
9 yr	18.5	77.1		35.4	41.7	3.1	0.7	0.6		64.0			1.074	
10 yr	18.7	76.9		34.9	42.0	3.1	0.7	0.6		64.5			1.075	

Note. B = data from Butte et al. 2000; CHO = carbohydrates; D_{FFM} = density of fat-free mass; ECW = extracellular water; F = data from Foman and Nelson 2002 and Fomon et al. 1982; ICW = intracellular water; FFM = fat-free mass; mo = month; MO = bone mineral; MS = soft tissue mineral; TBK = total body potassium; TBW = total body water; yr = year.

Table A.6 Atomic Level Composition (g) of Organs and Tissues of Reference Man

	Weight	O	C	H	N	Ca	P	S	K	Na	Cl
Total body	70,000	43,000	16,000	7,000	1,800	1,000	580	140	140	100	95
Total soft tissue	60,000	38,000	14,000	6,300	1,500	14	80	120	120	68	81
Adipose tissue	15,000	3,500	9,600	1,800	120	0.34	2.4	11	4.8	7.6	18
Substaneous subcutaneous	7,500	1,700	4,800	870	61	0.17	1.2	5.5	2.4	3.8	9.0
Other separable	5,000	1,200	3,200	580	40	0.11	0.8	3.7	1.6	2.5	6.0
Interstitial	1,000	230	640	120	8.0	0.022	0.16	0.72	0.26	0.50	0.12
Blood (whole)	5,500	4,100	540	550	160	0.31	1.9	10	8.8	10	15
Plasma	3,100	2,700	130	340	34	0.29	0.34	2.7	0.50	10	11
Connective tissue	3,400	2,200	650	320	190			20		19	8.7
CNS	1,430	1,000	180	150	18	0.12	4.8	2.4	4.2	2.5	3.2
Brain	1,400	1,000	170	150	18	0.12	4.8	2.4	4.2	2.5	3.2
GI tract	1,200	890	140	130	26	0.12	1.1	1.3	1.5	1.3	1.6
Heart	330	230	54	34	8.8	0.012	0.48	0.54	0.72	0.40	0.54
Kidneys	310	230	40	32	8.5	0.29	0.50		0.59	0.62	0.74
Liver	1,800	1,200	260	180	51	0.090	4.7	5.2	4.5	1.8	3.6
Lung	1,000	740	100	99	28	0.087	0.78	2.2	1.9	1.8	2.6
Muscle (skeletal)	28,000	21,000	3,000	2,800	770	0.87	50	67	84	21	22
Pancreas	100	67	13	9.7	2.1	0.0091	0.23		0.23	0.14	0.16
Skeleton	10,000	4,700	2,500	720	300	1,000	500	17	15	32	14
Bone	5,000	2,100	740	190	210		500				
Cortical	4,000	1,700	550	180	160	800	400	12			
Trabecular	1,000	260	130	43	38		100				
Red marrow	1,500	620	620	150	48						
Yellow marrow	1,500	340	950	170	9.6	0.029	0.21	1.1		0.66	1.6
Cartilage	1,100	800	110	110	29			6.6		6.0	2.8
Periarticular tissue	900	660	82	88	22			5.4		4.9	2.3
Skin	2,600	1,600	590	260	120	0.39	0.85	4.1	2.2	4.7	6.9
Spleen	180	130	20	18	5.6	0.012	0.40	0.29	0.56	0.22	0.29

Note. CNS = central nervous system; GI = gastrointestinal.

Table A.7 Molecular Level Composition (g) of Organs and Tissues of Reference Man

	Weight	Water	Ash	Fat	Protein	Density (g/cm³)
Total body	70,000	42,000	3,700	13,300	10,600	1.07
Total soft tissue	60,000	38,700	400	11,400	8,700	
Adipose tissue	15,000	2,300	30	12,000	750	0.92
Subcutaneous	7,500	1,100	15	6,000	380	0.92
Other separable	5,000	750	10	4,000	250	0.92
Interstitial	1,000	150	2.0	800	50	0.92
Blood (whole)	5,500	4,400	55	36	990	1.06
Plasma	3,100	2,900	29	23	210	1.03
Connective tissue	3,400	2,100	140	44	1,200	1.2
CNS	1,430	1,100	21	160	110	1.03
Brain	1,400	1,100	21	150	110	1.03
GI tract	1,200	950	10	74	160	1.04
Heart	330	240	3.6	33	55	1.03
Kidneys	310	240	3.4	16	53	1.05
Liver	1,800	1,300	23	120	320	
Lung	1,000	780	11	9.9	177	1.05
Muscle (skeletal)	28,000	22,000	340	620	4,800	1.04
Pancreas	100	71	1.2	8	13	1.05
Skeleton	10,000	3,300	2,800	1,900	1,900	1.4
Bone	5,000	850	2,700	50	1,300	2.2
Cortical	4,000	600	2,200	40	1,000	1.85
Trabecular	1,000	230	500	10	240	1.08
Red marrow	1,500	600	9	600	300	1.03
Yellow marrow	1,500	230	3	1,200	60	0.98
Cartilage	1,100	860	45	14	180	1.1
Periarticular tissue	900	570	37	12	140	1.1
Skin	2,600	1,600	18	260	750	1.10
Spleen	180	140	2.5	2.9	35	1.06

Note. CNS = central nervous system; GI = gastrointestinal.

Table A.8 Reference Caucasian Man and Woman

	Unit	Reference Man	Reference Woman
Body mass	kg	70	58
Body height	cm	170	160
Body surface area	cm²	18,000	16,000
Total body specific gravity	g/cm³	1.07	1.04
Water content	ml/kg BM	600	500
Extracellular water	ml/kg BM	260	200
Intracellular water	ml/kg BM	340	300
Total blood	ml	5,200	3,900
	g	5,500	4,100
Red blood cells	ml	2,200	1,350
	g	2,400	1,500
Plasma	ml	3,000	2,500
	g	3,100	2,600
Total body fat	kg	13.5	16.0
Total body adipose tissue	kg	15.0	19.0
Subcutaneous AT	kg	7.5	13.0
Separable AT	kg	5.0	4.0
Yellow marrow	kg	1.5	1.3
Interstitial AT	kg	1.0	0.7
Total body connective tissue	kg	5.05	4.10
Cartilage	kg	2.50	2.00
Tendons and fascia	kg	0.85	0.70
Other connective tissue	kg	1.70	1.40
Total body skin	kg	2.60	1.79
Total skin thickness	µm	1,300	1,300
Wet skeletal weight	kg	10.0	6.8
Fat-free wet skeletal weight	kg	8.0	5.8
Dry skeletal weight	kg	5.0	3.4
Percent ash of wet skeletal weight		28	28
Total bone marrow	kg	3.00	2.60
Red bone marrow	kg	1.50	1.30
Yellow bone marrow	kg	1.50	1.30

Table A.8 *(continued)*

	Unit	Reference Man	Reference Woman
Skeletal muscle	kg	28	17
Spleen	g	180	150
Heart	g	330	240
Stomach	g	150	140
Liver	g	1,800	1,400
Pancreas	g	100	85
Lung	g	1,000	800
Kidneys	g	310	275
Uterus	g	—	80
Both breasts	g	26	360
Brain	g	1,400	1,200
Spinal cord	g	30	28
Cerebrospinal fluid	ml	120	100
24-hr energy expenditure	kcal/day	3,000	2,100
Resting energy expenditure	kcal/day	1,680	1,340
Resting metabolic rate	cal/min · kg^{-1}	16.7	16
Oxygen inhaled	g/day	920	640

Abbreviations

ACT = adjuvant chemotherapy

ACE = angiotensin-converting enzyme

ADP = air displacement plethysmography

ADT = androgen deprivation therapy

AN = anorexia nervosa

AR = androgen receptor

AT = adipose tissue

ATFM = adipose tissue–free body mass

AVF = abdominal visceral fat

BCM = body cell mass

BEST = Bone, Estrogen and Strength Training study

BIA = bioelectric impedance analysis

BM = body mass

BMC = bone mineral content

BMD = bone mineral density

BMI = body mass index

BUA = broadband ultrasonic attenuation

BV = body volume

BW = body weight

CBF = cerebral blood flow

CHF = congestive heart failure

CHO = carbohydrates

Chol = cholesterol

CI = confidence interval

CMF = Cyclophosphamide, Methotrexate, and 5-Fluorouracil

CMR_{glc} = cerebral metabolic rate for glucose

$CMRO_2$ = cerebral metabolic rate for oxygen

CNS = central nervous system

COPD = chronic obstructive pulmonary disease

CR = coefficient of reliability

CRP = C-reactive protein

CT = computed axial tomography

CV = coefficient of variation

CVD = cardiovascular disease

DAT = deep adipose tissue

D_b = density of body

DBP = diastolic blood pressure

D_F = density of fat

D_{FFM} = density of fat-free mass

DGNA = delayed-γ neutron activation

DM = diabetes mellitus

DPA = dual-photon absorptiometry

DSAT = deep subcutaneous adipose tissue

D_w = density of water

DXA = dual-energy X-ray absorptiometry

DZ = dizygous

E = extremity

ECF = extracellular fluid

ECS = extracellular solids

ECW = extracellular water

ECW–ICW = extracellular water to intracellular water ratio

$ER\alpha$ = estrogen receptor-α

FFM = fat-free mass

FFMI = fat-free mass index

FM = fat mass

FMI = fat mass index

FT = fast-twitch fiber

G = glycogen

GH = growth hormone

GHRH = growth hormone-releasing hormone receptor gene

GI = gastrointestinal

GR = glucocorticoid receptor

HAART = highly active antiretroviral therapy

HbA_{1c} = glycosylated hemoglobin

HD = hydrodensitometry

HDL = high-density lipoprotein

HFEA = high-frequency energy absorption

HPA = hypothalamic–pituitary–adrenal

HU = Hounsfield unit

IAD = intra-abdominal depth

IBD = inflammatory bowel disease

ICF = intracellular fluid

ICW = intracellular water

IGF = insulin-like growth factor

IGF1R = insulin-like growth factor-1 receptor gene

IL = interleukin

IMAT = intermuscular adipose tissue

IMLC = intramyocellular lipid content

IOM = Institute of Medicine

IR = insulin resistance

IRS-1 = insulin receptor substrate-1

IVNA = in vivo neutron activation analysis

LBM = lean body mass

LDL = low-density lipoprotein

LFM = lipid-free body mass

LPL = lipoprotein lipase

LST = lean soft tissue

LTM = bone mineral–free lean tissue mass

M = mineral

MA = mass

MAC = *mycobacterium avium* complex

MCSA = muscle cross-sectional area

M/FFM = mineral fraction of the fat-free mass

Mo = bone mineral

MRI = magnetic resonance imaging

mRNA = messenger RNA

MRS = magnetic resonance spectroscopy

Ms = soft tissue mineral

MZ = monozygotic

NHANES = National Health and Nutrition Survey

NMR = nuclear magnetic resonance

NPY = neuropeptide Y gene

OGDH = oxoglutarate dehydrogenase

OI = osteogenic index

OMR = organ metabolic rate

1RM = 1 repetition maximum

P = protein

PE = pure error

%BF = percent body fat

PET = positron emission tomography

PFK = phosphofructokinase

PGNA = prompt-γ neutron activation

PPAR = peroxisome proliferator-activated receptor

pQCT = peripheral quantitative computed tomography

PRE = progressive resistance exercise

PRESS = prediction of the sum of squares

QFS = Québec Family Study

QTL = quantitative trait locus

R = resistance

RA = rheumatoid arthritis

REE = resting energy expenditure

rhGH = recombinant human growth hormone

RMSE = root mean square error

RPAT = retroperitoneal adipose tissue

RR = rate ratio

R_{ST} = ratio of the attenuation of photons at the low energy relative to the higher energy in soft tissue

RV = residual lung volume

S = stature

SAA = surface area artifact

SAT = subcutaneous adipose tissue

SBP = systolic blood pressure

SD = standard deviation

SEE = standard error of estimate

SFT = skinfold thickness

SHBG = sex hormone binding globulin

SI = stiffness index

SKF = skinfold

SLE = systemic lupus erythematosus

SM = skeletal muscle

SMM = skeletal muscle mass

SOS = speed of sound

SPA = single-energy photon absorptiometry

S^2/R = resistance index

SSAT = superficial subcutaneous adipose tissue

ST = slow-twitch fiber

STM = soft tissue mass

T = trunk

TBBA = total body bone ash

TBBM = total body bone mineral

TBC = total body carbon

TBCa = total body calcium

TBCl = total body chlorine

TBF = total body fat

TBH = total body hydrogen

TBK = total body potassium

TBN = total body nitrogen

TBNa = total body sodium

TBO = total body oxygen

TBP = total body phosphorus

TBPro = total body protein

TBW = total body water

TEM = technical error of measurement

TER = trunk–extremity skinfold ratio

3-MH = 3-methylhistidine

TLC = total lung capacity

TNF = tumor necrosis factor

TPN = total parenteral nutrition

TOBEC = total body electrical conductivity

UCP = uncoupling protein

UWW = underwater weighing

V = volume

VAT = visceral adipose tissue

VIF = variance inflation factor

V/S = visceral to subcutaneous

V_{TG} = thoracic gas volume

W_a = weight in air

WC = waist circumference

W/FFM = aqueous fraction of the fat-free mass

WHR = waist circumference to hip circumference ratio

WHtR = waist–height ratio

W_w = weight in water

Xc = reactance

YLL = years of life lost

References

Chapter 1

Anderson EC, Langham W (1959) Average potassium concentration of the human body as a function of age. *Science* 130:713-714.

Anderson J, Osborn SB, Tomlinson RWS, Newton D, Rundo J, Salmon L, Smith JW (1964) Neutron-activation analysis in man *in vivo*: A new technique in medical investigation. *Lancet* ii:1201-1205.

Anderwald C, Bernroider E, Krssak M, et al. (2002) Effects of insulin treatment in type 2 diabetic patients on intracellular lipid content in liver and skeletal muscle. *Diabetes* 51:3025-3032.

Behnke AR, Jr, Feen BG, Welham WC (1942) The specific gravity of healthy men. *Journal of the American Medical Association* 118:495-498.

Benjamin E (1914) Der Eiweissnährachaden des Säuglings. *Zeitschrift für Kinderheilkunde* 10:185-302.

Bermudez OI, Tucker KL (2001) Total and central obesity among elderly Hispanics and the association with type 2 diabetes. *Obesity Research* 9: 443-451.

Bischoff E (1863) Einige Gewichts—und Trocken—Bestimmungen der organe des menschlichen Körpers. *Zeitschrift für Rationelle Medizin* 20:75-118.

Björntorp P (2000) Metabolic difference between visceral fat and subcutaneous abdominal fat. *Diabetes and Metabolism* 26: 10-2.

Borkan GA, Hults DE, Gerzof SF, Burrows BA, Robbins AH (1983) Relationships between computed tomography tissue areas, thicknesses and total body composition. *Annals of Human Biology* 10:537-546.

Brozek J, Grande F, Anderson JT, Keys A (1963) Densitometric analysis of body composition: Revisions of some quantitative assumptions. *Annals of the New York Academy of Science* 110:113-140.

Camerer W, Söldner C (1900) Die chemische Zusammensetzung des Neugeborenen. *Zeitschrift für Biologie* 39:173-192.

Cathcart EP (1907) Über die Zusammensetzung des Hungerharns. *Biochemische Zeitschrift* 6:109-148.

Cheek DB (1968) Human Growth: Body Composition, Cell Growth, Energy, and Intelligence. Lea & Febiger, Philadelphia.

Clarys JP, Martin AD, Drinkwater DT (1985) Gross tissue weights in the human body by cadaver dissection. *Human Biology* 56:459-473.

Cohn SH, Dombrowski CS (1971) Measurement of total-body calcium, sodium, chlorine, nitrogen, and phosphorus in man by in vivo neutron activation analysis. *Journal of Nuclear Medicine* 12:499-505.

Cohn SH, Dombrowski CS, Pate HR, Robertson JS (1969) A whole-body counter with an invariant response to radionuclide distribution and body size. *Physics in Medicine and Biology* 14:645-658.

Colberg SR, Simoneau JA, Thaete FL, Kelley DE (1995) Skeletal muscle utilization of free fatty acids in women with visceral obesity. *Journal of Clinical Investigation* 95:1846-1853.

Comizio R, Pietrobelli A, Tan YX, et al. (1998) Total body lipid and triglyceride response to energy deficit: Relevance to body composition models. *American Journal of Physiology* 274:E860-E866.

Du Bois D, Du Bois EF (1916) Clinical calorimeter. A formula to estimate the approximate surface if height and weight be known. *Archives of Internal Medicine* 17:863-871.

Edelman IS, Leibman J, O'Meara MP, Birkenfeld LW (1958) Interrelationship between serum sodium concentration, serum osmolarity, and total exchangeable sodium, total exchangeable potassium and total body water. *Journal of Clinical Investigation* 37:1236-1256.

Ellis KJ (1990) Reference man and woman more fully characterized: Variations on the basis of body size, age, and race. *Biological Trace Element Research* 26-27:385-400.

Enevoldsen LH, Simonsen L, Stallknecht B, Calbolt L, Bulow J (2001) In vivo human lipolytic activity in preperitoneal and subdivisions of subcutaneous abdominal adipose tissue. *American Journal of Physiology Endocrinology and Metabolism* 281:E1110-E1114.

Fehling H (1876) Beitrage zur Physiologie des placentaren Stoffverkehrs. *Archiv für Gynaekologie* 11:523.

Fidanza F, Keys A, Anderson JT (1953) The density of body fat in man and other mammals. *Journal of Applied Physiology* 6:252-256.

Folin O (1905) Laws governing the chemical composition of urine. *American Journal of Physiology* 13:66-115.

Fomon SJ, Haschke F, Ziegler EE, Nelson SE (1982) Body composition of reference children from birth to age 10 years. *American Journal of Clinical Nutrition* 35:1169-1175.

Forbes GB, Hursh J, Gallup J (1961) Estimation of total body fat from potassium-40 content. *Science* 133:101-102.

Foster MA, Hutchison JMS, Mallard JR, Fuller M (1984) Nuclear magnetic resonance pulse sequence and discrimination of high- and low-fat tissues. *Magnetic Resonance Imaging* 2: 187-192.

Foy JM, Schneider H (1960) Estimation of total body water (virtual tritium space) in the rat, cat, rabbit, guinea pig and man, and of the biological half-life of tritium in man. *Journal of Physiology (London)* 154:169-176.

Gurr MI, Harwood JL (1991) *Lipid Biochemistry*. London: Chapman and Hall.

Heymsfield SB, Olafson RP, Kutner MH, Nixon DW (1979) A radiographic method of quantifying protein-calorie undernutrition. *American Journal of Clinical Nutrition* 32:693-702.

Heymsfield SB, Smith R, Aulet M, Bensen B, Lichtman S, Wang J, Pierson RN, Jr (1990) Appendicular skeletal muscle mass: Measurement by dual-photon absorptiometry. *American Journal of Clinical Nutrition* 52:214-218.

Hounsfield GN (1973) Computerized transverse axial scanning (tomography). 1. Description of system. *British Journal of Radiology* 46:1016-1022.

Iob V, Swanson WW (1938) Mineral growth. *Growth* 2:252-256.

Katz J (1896) Die mineralischen Bestandtheile des Muskelfleisches. *Archiv für die Gessamte Physiologie des Menschen und der Thiere* 63:1-85.

Kehayias JJ, Ellis KJ, Cohn SH, Weinlein JH (1987) Use of a high-repetition rate neutron generator for *in vivo* body composition measurements via neutron inelastic scattering. *Nuclear Instruments and Methods in Physics Research* B24/25:1006-1009.

Kehayias JJ, Heymsfield SB, LoMonte AF, Wang J, Pierson RN, Jr (1991) *In vivo* determination of body fat by measuring total body carbon. *American Journal of Clinical Nutrition* 53:1339-1344.

Keith NM, Rowntree LG, Geraghty JT (1915) A method for the determination of plasma and blood volume. *Archives of Internal Medicine* 16:547-576.

Keys A, Brožek J (1953) Body fat in adult man. *Physiological Reviews* 33:245-345.

Kleiber M (1932) Body size and metabolism. *Hilgardia* 6:315-353.

Kulwich R, Feinstein L, Anderson EC (1958) Correlation of potassium-40 concentration and fat-free lean content of hams. *Science* 127:338-339.

Lawes JB, Gilbert JH (1859) Experimental inquiry into the composition of some of the animals fed and slaughtered as human food. *Philosophical Transactions of the Royal Society of London* 149:493-680.

Lifson N, Gordon GB, McClintock R (1955) Measurement of total carbon dioxide production by means of $D_2{}^{18}O$. *Journal of Applied Physiology* 7:704-710.

Lohman TG (1981) Skinfolds and body density and their relation to body fatness: A review. *Human Biology* 53: 181-225.

Lohman TG (1986) Applicability of body composition techniques and constants for children and youths. *Exercise and Sport Sciences Reviews* 14: 325-357.

Lukaski HC, Mendez J, Buskirk ER, Cohn SH (1981) Relationship between endogenous 3-methylhistidine excretion and body composition. *American Journal of Physiology* 240:E302-E307.

Magnus-Levy A (1906) Physiologie des Stoffwechsels. In C von Noorden (ed), *Handbuch der Pathologie des Stoffwechsels*. Berlin: Hirschwald, pp. 446.

Marriott WM (1923) Anhydremia. *Physiological Reviews* 3: 275-294.

Matiegka J (1921) The testing of physical efficiency. *American Journal of Physical Anthropology* 4:223-230.

Mazess RB, Cameron JR, Sorenson JA (1970) Determining body composition by radiation absorption spectrometry. *Nature* 228:771-772.

McCance RA, Widdowson EM (1951) A method of breaking down the body weights of living persons into terms of extracellular fluid, cell mass and fat, and some applications of it to physiology and medicine. *Proceedings of the Royal Society of London, Series B* 138:115-130.

Mitchell HH, Hamilton TS, Steggerda FR, Bean HW (1945) The chemical composition of the adult human body and its bearing on the biochemistry of growth. *Journal of Biological Chemistry* 158:625-637.

Moleschott (1859) *Physiologie der Nahrungsmittel*. Giessen, Ferber'sche Universitätsbuchhandlung, Emil Roth, pp 224.

Moore FD (1946) Determination of total body water and solids with isotopes. *Science* 104:157-160.

Moore FD, Olesen KH, McMurray JD, Parker HV, Ball MR, Boyden CM (1963) *The Body Cell Mass and Its Supporting Environment: Body Composition in Health and Disease*. Philadelphia: Saunders.

Moulton CR (1923) Age and chemical development in mammals. *Journal of Biological Chemistry* 57:79-97.

Pace N, Rathbun EN (1945) Studies on body composition, III: The body water and chemically combined nitrogen content in relation to fat content. *Journal of Biological Chemistry* 158:685-691.

Pfeiffer L (1887) Über den Fettgehalt des Körpers und verschiedener Theile desselben bei mageren und fetten Thieren. *Zeitschrift für Biologie* 23:340-380.

Quetelet LAJ (1871) *Anthropometric ou Msure des Differentes Facultes de L'homme*. Brussels: C Marquardt, pp. 479.

Raji A, Seely EW, Arky RA, Simonson DC (2001) Body fat distribution and insulin resistance in healthy Asian Indians and Caucasians. *Journal of Clinical Endocrinology and Metabolism* 86:5366-5371.

Reynolds EL (1950) The distribution of subcutaneous fat in childhood and adolescence. *Monographs of the Society for Research in Child Development* 15 (serial no. 50).

Roche AF, Davila GH (1972) Late adolescent growth in stature. *Pediatrics* 50:874-880.

Ross R, Fortier L, Hudson R (1996) Separate associations between visceral and subcutaneous adipose tissue distribution, insulin and glucose levels in obese women. *Diabetes Care* 19:1404-1411.

Rubner M (1902) *Die Gesetze des Energieverbrauchs bei der Ernährung*. Leipzig and Vienna: Deutsch.

Rundo J, Bunce LJ (1966) Estimation of total hydrogen content of the human body. *Nature* 210:1023-1065.

Schultz SG (2002) William Harvey and the circulation of the blood: The birth of a scientific revolution and modern physiology. *News in Physiological Science* 17:175-180.

Shaffer PA, Coleman W (1909) Protein metabolism in typhoid fever. *Archives of Internal Medicine* 4:538-600.

Sheng HP, Huggins RA (1979) A review of body composition studies with emphasis on total body water and fat. *American Journal of Clinical Nutrition* 32:630-647.

Siri WE (1961) Body composition from fluid spaces and density. In J Brozek and A Henschel (eds.), *Techniques for Measuring Body Composition*. Washington, DC: National Academy of Sciences, pp 223-244.

Sjöström L, Kvist H, Cederblad A, Tylen U (1986) Determination of total adipose tissue and body fat in women by computed tomography, ^{40}K, and tritium. *American Journal of Physiology* 250:E736-E745.

Snyder WS, Cook MJ, Nasset ES, Karhausen LR, Howells GP, Tipton IH (1975) *Report of the Group on Reference Man*. Oxford, UK: Pergamon Press.

Stuart HC, Hill P, Shaw C (1940) The growth of bone, muscle and overlying tissue as revealed by studies of roentgenograms of the leg areas. *Monographs of Society for Research in Child Development* 5:1-190.

Sutcliffe JF, Smith AH, Barker MC, Smith MA (1993) A theoretical analysis using ratios of the major elements measured by neutron activation analysis to derive total body water, protein, and fat. *Medical Physics* 20:1129-1134.

Talbot NB (1938) Measurement of obesity by the creatinine coefficient. *American Journal of Diseases of Children* 55:42-50.

Thomasset A (1962) Bio-electrical properties of tissue impedance measurements. *Lyon Medical* 207:107-118.

Tokunaga K, Matsuzawa Y, Ishikawa K, Tarui S (1983) A novel technique for the determination of body fat by computed tomography. *International Journal of Obesity and Related Metabolic Disorders* 7:437-445.

Voit E (1901) Die Bedeutung des Körperfettes für die Eiweisszersetzung des hungernden Tieres. *Zeitschrift für Biologie* 41:502-549.

von Bezold A (1857) Untersuchngen Über die vertheilung von Wasser, organischer Materie und anorganischen Verbindungen im Thierreiche. *Zeitschrift für Wissenschaftliche Zoologie* 8:487-524.

von Hevesy G, Hofer E (1934) Die Verweilzeit des Wassers im menschlichen Körper, untersucht mit Hilfe von "schwerem" Wasser als Indicator. *Klinische Wochenschrift* 13:1524-1526.

Wajchenberg BL (2000) Subcutaneous and visceral adipose tissue: their relation to the metabolic syndrome. *Endocrine Reviews* 21:697-738.

Wang ZM, Deurenberg P, Wang W, et al. (1999) Hydration of fat-free body mass: Review and critique of a classic body-composition constant. *American Journal of Clinical Nutrition* 69:833-841.

Wang ZM, Pierson RNJ, Heymsfield SB (1992) The five level model: A new approach to organizing body composition research. *American Journal of Clinical Nutrition* 56:19-28.

Wang ZM, Wang ZM, Heymsfield SB (1999) History of the study of human body composition: A brief review. *American Journal of Human Biology* 11:157-165.

Widdowson EM, McCance RA, Spray CM (1951) The chemical composition of the human body. *Clinical Science* 10:113-125.

Zhu S, Wang ZM, Heshka S, Heo M, Faith MS, Heymsfield, SB (2002) Waist circumference and obesity-associated risk factors among whites in the third National Health and Nutrition Examination Survey: Clinical action thresholds. *American Journal of Clinical Nutrition* 76:743-749.

Chapter 2

Akers, R., & Buskirk, E.R. (1969). An underwater weighing system utilizing "force cube" transducers. *Journal of Applied Physiology, 26*, 649-652.

Baumgartner, R.N., Heymsfield, S.B., Lichtman, S., Wang, J., & Pierson R.N., Jr. (1991). Body composition in elderly people: Effect of criterion estimates on predictive equations. *American Journal of Clinical Nutrition, 53*, 1345-1353.

Behnke, A.R., & Wilmore, J.H. (1974). *Evaluation and regulation of body build and composition.* Englewood Cliffs, NJ: Prentice Hall.

Behnke, A.R., Feen, B.G., & Welham, W.C. (1942). The specific gravity of healthy men. *Journal of the American Medical Association, 118*, 495-498.

Biaggi, R.R., Vollman, M.W., Nies, M.A., Brener, C.E., Flakoll, P.J., Levenhagen, D.K., et al. (1999). Comparison of air-displacement plethysmography with hydrostatic weighing and bioelectrical impedance analysis for the assessment of body composition in healthy adults. *American Journal of Clinical Nutrition, 69*, 898-903.

Bland, J.M., & Altman, D.G. (1986). Statistical methods for assessing agreement between two methods of clinical measurement. *Lancet, 1*, 307-310.

Bonge, D., & Donnelly, J.E. (1989). Trials to criteria for hydrostatic weighing at residual volume. *Research Quarterly for Exercise and Sport, 60*, 176-179.

Brozek, J., Grande, F., & Anderson, J.T. (1963). Densitometric analysis of body composition: Revision of some quantitative assumptions. *Annals of the New York Academy of Sciences, 110*, 113-140.

Buskirk, E.R. (1961). Underwater weighing and body density: A review of procedures. In J. Brozek & A. Henschel (Eds.), *Techniques for Measuring Body Composition* (pp. 90-105). Washington, DC: National Academy of Sciences, National Research Council.

Clarys, J.P., Martin, A.D., & Drinkwater, D.T. (1984). Gross tissue weights in the human body by cadaver dissection. *Human Biology, 56*, 459-473.

Collins, M.A., Millard-Stafford, M.L., Sparling, P.B., Snow, T.K., Rosskopf, L.B., Webb, S.A., et al. (1999). Evaluation of the BOD POD for assessing body fat in collegiate football players. *Medicine and Science in Sports and Exercise, 31*, 1350-1356.

Cournand, A., Baldwin, E.D., Darling, R.C., & Richards, D.W. (1941). Studies on intrapulmonary mixture of gases, IV: The significance of the pulmonary emptying rate and a simplified open circuit measurement of residual air. *Journal of Clinical Investigation, 20*, 681-689.

Demerath, E.W., Guo, S.S., Chumlea, W.C., Towne, B., Roche, A.F., & Siervogel, R.M. (2002). Comparison of percent body fat estimates using air displacement plethysmography and hydrodensitometry in adults and children. *International Journal of Obesity and Related Metabolic Disorders, 26*, 389-397.

Dempster, P., & Aitkens, S. (1995). A new air displacement method for the determination of human body composition. *Medicine and Science in Sports and Exercise, 27*, 1692-1697.

Dewit, O., Fuller, N.J., Fewtrell, M.S., Elia, M., & Wells, J.C. (2000). Whole body air displacement plethysmography compared with hydrodensitometry for body composition analysis. *Archives of Disease in Childhood, 82*, 159-164.

Donnelly, J.E., Brown, T.E., Israel, R.G., Smith-Sintek, S., O'Brien, K.F., & Caslavka, B. (1988) Hydrostatic weighing without head submersion: Description of a method. *Medicine and Science in Sports and Exercise, 20*, 66-69.

DuBois, D., & DuBois, E.F. (1916). A formula to estimate the approximate surface area if height and weight be known. *Archives in Internal Medicine, 17*, 863-871.

Economos, C.D., Nelson, M.E., Fiatarone, M.A., Dallal, G.E., Heymsfield, S.B., Wang, J., et al. (1997). A multi-center comparison of dual energy X-ray absorptiometers: In vivo and in vitro soft tissue measurement. *European Journal of Clinical Nutrition, 51*, 312-317.

Faires, V.M. (1962). *Thermodynamics.* New York: Macmillan.

Fields, D.A., & Goran, M.I. (2000). Body composition techniques and the four-compartment model in children. *Journal of Applied Physiology, 89*, 613-620.

Fields, D.A., Goran, M.I., & McCrory, M.A. (2002). Body-composition assessment via air-displacement plethysmography in adults and children: A review. *American Journal of Clinical Nutrition,* 75, 453-467.

Fields, D.A., Hunter, G.R., & Goran, M.I. (2000). Validation of the BOD POD with hydrostatic weighing: Influence of body clothing. *International Journal of Obesity and Related Metabolic Disorders,* 24, 200-205.

Fields, D.A., Wilson, G.D., Gladden, L.B., Hunter, G.R., Pascoe, D.D., & Goran, M.I. (2001). Comparison of the BOD POD with the four-compartment model in adult females. *Medicine and Science in Sports and Exercise,* 33, 1605-1610.

Forsyth, R., Plyley, J.J., & Shephard, R.J. (1998). Residual volume as a tool in body fat prediction. *Annals of Nutrition and Metabolism,* 32:62-67.

Friedl, K.E., DeLuca, J.P., Marchitelli, L.J., & Vogel, J.A. (1992). Reliability of body-fat estimations from a four-compartment model by using density, body water, and bone mineral measurements. *American Journal of Clinical Nutrition,* 55, 764-770.

Girandola, R.N., Wiswell, R.A., Mohler, J.G., Romero, G.T., & Barnes, W.S. (1977). Effects of water immersion on lung volumes: Implications for body composition analysis. *Journal of Applied Physiology,* 43, 276-279.

Going, S.B. (1996). Densitometry. In A.F. Roche, S.B. Heymsfield, & T.G. Lohman (Eds.), *Human Body Composition.* Champaign, IL: Human Kinetics.

Going, S.B., Williams, D.P., & Lohman, T.G. Aging and body composition: Biological changes and methodological issues. *Exercise and Sport Sciences Reviews,* 23, 411-458.

Goldman, R.F., & Buskirk, E.R.. (1961). Body volume measurement by underwater weighing: Description of a method. In. J. Brozek & A. Henschel (Eds.), *Techniques for Measuring Body Composition* (pp. 78-89). Washington, DC: National Academy of Sciences, National Research Council.

Graedinger, R.H., Reineke, E.P., Pearson, A.M., Van Huss, W.D., Wessel, J.A., & Montoye, H.J. (1963). Determination of body density by air displacement, helium dilution, and underwater weighing. *Annals of the New York Academy of Sciences,* 110, 96-108.

Heymsfield, S.B., Wang, J., Kehayias, J., Heshka, S., Lichtman, S., & Pierson, R.N. Jr. (1989). Chemical determination of human body density in vivo: Relevance to hydrodensitometry. *American Journal of Clinical Nutrition,* 50, 1282-1289.

Higgins, P.B., Fields, D.A., Hunter, G.R., & Gower, B.A. (2001). Effect of scalp and facial hair on air displacement plethysmography estimates of percentage of body fat. *Obesity Research* 9, 326-330.

Iwaoka, H., Yokoyama, T., Nakayama, T., Matsumura, Y., Yoshitake, Y., Fuchi, T., Yoshiike, N., & Tanaka, H. (1998). Determination of percent body fat by the newly developed sulfur hexafluoride dilution method and air displacement plethysmography. *Journal of Nutritional Science and Vitaminology,* 44, 561-568.

Jackson, A.S., Pollock, M.L., Graves, J., & Mahar, M.T. (1988). Reliability and validity of bioelectrical impedance in determining body composition. *Journal of Applied Physiology,* 64, 529-534.

Katch, F.I. (1969). Practice curves and errors of measurement in estimating underwater weight by hydrostatic weighing. *Medicine and Science in Sports,* 1, 212-216.

Keys, A., & Brozek, J. (1953). Body fat in adult man. *Physiological Reviews,* 33: 245-325.

Koda, M., Tsuzuku, S., Ando, F., Niino, N., & Shimokata, H. (2000). Body composition by air displacement plethysmography in middle-aged and elderly Japanese. Comparison with dual-energy X-ray absorptiometry. *Annals of the New York Academy of Sciences,* 904, 484-488.

Levenhagen, D.K., Borel M.J., Welch, D.C., Piasecki, J.H., Piasecki, D.P., Chen, K.Y., et al. (1999). A comparison of air displacement plethysmography with three other techniques to determine body fat in healthy adults. *Parenteral and Enteral Nutrition,* 23, 293-299.

Lockner, D.W., Heyward, V.H., Baumgartner, R.N., & Jenkins, K.A. (2000). Comparison of air-displacement plethysmography, hydrodensitometry, and dual X-ray absorptiometry for assessing body composition of children 10 to 18 years of age. *Annals of the New York Academy of Sciences,* 904, 72-78.

Lohman, T.G. (1986). Applicability of body composition techniques and constants for children and youth. *Exercise and Sport Sciences Reviews,* 14, 325-357.

Lohman, T.G. (1992). *Advances in Body Composition Assessment.* Champaign, IL: Human Kinetics.

Lohman, T.G., & Chen, Z. (2004). Dual energy x-ray absorptiometry. In S. Heymsfield, T. Lohman, Z. Wang, & S. Going (Eds.), *Human Body Composition.* Champaign, IL: Human Kinetics.

Lohman, T.G., & Going, S.B. (1993). Multicomponent models in body composition research: Opportunities and pitfalls. In K. Ellis & J. Eastman (Eds.), *Human Body Composition: In Vivo Methods, Models and Assessment* (pp. 53-58). New York: Plenum.

Lohman, T.G., Roche, A.F., & Martorell, R. (Eds.) (1988). *Anthropometric standardization reference manual,* Champaign, IL: Human Kinetics.

McCrory, M.A., Gomez, T.D., Bernauer, E.M., & Mole, P.A. (1995). Evaluation of a new air displacement plethysmograph for measuring human body composition. *Medicine and Science in Sports and Exercise,* 27, 1686-1691.

McCrory, M.A., Mole, P.A., Gomez, T.D., Dewey, K.G., & Bernauer, E.M. (1998). Body composition by air-displacement plethysmography by using predicted and measured thoracic gas volumes. *Journal of Applied Physiology,* 84, 1475-1479.

Millard-Stafford, M.L., Collins, M.A., Evans, E.M., Snow, T.K., Cureton, K., & Rosskopf, L. (2001). Use of air displacement plethysmography for estimating body fat in a four-component model. *Medicine and Science in Sports and Exercise,* 33, 1331-1317.

Miyatake, N., Nonaka, K., & Fujii, M. (1999). A new air displacement plethysmograph for the determination of Japanese body composition. *Diabetes, Obesity and Metabolism,* 1, 347-351.

Nunez, C., Kovera, A.J., Pietrobelli, A., Heshka, S., Horlick, M., Kehayias, J.J., et al. (1999). Body composition in children and adults by air displacement plethysmography. *European Journal of Clinical Nutrition,* 53, 382-387.

Organ, L.W., Eklund, A.D., & Ledbetter, J.D. (1994). An automated real time underwater weighing system. *Medicine and Science in Sports and Exercise,* 26, 383-391.

Petty, D., Iwanski, R., & Gap, C. (1984). Total body plethysmography for body volume determination. *IEEE Frontiers Engineering Computing Health Care,* 6, 316-319.

Pierson, R.N., Jr., Wang, J., Heymsfield, S.B., Russel-Aulet, M., Mazariegos, M., Tierney, M., et al. (1991). Measuring body fat: calibrating the rulers. Intermethod comparisons in 389 normal Caucasian subjects. *American Journal of Physiology,* 261(1 Pt 1), E103-E108.

Rosenthal, M., Bain, S.H., Cramer, D., Helms, P., Denison, D., Bush, A., & Warner, J.O. (1993). Lung function in white children aged 4 to 19 years: I—Spirometry. *Thorax,* 48, 794-802.

Ruppell, G. (1994). *Manual of Pulmonary Function Testing.* St. Louis, MO: Mosby.

Sardinha, L.B., Lohman, T.G., Teixeira, P.J., Guedes, D.P., & Going, S.B. (1998). Comparison of air displacement plethysmography with dual-energy X-ray absorptiometry and 3 field methods for estimating body composition in middle-aged men. *American Journal of Clinical Nutrition,* 68, 786-793.

Selinger, A. (1977). *The Body As a Three Component System.* Unpublished doctoral dissertation, University of Illinois, Urbana.

Siri, W.E. (1956). The gross composition of the body. *Advances in biological and medical physics,* 4: 239-280.

Siri, W.E. (1961). Body composition from fluid spaces and density: Analysis of methods. In J. Brozek & A. Henschel (Eds.), *Techniques for measuring body composition* (pp. 223-224). Washington, DC: National Academy of Sciences, National Research Council.

Sly, P.D., Lanteri, C., & Bates, J.H. (1990). Effect of the thermodynamics of an infant plethysmograph on the measurement of thoracic gas volume. *Pediatric Pulmonology,* 8, 203-208.

Taylor, A., Aksoy, Y., Scopes, J.W., du Mont, G., & Taylor, B.A. (1985). Development of an air displacement method for whole body volume measurement of infants. *Journal of Biomedical Engineering,* 7, 9-17.

Thomas, T.R., & Etheridge, G.L. (1980). Hydrostatic weighing at residual volume and functional residual capacity. *Journal of Applied Physiology,* 49, 157-159.

Timson, B.F., & Coffman, J.L. (1984). Body composition by hydrostatic weighing at total lung capacity and residual volume. *Medicine and Science in Sports and Exercise,* 16, 411-414.

Van der Ploeg, G., Gunn, S. Withers, R., Modra, A., & Crockett, A. (2000). Comparison of two hydrodensitometric methods for estimating percent body fat. *Journal of Applied Physiology,* 88, 1175-1180.

Wagner, D.R., Heyward, V.H., & Gibson, A.L. (2000). Validation of air displacement plethysmography for assessing body composition. *Medicine and Science in Sports and Exericse,* 32, 1339-1344.

Welch, B.D., & Crisp, C.E. (1958). Effect of level of expiration on body density measurement. *Journal of Applied Physiology,* 12, 399-402.

Wells, J.C., Douros, I., Fuller, N.J., Elia, M., & Dekker, L. (2000). Assessment of body volume using three-dimensional photonic scanning. *Annals of the New York Academy of Sciences,* 904, 247-254.

Wells, J.C., & Fuller, N.J. (2001). Precision of measurement and body size in whole-body air-displacement plethysmography. *International Journal of Obesity and Related Metabolic Disorders,* 25, 1161-1167.

Weltman, A., & Katch, V. (1981). Comparison of hydrostatic weighing at residual volume and total lung capacity. *Medicine and Science in Sports and Exercise,* 13, 210-213.

Wilmore, J.H. (1969a). A simplified method for determination of residual lung volumes. *Journal of Applied Physiology,* 27, 96-100.

Wilmore, J.H., Vodak, P.A., Parr, R.B., Girandola, R.N., & Billing, J.E. (1980). Further simplification of a method for determination of residual lung volume. *Medicine and Science in Sports and Exercise,* 12, 216-218.

Yee, A.J., Fuerst, T., Salamone, L., Visser, M., Dockrell, M., Van Loan, M., & Kern, M. (2001). Calibration and validation of an air-displacement plethysmography method for estimating percentage body fat in an elderly population: A comparison among compartmental models. *American Journal of Clinical Nutrition,* 74, 637-642.

Zapletal, A., Paul, T., & Samanek, M. (1976). Normal values of static pulmonary volumes and ventilation in children and adolescents. *Cesk Pediatr,* 31(10), 532-539.

Chapter 3

Annegers, J. (1954). Total body water in rats and in mice. *Proceedings of the Society for Experimental Biology and Medicine,* 87, 454-456.

Babineau, L.-M., & Page, E. (1955). On body fat and body water in rats. *Journal of Biochemistry and Physiology,* 33, 970-979.

Barac-Nieto, M., Spurr, G.B., Lotero, H., Maksud, M.G., & Dahners, H.W. (1979). Body composition during nutritional repletion of severely undernourished men. *American Journal of Clinical Nutrition,* 32, 981-991.

Bartoli, W.P., Davis, J.M., Pate, R.R., Ward, D.S., & Watson, P.D. (1993). Weekly variability in total body water using ^2H_2O dilution in college-age males. *Medicine and Science in Sports and Exercise,* 25, 1422-1428.

Basalt, R.C. (1980). *Analytical procedures for therapeutic drug monitoring and emerging toxicity.* Davis, CA: Biomedical.

Beddoe, A.H., Streat, S.J., & Hill, G.L. (1985). Hydration of fat-free body in protein-depleted patients. *American Journal of Physiology,* 249, E227-E233.

Bell, E.F. (1985). Body water in infancy. In A.F. Roche (Ed.), *Body composition assessments in youth and adults: Sixth Ross conferences on medical research* (pp. 30-33). Columbus, OH: Ross Laboratories.

Blanc, S., Colligan, A.S., Trabulsi, J., Everhart, J., Bauer, D., Harris, T., & Schoeller, D.A. (2002). Influence of delayed isotopic equilibration on the accuracy of doubly labeled water in the elderly. *Journal of Applied Physiology,* 92, 1036-1044.

Boileau, R.A., Lohman, T.G., & Slaughter, M.H. (1984). Hydration of the fat-free body in children and during maturation. *Human Biology,* 56, 651-666.

Brodie, B.B., Brand, E., & Leshin, S. (1939). The use of bromide as a measure of extracellular fluid. *Journal of Biological Chemistry,* 130, 555-563.

Cheek, D.B. (1953). Estimation of the bromide space with a modification of Conway's method. *Journal of Applied Physiology,* 5, 639-645.

Coward, W.A. (1988). The doubly-labeled-water ($^2H_2^{18}0$) method: Principles and practice. *Proceedings of the Society for Nutrition,* 47, 209-218.

Coward, W.A. (1990). Calculation of pool sizes and flux rates. In A.M. Prentice (Ed.), *The doubly-labeled water method for measuring energy expenditure: Technical recommendations for use in humans (report)* (p. 48). Vienna: International Dietary Energy Consultancy Group.

Culebras, J.M., Fitzpatrick, G.F., Brennan, M.F., Doyden, C.M., & Moore, F.D. (1977). Total body water and the exchangeable hydrogen, II: A review of comparative data from animals based on isotope dilution and desiccation, with a report of new data from the rat. *American Journal of Physiology,* 232, R60-R65.

Culebras, J.M., & Moore, F.D. (1977). Total body water and the exchangeable hydrogen, I: Theoretical calculation of non-aqueous exchangeable hydrogen in man. *American Journal of Physiology,* 232, R54-R59.

Dansgaard, W. (1964). Stable isotopes in precipitation. *Telus,* 16, 436-468.

Dawson, N.J., Stephenson, S.K., & Fredline, D.K. (1972). Body composition of mice subjected to genetic selection for different body proportions. *Comparative Biochemistry and Physiology,* 42B, 679-691.

Denne, S.C., Patel, D., & Kalhan, S.C. (1990). Total body water measurement in normal and diabetic pregnancy: Evidence for maternal and amniotic fluid equilibrium. *Biology Neonate,* 57, 284-291.

Donnan, F.G., & Allmand, A.J. (1914). Ionic equilibria across semi-permeable membranes. *Journal of Chemistry Society,* 105, 1941-1963.

Drews, D., & Stein, T.P. (1992). Effect of bolus fluid intake on energy expenditure values as determined by the doubly labeled water method. *Journal of Applied Physiology,* 72, 82-86.

Dunning, M.F., Steele, J.M., & Berger, E.Y. (1951). Measurement of total body chloride. *Proceedings for the Society for Experimental Biology and Medicine,* 77, 854-858.

Edelman, I.S. (1952). Exchange of water between blood and tissues: Characteristics of deuterium oxide equilibration in body water. *American Journal of Physiology,* 171, 279-296.

Edelman, I.S., & Leibman, J. (1959). Anatomy of body water and electrolytes. *American Journal of Medicine, 27,* 256-277.

Edelman, I.S., Olney, J.M., James, A.H., Brooks, L. & Moore, F.D. (1952). Body composition: Studies in the human being by the dilution principle. *Science,* 115, 447-454.

Fjeld, C.R., Brown, K.H., & Schoeller, D.A. (1988). Validation of the deuterium oxide method for measuring average daily milk intake in infants. *American Journal of Clinical Nutrition,* 48, 671-679.

Fomon, S.J., Haschke, F., Ziegler, E.E., & Nelson, S.E. (1982). Body composition of reference children from birth to age 10 years. *American Journal of Clinical Nutrition,* 35, 1169-1175.

Forbes, G.B. (1962). Methods for determining composition of the human body. With a note on the effect of diet on body composition. *Pediatrics,* 29, 477-494.

Foy, J.M., & Schneider, H. (1960). Estimation of total body water (virtual tritium space) in the rat, cat, rabbit, guinea pig and man, and of the biological half-life of tritium in man. *Journal of Physiology (London),* 154, 169-176.

Gamble, J.L., Jr., Robertson, J.S., Hannigan, C.A., Foster, G.C., & Farr, L.E. (1953). Chloride, bromide, sodium and sucrose spaces in man. *Journal for Clinical Investigation,* 32, 483-487.

Gettinger, R.D. (1983). Use of doubly-labeled water ($^2HH^{18}0$) for determination of H_2O flux and CO_2 production by a mammal in a humid environment. *Oecologia (Berlin),* 59, 54-57.

Green, B., & Dunsmore, J.D. (1978). Turnover of tritiated water and sodium in captive rabbits (Oryctolagus cuniculus). *Journal of Mammalogy,* 59, 12-17.

Harrison, H.E., Darrow, D.C., & Yannet, H. (1936). The total electrolyte content of animals and its probable relation to the distribution of body water. *Journal of Biological Chemistry,* 113, 515-529.

Hewitt, M.J., Going, S.B., Williams, D.P., & Lohman, T.G. (1993). Hydration of the fat-free body mass in children and adults: Implications for body composition assessment. *American Journal of Physiology,* 265, E88-E95.

Holleman, D.F., & Dieterich, R.A. (1975). An evaluation of the tritiated water method for estimating body water in small rodents. *Canadian Journal of Zoology,* 53, 1376-1378.

Horvitz, M.A., & Schoeller, D. (2001) Natural abundance deuterium and 18-oxygen effects on the doubly labeled water method. *American Journal of Physiology,* 280, E965-E972.

Houseman, R.A., McDonald, I., & Pennie, K. (1973). The measurement of total body water in living pigs by deuterium oxide dilution and its relation to body composition. *British Journal of Nutrition,* 30, 149-156.

International Commission on Radiologic Protection. (1975). *Report of the Task Force on Reference Man* (p. 28). Oxford, UK: Pergamon.

Janghorbani, M., Davis, T.A., & Ting, B.T.G. (1988). Measurement of stable isotopes of bromine in biological fluids with inductively coupled plasma mass spectrometry. *Analyst,* 113, 403-411.

Johnson, J.A., Cavert, H.M., Lifson, N., & Visscher, M.B. (1951). Permeability of the bladder to water studied by means of isotopes. *American Journal of Physiology,* 165, 87-92.

Jones, P.J.H., Winthrop, A.L., Schoeller, D.A., Filler, R.M., & Heim, T. (1987). Validation of doubly labeled water for assessing energy expenditure in infants. *Pediatric Research,* 21, 242-246.

Kanto, U., & Clawson, A.J. (1980). Use of deuterium oxide for the in vivo prediction of body composition in female rats in various physiological states. *Journal of Nutrition,* 110, 1840-1848.

Keys, A., & Brozek, J. (1953). Body fat in adult man. *Physiological Reviews,* 33, 245-325.

Khaled, M.A., Lukaski, H.C., & Watkins, C.L. (1987). Determination of total body water by deuterium NMR. *American Journal of Clinical Nutrition,* 45, 1-6.

Knight, G.S., Beddoe, A.H., Streat, S.J., & Hill, G.L. (1986). Body composition of two human cadavers by neutron activation and chemical analysis. *American Journal of Physiology,* 250, E179-E185.

LaForgia, J., & Withers, R.T. (2002). Measurement of total body water using deuterium dilution: Impact of different calculations for determining body fat. *British Journal of Nutrition,* 88, 325-329.

Lesser, G.T., Deutsch, S., & Markofsky, J. (1980). Fat-free mass, total body water, and intracellular water in the aged rat. *American Journal of Physiology,* 238, R82-R90.

Lifson, N., Gordon, G.B., & McClintock, R. (1955). Measurement of total carbon dioxide production by means of $D_2^{18}O$. *Journal of Applied Physiology,* 7, 704-710.

Lohman, T.G., (1986). Applicability of body composition techniques and constants for children and youths. *Exercise Sports Sciences Reviews,* 14, 325-357.

Lohman, T.G. (1989). Assessment of body composition in children. *Pediatric Exercise Science,* 1, 19-30.

Lohman, T.G., Harris, M., Teixeira, P.J., & Weiss, L. (2000). DXA: Assessing body composition and body composition changes—another look. *Annals of the New York Academy of Sciences,* 904, 45-54.

Lukaski, H.C. (1987). Methods for the assessment of human body composition: Traditional and new. *American Journal of Clinical Nutrition, 46,* 537-556.

Lukaski, H.C., & Johnson, P.W. (1985). A simple, inexpensive method of determining total body water using a tracer dose of D_2O and infrared absorption of biological fluids. *American Journal of Clinical Nutrition, 41,* 363-370.

McCullough, A.J., Mullen, K.D., & Kalhan, S.C. (1991). Measurements of total body and extracellular water in cirrhotic patients with and without ascites. *Hepatology, 14,* 1102-1111.

Mendez, J., Prokop, E., Picon-Reategui, E., Akers, R., & Buskirk, E.R. (1970). Total body water by D_2O dilution using saliva samples and gas chromatography. *Journal of Applied Physiology, 28,* 354-357.

Modlesky, C.M., Cureton, K.J., Lewis, B.M., Prior, B.M., Sloniger, M.A., & Rowe, D.A. (1996). Density of fat-free mass and estimates of body compostion in male weight trainers. *Journal of Applied Physiology, 80,* 2085-2096.

Moore, F.D., Lister, J., Boyden, C.M., Ball, M.R., Sullivan, N., & Dagher, F.J. (1968). Skeleton as a feature of body composition. *Human Biology, 40,* 135-188.

Moore, F.D., Olesen, K.H., McMurray, J.D., Parker, H.V., Ball, M.R., & Boyden, C.M. (1963). *The body cell mass and its supporting environment.* Philadelphia: Saunders.

Nagy, K.A., & Costa, D. (1980). Water flux in animals: Analysis of potential errors in the doubly labeled water method. *American Journal of Physiology, 238,* R454-R465.

National Research Council and Subcommittee on the Tenth Edition. (1989). *Recommended daily allowances, tenth edition. Water and electrolytes* (pp. 247-249). Washington, DC: National Academy.

Pace, N., & Rathbun, E.N. (1945). Studies on body composition, III: The body water and chemically combined nitrogen content in relation to fat content. *Journal of Biological Chemistry, 158,* 685-691.

Penn, I.W., Wang, Z.M., Buhl, K.M, Allison, D.B., Burastro, S.E., & Heysmfield, S.B. (1994). Body composition and two-compartment model assumptions in male long distance runners. *Medicine and Science in Sports and Exercise, 26,* 392-397.

Pierson, R.N., Jr., Price, D.C., Wang, J., & Jain, R.K. (1978). Extracellular water measurements: Organ tracer kinetics of bromide and sucrose in rats and man. *American Journal of Physiology, 235,* F254-F264.

Racette, S.B., Schoeller, D.A., Luke, A.H., Shay, K., Hnilicka, J.H., & Kushner, R.F. (1994). Relative dilution spaces of ^2H- and ^{18}O-labeled water in humans. *American Journal of Physiology, 267,* E585-E590.

Reilly, J.J., & Fedak, M.A. (1990). Measurement of the body composition of living gray seals by hydrogen isotope dilution. *Journal of Applied Physiology, 69,* 885-891.

Ritz, P. (2000). Body water spaces and cellular hydration during healthy aging. *Annals of the New York Academy of Sciences, 904,* 474-483.

Roberts, S., Fjeld, C., Westerterp, K., & Goran, M. (1990). Use of the doubly-labelled water method under difficult circumstances. In A.M. Prentice (Ed.), *The doubly-labelled water method for measuring energy expenditure: Technical recommendations for use in humans* (pp 251-263). Vienna: International Dietary Energy Consultancy Group.

Rothwell, N.J., & Stock, M.J. (1979). In vivo determination of body composition by tritium dilution in the rat. *British Journal of Nutrition, 4,* 625-628.

Scatchard, G., Scheinberg, I.H., & Armstrong, S.H., Jr. (1950). Physical chemistry of protein solutions, IV: The combination of human serum albumin with chloride ion. *Journal of the American Chemistry Society, 72,* 535-540.

Schloerb, P.R., Friis-Hansen, B.J., Edelman, I.S., Solomon, A.K., & Moore, F.D. (1950). The measurement of total body water in the human subject by deuterium oxide dilution. With a consideration of the dynamics of deuterium distribution. *Journal of Clinical Investigation, 29,* 1296-1310.

Schloerb, P.R., Friis-Hansen, B.J., Edelman, I.S., Sheldon, D.B., & Moore, F.D. (1951). The measurement of deuterium oxide in body fluids by the falling drop method. *Journal of Laboratory and Clinical Medicine, 37,* 653-662.

Schober, O., Lehr, L., & Hundeshagen, H. (1982). Bromide space, total body water and sick cell syndrome. *European Journal of Nuclear Medicine, 7,* 14-15.

Schoeller, D.A., van Santen, E., Peterson, D.W., Dietz, W.H., Jaspan, J., & Klien, P.D. (1980). Total body water measurement in humans with ^{18}O and ^2H labeled water. *American Journal of Clinical Nutrition, 33,* 2686-2693.

Schoeller, D.A. (1988). Measurement of energy expenditure in free-living humans by using doubly labeled water. *Journal of Nutrition, 118,* 1278-1289.

Schoeller, D.A. (1989). Changes in total body water with age. *American Journal of Clinical Nutrition, 50,* 1176-1181.

Schoeller, D.A. (1991). Isotope dilution methods. In P. Bjorntorp & B.N. Brodoff (Eds.), *Obesity* (pp. 80-88). New York: Lippincott.

Schoeller, D.A., Kushner, R.F., Taylor, P., Dietz, W.H., & Bandini, L. (1985). Measurement of total body water: Isotope dilution techniques. In A.F. Roche (Ed.), *Body composition assessments in youth and adults: Sixth Ross conferences on medical research* (pp. 124-129). Columbus, OH: Ross Laboratories.

Schoeller, D.A., Leitch, C.A., & Brown, C. (1986a). Doubly labeled water method: In vivo oxygen and hydrogen isotope fractionation. *American Journal of Physiology, 251,* R1137-R1143.

Schoeller, D.A., Ravussin, E., Schutz, Y., Acheson, K.J., Baertschi, P., & Jequier, E. (1986b). Energy expenditure by doubly labeled water: Validation in humans and proposed calculation. *American Journal of Physiology, 250,* R823-R830.

Scholer, J.F., & Code, C.F. (1954). Rate of absorption of water from stomach and small bowel of human beings. *American Journal of Physiology, 27,* 565-577.

Sheng, H-P., & Huggins, R.A. (1979). A review of body composition studies with emphasis on total body water and fat. *American Journal of Clinical Nutrition, 32,* 630-647.

Singh, J., Prentice, A.M., Diaz, E., Coward, W.A., Ashford, J., Sawyer, M., & Whitehead, R.G. (1989). Energy expenditure of Gambian women during peak agricultural activity measured by the doubly-labelled water method. *British Journal of Nutrition, 62,* 315-329.

Sohlstrom, A., & Forsum, E. (1997). Changes in total body fat during the human reproductive cycle as assessed by magnetic resonance imaging, body water dilution, and skinfold thickness: comparison of methods. *American Journal of Clinical Nutrition 66,* 1315-1322.

Speakman, J.R., Nair, K.S., & Goran, M.I. (1993). Revised equations for calculating CO_2 production from doubly labeled water in humans. *American Journal of Physiology, 264,* E912-E917.

Spears, C.P., Hyatt, K.H., Vogal, J.M., & Lang, F.H. (1974). Unified method for serial study of body fluid compartments. *Aerospace Medicine, 45,* 274-278.

Thomas, L.D., Van der Velde, D., & Schloerb, P.R. (1991). Optimum doses of deuterium oxide and sodium bromide for the determination of total body water and extracellular fluid. *Journal of Pharmaceutical Biomedical Analysis, 9,* 581-584.

Thompson, W.O., & Yates, B.J. (1941). Venous afferent elicited muscle pumping of a new orthostatic vasopressor mechanism. *Physiologist, 26* (Suppl), S74-S75.

Thornton, W.E., Moore, T.P., & Pool, S.L. (1987). Fluid shifts in weightlessness. *Aviation, Space, and Environmental Medicine, 58,* A86-A90.

Tisavipat, A.S., Vibulsreth, H.P., Sheng, H.P., & Huggins, R.A. (1974). Total body water measured by desiccation and by tritiated water in adult rats. *Journal of Applied Physiology, 37,* 699-701.

Trapp, S.A., & Bell, E.F. (1989). An improved spectrophotometric bromide assay for the estimation of extracellular water volume. *Clinica Chimica Acta, 181,* 207-212.

Vaisman, N., Pencharz, P.B., & Koren, G. (1987). Comparison of oral and intravenous administration of sodium bromide for extracellular water measurements. *American Journal of Clinical Nutrition, 46,* 1-4.

Visser, M., Fuerst, T., Lang, T., Salamone, L., & Harris, T.B. (1999). Validity of fan-beam dual-energy x-ray absorptiometry for measuring fat-free mass and leg muscle mass. *Journal of Applied Physiology, 87,* 1513-1520.

Wallace, G.W.B., & Brodie, B.B. (1939). The distribution of iodide, thiocyanate, bromide and chlo-ride in the central nervous system and spinal fluid. *Journal of Pharmacology and Experimental Therapy, 65,* 214-219.

Wang, J., Pierson, R.N., Jr., & Kelly, W.G. (1973). A rapid method for the determination of deuterium oxide in urine: Application to the measurement of total body water. *Journal of Laboratory and Clinical Medicine, 82,* 170-178.

Wang, Z-M., Pierson, R.N., Jr., & Heymsfield, S.B. (1992). The five-level model: A new approach to organizing body-composition research. *American Journal of Clinical Nutrition, 56,* 19-28.

Wang, Z-M., Deurenberg, P., Wang, W., Pietrobelli, A., Baumgartner, R.N., & Heymsfield, S.B. (1999). Hydration of fat-free mass: New physiological modeling approach. *American Journal of Physiology, 276,* E995-E1003.

Weir, E.G., & Hastings, A.B. (1939). The distribution of bromide and chloride in tissues and body fluids. *Journal of Biological Chemistry, 129,* 547-558.

Wells, J.C., Fuller, N.J., Dewit, O., Fewtrell, M.S., Elia, M., & Cole, T.J. (1999). Four-compartment model of body composition in children: Density and hydration of fat-free mass and comparison with simpler models. *American Journal of Clinical Nutrition, 69,* 904-912.

Wentzel, A.D., Iacono J.M., Allen T.H., & Roberts, J.E. (1958). Determination of heavy water (HDO) in body fluids by direct introduction of water with a mass spectrometer: Measurement of total body water. *Physiological and Medical Biology, 3,* 1-6.

Whyte, R.K., Bayley, H.S., & Schwarcz, H.P. (1985). The measurement of whole body water by $H_2{}^{18}O$ dilution in newborn pigs. *American Journal of Clinical Nutrition, 41,* 801-809.

Wong, W.W., Butte, N.F, Smith, E.O., Garza, C., & Klein, P.D. (1989a). Body composition of lactating women determined by anthropometry and deuterium dilution. *British Journal of Nutrition, 61,* 25-33.

Wong, W.W., Cochran, W.J., Klish, W.J., Smith, E.O., Lee, L.S., & Klein, P.D. (1988). In vivo isotope-fractionation factors and the measurement of deuterium- and oxygen-18-dilution spaces from plasma, urine, saliva, respiratory water vapor, and carbon dioxide. *American Journal of Clinical Nutrition, 47,* 1-6.

Wong, W.W., Legg, H.J.L., Clark, L.L., & Klein, P.D. (1991). Rapid preparation of pyrogen-free $^2H_2{}^{18}O$ for human nutrition studies. *American Journal of Clinical Nutrition, 53,* 585-586.

Wong, W.W., Sheng, H-P., Morkenberg, J.C., Kosanovich, J.L., Clark, L.L., & Klein, P.D. (1989b). Measurement of extracellular water volume by bromide ion chromatography. *American Journal of Clinical Nutrition, 50,* 1290-1294.

Chapter 4

Ahlgren, L., Albertsson, M., Areberg, J., Kadar, L., Linden, M., Mattson, S., & McNeill, F. (1999) A 252 Cf-based instrument for in vivo body protein monitoring in cancer patients. *Acta Oncologica, 38,* 431-437.

Alpsten, M., & Mattsson, S., eds. (1998). International Symposium on In Vivo Body Composition Studies. *Applied Radiation Isotopes, 49,* 429-752.

Anderson, E.C., & Langham, W. (1959). Average potassium concentration of the human body as a function of age. *Science, 130,* 713-714.

Anderson, J., Osborn, S.B., Newton, D., Rundo, J., Salmon, L., & Smith, J.W. (1964). Neutron activation analysis in man in vivo. A new technique in medical investigation. *Lancet,* 1201-1205.

Baur, L.A., Allen, B.J., Rose A., Blagojevic, N., & Gaskin K.J. (1991). A total body nitrogen facility for pediatric use. *Physics in Medicine and Biology, 36,* 1363-1375.

Bewley, D.K. (1988). Anthropomorphic models for checking the calibration of whole body counters and activation analysis systems. *Physics in Medicine and Biology, 33,* 805-813.

Biggin, H.C., Chen, C.S., Ettinger, K.V., Fremlin, J.H., Morgan, W.D., Nowotny, R., & Chamberlain, M.J. (1972). Determination of nitrogen in living patients. *Nature, 236,* 187-189.

Borovnicar, D.J., Stroud, D.B., Wahlqvist, M.L., & Strauss, B.J. (1996) A neutron activation analysis facility for in vivo measurement of nitrogen and chlorine in children. *Australia's Physical Engineering Science Medicine, 19,* 252-263.

Burmeister, W. (1965). Potassium-40 content as a basis for the calculation of body cell mass in man. *Science, 148,* 1336-1344.

Chettle, D.R., & Fremlin, J.H. (1984). Techniques of in vivo neutron activation analysis. *Physics in Medicine and Biology, 29,* 1011-1143.

Chung, C., Wei, Y.Y., & Chen, Y.Y. (1993). Determination of whole body nitrogen and radiation assessment using in vivo prompt gamma acation technique. *Applied Radiation Isotopes, 44,* 941-948.

Cohn, S.H., & Dombrowski, C.S. (1971). Measurement of total-body calcium, sodium, chlorine, nitrogen, and phosphorus in man by in-vivo neutron activation. *Journal of Nuclear Medicine, 12,* 499-505.

Cohn, S.H., Dombrowski, C.S., Pate, H.R., & Robertson, J.S. (1969). A whole-body counter with an invariant response to radionuclide distribution and body size. *Physics in Medicine and Biology, 14,* 645-658.

Cohn, S.H., & Parr, R.M., eds. (1985). Nuclear-based techniques for the in vivo study of human body composition. *Clinical Physics and Physiological Measurements, 6,* 275-301.

Cohn, S.H., Vartsky, D., Yasumura, S., Sawitsky, A., Zanzi, I., Vaswani, A., & Ellis, K.J. (1980) Compartmental body composition based on total-body nitrogen, potassium, and calcium. *American Journal of Physiology,* 239, E524-E530.

Cordain, L., Johnson, J.E., Bainbridge, C.N., Wicker, R.E., & Stockler, J.M. (1989). Potassium content of the fat free body in children. *Journal of Sports Medicine and Physical Fitness,* 29, 170-176.

De Lorenzo, A., & Mohamed, E.I., eds. (2003). Sixth International Symposium on In Vivo Body Composition Studies. *Acta Diabetological* 40, suppl. 1, s1-s319.

Ellis, K.J. (1991). Planning in vivo body composition studies in humans. In: K.S. Subramanian, K. Okamoto, & G.V. Iyengar, eds. *Biological Trace Element Research: Multidisciplinary Perspectives.* Washington, DC: American Chemical Society, pp. 25-39.

Ellis, K.J. (1992). Measurement of whole-body protein content in vivo. In: S. Nissen, ed., *Methods in Protein Nutrition and Metabolism.* San Diego: Academic Press, pp. 195-223.

Ellis, K.J., & Eastman, J., eds. (1993a). *Human Body Composition: In Vivo Methods, Models and Assessment.* Plenum Press, New York.

Ellis, K.J., Shukla, K.K., & Cohn, S.H. (1974). A predictor for total body potassium in man based on height, weight, sex and age: application in metabolic disorders. *Journal of Laboratory and Clinical Medicine,* 83, 716-727.

Ellis, K.J., & Shypailo, R.J. (1992a). Multi-geometry [241]AmBe neutron irradiator: design and calibration for total-body neutron activation analysis. *Journal of Radioanalytical and Nuclear Chemistry,* 161, 51-60.

Ellis, K.J., & Shypailo, R.J. (1992b). [40]K measurements in the infant. *Journal of Radioanalytical and Nuclear Chemistry,* 161, 61-69.

Ellis, K.J., Shypailo, R.J., Sheng, H-P., & Pond, W.G. (1992). In vivo measurements of nitrogen, hydrogen, and carbon in genetically obese and lean pigs. *Journal of Radioanalytical and Nuclear Chemistry,* 160, 159-168.

Ellis, K.J., & Shypailo, R.J. (2005). Total body nitrogen measurements: comparison of the delayed-gamma, prompt-gamma, and inelastic neutron scattering methods using a D,T neutron generator. *Journal of Radioanalytical and Nuclear Chemistry,* in press.

Ellis, K.J., Yasumura, S., & Morgan, W.D., eds. (1987). *In Vivo Body Composition Studies.* London: Institute of Physical Sciences in Medicine.

Fenwick, J.D., McKenzie, A.L., & Boddy, K. (1991). Intercomparison of whole-body counters using a multinuclide calibration phantom. *Physics in Medicine and Biology,* 36, 191-198.

Fomon, S.J., Haschke, F., Ziegler, E.E., & Nelson, S.E. (1982). Body composition of reference children from birth to age 10 years. *American Journal of Clinical Nutrition,* 35, 1169-1175.

Forbes, G.B. (1968). A 4π plastic scintillation detector. *International Journal of Applied Radiation Isotopes,* 19, 535-541.

Forbes, G.B., Hursh, J., & Gallup. J. (1961) Estimation of total body fat from potassium-40 content. *Science,* 133, 101-102.

Forbes, G.B., & Lewis A. (1956). Total sodium, potassium, and chloride in adult man. *Journal of Clinical Investigation,* 35, 596-600.

Garrett, R., & Mitra, S. (1991) A feasibility study of in vivo 14-Mev neutron activation analysis using the associated particle technique. *Medical Physics,* 18, 916-920.

Gerace, L., Aliprantis, A., Russell, M., Allison, D.B., Buhl, K.M., Wang, J., Wang, Z., Pierson, R.N., Jr., & Heymsfield, S.B. (1994) Skeletal differences between black and white men and their relevance to body composition estimates. *American Journal of Human Biology,* 6, 255-262.

Haschke, F. (1989). Body composition during adolescence. In: W.J. Klish & N. Kretchner, eds. *Body Composition Measurements in Infants and Children.* 98th Ross Conference on Pediatric Research, Columbus, OH, pp. 76-82.

Hassager, C., Sorensen, S.S., Nielsen, B., & Christiansen, C. (1989) Body composition measurement by dual photon absorptiometry: comparison with body density and total body potassium measurements. *Clinical Physiology,* 9, 353-360.

Heymsfield, S.B., Wang, Z., Baumgartner, R.N., Dilmanian, F.A., Ma, R., & Yasumura, S. (1993) Body composition and aging: a study by in vivo neutron activation analysis. *Journal of Nutrition,* 123, 432-437.

International Atomic Energy Agency. (1970). *Director of Whole-Body Radioactivity Monitors.* IAEA STI/PUB/213, Vienna.

International Commission on Radiological Protection. (1984). *Report of the Task Group on Reference Man.* ICRP Report 23. New York: Pergamon Press.

Kehayias, J.J., Ellis, K.J., Cohn, S.H., & Weinlein, J.H. (1987). Use of a high-repetition rate neutron generator for in vivo body composition measurements via neutron inelastic scattering. *Nuclear Instruments Methodological Physics Research,* B24/25, 1006-1009.

Kehayias, J.J., Heymsfield, S.B., LoMonte, A.F., Wang, J., & Pierson, R.M. (1991). In vivo determination of body fat by measuring total body carbon. *American Journal of Clinical Nutrition,* 52, 1339-1344.

Kehayias, J.J., & Zhuang, H. (1993). Measurement of regional body fat in vivo in humans by simultaneous detection of regional carbon and oxygen, using neutron in elastic scattering at low radiation exposure. In: K.J. Ellis & J.D. Eastman, eds., *Human Body Composition: In Vivo Methods, Models and Assessment,* New York: Plenum Press, pp. 49-52.

Knight, G.S., Beddoe, A.H., Streat, S.J., & Hill, G.L. (1986). Body composition of two human cadavers by neutron activation and chemical analysis. *American Journal of Physiology,* 250, E179-E185.

Kulwich, R., Feinstein, L., & Anderson, E.C. (1958). Correlation of potassium-40 concentration and fat-free lean content of hams. *Science,* 127, 338-339.

Kvist, H., Chowdhury, B., Sjostrom, L., Tylen, U., & Cederblad, A. (1988) Adipose tissue volume determination in males by computed tomography and [40]K. *International Journal of Obesity,* 12, 249-266.

Kyere, K., Oldroyd, B., Oxby, C.B., Burkinshaw, L., Ellis, R.E., & Hill G.L. (1982). The feasibility of measuring total body carbon by counting neutron inelastic scatter gamma rays. *Physics in Medicine and Biology,* 27, 805-817.

Lohman, T.G. (1986). Applicability of body composition techniques and constraints for children and youth. *Exercise and Sport Sciences Reviews,* 14, 325-357.

Lykken, G.I., Lukaski, H.C., Bolonchuk, W.W., & Sandstead, H.H. (1983). Potential errors in body composition as estimated by whole body scintillation counting. *Journal of Laboratory and Clinical Medicine,* 4, 651-658.

Ma, R., Ellis, K.J., Yasumura, S., Shypailo, R.J., & Pierson, R.N., Jr. (1999) Total body calcium measurements: comparison of two delayed-gamma neutron activation facilities. *Physics in Medicine and Biology,* 44, 113-118.

Moore, F.D., Olesen, K.H., McMurray, J.D., Parker, H.V., Ball M.R., & Boyden, C.M. (1963). *The Body Cell Mass and Its Supporting Environment,* Philadelphia: Saunders.

Ortiz, O., Russell, M., Daley, T.L., Baumgartner, R.N., Waki, M., Lichtman, S., Wang, J., Pierson, R.N., Jr., & Heymsfield, S.B. (1992) Differences in skeletal muscle and bone mineral mass between black and white females and their relevance to estimates of body composition. *American Journal of Clinical Nutrition,* 55, 8-13.

Parr, R.M., ed. (1973). *In Vivo Neutron Activation Analysis.* Proceedings of IAEA Expert Panel, International Atomic Energy Agency, Vienna.

Penn, I-W., Wang, Z-M., Buhl, K.M., Allison, D.B., Burastero, S.E., & Heymsfield, S.B. (1994) Body composition and two-compartment model assumptions in male long distance runners. *Medicine and Science in Sports and Exercise,* 26, 392-397.

Pierson, R.N., Jr. ed. (2000). *Quality of the Body Cell Mass.* New York: Springer.

Pierson, R.N., Jr., & Wang, J. (1988). Body composition denominators for measurements of metabolism: what measurements can be believed? *Mayo Clinical Procedures,* 63, 947-949.

Pierson, R.N., Jr., Wang, J., Heymsfield, S.B., Russell-Aulet, M., Mazariegos, M., Tierney, M., Smith, R., Thornton, J.C., Kehayias, J., & Weber, D.A. (1991), Measuring body fat: calibrating the rulers. Intermethod comparisons in 389 normal Caucasian subjects. *American Journal of Physiology,* 261, E103-E108.

Pierson, R.N., Jr., Wang, J., Thornton, J.C., Van Itallie, T.E., & Colt, E.W.D. (1984) Body potassium by four-pi 40K counting: an anthropometric correction. *American Journal of Physiology,* 246, F24-F239.

Rundo, J., & Bunce, L.J. (1966). Estimation of total hydrogen content of the human body. *Nature,* 210, 1023-1065.

Sjostrom, L., Kvist, H., Cederblad, A., & Tylen, U. (1986) Determination of total adipose tissue and body fat in women by computed tomography, ^{40}K, and tritium. *American Journal of Physiology,* 250, E736-E745.

Stamatelatos, I.E.M., Chettle, D.R., Green, S., & Scott, M.C. (1992). Design studies related to an in vivo neutron activation analysis facility for measuring total body nitrogen. *Physics in Medicine and Biology,* 37, 1657-1674.

Stamatelatos, I.E., Dilmanian, F.A., Ma, R., Lidofsky, L.J., Weber, D.A., Pierson R.N., Jr., Kamen, Y., & Yasumura S. (1993). Calibration for measuring total body nitrogen with a newly upgraded prompt gamma neutron activation facility. *Physics in Medicine and Biology,* 38, 615-626.

Sutcliffe, J.F., Knight, G.S., Pinilla, J.C., & Hill, G. (1993b). New and simple equations to estimate the energy and fat contents and energy density of humans in sickness and health. *British Journal of Nutrition,* 69, 631-644.

Sutcliffe, J.F., Smith, A.H., Barker, M.C.J., & Smith, A. (1993a). A theoretical analysis using ratios of the major elements measured by neutron activation analysis to derive total body water protein, and fat. *Physics in Medicine and Biology,* 20, 1129-1134.

Wang, Z.M., Pi-Sunyer, F.X., Kotler, D.P., Wang, J., Pierson, R.N., Jr., & Heymsfield, S.B. (2001). Magnitude and variation of the ratio of total body potassium to fat-free mass: a cellular level modeling study. *American Journal of Physiology and Endocrinology Metabolism,* 281, E1-E7.

Wang, Z.M., Visser, M., Ma, R., Baumgartner, R.N., Kotler, D., Gallagher, D., & Heymsfield, S.B. (1996). Skeletal muscle mass: evaluation of neutron activation and dual-energy X-ray absorptiometry methods. *Journal of Applied Physiology,* 80, 824-831.

Watson, W.W. (1987). Total body potassium measurement—the effect of fallout from Chernobyl. *Clinical Physics and Physiological Measurements,* 8, 337-341.

Yasumura, S., Harrison, J.E., McNeill, K.G., Woodhead, A.D., & Dilmanian, F.A., eds. (1990). *In Vivo Body Composition Studies.* New York: Plenum Press.

Yasumura, S., Wang, J., & Pierson, R.N., Jr., eds. (2000). In vivo body composition studies. *Annals of the New York Academy of Sciences,* 904, 1-631.

Ziegler, E.E., O'Donnell, A.M., Nelson, S.E., & Fomon, S.J. (1976). Body composition of the reference fetus. *Growth,* 40, 329-341.

Chapter 5

Altman, D.G., & Bland, J.M. (1983). Measurement in medicine: the analysis of method comparison studies. *Statistician,* 32, 307-317.

Barthe, N., Braillon, P., Ducassou, D., & Basse-Cathalinat, B. (1997). Comparison of two Hologic DXA systems (QDR 1000 and QDR 4500/A). *British Journal of Radiology,* 70, 728-739.

Bartok-Olson, C.J., Schoeller, D.A., Sullivan, J.C., & Clark, R.R. (2000). The "B" in the Selinger four-compartment body composition formula should be body mineral instead of bone mineral. *Annals of the New York Academy of Sciences,* 904, 342-344.

Baumgartner, R.N, Heymsfield, S.B., Lichtman, S., Wang, J. & Pierson, R.N. (1991). Body composition in elderly people: effect of criterion estimates on predictive equations. *European Journal of Clinical Nutrition,* 53, 1345-1353.

Baumgartner, R.N., Koehler, K.M., Gallagher, D., Romero, L., Heymsfield, S.B., Ross, R.R., Garry, P.J., & Lindeman, R.D. (1998). Epidemiology of sarcopenia among the elderly in New Mexico. *American Journal of Epidemiology,* 147, 755-763.

Black, A., Tilmont, E.M., Baer, D.J., Rumpler, W.V., Ingram, D.K., Roth, G.S., & Lane, M.A. (2001). Accuracy and precision of dual-energy x-ray absorptiometry for body composition measurements in rhesus monkeys. *Journal of Medical Primatology,* 30, 94-99.

Brozek, J., Grande, F., Anderson, J.T., & Keys, A. (1963). Densiotometric analysis of body composition: revision of some quantitative assumptions. *Annals of the New York Academy of Sciences,* 110, 113-140.

Brunton, J.A., Bayley, H.S., & Atkinson, S.A. (1993). Validation and application of dual-energy x-ray absorptiometry to measure bone mass and body composition in small infants. *American Journal of Clinical Nutrition,* 58, 839-845.

Cameron, J.R., & Sorensen, J.A. (1963). Measurement of bone mineral in vivo: an improved method. *Science,* 142, 230-232.

Clark, R.R., Kuta, J.M., & Sullivan, J.C. (1993). Prediction of percent body fat in adult males using dual energy x-ray absorptiometry, skinfolds, and hydrostatic weighing. *Medicine and Science in Sports and Exercise,* 25, 528-535.

Cordero-MacIntyre, Z.R., Peters, W., Libanati, C.R., Espanna, R.C., Howell, W.H., & Lohman, T.G. (2000a). Effect of a weight-reduction program on total and regional body composition in obese postmenopausal women. *Annals of the New York Academy of Sciences,* 904, 526-535.

Cordero-MacIntyre, Z.R., Peters, W., Libanati, C.R., Espanna, R.C., Howell, W.H., & Lohman, T.G. (2000b). Reproducibility of body measurements in very obese postmenopausal women. *Annals of the New York Academy of Sciences, 904,* 536-538.

Cordero-MacIntyre, Z.R., Peters, W., Libanati, CR, Espanna, R.C., Abilaso, XX, Howell, W.H., & Lohman, T.G. (2002). Reproducibility of DXA in obese women. *Journal of Clinical Densitometry, 5,* 35-44.

Cullum, I.D., Ell, P.J., & Ryder, J.P. (1989). X-ray dual-photon absorptiometry: a new method for the measurement of bone density. *British Journal of Radiology, 62,* 587-592.

Ellis, K.J., & Shypailo, R.J. (1998). Bone mineral and body composition measurements: cross-calibration of pencil-beam and fan-beam dual-energy x-ray absorptiometers. *Journal of Bone Mineral Research, 13,* 1613-1618.

Ellis, K.J., Shypailo, R.J., Abrams, S.A., & Wong, W.W. (2000). The reference child and adolescent models of body composition: a contemporary comparison. *Annals of the New York Academy of Sciences, 904,* 374-382.

Evans, E.M., Saunders, M.J., Spano, M.A., Arngrimsson, S.A., Lewis, R.D., & Cureton, K.J. (1999). Body-composition changes with diet and exercise in obese women: a comparison of estimates from clinical methods and a 4-component model. *American Journal of Clinical Nutrition, 70,* 5-12.

Faulkner, K., Glueer, C., Engelke, K., & Genant, H.K. (1992). Cross calibration of QDR-1000/w and QDR-2000 scanners. *Journal of Bone Mineral Research, 7,* 518S.

Fields, D.A., & Goran, M.I. (2000). Body composition techniques and the four-compartment model in children. *Journal of Applied Physiology, 89,* 613-620.

Fields, D.A., Goran, M.I., & McCrory, M.A. (2002). Body-composition assessment via air-displacement plethysmography in adults and children: a review. *American Journal of Clinical Nutrition, 75,* 453-67.

Fomon, S.J., Haschke, F., Ziegler, E.E., & Nelson, S.E. (1982). Body composition of reference children from birth to age 10 years. *American Journal of Clinical Nutrition, 35,* 1169-1175.

Fuller, N.J., Jebb, S.A., Laskey, M.A., Coward, W.A., & Elia, M. (1992). Four-component model for the assessment of body composition in humans: comparison with alternative methods, and evaluation of the density and hydration of fat-free mass. *Clinical Science, 82,* 687-693.

Fuller, N.J., Laskey, M.A., & Elia, M. (1992). Assessment of the composition of major body regions by dual-energy X-ray absorptiometry (DEXA), with special reference to limb muscle mass. *Clinical Physiology, 12,* 253-266.

Genton, L.D., Didier, H., Kyle, U.G., & Pichard, C. (2002). Dual-energy absorptiometry and body composition: differences between devices and comparison with reference methods. *Nutrition, 18,* 66-70.

Geriffiths, M.R., Noakes, K.A., & Pocock, N.A. (1997). Correcting the magnification error of fan beam densitometers. *Journal of Bone Mineral Research, 12,* 119-123.

Going, S.B., Massett, M.P., Hall, M.C., Bare, L.A., Root, P.A., Williams, D.P., & Lohman, T.G. (1993). Detection of small changes in body composition by dual-energy x-ray absorptiometry. *American Journal of Clinical Nutrition, 57,* 845-850.

Going, S.B., Pamenter, R., & Lohman, T. (1990). Estimation of total body composition by regional dual photon absorptiometry. *American Journal of Human Biology, 2,* 703-710.

Goran, M.I., Driscoll, P., Johnson, R., Nagy, T.R., & Hunter, G. (1996). Cross-calibration of body-composition techniques against dual-energy x-ray absorptiometry in young children. *European Journal of Clinical Nutrition, 3,* 299-305.

Gotfredsen, A., Borg, J., Christiansen, C., & Mazess, R.B. (1984). Total body bone mineral in vivo by dual photon absorptiometry. II. Accuracy. *Clinical Physiology, 4,* 357-362.

Gotfredsen, A., Borg, J., Christiansen, C., & Mazess, R.B. (1986). Measurement of lean body mass and total body fat using dual photon absorptiometry. *Metabolism, 35,* 88-93.

Hansen, N.J., Lohman, T.G., Going, S.B., Hall, M.C., Pamenter, R.W., Bare, L.A., Boyden, T.W., & Houtkooper, L.B. (1993). Prediction of body composition in premenopausal females from dual-energy x-ray absorptiometry. *Journal of Applied Physiology, 75,* 1637-1641.

Harper, K., Lobaugh, K., King, S.T., & Drezner, M.K. (1992). Upgrading dual-energy x-ray absorptiometry scanners: do new models provide equivalent methods? *Journal of Bone Mineral Research, 7,* S191.

Haschke, F. (1989). Body composition during adolescence. In W.J. Klish & N. Kretchmer (Eds.), *Body Composition Measurements in Infants and Children* (pp. 76-83). Columbus, OH: Ross Laboratories.

Heymsfield, S.B., Lichtman, S., Baumgartner, R.N., Wang, J., Kamen, Y., Aliprantis, A., & Pierson, R.N. Jr. (1990a). Body composition of humans: comparison of two improved four-compartment models that differ in expense, technical complexity, and radiation exposure. *European Journal of Clinical Nutrition, 52,* 52-58.

Heymsfield, S.B., Smith, R., Aulet, M., Bensen, B., Lichtman, S., Wang, J., & Pierson, R.N. Jr. (1990b). Appendicular skeletal muscle mass: measurement by dual-photon absorptiometry. *American Journal of Clinical Nutrition, 52,* 214-218.

Heymsfield, S.B., Wang, J., Heshka, S., Kehayias, J.J., & Pierson, R.N. Jr. (1989). Dual-photon absorptiometry: comparison of bone mineral and soft tissue mass measurements in vivo with established methods. *American Journal of Clinical Nutrition, 49,* 1283-1289.

Heymsfield, S.B., Wang, Z., Wang, J., et al. (1994). Theoretical foundation of dual energy x-ray absorptiometry (DEXA) soft tissue estimates: Validation in situ and in vivo. *Federation of American Societies for Experimental Biology, 8* (part 1), A278.

Horber, F.F., Thomi, F., Casez, J.P., Fonteille, J., & Jaeger, P. (1992). Impact of hydration status on body composition as measured by dual energy X-ray absorptiometry in normal volunteers and patients on haemodialysis. *British Journal of Radiology, 65,* 895-900.

Houtkooper, L.B., Going, S.B., Sproul, J., Blew, R.M., & Lohman, T.G. (2000). Comparison of methods for assessing body composition changes over one year in post-menopausal women. *American Journal of Clinical Nutrition, 72,* 401-406.

Jebb, S.A., Goldberg, G.R., & Elia, M. (1993). DXA measurements of fat and bone mineral density in relation to depth and adiposity. In K.J. Ellis & J.D. Eastman (Eds.), *Human Body Composition* (pp. 115-119). New York: Plenum Press.

Johnson, J., & Dawson-Hughes, B. (1991). Precision and stability of dual-energy X-ray absorptiometry measurements. *Calcified Tissue International, 49,* 174-178.

Kelly, T.L., Slovik, D.M., Schoenfeld, D.A., & Neer, R.M. (1988). Quantitative digital radiography versus dual photon absorptiometry of the lumbar spine. *Journal of Clinical Endocrinology and Metabolism, 67,* 839-844.

Kiebzak, G.M., Leamy, L.J., Pierson, L.M., Nord, R.H., & Zhang, ZY. (2000). Measurement precision of body composition variables using the lunar DPX-L densitometer. *Journal of Clinical Densitometry, 3,* 35-41.

Kim, J., Wang, Z., Heymsfield, SB, Baumgartner, R.N., & Gallagher, D. (2002). Total-body skeletal muscle mass: estimation by a new dual-energy x-ray absorptiometry method. *American Journal of Clinical Nutrition, 76,* 378-383.

Kohrt, W.M. (1998). Preliminary evidence that DEXA provides an accurate assessment of body composition. *Journal of Applied Physiology, 84,* 372-377.

Kopp-Hoolihan, L.E., Van Loan, M.D., Wong, W.W. & King, J.C. (1999). Fat mass deposition during pregnancy using a four-component model. *Journal of Applied Physiology, 87,* 196-202.

Laskey, M.A., Lyttle, K.D., Flarman, M.E., & Barber, R.W. (1992). The influence of tissue depth and composition on the performance of the Lunar dual-energy X-ray absorptiometer whole-body scanning mode. *European Journal of Clinical Nutrition, 46,* 39-45.

Lockner, D.W., Heyward, V.H., Baumgartner, R.N., & Jenkins, KA. (2000). Comparison of air-displacement plethysmography, hydrodensitometry, and dual X-ray absorptiometry for assessing body composition of children 10 to 18 years of age. *Annals of the New York Academy of Sciences, 904,* 72-78.

Lohman, T.G. (1986). Applicability of body composition techniques and constants for children and youth. *Exercise and Sport Sciences Reviews, 14,* 325-357.

Lohman, T. (1989). Assessment of body composition in children. *Pediatric Exercise Science, 1,* 19-30.

Lohman, T.G. (1992). *Advances in Body Composition Assessment.* Champaign, IL: Human Kinetics.

Lohman, T.G. (1996). Dual energy x-ray absorptiometry. In A.F. Roche, S.B. Heymsfield, and T.G. Lohman (Eds.), *Human Body Composition* (63-78). Champaign, IL: Human Kinetics.

Lohman, T.G., & Going, S.B. (1993). Multicomponent models in body composition research: opportunities and pitfalls. *Basic Life Science, 60,* 53-58.

Lohman, T.G., Harris, M., Teixeira, P.J., & Weiss, L. (2000). Assessing body composition and changes in body composition. Another look at dual-energy X-ray absorptiometry. *Annals of the New York Academy of Sciences, 904,* 45-54.

Lukaski, H.C., Hall, C.B., Marchello, M.J., & Siders, W.A. (2001). Validation of dual x-ray absorptiometry for body-composition assessment of rats exposed to dietary stressors. *Nutrition, 17,* 607-613.

Makan, S., Bayley, H.S., & Webber, C.E. (1997). Precision and accuracy of total body bone mass and body composition measurements in the rat using x-ray-based dual photon absorptiometry. *Canadian Journal of Physiology and Pharmacology, 75,* 1257-1261.

Mazess, R.B., & Barden, H.S. (2000). Evaluation of differences between fan-beam and pencil-beam densitometers. *Calcified Tissue International, 67,* 291-296.

Mazess, R.B., Barden, H., Bisek, J., & Hanson, J. (1990). Dual-energy x-ray absorptiometry for total-body and regional bone-mineral and soft-tissue composition. *American Journal of Clinical Nutrition, 51,* 1106-1112.

Mazess, R., Cameron, J., & Miller, H. (1972). Direct readout of bone mineral content using radionuclide absorptiometry. *International Journal of Applied Radiation Isotopes, 23,* 471-479.

Mazess, R.B., Hanson, J.A., Payne, R., Nord, R., & Wilson, M. (2000). Axial and total-body bone densitometry using a narrow-angle fan-beam. *Osteoporosis International, 11,* 158-166.

Mazess, R.B., Peppler, W.W., Chestnut, C.H., III, Nelp, W.B., Cohn, S.H., & Zanzi, I. (1981). Total body bone mineral and lean body mass by dual-photon absorptiometry, II: Comparison with total body calcium by neutron activation analysis. *Calcified Tissue International, 33,* 361-363.

Mazess, R.B., Peppler, W.W., & Gibbons, M. (1984). Total body composition by dual-photon (153Gd) absorptiometry. *American Journal of Clinical Nutrition, 40,* 834-839.

Mazess, R.B., Trempe, J.A., Bisek, J.P., Hanson, J.A., & Herns, D. (1991). Calibration of dual-energy x-ray absorptiometry for bone density. *Journal of Bone Mineral Research, 6,* 799-806.

Milliken, L.A., Going, S.B., & Lohman, T.G. (1996). Effects of variations in regional composition on soft tissue measurements by dual-energy X-ray absorptiometry. *International Journal of Obesity Related Metabolic Disorders, 20,* 677-682.

Modlesky, C.M., Lewis, R.D., Yetman, K.A., Rose, B., Rosskopf, L.B., Snow, T.K., & Sparling, P.B. (1996). Comparison of body composition and bone mineral measurements from two DXA instruments in young men. *European Journal of Clinical Nutrition, 64,* 669-676.

Nord, R.H., Homuth, J.R., Hanson, J.A., & Mazess, R.B. (2000). Evaluation of a new DXA fan-beam instrument for measuring body composition. *Annals of the New York Academy of Sciences, 904,* 118-125.

Nord, R., & Payne, R.K. (1995). Body composition by DXA: a review of technology. *Asia Pacific Journal of Clinical Nutrition, 4,* 167-171.

Nunez, C., Kovera, A.J., Pietrobelli, A., Heshka, S., Horlick, M., Kehayias, J.J., Wang, Z., & Heymsfield, S.B. (1999). Body composition in children and adults by air displacement plethysmography. *European Journal of Clinical Nutrition, 53,* 382-387.

Patel, R., Blake, G.M., Rymer, J., & Fogelman, I. (2000). Long-term precision of DXA scanning assessed over seven years in forty postmenopausal women. *Osteoporosis International, 11,* 68-75.

Peppler, W.W., & Mazess, R.B. (1981). Total body bone mineral and lean body mass by dual-photon absorptiometry. I. Theory and measurement procedure. *Calcified Tissue International, 33,* 353-359.

Picaud, J.C., Rigo, J., Nyamugabo, K., Milet, J., & Senterre, J. (1996). Evaluation of dual-energy X-ray absorptiometry for body-composition assessment in piglets and term human neonates. *American Journal of Clinical Nutrition, 63,* 157-163.

Pietrobelli, A., Formica, C., Wang, Z., & Heymsfield, S.B. (1996). Dual-energy X-ray absorptiometry body composition model: review of physical concepts. *American Journal of Physiology, 271,* E941-E951.

Pietrobelli, A., Wang, Z., Formica, C., & Heymsfield, S.B. (1998). Dual-energy x-ray absorptiometry: fat estimation errors due to variation in soft tissue hydration. *American Journal of Physiology, 274,* E808-E816.

Prior, B.M., Cureton, K.J., Modlesky, C.M., Evans, E.M., Sloniger, M.A., Saunders, M., & Lewis, R.D. (1997). In vivo validation of whole body composition estimates from dual-energy X-ray absorptiometry. *Journal of Applied Physiology, 83,* 623-630.

Roemmich, J.N., Clark, P.A., Weltman, A., & Rogol, A.D. (1997). Alterations in growth and body composition during puberty. I. Comparing multicompartment body composition models. *Journal of Applied Physiology, 83,* 927-935.

Roubenoff, R., Kehayias, J.J., Dawson-Hughes, B., & Heymsfield, S.B. (1993). Use of dual energy x-ray absorptiometry in body composition studies: not yet a "gold standard." *American Journal of Clinical Nutrition, 58,* 589-591.

Salamone, L.M, Fuerst, T., Visser, M., Kern, M., Lang, T., Dockrell, M., Cauley, J.A., Nevitt, M., Tylavsky, F., & Lohman, T.G. (2000). Measurement of fat mass using DEXA: a validation study in elderly adults. *Journal of Applied Physiology, 89,* 345-352.

Schoeller, D.A., Tylavsky, F.A., Baer, D.J., Chumlea, W.C., Earthman, C.P., Fuerst, T., Harris, T.B., Heymsfield, S.B., Horlick, M., Lohman, T.G., Lukaski, H.C., Shepherd, J., Siervogel, R.M., & Borrud, L.G. (in press). QDR 4500A dual X-ray absorptiometer underestimates fat mass in comparison with criterion methods in adults. *American Journal of Clinical Nutrition,* in press.

Selinger, A. (1977). The body as a three component supplement. PhD thesis, University of Illinois–Urbana, Champaign, IL.

Siri, W.E. (1961). Body composition from fluid spaces and density: analysis of method. In J. Brozek & A. Henschel (Eds.), *Techniques for Measuring Body Composition* (pp.223-244). Washington, D.C.: National Academy of Sciences.

Svendsen, O.L., Haarbo, J., Hassager, C., & Christiansen, C. (1993a). Accuracy of measurements of total-body soft-tissue composition by dual energy X-ray absorptiometry in vivo. *American Journal of Clinical Nutrition, 57,* 605-608.

Svendsen, O.L., Haarbo, J., Hassager, C., & Christiansen, C. (1993b). Accuracy of measurements of total-body soft-tissue composition by dual energy X-ray absorptiometry in vivo. In K.J. Ellis & J.D. Eastman (Eds.), *Human Body Composition* (pp. 381-383). New York: Plenum Press.

Svendsen, O.L., Hassager, C., Bergmann, I., & Christiansen, C. (1993c). Measurement of abdominal and intra-abdominal fat in postmenopausal women by dual energy X-ray absorptiometry and anthropometry: comparison with computerized tomography. *International Journal of Obesity, 17,* 45-51.

Tataranni, P.A., & Ravussin, E. (1995). Use of dual energy x-ray absorptiometry in obese individuals. *American Journal of Clinical Nutrition, 62,* 730-734.

Testolin, C.G., Gore, R., Rivkin, T., Horlick, M., Arbo, J., Wang, Z., Chiumello, G., & Heymsfield, S.B. (2000). Dual-energy x-ray absorptiometry: analysis of pediatric fat estimate errors due to tissue hydration effects. *Journal of Applied Physiology, 89,* 2365-2372.

Thomsen, T.K., Jensen, V.J., & Henriksen, M.G. (1998). In vivo measurement of human body composition by dual-energy X-ray absorptiometry (DXA). *European Journal of Surgery, 164,* 133-137.

Tothill, P., & Avenell, A. (1994). Errors in dual-energy X-ray absorptiometry of the lumbar spine owing to fat distribution and soft tissue thickness during weight change. *British Journal of Radiology, 67,* 71-75.

Tothill, P., & Hannan, W.J. (2000). Comparisons between a Hologic QDR 1000W, QDR 4500A and Lunar Expert Dual Energy X-ray Absorptiometry scans used for measuring total body bone and soft tissue. *Annals of the New York Academy of Sciences, 904,* 63-71.

Tothill, P., Laskey, M.A., Orphanidou, C.I., & Wijk, M.V. (1999). Anomalies in dual energy x-ray absorptiometry measurements of total-body bone mineral during weight change using Lunar, Hologic and Norland instruments. *British Journal of Radiology, 72,* 661-669.

Tylavsky, F.A., Fuerst, T., Nevitt, M., Dockrell, M., Waer, J.Y., Cauley, J.A., & Harris, T.B. (2000). Measurement of changes in soft tissue mass and fat mass with weight change: pencil versus fan beam dual energy x-ray absorptiometry. *Annals of the New York Academy of Sciences, 904,* 94-97.

Tylavsky, F., Lohman, T., Blunt, B.A., Schoeller, D.A., Fuerst, T., Cauley, J.A., Nevitt, M.C., Visser, M., & Harris, T.B. (2003). QDR 4500 DXA overestimates fat-free mass compared with criterion methods. *Journal of Applied Physiology, 94,* 959-965.

Van Loan, M.D., Johnson, H.L., & Barbieri, T.F. (1998). Effect of weight loss on bone mineral content and bone mineral density in obese women. *European Journal of Clinical Nutrition, 67,* 734-738.

Van Loan, M., Thompson, J., Butterfield, G., Marcus, R., & Mayclin, P. (1994). Comparison of bone mineral content (BMC), bone mineral density (BMD), lean and fat measurements from two different bone densitometers. *Medicine and Science in Sports and Exercise, 26,* S40.

Visser, M., Fuerst, T., Lang, T., Salamone, L., & Harris T.B. (1999). Validity of fan-beam dual-energy x-ray absorptiometry for measuring fat-free mass and leg muscle mass. *Journal of Applied Physiology, 87,* 1513-1520.

Wang, J., Heymsfield, S.B., Aulet, M., Thorton, J.C., & Pierson, R.N. Jr. (1989). Body fat from body density: underwater weighing vs. dual-photon absorptiometry. *American Journal of Physiology, 256,* E829-E834.

Wang, W., Wang, Z.M., Faith, M.S., Cotler, D., Shih, R., & Heymsfield, S.B. (1999). Regional skeletal muscle measurement: evaluation of new dual energy x-ray absorptiometry model. *Journal of Applied Physiology, 87,* 1163-1171.

Wang, Z.M., Pi-Sunyer, F.X., Kotler, D.P., Wielopolski, L.W., Withers, R.T., Pierson, R.N., Jr., & Heymsfield, S.B. (2002). Multicomponent methods: Evaluation of new and traditional soft tissue mineral models by in vivo neutron activation analysis. *American Journal of Clinical Nutrition, 76,* 968-974.

Wang, Z.M., Visser, M., Ma, R., Baumgartner, R.N., Kotler, D., Gallagher, D., & Heymsfield, S.B. (1996). Skeletal muscle mass: evaluation of neutron activation and dual-energy X-ray absorptiometry methods. *Journal of Applied Physiology, 80,* 824-831.

Wells, J.C., Fuller, N.J., Dewit, O., Fewtrell, M.S., Elia, M., & Cole, T.J. (1999). Four-component model of body composition in children: density and hydration of fat-free mass and comparison with simpler models. *American Journal of Clinical Nutrition, 69,* 904-912.

Winston, W.K., Koo, M.H., & Hockman, E.M. (2002). Use of fan beam dual-energy x-ray absorptiometry to measure body composition of piglets. *Journal of Nutrition, 132,* 1380-1383.

Withers, R.T., LaForgia, J., Pillans, R.K., Shipp, N.J., Chatterton, B.E., Schultz, C.G., & Leaney, F. (1998). Comparisons of two-, three-, and four-compartment models of body composition analysis in men and women. *Journal of Applied Physiology, 85,* 238-245.

Withers, R.T., Smith, D.A., Chatterton, B.E., Schultz, C.G., & Gaffney, R.D. (1992). A comparison of four methods of estimating the body composition of male endurance athletes. *European Journal of Clinical Nutrition, 46,* 773-784.

Wong, J.C., & Griffiths, M.R. (2002). *Journal of Clinical Densitometry, 5,* 199-205.

Chapter 6

Andreoli, A., Melchiorri, G., DeLorenzo, A., Caruso, I.S., Salimei, P., & Guerrisi, M. (2002). Bioelectrical impedance measures in different position and vs dual-energy X-ray absorptiometry (DXA). *Journal of Sports Medicine, 42,* 186-189.

Azcue, M., Wesson, D., Neuman, M., & Pencharz, P. (1993). What does bioelectrical impedance spectroscopy (BIS) measure? *Basic Life Science, 60,* 121-123.

Baker, L. (1989). Principles of the impedance technique. *IEEE Engineering in Medicine and Biology Magazine, 3,* 11-15.

Barak, N., Wall-Alonso, E., & Sitrin, C.A. (2003). Use of bioelectrical impedance analysis to predict energy expenditure of hospitalized patients receiving nutrition support. *Journal of Parenteral and Enteral Nutrition, 27,* 43-46.

Barbosa-Silva, M.C., Barros, A.J., Post, C.L., Waitzberg, D.L., & Heymsfield, S.B. (2003). Can bioelectrical impedance analysis identify malnutrition in preoperative nutrition assessment. *Nutrition, 19,* 422-426.

Barnett, A. (1937). The basic factors in proposed electrical methods for measuring thyroid function, I. The effect of body size and shape. *Western Journal of Surgical Obstetrics and Gynecology, 45,* 322-326.

Barnett, A., & Bagno, S. (1936). The physiological mechanisms involved in the clinical measure of phase angle. *American Journal of Physiology, 114,* 366-382.

Bauer, J., Capra, S., Davies, P.S., Ash, S., & Davidson, W. (2002). Estimation of total body water from bioelectrical impedance analysis in patients with pancreatic cancer. *Journal of Human Nutrition, 15,* 185-188.

Baumgartner, R.N., Chumlea, W.C., & Roche, A.F. (1988). Bioelectric impedance phase angle and body composition. *American Journal of Clinical Nutrition, 48,* 16-23.

Baumgartner, R.N., Chumlea, W.C., & Roche, A.F. (1990). Bioelectric impedance for body composition. *Exercise and Sport Sciences Reviews, 18,* 193-224.

Bedogni, G., Malavolti, M., Severi, S., Poli, M., Mussi, C., Fantuzzi, A.L., & Battistini, N. (2002). Accuracy of an eight-point tactile-electrode impedance method in the assessment of total body water. *European Journal of Clinical Nutrition, 56,* 1143-1148.

Berneis, K., & Keller, U. (2000). Bioelectrical impedance analysis during acute changes of extracellular osmolality in man. *Clinical Nutrition, 19,* 361-366.

Bolot, J.F., Fournier, G., Bertoye, A., Lenior, J., Jenin, P., & Thomasset, A. (1977). Determination of lean body mass by the electrical impedance measure. *Nouvelle Presse Medicale, 6,* 2249-2251.

Brozek, J., Grande, F., Anderson, J., & Keys, A. (1963). Densitometric analysis of body composition: revision of some quantitative assumptions. *Annals of the New York Academy of Sciences, 110,* 113-140.

Cable, A., Nieman, D.C., Austin, M., Hogen, E., & Utter, A.C. (2001). Validity of leg-to-leg bioelectrical impedance measurement in males. *Journal of Sports Medicine and Physical Fitness, 41,* 411-414.

Cha, K., Brown, E., & Wilmore, D. (1994). A new bioelectrical impedance method for measurement of the erythrocyte sedimentation rate. *Physiological Measurements, 15,* 499-508.

Chamney, P.W., Kramer, M., Rode, C., Kleinekofort, W., & Wizemann, V. (2002). A new technique for establishing dry weight in hemodialysis patients via whole body bioimpedance. *Kidney International, 61,* 2250-2258.

Chanchairujira, T., & Mehta, R.L. (2001). Assessing fluid change in hemodialysis: whole body versus sum of segmental bioimpedance spectroscopy. *Kidney International, 60,* 2337-2342.

Chertow, G.M., Lazarus, J.M., Lew, N.L., Ma, L., & Lowrie, E.G. (1997). Development of a population-specific regression equation to estimate total body water in hemodialysis patients. *Kidney International, 51,* 1578-1582.

Chumlea, W.C., Baumgartner, R.N., & Roche, A.F. (1987). Segmental bioelectric impedance measures of body composition. In Ellis, K.J., Yasumura, S., Morgan, W.D., eds. *In Vivo Body Composition Studies.* London: The Institute Physical Science Medicine, 103-107.

Chumlea, W.C., Baumgartner, R.N., & Roche, A.F. (1988). Specific resistivity used to estimate fat-free mass from segmental body measures of bioelectric impedance. *American Journal of Clinical Nutrition, 48,* 7-15.

Chumlea, W.C., & Guo, S.S. (1994). Bioelectrical impedance and body composition: present status and future directions. *Nutrition Reviews, 52,* 123-131.

Chumlea, W.C., Guo, S.S., Kuczmarski, R.J., Johnson, C.L., & Leahy, C.K. (1990). Reliability for anthropometry in the Hispanic health and nutrition examination survey (HHANES). *American Journal of Clinical Nutrition, 51,* 902-907.

Chumlea, W.C., Guo, S.S., Bellisari, A., Baumgartner, R.N., & Siervogel, R.M. (1994). Reliability for multiple frequency bioelectric impedance. *American Journal of Human Biology, 6,* 195-202.

Chumlea, W.C., Guo, S.S., Kuczmarski, R.J., Flegal, K.M., Johnson, C.L., Heymsfield, S.B., Lukaski, H., Friedl, K., & Hubbard, V.S. (2002). Body composition estimates from NHANES III bioelectrical impedance data. *International Journal of Obesity and Related Metabolic Disorders, 26,* 1596-1609.

Conlisk, E., Haas, J., Martinez, E., Flores, R., Rivera, J., & Martorell, R. (1992). Predicting body composition from anthropometry and bioimpedance in marginally undernourished adolescents and young adults. *American Journal of Clinical Nutrition, 55,* 1051-1059.

Cornish, B., Thomas, B., & Ward, L. (1993). Improved prediction of extracellular and total body water using impedance loci generated by multiple frequency bioelectrical impedance analysis. *Physics in Medicine and Biology, 38,* 337-346.

Demura, S., Yamaji, S., Goshi, F., & Nagasawa, Y. (2002). The influence of transient change of total body water on relative body fats based on three bioelectrical impedance analyses methods. Comparison between before and after exercise with sweat loss, and after drinking. *Journal of Sports Medicine and Physical Fitness, 42,* 38-44.

Deurenberg, P., Andreoli, A., & Delorenzo, A. (1996). Multi-frequency bioelectrical impedance: a comparison between the Cole-Cole modeling and Hanai equations with the classical impedance index approach. *Annals of Human Biology, 23,* 31-40.

Deurenberg, P., & Deurenberg-Yap, M. (2002). Validation of skinfold thickness and hand-held impedance measurements for estimation of body fat percentage among Singaporean Chinese, Malay and Indian subjects. *Asia Pacific Journal of Clinical Nutrition, 11,* 1-7.

Deurenberg, P., & Schouten, F. (1992). Loss of total body water and extracellular water assessed by multifrequency impedance. *European Journal of Clinical Nutrition, 46,* 247-255.

Deurenberg, P., Kooy, K., Paling, A., & Withagen, P. (1989). Assessment of body composition in 8-11 year old children by bioelectrical impedance. *European Journal of Clinical Nutrition,* 43, 623-629.

Deurenberg, P., Kusters, C.S.L., & Smit, H.E. (1990). Assessment of body composition by bioelectrical impedance in children and young adults is strongly age-dependent. *European Journal of Clinical Nutrition,* 44, 261-268.

Deurenberg, P., Vanderkooy, K., Leenen, R., Weststrate, J., & Seidell, J. (1991). Sex and age specific prediction formulas for estimating body composition from bioelectrical impedance—a cross-validation study. *International Journal of Obesity,* 15, 17-25.

Dittmar, M., & Reber, H. (2001). New equations for estimating body cell mass from bio-impedance parallel models in healthy older Germans. *American Journal of Physiology, Endocrinology and Metabolism,* 281, E1005-E1014.

Ducrot, H., Thomasset, A., Joly, R., Jungers, P., Eyraud, C., & Lenoir, J. (1970). Determination of extracellular fluid volume in man. *La Presse Medicale,* 78, 2269-2272.

Dumler, F., & Kilates, C. (2003). Body composition analysis by bioelectrical impedance in chronic maintenance dialysis patients: comparisons to the National Health and Nutrition Examination Survey III. *Journal of Renal Nutrition,* 13, 166-172.

Ellis, K.J., Bell, S.J., Chertow, G.M., Chumlea, W.C., Knox, T., Kotler, D.P., Lukaski, H.C., & Schoeller, D.A. (1999). Bioelectrical impedance methods in clinical research: a follow-up to the NIH Technology Assessment Conference. *Nutrition,* 15, 874-880.

Forbes, G., Simon, W., & Amatruda, J. (1992). Is bioimpedance a good predictor of body-composition change. *American Journal of Clinical Nutrition,* 56, 4-6.

Fornetti, W.C., Pivarnik, J.M., Foley, J.M., & Fiechtner, J.J. (1999). Reliability and validity of body composition measures in female athletes. *Journal of Applied Physiology,* 87, 1114-1122.

Fredrix, E., Saris, W., Soeters, P., Wouters, E., Kester, A., & Westerterp, K. (1990). Estimation of body composition by bioelectrical impedance in cancer patients. *European Journal of Clinical Nutrition,* 44, 749-752.

Fuller, N., & Elia, M. (1989). Potential use of bioelectrical impedance of the whole body and of body segments for the assessment of body composition, comparison with densitometry and anthropometry. *European Journal of Clinical Nutrition,* 43, 779-791.

Geddes, L.A., & Baker, L.E. (1967). The specific resistance of biological material. *Medical and Biological Engineering,* 5, 271-293.

Gudivaka, R., Schoeller, D.A., Kushner, R.F., & Bolt, M.J. (1999). Single- and multifrequency models for bioelectrical impedance analysis of body water compartments. *Journal of Applied Physiology,* 87, 1087-1096.

Guo, S.S., Roche, A.F., Chumlea, W.C., Miles, D.C., & Pohlman, R.H. (1987). Body composition predictions from bioelectric impedance. *Human Biology,* 59, 221-234.

Guo, S.S., Roche, A.F., & Houtkooper, L.H. (1989). Fat-free mass in children and young adults from bioelectric impedance and anthropometry variables. *American Journal of Clinical Nutrition,* 50, 435-443.

Hannan, W., Cowen, S., Fearon, K., Plester, C., Falconer, J., & Richardson, R. (1994). Evaluation of multi-frequency bioimpedance analysis for the assessment of extracellular and total body water in surgical patients. *Clinical Science,* 86, 479-485.

Heitmann, B. (1990). Prediction of body water and fat in adult Danes from measurement of electrical impedance–A validation study. *International Journal of Obesity,* 14, 789-802.

Ho, L.T., Kushner, R.F., Schoeller, D.A., Gudivaka, R., & Spiegel, D.M. (1994). Bioimpedance analysis of total body water in hemodialysis patients. *Kidney International,* 46, 1438-1442.

Hodgdon, J.A., & Fitzgerald, P.I. (1987). Validity of impedance predictions at various levels of fatness. *Human Biology,* 59, 281-298.

Hoffer, E., Meador, C., & Simpson, D. (1969). Correlation of whole-body impedance with total body water volume. *Journal of Applied Physiology,* 27, 531-534.

Horlick, M., Arpadi, S.M., Bethel, J., Wang, J., Moye, J.J., Cuff, P., Pierson, R.N., Jr., & Kotler, D. (2002). Bioelectrical impedance analysis models for prediction of total body water and fat-free mass in healthy and HIV-infected children and adolescents. *American Journal of Clinical Nutrition,* 76, 991-999.

Houtkooper, L.B., Going, S.B., Lohman, T.G., Roche, A.F., & VanLoan, M. (1992). Bioelectrical impedance estimation of fat-free body mass in children and youth: a cross-validation study. *Journal Applied Physiology,* 71, 366-373.

Houtkooper, L.B., Lohman, T.G., Going, S.B., & Howell, W.H. (1996). Why bioelectrical impedance analysis should be used for estimating adiposity. *American Journal of Clinical Nutrition,* 64, 436S-448S.

Jenin, P., Lenoir, J., Roullet, C., Thomasset, A., & Ducrot, H. (1975). Determination of body fluid compartments by electrical impedance measurements. *Aviation Space and Environmental Medicine,* 46, 152-155.

Kanai, H., Haeno, M., & Sakamoto, K. (1987). Electrical measurement of fluid distribution in legs and arms. *Medical Progress Through Technology,* 12, 159-170.

Kay, C.F., Bothwell, P.T., & Foltz, E.L. (1954). Electrical resistivity of living body tissues at low frequencies. *Journal of Physiology,* 13, 131-136.

Konings, C.J., Kooman, J.P., Schonck, M., van Kreel, B., Heidendal, G.A., Cheriex, E.C., van der Sande, F.M., & Leunissen, K.M. (2003). Influence of fluid status on techniques used to assess body composition in peritoneal dialysis patients. *Peritoneal Dialysis International,* 23, 184-190.

Kotler, D.P., Burastero, S., Wang, J., & Pierson, R.N., Jr. (1996). Prediction of body cell mass, fat-free mass, and total body water with bioelectrical impedance analysis: effects of race, sex, and disease. *American Journal of Clinical Nutrition,* 64, 489S-497S.

Kushner, R. (1992). Bioelectrical impedance analysis—a review of principles and applications. *Journal of the American College of Nutrition,* 11, 199-209.

Kushner, R.F., & Roxe, D.M. (2002). Bipedal bioelectrical impedance analysis reproducibly estimates total body water in hemodialysis patients. *American Journal of Kidney Disease,* 39, 154-158.

Kushner, R., & Schoeller, D. (1986). Estimation of total body water by bioelectrical impedance analysis. *American Journal of Clinical Nutrition,* 44, 417-424.

Kyle, U.G., Genton, L., Slosman, D.O., & Pichard, C. (2001). Fat-free and fat mass percentiles in 5225 healthy subjects aged 15 to 98 years. *Nutrition,* 17, 534-541.

Lafarque, A.L., Cabrales, L.B., & Larramendi, R.M. (2002). Bioelectrical parameters of the whole human body obtained through bioelectrical impedance analysis. *Bioelectromagnetics,* 23, 450-454.

Lee, S.W., Song, J.H., Kim, G.A., Lee, K.J., & Kim, M.J. (2001). Assessment of total body water from anthropometry-based equations using bioelectrical impedance as reference in Korean adult control and haemodialysis subjects. *Nephrology Dialysis Transplant,* 16, 91-97.

Lichtenbelt, W.D.V.M., Snel, Y.E.M., Brummer, R.-J.M., & Koppeschaar, H.P.F. (1997). Deuterium and bromide dilution, and bioimpedance spectrometry independently show that growth hormone-deficiency adults have an enlarged extracellular water compartment related to intracellular water. *Journal of Clinical Endocrinology and Metabolism,* 82, 907-911.

Lohman, T. (1986). Applicability of body composition techniques and constants for children and youths. *Exercise and Sport Sciences Reviews,* 14, 325-357.

Lohman, T.G., Caballero, B., Himes, J.H., Davis, C.E., Stewart, D., Houtkooper, L.H., Going, S.B., Hunsberger, S., Weber, J.L., Reid, R., & Stephenson, L. (2000). Estimation of body fat from anthropometry and bioelectrical impedance in Native American children. *International Journal of Obesity and Related Metabolic Disorders,* 24, 82-88.

Lohman, T.G., Thompson, J., Going, S., Himes, J.H., Caballero, B., Norman, J., Cano, S., & Ring, K. (2003). Indices of changes in adiposity in American Indian children. *Preventive Medicine,* 37(Suppl 1), S91-S96.

Lukaski, H. (1987). Methods for the assessment of human body composition: traditional and new. *American Journal of Clinical Nutrition,* 46, 537-556.

Lukaski, H.C., & Bolonchuk, W.W. (1987). Theory and validation of the tetrapolar bioelectrical impedance method to assess human body composition. In *In Vivo Body Composition Studies* (pp. 49-60). London: The Institute of Physical Sciences in Medicine.

Lukaski, H., Bolonchuk, W., Hall, C., & Siders, W. (1986). Validation of tetrapolar bioelectrical impedance method to assess human body composition. *Journal of Applied Physiology,* 60, 1327-1332.

Lukaski, H., Johnson, P., Bolonchuk, W., & Lykken, G. (1985). Assessment of fat-free mass using bioelectrical impedance measurements of the human body 1,2. *American Journal of Clinical Nutrition,* 41, 810-817.

Mast, M., Sonnichsen, A., Langnase, K., Labitzke, K., Bruse, U., Preub, U., & Muller, M.J. (2002). Inconsistencies in bioelectrical impedance and anthropometric measurements of fat mass in a field study of prepubertal children. *British Journal of Nutrition,* 87, 163-175.

Matthie, J.R., & Withers, P.O. (1998). Bioimpedance: 50 kHz parallel reactance and the prediction of body cell mass. *American Journal of Clinical Nutrition,* 68, 403-404.

Matthie, J., Zarowitz, B., De Lorenzo, A., Andreoli, A., Katzarski, K., Pan, G., & Withers, P. (1998). Analytic assessment of the various biompedance methods used to estimate body water. *Journal of Applied Physiology,* 84, 1801-1816.

Moore, F.D. (1963). *The Body Cell Mass and its Supporting Environment.* Philadelphia: Saunders.

NIH (1996). Bioelectrical impedance analysis in body composition measurement. *American Journal of Clinical Nutrition,* 64, 3875-4855.

Nyboer, J. (1959). *Electrical impedance plethysmography.* Springfield, IL: Charles C Thomas.

Nyboer, J. (1970). Electrorheometric properties of tissues and fluids. *Annals of the New York Academy of Sciences,* 170, 410-420.

O'Brien, C., Young, A.J., & Sawka, M.N. (2002). Bioelectrical impedance to estimate changes in hydration status. *International Journal of Sports Medicine,* 23, 361-366.

Patterson, R. (1989). Body fluid determinations using multiple impedance measurements. *IEEE Engineering in Medicine and Biology Magazine,* 8, 16-18.

Phillips, S.M., Bandini, L.G., Compton, D.V., Naumova, E.N., & Must, A. (2003). A longitudinal comparison of body composition by total body water and bioelectrical impedance in adolescent girls. *Journal of Nutrition,* 133, 1419-1425.

Pietrobelli, A., Nunez, C., Zingaretti, G., Battistini, N., Morini, P., Wang, Z.M., Yasumura, S., & Heymsfield, S.B. (2002). Assessment by bioimpedance of forearm cell mass: a new approach to calibration. *European Journal of Clinical Nutrition,* 56, 723-728.

Rising, R., Swinburn, B., Larson, K., & Ravussin, E. (1991). Body composition in Pima Indians—validation of bioelectrical resistance. *American Journal of Clinical Nutrition,* 53, 594-598.

Roche, A.F., Chumlea, W.C., & Guo, S.S. (1986). *Identification/Validation of New Anthropometric Techniques for Quantifying Body Composition.* TR-86/058. Natick, MA: U.S. Army Natick Research and Development Center.

Roubenoff, R. (1996). Applications of bioelectrical impedance analysis for body composition to epidemiologic studies. *American Journal of Clinical Nutrition,* 64(3 suppl), 459s-462s.

Rush, S., Abildskov, J.A., & McFee, R. (1963). Resistivity of body tissues at low frequencies. *Circulation Research,* 12, 40-50.

Salinari, S., Bertuzzi, A., Mingrone, G., Capristo, E., Pietrobelli, A., Campioni, P., Greco, A.V., & Heymsfield, S.B. (2002). New bioimpedance model accurately predicts lower limb muscle volume: validation by magnetic resonance imaging. *American Journal of Endocrinology and Metabolism,* 282, E960.

Scheltinga, M.R., Jacobs, D.O., Kimbrough, T.D., & Wilmore, D.W. (1991). Alterations in body fluid content can be detected by bioelectrical impedance analysis. *Journal of Surgical Research,* 50, 461-468.

Schoeller, D.A., & Luke, A. (2000). Bioelectrical impedance analysis prediction equations differ between African Americans and Caucasians but it is not clear why. *Annals of the New York Academy of Sciences,* 904, 225-226.

Schols, A., Wouters, E., Soeters, P., & Westerterp, K. (1991). Body composition by bioelectrical-impedance analysis compared with deuterium dilution and skinfold anthropometry in patients with chronic obstructive pulmonary disease. *American Journal of Clinical Nutrition,* 53, 421-424.

Schwan, H., & Kay, C. (1956). The conductivity of living tissues. *Annals of the New York Academy of Sciences,* 65, 1007-1013.

Schwan, H., & Li, K. (1953). Capacity and conductivity of body tissues at ultrahigh frequencies. *Proceed IRE,* 4, 1735-1740.

Schwenk, A., Breuer, P., Kremer, G., & Ward, L. (2001). Clinical assessment of HIV-associated lipodystrophy syndrome: bioelectrical impedance analysis, anthropometry and clinical scores. *Clinical Nutrition,* 20, 243-249.

Segal, K.R, Gutin, B., Presta, E., Wang, J., & Van Itallie, T.B. (1985). Estimation of human body composition by electrical impedance methods: a comparative study. *Journal of Applied Physiology,* 58, 1565-1571.

Segal, K., Burastero, S., Chun, A., Coronel, P., Pierson, R., & Wang, J. (1991). Estimation of extracellular and total body water by multiple-frequency bioelectrical-impedance measurement. *American Journal of Clinical Nutrition,* 54, 26-29.

Sergi, G., Bussolotto, M., Perini, P., Calliari, I., Giantin, V., Ceccon, A., Scanferla, F., Bressan, M., Moschini, G., & Enzi, G. (1994). Accuracy of bioelectrical impedance analysis in estimation of extracellular space in healthy subjects and in fluid retention states. *Annals of Nutrition and Metabolism,* 38, 158-165.

Settle, R., Foster, K., Epstein, B., & Mullen, J. (1980). Nutritional assessment: whole body impedance and body fluid compartments. *Nutrition and Cancer, 2,* 72-80.

Shirreffs, S., & Maughan, R. (1994). The effect of posture change on blood volume, serum potassium and whole body electrical impedance. *European Journal of Applied Physiology and Occupational Physiology,* 69, 461-463.

Shirreffs, S., Maughan, R., & Bernardi, M. (1994). Effect of posture change on blood volume, serum potassium and body water as estimated by bioelectrical impedance analysis (BIA). *Clinical Science (London),* 87, 21.

Simpson, J.A., Lobo, D.N., Anderson, J.A., Macdonald, I.A., Perkins, A.C., Neal, K.R., Allison, S.P., & Rowlands, B.J. (2001). Body water compartment measurements: a comparison of bioelectrical impedance analysis with tritium and sodium bromide dilution techniques. *Clinical Nutrition,* 20, 339-343.

Siri, W. (1961). Body composition from fluid spaces and density analysis of methods. In J. Brozek & A. Henschel (Eds.), *Techniques for Measuring Body Composition* (Vol. 61, pp. 223-244). Washington, DC: National Academy Press.

Slinde, F., Bark, A., Jansson, J., & Rossander-Hulthen, L. (2003). Bioelectrical impedance variation in healthy subjects during 12 in the supine position. *Clinical Nutrition,* 22, 153-157.

Slinde, F., & Rossander-Hulthen, L. (2001). Bioelectrical impedance: effect of 3 identical meals on diurnal impedance variation and calculation of body composition. *American Journal of Clinical Nutrition,* 74, 474-478.

Spence, J., Baliga, R., Nyboer, J., Seftick, J., & Fleischmann, L. (1979). Changes during hemodialysis in total body water cardiac output and chest fluid as detected by bioelectric impedance analysis. *Transactions American Society Artificial Internal Organs,* 25, 51-55.

Stolarczyk, L., Heyward, V., Hicks, V., & Baumgartner, R. (1994). Predictive accuracy of bioelectrical impedance in estimating body composition of native american women. *American Journal of Clinical Nutrition,* 59, 964-970.

Subramanyan, R., Manchanda, S.C., Nyboer, J., & Bhatia, M.L. (1980). Total body water in congestive heart failure. A pre- and post-treatment study. *Journal of Association of Physicians of India,* 28, 257-262.

Sun, S.S., Chumlea, W.C., Heymsfield, S.B., Lukaski, H., Da, S., Friedl, K., Kuczmarski, R.J., Flegal, K.M., Johnson, C.L., & Hubbard, V. (2003). Development of bioelectrical impedance analysis prediction equations for body composition with the use of a multicomponent model for use in epidemiological surveys. *American Journal of Clinical Nutrition,* 77, 331-340.

Tagliabue, A., Andreoli, A., Comelli, M., Bertoli, S., Testolin, G., Oriani, G., & DeLorenzo, A. (2001). Prediction of lean body mass from multifrequency segmental impedance: influence of adiposity. *Acta Diabetologica,* 38, 93-97.

Thomasset, A. (1962). Bioelectrical properties of tissue impedance measurements. *Lyon Medical,* 207, 107-118.

Tyrrell, V.J., Richards, G., Hofman, P., Gillies, G.F., Robinson, E., & Cutfield, W.S. (2001). Foot-to-foot bioelectrical impedance analysis: a valuable tool for the measurement of body composition in children. *International Journal of Related Metabolic Disorders,* 25, 273-278.

Van Loan, M.D., & Mayclin, P.L. (1992). Use of multi-frequency bioelectrical impedance analysis for the estimation of extracellular fluid. *European Journal of Clinical Nutrition,* 46, 117-124.

van Marken Lichtenbelt, W.D., Westerterp, K.R., Wouters, L., & Luijendijk, S. (1994). Validation of bioelectrical-impedance measurements as a method to estimate body-water compartments. *American Journal of Clinical Nutrition,* 60, 159-166.

VanderJagt, D.J., Huang, Y.S., Chuang, L.T., Bonnett, C., & Glew, R.H. (2002). Phase angle and n-3 polyunsaturated fatty acids in sickle cell disease. *Archives of Disease in Childhood,* 87, 252-254.

Varlet-Marie, E., Gaudard, A., Mercier, J., Bressole, F., & Brun, J.F. (2003). Is whole body impedance a predictor of blood viscosity? *Clinical Hemorheology and Microcirculation,* 28, 129-137.

Visser, M., Deurenberg, P., & Van Staveren, W.A. (1995). Multi-frequency bioelectrical impedance for assessing total body water and extracellular water in the elderly. *European Journal of Clinical Nutrition,* 49, 256-266.

Wagner, D.R., Heyward, V.H., Kocina, P.S., Stolarczyk, L., & Wilson, W.L. (1997). Predictive accuracy of BIA equations for estimating fat-free mass of black men. *Medicine and Science in Sports and Exercise,* 29, 69-74.

Ward, L., Bunce, I., Cornish, B., Mirolo, B., Thomas, B., & Jones, L. (1992). Multi-frequency bioelectrical impedance augments the diagnosis and management of lymphoedema in post-mastectomy patients. *European Journal of Clinical Investigation,* 22, 751-754.

Withers, R.T., LaForgia, J., & Heymsfield, S.B. (1999). Critical appraisal of the estimation of body composition via two, three, and four-compartment models. *American Journal of Human Biology,* 11, 175-185.

Zillikens, M., & Conway, J.M. (1991). Estimation of total body water by bioelectrical impedance analysis in Blacks. *American Journal of Human Biology,* 3, 25-32.

Chapter 7

Abate, N., Burns, D., Peshock, R.M., Garg, A., & Grundy, S.M. (1994): Estimation of adipose tissue mass by magnetic resonance imaging: validation against dissection in human cadavers. *Journal of Lipid Research* 35, 1490-6.

Abate, N., and Garg, A. (1995): Heterogeneity in adipose tissue metabolism: causes, implications and management of regional adiposity. *Progress in Lipid Research* 34, 53-70.

Abate, N., Garg, A., Coleman, R., Grundy, S.M., and Peshock, R.M. (1997): Prediction of total subcutaneous abdominal, intraperitoneal, and retroperitoneal adipose tissue masses in men by a single axial magnetic resonance imaging slice. *American Journal of Clinical Nutrition* 65, 403-8.

Abate, N., Garg, A., Peshock, R.M., Stray-Gundersen, J., Adams-Huet, B., and Grundy, S.M. (1996): Relationship of generalized and regional adiposity to insulin sensitivity in men with NIDDM. *Diabetes* 45, 1684-93.

Abate, N., Garg, A., Peshock, R.M., Stray-Gundersen, J., and Grundy, S.M. (1995): Relationships of generalized and regional adiposity to insulin sensitivity in men. *Journal of Clinical Investigation* 96, 88-98.

Abe, T., Kawakami, Y., Suzuki, Y., Gunji, A., and Fukunaga, T. (1997): Effects of 20 days bed rest on muscle morphology. *Journal of Gravitational Physiology* 4, S10-4.

Banerji, M.A., Buckley, M.C., Chaiken, R.L., Gordon, D., Lebovitz, H.E., and Kral, J.G. (1995): Liver fat, serum triglycerides and visceral adipose tissue in insulin-sensitive and insulin-resistant black men with NIDDM. *International Journal of Obesity and Related Metabolic Disorders* 19, 846-50.

Baumgartner, R.N., Koehler, K.M., Gallagher, D., Romero, L., Heymsfield, S.B., Ross, R.R., Garry, P.J., and Lindeman, R.D. (1998): Epidemiology of sarcopenia among the elderly in New Mexico. *American Journal of Epidemiology* 147, 755-63.

Baumgartner, R.N., Rhyne, R.L., and Garry, P.J. (1993): Body composition in the elderly from MRI: associations with cardiovascular disease risk factors, pp. 35-38. In K.J. Ellis, and J. Eastman (Eds): *In Vivo Methods, Models, and Assessment*, Plenum Press, New York.

Bjorntorp, P. (1990): "Portal" adipose tissue as a generator of risk factors for cardiovascular disease and diabetes. *Arteriosclerosis* 10, 493-6.

Bjorntorp, P. (1991): Metabolic implications of body fat distribution. *Diabetes Care* 14, 1132-43.

Boada, F.E., Ross, R., Stenger, V.A., Noll, D.C., Goodpaster, B.H., and Kelley, D.E. (1999): Absolute quantification of skeletal muscle lipid content with MRI, pp. 21-51: *Proceedings of the 7th Scientific Meeting of the International Society for Magnetic Resonance in Medicine*, Philadelphia.

Boesch, C., and Kreis, R. (2000): Observation of intramyocellular lipids by 1H-magnetic resonance spectroscopy. *Annals of the New York Academy of Sciences* 904, 25-31.

Boesch, C., Slotboom, J., Hoppeler, H., and Kreis, R. (1997): In vivo determination of intra-myocellular lipids in human muscle by means of localized 1H-MR-spectroscopy. *Magnetic Resonance in Medicine* 37, 484-93.

Borkan, G.A., Gerzof, S.G., Robbins, A.H., Hults, D.E., Silbert, C.K., and Silbert, J.E. (1982): Assessment of abdominal fat content by computed tomography. *American Journal of Clinical Nutrition* 36, 172-7.

Borkan, G.A., Hults, D.E., Gerzof, S.G., Robbins, A.H., and Silbert, C.K. (1983): Age changes in body composition revealed by computed tomography. *Journal of Gerontology* 38, 673-7.

Brochu, M., Starling, R.D., Tchernof, A., Matthews, D.E., Garcia-Rubi, E., and Poehlman, E.T. (2000): Visceral adipose tissue is an independent correlate of glucose disposal in older obese postmenopausal women. *Journal of Clinical Endocrinology and Metabolism* 85, 2378-84.

Brooks, S.V., and Faulkner, J.A. (1994): Skeletal muscle weakness in old age: underlying mechanisms. *Medicine and Science in Sports and Exercise* 26, 432-9.

Brummer, R.J., Lonn, L., Kvist, H., Grangard, U., Bengtsson, B.A., and Sjostrom, L. (1993): Adipose tissue and muscle volume determination by computed tomography in acromegaly, before and 1 year after adenomectomy. *European Journal of Clinical Investigation* 23, 199-205.

Busetto, L., Tregnaghi, A., De Marchi, F., Segato, G., Foletto, M., Sergi, G., Favretti, F., Lise, M., and Enzi, G. (2002): Liver volume and visceral obesity in women with hepatic steatosis undergoing gastric banding. *Obesity Research*, 10, 408-11.

Carey, G.B. (1997): The swine as a model for studying exercise-induced changes in lipid metabolism. *Medicine and Science in Sports and Exercise* 29, 1437-43.

Chowdhury, B., Sjostrom, L., Alpsten, M., Kostanty, J., Kvist, H., and Lofgren, R. (1994): A multicompartment body composition technique based on computerized tomography. *International Journal of Obesity and Related Metabolic Disorders,* 18, 219-34.

Cohn, S.H., Vartsky, D., Yasumura, S., Sawitsky, A., Zanzi, I., Vaswani, A., and Ellis, K.J. (1980): Compartmental body composition based on total-body nitrogen, potassium, and calcium. *American Journal of Physiology* 239, E524-30.

Conley, M.S., Stone, M.H., Nimmons, M., and Dudley, G.A. (1997): Specificity of resistance training responses in neck muscle size and strength. *European Journal of Applied Physiology and Occupational Physiology* 75, 443-8.

Constantinides, C.D., Gillen, J.S., Boada, F.E., Pomper, M.G., and Bottomley, P.A. (2000): Human skeletal muscle: sodium MR imaging and quantification-potential applications in exercise and disease. *Radiology* 216, 559-68.

Cruz, M.L., Bergman, R.N., and Goran, M.I. (2002): Unique effect of visceral fat on insulin sensitivity in obese Hispanic children with a family history of type 2 diabetes. *Diabetes Care* 25, 1631-6.

Cunningham, J.J. (1980): A reanalysis of the factors influencing basal metabolic rate in normal adults. *American Journal of Clinical Nutrition* 33, 2372-4.

de Ridder, C.M., de Boer, R.W., Seidell, J.C., Nieuwenhoff, C.M., Jeneson, J.A., Bakker, C.J., Zonderland, M.L., and Erich, W.B. (1992): Body fat distribution in pubertal girls quantified by magnetic resonance imaging. *International Journal of Obesity and Related Metabolic Disorders* 16, 443-9.

Deans, H.E., Smith, F.W., Lloyd, D.J., Law, A.N., and Sutherland, H.W. (1989): Fetal fat measurement by magnetic resonance imaging. *British Journal of Radiology* 62, 603-7.

DeNino, W.F., Tchernof, A., Dionne, I.J., Toth, M.J., Ades, P.A., Sites, C.K., and Poehlman, E.T. (2001): Contribution of abdominal adiposity to age-related differences in insulin sensitivity and plasma lipids in healthy nonobese women. *Diabetes Care* 24, 925-32.

Despres, J.P., Moorjani, S., Ferland, M., Tremblay, A., Lupien, P.J., Nadeau, A., Pinault, S., Theriault, G., and Bouchard, C. (1989): Adipose tissue distribution and plasma lipoprotein levels in obese women. Importance of intra-abdominal fat. *Arteriosclerosis* 9, 203-10.

Despres, J.P., Moorjani, S., Lupien, P.J., Tremblay, A., Nadeau, A., and Bouchard, C. (1990): Regional distribution of body fat, plasma lipoproteins, and cardiovascular disease. *Arteriosclerosis* 10, 497-511.

Doherty, T.J., Vandervoort, A.A., & Brown, W.F. (1993): Effects of ageing on the motor unit: a brief review. *Can J Appl Physiol* 18, 331-58.

Eliakim, A., Scheett, T., Allmendinger, N., Brasel, J.A., and Cooper, D.M. (2001): Training, muscle volume, and energy expenditure in nonobese American girls. *Journal of Applied Physiology* 90, 35-44.

Engstrom, C.M., Loeb, G.E., Reid, J.G., Forrest, W.J., and Avruch, L. (1991): Morphometry of the human thigh muscles. A comparison between anatomical sections and computer tomographic and magnetic resonance images. *Journal of Anatomy* 176, 139-56.

Evans, W. (1997): Functional and metabolic consequences of sarcopenia. *Journal of Nutrition* 127, 998S-1003S.

Figueroa-Colon, R., Mayo, M.S., Treuth, M.S., Aldridge, R.A., Hunter, G.R., Berland, L., Goran, M.I., and Weinsier, R.L. (1998): Variability of abdominal adipose tissue measurements using computed tomography in prepubertal girls. *International Journal of Obesity and Related Metabolic Disorders* 22, 1019-23.

Foster, M.A., Hutchison, J.M., Mallard, J.R., and Fuller, M. (1984): Nuclear magnetic resonance pulse sequence and discrimination of high- and low-fat tissues. *Magnetic Resonance Imaging* 2, 187-92.

Fowler, P.A., M.F. Fuller, C.A. Glasby, M.A. Foster, G.G. Cameron, G. McNiel, and Maughan, R.J. (1991). Total and subcutaneous adipose tissue distribution in women: the measurement of distribution and accurate prediction of quantity by using magnetic resonance imaging. *American Journal of Clinical Nutrition* 54: 18-25.

Frayne, K.N. (2000): Visceral fat and insulin resistance—causiative or correlative? *British Journal of Nutrition* 83, S71-S77.

Frontera, W.R., Hughes, V.A., Fielding, R.A., Fiatarone, M.A., Evans, W.J., and Roubenoff, R. (2000): Aging of skeletal muscle: a 12-yr longitudinal study. *Journal of Applied Physiology* 88, 1321-6.

Frontera, W.R., Hughes, V.A., Lutz, K.J., and Evans, W.J. (1991): A cross-sectional study of muscle strength and mass in 45- to 78-yr-old men and women. *Journal of Applied Physiology* 71, 644-50.

Gallagher, D., Belmonte, D., Deurenberg, P., Wang, Z., Krasnow, N., Pi-Sunyer, F.X., and Heymsfield, S.B. (1998): Organ-tissue mass measurement allows modeling of REE and metabolically active tissue mass. *American Journal of Physiology* 275, E249-58.

Gallagher, D., and Heymsfield, S.B. (1998): Muscle distribution: variations with body weight, gender, and age. *Applied Radiation and Isotopes* 49, 733-4.

Gallagher, D., Visser, M., De Meersman, R.E., Sepulveda, D., Baumgartner, R.N., Pierson, R. N., Harris, T., and Heymsfield, S.B. (1997): Appendicular skeletal muscle mass: effects of age, gender, and ethnicity. *Journal of Applied Physiology* 83, 229-39.

Gerard, E.L., Snow, R.C., Kennedy, D.N., Frisch, R.E., Guimaraes, A.R., Barbieri, R.L., Sorensen, A.G., Egglin, T.K., and Rosen, B.R. (1991): Overall body fat and regional fat distribution in young women: quantification with MR imaging. *AJR American Journal of Roentgenology*, 157, 99-104.

Goodpaster, B.H., Carlson, C.L., Visser, M., Kelley, D.E., Scherzinger, A., Harris, T.B., Stamm, E., and Newman, A.B. (2001): Attenuation of skeletal muscle and strength in the elderly: the Health ABC Study. *Journal of Applied Physiology* 90, 2157-65.

Goodpaster, B.H., Kelley, D.E., Thaete, F.L., He, J., and Ross, R. (2000a): Skeletal muscle attenuation determined by computed tomography is associated with skeletal muscle lipid content. *Journal of Applied Physiology* 89, 104-10.

Goodpaster, B.H., Kelley, D.E., Wing, R.R., Meier, A., and Thaete, F.L. (1999): Effects of weight loss on regional fat distribution and insulin sensitivity in obesity. *Diabetes* 48, 839-47.

Goodpaster, B.H., Krishnaswami, S., Resnick, H., Kelley, D.E., Haggerty, C., Harris, T.B., Schwartz, A.V., Kritchevsky, S., and Newman, A.B. (2003): Association between regional adipose tissue distribution and both type 2 diabetes and impaired glucose tolerance in elderly men and women. *Diabetes Care* 26, 372-9.

Goodpaster, B.H., Thaete, F.L., and Kelley, D.E. (2000b): Thigh adipose tissue distribution is associated with insulin resistance in obesity and in type 2 diabetes mellitus. *American Journal of Clinical Nutrition* 71, 885-92.

Goodpaster, B.H., Thaete, F.L., Simoneau, J.A., and Kelley, D.E. (1997): Subcutaneous abdominal fat and thigh muscle composition predict insulin sensitivity independently of visceral fat. *Diabetes* 46, 1579-85.

Goran, M.I., Kaskoun, M., and Shuman, W.P. (1995): Intra-abdominal adipose tissue in young children. *International Journal of Obesity and Related Metabolic Disorders* 19, 279-83.

Goto, T., Onuma, T., Takebe, K., and Kral, J.G. (1995): The influence of fatty liver on insulin clearance and insulin resistance in non-diabetic Japanese subjects. *International Journal of Obesity and Related Metabolic Disorders* 19, 841-5.

Gower, B.A., Nagy, T.R., and Goran, M.I. (1999): Visceral fat, insulin sensitivity, and lipids in prepubertal children. *Diabetes* 48, 1515-21.

Gray, D.S., Fujioka, K., Colletti, P.M., Kim, H., Devine, W., Cuyegkeng, T., and Pappas, T. (1991): Magnetic-resonance imaging used for determining fat distribution in obesity and diabetes. *American Journal of Clinical Nutrition* 54, 623-7.

Greenfield, J.R., Samaras, K., Chisholm, D.J., and Campbell, L.V. (2002): Regional intra-subject variability in abdominal adiposity limits usefulness of computed tomography. *Obesity Research* 10, 260-5.

Grimby, G. (1995): Muscle performance and structure in the elderly as studied cross-sectionally and longitudinally. *Journal of Gerontology A Biological Sciences and Medical Sciences* 50 Spec No, 17-22.

Han, T.S., Kelly, I.E., Walsh, K., Greene, R.M., and Lean, M.E. (1997): Relationship between volumes and areas from single transverse scans of intra-abdominal fat measured by magnetic resonance imaging. *International Journal of Obesity and Related Metabolic Disorders*, 21, 1161-6.

Hayes, P.A., Sowood, P.J., Belyavin, A., Cohen, J.B., and Smith, F.W. (1988): Sub-cutaneous fat thickness measured by magnetic resonance imaging, ultrasound, and calipers. *Medicine and Science in Sports and Exercise* 20, 303-9.

Herd, S.L., Gower, B.A., Dashti, N., and Goran, M.I. (2001): Body fat, fat distribution and serum lipids, lipoproteins and apolipoproteins in African-American and Caucasian-American prepubertal children. *International Journal of Obesity and Related Metabolic Disorders*, 25, 198-204.

Heymsfield, S.B., Fulenwider, T., Nordlinger, B., Barlow, R., Sones, P., and Kutner, M. (1979): Accurate measurement of liver, kidney, and spleen volume and mass by computerized axial tomography. *Annals of Internal Medicine* 90, 185-7.

Heymsfield, S.B., Gallagher, D., Kotler, D.P., Wang, Z., Allison, D.B., and Heshka, S. (2002): Body-size dependence of resting energy expenditure can be attributed to nonenergetic homogeneity of fat-free mass. *American Journal of Physiology-Endocrinology and Metabolism*, 282, E132-8.

Heymsfield, S.B., McManus, C., Stevens, V., and Smith, J. (1982): Muscle mass: reliable indicator of protein-energy malnutrition severity and outcome. *American Journal of Clinical Nutrition* 35, 1192-9.

Hudash, G., Albright, J.P., McAuley, E., Martin, R.K., and Fulton, M. (1985): Cross-sectional thigh components: computerized tomographic assessment. *Medicine and Science in Sports and Exercise* 17, 417-21.

Illner, K., Brinkmann, G., Heller, M., Bosy-Westphal, A., and Muller, M.J. (2000): Metabolically active components of fat free mass and resting energy expenditure in nonobese adults. *American Journal of Physiology: Endocrinology and Metabolism*, 278, E308-15.

Ivey, F.M., Tracy, B.L., Lemmer, J.T., NessAiver, M., Metter, E.J., Fozard, J.L., and Hurley, B.F. (2000): Effects of strength training and detraining on muscle quality: age and gender comparisons. *Journal of Gerontology* 55, B152-9.

Jacob, S., Machann, J., Rett, K., Brechtel, K., Volk, A., Renn, W., Maerker, E., Matthaei, S., Schick, F., Claussen, C.D., and Haring, H.U. (1999): Association of increased intra-myocellular lipid content with insulin resistance in lean nondiabetic offspring of type 2 diabetic subjects. *Diabetes* 48, 1113-9.

Janssen, I., Fortier, A., Hudson, R., and Ross, R. (2002a): Effects of an energy-restrictive diet with or without exercise on abdominal fat, intermuscular fat, and metabolic risk factors in obese women. *Diabetes Care* 25, 431-8.

Janssen, I., Heymsfield, S.B., Allison, D.B., Kotler, D.P., and Ross, R. (2002b): Body mass index and waist circumference independently contribute to the prediction of nonabdominal, abdominal subcutaneous, and visceral fat. *American Journal of Clinical Nutrition* 75, 683-8.

Janssen, I., Heymsfield, S.B., Baumgartner, R.N., and Ross, R. (2000a): Estimation of skeletal muscle mass by bioelectrical impedance analysis. *Journal of Applied Physiology* 89, 465-71.

Janssen, I., Heymsfield, S.B., and Ross, R. (2002c): Low relative skeletal muscle mass (sarcopenia) in older persons is associated with functional impairment and physical disability. *Journal of the American Geriatric Society*, 50, 889-96.

Janssen, I., Heymsfield, S.B., Wang, Z.M., and Ross, R. (2000b): Skeletal muscle mass and distribution in 468 men and women aged 18-88 yr. *Journal of Applied Physiology* 89, 81-8.

Janssen, I., and Ross, R. (1999): Effects of sex on the change in visceral, subcutaneous adipose tissue and skeletal muscle in response to weight loss. *International Journal of Obesity and Related Metabolic Disorders* 23, 1035-46.

Jensen, M.D. (1997): Lipolysis: contribution from regional fat. *Annual Review of Nutrition*, 17, 127-39.

Kelley, D.E., Thaete, F.L., Troost, F., Huwe, T., and Goodpaster, B.H. (2000): Subdivisions of subcutaneous abdominal adipose tissue and insulin resistance. *American Journal of Physiology and Endocrinology and Metababolism* 278, E941-8.

Kleiber, M. (1932): Body size and metabolism. *Hilgradia* 6, 315-53.

Klein, C.S., Allman, B.L., Marsh, G.D., and Rice, C.L. (2002): Muscle size, strength, and bone geometry in the upper limbs of young and old men. *Journal of Gerontology* 57, M455-9.

Kvist, H., Sjostrom, L., and Tylen, U. (1986): Adipose tissue volume determinations in women by computed tomography: technical considerations. *International Journal of Obesity Research* 10, 53-67.

LeBlanc, A., Lin, C., Shackelford, L., Sinitsyn, V., Evans, H., Belichenko, O., Schenkman, B., Kozlovskaya, I., Oganov, V., Bakulin, A., Hedrick, T., and Feeback, D. (2000): Muscle volume, MRI relaxation times (T2), and body composition after spaceflight. *Journal of Applied Physiology* 89, 2158-64.

LeBlanc, A., Rowe, R., Schneider, V., Evans, H., and Hedrick, T. (1995): Regional muscle loss after short duration spaceflight. *Aviation, Space and Environmental Medicine*, 66, 1151-4.

Lee, R.C., Wang, Z., Heo, M., Ross, R., Janssen, I., and Heymsfield, S.B. (2000): Total-body skeletal muscle mass: development and cross-validation of anthropometric prediction models. *American Journal of Clinical Nutrition* 72, 796-803.

Levine, J.A., Abboud, L., Barry, M., Reed, J.E., Sheedy, P.F., and Jensen, M.D. (2000): Measuring leg muscle and fat mass in humans: comparison of CT and dual-energy X-ray absorptiometry. *Journal of Applied Physiology* 88, 452-6.

Liu, M., Chino, N., and Ishihara, T. (1993): Muscle damage progression in Duchenne muscular dystrophy evaluated by a new quantitative computed tomography method. *Archives Physical Medicine Rehabilitation* 74, 507-14.

Mahmood, S., Taketa, K., Imai, K., Kajihara, Y., Imai, S., Yokobayashi, T., Yamamoto, S., Sato, M., Omori, H., and Manabe, K. (1998): Association of fatty liver with increased ratio of visceral to subcutaneous adipose tissue in obese men. *Acta Medica Okayama* 52, 225-31.

Marin, P., Andersson, B., Ottosson, M., Olbe, L., Chowdhury, B., Kvist, H., Holm, G., Sjostrom, L., and Bjorntorp, P. (1992): The morphology and metabolism of intraabdominal adipose tissue in men. *Metabolism* 41, 1242-8.

Marks, S.J., Moore, N.R., Ryley, N.G., Clark, M.L., Pointon, J.J., Strauss, B.J., & Hockaday, T.D. (1997): Measurement of liver fat by MRI and its reduction by dexfenfluramine in NIDDM. *Int J Obes Relat Metab Disord* 21, 274-9.

Mersmann, H.J., and Leymaster, K.A. (1984): Differential deposition and utilization of backfat layers in swine. *Growth* 48, 321-30.

Metter, E.J., Lynch, N., Conwit, R., Lindle, R., Tobin, J., and Hurley, B. (1999): Muscle quality and age: cross-sectional and longitudinal comparisons. *Journal of Gerontology* 54, B207-18.

Misra, A., Garg, A., Abate, N., Peshock, R.M., Stray-Gundersen, J., and Grundy, S.M. (1997): Relationship of anterior and posterior subcutaneous abdominal fat to insulin sensitivity in nondiabetic men. *Obesity Research* 5, 93-9.

Mitsiopoulos, N., Baumgartner, R.N., Heymsfield, S.B., Lyons, W., Gallagher, D., and Ross, R. (1998): Cadaver validation of skeletal muscle measurement by magnetic resonance imaging and computerized tomography. *Journal of Applied Physiology* 85, 115-22.

Monzon, J.R., Basile, R., Heneghan, S., Udupi, V., and Green, A. (2002): Lipolysis in adipocytes isolated from deep and superficial subcutaneous adipose tissue. *Obesity Research* 10, 266-9.

Mourier, A., Gautier, J.F., De Kerviler, E., Bigard, A.X., Villette, J.M., Garnier, J.P., Duvallet, A., Guezennec, C.Y., and Cathelineau, G. (1997): Mobilization of visceral adipose tissue related to the improvement in insulin sensitivity in response to physical training in NIDDM. Effects of branched-chain amino acid supplements. *Diabetes Care* 20, 385-91.

Nguyen-Duy, T.B., Nichaman, M.Z., Church, T.S., Blair, S.N., and Ross, R. (2003): Visceral fat and liver fat are independent predictors of metabolic risk factors in men. *American Journal of Physiology-Endocrinology and Metabolism* 284, E1065-E1071.

Nordal, H.J., Dietrichson, P., Eldevik, P., and Gronseth, K. (1988): Fat infiltration, atrophy and hypertrophy of skeletal muscles demonstrated by X-ray computed tomography in neurological patients. *Acta Neurologica Scandinavica* 77, 115-22.

Overend, T.J., Cunningham, D.A., Kramer, J.F., Lefcoe, M.S., and Paterson, D.H. (1992a): Knee extensor and knee flexor strength: cross-sectional area ratios in young and elderly men. *Journal of Gerontology* 47, M204-10.

Overend, T.J., Cunningham, D.A., Paterson, D.H., and Lefcoe, M.S. (1992b): Thigh composition in young and elderly men determined by computed tomography. *Clinical Physiology* 12, 629-40.

Perseghin, G., Scifo, P., De Cobelli, F., Pagliato, E., Battezzati, A., Arcelloni, C., Vanzulli, A., Testolin, G., Pozza, G., Del Maschio, A., and Luzi, L. (1999): Intramyocellular triglyceride content is a determinant of in vivo insulin resistance in humans: a 1H-13C nuclear magnetic resonance spectroscopy assessment in offspring of type 2 diabetic parents. *Diabetes* 48, 1600-6.

Piekarski, J., Goldberg, H.I., Royal, S.A., Axel, L., and Moss, A.A. (1980): Difference between liver and spleen CT numbers in the normal adult: its usefulness in predicting the presence of diffuse liver disease. *Radiology* 137, 727-9.

Randle, P., Hales, C., Garland, P., and Newsholme, E. (1963): The glucose fatty acid cycle. *Lancet* 1, 785-9.

Rebuffe-Scrive, M., Bronnegard, M., Nilsson, A., Eldh, J., Gustafsson, J.A., and Bjorntorp, P. (1990): Steroid hormone receptors in human adipose tissues. *Journal of Clinical Endocrinology and Metabolism* 71, 1215-9.

Ricci, C., Longo, R., Gioulis, E., Bosco, M., Pollesello, P., Masutti, F., Croce, L.S., Paoletti, S., de Bernard, B., Tiribelli, C., and Dalla Palma, L. (1997): Noninvasive in vivo quantitative assessment of fat content in human liver. *Journal of Hepatology* 27, 108-13.

Rice, B., Janssen, I., Hudson, R., and Ross, R. (1999): Effects of aerobic or resistance exercise and/or diet on glucose tolerance and plasma insulin levels in obese men. *Diabetes Care* 22, 684-91.

Rice, C.L., Cunningham, D.A., Paterson, D.H., and Lefcoe, M.S. (1989): Arm and leg composition determined by computed tomography in young and elderly men. *Clinical Physiology* 9, 207-20.

Rissanen, J., Hudson, R., and Ross, R. (1994): Visceral adiposity, androgens, and plasma lipids in obese men. *Metabolism* 43, 1318-23.

Roos, M.R., Rice, C.L., Connelly, D.M., & Vandervoort, A.A. (1999): Quadriceps muscle strength, contractile properties, and motor unit firing rates in young and old men. *Muscle Nerve* 22, 1094-103.

Ross, R. (1996): Magnetic resonance imaging provides new insights into the characterization of adipose and lean tissue distribution. *Canadian Journal of Physiology and Pharmacology* 74, 778-85.

Ross, R. (1997): Effects of diet- and exercise-induced weight loss on visceral adipose tissue in men and women. *Sports Medicine* 24, 55-64.

Ross, R., Aru, J., Freeman, J., Hudson, R., and Janssen, I. (2002): Abdominal adiposity and insulin resistance in obese men. *American Journal of Physiology-Endocrinology and Metabolism* 282, E657-63.

Ross, R., Dagnone, D., Jones, P.J., Smith, H., Paddags, A., Hudson, R., and Janssen, I. (2000): Reduction in obesity and related comorbid conditions after diet-induced weight loss or exercise-induced weight loss in men. A randomized, controlled trial. *Annals of Internal Medicine* 133, 92-103.

Ross, R., Fortier, L., and Hudson, R. (1996a): Separate associations between visceral and subcutaneous adipose tissue distribution, insulin and glucose levels in obese women. *Diabetes Care* 19, 1404-11.

Ross, R., Leger, L., Morris, D., de Guise, J., and Guardo, R. (1992): Quantification of adipose tissue by MRI: relationship with anthropometric variables. *Journal of Applied Physiology,* 72, 787-95.

Ross, R., Pedwell, H., and Rissanen, J. (1995): Effects of energy restriction and exercise on skeletal muscle and adipose tissue in women as measured by magnetic resonance imaging. *American Journal of Clinical Nutrition* 61, 1179-85.

Ross, R., and Rissanen, J. (1994): Mobilization of visceral and subcutaneous adipose tissue in response to energy restriction and exercise. *American Journal of Clinical Nutrition* 60, 695-703.

Ross, R., Rissanen, J., Pedwell, H., Clifford, J., and Shragge, P. (1996b): Influence of diet and exercise on skeletal muscle and visceral adipose tissue in men. *Journal of Applied Physiology* 81, 2445-55.

Rössner, S., Bo, W.J., Hiltbrandt, E., Hinson, W., Karstaedt, N., Santago, P., Sobol, W.T., and Crouse, J.R. (1990): Adipose tissue determinations in cadavers—a comparison between cross-sectional planimetry and computed tomography. *International Journal of Obesity Research* 14, 893-902.

Roth, S.M., Martel, G.F., Ivey, F.M., Lemmer, J.T., Tracy, B.L., Metter, E.J., Hurley, B.F., and Rogers, M.A. (2001): Skeletal muscle satellite cell characteristics in young and older men and women after heavy resistance strength training. *Journal of Gerontology* 56, B240-7.

Saint-Marc, T., Partisani, M., Poizot-Martin, I., Rouviere, O., Bruno, F., Avellaneda, R., Lang, J.M., Gastaut, J.A., and Touraine, J.L. (2000): Fat distribution evaluated by computed tomography and metabolic abnormalities in patients undergoing antiretroviral therapy: preliminary results of the LIPOCO study. *AIDS* 14, 37-49.

Schick, F., Machann, J., Brechtel, K., Strempfer, A., Klumpp, B., Stein, D.T., and Jacob, S. (2002): MRI of muscular fat. *Magnetic Resonance in Medicine* 47, 720-7.

Seidell, J.C., Bakker, C.J., and van der Kooy, K. (1990): Imaging techniques for measuring adipose-tissue distribution—a comparison between computed tomography and 1.5-T magnetic resonance. *American Journal of Clinical Nutrition* 51, 953-7.

Seidell, J.C., and Bouchard, C. (1997): Visceral fat in relation to health: is it a major culprit or simply an innocent bystander? *International Journal of Obesity Research* 21, 626-31.

Seidell, J.C., Oosterlee, A., Thijssen, M.A., Burema, J., Deurenberg, P., Hautvast, J.G., and Ruijs, J.H. (1987): Assessment of intra-abdominal and subcutaneous abdominal fat: relation between anthropometry and computed tomography. *American Journal of Clinical Nutrition* 45, 7-13.

Selberg, O., Burchert, W., Graubner, G., Wenner, C., Ehrenheim, C., and Muller, M.J. (1993): Determination of anatomical skeletal muscle mass by whole body nuclear magnetic resonance. *Basic Life Science* 60, 95-7.

Shen, W., Wang, Z., Tang, H., Heshka, S., Punyanitya, M., Zhu, S., Lei, J., and Heymsfield, S. B. (2003): Volume estimates by imaging methods: model comparisons with visible women as the reference. *Obesity Research,* 11, 217-225.

Shulman, G.I. (2000): Cellular mechanisms of insulin resistance. *Journal of Clinical Investigation* 106, 171-6.

Simoneau, J.A., Colberg, S.R., Thaete, F.L., and Kelley, D.E. (1995): Skeletal muscle glycolytic and oxidative enzyme capacities are determinants of insulin sensitivity and muscle composition in obese women. *FASEB Journal* 9, 273-8.

Sjöström, L. (1991): A computer-tomography based multicompartment body composition technique and anthropometric predictions of lean body mass, total and subcutaneous adipose tissue. *International Journal of Obesity Research* 15 Suppl 2, 19-30.

Sjöström, L., Kvist, H., Cederblad, A., and Tylen, U. (1986): Determination of total adipose tissue and body fat in women by computed tomography, 40K, and tritium. *American Journal of Physiology* 250, E736-45.

Smith, S.R., Lovejoy, J.C., Greenway, F., Ryan, D., deJonge, L., de la Bretonne, J., Volafova, J., and Bray, G.A. (2001): Contributions of total body fat, abdominal subcutaneous adipose tissue compartments, and visceral adipose tissue to the metabolic complications of obesity. *Metabolism* 50, 425-35.

Snyder, W.S., Cooke, M.J., Manssett, E.S., Larhansen, L.T., Howells, G.P., and Tipton, I.H. (1975): *Report of the Task Group on Reference Man.* Pergamon. Oxford, UK.

Sobol, W., Rossner, S., Hinson, B., Hiltbrandt, E., Karstaedt, N., Santago, P., Wolfman, N., Hagaman, A., and Crouse, J.R., III (1991): Evaluation of a new magnetic resonance imaging method for quantitating adipose tissue areas. *International Journal of Obesity Research* 15, 589-99.

Sohlstrom, A., Wahlund, L.O., and Forsum, E. (1993): Adipose tissue distribution as assessed by magnetic resonance imaging and total body fat by magnetic resonance imaging, underwater weighing, and body-water dilution in healthy women. *American Journal of Clinical Nutrition* 58, 830-8.

Sprawls, P. (1977): *The Physical Principles of Diagnostic Radiology.* University Park Press. Baltimore.

Staten, M.A., Totty, W.G., and Kohrt, W.M. (1989): Measurement of fat distribution by magnetic resonance imaging. *Investigative Radiology* 24, 345-9.

Thaete, F.L., Colberg, S.R., Burke, T., and Kelley, D.E. (1995): Reproducibility of computed tomography measurement of visceral adipose tissue area. *International Journal of Obesity and Related Metabolic Disorders* 19, 464-7.

Thomas, E.L., and Bell, J.D. (2003): Influence of undersampling on magnetic resonance imaging measurements of intra-abdominal adipose tissue. *International Journal of Obesity and Related Metabolic Disorders* 27, 211-8.

Thomas, E.L., Saeed, N., Hajnal, J.V., Byrnes, A., Goldstone, A.P., Frost, G., and Bell, J.D. (1998): Magnetic resonance imaging of total body fat. *Journal of Applied Physiology* 85, 1778-85.

Tokunaga, K., Matsuzawa, Y., Ishikawa, T., and Tarui, S. (1983): A novel technique for the determination of body fat by computed tomography. *International Journal of Obesity Research* 7, 437-445.

Tomlinson, B.E., & Irvin, D. (1977). The numbers of limb motor neurons in the human lumbosacral cord throughout life. *J Neurol Sci* 34, 213-9.

Toth, M.J., Sites, C.K., Cefalu, W.T., Matthews, D.E., and Poehlman, E.T. (2001): Determinants of insulin-stimulated glucose disposal in middle-aged, premenopausal women. *American Journal of Physiology-Endocrinology and Metabolism* 281, E113-21.

Tracy, B.L., Ivey, F.M., Hurlbut, D., Martel, G.F., Lemmer, J.T., Siegel, E.L., Metter, E.J., Fozard, J.L., Fleg, J.L., and Hurley, B.F. (1999): Muscle quality. II. Effects of strength training in 65- to 75-yr-old men and women. *Journal of Applied Physiology* 86, 195-201.

Tsubahara, A., Chino, N., Akaboshi, K., Okajima, Y., and Takahashi, H. (1995): Age-related changes of water and fat content in muscles estimated by magnetic resonance (MR) imaging. *Disability Rehabilitation* 17, 298-304.

Tzankoff, S.P., and Norris, A.H. (1977): Effect of muscle mass decrease on age-related BMR changes. *Journal of Applied Physiology* 43, 1001-6.

U.S. National Library of Medicine (1995): *Visible Human Project. Visible Human CD-ROM, Version 1.1.*

Visser, M., Kritchevsky, S.B., Goodpaster, B.H., Newman, A.B., Nevitt, M., Stamm, E., and Harris, T.B. (2002): Leg muscle mass and composition in relation to lower extremity performance in men and women aged 70 to 79: the health, aging and body composition study. *Journal of American Geriatric Society*, 50, 897-904.

Wang, Z.M., Gallagher, D., Nelson, M.E., Matthews, D.E., and Heymsfield, S.B. (1996): Total-body skeletal muscle mass: evaluation of 24-h urinary creatinine excretion by computerized axial tomography. *American Journal of Clinical Nutrition* 63, 863-9.

Weinsier, R.L., Schutz, Y., and Bracco, D. (1992): Reexamination of the relationship of resting metabolic rate to fat-free mass and to the metabolically active components of fat-free mass in humans. *American Journal of Clinical Nutrition* 55, 790-4.

Chapter 8

Abate, N., Burns D., Peshock, R.M., Garg A., & Grundy, S.M. (1994). Estimation of adipose tissue mass by magnetic resonance imaging: validation against dissection in human cadavers. *Journal of Lipid Research, 35,* 1490-1496.

Abdel-Malek, A.K., Mukherjee, D., & Roche, A.F. (1985). A method of constructing an index of obesity. *Human Biology, 57,* 415-430.

Abe, T., Kawakami, Y., Sugita, M., Yoshikawa, K., & Fukunaga, T. (1995). Use of B-mode ultrasound for visceral fat mass evaluation: comparisons with magnetic resonance imaging. *Applied Human Science: Journal of Physiological Anthropology, 14,* 133-139.

Abe, T., Kondo, M., Kawakami, Y., & Fukunaga, T. (1994). Prediction equations for body composition of Japanese adults by B-mode ultrasound. *American Journal of Human Biology, 6,* 161-170.

Abe, T., Tanaka, F., Kawakami, Y., Yoshikawa, K., & Fukunaga, T. (1996). Total and segmental subcutaneous adipose tissue volume measured by ultrasound. *Medicine and Science in Sports and Exercise, 28,* 908-912.

Adams, W.C., Deck-Côté, K., & Winters, K.M. (1992). Anthropometric estimation of bone mineral content in young adult females. *American Journal of Human Biology, 4,* 767-774.

Alexander, H.G., & Dugdale, A.E. (1992). Fascial planes within subcutaneous fat in humans. *European Journal of Clinical Nutrition, 46,* 903-906.

Aloia, J.F., Vaswani, A., Delerme-Pagan, C., & Flaster, E. (1998). Discordance between ultrasound of the calcaneus and bone mineral density in black and white women. *Calcified Tissue International, 62,* 481-485.

Armellini, F., Zamboni, M., Castelli, S., Micciolo, R., Mino, A., Turcato, E., Rigo, L., Bergamo-Andreis, IA., & Bosello, O. (1994). Measured and predicted total and visceral adipose tissue in women—correlations with metabolic parameters. *International Journal of Obesity, 18,* 641-647.

Armellini, F., Zamboni, M., Harris, T., Micciolo, R., & Bosello, O. (1997). Sagittal diameter minus subcutaneous thickness. An easy-to-obtain parameter that improves visceral fat prediction. *Obesity Research, 5,* 315-320.

Armellini, F., Zamboni, M., Rigo, L., Todesco, T., Bergamo-Andreis, I.A., Procacci, C., & Bosello, O. (1990). The contribution of sonography to the measurement of intra-abdominal fat. *Journal of Clinical Ultrasound, 18,* 563-567.

Armellini, F., Zamboni, M., Rigo, L., Robbi, R., Todesco, T., Castelli, S., Mino, A., Bissoli, L, Turcato, E., and Bosello, O. (1993a). Measurements of intra-abdominal fat by ultrasound and computed tomography: Predictive equations in women. In K.L. Ellis & J.D. Eastman (Eds.), *Human Body Composition* (pp. 75-77). New York: Plenum Press.

Armellini, F., Zamboni, M., Robbi, R., Todesco, T., Rigo, L., Bergamo-Andreis, I.A., & Bosello, O. (1993b). Total and intra-abdominal fat measurements by ultrasound and computerized-tomography. *International Journal of Obesity, 17,* 209-214.

Ashwell, M., Cole, T.J., & Dixon, A.K. (1985). Obesity: New insight into the anthropometric classification of fat distribution shown by computed tomography. *British Medical Journal, 290,* 1692-1694.

Ashwell, M., McCall, S.A., Cole, T.J., & Dixon, A.K. (1987). Fat distribution and its metabolic complications: Interpretations. In N.G. Norgan (Ed.), *Human Body Composition and Fat Distribution. Report of an EC Workshop* (pp. 227-243). The Hague: CIP-gegenvens Koninklijke Bibliothekl (Euro-Nut Conference Series).

Baker, G.L. (1969). Human adipose tissue composition and age. *American Journal of Clinical Nutrition, 22,* 829-835.

Bakker, H.K., & Struikenkamp, R.S. (1977). Biological variability and lean body mass estimates. *Human Biology, 49,* 187-202.

Ballor, D.L., & Katch, V.L. (1989). Validity of anthropometric regression equations for predicting changes in body fat of obese females. *American Journal of Human Biology, 1,* 97-101.

Baran, D.T. (1995). Quantitative ultrasound: a technique to target women with low bone mass for preventive therapy. *American Journal of Medicine, 98* (Suppl 2A), 48S-51S.

Baumgartner, R.N., Heymsfield, S.B., Lichtman, S., Wang, J., & Pierson, R.N. (1991). Body composition in elderly people: effect of criterion estimates on predictive equations. *American Journal of Clinical Nutrition, 53,* 1345-1349.

Baumgartner, R.N., Heymsfield, S.B., Roche, A.F., & Bernadino, M. (1988). Quantification of abdominal composition by computed tomography. *American Journal of Clinical Nutrition, 48,* 936-945.

Baumgartner, R.N., Rhyne, R.L., Garry, P.J., & Chumlea, W.C. (1993). Body composition in the elderly from magnetic resonance imaging: associations with cardiovascular disease risk factors. In K.I. Ellis & I.D. Eastman (Eds.), *Human Body Composition* (pp. 35-38). New York: Plenum Press.

Baumgartner, R.N., Rhyne, R.L., Troup, C., Wayne, S., & Garry, P.J. (1992). Appendicular skeletal muscle areas assessed by magnetic resonance imaging in older persons. *Journal of Gerontology, 47,* M67-M72.

Baumgartner, R.N., Roche, A.F., Guo, S., Chumlea, W.C., Ryan, A.S., & Kuczmarski, R.J. (1990). Fat patterning and centralized obesity in Mexican-American children in the Hispanic Health and Nutrition Examination Survey (HHANES 1982-1984). *American Journal of Clinical Nutrition, 51,* 936S-943S.

Bellisari, A. (1993). Sonographic measurement of adipose tissue. *Journal of Diagnostic Medical Sonography, 9,* 11-18.

Bellisari, A. (1997). Validation of ultrasonic measurements of subcutaneous adipose tissue using magnetic resonance imaging. *American Journal of Human Biology, 9,* 123-124.

Bellisari, A., Roche, A.F., & Siervogel, R.M. (1993). Reliability of B-mode ultrasonic measurements of subcutaneous and intra-abdominal adipose tissue. *International Journal of Obesity, 17,* 475-480.

Bellisari, A., Wellens, R., Roche, A.F., Boska, M., Guo, S., Chumlea, W.C., & Siervogel, R.M. (1994). Validation of ultrasonic and skinfold measurements of subcutaneous adipose tissue using magnetic resonance imaging. *American Journal of Human Biology, 6,* 116.

Björntorp, P. (1990). "Portal" adipose tissue as a generator of risk factors for cardiovascular disease and diabetes. *Arteriosclerosis, 10,* 493-496.

Bliznak, I., & Staple, T.W. (1975). Roentgenographic measurement of skin thickness in normal individuals. *Radiology, 118,* 55-60.

Boileau, R.A., Lohman, T.G., Slaughter, M.H., Ball, T.E., Going, S.B., & Hendrix, M.K. (1984). Hydration of the fat-free body in children during maturation. *Human Biology, 56,* 651-666.

Boileau, R.A., Wilmore, J.H., Lohman, T.G., Slaughter, M., & Riner, W. (1981). Estimation of body density from skinfold thicknesses, body circumferences and skeletal widths in boys aged 8 to 11 years: comparison of two samples. *Human Biology, 53,* 575-592.

Bolinder, J., Kager, L., Ostman, J., & Arner, P. (1983). Differences at the receptor and post-receptor levels between human omental and subcutaneous adipose tissue in the action of insulin on lipolysis. *Diabetes, 32,* 117-123.

Booth, R.A.D., Goddard, B.A., & Paton, A. (1966). Measurement of fat thickness in man: A comparison of ultrasound, Harpenden calipers and electrical conductivity. *British Journal of Nutrition, 20,* 719-725.

Borkan, G.A., Hults, D.E., Gerzof, S.G., Burrows, B.A., & Robbins, A.H. (1983). Relationships between computed tomograpy tissue areas, thicknesses and total body composition. *Annals of Human Biology, 10,* 537-546.

Bosello, O., Zamboni, M., Armellini, F., & Todesco, T. (1993). Biological and clinical aspects of regional body fat distribution. *Diabetes Nutrition and Metabolism, Clinical and Experimental, 6,* 163-171.

Bouchard, C. (1988). Introductory notes on the topic of fat distribution. In C. Bouchard & F.E. Johnston (Eds.), *Fat Distribution During Growth and Later Health Outcomes* (pp. 1-8). New York: Liss.

Bradfield, R.B., Schultz, Y., & Lechtig, A. (1979). Skinfold changes with weight loss. *American Journal of Clinical Nutrition, 32,* 1756.

Braillon, P.M., Salle, B.L., Brunet, J., Glorieux, F.H., Delmas, P.D., and Meunier, P.J. (1992). Dual energy x-ray absorptiometry measurement of bone mineral content in newborns: validation of the technique. *Pediatric Research, 32,* 77-89.

Brans, Y.W., Sumners, J.E., Dweck, H.S., & Cassady, G. (1974). A noninvasive approach to body composition in the neonate: dynamic skinfold measurements. *Pediatric Research, 8,* 215-222.

Bray, G.A., Greenway, F.L., Molitch, M.E., Dahms, W.T., Atkinson, R.L., & Hamilton, K. (1978). Use of anthropometric measures to assess weight loss. *American Journal of Clinical Nutrition, 31,* 769-773.

Bunt, J.C., Going, S.B., Lohman, T.G., Heinrich, C.H., Perry, C.D., and Parmenter, R.W. (1990). Variation in bone mineral content and estimated body fat in young adult females. *Medicine and Science in Sports and Exercise, 22,* 564-569.

Bunt, J.C., Lohman, T.G., & Boileau, R.A. (1989). Impact of total body water fluctuation on estimating of body fat from body density. *Medicine and Science in Sports and Exercise, 21,* 96-100.

Bushberg, J.T., Seibert, J.A., Leidholdt, E.M., & Boone, J.M. (2002). *The Essential Physics of Medical Imaging* (2nd edition). Philadelphia: Lippincott Williams & Wilkins.

Carlyon, R.G., Bryant, R.W., Gore, C.J., & Walker, R.E. (1998) Apparatus for precision calibration of skinfold calipers. *American Journal of Human Biology, 10,* 689-697.

Çetin, A., Ertürk, H., Çeliker, R., Sivri, A., & Hascelik, Z. (2001) The role of quantitative ultrasound in predicting osteoporosis defined by dual X-ray absorptiometry. *Rheumatology International, 20,* 55-59.

Chan, G.M. (1992). Performance of dual-energy x-ray absorptiometry in evaluating bone, lean body mass, and fat in pediatric subjects. *Journal of Bone Mineral Research, 7,* 369-374.

Chestnut, C.H., III, Manske, E., Baylink, D., & Nelp, W.B. (1973). Preliminary report: correlation of total body calcium (bone mass), as determined by neutron activation analysis with regional bone mass as determined by photon absorption. In R.B. Mazess (Ed.), *International Conference on Bone Mineral Measurement* (NIH Publication No.75-683, pp. 34-38). Washington, DC: U.S. Department of Health, Education, and Welfare.

Chinn, K.S.K., & Allen, T.H. (1960). *Body Fat in Men From Two Skinfolds, Weight, Height, and Age* (U.S. Army Medical Research and Nutrition Laboratory, Report No. 248). Washington, DC: U.S. Government Printing Office.

Chumlea, W.C., Roche, A.F., & Steinbaugh, M.L. (1985a). Estimating stature from knee height for persons 60 to 90 years of age. *Journal of American Geriatric Society, 33,* 116-120.

Chumlea, W.C., Steinbaugh, M.L., Roche, A.F., Mukherjee, D., & Gopalaswamy, N. (1985b). Nutritional anthropometric assessment in elderly persons 65 to 90 years of age. *Journal of Nutrition for the Elderly, 4,* 39-51.

Cignarelli, M., DePergola, G., Picca, G., Sciaraffia, M., Pannaciulli, N., Tarallo, M., Laudadio, E., Turrisi, E., & Giorgino, R. (1996). Relationship of obesity and body fat distribution with ceruloplasmin serum levels. *International Journal of Obesity, 20,* 809-813.

Cohn, S.H., Ellis, K.J., Vartsky, D., et al. (1981). Comparison of methods of estimating body fat in normal subjects and cancer patients. *American Journal of Clinical Nutrition, 34,* 2839-2847.

Cooke, R.W.I., Lucas, A., Udkin, P.L.N., & Pryse-Davies, J. (1977). Head circumference as an index of brain weight in the fetus and newborn. *Early Human Development, 1,* 145-149.

Côté, K.D., & Adams, W.C. (1993). Effect of bone density on body composition estimates in young adult black and white women. *Medicine and Science in Sports & Exercise, 25,* 290-296.

Crenier, E.J. (1966). La prédiction du poids corporel "normal" [The prediction of normal body weight]. *Révue de la Societé de Biometrie Humaine, 1,* 10-24.

Damon, A. (1965). Note on anthropometric technique: II. Skinfolds, right and left sides, held by one or two hands. *American Journal of Physical Anthropology, 23,* 305-306.

Dawson-Hughes, B., Shipp, C., Sadowski, L., & Dallal, G. (1987). Bone density of the radius, spine, and hip in relation to percent of ideal body weight in postmenopausal women. *Calcified Tissue International, 40,* 310-314.

DeNino, W.F., Tchernof, A., Dionne, I.J., Toth, M.J., Ades, P.A., Sites, C.K., & Poehlman, E.T. (2001). Contribution of abdominal adiposity to age-related differences in insulin sensitivity and plasma lipids in healthy nonobese women. *Diabetes Care, 24,* 925-932.

Després, J.P., Prud'Homme, D., Pouilot, M.C., Tremblay, A., & Bouchard, C. (1991). Estimation of deep abdominal adipose tissue accumulation from simple anthropometric measurements in men. *American Journal of Clinical Nutrition, 54,* 471-477.

Deurenberg, P., van der Kooy, K., & Hautvast, J.G.A.J. (1990). The assessment of body composition in the elderly by densitometry, anthropometry and bioelectrical impedance. In S. Yasumura, J.E. Harrison, K.G. McNeill, A.D. Woodhead, & F.A. Dilmanian (Eds.), *In Vivo Body Composition Studies. Recent Advances* (pp. 391-393). New York: Plenum Press.

Deurenberg, P., van der Kooy, K., Hulshof, T., & Eyers, P. (1989). Body mass index as a measure of body fatness in the elderly. *European Journal of Clinical Nutrition, 43,* 231-236.

Deutsch, M.I., Mueller, W.H., & Malina, R.M. (1985). Androgyny in fat patterning is associated with obesity in adolescents and young adults. *Annals of Human Biology, 12,* 275-286.

Durnin, J., & Womersley, J. (1974). Body fat assessed from total body density and its estimation from skinfold thickness: measurements on 481 men and women aged from 16 to 72 years. *British Journal of Nutrition, 32,* 77-97.

Edholm, O.G., Adam, J.M., & Best, T.W. (1974). Day-to-day weight changes in young men. *Annals of Human Biology, 3,* 3-12.

Edwards, D.A.W., Hammond, W.H., Healy, M.J.R., Tanner, J.M., & Whitehouse, R.H. (1955). Design and accuracy of calipers for measuring subcutaneous tissue thickness. *British Journal of Nutrition, 2,* 133-143.

Eston, R., Evans, R., & Fu, F. (1994). Estimation of body-composition in Chinese and British men by ultrasonographic assessment of segmental adipose tissue volume. *British Journal of Sports Medicine, 28,* 9-13.

Evens, R.G. (1991). The future of ultrasonography: report of the Ultrasonography Task Force. *Journal of the American Medical Association, 266,* 406-409.

Fanelli, M.T., Kuczmarski, R.J., & Hirsch, M. (1988). Estimation of body fat from ultrasound measures of subcutaneous fat and circumferences in obese women. *International Journal of Obesity, 12,* 125-132.

Ferland, M., Després, J.P., Tremblay, A., Pinault, S., Nadeau, A., Moorjani, S., Lupien, P.J., Theriault, G., & Bouchard, C. (1989). Assessment of adipose tissue distribution by computed axial tomography in obese women: association with body density and anthropometric measurements. *British Journal of Nutrition, 61,* 139-148.

Flygare, A., Valentin, L., Karsland-Akeson, P., Flodmark, P., Ivarson, S.A., & Axelsson, I. (1999). Ultrasound measurements of subcutaneous adipose tissue in infants are reproducible. *Journal of Pediatric Gastroenterology and Nutrition, 28,* 492-494.

Forbes, G.B. (1974). Stature and lean body mass. *American Journal of Clinical Nutrition, 27,* 595-602.

Forbes, G.B., Brown, M.R., & Griffiths, H.J.L. (1988). Arm muscle plus bone area: anthropometry and CAT scan compared. *American Journal of Clinical Nutrition, 47,* 929-931.

Forsberg, A.M., Nilsson, E., Werneman, J, Bergstrom, J., & Hultman, E. (1991). Muscle composition in relation to age and sex. *Clinical Science, 81,* 249-256.

Frantzell, A., & Ingelmark, B.E. (1951). Occurrence and distribution of fat in human muscles at various age levels: a morphologic and roentgenologic investigation. *Acta Societatis Medicorum Upsaliensis, 56,* 59-87.

Frerichs, R.R., Harsha, D.W., & Berenson, G.S. (1979). Equations for estimating percentage of body fat in children 10-14 years old. *Pediatric Research, 13,* 170-174.

Fuchs, R.J., Theis, C.F., & Lancaster, M.C. (1978). A nomogram to predict lean body mass in men. *American Journal of Clinical Nutrition, 31,* 673-678.

Fukagawa, N.K., Bandini, L.G., & Young, J.B. (1990). Effect of age on body composition and resting metabolic rate. *American Journal of Physiology, 259,* E233-E238.

Fülöp, T., Jr., Worum, I., Csongor, J., Foris, G., & Leovay, A. (1985). Body composition in elderly people. *Gerontology, 31,* 150-157.

Gerver, W.J.M., & deBruin, R. (2001). *Paediatric Morphometrics: A Reference Manual.* (2nd ed.). Utrecht, The Netherlands: Wetenschappelijke Uitgeverij Bunge.

Goodpaster, B.H., Thaete, F.L., Simoneau, J.A., & Kelley, D.E. (1997). Subcutaneous abdominal fat and thigh muscle composition predict insulin sensitivity independently of visceral fat. *Diabetes, 46,* 1579-1585.

Grauer, W.O., Moss, A.A., Cann, C.E., & Goldberg, H.I. (1984). Quantification of body fat distribution in the abdomen using computed tomography. *American Journal of Clinical Nutrition, 39,* 631-637.

Greenspan, S.L., Cheng, S., Miller, P.D., & Orwoll, E.S. (2001). Clinical performance of a highly portable, scanning calcaneal ultrasonometer. *Osteoporosis International, 12,* 391-398.

Guo, S., Roche, A.F., & Houtkooper, L. (1989). Fat-free mass in children and young adults predicted from bioelectric impedance and anthropometric variables. *American Journal of Clinical Nutrition, 50,* 435-443.

Haffner, S.M., Stern, M.P., Hazunda, H.P., Pugh, J., & Patterons, J.K. (1987). Do upper-body and centralized adiposity measure different aspects of regional body-fat distribution? Relationship to non-insulin-dependent diabetes mellitus, lipids, and lipoproteins. *Diabetes, 36,* 43-51.

Hagen-Ansert, S.L. (2001). *Textbook of Diagnostic Ultrasonography* (5th ed.). St. Louis: Mosby.

Haisman, M.F. (1970). The assessment of body fat content in young men from measurements of body density and skinfold thickness. *Human Biology, 42,* 679-688.

Hammer, L.D., Wilson, D.M., Litt, I.F., Killen, J.D., Hayward, C., Miner, B., Vosti, C., & Taylor, C.B. (1991). Impact of pubertal development on body fat distribution among white, Hispanic, and Asian female adolescents. *Journal of Pediatrics, 118,* 975-980.

Harsha, D.W., Frerichs, R.R., & Berenson, G.S. (1978). Densitometry and anthropometry of black and white children. *Human Biology, 50,* 261-280.

Harsha, D.W., Voors, A.W., & Berenson, G.S. (1980). Racial differences in subcutaneous fat patterns in children aged 7-15 years. *American Journal of Physical Anthropology, 53,* 333-337.

Hewitt, D. (1958). Sib resemblance in bone, muscle, and fat. Measurement of the human calf. *Annals of Human Genetics, 22,* 213-221.

Hewitt, G., Withers, R.T., Brooks, A.G., et al. (2002). An improved rig for dynamically calibrating skinfold calipers: comparison between Harpenden and Slim Guide instruments. *American Journal of Human Biology, 14,* 721-727.

Heymsfield, S.B., Arteaga, C., McManus, C., Smith, J., & Moffitt, S. (1983). Measurement of muscle mass in humans: validity of the 24-hour urinary creatinine method. *American Journal of Clinical Nutrition, 37,* 478-494.

Heymsfield, S.B., McManus, C., Smith, J., Stevens, V., & Nixon, D.W. (1982). Anthropometric measurement of muscle mass: revised equations for calculating bone-free arm muscle area. *American Journal of Clinical Nutrition, 36,* 680-690.

Heymsfield, S.B., Smith, R., Aulet, M., Bensen, B., Litchman, S., Wang, J., & Pierson, R.N. (1990). Appendicular skeletal muscle mass: measurement by dual-photon absorptiometry. *American Journal of Clinical Nutrition, 52,* 214-218.

Heymsfield, S.B., Wang, J., Lichtman, S., Kamm, Y., Kchayias, J., & Pierson, R.N. (1989). Body composition in elderly subjects: a critical appraisal of clinical methodology. *American Journal of Clinical Nutrition, 50,* 1167-1175.

Himes, J.H. (1991). Considering frame size in nutritional assessment. In J.H. Himes (Ed.), *Anthropometric Assessment of Nutritional Status* (pp. 141-150). New York: Wiley-Liss.

Himes, J.H. (2001). Prevalence of individuals with skinfolds too large to measure. *American Journal of Public Health, 91,* 154-155.

Himes, J.H., & Bouchard, C. (1989). Validity of anthropometry in classifying youths as obese. *International Journal of Obesity, 13,* 183-193.

Himes, J.H., Roche, A.F., & Webb, P. (1980). Fat areas as estimates of total body fat. *American Journal of Clinical Nutrition, 33,* 2093-2100.

Imbeault, P., Prud'Homme, D., Tremblay, A., Després, J.P., & Mauriege, P. (2000). Adipose tissue metabolism in young and middle-aged men after control for total body fatness. *Journal of Clinical Endocrinology and Metabolism, 85,* 2455-2462.

Ishida, Y., Carroll, M.L., Pollock, J.E., Graves, J.E., & Leggett, S.H. (1992). Reliability of B-mode ultrasound for the measurement of body fat and muscle thickness. *American Journal of Human Biology, 4,* 511-520.

Ishida, Y., Kanehisa, H., Carroll, J.F., Pollock, M.L., Graves, J.E., & Ganzarella, L. (1997). Distribution of subcutaneous fat and muscle thicknesses in young and middle-aged women. *American Journal of Human Biology, 9,* 247-255.

Jackson, A.S., & Pollock, M.L. (1976). Factor analysis and multivariate scaling of anthropometric variables for the assessment of body composition. *Medicine and Science in Sports and Exercise, 8,* 196-203.

Jackson, A.S., & Pollock, M.L. (1978). Generalized equations for predicting body density of men. *British Journal of Nutrition, 40,* 497-504.

Jackson, A.S., Pollock, M.L., & Ward, A. (1980). Generalized equations for predicting body density of women. *Medicine and Science in Sports and Exercise, 12,* 175-182.

Johnson, D., Cormack, G.C., Abrahams, P.H., & Dixon, A.K. (1996). Computed tomographic observations on subcutaneous fat: implications for liposuction. *Plastic and Reconstructive Surgery, 97,* 387-396.

Kabir, N., & Forsum, E. (1993). Estimation of total body fat and subcutaneous adipose tissue in full-term infants less than 3 months old. *Pediatric Research, 34,* 448-454.

Kashiwazaki, H., Dejima, Y., Orias-Rivera, J., & Coward, W.A. (1996). Prediction of total body water and fatness from anthropometry. *American Journal of Human Biology, 8,* 331-340.

Katch, V.L., & Freedson, P.S. (1982). Body size and shape: derivation of the "HAT" frame size model. *American Journal of Clinical Nutrition, 36,* 669-675.

Katch, F.I., & McArdle, W.D. (1973). Prediction of body density from simple anthropometric measurements in college-age men and women. *Human Biology, 45,* 445-454.

Katch, F.I., & Michael, E.D., Jr. (1968). Prediction of body density from skin-fold and girth measurements of college females. *Journal of Applied Physiology, 25,* 92-94.

Katzman, D.K., Bachrach, L.K., Carter, D.R., & Marcus, R. (1991). Clinical and anthropometric correlates of bone mineral acquisition in healthy adolescent girls. *Journal of Clinical Endocrinology and Metabolism, 73,* 1332-1339.

Kaufman, J.J., & Einhorn, T.A. (1993). Perspectives: ultrasound assessment of bone. *Journal of Bone and Mineral Research, 8,* 517-525.

Kelley, D.E., Thaete, F.L., Troost, F., Huwe, T., & Goodpaster, T.H. (2000). Subdivisions of subcutaneous abdominal adipose tissue and insulin resistance. *American Journal of Physiology, Endocrinology and Metabolism, 278,* E941-E948.

Kissebah, A.H., Vydelingum, N., Murray, R., Evans, R.W., Hartz, D.J., Kalkhoff, R.K., & Adams, P.W. (1982). Relation of body fat distribution to metabolic complications of obesity. *Journal of Clinical Endocrinology and Metabolism, 54,* 254-260.

Koester, R.S., Hunter, G.R., Snyder, S., Khaled, M., & Berland, L.L. (1992). Estimation of computerized tomography derived abdominal fat distribution. *International Journal of Obesity, 16,* 543-554.

Kral, J.G., Mazariegos, M., McKeon, E.W., Pierson, R.N. Jr., & Wang, J. (1993). Body composition studies in severe obesity. In J.G. Kral & T.B. Van Itallie (Eds.), *Recent Developments in Body Composition Analysis: Methods and Applications* (pp. 137-146). London: Smith-Gordon.

Kuczmarkski, R.J., Fanelli, M.T., & Koch, G.G. (1987). Ultrasonic assessment of body composition in obese adults: overcoming the limitations of the skinfold caliper. *American Journal of Clinical Nutrition, 45,* 717-724.

Kuczmarski, R.J., Ogden, C.L., Guo, S.S., Grummer-Shawn, L.M., Flegal, K.M., Mei, Z., Wei, R., Curtin, L.R., Roche, A.F., & Johnson, C.L. (2002). *2002 CDC Growth Charts for the United States: Methods and Development,* (National Center for Health Statistics, Vital and Health Statistics Series 11, No. 246). Hyattsville, MD: U.S. Department of Health and Human Services, Centers for Disease Control and Prevention.

Kvist, H., Chowdhury, B., Grangard, V., Tylén, U., & Sjöström, L. (1988). Total and visceral adipose-tissue volumes derived from measurements with computed tomography in adult men and women: predictive equations. *American Journal of Clinical Nutrition, 48,* 1351-1361.

Kwok, T., & Whitelaw, M.N. (1991). The use of armspan in nutritional assessment of the elderly. *Journal of the American Geriatric Society, 39,* 492- 496.

Lee, R.C., Wang, Z., Heo, M., Ross, R., Jansen, I., & Heymsfield, S.B. (2000). Total-body skeletal muscle mass: development and cross-validation of anthropometric prediction models. *American Journal of Clinical Nutrition, 72,* 796-803.

Lemiux, S., Prud'Homme, D., Tremblay, A., Bouchard, C., & Després, J.P. (1996). Anthropometric correlates to changes in visceral adipose tissue over 7 years in women. *International Journal of Obesity, 20,* 618-624.

Lemieux, S., Després, J.P., Moorjani, S., Nadeau, A., Theriault, G., Prud'Homme, D., Tremblay, A., Bouchard, C., & Lupien,

P.J. (1994). Are gender differences in cardiovascular-disease risk-factors explained by the level of visceral adipose tissue? *Diabetologica, 37,* 757-764.

Lohman, T.G. (1981). Skinfolds and body density and their relation to body fatness: a review. *Human Biology, 53,* 181-225.

Lohman, T.G. (1988). Anthropometry and body composition. In T.G. Lohman, A.F. Roche, & R. Martorell (Eds.), *Anthropometric Standardization Reference Manual* (pp. 125-129). Champaign, IL: Human Kinetics.

Lohman, T.G. (1991). Anthropometric assessment of fat-free body mass. In J.H. Himes (Ed.), *Anthropometric Assessment of Nutritional Status* (pp. 173-183). New York: Wiley-Liss.

Lohman, T.G. (1992). *Advances in Body Composition Assessment* (Current Issues in Exercise Science. Monograph No.3). Champaign, IL: Human Kinetics.

Lohman, T.G., Boileau, R.A., & Massey, B.H. (1975). Prediction of lean body weight in young boys from skinfold thickness and body weight. *Human Biology, 47,* 245-262.

Lohman, T.G., Pollock, M.L., Slaughter, M.H., Brandon, L.J., & Boileau, R.A. (1984). Methodological factors and the prediction of body fat in female athletes. *Medicine and Science in Sports & Exercise, 16,* 92-96.

Lohman, T.G., Roche, A.F., & Martorell, R., (Eds.). (1988). *Anthropometric Standardization Reference Manual.* Champaign, IL: Human Kinetics.

Malina, R.M. (1969). Quantification of fat, muscle, and bone in man. *Clinical Orthopaedics and Related Research, 65,* 9-38.

Malina, R.M. (1986). Growth of muscle tissue and muscle mass. In F. Falkner & J.M. Tanner (Eds.), *Human Growth: A Comprehensive Treatise. Vol. 2: Postnatal Growth Neurobiology* (2nd ed., pp. 77-99). New York: Plenum Press.

Malina, R.M., Brown, K.H., & Zavaleta, A.N. (1987). Relative lower extremity length in Mexican American and in American black and white youth. *American Journal of Physical Anthropology, 72,* 89-94.

Martin, A.D., Ross, W.D., Drinkwater, D.T., & Clarys, J.P. (1985). Prediction of body fat by skin-fold caliper: assumptions and cadaver evidence. *International Journal of Obesity, 9,* 31-39.

Mayhew, J.L., Clark, B.A., McKeown, B.C., & Montaldi, D.H. (1985). Accuracy of anthropometric equations for estimating body composition of female athletes. *Journal of Sports Medicine and Physical Fitness, 25,* 120-126.

Mazess, R.B., Peppler, W.W., Chestnut, C.H., III, Nelp, W.B., Cohn, S.H., & Zanzi, I. (1981). Total body bone mineral and lean body mass by dual-photon absorptiometry, II: Comparison with total body calcium by neutron activation analysis. *Calcified Tissue International, 33,* 361-363.

Miller, J.Z., Slemenda, C.W., Meaney, F.J., Reister, T.K., Hui, S., & Johnston, C.C. (1991). The relationship of bone mineral density and anthropometric variables in healthy male and female children. *Bone Mineral, 14,* 137-152.

Moeller, S.J., & Christian, L.L. (1998). Evaluation of the accuracy of real-time ultrasonic measurements of backfat and loin muscle area in swine using multiple statistical analysis procedures. *Journal of Animal Science, 76,* 2503-2514.

Moris, M., Peretz, A., Tjeka, R., Negaben, N., Wouters, M., & Bergmann, P. (1995). Quantitative ultrasound bone measurements: normal values and comparison with bone mineral density by dual X-ray absorptiometry. *Calcified Tissue International, 57,* 6-10.

Mueller, W.H., & Malina, R.M. (1987). Relative reliability of circumferences and skinfolds as measures of body fat distribution. *American Journal of Physical Anthropology, 72,* 437-439.

Müeller, W.H., Marbella, A., Harrist, R.B., Kaplowitz, H.J., Grunbaum, J.A., & Labarthe, D.M. (1989). Body circumferences as alternatives to skinfold measures of body fat distribution in children. *Annals of Human Biology, 16,* 495-506.

Mueller, W., & Reid, R. (1979). A multivariate analysis of fatness and relative fat patterning. *American Journal of Physical Anthropology, 50,* 199-208.

Nagamine, S., & Suzuki, S. (1964). Anthropometry and body composition of Japanese young men and women. *Human Biology, 36,* 8-15.

Najjar, M.P., & Rowland, M. (1987). *Anthropometric Reference Data and Prevalence of Overweight. United States, 1976-1980* (Vital and Health Statistics, Series 11, No. 238, National Center for Health Statistics). Washington, DC: U.S. Government Printing Office.

Norgan, N.G., & Ferro-Luzzi, A. (1985). The estimation of body density in men: are general equations general? *Annals of Human Biology, 12,* 1-15.

Norgan, N.G., & Ferro-Luzzi, A. (1986). Simple indices of subcutaneous fat patterning. *Ecology of Food and Nutrition, 18,* 117-123.

Orphanidou, C., McCargar, L., Birmingham, C.L., Mathieson, J., & Godner, E. (1994). Accuracy of subcutaneous fat measurement—comparison of skinfold calipers, ultrasound, and computed-tomography. *Journal of the American Dietetic Association, 94,* 855-858.

Parízková, J. (1977). *Body Fat and Physical Fitness.* The Hague, The Netherlands: Martinus Nijhoff.

Pawan, G.E.S., & Clode, M. (1960). The gross chemical composition of subcutaneous adipose tissue in the lean and obese human subject. *Journal of Biochemistry, 74,* 9.

Pérusse, L., Rice, T., Chagnon, Y.C., Despés, J.P., Lemieux, S., Roy, S., Lacaille, M., Ho-Kim, M.A., Chagnon, M., Province, M.A., & Rao, D.C. (2001). A genome-wide scan for abdominal fat assessed by computed tomography in the Québec Family Study. *Diabetes, 50,* 614-621.

Pham, C.L., Mueller, W.H., & Wear, M.L. (1995). Precision of the one- versus two-handed method of skinfold measurement in the obese. *American Journal of Human Biology, 7,* 617-621.

Pollock, M.L., Hickman, T., Kendrick, Z., Jackson, A., Linnerud, A.C., & Dawson, G. (1976). Prediction of body density in young and middle-aged men. *Journal of Applied Physiology, 40,* 300-304.

Pollock, M.L., Laughridge, E.E., Coleman, B., Linnerud, A.C., & Jackson, A. (1975). Prediction of body density in young and middle-aged women. *Journal of Applied Physiology, 38,* 745-749.

Pouliot, M.C., Després, J.P., Lemieux, S., Moorjani, S., Bouchard, C., Tremblay, A., Nadeau, A., & Lupien, P.J. (1994). Waist circumference and abdominal sagittal diameter: best simple anthropometric indexes of abdominal visceral adipose tissue accumulation and related cardiovascular risk in men and women. *The American Journal of Cardiology, 73,* 460-468.

Ramirez, M.E. (1993). Subcutaneous fat distribution in adolescents. *Human Biology, 65,* 771-782.

Rasmussen, M.H., Andersen, T., Breum, L., Hilsted, J., & Gotzsche, P.C. (1993). Observer variation in measurements of waist-hip ratio and the abdominal sagittal diameter. *International Journal of Obesity, 17,* 323-327.

Reid, I.R., Evans, M.C., & Ames, R. (1992). Relationships between upper-arm anthropometry and soft-tissue composition in postmenopausal women. *American Journal of Clinical Nutrition, 56,* 463-466.

Ribiero-Filho, F.F., Faria, A.N., Kohlmann, N.E.B., Azjen, S., Zanella, M.T., & Ferreira, S.R.G. (2001). Ultrasonography for the diagnosis of visceral obesity. *Obesity Research, 9* (Suppl 3), 173S.

Rico, H., Revilla, M., Hernandez, E.R., Villa, L.F., DelBuergo, M.A., & Alonso, A.L. (1991). Age-related and weight-related changes in total body bone mineral in men. *Mineral and Electrolyte Metabolism, 17,* 321-323.

Roche, A.F. (1994). Sarcopenia: a critical review of its measurement and health-related significance in the middle-aged and elderly. *American Journal of Human Biology, 6,* 33-42.

Roche, A.F., Abdel-Malek, A.K., & Mukerjee, D. (1985). New approaches to clinical assessment of adipose tissue. In A.F. Roche (Ed.), *Body Composition Assessments in Youth and Adults* (pp. 14-19). Proceedings of Sixth Ross Conference on Medical Research. Columbus, OH: Ross Laboratories.

Roche, A.F., Siervogel, R.M., Chumlea, W.C., & Webb, P. (1981). Grading body fatness from limited anthropometric data. *American Journal of Clinical Nutrition, 34,* 2831-2838.

Ross, R., Léger, L., Morris, E.B., DeGuise, J., & Guardo, R. (1992). Quantification of adipose tissue by MRI: relationship with anthropometric variables. *Journal of Applied Physiology, 72,* 787-795.

Rössner, S., Bo, W.J., Hiltbrandt, E., Hinson, W., Karstaedt, N., Santago, P., Sobol, W.T., & Crouse, J.R. (1990). Adipose tissue determinations in cadavers: a comparison between cross-sectional planimetry and computed tomography. *International Journal of Obesity, 14,* 893-902.

Ruiz, L., Colley, J.R.T., & Hamilton, P.J.S. (1971). Measurement of triceps skinfold thickness: an investigation of sources of variation. *British Journal of Preventive and Social Medicine, 25,* 165-167.

Russell-Aulet, M., Wang, J., Thornton, J.C., Colt, E.W., & Pierson, R.N., Jr. (1991). Bone mineral density and mass by total body dual-photon absorptiometry in normal white and Asian men. *Journal of Bone Mineral Research, 6,* 1109-1113.

Russell-Aulet, M., Wang, J., Thornton, J.C., Colt, E.W., & Pierson, R.N., Jr. (1993). Bone mineral density and mass in a cross-sectional study of white and Asian women. *Journal of Bone Mineral Research, 8,* 575-582.

Sanders, R.C., & Maggio, M. (1984). Artifacts. In *Clinical Sonography: A Practical Guide* (pp. 355-367). Boston: Little, Brown.

Satwanti, K., Bharadwaj, H., & Singh, I.P. (1977). Relationship of body density to body measurements in young Punjabi women: applicability of body composition prediction equations developed for women of European descent. *Human Biology, 49,* 203-213.

Scherf, J., Franklin, B.A., Lucas, C.P., Stevenson, D., & Rubenfire, M. (1986). Validity of skinfold thickness measures of formerly obese adults. *American Journal of Clinical Nutrition, 43,* 128-135.

Schlemmer, A., Hassager, C., Haarbo, J., & Christiansen, C. (1990). Direct measurement of abdominal fat by dual photon absorptiometry. *International Journal of Obesity, 14,* 603-611.

Segal, K.R., Van Loan, M., Fitzgerald, P.I., Hogdon, J.A., & Van Itallie, T.B. (1988). Lean body mass estimation by bioelectrical impedance analysis: a four-site cross-validation study. *American Journal of Clinical Nutrition, 47,* 7-14.

Seidell, J.C., Cigonili, M., Charzewski, J., Contaldo, F., Ellsinger, B.M., & Björntörp, P. (1988a). Measurement of regional distribution of adipose tissue. In P Björntörp & S. Rössner (Eds.) *Obesity in Europe* (pp. 351-357). London: John Libbey.

Seidell, J.C., Oosterlee, A., Deurenberg, P., Hautvast, J.G.A.J., & Ruijs, J.H.J. (1988b). Abdominal fat depots measured with computed tomography: effects of degree of obesity, sex and age. *European Journal of Clinical Nutrition, 42,* 805-815.

Seidell, J., Oosterlee, A., Thijssen, M.A.O., Burrema, J., Deurenberg, P., Hautvast, J.G.A.J., & Ruijs, J.H.J. (1987). Assessment of intra abdominal and subcutaneous abdominal fat: relation between anthropometry and computed tomography. *American Journal of Clinical Nutrition, 45,* 7-13.

Seip, R.L., Snead, D., & Weltman, A. (1993). Validity of anthropometric techniques for estimating percentage of body fat in obese females before and after sizable weight loss. *American Journal of Human Biology, 5,* 549-557.

Shephard, R.J., Jones, J., Ishii, K., Kaneko, M., & Olbrecht, A.J. (1969). Factors affecting body density and thickness of subcutaneous fat: data on 518 Canadian city dwellers. *American Journal of Clinical Nutrition, 22,* 1175-1189.

Singer, D.B., Sung, C.J., & Wigglesworth, J.S. (1991). Fetal growth and maturation: With standards for body and organ development. In J. Wigglesworth & D. Singer (Eds.), *Textbook of Fetal and Perinatal Pathology* (pp. 11-47). Chicago: Blackwell Scientific.

Sinning, W.E., Dolny, D.G., Little, K.D., Cunningham, L.N., Racaniello, A., Siconoifi, S.F., & Sholes, J.L. (1985). Validity of "generalized" equations for body composition analysis in male athletes. *Medicine and Science in Sports and Exercise, 17,* 124-130.

Sinning, W.E., & Wilson, J.R. (1984). Validity of "generalized" equations for body composition analysis in women athletes. *Research Quarterly for Exercise and Sport, 55,* 153-160.

Sjöström, L. (1987). New aspects of weight-for-height indices and adipose tissue distribution in relation to cardiovascular risk and total adipose tissue volume. In E. Berry, S. Blondheim, H. Elihau, & E. Shafrir (Eds.), *Recent Advances In Obesity Research, V: Proceedings of the 5th International Congress on Obesity* (pp. 66-76). London: Libbey.

Slaughter, M., & Lohman, T. (1980). An objective method for measurement of the musculoskeletal size to characterize body physique with application to the athletic population. *Medicine and Science in Sports and Exercise, 12,* 170-174.

Slemenda, C.W., Hui, S.L., Longcope, C., Wellman, H., & Johnston, C.C. (1990). Predictors of bone mass in perimenopausal women. A prospective study of clinical data using photon absorptiometry. *Annals of Internal Medicine, 112,* 96-101.

Sloan, A.W. (1967). Estimation of body fat in young men. *Journal of Applied Physiology, 23,* 311-315.

Sloan, A.W., Burt, J.J., & Blyth, C.S. (1962). Estimation of body fat in young women. *Journal of Applied Physiology, 17,* 967-970.

Snyder, W.S., Cook, M.J., Nasset, E.S., Karhausen, L.R., Howells, G.P., & Tipton, I.H. (1984). *Report No. 23 of the Task Group on Reference Man.* Oxford, UK: International Commission on Radiological Protection.

Sprawls, P. (1987). *Physical Principles of Medical Imaging.* Rockville, MD: Aspen.

Steinkamp, R.C., Cohen, N.L., Gaffey, W.R., McKey, T., Bron, G., Siri, W.E., Sargent, T.W., & Isaacs, E. (1965). Measures of body fat and related factors in normal adults, II: A simple clinical method to estimate body fat and lean body mass. *Journal of Chronic Diseases, 18,* 1291-1307.

Stevens-Simon, C., Thureen, P., Barrett, J., & Stamm, E. (2001). Skinfold caliper and ultrasound assessments of change in the distribution of subcutaneous fat during adolescent pregnancy. *International Journal of Obesity, 25,* 1340-1345.

Stolk, R.P., Wink, O., Zelissen, P.M.J., Meijer, R., van Gils, A.P.G., & Grobbee, D.E. (2001). Validity and reproducibility of ultrasonography for the measurement of intra-abdominal adipose tissue. *International Journal of Obesity, 25,* 1346-1351.

Suzuki, R., Watanabe, S., Hirai, Y., Akiyama, K., Nishide, T., Matsushima, Y., Murayama, H., Ohshima, H., Shinomiya, M., Shirai, K., Saito, Y., Yoshida, S., Saisho, H., & Ohto, M. (1993). Abdominal wall fat index, estimated by ultrasonography, for assessment of the ration of visceral fat to subcutaneous fat in the abdomen. *American Journal of Medicine, 95,* 309-314.

Takeda, N., Miyake, M., Kita, S., Tomomitsu, T., & Fukunaga, M. (1996). Sex and age patterns of quantitative ultrasound densitometry of the calcaneus in normal Japanese subjects. *Calcified Tissue International, 59,* 84-88.

Targher, G., Tonoli, M., Agostino, G., Rigo, L., Boschini, K., Muggeo, M., DeSandre, G., Cigolini, M. (1996). Ultrasonographic intra-abdominal depth and its relation to haemostatic factors in healthy males. *International Journal of Obesity, 20,* 882-885.

Teran, J.C., Sparks, K.E., Quinn, L.M., Fernandex, B.S., Krey, S.H., & Steffee, W.P. (1991). Percent body fat in obese white females predicted by anthropometric measurements. *American Journal of Clinical Nutrition, 53,* 7-13.

Thomas, L.W. (1962). The chemical composition of adipose tissue of man and mice. *Quarterly Journal of Experimental Physiology, 47,* 179-188.

Thorland, W.G., Johnson, G.O., Tharp, G.D., Fagot, T.G., & Hammer, R.W. (1984). Validity of anthropometric equations for the estimation of body density in adolescent athletes. *Medicine and Science in Sports and Exercise, 16,* 77-81.

Tornaghi, G., Raiteri, R., Pozzato, C., Rippoli, A., Bramani, M., Cipolat, M., & Craveri, A. (1994). Anthropometric or ultrasonic measurements in assessment of visceral fat—a comparative study. *International Journal of Obesity, 18,* 771-775.

Toth, M.J., Tchernof, A., Rosen, C.J., Matthews, D.E., & Poehlman, E.T. (2000). Regulation of protein metabolism in middle-aged, premenopausal women: roles of adiposity and estradiol. *Journal of Clinical Endocrinology and Metabolism, 85,* 1382-1389.

van der Kooy, K., Leenen, R., Seidell, J.C., Deurenberg, P., & Visser, M. (1993). Abdominal diameters as indicators of visceral fat: comparison between magnetic resonance imaging and anthropometry. *British Journal of Nutrition, 70,* 47-58.

Vu Tran, Z., & Weltman, A. (1988). Predicting body composition of men from girth measurements. *Human Biology, 60,* 167-175.

Wang, J., Robinowitz, D., Aulet, M., Smith, R.P., Tierney, M., Greene, A., Thornton, J., Heymsfield, S.B., & Pierson, R.N. (1988). Bone density is a function of age, sex, race and body weight, but not of height, water, fat, or body cell mass in normals. *American Journal of Clinical Nutrition, 47,* 773.

Weiss, L.W., & Clark, F.C. (1987). Three protocols for measuring subcutaneous fat thickness on the upper extremities. *European Journal of Applied Physiology, 56,* 217-221.

Weits, T., van der Bek, E.J., Wedel, M., & ter Haar Romeny, B.M. (1988). Computed tomography measurement of abdominal fat deposition in relation to anthropometry. *International Journal of Obesity, 23,* 217-225.

Wilmore, J.H., & Behnke, A.R. (1969). Anthropometric estimation of body density and lean body weight in young men. *Journal of Applied Physiology, 27,* 25-31.

Wilmore, J.H., & Behnke, A.R. (1970). An anthropometric estimation of body density and lean body weight in young women. *American Journal of Clinical Nutrition, 23,* 267-274.

Wirth, A., & Steinmetz, B. (1998). Gender differences in changes in subcutaneous and intra-abdominal fat during weight reduction: an ultrasound study. *Obesity Research, 6,* 393-399.

Withers, R.T., Norton, K.I., Craig, N.P., Hartland, M.C., & Venables, W. (1987). The relative body fat and anthropometric prediction of body density of South Australian females aged 17-35 years. *European Journal of Applied Physiology, 56,* 181-190.

Womersley, J., & Durnin, J.V.G.A. (1973). An experimental study of variability of measurements of skinfold thicknesses in young adults. *Human Biology, 45,* 281-292.

Womersley, J., & Durnin, J.V.G.A. (1977). A comparison of the skinfold method with extent of overweight and various weight-height relationships in the assessment of obesity. *British Journal of Nutrition, 38,* 271-284.

Young, C.M., & Tensuan, R.S. (1963). Estimating the lean body mass of young women. *Journal of the American Dietetic Association, 42,* 46-51.

Chapter 9

Arpadi, S.M., Horlick, M.N., Wang, J., Cuff, P., Bamji, M., & Kotler, D.P. (1998). Body composition in prepubertal children with human immunodeficiency virus type 1 infection. *Arch Pediatr Adolesc Med* 152: 688-93.

Barera, G., Mora, S., Brambilla, P., Menni, L., Beccio, S., & Bianchi, C. (2000). Body composition in children with celiac disease and the effects of a gluten-free diet: a prospective case-control study. *Am J Clin Nutr* 72: 71-5.

Barlow, S.E., & Dietz, W.H. (1998). Obesity evaluation and treatment: Expert Committee recommendations. The Maternal and Child Health Bureau, Health Resources and Services Administration and the Department of Health and Human Services. *Pediatrics* 102: E29.

Baur, L.A. (1995). Body composition measurement in normal children: ethical and methodological limitations. *Asia Pacific J Clin Nutr* 4: 35-8.

Benn, R.T. (1971). Some mathematical properties of weight-for-height indexes used as measures of adiposity. *Br J Prev Soc Med* 25: 42-50.

Boileau, R.A. (1988). Utilization of total body electrical conductivity in determining body composition. In N. York (Ed), *Designing Foods* (251-327). New York: National Academy of Science.

Brook, C.G.D. (1971). Determination of body composition in children from skinfold measurements. *Arch Dis Child* 46: 182-4.

Butte, N., Heinz, C., Hopkinson, J., Wong, W., Shypailo, R., & Ellis, K. (1999). Fat mass in infants and toddlers: comparability of total body water, total body potassium, total body electrical conductivity, and dual-energy X-ray absorptiometry. *J Pediatr Gastroenterol Nutr* 29: 184-9.

Butte, N.F., Hopkinson, J.M., Wong, W.W., Smith, E.O., & Ellis, K.J. (2000). Body composition during the first 2 years of life: an updated reference. *Pediatr Res* 47: 578-85.

Cacciari, E., Milani, S., Balsamo, A., Dammacco, F., De Luca, F., Chiarelli, F., Pasquino, A.M., Tonini, G., & Vanelli, M. (2002). Italian cross-sectional growth charts for height, weight and BMI (6-20 y). *Eur J Clin Nutr* 56: 171-80.

Caprio, S., Hyman, L.D., McCarthy, S., Lange, R., Bronson, M., & Tamborlane, W.V. (1996). Fat distribution and cardiovascular risk factors in obese adolescent girls: importance of the intraabdominal fat depot. *Am J Clin Nutr* 64: 12-7.

Chinn, S., Rona, R.J., Gulliford, M.C., & Hammond, J. (1992). Weight for height in children aged 4-12 years. A new index compared to the normalized body mass index. *Eur J Clin Nutr* 46: 489-500.

Cole, T.J. (1993). The use and construction of anthropometric growth reference standards. *Nutr Res Rev* 6: 19-50.

Cole, T.J., Bellizzi, M.C., Flegal, K.M., & Dietz, W.H. (2000). Establishing a standard definition for child overweight and obesity worldwide: international survey. *BMJ*, 320:1240-3.

Cole, T.J., Freeman, J.V., & Preece, M.A. (1995). Body mass index reference curves for the UK, 1990. *Arch Dis Child* 73: 25-9.

Cole, T.J., & Roede, M.J. (1999). Centiles of body mass index for Dutch children aged 0-20 years in 1980—a baseline to assess recent trends in obesity. *Ann Hum Biol* 26: 303-8.

Cole, T.J., & Rolland-Cachera, M.F. (2002). Measurement and definition. In W. Burniat, T. Cole, I. Lissau, & E. Poskitt (Eds), *Child and Adolescent Obesity* (pp 3-27). Cambridge, UK: Cambridge University Press.

Crapo, R.O., Morris, A.H., & Gardner, R.M. (1982). Reference values for pulmonary tissue volume, membrane diffusing capacity, and pulmonary capillary blood volume. *Bull Eur Physiopathol Respir* 18: 893-9.

Daniels, S.R., Khoury, P.R., & Morrison, J.A. (1997). The utility of body mass index as a measure of body fatness in children and adolescents: differences by race and gender. *Pediatrics* 99: 804-7.

Deans, H.E., Smith, F.W., Lloyd, D.J., Law, A.N., & Sutherland, H.W. (1989). Fetal fat measurement by magnetic resonance imaging. *Br J Radiol* 62: 603-7.

Demerath, E.W., Guo, S.S., Chumlea, W.C., Towne, B., Roche, A.F., & Siervogel, R.M. (2002). Comparison of percent body fat estimates using air displacement plethysmography and hydrodensitometry in adults and children. *Int J Obes Relat Metab Disord* 26: 389-97.

Deurenberg, P., & Deurenberg-Yap, M. (2001). Differences in body-composition assumptions across ethnic groups: practical consequences. *Cr Opin CLin Nutr Metab Care* 4:377-83.

Dewit, O., Fuller, N.J., Fewtrell, M.S., Elia, M., & Wells, J.C. (2000). Whole body air displacement plethysmography compared with hydrodensitometry for body composition analysis. *Arch Dis Child* 82: 159-64.

Dietz, W.H., & Bellizzi, M.C. (1999). Introduction: the use of body mass index to assess obesity in children. *Am J Clin Nutr* 70: 123S-5S.

Durnin, J.V.G.A., & Rahaman, M.M. (1967). The assessment of the amount of fat in the human body from measurements of skinfold thickness. *Br J Nutr* 21: 681-9.

Dwyer, T., & Blizzard, C.L. (1996). Defining obesity in children by biological endpoint rather than population distribution. *Int J Obes Relat Metab Disord* 20: 472-80.

Eid, E.E. (1970). Follow-up study of physical growth of children who had excessive weight gain in the first six months of life. *Br Med J* 2: 74-76.

Elbers, J.M., Haumann, G., Asscheman, H., Seidell, J.C., & Gooren, L.J. (1997). Reproducibility of fat area measurements in young, non-obese subjects by computerized analysis of magnetic resonance images. *Int J Obes Relat Metab Disord* 21: 1121-9.

Ellis, K.J. (1998). Body composition of the neonate. In R.M. Cowett (Ed), *Principles of Perinatal-Neonatal Metabolism* (pp 1077-1096). New York: Springer-Verlag.

Ellis, K.J., Abrams, S.A., & Wong, W.W. (1999). Monitoring childhood obesity: assessment of the weight/height index. *Am J Epidemiol* 150: 939-46.

Fernandez, J.R., Pietrobelli, A., Redden, D.T., & Allison, D.B. (2004). Waist circumference percentile in nationally representative samples of black, white and Hispanic children. *J Pediatrics* 145:439-44.

Fields, D.A., & Goran, M.I. (2000). Body composition techniques and the four-compartment model in children. *J Appl Physiol* 89: 613-20.

Fields, D.A., Goran, M.I., & McCrory, M.A. (2002). Body-composition assessment via air-displacement plethysmography in adults and children: a review. *Am J Clin Nutr* 75: 453-67.

Fiorotto, M.L., Cochran, W.J., Funk, R.C., Sheng, H.P., & Klish, W.J. (1987). Total body electrical conductivity measurements: effects of body composition and geometry. *Am J Physiol* 252: R794-800.

Fomon, S.J., Haschke, F., Ziegler, E.E., & Nelson, S.E. (1982). Body composition of reference children from birth to age 10 years. *Am J Clin Nutr* 35: 1169-75.

Fomon, S.J., & Nelson, S.E. (2002). Body composition of the male and female reference infants. *Annu Rev Nutr* 22: 1-17.

Forbes, G.B. (Ed.) (1987). *Human Body Composition: Growth, Aging, Nutrition, and Activity.* New York: Springer-Verlag.

Freedman, D.S. (2002). Clustering of coronary heart disease risk factors among obese children. *J Pediatr Endocrinol Metab* 15: 1099-108.

Gallagher, D., Belmonte, D., Deurenberg, P., Wang, Z., Krasnow, N., Pi-Sunyer, F.X., & Heymsfield, S.B. (1998). Organ-tissue mass measurement allows modeling of REE and metabolically active tissue mass. *Am J Physiol* 275: E249-58.

Gallagher, D., Heymsfield, S.B., Heo, M., Jebb, S.A., Murgatroyd, P.R., & Sakamoto, Y. (2000). Healthy percentage body fat ranges: an approach for developing guidelines based on body mass index. *Am J Clin Nutr* 72: 694-701.

Garn, S.M., Leonard, W.R., & Hawthorne, V.M. (1986). Three limitations of the body mass index. *Am J Clin Nutr* 44: 996-7.

Gaskin, P.S., & Walker, S.P. (2003). Obesity in a cohort of black Jamaican children as estimated by BMI and other indices of adiposity. *Eur J Clin Nutr* 57:420-6.

Giampietro, O., Virgone, E., Carneglia, L., Griesi, E., Calvi, D., & Matteucci, E. (2002). Anthropometric indices of school children and familiar risk factors. *Prev Med* 35: 492-8.

Goran, M.I. (1998). Measurement issues related to studies of childhood obesity: assessment of body composition, body fat distribution, physical activity, and food intake. *Pediatrics* 101: 505-18.

Goran, M.I., & Gower, B.A. (2001). Longitudinal study on pubertal insulin resistance. *Diabetes* 50: 2444-50.

Hara, M., Saitou, E., Iwata, F., Okada, T., & Harada, K. (2002). Waist-to-height ratio is the best predictor of cardiovascular disease risk factors in Japanese schoolchildren. *J Atheroscler Thromb* 9: 127-32.

Harrington, T.A., Thomas, E.L., Modi, N., Frost, G., Coutts, G.A., & Bell, J.D. (2002). Fast and reproducible method for the direct quantitation of adipose tissue in newborn infants. *Lipids* 37: 95-100.

Harsha, D.W., Frerichs, R.R., & Berenson, G.S. (1978). Densitometry and anthropometry of black and white children. *Hum Biol* 50: 261-80.

Haschke, F. (1983). Body composition of adolescent males Part 2. Body composition of male reference adolescents. *Acta Pediatr Scand* 307(Suppl): 13-23.

Hashimoto, K., Wong, W.W., Thomas, A.J., Uvena-Celebrezze, J., Huston-Pressley, L., Amini, S.B., & Catalano, P.M. (2002). Estimation of neonatal body composition: isotope dilution versus total-body electrical conductivity. *Biol Neonate* 81: 170-5.

He, Q., Albertsson-Wikland, K., & Karlberg, J. (2000). Population-based body mass index reference values from Goteborg, Sweden: birth to 18 years of age. *Acta Paediatr* 89: 582-92.

Heymsfield, S.B., Wang, Z., Baumgartner, R.N., & Ross, R. (1997). Human body composition: advances in models and methods. *Annu Rev Nutr* 17: 527-58.

Higgins, P.B., Gower, B.A., Hunter, G.R., & Goran, M.I. (2001). Defining health-related obesity in prepubertal children. *Obes Res* 9: 233-40.

Himes, J.H., & Dietz, W.H. (1994). Guidelines for overweight in adolescent preventive services: recommendations from an expert committee. The Expert Committee on Clinical Guidelines for Overweight in Adolescent Preventive Services. *Am J Clin Nutr* 59: 307-16.

Houtkooper, L.B., Going, S.B., Lohman, T.G., Roche, A.F., & Van Loan, M. (1992). Bioelectrical impedance estimation of fat-free body mass in children and youth: a cross-validation study. *J Appl Physiol* 72: 366-73.

Kabir, N., & Forsum, E. (1993). Estimation of total body fat and subcutaneous adipose tissue in full-term infants less than 3 months old. *Pediatr Res* 34: 448-54.

Kang, H.S., Gutin, B., Barbeau, P., Litaker, M.S., Allison, J., & Le, N.A. (2002). Low-density lipoprotein particle size, central obesity, cardiovascular fitness, and insulin resistance syndrome markers in obese youths. *Int J Obes Relat Metab Disord* 26: 1030-5.

Kuczmarski, R.J., Ogden, C.L., Grummer-Strawn, L.M., Flegal, K.M., Mei, Z., Wei, R., Curtin, L.R., Roche, A.F., & Johnson, C.L. (2000). CDC growth charts for the United States: methods and development. *Vital Health Stat* 11:1-190.

Kushner, R.F., Schoeller, D.A., Fjeld, C.R., & Danford, L. (1992). Is the impedance index (ht^2/R) significant in predicting total body water? *Am J Clin Nutr* 56: 835-9.

Laskey, M.A., Flaxman, M.E., Barber, R.W., Trafford, S., Hayball, M.P., Lyttle, K.D., Crisp, A.J., & Compston, J.E. (1991). Comparative performance in vitro and in vivo of Lunar DPX and Hologic QDR-1000 dual energy X-ray absorptiometers. *Br J Radiol* 64: 1023-9.

Lee, R.C., Wang, Z., Heo, M., Ross, R., Janssen, I., & Heymsfield, S.B. (2000). Total-body skeletal muscle mass: development and cross-validation of anthropometric prediction models. *Am J Clin Nutr* 72: 796-803.

Leger, J., Garel, C., Fjellestad-Paulsen, A., Hassan, M., & Czernichow, P. (1998). Human growth hormone treatment of short-stature children born small for gestational age: effect on muscle and adipose tissue mass during a 3-year treatment period and after 1 year's withdrawal. *J Clin Endocrinol Metab* 83: 3512-6.

Leung, S.S., Cole, T.J., Tse, L.Y., & Lau, J.T. (1998). Body mass index reference curves for Chinese children. *Ann Hum Biol* 25: 169-74.

Lewy, V.D., Danadian, K., & Arslanian, S. (1999). Determination of body composition in African-American children: validation of bioelectrical impedence with dual energy X-ray absorptiometry. *J Pediatr Endocrinol Metab* 12: 443-8.

Lockner, D.W., Heyward, V.H., Baumgartner, R.N., & Jenkins, K.A. (2000). Comparison of air-displacement plethysmography, hydrodensitometry, and dual X-ray absorptiometry for assessing body composition of children 10 to 18 years of age. *Ann N Y Acad Sci* 904: 72-8.

Lohman, T.G. (1986). Applicability of body composition techniques and constants for children and youths. *Exerc Sport Sci Rev* 14: 325-57.

Lohman, T.G. (1992). *Advances in Body Composition Assessment*. Champaign, IL: Human Kinetics.

Lohman, T.G., & Going, S.B. (1998). Assessment of body composition and energy balance. In D. Lamb & R. Murray (Eds), *Perspectives in Exercise Science and Sports Medicine, 11, Exercise, Nutrition and the Control of Body Weight* (pp 61-105) Carmel, IN: Cooper.

Maffeis, C., Pietrobelli, A., Grezzani, A., Provera, S., & Tato, L. (2001). Waist circumference and cardiovascular risk factors in prepubertal children. *Obes Res* 9: 179-87.

Magarey, A.M., Daniels, L.A., Boulton, T.J., & Cockington, R.A. (2001). Does fat intake predict adiposity in healthy children and adolescents aged 2-15 y? A longitudinal analysis. *Eur J Clin Nutr* 55: 471-81.

Marcus, M.A., Wang, J., Thornton, J.C., Ma, R., Burastero, S., & Pierson, R.N.J. (1997). Anthropometrics do not influence dual X-ray absorptiometry (DXA) measurement of fat in normal to obese adults: a comparison with in vivo neutron activation analysis (IVNA). *Obes Res* 5: 122-30.

Mast, M., Langnase, K., Labitzke, K., Bruse, U., Preuss, U., & Muller, M. J. (2002). Use of BMI as a measure of overweight and obesity in a field study on 5-7 year old children. *Eur J Nutr* 41: 61-7.

Maynard, L.M., Wisemandle, W., Roche, A.F., Chumlea, W.C., Guo, S.S., & Siervogel, R.M. (2001). Childhood body composition in relation to body mass index. *Pediatrics* 107: 344-50.

McCrory, M.A., Fuss, P.J., Saltzman, E., Hays, N.P., & Roberts, S.B. (2000). Body composition measurement by air-displacement plethysmography and underwater weighing: effects of gas-producing and gas-containing foods *FASEB J* 14: A498.

Moulton, C.R. (1923). Age and chemical development in mammals. *J Biol Chem* 57: 79-97.

Must, A., Jacques, P.F., Dallal, G.E., Bajema, C.J., & Dietz, W.H. (1992). Long-term morbidity and mortality of overweight adolescents. A follow-up of the Harvard Growth Study of 1922 to 1935. *N Engl J Med* 327: 1350-5.

Must, A., Spadano, J., Coakley, E.H., Field, A.E., Colditz, G., & Dietz, W.H. (1999). The disease burden associated with overweight and obesity. *JAMA* 282: 1523-9.

Nunez, C., Kovera, A.J., Pietrobelli, A., Heshka, S., Horlick, M., Kehayias, J.J., Wang, Z., & Heymsfield, S.B. (1999). Body composition in children and adults by air displacement plethysmography. *Eur J Clin Nutr* 53: 382-7.

Nysom, K., Holm, K., Michaelsen, K.F., Hertz, H., Muller, J., & Molgaard, C. (1999). Degree of fatness after treatment for acute lymphoblastic leukemia in childhood. *J Clin Endocrinol Metab* 84: 4591-6.

Olhager, E., Thomas, K.A., Wigstrom, L., & Forsum, E. (1998). Description and evaluation of a method based on magnetic resonance imaging to estimate adipose tissue volume and total body fat in infants. *Pediatr Res* 44: 572-7.

Pietrobelli, A., Faith, M.S., Allison, D.B., Gallagher, D., Chiumello, G., & Heymsfield, S.B. (1998a). Body mass index as a measure of adiposity among children and adolescents: a validation study. *J Pediatrics* 132: 204-10.

Pietrobelli, A., Formica, C., Wang, Z., & Heymsfield, S. B. (1996). Dual-energy X-ray absorptiometry body composition model: review of physical concepts. *Am J Physiol* 271: E941-51.

Pietrobelli, A., Wang, Z.M., Formica, C., & Heymsfield, S.B. (1998b). Dual energy X-ray absorptiometry: fat estimation errors due to variation in soft tissue hydration. *Am J Physiol* 274: 808-16.

Pintauro, S.J., Nagy, T.R., Duthie, C.M., & Goran, M. I. (1996). Cross-calibration of fat and lean measurements by dual-energy X-ray absorptiometry to pig carcass analysis in the pediatric body weight range. *Am J Clin Nutr* 63: 293-8.

Quetelet, A. (1869) *Physique Sociale ou Essay Sur le Development des Facultes de L'homme*. Bruxelles: C. Muquardt.

Reilly, J.J., Dorosty, A.R., & Emmett, P.M. (2000). Identification of the obese child: adequacy of the body mass index for clinical practice and epidemiology. *Int J Obes Relat Metab Disord* 24: 1623-7.

Roemmich, J.N., Clark, P.A., Weltman, A., & Rogol, A.D. (1997). Alterations in growth and body composition during puberty. I. Comparing multicompartment body composition models. *J Appl Physiol* 83: 927-35.

Rohrer, F. (1921). Der index der Korperfulle als des Ernahrungszustandes. *Munch Med Wochenschr* 68:580-82.

Rolland-Cachera, M.F., Sempe, M., Guilloud-Bataille, M., Patois, E., Pequignot-Guggenbuhl, F., & Fautrad, V. (1982). Adiposity indices in children. *Am J Clin Nutr* 36: 178-84.

Russell-Aulet, M., Wang, J., Thornton, J., & Pierson, R.N.J. (1991). Comparison of dual-photon absorptiometry systems for total-body bone and soft tissue measurements: dual-energy X-rays versus gadolinium 153. *J Bone Miner Res* 6: 411-5.

Schaefer, F., Georgi, M., Wuhl, E., & Scharer, K. (1998). Body mass index and percentage fat mass in healthy German schoolchildren and adolescents. *Int J Obes Relat Metab Disord* 22: 461-9.

Schneider, S., Kolesnik, J.A., Wang, J., & Pierson, R.N.J. (1998). *Total Body Potassium (TBK) Measurement: Accuracy (A), Efficiency (E), and Reproducibility (R)*. San Francisco: Experimental Biology.

Sentongo, T.A., Semeao, E.J., Piccoli, D.A., Stallings, V.A., & Zemel, B.S. (2000). Growth, body composition, and nutritional status in children and adolescents with Crohn's disease. *J Pediatr Gastroenterol Nutr* 31: 33-40.

Shen, W., Wang, Z.M., Punyanita, M., Lei, J., Sinav, A., Kral, J.G., Imielinska, C., Ross, R., & Heymsfield, S.B. (2003). Adipose tissue quantification by imaging methods: a proposed classification. *Obes Res* 11: 5-16.

Sjostrom, L. (1991). A computer-tomography based multicompartment body composition technique and anthropometric predictions of lean body mass, total and subcutaneous adipose tissue. *Int J Obes Relat Metab Disord.* 15: 19-30.

Slaughter, M.H., Lohman, T.G., Boileau, R.A., Horswill, C.A., & Stillman, R.J. (1988). Skinfold equations for estimation of body fatness in children and youth. *Hum Biol* 60:709-23.

Snyder, W.M., Cook, M.J., Nasset, E.S., Karhausen, L.R., Howells, G.P., & Tipton, I.H. (1975). *Report of the Task Group on Reference Man* (ICRP publication 23). Oxford, UK: Pergamon Press.

Sopher, A.B., Thornton, J.C., Wang, J., Pierson, R.N., Jr, Heymsfield, S.B., & Horlick, M. (2004). Measurement of percent body fat in 411 children and adolescents: a comparison of dual-energy x-ray absorptiometry to a four-compartment model. *Pediatrics* 113:1285-90.

Stettler, N., Kawchak, D.A., Boyle, L.L., Propert, K.J., Scanlin, T.F., Stallings, V.A., & Zemel, B.S. (2000). Prospective evaluation of growth, nutritional status, and body composition in children with cystic fibrosis. *Am J Clin Nutr* 72: 407-13.

Tanner, J.M. (1951). The assessment of growth and development in children. *Arch Dis Child* 10: 33-6.

Teixeira, P.J., Sardinha, L.B., Going, S.B., & Lohman, T.G. (2001). Total and regional fat and serum cardiovascular disease risk factors in lean and obese children and adolescents. *Obes Res* 9: 432-42.

Treuth, M.S., Butte, N.F., Wong, W.W., & Ellis, K.J. (2001). Body composition in prepubertal girls: comparison of six methods. *Int J Obes Relat Metab Disord* 25: 1352-9.

Urlando, A., Dempster, P., & Aitkens, S. (2003). A new air displacement plethysmograph for the measurement of body composition in infants. *Pediatr Res* 53: 486-92.

Van Loan, M. (1990). Assessment of fat-free mass in teen-agers: use of TOBEC methodology. *Am J Clin Nutr* 52: 586-90.

Wagner, D.R., & Heyward, V.H. (1999). Techniques of body composition assessment: a review of laboratory and field methods. *Res Q Exerc Sport* 70: 135-49.

Wang, Z.M., Shen, W., Kotler, D.P., Heshka, S., Wielopolski, L., Aloia, J.F., Nelson, M.E., Pierson, R.N., Jr., & Heymsfield, S.B. (2004) Total-body protein: a new cellular level mass and distribution prediction model. *Am J Clin Nutr.* In Press.

Wang, Z., Zhu, S., Wang, J., Pierson, R.N.J., & Heymsfield, S.B. (2003). Whole-body skeletal muscle mass: development and validation of total-body potassium prediction models. *Am J Clin Nutr* 77: 76-82.

Warner, M.M., Guo, J., & Zhao, Y. (2001). The relationship between plasma apolipoprotein A-IV levels and coronary heart disease. *Chin Med J (Engl)* 114: 275-9.

Waterloo, J.C., Buzina, R., Keller, W., Lane, J.M., Nichaman, M.Z., & Tanner, J.M. (1977). The presentation and use of height and weight data for comparing the nutritional status of groups of children under the age of 10 years. *Bull World Health Org* 55:489-98.

Wells, J.C., Coward, W.A., Cole, T.J., & Davies, P.S. (2002). The contribution of fat and fat-free tissue to body mass index in contemporary children and the reference child. *Int J Obes Relat Metab Disord* 26: 1323-8.

Wells, J.C., Fuller, N.J., Dewit, O., Fewtrell, M.S., Elia, M., & Cole, T.J. (1999). Four-component model of body composition in children: density and hydration of fat-free mass and comparison with simpler models. *Am J Clin Nutr* 69: 904-12.

Weststrate, J.A., & Deurenberg, P. (1989). Body composition in children: proposal for a method for calculating body fat percentage from total body density or skinfold-thickness measurements. *Am J Clin Nutr* 50: 1104-15.

White, D.R., Widdowson, E.M., Woodard, H.Q., & Dickerson, J.W. (1991). The composition of body tissues (II). Fetus to young adult. *Br J Radiol* 64: 149-59.

Williams, D.P., Going, S.B., Lohman, T.G., Harsha, D.W., Srinivasan, S.R., Webber, L.S., & Berenson, G.S. (1992). Body fatness and risk for elevated blood pressure, total cholesterol, and serum lipoprotein ratios in children and adolescents. *Am J Public Health* 82: 358-63.

Wong, W.W., Hergenroeder, A.C., Stuff, J.E., Butte, N.F., Smith, E.O., & Ellis, K.J. (2002). Evaluating body fat in girls and female adolescents: advantages and disadvantages of dual-energy X-ray absorptiometry. *Am J Clin Nutr* 76: 384-9.

Yanovski, S.Z., Van Hubbard, S., Heymsfield, S.B., & Lukaski, H.C. (Eds). (1996). Bioelectrical impedance analysis in body composition measurement. Proceedings of a National Institutes of Health Technology Assessment Conference. Bethesda, Maryland, December 12-14, 1994. *Am J Clin Nutr* 64: 387S-532S.

Chapter 10

Acquarone, C., Cucco, M., Cauli, S.L., & Malacarne, G. (2002). Effects of food abundance and predictability on body composition and health parameters: experimental tests with the Hooded Crow. *Ibis, 144,* E155-E163.

Acquarone, C., Cucco, M., & Malacarne, G. (2001). Short-term effects on body condition and size of immunocompetent organs in the hooded crow. *Italian Journal of Zoology, 68,* 195-199.

Andersson, N., Lindberg, M.K., Ohlsson, C., Andersson, K., & Ryberg, B. (2001). Repeated *in vivo* determinations of bone mineral density during parathyroid hormone treatment in ovariectomized mice. *Journal of Endocrinology, 170,* 529-537.

Angilletta, M.J. (1999). Estimating body composition of lizards from total body electrical conductivity and total body water. *Copeia, 3,* 587-595.

Arnould, J.P.Y., Boyd, I.L., & Speakman, J.R. (1996). Measuring the body composition of Antarctic fur seals (*Arctocephalus gazella*): validation of hydrogen isotope dilution. *Physiological Zoology, 69,* 93-116.

Baer, D.J., Rumpler, W.V., Barnes, R.E., Kressler, L.L., Howe, J.C., & Haines, T.E. (1993). Measurement of body composition of live rats by electromagnetic conductance. *Physiology and Behavior, 53,* 1195-1199.

Baffy, G., Zhang, C.-Y., Glickman, J.N., & Lowell, B.B. (2002). Obesity-related fatty liver is unchanged in mice deficient for mitochondrial uncoupling protein 2. *Hepatology, 35,* 753-761.

Barzilai, N., Banerjee, S., Hawkins, M., Chen, W., & Rossetti, L. (1998). Caloric restriction reverses hepatic insulin resistance in aging rats by decreasing visceral fat. *Journal of Clinical Investigation, 101,* 1353-1361.

Barziza, D.E. & Gatlin, D.M. (2000). An evaluation of total body electrical conductivity to estimate body composition of largemouth bass, *Micropterus salmoides. Aquatic Living Resources, 13,* 439-447.

Beckmann, N., Gentsch, C., Baumann, D., Bruttel, K., Vassout, A., Schoeffter, P., Loetscher, E., Bobadilla, M., Perentes, E., & Rudin, M. (2001). Non-invasive, quantitative assessment of the anatomical phenotype of corticotropin-releasing factor-overexpressing mice by MRI. *NMR in Biomedicine, 14,* 210-216.

Bell, R.C., Lanou, A.J., Frongillo, E.A., Jr., Levitsky, D.A., & Campbell, T. C. (1994). Accuracy and reliability of total body electrical conductivity (TOBEC) for determining body composition of rats in experimental studies. *Physiology and Behavior, 56,* 767-773.

Bellinger, L.L. & Williams, F.E. (1993). Validation study of a total body electrical conductive (TOBEC) instrument that measures fat-free body mass. *Physiology and Behavior, 53,* 1189-1194.

Bland, J.M., & Altman, D.G. (1986). Statistical methods for assessing agreement between two methods of clinical measurement. *Lancet, 1(8476),* 307-310.

Brix, A.E., Elgavish, A., Nagy, T.R., Gower, B.A., Rhead, W.J., & Wood, P.A. (2002). Evaluation of liver fatty acid oxidation in the leptin-deficient obese mouse. *Molecular Genetics and Metabolism, 75,* 219-226.

Bronson, F.H. (1987). Susceptibility of the fat reserves of mice to natural challenges. *Journal of Comparative Physiology B, 157,* 551-554.

Castro, G., Wunder, B.A., & Knopf, F.L. (1990). Total body electrical conductivity (TOBEC) to estimate total body fat of free-living birds. *Condor, 92,* 496-499.

Changani, K.K., Nicholson, A., White, A., Katcham, J.K., Reid, D.G., & Clapham, J.C. (2003). A longitudinal magnetic resonance imaging (MRI) study of differences in abdominal fat distribution between normal mice, and lean overexpressors of mitochondrial uncoupling protein-3 (UCP-3). *Diabetes, Obesity and Metabolism, 5,* 99-105.

Coleman, T.G., Manning, R.D.J., Norman, R.A.J., & Guyton, A.C. (1972). Dynamics of water-isotope distribution. *American Journal of Physiology, 223,* 1371-1375.

Crum, B.G., Williams, J.B., & Nagy, K.A. (1985). Can tritiated water-dilution space accurately predict total body water in chukar partridges? *Journal of Applied Physiology, 59,* 1383-1388.

Culebras, J.M., Fitzpatrick, G.F., Brennan, M.F., Boyden, C.M., & Moore, F.D. (1977). Total body water and the exchangeable hydrogen. II. A review of comparative data from animals based on isotope dilution and desiccation, with a report of new data from the rat. *American Journal of Physiology, 232,* R60-R65.

de Souza, C.J., Eckhardt, M., Gagen, K., Dong, M., Chen, W., Laurent, D. & Burkey, B.F. (2001). Effects of pioglitazone on adipose tissue remodeling within the setting of obesity and insulin resistance. *Diabetes, 50,* 1863-1871.

Degen, A.A., Pinshow, B., Alkon, P.U., & Arnon, H. (1981). Tritiated water for estimating total body water and water turnover rate in birds. *Journal of Applied Physiology, 51,* 1183-1188.

Denny, M.J.S. & Dawson, T.J. (1975). Comparative metabolism of tritiated water by macropodid marsupials. *American Journal of Physiology, 228,* 1794-1799.

Dickinson, K., North, T.J., Telford, G., Smith, S., Brammer, R., Jones, R.B. & Heal, D.J. (2001). Determination of body composition in conscious adult female Wistar utilising total body electrical conductivity. *Physiology and Behavior, 74,* 425-433.

Dobush, G.R., Ankney, C.D., & Krementz, D.G. (1985). The effect of apparatus, extraction time, and solvent type on lipid extractions of snow geese. *Canadian Journal of Zoology, 63,* 1917-1920.

Elliot, D.A., Backus, R.C., Van Loan, M.D., & Rogers, Q.R. (2002). Evaluation of multifrequency bioelectrical impedance analysis for the assessment of extracellular and total body water in healthy cats. *Journal of Nutrition, 132,* 1757S-1759S.

Ferrier, L., Robert, P., Dumon, H., Martin, L., & Nguyen, P. (2002). Evaluation of body composition in dogs by isotopic dilution using a low-cost technique, Fourier-transform infrared spectroscopy. *Journal of Nutrition, 132,* 1725S-1727S.

Fink, C., Cooper, H.J., Heubner, J.L., Guilak, F., & Kraus, V.B. (2002). Precision and accuracy of a transportable dual-energy X-ray absorptiometry unit for bone mineral measurements in Guinea pigs. *Calcified Tissue International, 70,* 164-169.

Fortun-Lamothe, L., Lamboley-Gauzere, B., & Bannelier, C. (2002). Prediction of body composition in rabbit females using total body electrical conductivity (TOBEC). *Livestock Production Science, 78,* 133-142.

Foy, J.M. & Schnieden, H. (1960). Estimation of total body water (virtual tritium space) in the rat, cat, rabbit, guinea-pig and man, and of the biological half-life of tritium in man. *Journal of Physiology, 154,* 169-176.

Frank, C.L. (2002). Short-term variations in diet fatty acid composition and torpor by ground squirrels. *Journal of Mammalogy, 83,* 1013-1019.

Golet, G.H., & Irons, D.B. (1999). Raising young reduces body condition and fat stores in black-legged kittiwakes. *Oecologia, 120,* 530-538.

Goto, T., Onuma, T., Takebe, K., & Kral, J.G. (1995). The influence of fatty liver on insulin clearance and insulin resistance in non-diabetic Japanese subjects. *International Journal of Obesity, 19,* 841-845.

Grier, S.J., Turner, A.S., & Alvis, M.R. (1996). The use of dual-energy X-ray absorptiometry in animals. *Investigative Radiology, 31,* 50-62.

Guo, Z., Mishra, P., & Macura, S. (2001). Sampling the intramyocellular triglycerides from skeletal muscle. *Journal of Lipid Research, 42,* 1041-1048.

Harris, R.B., Kasser, T.R., & Martin, R.J. (1986). Dynamics of recovery of body composition after overfeeding, food restriction or starvation of mature female rats. *Journal of Nutrition, 116,* 2536-2546.

Harris, R.B., & Martin, R.J. (1984). Recovery of body weight from below "set point" in mature female rats. *Journal of Nutrition, 114,* 1143-1150.

Heiman, M.L., Tinsley, F.C., Mattison, J.A., Hauck, S., & Bartke, A. (2003). Body composition of prolactin-, growth hormone-, and thyrotropin-deficient Ames dwarf mice. *Endocrine, 20,* 149-154.

Hockings, P.D., Roberts, T., Campbell, S.P., Reid, D.G., Greenhill, R.W., Polley, S.R., Nelson, P., Bertram, T.A., & Kramer, K. (2002). Longitudinal magnetic resonance imaging quantitation of rat liver regeneration after partial hepatectomy. *Toxicologic Pathology, 30,* 606-610.

Hunter, H.L., & Nagy, T.R. (2002). Body composition in a seasonal model of obesity: longitudinal measures and validation of dual-energy X-ray absorptiometry. *Obesity Research, 10,* 1180-1187.

Ishikawa, M., & Koga, K. (1998). Measurement of abdominal fat by magnetic resonance imaging of Oletf rats, an animal model of NIDDM. *Magnetic Resonance Imaging, 16,* 45-53.

Johnson, G.A., Cofer, G.P., Gewalt, S.L., & Hedlund, L.W. (2002). Morphologic phenotyping with MR microscopy: the visible mouse. *Radiology, 222,* 789-793.

Johnson, R.J., & Farrell, D J. (1988). The prediction of body composition in poultry by estimation in vivo of total body water with tritiated water and deuterium oxide. *British Journal of Nutrition, 59,* 109-124.

Katoh, S., Hata, S., Matsushima, M., Ikemoto, S., Inoue, Y., Yokoyama, J., & Tajima, N. (2001). Troglitazone prevents the rise in visceral adiposity and improves fatty liver associated with sulfonylurea therapy—a randomized controlled trial. *Metabolism, 50,* 414-417.

Kodama, A.M. (1971). In vivo and in vitro determinations of body fat and body water in the hamster. *Journal of Applied Physiology, 31,* 218-222.

Koteja, P. (1996). The usefulness of a new TOBEC instrument (ACAN) for investigating body composition in small mammals. *Acta Theriologica, 41,* 107-112.

Koteja, P., Jurczyszyn, M., & Woloszyn, B.W. (2001). Energy balance of hibernating mouse-eared bat *Myotis myotis*: a study with a TOBEC instrument. *Acta Theriologica, 46,* 1-12.

Krol, E., & Speakman, J.R. (1999). Isotope dilution spaces of mice injected simultaneously with deuterium, tritium and oxygen-18. *Journal of Experimental Biology, 202,* 2839-2849.

Lantry, B.F., & Stewart, D.J. (1999). Evaluation of total-body electrical conductivity to estimate whole-body water content of yellow perch, *Perca flavescens*, and alewife, *Alosa pseudoharengus*. *Fishery Bulletin, 97,* 71-79.

Moore, F.D., Muldowney, F.P., Haxhe, J.J., Marczynska, A.W., Ball, M.R., & Boyden, C.M. (1962). Body composition in the dog. I. Findings in the normal animal. *Journal of Surgical Research, 2,* 245-253.

Mystkowski, P., Shankland, E., Schreyer, S.A., LeBoeuf, R.C., Schwartz, R.C., Cummings, D.E., Kushmerick, M., & Schwartz, M.W. (2000). Validation of whole-body magnetic resonance spectroscopy as a tool to assess murine body composition. *International Journal of Obesity, 24,* 719-724.

Nagy, T.R. (2001). The use of dual-energy X-ray absorptiometry for the measurement of body composition. In J.R. Speakman (Ed.), *Body Composition Analysis of Animals: A Handbook of Non-Destructive Methods* (pp. 211-229). Cambridge, UK: Cambridge University Press.

Nagy, T.R., & Clair, A.L. (2000). Precision and accuracy of dual-energy X-ray absorptiometry for determining in vivo body composition of mice. *Obesity Research, 8,* 392-398.

Nagy, T.R., & Johnson, M.S. (2003). Measurement of body and liver fat in small animals using peripheral quantitative computed tomography. *International Journal of Body Composition Research, 4,* 155-160.

Nagy, T.R., Prince, C.W., & Li, J. (2001). Validation of peripheral dual-energy X-ray absorptiometry for the measurement of bone mineral in intact and excised long bones of rats. *Journal of Bone and Mineral Research, 16,* 1682-1687.

Nunes, S., Mueke, E.M., & Holekamp, K. (2002). Seasonal effects of food provisioning on body fat, insulin and corticosterone in free-living juvenile Belding's ground squirrels (*Spermophilus beldingi*). *Canadian Journal of Zoology, 80,* 366-371.

Pietrobelli, A., Formica, C., Wang, Z., & Heymsfield, S. (1996). Dual-energy X-ray absorptiometry body composition model: review of physical concepts. *American Journal of Physiology, 271,* E941-E951.

Pulawa, L.K., & Florant, G.L. (2000). The effects of caloric restriction on the body composition and hibernation of the golden-mantled ground squirrel (*Spermophilus lateralis*). *Physiological and Biochemical Zoology, 73,* 538-546.

Razani, B., Combs, T.P., Wang, X.B., Frank, P.G., Park, D.S., Russell, R.G., Li, M., Tang, B., Jelicks, L.A., Scherer, P.E., & Lisanti, M.P. (2002). Caveolin-1-deficient mice are lean, resistant to diet-induced obesity, and show hypertriglyceridemia with adipocyte abnormalities. *Journal of Biological Chemistry, 277,* 8635-8647.

Robin, J.P., Heitz, A., Le Maho, Y., & Lignon, J. (2002). Physical limitations of the TOBEC method: accuracy and long-term stability. *Physiology and Behavior, 75,* 105-118.

Rosen, H.N., Tollin, S., Balena, R., Middlebrooks, V.L., Beamer, W.G., Donohue, L.R., Rosen, C., Turner, A., Holick, M., & Greenspan, S.L. (1995). Differentiating between orcheicto-mized rats and controls using measurements of trabecular bone density: a comparison among DXA, histomorphometry, and peripheral quantitative computerized tomography. *Calcified Tissue International, 57,* 35-39.

Ross, R., Léger, L., Guardo, R., De Guise, J., & Pike, B.G. (1991). Adipose tissue volume measured by magnetic resonance imaging and computerized tomography in rats. *Journal of Applied Physiology, 70,* 2164-2172.

Rudin, M., Beckmann, N., Porszasz, R., Resse, T., Bochelen, D., & Sauter, A. (1999). In vivo magnetic resonance methods in pharmaceutical research: current status and perspectives. *NMR in Biomedicine, 12,* 69-97.

Sbarbati, A., Baldassarri, A.M., Zancanaro, C., Boicelli, A., & Osculati, F. (1991). In vivo morphometry and functional morphology of brown adipose tissue by magnetic resonance imaging. *Anatomical Record, 231,* 293-297.

Sheng, H.P., & Huggins, R.A. (1971). Direct and indirect measurement of total body water in the growing beagle. *Proceedings of the Society of Experimental Biology and Medicine, 137,* 1093-1099.

Shields, R.G., Jr., Mahan, D.C., & Byers, F.M. (1984). In vivo body composition estimation in nongravid and reproducing first-litter sows with deuterium oxide. *Journal of Animal Science, 59,* 1239-1246.

Simoneau, J.-A., Colberg, S.R., Thaete, F.L., & Kelley, D.E. (1995). Skeletal muscle glycolytic and oxidative enzyme capacities are determinants of insulin sensitivity and muscle composition in obese women. *FASEB Journal, 9,* 273-278.

Speakman, J.R. (1997). *Doubly Labelled Water. Theory and Practice.* London: Chapman and Hall.

Speakman, J.R., Visser, G.H., Ward, S., & Krol, E. (2001). The isotope dilution method for the evaluation of body composition. In J.R. Speakman (Ed.), *Body Composition Analysis of Animals. A Handbook of Non-Destructive Methods* (pp. 56-98). Cambridge, UK: Cambridge University Press.

Stein, D.T., Babcock, E.E., Malloy, C.R., & McGarry, J.D. (1995). Use of proton spectroscopy for detection of homozygous fatty ZDF-drt rats before weaning. *International Journal of Obesity, 19,* 804-810.

Stenger, J., & Bielajew, C. (1995). Comparison of TOBEC-derived total body fat with fat pad weights. *Physiology and Behavior, 57,* 319-323.

Tang, H., Vasselli, J.R., Wu, E.X., Boozer, C.N., & Gallagher, D. (2002). High-resolution magnetic resonance imaging tracks changes in organ and tissue mass in obese and aging rats. *American Journal of Physiology, 282,* R890-R899.

Tisavipat, A., Vibulsreth, S., Sheng, H.P., & Huggins, R.A. (1974). Total body water measured by desiccation and by tritiated water in adult rats. *Journal of Applied Physiology, 37,* 699-701.

Tobin, B.W., & Finegood, D.T. (1995). Estimation of rat body composition by means of electromagnetic scanning is altered by duration of anesthesia. *Journal of Nutrition, 125,* 1512-1520.

Trocki, O., Baer, D.J., & Castonguay, T.W. (1995). An evaluation of the use of total body electrical conductivity for the estimation of body composition in adult rats: effect of dietary obesity and adrenalectomy. *Physiology and Behavior, 57,* 765-772.

Unangst, E.T., Jr., & Wunder, B.A. (2001). Need for species-specific models for body-composition estimates of small mammals using EM-SCAN (R). *Journal of Mammalogy, 82,* 527-534.

Unangst, E.T., Jr., & Wunder, B.A. (2002). Effects of trap retention on body composition of live meadow voles. *Physiological and Biochemical Zoology, 75,* 627-634.

Walsberg, G.E. (1988). Evaluation of a nondestructive method for determining fat stores in small birds and mammals. *Physiological Zoology, 61,* 153-159.

West, D.B., York, D.A., Goudey-Lefevre, J., & Truett, G.E. (1994). Genetics and physiology of dietary obesity in the mouse. In G.A. Bray & D. Ryan (Eds.), *Molecular and Genetic Aspects of Obesity.* Baton Rouge: Louisiana State University Press.

Wirsing, A.J., Steury, T.D., & Murray, D.L. (2002). Noninvasive estimation of body composition in small mammals: a comparison of conductive and morphometric techniques. *Physiological and Biochemical Zoology, 75,* 489-497.

Wolf, T.J., Ellington, C.P., Davis, S., & Feltham, M.J. (1996). Validation of the doubly labelled water technique for bumblebees *Bombus terrestris* (L.). *Journal of Experimental Biology, 199,* 959-972.

Yasui, T., Ishiko, O., Sumi, T., Honda, K., Hirai, K., Nishimura, S. Matsumoto, Y., & Ogita, S. (1998). Body composition analysis of cachectic rabbits by total body electrical conductivity. *Nutrition and Cancer, 32,* 190-193.

Zuercher, G.L., Roby, D.D., & Rexstad, E.A. (2003). Validation of two new total body electrical conductivity (TOBEC) instruments for estimating body composition of live northern red-backed voles *Clethrionomys rutilus. Acta Theriologica, 42,* 387-397.

Chapter 11

Beaton, A.E., & Tukey, J.W. (1979). The fitting of power series, mean polynomials, illustrated on band spectroscopic data. *Technometrics,* 21:215-223.

Belsley, D.A., Kuh, E., & Welsch, R.E. (1980). Regression diagnostics: Identifying influential data and sources of collinearity. New York: Wiley.

Chatterjee, S., & Price, B. (1979). Regression analysis by example. New York: Wiley.

Chumlea, W.C., & Guo, S.S. (2000). Assessment and prevalence of obesity: application of new methods to a major problem. *Endocrine,* 13:135-142.

Chumlea, W.C., & Kuczmarski, R.J. (1995). Using a bony landmark to measure waist circumference. *Journal of American Dietetic Association,* 95:12.

Chumlea, W.C., Guo, S.S., Kuczmarski, R.J., Flegal, K.M., Johnson, C.L., Heymsfield, S.B., Lukaski, H.C., Friedl, K., & Hubbard, V.S. (2002). Body composition estimates from NHANES III bioelectrical impedance data. *International Journal of Obesity and Related Metabolic Disorders,* 26:1596-1609.

Chumlea, W.C., Guo, S.S., Kuczmarski, R.J., Johnson, C.L., & Leahy, C.K. (1990). Reliability for anthropometry in the Hispanic Health and Nutrition Examination Survey (HHANES). *American Journal of Clinical Nutrition,* 51:902-907.

Chumlea, W.C., Guo, S.S., Siervogel, R.M., Garry, P.J., Wang, J., Pierson, R.N., & Heymsfield, S.B. (1999). Total body water values for adults 20 to 80 years of age. Workshop, Nutrition & Successful Ageing Abstracts Danone Res Center. Conference paper.

Chumlea, W.C., Guo, S.S., Zeller, C.M., Reo, N.V., Baumgartner, R.N., Garry, P., Wang, J., Pierson, P.N. Jr., Heymsfield, S.B., & Siervogel, R.M. (2001). Total body water reference values and prediction equations for adults. *Kidney International,* 59:2250-2258.

Conlisk, E., Haas, J., Martinez, E., Flores, R., Rivera, J., & Martorell, R. (1992). Predicting body composition from anthro-

pometry and bioimpedance in marginally undernourished adolescents and young adults. *American Journal of Clinical Nutrition,* 55:1051-1059.

Deurenberg, P., Vanderkooy, K., Leenen, R., Weststrate, J., & Seidell, J. (1991). Sex and age specific prediction formulas for estimating body composition from bioelectrical impedance—a cross-validation study. *International Journal of Obesity,* 15:17-25.

Deurenberg, P., Kooij, K., Evers, P., & Hulshof, T. (1990a). Assessment of body composition by bioelectrical impedance in a population age >60 y. *American Journal of Clinical Nutrition,* 51:3-6.

Deurenberg, P., Kusters, C., & Smit, H. (1990b). Assessment of body composition by bioelectrical impedance in children and young adults is strongly age-dependent. *European Journal of Clinical Nutrition,* 44:261-268.

Deurenberg, P., Smit, H.E., & Kusters, C.S.L. (1989a). Is the bioelectric impedance method suitable for epidemiologic field studies? *European Journal of Clinical Nutrition,* 43: 647-654.

Deurenberg, P., Vanderkooy, K., Paling, A., & Withagen, P. (1989b). Assessment of body composition in 8-11 year old children by bioelectrical impedance. *European Journal of Clinical Nutrition,* 43:623-629.

Dittmar, M., & Reber, H. (2001). New equations for estimating body cell mass from bioimpedance parallel models in healthy older Germans. *American Journal of Physiology Endocrinology and Metabolism,* 281:E1005-E1014.

Duncan, G.T. (1978). An empirical study of jackknife constructed confidence regions in non-linear regression. *Technometrics,* 20:123-129.

Geisser, S.A. (1974). Predictive approach to the random effect model. *Biometrika,* 61:101-107.

Guo, S.S., Khoury, P., Specker, B., Heubi, J., Chumlea, W.C., Siervogel, R., & Morrison, J. (1993). Prediction of fat-free mass in black and white pre-adolescent and adolescent girls from anthropometry and impedance. *American Journal of Human Biology,* 5:735-745.

Guo, S.S., Roche, A.F., & Houtkooper, L.H. (1989). Fat-free mass in children and young adults from bioelectric impedance and anthropometry variables. *American Journal of Clinical Nutrition,* 50:435-443.

Guo, S.S., Wu, W., Chumlea, W.C., & Roche, A.F. (2002). Predicting overweight and obesity in adulthood from body mass index values in childhood and adolescence. *American Journal of Clinical Nutrition,* 76:653-658.

Hassager, C., Gotfredsen, A., Jensen, J., & Christiansen, C. (1986). Prediction of body composition by age, height, weight, and skinfold thickness in normal adults. *Metabolism,* 35:1081-1084.

Heitmann, B. (1990). Prediction of body water and fat in adult Danes from measurement of electrical impedance—a validation study. *International Journal of Obesity,* 14: 789-802.

Heymsfield, S.B., Lichtman, S., Baumgartner, R.N., Wang, J., Kamen, Y., Aliprantis, A., & Pierson, R.N., Jr. (1990). Body composition of humans: Comparison of two improved four-compartment models that differ in expense, technical complexity, and radiation exposure. *American Journal of Clinical Nutrition,* 52:52-58.

Huber, P. (1963). Robust estimation of a location parameter. *Annals of Mathematical Statistics,* 35:73-100.

Kwok, T., Woo, J., & Lau, E. (2000). Prediction of body fat by anthropometry in older Chinese people. *Obesity Research,* 9:97-101.

Lohman, T. (1986). Applicability of body composition techniques and constants for children and youths. *Exercise and Sport Sciences Reviews,* 14:325-357.

Lukaski, H. (1987). Methods for the assessment of human body composition: Traditional and new. *American Journal of Clinical Nutrition,* 46:537-556.

Lukaski, H., Bolonchuk, W., Hall, C., & Siders, W. (1986). Validation of tetrapolar bioelectrical impedance method to assess human body composition. *Journal of Applied Physiology,* 60:1327-1332.

Mallows, C.L. (1973). Some comments on Cp. *Technometrics,* 15:661-675.

Mason, R., & Gunst, R. (1985). Outlier-induced collinearities. *Technometrics,* 27:401-407.

Maynard, L.M., Wisemandle, W.A., Roche, A.F., Chumlea, W.C., Guo, S.S., & Siervogel, R.M. (2001). Childhood body composition in relation to body mass index: The Fels Longitudinal Study. *Pediatrics,* 107:344-350.

Montgomery, D.C., & Peck, E.A. (1981). Introduction to linear regression analysis. New York: Wiley.

Morrison, J.A., Guo, S.S., Specker, B., Chumlea, W.C., Yanovski, S., & Yanovski, J.A. (2001). Assessing body composition of 6-17 year-old Black and White girls in field studies. *American Journal of Human Biology,* 13:249-254.

Myers, R.H. (1986). Classical & modern regression with applications. Boston: Duxbury.

Roubenoff, R., Kehayias, J., Dawson-Hughes B., & Heymsfield, S.B. (1993). Use of dual-energy x-ray absorptiometry in body-composition studies—not yet a gold standard. *American Journal of Clinical Nutrition,* 58:589-591.

Schoeller, D.A., & Luke, A. (2000). Bioelectrical impedance analysis prediction equations differ between African Americans and Caucasians but it is not clear why. *Annals of the New York Academy of Sciences,* 225-226.

Segal, K., Gutin, B., Presta, E., Wang, J., & Van Itallie, T.B. (1985). Estimation of human body composition by electrical impedance methods: a comparative study. *Journal of Applied Physiology,* 58:1565-1571.

Segal, K., Van Loan, M., Fitzgerald, P., Hodgdon, J., & Van Itallie, T.B. (1988). Lean body mass estimation by bioelectrical impedance analysis: a four-site cross-validation study. *American Journal of Clinical Nutrition,* 47:7-14.

Shapiro, S.S., & Wilks, M.B. (1965). An analysis of variance test for normality (complete samples). *Biometrika,* 115-124.

Stone, M. (1974). Cross-validatory choice and assessment of statistical predictions. *Journal of the Royal Statistical Society,* 36:111-133.

Sun, S.S., Chumlea, W.C., Heymsfield, S.B., Lukaski, H., Schoeller, D., Friedl, K., Kuczmarksi, R., Flegal, K., Johnson, C., & Hubbard, V. (2003). Development of bioelectrical impedance analysis prediction equations for body composition with the use of a multicomponent model for use in epidemiological surveys. *American Journal of Clinical Nutrition,* 77: 331-340.

Tukey, J.W. (1977). Exploratory data analysis. Reading MA: Addison-Wesley.

Zillikens, M., & Conway, J. (1990). Anthropometry in blacks—applicability of generalized skinfold equations and differences in fat patterning between blacks and whites. *American Journal of Clinical Nutrition,* 52:45-51.

Chapter 12

Allen, T.H., Krzywicki, H.J., & Roberts, J.E. (1959). Density, fat, water and solids in freshly isolated tissues. *Journal of Applied Physiology,* 14, 1005-1008.

Bartoli, W.P., Davis, J.M., Pate, R.R., Ward, D.S., & Watson, P.D. (1993). Weekly variability in total-body water using ^2H_2O dilution in college-age males. *Medicine and Science in Sports and Exercise,* 25, 1422-1428.

Baumgartner, R.N., Heymsfield, S.B., Lichtman, S., Wang, J., & Pierson, R.N., Jr. (1991). Body composition in elderly people: Effect of criterion estimates on predictive equations. *American Journal of Clinical Nutrition,* 53, 1345-1353.

Beddoe, A.H., Streat, S.J., & Hill, G.L. (1984). Evaluation of an in vivo prompt gamma neutron activation facility for body composition studies in critically ill intensive care patients: Results on 41 normals. *Metabolism,* 33, 270-280.

Behnke, A.R., Jr., Feen, B.G., & Welham, W.C. (1942). The specific gravity of healthy men. *Journal of the American Medical Association,* 118, 495-498.

Biltz, R.M., & Pellegrino, E.D. (1969). The chemical anatomy of bone. *Journal of Bone and Joint Surgery [AM],* 51A, 456-466.

Brožek, J., Grande, F., Anderson, J.T., & Keys, A. (1963). Densitometric analysis of body composition: Revision of some quantitative assumptions. *Annals of the New York Academy of Sciences,* 110, 113-140.

Burkinshaw, L., Hill, G.L., & Morgan, D.B. (1979). Assessment of the distribution of protein in the human body by in-vivo neutron activation analysis. In *Nuclear Activation Techniques in the Life Sciences 1978* (pp. 787-798). Vienna: International Atomic Energy Agency.

Burnell, J.M., Baylink, D.J., Chestnut, C.H., III, Mathews, M.W., & Teubner, E.J. (1982). Bone matrix and mineral abnormalities in postmenopausal osteoporosis. *Metabolism, Clinical and Experimental,* 31, 1113-1120.

Burton, A.C. (1935). Human calorimetry. II. The average temperature of the tissues of the body. *Journal of Nutrition,* 9, 261-280.

Cohn, S.H., Vartsky, D., Yasumura, S., Sawitsky, A., Zanzi, I., Vaswani, A., & Ellis, K.J. (1980). Compartmental body composition based on total-body nitrogen, potassium, and calcium. *American Journal of Physiology (Endocrinology and Metabolism),* 239, E524-E530.

Cohn, S.H., Vaswani, A.N., Yasumura S., Yuen, K., & Ellis, K.J. (1984). Improved models for determination of body fat by in vivo neutron activation. *American Journal of Clinical Nutrition,* 40, 255-259.

Culebras, J.M., & Moore, F.D. (1977). Total-body water and the exchangeable hydrogen. 1. Theoretical calculation of nonaqueous exchangeable hydrogen in man. *American Journal of Physiology (Regulatory, Integrative and Comparative Physiology),* 232, R54-R59.

Cunningham, J.J. (1994). N × 6.25: Recognizing a bivariate expression for protein balance in hospitalized patients. *Nutrition,* 10, 124-127.

Dallemagne, M.J., & Melon, J. (1945). Le poids spécifique et l'indice de réfraction de l'os, de l'émail, de la dentine et du cément. *Bulletin de la Societe de Chimique Biologique,* 27, 85-89.

Diem, K. (Ed.) (1962). *Documenta Geigy Scientific Tables*, Ardsley, NY: Geigy Pharmaceuticals.

Dilmanian, F.A., Weber, D.A., Yasumura, S., Kamen, Y., Kindofsky, L., Heymsfield, S.B., Pierson, R.N., Jr., Wang, J., Kehayias, J.J., & Ellis, K.J. (1990). The performance of the BNL delayed and prompt-gamma neutron activation systems at Brookhaven National Laboratory. In S. Yasumura, J.E. Harrison, K.G. McNeill, et al. (eds.), *Advances in In Vivo Body Composition Studies* (pp. 309-315). New York: Plenum Press.

Dutton, J. (1991). *In vivo* analysis of body elements and body composition. *University of Wales Science and Technology Review*, 8, 19-30.

Fidanza, F., Keys, A., & Anderson, J.T. (1953). Density of body fat in man and other mammals. *Journal of Applied Physiology*, 6, 252-256.

Forbes, G.B. (1987). *Human Body Composition: Growth, Aging, Nutrition, and Activity*. New York: Springer-Verlag.

Forslund, A.H., Johansson, A.G., Sjodin, A., Bryding, G., Ljunghall, S., & Hambraeus, L. (1996). Evaluation of modified multicomponent models to calculate body composition in healthy males. *American Journal of Clinical Nutrition*, 63, 856-862.

Friedl, K.E., DeLuca, J.P., Marchitelli, L.J., & Vogel, J.A. (1992). Reliability of body-fat estimations from a four-compartment model by using density, body water, and bone mineral measurements. *American Journal of Clinical Nutrition*, 55, 764-770.

Fuller, N.J., Jebb, S.A., Laskey, M.A., & Coward, W.A. (1992). Four compartment model for the assessment of body composition in humans: Comparison with alternative methods, and evaluation of the density and hydration of fat-free mass. *Clinical Science*, 82, 687-693.

Fuller, N.J., Wells, J.C.K., & Elia, M. (2001). Evaluation of a model for total-body protein mass based on dual energy X-ray absorptiometry: Comparison with a reference four component model. *British Journal of Nutrition*, 86, 45-52.

Goran, M.I., Poehlman, E.T., Nair, K.S., & Danforth, E., Jr. (1992). Effect of gender, body composition, and equilibration time on the ^{3}H-to-^{18}O dilution space ratio. *American Journal of Physiology (Endocrinology and Metabolism)*, 263, E1119-E1124.

Gurr, M.I., & Harwood, J.L. (1991) *Lipid Biochemistry*. 4th Ed. London: Chapman and Hall.

Haurowitz, F. (1963). *The Chemistry and Function of Proteins*. New York: Academic Press.

Heymsfield, S.B., Lichtman, S., Baumgartner, R.N., Wang, J., Kamen, Y., Aliprantis, A., & Pierson, R.N., Jr. (1990). Body composition of humans: Comparison of two improved four-compartment models that differ in expense, technical complexity, and radiation exposure. *American Journal of Clinical Nutrition*, 52, 52-58.

Heymsfield, S.B., & Matthews, D. (1994). Body composition: Research and clinical advances. *Journal of Parenteral and Enteral Nutrition*, 18, 91-103.

Heymsfield, S.B., & Waki, M. (1991). Body composition in humans: Advances in the development of multicompartment chemical models. *Nutrition Reviews*, 49, 97-108.

Heymsfield, S.B., Waki, M., Kehayias, J., Lichtman, S., Dilmanian, F.A., Kamen, Y., Wang, J., & Pierson, R.N., Jr. (1991). Chemical and elemental analysis of humans in vivo using improved body composition models. *American Journal of Physiology (Endocrinology and Metabolism)*, 261, E190-E198.

Heymsfield, S.H., Wang, J., Funfar, J., Kehayias, J.J., & Pierson, R.N., Jr. (1989). Dual photo absorptiometry: Accuracy of bone mineral and soft tissue mass measurement in vivo. *American Journal of Clinical Nutrition*, 49, 1283-1289.

Heymsfield, S.B., Wang, Z.M., & Withers, R.T. (1996) Multicomponent molecular level model of body composition analysis. In A.F. Roche, S.B. Heymsfield, & T.G. Lohman (eds.), *Human Body Composition* (pp 129-147). Champaign, IL: Human Kinetics.

Hulmes, D.J.S., & Miller, A. (1979). Quasi-hexagonal molecular packing in collagen fibrils. *Nature*, 282, 878-880.

James, H.M., Dabek, J.T., Chettle, D.R., Dykes, P.W., Fremlin, J.H., Hardwicke, J., Thomas, B.J., & Vartsky, D. (1984). Whole body cellular and collagen nitrogen in healthy and wasted men. *Clinical Science*, 67, 73-82.

Jue, T., Rothman, D.L., Shulman, G.I., Tavitian, B.A., DeFronzo, R.A., & Shulman, R.G. (1989). Direct observation of glycogen synthesis in human muscle with ^{13}C NMR. *Proceedings of the National Academy of Sciences USA*, 86, 4489-4491.

Kehayias, J.J., Ellis, K.J., Cohn, S.H., Yasumura, S., & Weinlein, J. (1987). Use of a pulsed neutron generator for in vivo measurement of body carbon. In K.J. Ellis, S. Yasumura, & W.D. Morgan (eds.), *In Vivo Body Composition Studies* (pp 427-435). London: The Institute of Physical Sciences in Medicine.

Kehayias, J.J., Heymsfield, S.B., LoMonte, A.F., Wang, J., & Pierson, R.N., Jr. (1991). In vivo determination of body fat by measuring total-body carbon. *American Journal of Clinical Nutrition*, 53, 1339-1344.

Kleiber, M. (1975). *The Fire of Life*. Huntington, NY: Kreiger.

Knight, G.S., Beddoe, A.H., Streat, S.J., & Hill, G.L. (1986). Body composition of two human cadavers by neutron activation and chemical analysis. *American Journal of Physiology (Endocrinology and Metabolism)*, 250, E179-E185.

Kyere, K., Oldroyd, B., Oxby, C.B., Burkinshaw, L., Ellis, R.E., & Hill, G.L. (1982). The feasibility of measuring total body carbon by counting neutron inelastic scatter gamma rays, *Physics in Medicine and Biology*, 27, 805-817.

Lohman, T.G. (1981). Skinfolds and body density and their relation to body fatness: A review. *Human Biology*, 53, 181-225.

Lohman, T.G. (1986). Applicability of body composition techniques and constants for children and youths. *Exercise and Sport Sciences Reviews*, 14, 325-357.

Lohman, T.G. (1992). *Advances in Body Composition Assessment*. Champaign, IL: Human Kinetics.

Lohman, T.G., & Going, S.B. (1993). Multicomponent models in body composition research: Opportunities and pitfalls. In K.J. Ellis & J.D. Eastman (eds.), *Human Body Composition* (pp. 53-58). New York: Plenum Press.

Maffy, R.H. (1976). The body fluids: volume, composition, and physical chemistry. In B.M. Brenner & F.C. Rector (eds.), *The Kidney* (Vol 1, pp. 65-103). Philadelphia: Saunders.

Méndez, J., Keys, A., Anderson, J.T., & Grande F. (1960). Density of fat and bone mineral of the mammalian body. *Metabolism*, 9, 472-477.

Mernagh, J.R., Harrison, J.E., & McNeill, K.G. (1977). *In vivo* determination of nitrogen using Pu-Be sources. *Physics in Medicine and Biology*, 22, 831-835.

Modlesky, C.M., Evans, E.M., Millard-Stafford, M.L., Collins, M.A., Lewis, R.D., & Cureton, K.J. (1999). Impact of bone mineral estimates on percent fat estimates from a four-component model. *Medicine and Science in Sports and Exercise*, 31, 1861-1868.

Modlesky, C.M., Lewis, R.D., Yetman, K.A., Rose, B., Rosskopf, L.B., Snow, T.K., & Sparling, P.B. (1996). Comparison of body composition and bone mineral measurements from two DXA instruments in young men. *American Journal of Clinical Nutrition, 64,* 669-676.

Mueller, S.H., & Martorell, R. (1988). Reliability and accuracy of measurement. In T.G. Lohman, A.F. Roche, & R. Martorell (eds.), *Anthropometric Standardization Reference Manual* (pp. 83-86). Champaign, IL: Human Kinetics.

Pace, N., & Rathbun, E.N. (1945). Studies on body composition. III. The body water and chemically combined nitrogen content in relation to fat content. *Journal of Biological Chemistry, 158,* 685-691.

Pierson, R.N., Jr., Wang, J., Heymsfield, S.B., Dilmanian, F.A., & Weber, D.A. (1990). High precision in-vivo neutron activation analysis: A new era for compartmental analysis on body composition. In S. Yasumura, J. Harrison, K.G. McNeil, et al. (eds.), *Advances in In Vivo Body Composition Studies* (pp. 317-325). New York: Plenum Press.

Ryde, S.J.S., Birks, J.L., Morgan, W.D., Evans, C.J., & Dutton, J. (1993). A five-compartment model of body composition of healthy subjects assessed using *in vivo* neutron activation analysis. *European Journal of Clinical Nutrition, 47,* 863-874.

Schoeller, D.A., & Jones, P.J.H. (1987). Measurement of total body water by isotope dilution: a unified approach to calculations. In K.J. Ellis, S. Yasumura, & W.D. Morgan (eds.), *In Vivo Body Composition Studies* (pp. 131-137). London: Institute of Physical Sciences in Medicine.

Schoeller, D.A., Van Santen, E., Peterson, D.W., Jaspan, J., & Klein, P.D. (1980). Total body water measurement in humans with ^{18}O and ^{2}H labeled water. *American Journal of Clinical Nutrition, 33,* 2686-2693.

Selinger, A. (1977). The body as a three component system. Unpublished doctoral dissertation, University of Illinois, Urbana.

Sheng, H.P., & Huggins, R.A. (1979). A review of body composition studies with emphasis on total-body water and fat. *American Journal of Clinical Nutrition, 32,* 630-647.

Siconolfi, S.F., Gretebeck, R.J., & Wong, W.W. (1995). Assessing total body protein, mineral, and bone mineral content from total body water and body density. *Journal of Applied Physiology, 79,* 1837-1843.

Siri, W.E. (1961). Body composition from fluid spaces and density: Analysis of methods. In J. Brožek & A. Henschel (eds.), *Techniques for Measuring Body Composition* (pp. 223-244). Washington, DC: National Academy of Sciences/National Research Council.

Snyder, W.S., Cook, M.J., Nasset, E.S., Karhausen, L.R., Howells, G.P., & Tipton, I.H. (1975). *Report of the Task Group on Reference Man.* Elmsford, NY: Pergamon Press.

Taylor, R., Price, T.B., Rothman, D.L., Shulman, R.G., & Shulman, G.I. (1992). Validation of ^{13}C NMR measurement of human skeletal muscle glycogen by direct biochemical assay of needle biopsy samples. *Magnetic Resonance in Medicine, 27,* 13-20.

Vartsky, D., Ellis, K.J., & Cohn, S.H. (1979). *In vivo* measurement of body nitrogen by analysis of prompt gammas from neutron capture. *Journal of Nuclear Medicine, 20,* 1158-1165.

Wang, J., Pierson, R.N., & Kelly, W.G. (1973). A rapid method for the determination of deuterium oxide in urine: Application to the measurement of total-body water. *Journal of Laboratory and Clinical Medicine, 82,* 170-178.

Wang, Z.M., Deurenberg, P., Wang, W., Pietrobelli, A., Baumgartner, R.N., & Heymsfield, S.B. (1999). Hydration of fat-free body mass: Review and critique of a classic body-composition constant. *American Journal of Clinical Nutrition, 69,* 833-841.

Wang, Z.M., Heshka, S., Pierson, R.N., Jr., & Heymsfield, S.B. (1995). Systematic organization of body-composition methodology: An overview with emphasis on component-based methods. *American Journal of Clinical Nutrition, 61,* 457-465.

Wang, Z.M., Ma, R., Pierson, R.N., Jr., & Heymsfield, S.B. (1993). Five-level model: Reconstruction of body weight at atomic, molecular, cellular, and tissue-system levels from neutron activation analysis. In K.J. Ellis & J.D. Eastman (eds.), *Human Body Composition, In Vivo Methods, Models, and Assessment* (pp.125-128). New York: Plenum Press.

Wang, Z.M., Pierson, R.N., Jr., & Heymsfield, S.B. (1992). The five-level model: A new approach to organizing body-composition research. *American Journal of Clinical Nutrition, 56,* 19-28.

Wang, Z.M., Pi-Sunyer, F.X., Kotler, D.P., Wielopolski, L.W., Withers, R.T., Pierson, R.N., Jr., & Heymsfield, S.B. (2002). Multicomponent methods: Evaluation of new and traditional soft tissue mineral models by in vivo neutron activation analysis. *American Journal of Clinical Nutrition, 76,* 968-974.

Wang, Z.M., Shen, W., Kotler, D., Heshka, S., Wielopolski, L., Aloia, J.F., Nelson, M.E., Pierson, R.N., Jr., & Heymsfield, S.B. (2003). Total-body protein: a new cellular level mass and distribution prediction model. *American Journal of Clinical Nutrition, 78,* 979-984.

Wang, Z.M., Visser, M., Ma, R., Baumgartner, R.N., Kotler, D., Gallagher, D., & Heymsfield, S.B. (1996). Skeletal muscle mass: Evaluation of neutron activation and dual-energy X-ray absorptiometry methods. *Journal of Applied Physiology, 80,* 824-831.

Werner, J., & Buse, M. (1988). Temperature profiles with respect to inhomogeneity and geometry of the human body. *Journal of Applied Physiology, 65,* 1110-1118.

Withers, R.T., LaForgia, J., & Heymsfield, S.B. (1999). Critical appraisal of the estimation of body composition via two-, three-, and four-compartment models. *American Journal of Human Biology, 11,* 175-185.

Withers, R.T., LaForgia, J., Pillans, R.K., Shipp, N.J., Chattertan, B.E., Schultz, C.G., & Leaney, F. (1998). Comparisons of two-, three-, and four-compartment models of body composition analysis in men and women. *Journal of Applied Physiology, 85,* 238-245.

Withers, R.T., Smith, D.A., Chatterton, B.E., Schultz, C.G., & Gaffney, R.D. (1992). A comparison of four methods of estimating the body composition of male endurance athletes. *European Journal of Clinical Nutrition, 46,* 773-784.

Wong, W.W., Cochran, W.J., Klish, W.J., Smith, E.O., Lee, L.S., & Klein, P.D. (1988). *In vivo* isotope-fractionation factors and the measurement of deuterium-and oxygen-18 dilution spaces from plasma, urine, saliva, respiratory water vapor, and carbon dioxide. *American Journal of Clinical Nutrition, 47,* 1-6.

Woodard, H.Q. (1962). The elementary composition of human cortical bone. *Health Physics, 8,* 513-517.

Woodard, H.Q. (1964). The composition of human cortical bone: Effect of age and of some abnormalities. *Clinical Orthopaedics and Related Research, 37,* 187-193.

Chapter 13

Abate, N., & Garg, A. (1995). Heterogeneity in adipose tissue metabolism: causes, implications and management of regional adiposity. *Prog Lipid Res, 34,* 53-70.

Abate, N., Garg, A., Peshock, R.M., Stray-Gundersen, J., Adams-Huet, B., & Grundy, S.M. (1996). Relationship of generalized and regional adiposity to insulin sensitivity in men with NIDDM. *Diabetes, 45,* 1684-1693.

Abate, N., Garg, A., Peshock, R.M., Stray-Gunderson, J., & Grundy, S.M. (1995). Relationships of generalized and regional adiposity to insulin sensitivity in men. *J Clin Invest, 96,* 88-98.

Ahima, R.S., & Flier, J. (2000). Adipose tissue as an endocrine organ. *Trens Endocr Metab, II,* 327-329.

Allison, D.B., Paultre, F., Goran, M I., Poehlman, E.T., & Heymsfield, S.B. (1995). Statistical considerations regarding the use of ratios to adjust data. *Int J Obes Relat Metab Disord, 19,* 644-652.

Anderson, P.J., Chan, J.C.N., Chan, Y.L., Tomlinson, B., Young, R.P., Lee, Z.S., Lee, K.K., Metreweli, C., Cockram, C.S., & Critchley, J.A. (1997). Visceral fat and cardiovascular risk factors in Chinese NIDDM patients. *Diabetes Care, 20,* 1854–1858.

Ardern, C.I., Janssen, I., Ross, R., & Katzmarzyk, P.T. (2004). Development of health-related waist circumference thresholds within BMI categories. *Obes Res, 12,* 1094-1103.

Armellini, F., Zamboni, M., Robbi, R., Todesco, T., Rigo, L., Bergamo-Andreis, I.A., & Bosello, O. (1993). Total and intra-abdominal fat measurements by ultrasound and computerized tomography. *Int J Obes Relat Metab Disord, 17,* 209-214.

Aronne, L.J., & Segal, K.R. (2002). Adiposity and fat distribution outcome measures: assessment and clinical implications. *Obes Res, 10 Suppl 1,* 14S-21S.

Asayama, K., Oguni, T., Hayashi, K., Dobashi, K., Fukunaga, Y., Kodera, K., Tamai, H., & Nakazawa, S. (2000). Critical value for the index of body fat distribution based on waist and hip circumferences and stature in obese girls. *Int J Obes Relat Metab Disord, 24,* 1026-1031.

Asayama, K., Hayashibe, H., Dobashi, K., Uchida, N., Kawada, Y., & Nakazawa, S. (1995). Relationships between biochemical abnormalities and anthropometric indices of overweight, adiposity and body fat distribution in Japanese elementary school children. *Int J Obes Relat Metab Disord, 19,* 253-259.

Ashwell, M., Lejeune, S., & McPherson, K. (1996). Ratio of waist circumference to height may be better indicator of need for weight management. *Br Med J, 312,* 377.

Bao, W., Srinivasan, S.R., Wattigney, W.A., Bao, W., & Berenson, G.S. (1996). Usefulness of childhood low-density lipoprotein cholesterol level in predicting adult dyslipidemia and other cardiovascular risks—The Bogalusa Heart Study. *Arch Intern Med, 156,* 1315-1320.

Bellu, R., Ortisi, M.T., Scaglioni, S., Agostini, C., Salanitri, V.S., Riva, E., & Giovannini, M. (1993). Lipid and apoprotein A-I and B levels in obese school-age children: results of a study in the Milan area. *J Pediatr Gastroenterol Nutr, 16,* 446-450.

Bjorntorp, P. (1985). Regional patterns of fat distribution. *Ann Intern Med, 103,* 994-995.

Bjorntorp, P. (1990). "Portal" adipose tissue as a generator of risk factors for cardiovascular disease and diabetes. *Arteriosclerosis, 10,* 493-496.

Bjorntorp, P. (1997). Body fat distribution, insulin resistance, and metabolic diseases. *Nutrition, 13,* 795-803.

Blake, G.J., & Ridker, P.M. (2001). Novel clinical markers of vascular wall inflammation. *Circ Res, 89,* 763-771.

Blew, R.M., Sardinha, L.B., Milliken, L.A., Teixeira, P.J., Going, S.B., Ferreira, D.L., Harris, M.M., Houtkooper, L.B., & Lohman, T.G. (2002). Assessing the validity of body mass index standards in early postmenopausal women. *Obes Res, 10,* 799-808.

Boesch, C., Slotboom, J., Hoppeler, H., & Kreis, R. (1997). In vivo determination of intra-myocellular lipids in human muscle by means of localized 1H-MR-spectroscopy. *Magn Reson Med, 37,* 484-493.

Brambilla, P., Manzoni, P., Sironi, S., Simone, P., Del Maschio, A., di Natale, B., & Chiumello, G. (1994). Peripheral and abdominal adiposity in childhood obesity. *Int J Obes Relat Metab Disord, 18,* 795-800.

Bray, G.A. (1987). Overweight is risking fate. Definition, classification, prevalence, and risks. *Ann NY Acad Sci, 499,* 14-28.

Bray, G.A. (2004). Don't throw the baby out with the bath water. *Am J Clin Nutr 79,* 347-349.

Brook, C.G.D. (1971). Determination of body composition of children from skinfold measurements. *Arch Diseas Child, 46,* 182-184.

Brozek, J., Grande, F., & Anderson, J.T. (1963). Densitometry analysis of body composition: revision of some quantitative assumptions. *Ann NY Acad Sci ,110,* 113-140.

Busetto, L., Baggio, M.B., Zurlo, F., Carraro, R., Digito, M., & Enzi, G. (1992). Assessment of abdominal fat distribution in obese patients: anthropometry versus computerized tomography. *Int J Obes Relat Metab Disord, 16,* 731-736.

Cacciari, E., Milani, S., Balsamo, A., Dammacco, F., De Luca, F., Chiarelli, F., Pasquino, A.M., Tonini, G., & Vanelli, M. (2002). Italian cross-sectional growth charts for height, weight and BMI (6-20 y). *Eur J Clin Nutr, 56,* 171-180.

Caprio, S. (1999). Relationship between abdominal visceral fat and metabolic risk factors in obese adolescents. *Am J Hum Biol, 11,* 259-266.

Caprio, S., Hyman, L.D., McCarthy, S., Lange, R., Bronson, M., & Tamborlane, W.V. (1996). Fat distribution and cardiovascular risk factors in obese adolescent girls: importance of the intraabdominal fat depot. *Am J Clin Nutr, 64,* 12-17.

Caprio, S., Hyman, L.D., Limb, C., McCarthy, S., Lange, R., Sherwin, R.S., Shulman, G., & Tamborlane, W.V. (1995). Central adiposity and its metabolic correlates in obese adolescent girls. *Am J Physiol, 269,* E118-E126.

Carey, D.G., Jenkins, A.B., Campbell, L.V., Freund, J., & Chisholm, D.J. (1996). Abdominal fat and insulin resistance in normal and overweight women: direct measurements reveal a strong relationship in subjects at both low and high risk of NIDDM. *Diabetes, 45,* 633-638.

Carroll, S., Cooke, C.B., Butterly, R J., Moxon, J.W.D., Moxon, J.W.A., & Dudfield, M. (2000). Waist circumference in the assessment of obesity and associated risk factors in coronary artery disease patients. *Coronary Health Care, 4,* 179-186.

Cassano, P.A., Rosner, B., Vokonas, P.S., & Weiss, S.T. (1992). Obesity and body fat distribution in relation to the incidence of non-insulin-dependent diabetes mellitus. A prospective cohort study of men in the normative aging study. *Am J Epidemiol, 136,* 1474-1486.

Chang, C.J., Wu, C.H., Yao, W.J., Yang, Y.C., Wu, J.S., & Lu, F.H. (2000). Relationships of age, menopause and central obesity on cardiovascular disease risk factors in Chinese women. *Int J Obes Relat Metab Disord, 24,* 1699-1704.

Chu, N.F., Chang, J.B., & Shieh, S.M. (2003). Plasma leptin, fatty acids, and tumor necrosis factor-receptor and insulin resistance in children. *Obes Res, 11,* 532-540.

Chu, N.-F., Wang, D.-J., Shieh, S.-M., & Rimm, E.B. (2000). Plasma leptin concentrations and obesity in relation to insulin resistance syndrome components among school children in Taiwan—the Taipei children heart study. *Int J Obes Relat Metab Disord, 24,* 1265-1271.

Chumlea, W.C., Baumgartner, R.N., Garry, P.J., Rhyne, R.L., Nicholson, C., & Wayne, S. (1992). Fat distribution and blood lipids in a sample of healthy elderly people. *Int J Obes Relat Metab Disord, 16,* 125-133.

Cole, T.J., Bellizzi, M.C., Flegal, K.M., & Dietz, W.H. (2000). Establishing a standard definition for child overweight and obesity worldwide: international survey. *Br Med J, 320,* 1240-1243.

Cole, T.J., Freeman, J.V., & Preece, M.A. (1990). Body mass index reference curves for the UK. *Arch Dis Child, 73,* 25-29.

Cook, D.G., Mendall, M.A., Whincup, P.H., Carey, I.M., Ballam, L., Morris, J.E., Miller, G.J., & Strachan, D.P. (2000). C-reactive protein concentration in children: relationship to adiposity and other cardiovascular risk factors. *Atherosclerosis, 149,* 139-150.

Cox, B.D., & Whichelow, M. (1996). Ratio of waist circumference to height is better predictor of death than body mass index. *Br Med J, 313,* 1487.

Daniels, S.R., Morrison, J.A., Sprecher, D.L., Khoury, P.R., & Kimball, T.R. (1999). Association of body fat distribution and cardiovascular risk factors in children and adolescents. *Circulation, 99,* 541-545.

Das, U.N. (2002). Obesity, metabolic syndrome X, and inflammation. *Nutrition, 18,* 430-438.

Després, J.P., & Lamarche, B. (1993). Effects of diet and physical activity on adiposity and body fat distribution: Implications for the prevention of cardiovascular disease. *Nutr Res Rev, 6,* 137–159.

Després, J.P., Lemieux, I., & Prud'homme, D. (2001). Treatment of obesity: need to focus on high risk abdominally obese patients. *Br Med J, 322,* 716-720.

Després, J.P., Prud'homme, D., Pouliot, M.C., Tremblay, A., & Bouchard, C. (1991). Estimation of deep abdominal adipose-tissue accumulation from simple anthropometric measurements in men. *Am J Clin Nutr, 54,* 471-477.

Deurenberg, P. (2001). Universal cut-off points for obesity are not appropriate. *Brit J Nutr, 85,* 135-136.

Dobbelsteyn, C.J., Joffres, M.R., MacLean, D.R., & Flowerdew, G. (2001). A comparative evaluation of waist circumference, waist-to-hip ratio and body mass index as indicators of cardiovascular risk factors. The Canadian Heart Health Surveys. *Int J Obes Relat Metab Disord, 25,* 652-661.

Durnin, J.V., & Rahaman, M.M. (1967). The assessment of the amount of fat in the human body from measurements of skinfold thickness. *Br J Nutr, 21(3),* 681-689.

Dwyer, T., & Blizzard, C.L. (1996). Defining obesity in children by biological endpoint rather than population distribution. *Int J Obes Relat Metab Disord, 20,* 472-480.

Eckel, R.H., & Krauss, R.M. (1998). American Heart Association call to action: obesity as a major risk factor for coronary heart disease. AHA Nutrition Committee. *Circulation, 21,* 2099-2100.

Ellis, K.J. (2001). Selected body composition methods can be used in field studies. *J Nutr, 131,* 1589S-1595S.

Expert Panel on the Identification, Evaluation, and Treatment of Overweight and Obesity in Adults. (1988). Executive summary of the clinical guidelines on the identification, evaluation and treatment of overweight and obesity in adults. *Arch Int Med, 158,* 1855-1667.

Ferland, M., Després, J.-P., Tremblay, A., Pinault, S., Nadeau, A., Moorjani, S., Lupien, P.J., Thériault, G., & Bouchard, C. (1989). Assessment of adipose tissue distribution by computed axial tomography in obese women: association with body density and anthropometric measurements. *Br J Nutr, 61,* 139-148.

Flodmark, C.E., Sveger, T., & Nilsson-Ehle, P. (1994). Waist measurement correlates to a potentially atherogenic lipoprotein profile in obese 12-14-year-old children. *Acta Paediatr, 83,* 941-945.

Folsom, A., Kaye, S., & Sellers, T. (1993). Body fat distribution and 5-year risk of death in older women. *JAMA, 269,* 483-487.

Folsom, A.R., Kushi, L.H., Anderson, K.E., Mink, P.J., Olson, J.E., Hong, C.P., Sellers, T.A., Lazovich, D., & Prineas, R.J. (2000). Associations of general and abdominal obesity with multiple health outcomes in older women: the Iowa Women's Health Study. *Arch Intern Med, 160,* 2117-2128.

Folsom, A.R., Prineas, R.J., Kaye, S., & Munger, R.G. (1990). Incidence of hypertension and stroke in relation to body fat distribution and other risk factors in older women. *Stroke, 21,* 701-706.

Folsom, A.R., Prineas, R.J., Kaye, S.A., & Soler, J.T. (1989). Body fat distribution and self-reported prevalence of hypertension, heart attack, and other heart disease in older women. *Int J Epidemiol, 18,* 361-367.

Fox, K., Peters, D., Armstrong, N., Sharpe, P., & Bell, M. (1993). Abdominal fat deposition in 11-year-old children. *Int J Obes Relat Metab Disord, 17,* 11-16.

Fox, K.R., Peters, D.M., Sharpe, P., & Bell, M. (2000). Assessment of abdominal fat development in young adolescents using magnetic resonance imaging. *Int J Obes Relat Metab Disord, 24,* 1653-1659.

Freedman, D.S., & Perry, G. (2000). Body composition and health status among children and adolescents. *Prev Med, 31,* S34-S53.

Freedman, D.S., Williamson, D.F., Croft, J.B., Ballew, C., & Byers, T. (1995). Relation of body distribution to ischemic heart disease. *Am J Epidemiol, 142,* 53-63.

Freedman, D.S., Srinivasan, S.R., Harsha, D.W., Webber, L.S., & Berenson, G.S. (1989). Relation of body fat patterning to lipid and lipoprotein concentrations in children and adolescents: the Bogalusa Heart Study. *Am J Clin Nutr, 50,* 930-939.

Freedman, D.S., Srinivasan, S.R., Burke, G.L., Shear, C.L., Smoak, C.G., Harsha, D.W., Webber, L.S., & Berenson, G.S. (1987). Relation of body fat distribution to hyperinsulinemia in children and adolescents: the Bogalusa Heart Study. *Am J Clin Nutr, 46,* 403-410.

Freedman, D.S., Srinivasan, S.R., Burke, G.L., Shear, C.L., Smoak, C.G., Harsha, D.W., Webber, L.S., & Berenson, G.S. (1987). Relation of body fat distribution to hyperinsulinemia in children and adolescents: the Bogalusa Heart Study. *Am J Clin Nutr, 46,* 403-410.

Frontini, M G., Bao, W., Elkasabany, A., Srinivasan, S.R., & Berenson, G. (2001). Comparison of weight-for-height indices as a measure of adiposity and cardiovascular risk from childhood to young adulthood: the Bogalusa heart study. *J Clin Epidemiol, 54,* 817-822.

Fujioka, S., Matsuzawa, Y., Tokunaga, K., & Tarui, S. (1987). Contribution of intra-abdominal fat accumulation to the impairment of glucose and lipid metabolism in human obesity. *Metabolism, 36,* 54-59.

Gallagher, D., Heymsfield, S.B., Heo, M., Jebb, S.A., Murgatroyd, P.R., & Sakamoto, Y. (2000). Healthy percentage body fat ranges: an approach for developing guidelines based on body mass index. *Am J Clin Nutr, 72,* 694-701.

Geib, H.C., Parhofer, K.G., & Schwandt, P. (2001). Parameters of childhood obesity and their relationship to cardiovascular risk factors in healthy prepubescent children. *Int J Obes Relat Metab Disord, 25,* 830-837.

Gill, T.P. (2001). Cardiovascular risk in the Asia-Pacific region from a nutrition and metabolic point of view: abdominal obesity. *Asia Pacific J Clin Nutr, 10,* 85-89.

Goodpaster, B.H. (2002). Measuring body fat distribution and content in humans. *Curr Opin Clin Nutr Metab Care, 5,* 481-487.

Goodpaster, B.H., Kelley, D.E., Thaete, F.L., He, J., & Ross, R. (2000). Skeletal muscle attenuation determined by computed tomography is associated with skeletal muscle lipid content. *J Appl Physiol, 89,* 104-110.

Goodpaster, B.H., Thaete, F.L., Simoneau, J.A., & Kelley, D.E. (1997). Subcutaneous abdominal fat and thigh muscle composition predict insulin sensitivity independently of visceral fat. *Diabetes, 46,* 1579-1585.

Goran, M.I. (1998). Measurement issues related to studies of childhood obesity: assessment of body composition, body fat distribution, physical activity, and food intake. *Pediatrics, 101,* 505-518.

Goran, M.I., Bergman, R.N., & Gower, B.A. (2001). Influence of total vs. visceral fat on insulin action and secretion in African American and white children. *Obes Res, 9,* 423-431.

Goran, M.I., & Malina, R.M. (1999). Fat distribution during childhood and adolescence: implications for later health outcomes. *Am J Hum Biol, 11,* 187-188.

Goran, M.I., Nagy, T.R., Treuth, M.S., Trowbridge, C., Dezenberg, C., McGloin, A., & Gower, B.A. (1997). Visceral fat in white and African American prepubertal children. *Am J Clin Nutr, 65,* 1703-1708.

Gower, B. (1999). Syndrome X in children: influence of ethnicity and visceral fat. *Am J Hum Biol, 11,* 249-257.

Gower, B.A., Nagy, T.R., & Goran, M.I. (1999). Visceral fat, insulin sensitivity, and lipids in prepubertal children. *Diabetes, 48,* 1515-1521.

Gower, B.A., Nagy, T.R., Trowbridge, C.A., Dezenberg, C., & Goran, M.I. (1998). Fat distribution and insulin response in prepubertal African American and white children. *Am J Clin Nutr, 67,* 821-827.

Haarbo, J., Hassager, C., Riis, B.J., & Christiansen, C. (1989). Relation of body fat distribution to serum lipids and lipoproteins in elderly women. *Atherosclerosis, 80,* 57-62.

Han, T.S., Lean, M.E., & Seidell, J.C. (1996). Waist circumference remains useful predictor of coronary heart disease. *BMJ, 312,* 1227-1228.

Han, T.S., van Leer, E.M., Seidell, J.C., & Lean, M.E. (1995). Waist circumference action levels in the identification of cardiovascular risk factors: prevalence study in a random sample. *BMJ, 311,* 1401-1405.

Han, T.S., van Leer, E.M., Seidell, J.C., & Lean, M.E. (1996). Waist circumference as a screening tool for cardiovascular risk factors: evaluation of receiver operating characteristics (ROC). *Obes Res, 4,* 533-547.

Harris, T.B., Visser, M., Everhart, J., Cauley, J., Tylavsky, F., Fuerst, T., Zamboni, M., Taaffe, D., Resnick, H.E., Scherzinger, A., & Nevitt, M. (2000). Waist circumference and sagittal diameter reflect total body fat better than visceral fat in older men and women. The Health, Aging and Body Composition Study. *Ann N Y Acad Sci, 904,* 462-473.

Heiat, A., Vaccarino, V., & Krumholz, H.M. (2001). An evidence-based assessment of federal guidelines for overweight and obesity as they apply to elderly persons. *Arch Intern Med, 161,* 1194-1203.

Herd, S.L., Gower, B.A., Dashti, N., & Goran, M.I. (2001). Body fat, fat distribution and serum lipids, lipoproteins and apolipoproteins in African-American and Caucasian-American prepubertal children. *Int J Obes Relat Metab Disord, 25,* 198-204.

Higgins, P.B., Gower, B.A., Hunter, G.R., & Goran, M.I. (2001). Defining health-related obesity in prepubertal children. *Obes Res, 9,* 233-240.

Hill, J.O., Sidney, S., Lewis, C.E., Tolan, K., Scherzinger, A.L., & Stamm, E.R. (1999). Racial differences in amounts of visceral adipose tissue in young adults: the CARDIA (Coronary Artery Risk Development in Young Adults) study. *Am J Clin Nutr, 69,* 381-387.

Himes, J.H., & Dietz, W.H. (1994). Guidelines for overweight in adolescent preventive services: recommendations from an expert committee. *Am J Clin Nutr, 59,* 307-316.

Ho, S.C., Chen, Y.M., Woo, J.L., Leung, S.S., Lam, T.H., & Janus, E.D. (2001). Association between simple anthropometric indices and cardiovascular risk factors. *Int J Obes Relat Metab Disord, 25,* 1689-1697.

Hsieh, S.D., & Yoshinaga, H. (1995). Abdominal fat distribution and coronary heart disease risk factors in men—waist/height ratio as a single and useful predictor. *Int J Obes Relat Metab Disord, 19,* 585-589.

Huang, T.T., Johnson, M.S., Gower, B.A., & Goran, M.I. (2002). Effect of changes in fat distribution on the rates of change of insulin response in children. *Obes Res, 10,* 978-984.

Hunter, G.R., Kekes-Szabo, T., Snyder, S.W., Nicholson, C., Nyikos, I., & Berland, L. (1997). Fat distribution, physical activity, and cardiovascular risk factors. *Med Sci Sports Exerc, 29,* 362-369.

Hunter, G.R., Snyder, S.W., Kekes-Szabo, T., Nicholson, C., & Berland, I. (1994). Intra-abdominal adipose tissue variables associated with risk of possessing elevated blood lipids and blood pressure. *Obes Res, 2,* 563–569.

Ito, H., Nakasuga, K., Ohshima, A., Maruyama, T., Kaji, Y., Harada, M., Fukunaga, M., Jingu, S., & Sakamoto, M. (2003). Detection of cardiovascular risk factors by indices of obesity obtained from anthropometry and dual-energy X-ray absorptiometry in Japanese individuals. *Int J Obes Relat Metab Disord, 27,* 232-237.

Iwao, S., Iwao, N., Muller, D.C., Elahi, D., Shimokata, H., & Andres, R. (2000). Effect of aging on the relationship between multiple risk factors and waist circumference. *J Am Geriatr Soc, 48,* 788-794.

Jacob, S., Machann, J., Rett, K., Brechtel, K., Volk, A., Renn, W., Maerker, E., Matthaei, S., Schick, F., Claussen, C.D., & Haring,

H. U. (1999). Association of increased intramyocellular lipid content with insulin resistance in lean nondiabetic offspring of type 2 diabetic subjects. *Diabetes, 48,* 1113-1119.

Janssen, I., Katzmarzyk, P.T., & Ross, R. (2002). Body mass index, waist circumference, and health risk: evidence in support of current National Institutes of Health guidelines. *Arch Intern Med, 162,* 2074-2079.

Janssen, I., Katzmarzyk, P.T., & Ross, R. (2004). Waist circumference and not body mass index explains obesity-related health risk. *Am J Clin Nutr, 79, 3,* 379-384.

Jebb, S.A., & Prentice, A.M. (2001). Single definition of overweight and obesity should be used. *Br Med J, 323,* 999.

Jensen, M.D., Kanaley, J.A., Reed, J.E., & Sheedy, P.F. (1995). Measurement of abdominal and visceral fat with computed tomography and dual-energy x-ray absorptiometry. *Am J Clin Nutr, 61,* 274-278.

Kahn, H.S. (1993). Choosing an index for abdominal obesity: an opportunity for epidemiologic clarification. *J Clin Epidemiol, 46,* 491-494.

Karlberg, J., Luo, Z.C., & Albertsson-Wikland, K. (2001). Body mass index reference values (mean and SD) for Swedish children. *Acta Paediatr Scand, 90,* 1427-1434.

Katzmarzyk, P.T., Tremblay, A., Pérusse, L., Després, J.P., & Bouchard, C. (2003). The utility of international child and adolescent overweight guidelines for predicting coronary heart disease risk factors. *J Clin Epidemiol, 56,* 456-462.

Kelley, D.E., Goodpaster, B.H., & Storlien, L. (2002). Muscle triglyceride and insulin resistance. *Annu Rev Nutr, 22,* 325-346.

Kelley, D.E., Thaete, F.L., Troost, F., Huwe, T., & Goodpaster, B.H. (2000). Subdivisions of subcutaneous abdominal adipose tissue and insulin resistance. *Am J Physiol (Endocrinol Metab), 278,* E941-E948.

Kiernan, M., & Winkleby, M.A. (2000). Identifying patients for weight-loss treatment: an empirical evaluation of the NHLBI obesity education initiative expert panel treatment recommendations. *Arch Intern Med, 160,* 2169-2176.

Kikuchi, D.A., Srinivasan, S.R., Harsha, D.W., Webber, L.S., Sellers, T.A., & Berenson, G.S. (1992). Relation of serum lipoprotein lipids and apolipoproteins to obesity in children: the Bogalusa Heart Study. *Prev Med, 21,* 177-190.

Kissebah, A.H., & Krakower, G.R. (1994). Regional adiposity and morbidity. *Physiol Rev, 74,* 761-809.

Ko, G., Chan, J., Cockram, C., & Woo, J. (1999). Prediction of hypertension, diabetes, dyslipidaemia or albuminuria using simple anthropometric indexes in Hong Kong Chinese. *Int J Obes Relat Metab Disord, 23,* 1136-1142.

Kuczmarski, R.J., & Flegal, K.M. (2000). Criteria for definition of overweight in transition: background and recommendations for the United States. *Am J Clin Nutr, 72,* 1074-1081.

Kvist, H., Chowdhury, B., Grangard, U., Tylen, U., & Sjostrom, L. (1988). Total and visceral adipose-tissue volumes derived from measurements with computed tomography in adult men and women: predictive equations. *Am J Clin Nutr, 48,* 1351-1361.

Lapidus, L., Bengtsson, C., Hallstrom, T., & Bjorntorp, P. (1989). Obesity, adipose tissue distribution and health in women—results from a population study in Gothenburg, Sweden. *Appetite, 13,* 25-35.

Lapidus, L., Bengtsson, C., Larsson, B., Pennert, K., Rybo, E., & Sjostrom, L. (1984). Distribution of adipose tissue and risk of cardiovascular disease and death: a 12 year follow up of participants in the population study of women in Gothenburg, Sweden. *Br Med J, 289,* 1257-1261.

Larsson, B., Svardsudd, K., Welin, L., Wilhelmsen, L., Bjorntorp, P., & Tibblin, G. (1984). Abdominal adipose tissue distribution, obesity, and risk of cardiovascular disease and death: 13 year follow up of participants in the study of men born in 1913. *Br Med J, 288,* 1401-1404.

Lean, M.E., Han, T.S., & Morrison, C.E. (1995). Waist circumference as a measure for indicating need for weight management. *Br Med J, 311,* 158-161.

Lemieux, S., Prud'homme, D., Bouchard, C., Tremblay, A., & Després, J. (1996). A single threshold value of waist girth identifies normal-weight and overweight subjects with excess visceral adipose tissue. *Am J Clin Nutr, 64,* 685-693.

Libby, P. (2002). Inflammation in atherosclerosis. *Nature, 420,* 868-874.

Lin, W.Y., Lee, L.T., Chen, C.Y., Lo, H., Hsia, H.H., Liu, I.L., Lin, R.S., Shau, W.Y., & Huang, K.C. (2002). Optimal cut-off values for obesity: using simple anthropometric indices to predict cardiovascular risk factors in Taiwan. *Int J Obes Relat Metab Disord, 26,* 1232-1238.

Lindsay, R.S., Hanson, R.L., Roumain, J., Ravussin, E., Knowler, W.C., & Tataranni, P.A. (2001). Body mass index as a measure of adiposity in children and adolescents: relationship to adiposity by dual energy X-ray absorptiometry and to cardiovascular risk factors. *J Clin Endocrinol Metabol, 86,* 4061-4067.

Lohman, T.G. (1989). Assessment of body composition in children. *Pediatr Exerc Sci, 1,* 19-30.

Lottenberg, S.A., Giannella-Neto, D., Derendorf, H., Rocha, M., Bosco, A., Carvalho, S.V., Moretti, A.E., Lerario, A.C., & Wajchenberg, B.L. (1998). Effect of fat distribution on pharmacokinetics of cortisol in obesity. *Int J Clin Pharmacol Ther 36,* 501–505.

Maffeis, C., Pietrobelli, A., Grezzani, A., Provera, S., & Tato, L. (2001). Waist circumference and cardiovascular risk factors in prepubertal children. *Obes Res, 9,* 179-187.

Marin, P., Andersson, B., Ottosson, M., Olbe, L., Chowdhury, B., Kvist, H., Holm, G., Sjostrom, L., & Bjorntorp, P. (1992). The morphology and metabolism of intraabdominal adipose tissue in men. *Metabolism, 41,* 1242-1248.

Matsuzawa, Y., Nakamura, T., Shimomura, I., & Kotami, K.(1996). Visceral fat accumulation and cardiovascular disease. In: A. Angel, H. Anderson, C. Bouchard, D. Lau, L. Leiter, & R. Mendelson. (eds) *Progress in Obesity Research: Proceedings of the Seventh International Congress on Obesity.* John Libbey & Company, London, vol 7, pp.569–572.

Maynard, L.M., Wisemandle, W., Roche, A.F., Chumlea, C., Guo, S.S., & Siervogel, R.M. (2001). Childhood body composition in relation to body mass index. *Pediatrics, 107,* 344-350.

Misra, A., Garg, A., Abate, N., Peshock, R.M., Stray-Gundersen, J., & Grundy, S.M. (1997). Relationship of anterior and posterior subcutaneous abdominal fat to insulin sensitivity in nondiabetic men. *Obes Res, 5,* 93-99.

Mokdad, A.H., Ford, E.S., Bowman, B.A., Dietz, W.H., Vinicor, F., Bales, V.S., & Marks, J.S. (2003). Prevalence of obesity, diabetes, and obesity-related health risk factors, 2001. *JAMA, 289,* 76-79.

Molarius, A., & Seidell, J. C. (1998). Selection of anthropometric indicators for classification of abdominal fatness—a critical review. *Int J Obes Relat Metab Disord, 22,* 719-727.

Molarius, A., Seidell, J.C., Sans, S., Tuomilehto, J., & Kuulasmaa, K. (1999). Varying sensitivity of waist action levels to identify subjects with overweight or obesity in 19 populations of the WHO MONICA Project. *J Clin Epidemiol, 52,* 1213-1224.

Molarius, A., Seidell, J.C., Visscher, T.L., & Hofman, A. (2000). Misclassification of high-risk older subjects using waist action levels established for young and middle-aged adults—results from the Rotterdam Study. *J Am Geriatr Soc, 48,* 1638-1645.

Mueller, W.H., & Stallones, L. (1981). Anatomical distribution of subcutaneous fat: skinfold site choice and construction of indices. *Hum Biol, 53,* 321-335.

Must, A., Dallal, G.E., & Dietz, W.H. (1991). Reference data for obesity: 85th and 95th percentiles of body mass index (wt/ht2) and triceps skinfold thickness. *Am J Clin Nutr, 53,* 839-846.

Must, A., Spadano, J., Coakley, E.H., Field, A.E., Colditz, G., & Dietz, W.B. (1999). The disease burden associated with overweight and obesity. *JAMA, 282,* 1523-1529.

Mykkanen, L., Laakso, M., & Pyorala, K. (1992). Association of obesity and distribution of obesity with glucose tolerance and cardiovascular risk factors in the elderly. *Int J Obes Relat Metab Disord, 16,* 695-704.

NHLBI. (1998). *Clinical guidelines on the identification, evaluation, and treatment of overweight and obesity in adults: The Evidence Report.* Washington, DC: NIH, National Heart, Lung, and Blood Institute.

Ogden, C.L., Flegal, K.M., Carroll, M.D., & Johnson, C.L. (2002a). Prevalence and trends in overweight among US children and adolescents, 1999-2000. *JAMA, 288,* 1728-1732.

Ogden, C.L., Kuczmarski, R.J., Flegal, K.M., Mei, Z., Guo, S., Wei, R., Grummer-Strawn, L.M., Curtin, L.R., Roche, A.F., & Johnson, C.L. (2002b). Centers for Disease Control and Prevention 2000 growth charts for United States: improvements to the 1977 National Center for Health Statistics version. *Pediatrics, 109,* 45-60.

Ohrvall, M., Berglund, L., & Vessby, B. (2000). Sagittal abdominal diameter compared with other anthropometric measurements in relation to cardiovascular risk. *Int J Obes Relat Metab Disord, 24,* 497-501.

Okosun, I.S., Liao, Y., Rotimi, C.N., Prewitt, T.E., & Cooper, R.S. (2000). Abdominal adiposity and clustering of multiple metabolic syndrome in white, black and Hispanic Americans. *Ann Epidemiol, 10,* 263-270.

Owens, S., Gutin, B., Ferguson, M., Allison, J., Karp, W., & Le, N.A. (1998). Visceral adipose tissue and cardiovascular risk factors in obese children. *J Pediatr, 133,* 41-45.

Pietrobelli, A., Faith, M.S., Allison, D.B., Gallagher, D., Chiumello, G., & Heymsfield, S.B. (1998). Body mass index as a measure of adiposity among children and adolescents: A validation study. *J Pediatr, 132,* 204-210.

Pintauro, S., Nagy, T.R., Duthie, C., & Moran, M.I. (1996). Cross validation of fat and lean measurement by dual energy X-ray absorptiometry to pig carcass assessments in paediatric body weight range. *Am J Clin Nutr, 63,* 293-298.

Poehlman, E.T., Toth, M.J., & Gardner, A.W. (1995). Changes in energy balance and body composition at menopause: a controlled longitudinal study. *Ann Intern Med, 123,* 673-675.

Pouliot, M.C., Després, J.P., Lemieux, S., Moorjani, S., Bouchard, C., Tremblay, A., Nadeau, A., & Lupien, P.J. (1994). Waist circumference and abdominal sagittal diameter: best simple anthropometric indexes of abdominal visceral adi-

pose tissue accumulation and related cardiovascular risk in men and women. *Am J Cardiol, 73,* 460-468.

Pouliot, M.C., Després, J.P., Nadeau, A., Moorjani, S., Prud'homme, D., Lupien, P.J., Tremblay, A., & Bouchard, C. (1992). Visceral obesity in men. Associations with glucose tolerance, plasma insulin, and lipoprotein levels. *Diabetes, 41,* 826-834.

Power, C., Lake, J.K., & Cole, T.J. (1997). Measurement and long-term health risks of child and adolescent fatness. *Int J Obes Relat Metab Disord, 21,* 507-526.

Rankinen, T., Kim, S.Y., Perusse, L., Després, J.P., & Bouchard, C. (1999). The prediction of abdominal visceral fat level from body composition and anthropometry: ROC analysis. *Int J Obes Relat Metab Disord, 23,* 801-809.

Reilly, J.J. (2002). Assessment of childhood obesity: national reference data or international approach? *Obes Res, 10,* 838-840.

Richelson, B., & Pederson, S.B. (1995). Associations between different anthropometric measurements of fatness and metabolic risk parameters in non-obese, healthy, middle-aged men. *Int J Obes Relat Metab Disord, 19,* 169-174.

Rimm, E.B., Stampfer, M.J., Giovannucci, E., Ascherio, A., Spiegelman, D., Colditz, G.A., & Willett, W.C. (1995). Body size and fat distribution as predictors of coronary heart disease among middle-aged and older US men. *Am J Epidemiol, 141,* 1117-1127.

Rissanen, P., Hamalainen, P., Vanninen, E., Tenhunen-Eskelinen, M., & Uusitupa, M. (1997). Relationship of metabolic variables to abdominal adiposity measured by different anthropometric measurements and dual-energy X-ray absorptiometry in obese middle-aged women. *Int J Obes Relat Metab Disord, 21,* 367-371.

Rolland-Cachera, M.F., Bellisle, F., Deheeger, M., Pequignot, F., & Sempe, M. (1990). Influence of body fat distribution during childhood on body fat distribution in adulthood: a two-decade follow-up study. *Int J Obes Relat Metab Disord, 14,* 473-481.

Ross, R., Shaw, K.D., Martel, Y., de Guise, J., & Avruch, L. (1993). Adipose tissue distribution measured by magnetic resonance imaging in obese women. *Am J Clin Nutr, 57,* 470-475.

Saito, Y., Kobayashi, J., Seimiya, K., Hikita, M., Takahashi, K., Murano, S., Bujo, H., & Morisaki, N. (1998). Contribution of visceral fat accumulation to postprandial hyperlipidemia in human obesity. *Int J Obes Relat Metab Disord,* Suppl 3, S226.

Sakamoto, Y., Nisizawa, M., Sato, T., Ohno, M., & Ikeda, Y. (1993). Measurement of body fat by bioelectrical impedance analysis. *Health Med, 8,* 38-41.

Sangi, H., & Mueller, W.H. (1991). Which measure of body fat distribution is best for epidemiologic research among adolescents? *Am J Epidemiol, 133,* 870-883.

Sangi, H., Mueller, W.H., Harrist, R.B., Rodriguez, B., Grunbaum, J.G., & Labarthe, D.R. (1992). Is body fat distribution associated with cardiovascular risk factors in childhood? *Ann Hum Biol, 19,* 559-578.

Sardinha, L.B., Going, S.B., Teixeira, P.J., & Lohman, T.G. (1999). Receiver operating characteristic analysis of body mass index, triceps skinfold thickness, and arm girth for obesity screening in children and adolescents. *Am J Clin Nutr, 70,* 1090-1095.

Sardinha, L.B., & Teixeira, P.J. (2000). Obesity screening in older women with the body mass index: a receiver operating characteristic (ROC) analysis. *Science and Sports, 15,* 212-219.

Sardinha, L.B., Teixeira, P.J., Guedes, D.P., Going, S.B., & Lohman, T.G. (2000). Subcutaneous central fat is associated with cardiovascular risk factors in men independently of total fatness and fitness. *Metabolism, 49,* 1379-1385.

Savva, S.C., Tornaritis, M., Savva, M.E., Kourides, Y., Panagi, A., Silikiotou, N., Georgiou, C., & Kafatos, A. (2000). Waist circumference and waist-to-height ratio are better predictors of cardiovascular disease risk factors in children than body mass index. *Int J Obes Relat Metab Disord, 24,* 1453-1458.

Schreiner, P.J., Terry, J.G., Evans, G.W., Hinson, W.H., Crouse, J.R., III, & Heiss, G. (1996). Sex-specific associations of magnetic resonance imaging-derived intra-abdominal and subcutaneous fat areas with conventional anthropometric indices. The Atherosclerosis Risk in Communities Study. *Am J Epidemiol, 144,* 335-345.

Schutz, Y., Kyle, U.U.G., & Pichard, C. (2002). Fat-free mass index and fat mass index percentiles in Caucasians aged 18-98 y. *Int J Obes Relat Metab Disord, 26,* 953-960.

Seidell, J.C., Andres, R., Sorkin, J.D., & Muller, D.M. (1994). The sagittal waist diameter and mortality in men: the Baltimore Longitudinal Study of Aging. *Int J Obes Relat Metab Disord, 18,* 61-67.

Seidell, J.C., Bakker, C.J.G., & Van Der Kooy, K. (1990). Imaging techniques for measuring adipose-tissue distribution—a comparison between computed tomography and 1.5-T magnetic resonance. *Am J Clin Nutr, 51,* 953-957.

Seidell, J.C., & Bouchard, C. (1997). Visceral fat in relation to health: is it a major culprit or simply an innocent bystander? *Int J Obes Relat Metab Disord, 21,* 626-631.

Seidell, J.C., & Bouchard, C. (1999). Abdominal adiposity and risk of heart disease. *JAMA, 281,* 2284-2285.

Seidell, J.C., Perusse, L., Després, J.P., & Bouchard, C. (2001). Waist and hip circumferences have independent and opposite effects on cardiovascular disease risk factors: the Quebec Family Study. *Am J Clin Nutr, 74,* 315-321.

Shen, W., Wang, Z., Punyanita, M., Lei, J., Sinav, A., Kral, J.G., Imielinska, C., Ross, R., & Heymsfield, S.B. (2003). Adipose tissue quantification by imaging methods: a proposed classification. *Obes Res, 11,* 5-16.

Sinaiko, A.R., Steinberger, J., Moran, A., Prineas, R.J., & Jacobs, D. (2002). Relation of insulin resistance to blood pressure in childhood. *J Hypertens, 20,* 509-517.

Sinha, R., Fisch, G., Teague, B., Tamborlane, W.V., Banyas, B., Allen, K., Savoye, M., Rieger, V., Taksali, S., Barbetta, G., Sherwin, R.S., & Caprio, S. (2002). Prevalence of impaired glucose tolerance among children and adolescents with marked obesity. *N Engl J Med, 346,* 802-810.

Siri, W.E. (1961). Body composition from fluid spaces and density: analysis of method. In: J. Brozek & A. Henschel (eds). *Techniques for measuring body composition.* Washington, DC, National Academy of Sciences, pp.223-244.

Sjostrom, L. (1991). A computer-tomography based multicompartment body composition technique and anthropometric predictions of lean body mass, total and subcutaneous adipose tissue. *Int J Obes Relat Metab Disord, 15 Suppl 2,* 19-30.

Snijder, M.B., Zimmet, P.Z., Visser, M., Dekker, J.M., Seidell, J.C., & Shaw, J.E. (2004). Independent and opposite associations of waist and hip circumferences with diabetes, hypertension and dyslipidemia: the AusDiab Study. *Int J Obes Relat Metab Disord, 28,* 402-429.

Sparrow, D., Borkan, G.A., Gerzof, S.G., Wisniewski, C., & Silbert, C.K. (1986). Relationship of fat distribution to glucose tolerance. Results of computed tomography in male participants of the Normative Aging Study. *Diabetes, 35,* 411-415.

Stevens, J. (2003). Ethnic-specific cutpoints for obesity vs country-specific guidelines for action. *Int J Obes Relat Metab Disord, 27,* 287-288.

Stevens, J., Cai, J., Pamuk, E.R., Williamson, D.F., Thun, M.J., & Wood, J.L. (1998). The effect of age on the association between body-mass index and mortality. *N Engl J Med, 338,* 1-7.

Stevens, J., & Nowicki, E. (2003). Body mass index and mortality in Asian populations: implications for obesity cutpoints. *Nutr Rev, 16,* 104-107.

Suter, E., & Hawes, M.R. (1993). Relationship of physical activity, body fat, diet, and blood lipid profile in youths 10-15 yr. *Med Sci Sports Exerc, 25,* 748-754.

Suwaidi, J.A., Higano, S.T., Hamasaki, S., Holmes, D.R., & Lerman, A. (2001). Association between obesity and coronary atherosclerosis and vascular remodeling. *Am J Cardiol, 88,* 1300-1303.

Svendsen, O.L., Haarbo, J., Hassager, C., & Christiansen, C. (1993). Accuracy of measurements of body composition by dual-energy x-ray absorptiometry in vivo. *Am J Clin Nutr, 57,* 605-608.

Svendsen, O.L., Hassager, C., Bergmann, I., & Christiansen, C. (1993). Measurement of abdominal and intra-abdominal fat in postmenopausal women by dual energy X-ray absorptiometry and anthropometry: comparison with computerized tomography. *Int J Obes Relat Metab Disord, 17,* 45-51.

Tanaka, S., Togashi, K., Rankinen, T., Pérusse, L., Leon, A.S., Rao, D.C., Skinner, J.S., Wilmore, J.H., & Bouchard, C. (2002). Is adiposity at normal body weight relevant for cardiovascular disease risk? *Int J Obes Relat Metab Disord, 26,* 176-183.

Taylor, R.W., Jones, I.E., Williams, S.M., & Goulding, A. (2000). Evaluation of waist circumference, waist-to-hip ratio, and the conicity index as screening tools for high trunk fat mass, as measured by dual-energy X-ray absorptiometry, in children aged 3-19 y. *Am J Clin Nutr, 72,* 490-495.

Teixeira, P.J., Sardinha, L.B., Going, S.B., & Lohman, T.G. (2001). Total and regional fat and serum cardiovascular disease risk factors in lean and obese children and adolescents. *Obes Res, 9,* 432-442.

Terry, R.B., Stefanick, M.L., Haskell, W.L., & Wood, P.D. (1991). Contributions of regional adipose tissue depots to plasma lipoprotein concentrations in overweight men and women: possible protective effects of thigh fat. *Metabolism, 40,* 733-740.

Thaete, F.L., Colberg, S.R., Burke, T., & Kelley, D.E. (1995). Reproducibility of computed tomography measurement of visceral adipose tissue area. *Int J Obes Relat Metab Disord, 19,* 464-467.

Tokunaga, K., Matsuzawa, Y., Ishikawa, K., & Tarui, S. (1983). A novel technique for the determination of body fat by computed tomography. *Int J Obes Relat Metab Disord, 7,* 437-445.

Treuth, M.S., Hunter, G.R., & Kekes-Szabo, T. (1995). Estimating intraabdominal adipose tissue in women by dual-energy X-ray absorptiometry. *Am J Clin Nutr, 62,* 527-532.

Turcato, E., Bosello, O., Di Francesco, V., Harris, T.B., Zoico, E., Bissoli, L., Fracassi, E., & Zamboni, M. (2000). Waist circumference and abdominal sagittal diameter as surrogates of body fat distribution in the elderly: their relation with cardiovascular risk factors. *Int J Obes Relat Metab Disord, 24,* 1005-1010.

USDA. (1990). *Dietary guidelines for Americans* (Publication 261-495/20124). Washington, DC: U.S. Department of Agriculture.

USDHHS. (2000). *The practical guide to the identification, evaluation, and treatment of overweight and obesity in adults.* Bethesda, MD: NIH.

Valdez, R., Seidell, J.C., Ahn, Y.I., & Weiss, K.M. (1993). A new index of abdominal adiposity as an indicator of risk for cardiovascular disease. A cross population study. *Int J Obes Relat Metab Disord, 17,* 77-82.

Van der Kooy, K., Leenen, R., Seidell, J.C., Deurenberg, P., & Visser, M. (1993). Abdominal diameters as indicators of visceral fat: comparison between magnetic resonance imaging and anthropometry. *Br J Nutr, 70,* 47-58.

Van der Kooy, K., & Seidell, J. C. (1993). Techniques for the measurement of visceral fat: a practical guide. *Int J Obes Relat Metab Disord, 17,* 187-196.

Van Itallie, T.B., Yang, M.U., Heysmsfield, S.B., Funk, R.C., & Boileau, R.A. (1990). Height-normalized indices of the body's fat-free mass and fat mass: potentially useful indicators of nutritional status. *Am J Clin Nutr, 52,* 953-959.

Van Loan, M.D. (1996). Body fat distribution from subcutaneous to intraabdominal: a perspective. *Am J Clin Nutr, 64,* 787-788.

Visscher, T.L., Seidell, J.C., Molarius, A., van der Kuip, D., Hofman, A., & Witteman, J.C. (2001). A comparison of body mass index, waist-hip ratio and waist circumference as predictors of all-cause mortality among the elderly: the Rotterdam study. *Int J Obes Relat Metab Disord, 25,* 1730-1735.

Visser, M. (2001). Higher levels of inflammation in obese children. *Nutrition, 17,* 480-484.

Wajchenberg, B.L. (2000). Subcutaneous and visceral adipose tissue: their relation to the metabolic syndrome. *Endocr Rev, 21,* 697-738.

Walton, C., Lees, B., Crook, D., Worthington, M., Godsland, I.F., & Stevenson, J.C. (1995). Body fat distribution, rather than overall adiposity, influences serum lipids and lipoproteins in healthy men independent of age. *Am J Med, 99,* 459-464.

Wang, Z.-M., Pierson, R.N., Jr, & Heymsfield, S.B. (1992). The five-level model: a new approach to organizing body-composition research. *Am J Clin Nutr, 56,* 19-28.

Ward, K.D., Sparrow, D., Vokonas, P.S., Willett, W.C., Landsberg, L., & Weiss, S.T. (1994). The relationships of abdominal obesity, hyperinsulinemia and saturated fat intake to serum lipid levels: the Normative Aging Study. *Int J Obes Relat Metab Disord, 18,* 137-144.

Washino, K., Takada, H., Nagashima, M., & Iwata, H. (1999). Significance of the atheroclerogenic index and body fat in children as markers for future, potential coronary heart disease. *Pediatr Intern, 41,* 260-265.

Wassertheil-Smoller, S., Fann, C., Allman, R.M., Black, H.R., Camel, G.H., Davis, B., Masaki, K., Pressel, S., Prineas, R.J., Stamler, J., & Vogt, T.M. (2000). Relation of low body mass to death and stroke in the systolic hypertension in the elderly program. *Arch Intern Med, 160,* 494-500.

Wattigney, W.A., Harsha, D.W., Srinivasan, S.R., Webber, L.S., & Berenson, G.S. (1991). Increasing impact of obesity on serum lipids and lipoproteins in young adults. The Bogalusa Heart Study. *Arch Intern Med, 151,* 2017-2022.

Wellens, R., Roche, A., Khamis, H., Jackson, A., Pollock, M., & Siervogel, R. (1996). Relationship between the body mass index and body composition. *Obes Res, 4,* 35-44.

Wells, J.C.K. (2001). A critique of the expression of paediatric body composition data. *Arch Dis Child, 85,* 67-72.

Wells, J.C.K., & Cole, T.J. (2002). Adjustment of fat-free mass and fat mass for height in children aged 8 y. *Int J Obes Relat Metab Disord, 26,* 947-952.

Wells, J.C.K., Coward, W.A., Cole, T.J., & Davies, P.S.W. (2002). The contribution of fat and fat-free tissue to body mass index in contemporary children and the reference child. *Int J Obes Relat Metab Disord, 26,* 1323-1328.

WHO. (1995). *Physical status: the use and interpretation of anthropometry.* Geneva: World Health Organization.

WHO. (1997). *Obesity: preventing and managing the global epidemic* (WHO/NUT/NCD/98.1). Report of a WHO Consultation prevented at the World Health Organization, Geneva.

WHO. (1998). *Obesity: preventing and managing the global epidemic.* Geneva: World Health Organization.

Williams, D.P., Going, S.B., Lohman, T., Harsha, D.W., Srinivasan, S.R., Webber, L.S., & Berenson, G.S. (1992). Body fatness and risk for elevated blood pressure, total cholesterol, and serum lipoprotein ratios in children and adolescents. *Am J Public Health, 82,* 358-363.

Williams, M.J., Hunter, G.R., Kekes-Szabo, T., Treuth, M.S., Snyder, S., Berland, L., & Blaudeau, T. (1996). Intra-abdominal adipose tissue cutpoints related to elevated cardiovascular risk in women. *Int J Obes Relat Metab Disord, 20,* 613–617.

Wilson, P.W.F., D'Agostino, R.B., Sullivan, L., Parise, H., & Kannel, W.B. (2002). Overweight and obesity as determinants of cardiovascular risk. *Arch Intern Med, 162,* 1867-1872.

Woo, J., Ho, S.C., Yu, A.L., & Sham, A. (2002). Is waist circumference a useful measure in predicting health outcomes in the elderly? *Int J Obes Relat Metab Disord, 26,* 1349-1355.

Wright, C.M., Booth, I.W., Buckler, J.M., Cameron, N., Cole, T.J., Healy, M.J., Hulse, J.A., Preece, M.A., Reilly, J.J., & Williams, A.F. (2002). Growth reference charts for use in the United Kingdom. *Arch Dis Child, 86,* 11-14.

Wu, C.H., Yao, W.J., Lu, F.H., Yang, Y.C., Wu, J.S., & Chang, C.J. (2001). Sex differences of body fat distribution and cardiovascular dysmetabolic factors in old age. *Age Ageing, 30,* 331-336.

Wu, D.M., Chu, N.F., Shen, M.H., & Chang, J.B. (2003). Plasma C-reactive protein levels and their relationship to anthropometric and lipid characteristics among children. *Int J Epidemiol, 56,* 94-100.

Zamboni, M., Armellini, F., Harris, T., Turcato, E., Micciolo, R., Bergamo-Andreis, I.A., & Bosello, O. (1997). Effects of age on body fat distribution and cardiovascular risk factors in women. *Am J Clin Nutr, 66,* 111-115.

Zhu, S., Heshka, S., Wang, Z., Shen, W., Allison, D.B., Ross, R., & Heymsfield, S.B. (2004). Combination of BMI and waist circumference for identifying cardiovascular risk factors in whites. *Obes Res, 12,* 633-645.

Chapter 14

Boddy, K., Holloway, I., & Elliot, A. (1973). A simple facility for total body in vivo neutron activation analysis. *International Journal of Applied Radiation, 24,* 428-430.

Boling, E.A., Taylor, W.L., Entenman, C., & Behnke, A.R. (1962). Total exchangeable potassium and chloride, and total body water in healthy men of varying fat content. *Journal of Clinical Investigation, 41,* 1840-1849.

Borsook, H., & Dubnoff, J.W. (1947). The hydrolysis of phosphocreatine and the origin of urinary creatinine. *Journal of Biological Chemistry, 168*, 493-510.

Brown, B.H., Karatzas, T., Nakielny, R., & Clark, R.G. (1988). Determination of upper arm muscle and fat areas using electrical impedance measurements. *Clinical Physics and Physiological in Measurements, 9*, 47-55.

Bulcke, J.A.L., Termote, J.-L., Palmers, Y., & Crolla, D. (1979). Computed tomography of the human skeletal muscular system. *Neuroradiology, 17*, 127-136.

Burkinshaw, L. (1987). Models of the distribution of protein in the human body. In K.J. Ellis, S. Yasumura, & W.D. Morgan (Eds.), *In vivo body composition studies* (pp. 15-24). London: Institute of Physical Sciences in Medicine.

Burkinshaw, L., Hedge, A.P., King, R.F.J.G., & Cohn, S.H. (1990). Models of the distribution of protein, water and electrolytes in the human body. *Infusiontherapie, 17* (*suppl 3*), 21-25.

Burkinshaw, L., Hill, G.L., & Morgan, D.B. (1978). Assessment of the distribution of protein in the human body by in vivo neutron activation analysis. In *International symposium on nuclear activation techniques in the life sciences* (pp. 787-798). Vienna: International Atomic Energy Association (IAEA-SM-227/39).

Calloway, D.H., & Margen, S. (1971). Variation in endogenous nitrogen excretion and dietary nitrogen utilization as determinants of human protein requirement. *Journal of Nutrition, 101*, 205-216.

Cheek, D.B. (1968). *Human growth: body composition, cell growth, energy and intelligence*. Philadelphia: Lea & Febiger.

Clarys, J.P., Martin, A.D., & Drinkwater, D.T. (1984). Gross tissue weights in the human body by cadaver dissection. *Human Biology, 56*, 459-473.

Cohn, S.H., & Dombrowski, C.S. (1971). Measurement of total body calcium, sodium, chlorine, nitrogen, and phosphorus in man by neutron activation analysis. *Journal of Nuclear Medicine, 12*, 499-505.

Cohn, S.H., Gartenhaus, W., Sawitsky, A., Rai, K., Zanzi, I., Vaswani, A., Ellis, K.J., Yasumura, S., Cortes, E., & Vartsky, D. (1980a). Compartmental body composition of cancer patients by measurement of total body nitrogen, potassium and water. *Metabolism, 30*, 222-229.

Cohn, S.H., Vartsky, D., Yasumura, S., Sawitsky, A., Zanzi, I., Vaswani, A., & Ellis, K.J. (1980b). Compartmental body composition based on total body nitrogen, potassium and calcium. *American Journal of Physiology, 239*, E524-E530.

Crim, M.C., Calloway, D.H., & Margen, S. (1975). Creatine metabolism in men: urinary creatine and creatinine excretions with creatine feedings. *Journal of Nutrition, 105*, 428-438.

Dixon, A.K. (1991). Imaging techniques in nutrition and the assessment of bone status: computed tomography. In R.G. Whitehead & A. Prentice (Eds.), *New techniques in nutritional research* (pp. 361-377). San Diego: Academic Press.

Elia, M., Carter, A., & Smith, R. (1979). The 3-methylhistidine content of human tissues. *British Journal of Nutrition, 42*, 567-570.

Fiatarone, M.A., Marks, E.C., Ryan, N.D., Meredith, C.N., Lipsitz, L.A., & Evans, W.E. (1990). High-intensity strength training in nonagenarians. *Journal of the American Medical Association, 263*, 3029-3034.

Folin, O. (1905). Laws governing the chemical composition of urine. *American Journal of Physiology, 13*, 66-115.

Forbes, G.B. (1987). *Human body composition*. New York: Springer-Verlag.

Forbes, G.B., & Bruining, G.J. (1976). Urinary creatinine excretion and lean body mass. *American Journal of Clinical Nutrition, 29*, 1359-1366.

Fuller, M.F., Fowler, P.A., McNeill, G., & Foster, M.A. (1994). Imaging techniques for the assessment of body composition. *Journal of Nutrition, 124*, 1546S-1550S.

Fuller, N.J., Hardington, C.R., Graves, M., Screaton, N., Dixon, A.K., Ward, L.C., & Elia, M. (1999). Predicting composition of leg sections with anthropometry and bioelectrical impedance analysis using magnetic resonance imaging as a reference. *Clinical Science, 96*, 647-657.

Greenblatt, D.C., Ransil, B.J., Harmatz, J.S., Smith, T.W., Dehme, D.W., & Koch-Weser, J. (1976). Variability of 24-hr urinary creatinine excretion by normal subjects. *Journal of Clinical Pharmacology, 16*, 321-328.

Gurney, L.M, & Jelliffe, D.B. (1973). Arm anthropometry in nutritional assessment: nomogram for rapid calculation of muscle circumference and cross-sectional muscle over fat areas. *American Journal of Clinical Nutrition, 26*, 912-915.

Hansen, G., Crooks, L.E., & Marguilis, A.R. (1980). In vivo imaging of the rat with nuclear magnetic resonance. *Radiology, 136*, 695-700.

Haus, A.G. (1979). *The physics of medical imaging: recording system measurements and techniques*. New York: American Institute of Physics.

Heymsfield, S.B. (1987). Human body composition: analysis by computerized axial tomography and nuclear magnetic resonance. In O.E. Levander (Ed.), *AIN symposium proceedings: nutrition '87* (pp. 92-96). Bethesda, MD: American Institute of Nutrition.

Heymsfield, S.B., McMannus, C., Smith, J., Stevens, V., & Nixon, D.W. (1982a). Anthropometric measurements of muscle mass: revised equations for calculating bone-free arm muscle area. *American Journal of Clinical Nutrition, 36*, 680-690.

Heymsfield, S.B., McMannus, C., Stevens, V., & Smith, J. (1982b). Muscle mass: reliable indicator of protein energy malnutrition severity and outcome. *American Journal of Clinical Nutrition, 35*, 1192-1199.

Heymsfield, S.B., Olafson, R.P., Kutner, M.H., & Nixon, D.W. (1979). A radiographic method of quantifying protein-calorie undernutrition. *American Journal of Clinical Nutrition, 32*, 693-702.

Heymsfield, S.B., Smith, R., Aulet, M., Bensen, B., Lichtman, S., Wang, J., & Pierson, R.N. (1990). Appendicular skeletal muscle mass: measurement by dual-photon absorptiometry. *American Journal of Clinical Nutrition, 52*, 214-218.

Hoberman, H.D., Sims, E.A.H., & Peters, J.H. (1948). Creatine and creatinine metabolism in the normal male studied with the aid of isotopic nitrogen. *Journal of Biological Chemistry, 172*, 45-58.

Horber, F.F., Thomi, F., Casez, J.P., Fonteille, J., & Jaeger, P. (1992). Impact of hydration status on body composition as measured by dual energy x-ray absorptiometry in normal volunteers and patients on haemodialysis. *British Journal of Radiology, 65*, 895-900.

International Commission on Radiological Protection (ICRP). (1975). *Report of the task group on reference man* (pp. 108-112). Oxford: Pergamon Press.

James, H.M., Dabek, J.T., Chettle, D.R., Dykes, P.W., Fremlin, J.H., Hardwick, J., Thomas, B.J., & Vartsky, D. (1984). Whole body cellular and collagen nitrogen in healthy and wasted man. *Clinical Science*, 67, 73-82.

Janssen, I. Heymsfield, S.B., Baumgartner, R.N., Ross, R. (2000). Estimation of skeletal muscle mass by bioelectrical impedance analysis. *Journal of Applied Physiology*, 89, 465-471.

Jelliffe, D.B. (1966). *The assessment of the nutritional status of the community* (WHO monograph series no. 53). Geneva: World Health Organization.

Jelliffe, E.F.P., & Jelliffe, D.B. (1969). The arm circumference as a public health index of protein-calorie malnutrition of early childhood. *Journal of Tropical Pediatrics, 15,* 225-230.

Johnson, P., Harris, C.I., & Perry, S.V. (1997). 3-Methylhistidine in actin and other muscle proteins. *Biochemical Journal,* 105, 361-370.

Kellie, S.E. (1992). Measurement of bone density with dual-energy x-ray absorptiometry (DXA). *Journal of the American Medical Association, 267,* 286-294.

Lee, R.C., Wang, Z.M., Heo, M., Ross, R., Janssen, I., & Heymsfield, S.B. (2000). Total-body skeletal muscle mass: development and cross-validation of anthropometric prediction models. *American Journal of Clinical Nutrition, 72,* 796-803.

Lewis, D.S., Rollwitz, W.L., Bertrand, H.A., & Masoro, E.J. (1986). Use of NMR for measurement of total body water and fat. *Journal of Applied Physiology, 60,* 836-840.

Lukaski, H.C. (1987). Methods for the assessment of human body composition: traditional and new. *American Journal of Clinical Nutrition, 46,* 537-556.

Lukaski, H.C. (1991). Assessment of body composition using tetrapolar bioelectrical impedance analysis. In R.G. Whitehead & A. Prentice (Eds.), *New techniques in nutritional research* (pp. 303-315). San Diego: Academic Press.

Lukaski, H.C. (1992). Methodology of body composition studies. In J.C. Watkins, R. Roubenoff, & I.H. Rosenberg (Eds.), *Body composition: the measure and meaning of changes with aging* (pp. 13-24). Boston: Foundation for Nutritional Advancement.

Lukaski, H.C. (1993). Soft tissue composition and bone mineral status: evaluation by dual energy x-ray absorptiometry. *Journal of Nutrition, 123,* 438-443.

Lukaski, H.C. (2000). Assessing regional muscle mass with segmental measurements of bioelectrical impedance in obese women during weight loss. *Annals New York Academy of Science*, 904, 154-158.

Lukaski, H.C., & Mendez, J. (1980). Relationship between fat-free weight and urinary 3-methylhistidine in man. *Metabolism, 29,* 758-761.

Lukaski, H.C., Mendez, J., Buskirk, E.R., & Cohn, S.H. (1981a). A comparison of methods of assessment of body composition including neutron activation analysis of total body nitrogen. *Metabolism,* 30, 777-782.

Lukaski, H.C., Mendez, J., Buskirk, E.R., & Cohn, S.H. (1981b). Relationship between endogenous 3-methylhistidine excretion and body composition. *American Journal of Physiology,* 240, E302-E307.

Lykken, G.I., Jacob, R.A., Munoz, J.M., & Sandstead, H.H. (1980). A mathematical model of creatine metabolism in normal males—comparison between theory and experiment. *American Journal of Clinical Nutrition,* 33, 2674-2685.

Ma, R., Wang, Z., Gallagher, D., Yasumura, S., Pierson, R.N., Wang, J., Kotler, D., & Heymsfield, S.B. (1994). Measurement of skeletal muscle in vivo: comparison of results from radiographic and neutron activation methods. *Federation of American Societies for Experimental Biology,* 8, A279 (abstract 1611).

Martin, A.D., Spenst, L.F., Drinkwater, D.T., & Clarys, J.P. (1990). Anthropometric estimation of muscle mass. *Medicine and Science in Sports and Exercise,* 22, 729-733.

Mategrano, V.C., Petasnick, J.P., Clark, J.W., Bin, A.C., & Weinstein, R. (1977). Attenuation values in computed tomography of the abdomen. *Radiology,* 125, 135-140.

Mateiga, J. (1921). The testing of physical efficiency. *American Journal of Physical Anthropology,* 4, 223-230.

Michaelsen, K.F., Wellens, R., Roche, A.F., Northeved, A., Culmsee, J., Boska, M., Guo, S., & Siervogel, R.M. (1993). The use of high frequency energy absorption to measure limb musculature. In K.J. Ellis & J.D. Eastman (Eds.), *Human body composition in vivo methods, models and assessment* (pp. 359-362). New York: Plenum Press.

Miyatani, M., Kanehisa, H., & Fukunaga, T. (2000). Validity of bioelectrical impedance and ultrasonographic methods for estimating the muscle volume of the upper arm. *European Journal of Applied Physiology,* 82, 391-396.

Miyatani, M., Kanehisa, H., Masuo, Y., Ito, M., & Fukunaga, T. (2001). Validity of estimating limb muscle volume by bioelectrical impedance. *Journal of Applied Physiology,* 91, 386-394.

Nord, R.H., & Payne, R.K. 1990. Standards for body composition calibration in DEXA. In E.F.J. Ring (Ed.), *Current research in osteoporosis and bone mineral measurement* (pp. 27-28). London: British Institute of Radiology.

Nunez, C., Gallagher, D., Grammes, J., Baumgartner, R.N., Ross, R., Wang, Z.M., Thornton, J. & Heymsfield, S.B. (1999). Bioimpedance analysis: potential for measuring lower limb skeletal muscle mass. *Journal of Parenteral and Enteral Nutrition,* 23, 96-103.

Nunez, C., Gallagher, D., Visser, M., Pi-Sunyer, F.X., Wang, Z., & Heymsfield, S.B. (1997). Bioimpedance analysis: evaluation of leg-to-leg system based on pressure contact foot-pad electrodes. *Medicine and Science in Sports and Exercise,* 29, 524-531.

Ransil, B.J., Greenblatt, D.J., & Koch-Weser, J. (1977). Evidence for systematic temporal variation in 24-hr urinary creatinine excretion. *Journal of Clinical Pharmacology,* 17, 108-119.

Rennie, M.J., & Millward, D.J. (1983). 3-Methylhistidine excretion and the urinary 3-methylhistidine/creatinine ratio are poor indicators of skeletal muscle protein breakdown. *Clinical Science,* 65, 217-225.

Ross, R., Leger, L., Morris, D., de Guise, J., & Guardo, R. (1992). Quantification of adipose tissue by MRI: relationship with anthropometric variables. *Journal of Applied Physiology,* 72, 787-795.

Ross, R., Shaw, K.D., Rissanen, J., Martel, Y., de Guise, J., & Avruch, L. (1994). Sex differences in lean and adipose tissue distribution by magnetic resonance imaging: anthropometric relationships. *American Journal of Clinical Nutrition,* 59, 1277-1285.

Talbot, N.B. (1938). Measurement of obesity by the creatinine coefficient. *American Journal of Diseases of Children,* 55, 42-50.

Tomas, F.M., Ballard, F.J., & Pope, L.M. (1979). Age-dependent changes in the rate of myofibrillar protein degradation in humans as assessed by 3-methylhistidine and creatinine excretion. *Clinical Science,* 56, 341-346.

Utter, A.C., Nieman, D.C., Ward, A.N., & Butterworth, D.E. (1999). Use of the leg-to-leg bioelectrical impedance method in assessing body composition change in obese women. *American Journal of Clinical Nutrition, 69*, 603-607.

Vartsky, D., Ellis, K.J., Vaswani, A.N., Yasumura, S., & Cohn, S.H. (1984). An improved calibration for the in vivo determination of body nitrogen, hydrogen, and fat. *Physics in Medicine and Biology, 29*, 209-218.

Wang, Z.M., Elam, R., Ma, R., Matthews, D., Nelson, M., Pierson, R.N., & Heymsfield, S.B. (1993). Validation of indirect skeletal muscle mass methods by computerized axial tomography. *Federation of American Societies for Experimental Biology, 7*, A83 (abstract 477).

Wassner, S.J., & Li, J.B. (1982). N-τ-Methylhistidine release: contributions of rat skeletal muscle, GI tract, and skin. *American Journal of Physiology, 243*, E293-E297.

Young, V.R., & Munro, H.N. (1978). N-τ-Methylhistidine (3-methylhistidine) and muscle protein turnover: an overview. *Federation Proceedings, 37*, 2291-2300.

Chapter 15

Abt I. Growth and development. Pediatrics, vol. I. London: Saunders, 1923.

Albu J, Shur M, Curi M, Murphy L, Heymsfield S, Pi-Sunyer FX. Resting metabolic rate in African-American Women. Am J Clin Nutr 1997;66:531-538.

Allison DB, Paultre F, Goran MI, Poehlman ET, Heymsfield SB. Statistical considerations regarding the use of ratios to adjust data. Int J Obes Relat Metab Disord 1995;19: 644-652.

Altman P, Dittmer D. Growth—including reproduction and morphological development. Washington DC: Federation of American Societies for Experimental Biology, 1962.

Amador M, Bacallao J, Hermelo M. Adiposity and growth: relationship of stature at fourteen years with relative body weight at different ages and serial measures of adiposity and body bulk. Euro J Clin Nutr 1992;46:213-219.

Bean BR. Composite study of weight of vital organs in man. Am J Phys Anthropol 1926;9:293-300.

Bertalanfey von L, Pirozynski WJ. Tissue respiration and body size. Science 1951;113:559-600.

Blanc S, Schoeller D, Kemnitz J, Weindruch R, Colman R, Newton W, Wink K, Baum S, Ramsey J. Energy expenditure of rhesus monkeys subjected to 11 years of dietary restriction. J Clin Endocrinol Metabol 2003;88:16-23.

Bogardus C, Lillioja S, Ravussin E, Abbott W, Zawadzki JK, Young A, Knowler WC, Jacobowitz R, Moll PP. Familial dependence of the resting metabolic rate. New Engl J Med 1986;315:96-100.

Boothby W, Berkson MJ, Dunn HL. Studies of the energy of normal individuals: a standard for basal metabolism, with a nomogram for clinical application. Am J Physiol 1936;116: 468-484.

Boxt LM, Katz J, Kolb T, Czegledy FP, Barst RJ. Direct quantitation of right and left ventricular volumes with nuclear magnetic resonance imaging in patients with primary pulmonary hypertension. J Am Coll Cardiol. 1992;19: 1508-1515.

Boyd E. In: Altman P, Dittmer D, eds. Growth, including reproduction and morphological development. Washington, DC: Biological Handbooks, FASEB, 1962, pp 346-348.

Carpenter WH, Fonong T, Toth MJ, Ades PJ, Calles-Escandon J, Walston JD, Poehlman ET. Total daily energy expenditure in free-living older African-Americans and Caucasians. Am J Physiol 1998;274:E96-E101.

Caviness VS, Kennedy DN, Richelme C, Rademacher J, Filipek PA. The human brain age 7-11 years: a volumetric analysis based on magnetic resonance images. Cerebral Cortex 1996;6:726-736

Creasey H, Rapoport SI. The aging human brain. Ann Neurol 1985;17:2-10.

Dekaban AS, Sadowsky D. Changes in brain weights during the span of human life: relation of brain weights to body heights and body weights. Ann Neurol 1978;4:345-356.

DeLany JP, Hansen BC, Bodkin NL, Hannah J, Bray GA. Long-term calorie restriction reduces energy expenditure in aging monkeys. J Gerontol Biol Sci 1999;54A:B5-B11.

DeSimone M, Farello G, Palumbo M, Gentile T, Ciuffreda M, Olioso P, Cirque M, DeMateis F. Growth charts, growth velocity, and bone measurement in childhood obesity. Int J Obes Related Metab Disord 1995;19:851-857.

Devereux RB, Reichek N. Echocardiographic determination of left ventricular mass in man. Anatomic validation of the method. Circulation 1977;55:613-618.

Dietze G, Wicklmayr M, Scifman R, Hellmut M. Metabolic fuels in fasting. Excerpta Medica International Congress Series 1980;314-320.

Dolan RJ, Mitchell J, Wakeling A. Structural brain changes in patients with anorexia nervosa. Psychol Med 1988;18:349-353.

Duara R, Margolin RA, Robertson-Tchabo EA, London ED, Schwartz M, Renfrew JW, Koziarz BJ, Sundaram M, Grady C, Moore AM, Ingvar DH, Sokoloff L, Weingartner H, Kessler RM, Manning RG, Channing MA, Cutler NR, Rapoport SI. Cerebral glucose utilization, as measured with positron emission tomography in 21 resting healthy men between the ages of 21 and 83 years. Brain 1983;106:761-775.

Dulloo AG, Jacquet J, Girardier L. Autoregulation of body composition during weight recovery in human: the Minnesota experiment revisited. Int J Obes 1996;20:393-405.

Elia M. Organ and tissue contribution to metabolic rate. In: Kinney JM, Tucker HN, eds. Energy metabolism. Tissue determinants and cellular corollaries. New York: Raven Press, 1992a, pp 61-77.

Elia M. Effect of starvation and very low calorie diets on protein energy interrelationships in lean and obese subjects. In: Scrimshaw NS and Schürch B, eds. Protein: Energy Interactions IDECG, Lausanne, 1992b, pp 249-284.

Elia M. Metabolism and nutrition of the gastrointestinal tract. In: Bindels JG, Goedhart AC, Visser HKA, eds. Recent developments in infant nutrition. Tenth Nutricia Symposium. Boston: Kluwer Academic, 1996, pp 318-348.

Elia M. Tissue distribution and energetics in weight loss and undernutrition. In: Kinney JM, Tucker HN, eds. Physiology, stress, and malnutrition. Philadelphia: Raven Press, 1997, pp 383-411.

Elia M. Metabolic response to starvation, injury, and sepsis. In: Payne-James J, Grimble G, Silk D, eds. Artificial nutritional support in clinical practices. Edward Arnold: London, 2001, pp 1-24.

Elia M. Special nutritional problems and the use of enteral and parenteral nutrition. In: Warrell DA, Cox TM, Firth JD, Benz EJ, eds. The Oxford textbook of medicine. Oxford, UK: Oxford University Press, 2003, pp 1073-1085.

Elia M, Livesey G. Energy expenditure and fuel selection in biological systems: the theory and practice of calculations based on indirect calorimetry and tracer methods. World Rev Nutr Diet 1992;70:68-131.

Forbes GB. Nutrition and growth. J Pediatr 1977;91:40-92.

Forbes G. Human body composition. New York: Springer-Verlag, 1987.

Foster GD, Wadden TA, Vogt RA. Resting energy expenditure in obese African American and Caucasian women. Obes Res 1997;5:1-6.

Fox PT, Raichle ME. Focal uncoupling of cerebral blood flow and oxidative metabolism during somato-sensory stimulation in human subjects. Proc Natl Acad Sci USA 1986;83:1140-1144.

Fox PT, Raichle ME, Mintnar MA, Dence C. Non-oxidative glucose consumption during focal physiologic activity. Science 1988;41:462-464.

Frackowiak RS, Wise RJ, Gibbs JM, Jones T, Leenders N. Oxygen extraction in the aging brain. Monogr Neural Sci 1984;11:118-122.

Fukagawa NK, Bandini LG, Young JB. Effect of age on body composition and resting metabolic rate. Am J Physiol 1990;259:E233-E238.

Gallagher D, Albu J, Kovera A, Janumala F, Wong S, Heshka S. Smaller mass of high metabolically active organ-tissues in African-American women. Obes Res 2000;8:Suppl;O13.

Gallagher D, Belmonte D, Deurenberg P, Wang, Z, Krasnow, N, Pi-Sunyer, FX, Heymsfield, SB. Organ-tissue mass measurement allows modeling of REE and metabolically active tissue mass. Am J Physiol 1998;275:E249-E258.

Gallagher D, Kovera AJ, Clay-Williams G, Agin D, Albu LJ, Matthews DE, et al. Weight loss in post-menopausal women: no evidence of adverse alterations in body composition. Am J Physiol Endocrinol Metab 2000;279:124-131.

Garby L, Lammert O, Kock K, Thobo-Carlsen B. Weights of brain, heart, liver, kidneys, and spleen in healthy and apparently healthy Danish subjects. Am J Hum Biol 1993;5:291-296.

Giedd J. Brain development, IX: human brain growth. Am J Psychiatry 1999;156:4.

Golden NH, Ashtari M, Kohn MR, Patel M, Jacobson MS, Fletcher A, Shenker IR. Reversibility of cerebral ventricular enlargement in anorexia nervosa, demonstrated by quantitative magnetic resonance imaging. J Pediatr 1996 128;2:296-301.

Gonzales-Pacheco DM, Buss WC, Koehler KM, Woodside WF, Alpert SS. Energy restriction reduces metabolic rate in adult male Fisher-344 rats. J Nutr 1993;123:90-97.

Grande F. Energy expenditure of organs and tissues. In: Kinney JN, ed. Assessment of energy metabolism in health and disease. Columbus, OH: Ross Laboratories, 1980, pp 88-92.

Haddad S, Restieri, C, Krishnan, K. Characterization of age-related changes in body weight and organ weights from birth to adolescence in humans. J Toxicol Environ Health 2001;64:453-464.

Harper C, Mina L. A comparison of Australian Caucasian and Aboriginal brain weights. Clin Exp Neurol 1981;18:44-51.

Heshka S, Yang MU, Wang J, Burt P, Pi-Sunyer FX. Weight loss and change in resting metabolic rate. Am J Clin Nutr 1990;52:981-986.

Heymsfield SB, Allison DB, Pi-Sunyer FX, and Sun Y. Columbia respiratory chamber indirect calorimeter: a new approach to air modelling. Med Biol Eng Comput 32:406-410, 1994.

Holliday MA, Potter D, Jarrah A, Bearg S. The relation of metabolic rate to body weight and organ size. Pediatr Res 1967;1:185-195.

Holliday MA. Metabolic rate and organ size during growth from infancy to maturity and during late gestation and early infancy. Pediatrics 1971;47:169-179.

Holliday MA. Body composition and energy needs during growth. In: Fulkner F, Tanner JM, eds. Human growth: a comprehensive treatise. New York: Plenum Press, 1986, pp 101-117.

Hsu A, Heshka S, Janumala I, Song M, Horlick M, Krasnow N, Gallagher D. Larger mass of high metabolic rate organs does not explain higher REE in children. Am J Clin Nutr 2003;77:1506-11.

Hunter GR, Weinsier RL, Darnell BE, Zuckerman P, Goran MI. Racial differences in energy expenditure and aerobic fitness in premenopausal women. Am J Clin Nutr 2000;71:500-506.

Illner K, Brinkman G, Heller M, Bosy-Westphal A, Muller M. Metabolically active components of fat free mass and resting energy expenditure in nonobese adults. Am J Physiol Endocrinol Metab 2000;278:E308-E315.

Inoue T, Otsu S. Statistical analysis of the organ weight in 1,000 autopsy cases of Japanese aged over 60 years. Acta Pathol Jpn 1987;37:343-359.

International Commission on Radiological Protection. Reference Man: Anatomical, physiological, and metabolic characteristics (ICRP Publication 23). New York: Pergamon Press, 1975.

Jahn DJ, DeMaria A, Kisslo J, Weyman A. Recommendations regarding quantification in M-mode echocardiography: results of a survey of echocardiographic measurements. Circulation 58:1072-1083, 1978.

Jain SC, Metha S, Kumar B, Reddy A, Nagaratnam A. Formulation of the reference Indian adult: anatomic and physiological data. Health Phys 1995;68:509-522.

Jernigan TL, Tallal P. Late childhood changes in brain morphology observable with MRI. Dev Med Child Neurol 1990;32:379-385.

Kaplan AS, Zemel BS, Stallings VA. Differences in resting energy expenditure in prepubertal black children and white children. J Pediatr 1996;129:643-647.

Katz J, Whang J, Boxt LM, Barst RJ. Estimation of right ventricular mass in normal subjects and in patients with primary pulmonary hypertension by nuclear magnetic resonance imaging. J Am Coll Cardiol 1993;21:1475-1481.

Katzman DK, Lambe EK, Mikulis DJ, Ridgley JN, Goldbloom DS, Zipursky RB. Cerebral gray matter and white matter volume deficits in adolescent girls with anorexia nervosa. J Pediatr1996;129:794-803.

Kennedy C, Sokoloff L. An adaptation of the nitrous oxide to the study of the cerebral circulation of children: normal values for cerebral blood flow and cerebral metabolic rate in childhood. J Clin Invest 1957;36:1130-1137.

Keys A, Brozek J, Henschel A, Mickelson O, Taylor HL. The biology of human starvation. Minneapolis: University of Minnesota Press, 1950, pp 81-535.

Keys A, Taylor HL, Grange F. Basal metabolism and age of adult man. Metabolism 1973;22:579-587.

Kingston K, Szmukler G, Andrewes D, Tress B, Desmond P. Neuropsychological and structural brain changes in anorexia nervosa before and after refeeding. Psychol Med 1996;26:15-28.

Kinney JM, Lister J, Moore FD. Relationship of energy expenditure to total exchangeable potassium. Ann New York Acad Sci 1963;110:711-722.

Klausen B, Toubro S, Astrup A. Age and sex effects on energy expenditure. Am J Clin Nutr 1997;65:895-907.

Kleiber M. Body size and metabolism. Hilgardia 1932;6:315.

Kleiber M. Body size and metabolic rate. Physiol Rev 1947;27:511.

Kleiber M. The fire of life, and introduction to animal energetics. New York: Wiley, 1961.

Kohn MR, Ashtari M, Golden NH, Schebendach J, Patel M, Jacobson MS, Shenker IR. Structural brain changes and malnutrition in anorexia nervosa. Ann NY Acad Sci 1997;817:398-399.

Krahn DD, Rock C, Dechert RE, Nairn KK, Hasse SA. Changes in resting energy expenditure and body composition in anorexia nervosa patients during refeeding. J Am Diet Assoc 1993;93:434-438.

Krieg JC, Pirke KM, Lauer C, Backmund H. Endocrine, metabolic, and cranial computed tomographic findings in anorexia nervosa. Biol Psychiatry 1988;23:377-387.

Krogh, A. The respiratory exchange of animals and man. Monographs in biochemistry. London: Longmans Green, 1916, p 133.

Kuhl DE, Metter EJ, Riege WH, Phelps ME. Effects of human aging on patterns of local cerebral glucose utilization determined by the [18F]flurodoxyglucose method. J Cereb Blood Flow Metabol 1982;2:163-171.

Kushner RF, Racette SB, Neil K, Schoeller DA. Measurement of physical activity among black and white obese women. Obes Res 1995;3:261S-265S.

Lambe EK, Katzman DK, Mikulis DJ, Kennedy SH, Zipursky RB. Cerebral gray matter volume deficits after weight recovery from anorexia nervosa. Arch Gen Psychiatry 1997;54:6:537-542.

Lane MA, Baer DJ, Rumpler WV, Weindruch R, Ingram DK, Tilmont EM, Cutler RG, Roth GS. Caloric restriction lowers body temperature in rhesus monkeys, consistent with a postulated anti-aging mechanism in rodents. Proc Natl Acad Sci USA 1996;93:4159-4164.

Leenders KL, Perani D, Lammertsma AA, Heather JD, Buckingham P, Healy MJR, Gibbs JM, Wise RJS, Hatazawa J, Herold S, Beaney RP, Brooks DJ, Spinks T, Rhodes C, Frackowiak RSJ, Jones T. Cerebral blood flow, blood volume and oxygen utilization. Brain 1990;113:27-47.

Leibel R, Rosenbaum M, Hirsch J. Changes in energy expenditure resulting from altered body weight. New Engl J Med 1995;332:621-628.

Marchal G, Rioux P, Petit-Taboue MC, Sette G, Travere JM, Le Poec C, Courtheoux P, Derlon JM, Baron JC. Regional cerebral oxygen consumption, blood flow, and blood volume in healthy human aging. Arch Neurol 1992;49:1013-1020.

Martin AJ, Friston KJ, Colebatch JG, Frackowiak RSJ. Decreases in regional cerebral blood flow with normal aging. J Cereb Blood Flow Metab 1991;11:684-689.

Mayer L, Walsh BT, Gallagher D, Heymsfield S, Killory E. Vital organ sizes in anorexia nervosa. International Conference on Eating Disorders, April 26-28, 2002, Boston. Abstract 353, P108.

McCarter RJ, Palmer J. Energy metabolism and aging: a lifelong study of Fischer 344 rats. Am J Physiol 1992;263:E448-E452.

Melchior JC, Rigaud D, Rozen R, Malon M, Apfelbaum M. Energy expenditure exonomy induced by decrease in lean body mass in anorexia nervosa. Eur J Clin Nutr 1989;43:793-799.

Moeller JR, Eidelberg D. Divergent expression of regional metabolic topographies in Parkinson's disease and normal aging. Brain 1997; 120:2197-2206.

Moeller JR, Ishikawa T, Dhawan V, Spetsieris P, Mandel F, Alexander GE, Grady C, Pietrini P, Eidelberg D. The metabolic topography of normal aging. J Cereb Blood Flow Metab 1996;16:385-398.

Moore FD, Olsen KH, McMurray JD, Parker HV, Ball MR, Boyden CM. The body cell mass and its supporting environment: body composition in health and disease. Philadelphia: Saunders, 1963.

Moossy J. Cerebral artherosclerosis: intracranial and extracranial lesions. In: Minckler J, ed. Pathology of the nervous system. New York: McGraw-Hill, 1971, pp 1423-1432.

Morrison JA, Alfaro MP, Khoury P, Thornton BB, Daniels SR. Determinants of resting energy expenditure in young black girls and young white girls. J Pediatr 1996;129:637-642.

Obarzanek E, Lesem MD, Jimerson DC. Resting metabolic rate of anorexia nervosa patients during weight gain. Am J Clin Nutr 1994;60:666-675.

Ogiu N, Nakamura Y, Ijiri I, Hiraiwa K, Ogiu T. A statistical analysis of the internal organ weights of normal Japanese people. Health Phys 1997;72:368-383.

Owen OE, Morgan AP, Kemp HG, Sullivan JM, Herrera MG, Cahill GF. Brain metabolism during fasting. J Clin Invest 1967;46:1589-1595.

Owen OE, Reichard A. Human forearm metabolism during progressive starvation. J Clin Invest 1971;50:1536-1545.

Pantano P, Baron JC, Lebrun-Grandie P, Duquesnoy N, Bousser MG, Comar D. Regional cerebral blood flow and oxygen consumption in human aging. Stroke 1984;15:635-641.

Pfefferbaum A, Mathalon DH, Sullivan EV, Rawles JM, Zipursky RB, Lim KO. A quantitative magnetic resonance imaging study of changes in brain morphology from infancy to late adulthood. Arch Neurol 1994;51:874-887.

Piers LS, Soars MJ, McCormack LM, O'Dea K. Is there evidence for an age-related reduction in metabolic rate? J Appl Physiol 1998;85:2196-2204.

Poehlman ET, Berke EM, Joseph JR, Gardner AW, Katzman-Rooks SM, Goran MI. Influence of aerobic capacity, body composition, and thyroid hormones on the age-related decline in resting metabolic rate. Metabolism 1992;41:915-921.

Poehlman ET, Goran MI, Gardner AW, Ades PA, Arciero PJ, Katzman-Rooks SM, Montgomery SM, Toth MJ, Sutherland PT. Determinants of decline in resting metabolic rate in aging females. Am J Physiol 1993;264:E450-E455.

Puggaard L, Bjornsbo K, Kock K, Luders K, Thobo-Carlsen B, Lammert O. Age-related decrease in energy expenditure at rest parallels reductions in mass of internal organs. Am J Hum Biol 2002;14:486-493.

Ravussin E, Bogardus C. Relationship of genetics, age, and physical fitness to daily energy expenditure and fuel utilization. Am J Clin Nutr 1989;49:968-975.

Ravussin E, Lillioja S, Knowler WC, Christin L, Freymond D, Abbott WGH, Boyce V, Howard BV, Bogardus C. Reduced rate of energy expenditure as a risk factor for body-weight gain. New Engl J Med 1988;318:467-472.

Redies C, Hoffer J, Beil C, Marliss EB, Evans AC, Lariviere F, Marrett S, Meyer E, Diksic M, Gjedde A. Generalized decrease in brain glucose metabolism during fasting in humans studied by PET. Am J Phys 1989;256:E805-E810.

Reiss AL, Abrams MT, Singer HS, Ross JL, Denckla MB. Brain development, gender, and IQ in children. A volumetric imaging study. Brain 1996;119:1763-1774.

Richet, C. La Chaleur Animale. Bibliothique Scientifique Internationale. Paris: Felix Alcan, 1889.

Rigaud D, Hassid J, Meulemans A, Poupard AT, Boulier A. A paradoxical increase in resting energy expenditure in malnourished patients near death: the king penguin syndrome. Am J Clin Nutr 2000;72:355-360.

Rissanen P, Franssila-Kullunki A, Rissanen A. Cardiac parasympathetic activity is increased by weight loss in healthy obese women. Obes Res 2001;9:637-643.

Ross, R. Magnetic resonance imaging provides new insights into the characterization of adipose and lean tissue distribution. Can J Clin Pharmacol 1996;74:778-785.

Rubner M. Die Gesetze des Energiever Brauchs bei der Ernahrung. Leipzig und Wien: Deutiche, 1902.

Rubner M. Probleme des wachatuma und der lebensdauer. Mitteilungen der Gesellschaft fur Innere Medizin' und Kinderheilkunde. 1908;7:58-72.

Sanchez R, Morales M, Cardozo J. Normal brain weight in Venezuelan adults related to sex and age. Invest Clin 1997;38:83-93.

Scalfi L, Di Biase G, Coltorti A, Contaldo F. Bioimpedance analysis and resting energy expenditure in undernourished and refed anorectic patients. Eur J Clin Nutr 1993;47:61-67.

Schebendach JE, Golden NH, Jacobson MS, Hertz S, Shenker IR. The metabolic responses to starvation and refeeding in adolescents with anorexia nervosa. Ann NY Acad Sci 1997;817:110-119.

Seo JS, Lee SY, Won KJ, Kim DJ, Sohn DS, Yang KM, Cho SH, Park JD, Lee KH, Kim HD. Relationship between normal heart size and body indices in Koreans. J Korean Med Sci 2000;15:641-646.

Simon G, Tanner JM. Radiographic centiles of lung and heart growth. Patterns of growth. Thorax 1972;27:261.

Snyder, WS, Cook MJ, Nasset ES, Karhausen LR, Howells GP, Tipton IH. Report of the Task Group on Reference Man International Commission on Radiological Protection (No. 23). Oxford, UK: Pergamon, 1975.

Sokoloff L. Circulation and energy metabolism of the brain. In: GJ Siegel, RW Albers, R. Katzman, BW Agranoff, eds. Basic Neurochemistry, Second Edition. Boston: Little Brown, 1976, pp 388-413.

Sparti A, DeLany JP, de la Bretonne JA, Sander GE, Bray GA. Relationship between resting metabolic rate and the composition of the fat-free mass. Metabolism 1997;46:1225-1230.

Sun M, Gower BA, Bartolucci AA, Hunter GR, Figueroa-Colon R, Goran MI. A longitudinal study of resting energy expenditure relative to body composition during puberty in African American and White children. Am J Clin Nutr 2001;73:308-315.

Swayze VW, Andersen AE, Andreasen NC, Arndt S, Sato Y, Ziebell S. Brain tissue volume segmentation in patients with anorexia nervosa before and after weight normalization. Int J Eat Disord 2003;33:33-44.

Swayze VW, Andersen A, Arndt S, Rajarethinam R, Fleming F, Sato Y, Andreasen NC. Reversibility of brain tissue loss in anorexia nervosa assessed with a computerized Talairach 3-D proportional grid. Psychol Med 1996;26:381-390.

Tanaka G, Nakahara Y, Nakazima Y. Japanese reference man—IV. Studies on the weight and size of internal organs of Normal Japanese. Nippon Igaku Hoshasen Gakkai Zasshi. 1989;49:344-365.

Van Bogaert P, Wikler D, Damhaut P, Szliwowski HB, Goldman S. Regional changes in glucose metabolism during brain development from the age of 6 years. Neuroimage 1998;8:62-68.

Vaughan L, F Zurlo, and E Ravussin. Aging and energy expenditure. Am J Clin Nutr. 53:821-5, 1991.

Visser M, Deurenberg P, van Staveren WA, Hautvast JGAJ. Resting metabolic rate and diet-induced thermogenesis in young and elderly subjects: relationship with body composition, fat distribution, and physical activity level. Am J Clin Nutr 1995;61:772-778.

Voigt J, Pakkenberg H. Brain weight of Danish children. Acta Anat 1983;116:290-301.

Wang Z, O'Connor TP, Heshka S, Heymsfield S. The reconstruction of Kleiber's law at the organ-tissue level. J Nutr 2001;131:2967-2970.

Weindruch R, Sohal RS. 1997. Seminars in medicine of the Beth Israel Deaconess Medical Center. Caloric intake and aging. N Engl J Med 337:986-994.

Weindruch R, Walford RL. The retardation of aging and disease by dietary restriction. Springfield, IL: Charles C Thomas, 1988.

Weinsier RL, Schutz Y, Bracco D. Reexamination of the relationship of resting metabolic rate to fat-free mass and to the metabolically active components of fat-free mass in humans. Am J Clin Nutr 1992;55:790-794.

Weinsier RL, Nagy TR, Hunter GR, Darnell BE, Hensrud DD, Weiss HL. Do adaptive changes in metabolic rate favor weight regain in weight-reduced individuals? An examination of the set-point theory. Am J Clin Nutr 2000;72:1088-1094.

Weir JB. New methods for calculating metabolic rate with special reference to protein metabolism. J Physiol 1949;109:1-9.

Weyer C, Snitker S, Bogardus C, Ravussin E. Energy metabolism in African Americans: potential risk factors for obesity. Am J Clin Nutr 1999;70:13-20.

Womack H. The relationship between human body weight, subcutaneous fat, heart weight and epicardial fat. Human Biol 1983;55:667-676.

Wong WW, Butte NF, Ellis KJ, Hergenroeder AC, Hill RB, Stuff JE, Smith EO. Pubertal African-American girls expend less energy at rest and during physical activity than Caucasian girls. J Clin Endocrinol Metab 1999;84:906-911.

World Health Organization (WHO). Energy and protein requirements. Technical Reports Series 724. Geneva: World Health Organization, 1985.

Yamaura H, Ito M, Kubota K, Matsuzawa T. Brain atrophy during aging: a quantitative study with computed tomography. J Gerontol 1980;35:492-498.

Yanovski SZ, Reynolds JC, Boyle AJ, Yanovski JA. Resting metabolic rate in African-American and Caucasian girls. Obes Res 1997;5:321-325.

Zeek P. Heart weight: I. The weight of the normal human heart. Arch Pathol Lab Med 1942;34:820-832.

Chapter 16

Ama, P.F.M., Simoneau, J.A., Boulay, M.R., Serresse, O., Thériault, G., & Bouchard, C. (1986). Skeletal muscle characteristics in sedentary Black and Caucasian males. *Journal of Applied Physiology, 61,* 1758-1761.

Austin, M.A., Friedlander, Y., Newman, B., Edwards, K., Mayer-Davis, E.J., & King, M.C. (1997). Genetic influences on changes in body mass index: a longitudinal analysis of women twins. *Obesity Research, 5,* 326-331.

Berg, K. (1981). Twin research in coronary heart disease. In L. Gedda, P. Parisi, & W.E. Nance (Eds.), *Twin research 3: Part C, epidemiological and clinical studies* (pp. 117-130). New York: Liss.

Berg, K. (1990). Molecular genetics and nutrition. In A.P. Simopoulos & B. Childs (Eds.), *Genetic variation and nutrition* (pp. 49-59). Basel: Karger.

Beunen, G., Maes, H.H., Vlietinck, R., Malina, R.M., Thomis, M., Feys, E., Loos, R., & Derom, C. (1998). Univariate and multivariate genetic analysis of subcutaneous fatness and fat distribution in early adolescence. *Behavioral Genetics, 28,* 279-288.

Blomstrand, E., & Ekblom, B. (1982). The needle biopsy technique for fiber type determination in human skeletal muscle. A methodological study. *Acta Physiologica Scandinavica, 116,* 437-442.

Borecki, I.B., Blangero, J., Rice, T., Pérusse, L., Bouchard, C., & Rao, D.C. (1998). Evidence for at least two major loci influencing human fatness. *American Journal of Human Genetics, 63,* 831-838.

Borecki, I.B., Rice, T., Pérusse, L., Bouchard, C., & Rao, D.C. (1995). Major gene influence on the propensity to store fat in trunk versus extremity depots: Evidence from the Québec Family Study. *Obesity Research, 3,* 1-8.

Borecki, I.B., & Suarez, B.K. (2001). Linkage and association: Basic concepts. In D.C. Rao & M.A. Province (Eds.), *Genetic dissection of complex traits.* (pp. 45-66), San Diego: Academic Press.

Bouchard, C. (1988). Inheritance of human fat distribution. In C. Bouchard & F.E. Johnson (Eds.), *Fat distribution during growth and later health outcomes* (pp. 103-125). New York: Liss.

Bouchard, C. (1990). Variation in human body fat: The contribution of the genotype. In G.A. Bray, D. Ricquier, & B.M. Spiegelman (Eds.), *Obesity: Towards a molecular approach* (pp. 17-28). New York: Liss.

Bouchard, C. (1991a). Current understanding of the etiology of obesity: Genetic and nongenetic factors. *American Journal of Clinical Nutrition, 53,* 1561S-1565S.

Bouchard, C. (1991b). Genetic aspects of anthropometric dimensions relevant to assessment of nutritional status. In J. Himes (Ed.), *Anthropometric assessment of nutritional status* (pp. 213-231). New York: Liss.

Bouchard, C. (1994a). Genetics of human obesities: Introductory notes. In C. Bouchard (Ed.), *The genetics of obesity* (pp. 1-15). Boca Raton, FL: CRC Press.

Bouchard, C. (1994b). Genetics of obesity: Overview and research directions. In C. Bouchard (Ed.), *The genetics of obesity* (pp. 223-233). Boca Raton, FL: CRC Press.

Bouchard, C., Chagnon, M., Thibault, M.C., Boulay, M.R., Marcotte, M., & Simoneau, J.A. (1988a). Absence of charge variants in human skeletal muscle enzymes of the glycolytic pathways. *Human Genetics, 78,* 100.

Bouchard, C., Després, J.P., Mauriège, P., Marcotte, M., Chagnon, M., Dionne, F.T., & Bélanger, A. (1991). The genes in the constellation of determinants of regional fat distribution. *International Journal of Obesity, 15,* 9-18.

Bouchard, C., & Pérusse, L. (1988). Heredity and body fat. *Annual Review of Nutrition, 8,* 259-277.

Bouchard, C., & Pérusse, L. (1993). Genetics of obesity. *Annual Review of Nutrition, 13,* 337-354.

Bouchard, C., & Pérusse, L. (1994). Genetics of obesity: Family studies. In C. Bouchard (Ed.), *The genetics of obesity* (pp. 79-92). Boca Raton, FL: CRC Press.

Bouchard, C., & Pérusse, L. (1996). Current status of the human obesity gene map. *Obesity Research, 4,* 81-90.

Bouchard, C., Pérusse, L., Dériaz, O., Després, J.P., & Tremblay, A. (1993). Genetic influences on energy expenditure in humans. *Critical Reviews in Food Science and Nutrition, 33,* 345-350.

Bouchard, C., Pérusse, L., Leblanc, C., Tremblay, A., & Thériault, G. (1988b). Inheritance of the amount and distribution of human body fat. *International Journal of Obesity, 12,* 205-215.

Bouchard, C., Pérusse, L., Rice, T., & Rao, D.C. (2004). The genetics of human obesity. In G.A. Bray & C. Bouchard (Eds.), *Handbook of obesity: Etiology and pathophysiology* (2nd ed., pp. 157-200). New York: Dekker.

Bouchard, C., Rice, T., Lemieux, S., Després, J.-P., Pérusse, L., & Rao, D.C. (1996). Major gene for abdominal visceral fat area in the Québec Family Study. *International Journal of Obesity, 20,* 420-427.

Bouchard, C., Simoneau, J.A., Lortie, G., Boulay, M.R., Marcotte, M., & Thibault, M.C. (1986). Genetic effects in human skeletal muscle fiber type distribution and enzyme activities. *Canadian Journal of Physiology and Pharmacology, 64,* 1245-1251.

Bouchard, C., Tremblay, A., Després, J.P., Nadeau, A., Lupien, P.J., Thériault, G., Dussault, J., Moorjani, S., Pineault, S., & Fournier, G. (1990). The response to long-term overfeeding in identical twins. *New England Journal of Medicine, 322,* 1477-1482.

Bouchard, C., Tremblay, A., Després, J.P., Poehlman, E.T., Thériault, G., Nadeau, A., Lupien, P.J., Moorjani, S., & Dussault, J. (1988c). Sensitivity to overfeeding: The Quebec experiment with identical twins. *Progress in Food and Nutrition Science, 12,* 45-72.

Bouchard, C., Tremblay, A., Després, J.P., Thériault, G., Nadeau, A., Lupien, P.J., Moorjani, S., Prud'homme, D., & Fournier, G. (1994). The response to exercise with constant energy intake in identical twins. *Obesity Research, 2,* 400-410.

Bray, G.A. (1981). The inheritance of corpulence. In L.A. Cioffi, W.P.T. James, & T.B. Van Itallie (Eds.), *The body weight regulatory system: Normal and disturbed mechanisms* (pp. 185-195). New York: Raven Press.

Brožek, J. (1961). Body measurements including skinfold thickness, as indicators of body composition. In J. Brožek & A. Henschel (Eds.), *Techniques for measuring body composition* (pp. 3-35). Washington, DC: National Research Council.

Cardon, L.R., Carmelli, D., Fabsitz, R.R., & Reed, T. (1994). Genetic and environmental correlations between obesity and body fat distribution in adult male twins. *Human Biology, 66,* 465-479.

Carmelli, D., Cardon, L.R., & Fabsitz, R. (1994). Clustering of hypertension, diabetes, and obesity in adult male twins: Same genes or same environments? *American Journal of Human Genetics, 55,* 566-573.

Chagnon, Y.C., Borecki, I.B., Pérusse, L., Roy, S., Lacaille, M., Chagnon, M., Ho-Kim, M.A., Rice, T., Province, M.A., Rao, D.C., & Bouchard, C. (2000a). Genome-wide search for genes related to the fat-free body mass in the Quebec Family Study. *Metabolism, 49,* 203-207.

Chagnon, Y.C., Pérusse, L., & Bouchard, C. (1998). The human obesity gene map: The 1997 update. *Obesity Research, 6,* 76-92.

Chagnon, Y.C., Pérusse, L., Weisnagel, S.J., Rankinen, T., & Bouchard, C. (2000b). The human obesity gene map: The 1999 update. *Obesity Research, 8,* 89-117.

Chagnon, Y.C., Rankinen, T., Snyder, E.E., Weisnagel, S.J., Pérusse, L., & Bouchard, C. (2003). The human obesity gene map: The 2002 update. *Obesity Research, 11,* 313-367.

Choh, A.C., Gage, T.B., McGarvey, S.T., & Comuzzie, A.G. (2001). Genetic and environmental correlations between various anthropometric and blood pressure traits among adult Samoans. *American Journal of Physical Anthropology, 115,* 304-311.

Comuzzie, A.G., Blangero, J., Mahaney, M.C., Mitchell, B.D., Hixson, J.E., Samollow, P.B., Stern, M.P., & MacCluer, J.W. (1995). Major gene with sex-specific effects influences fat mass in Mexican Americans. *Genetic Epidemiology, 12,* 475-488.

Comuzzie, A.G., Blangero, J., Mahaney, M.H., Mitchell, B.D., Stern, M.P., & MacCluer, J.W. (1994). Genetic and environmental correlations among skinfold measures. *International Journal of Obesity, 18,* 413-418.

Comuzzie, A.G., Blangero, J., Mitchell, B.D., Stern, M.P., & MacCluer, J.W. (1993). Segregation analysis of fat mass and fat free mass. *Genetic Epidemiology, 10,* 340 (Abstract).

Cui, J., Hopper, J.L., & Harrap, S.B. (2002). Genes and family environment explain correlations between blood pressure and body mass index. *Hypertension, 40,* 7-12.

Després, J.P., Bouchard, C., Savard, R., Prud'homme, D., Bukowiecki, L., & Thériault, G. (1984). Adaptive changes to training in adipose tissue lipolysis are genotype dependent. *International Journal of Obesity, 8,* 87-95.

Donahue, R.P., Prineas, R.J., Gomez, O., & Hong, C.P. (1992). Familial resemblance of body fat distribution: The Minneapolis Children's blood pressure study. *International Journal of Obesity, 16,* 161-167.

Enzi, G., Gasparo, M., Biondetti, P.R., Flor, D., Semisa, M., & Zurlo, F. (1986). Subcutaneous and visceral fat distribution according to sex, age, and overweight, evaluated by computed tomography. *American Journal of Clinical Nutrition, 44,* 739-746.

Fabsitz, R., Feinleib, M., & Hrubec, Z. (1980). Weight changes in adult twins. *Acta Geneticae Medicae Gemellologiae, 29,* 273-279.

Fabsitz, R.R., Sholinsky, P., & Carmelli, D. (1994). Genetic influences on adult weight gain and maximum body mass index in male twins. *American Journal of Epidemiology, 140,* 711-720.

Faith, M.S., Pietrobelli, A., Nunez, C., Heo, M., Heymsfield, S.B., & Allison, D.B. (1999). Evidence for independent genetic influences on fat mass and body mass index in a pediatric twin sample. *Pediatrics, 104,* 61-67.

Falconer, D.S. (1960). *Introduction to quantitative genetics.* New York: Ronald Press.

Feitosa, M.F., Rice, T., Nirmala-Reddy, A., Reddy, P.C., & Rao, D.C. (1999). Segregation analysis of regional fat distribution in families from Andhra Pradesh, India. *International Journal of Obesity and Related Metabolic Disorders, 23,* 874-880.

Ferland, M., Després, J.P., Tremblay, A., Pinault, S., Nadeau, A., Moorjani, S., Lupien, P.J., Thériault, G., & Bouchard, C. (1989). Assessment of adipose tissue distribution by computed axial tomography in obese women: Association with body density and anthropometric measurements. *British Journal of Nutrition, 61,* 139-148.

Garn, S.M., Sullivan, T.V., & Hawthorne, V.M. (1989). Fatness and obesity of the parents of obese individuals. *American Journal of Clinical Nutrition, 50,* 1308-1313.

Gollnick, P.D. (1982). Relationship of strength and endurance with skeletal muscle structure and metabolic potential. *International Journal of Sports Medicine, 3,* 26-32.

Gollnick, P.D., Armstrong, R.B., Saltin, B., Saubert, C.W. IV, Sembrovich, W.L., & Shepherd, R.E. (1973). Effect of training on enzyme activity and fiber type composition of human skeletal muscle. *Journal of Applied Physiology, 34,* 107-111.

Greenberg, D.A. (1993). Linkage analysis of "necessary" disease loci versus "susceptibility" loci. *American Journal of Human Genetics, 52,* 135-143.

Grilo, C.M., & Pogue-Geile, M.F. (1991). The nature of environmental influences on weight and obesity: A behavior genetic analysis. *Psychology Bulletin, 110,* 520-537.

Haseman, J.K., & Elston, R.C. (1972): The investigation of linkage between a quantitative trait and a marker locus. *Behavior Genetics, 2,* 3-19.

Hasstedt, S.J., Ramirez, M.E., Kuida, H., & Williams, R.R. (1989). Recessive inheritance of a relative fat pattern. *American Journal of Human Genetics, 45,* 917-925.

Heller, R., Garrison, R.J., Havlik, R.J., Feinleib, M., & Padgett, S. (1984). Family resemblances in height and relative weight in the Framingham Heart Study. *International Journal of Obesity, 8,* 399-405.

Hong, Y., Després, J.-P., Rice, T., Nadeau, A., Province, M.A., Gagnon, J., Leon, A.S., Skinner, J.S., Wilmore, J.H., Bouchard, C., & Rao, D.C. (2000). Evidence of pleiotropic loci for fasting insulin, total fat mass, and abdominal visceral fat in a sedentary population: The HERITAGE Family Study. *Obesity Research, 8,* 151-159.

Hong, Y., Pedersen, N.L., Brismar, K., & de Faire, U. (1997). Genetic and environmental architecture of the features of the insulin-resistance syndrome. *American Journal of Human Genetics, 60,* 143-152.

Hong, Y., Rice, T., Gagnon, J., Després, J.P., Nadeau, A., Pérusse, L., Bouchard, C., Leon, A.S., Skinner, J.S., Wilmore, J.H., & Rao, D.C. (1998). Familial clustering of insulin and abdominal visceral fat: The HERITAGE Family Study. *Journal of Clinical Endocrinology and Metabolism, 83,* 4239-4245.

Hunt, M., Katzmarzyk, P.T., Pérusse, L., Rice, T., Rao, D.C., & Bouchard, C. (2002). Familial resemblance of 7-year changes in body mass and adiposity. *Obesity Research, 10,* 507-517.

Ikoma, E., & Murotani, N. (1976). A genetic study on the length of tibia. *Annals of Human Genetics, 39,* 475-483.

Katzmarzyk, P.T., Pérusse, L., & Bouchard, C. (1999). Genetics of abdominal visceral fat levels. *American Journal of Human Biology, 11,* 225-235.

Korkeila, M., Kaprio, J., Rissanen, A., & Koskenvuo, M. (1995). Consistency and change of body mass index and weight. A study on 5967 adult Finnish twin pairs. *International Journal of Obesity and Related Metabolic Disorders, 19,* 310-317.

Lecomte, E., Herbeth, B., Nicaud, V., Rakotovao, R., Artur, Y., & Tiret, L. (1997). Segregation analysis of fat mass and fat-free mass with age- and sex-dependent effects: The Stanislas Family Study. *Genetic Epidemiology, 14,* 51-62.

Luke, A., Guo, X., Adeyemo, A.A., Wilks, R., Forrester, T., Lowe, W., Jr., Comuzzie, A.G., Martin, L.J., Zhu, X., Rotimi, C.N., & Cooper, R.S. (2001). Heritability of obesity-related traits among Nigerians, Jamaicans and US black people. *International Journal of Obesity and Related Metabolic Disorders, 25,* 1034-1041.

MacCluer, J.W. (1992). Biometrical studies to detect new genes with major effects on quantitative risk factors for atherosclerosis. *Current Opinion in Lipidology, 3,* 114-121.

MacDonald, A., & Stunkard, A.J. (1990). Body mass indexes of British separated twins. *New England Journal of Medicine, 322,* 1530.

Maes, H.H.M., Neale, M.C., & Eaves, L.J. (1997). Genetic and environmental factors in relative body weight and human adiposity. *Behavioral Genetics, 27,* 325-351.

Mahaney, M.H., Blangero, J., Comuzzie, A.G., VandeBerg, J.L., Stern, M., & MacCluer, J.W. (1995). Plasma HDL, cholesterol, triglycerides, and adiposity: A quantitative genetic test of the conjoint trait hypothesis in the San Antonio Heart Study. *Circulation, 92,* 3240-3248.

Marcotte, M., Chagnon, M., Côté, C., Thibault, M.C., Boulay, M.R., & Bouchard, C. (1987). Lack of genetic polymorphism in human skeletal muscle enzymes of the tricarboxylic acid cycle. *Human Genetics, 77,* 200.

Mauriège, P., Després, J.P., Marcotte, M., Tremblay, A., Nadeau, A., Moorjani, S., Lupien, P.J., Dussault, J., Fournier, G., Thériault, G., & Bouchard, C. (1992). Adipose tissue lipolysis after long-term overfeeding in identical twins. *International Journal of Obesity, 16,* 219-225.

Mitchell, B.D., Kammerer, C.M., Mahaney, M.C., Blangero, J., Comuzzie, A.G., Atwood, L.D., Haffner, S.M., Stern, M.P., & MacCluer, J.W. (1996). Genetic analysis of the IRS. Pleiotropic effects of genes influencing insulin levels on lipoprotein and obesity measures. *Arteriosclerosis, Thrombosis and Vascular Biology, 16,* 281-288.

Mueller, W.H. (1983). The genetics of human fatness. *Yearbook of Physiology and Anthropology, 26,* 215-230.

Nimmo, M.A., Wilson, R.H., & Snow, D.H. (1985). The inheritance of skeletal muscle fibre composition in mice. *Comparative Biochemistry and Physiology, 81A,* 109-115.

Olson, J.E., Atwood, L.D., Grabrick, D.M., Vachon, C.M., & Sellers, T.A. (2001). Evidence for a major gene influence on abdominal fat distribution: The Minnesota Breast Cancer Family Study. *Genetic Epidemiology, 20,* 458-478.

Pérusse, L., & Bouchard, C. (1994). Genetics of energy intake and food preferences. In C. Bouchard (Ed.), *The genetics of obesity* (pp. 125-134). Boca Raton, FL: CRC Press.

Pérusse L., Chagnon, Y.C., Dionne, F.T., & Bouchard, C. (1997). The human obesity gene map: The 1996 update. *Obesity Research, 5,* 49-61.

Pérusse, L., Chagnon, Y.C., Weisnagel, S.J., & Bouchard, C. (1999). The human obesity gene map: The 1998 update. *Obesity Research, 7,* 111-129.

Pérusse, L., Chagnon, Y.C., Weisnagel, S.J., Rankinen, T., Snyder, E., Sands, J., & Bouchard, C. (2001a). The human obesity gene map: The 2000 update. *Obesity Research, 9,* 135-169.

Pérusse, L., Després, J.-P., Lemieux, S., Rice, T., Rao, D.C., & Bouchard, C. (1996). Familial aggregation of abdominal visceral fat level: Results from the Québec Family Study. *Metabolism 45,* 378-382.

Pérusse, L., Leblanc, C., & Bouchard, C. (1988). Inter-generation transmission of physical fitness in the Canadian population. *Canadian Journal of Sport Sciences, 13,* 8-14.

Pérusse, L., Rice, T., Chagnon, Y.C., Després, J.-P., Lemieux, S., Roy, S., Lacaille, M., Ho-Kim, M.A., Chagnon, M., Province, M.A., Rao, D.C., & Bouchard, C. (2001b), A genome-wide scan for abdominal fat assessed by computed tomography in the Québec Family Study. *Diabetes, 50,* 614-21.

Pérusse, L., Rice, T., Després, J.-P., Rao, D.C., & Bouchard, C. (1997). Cross-trait familial resemblance for body fat and blood lipids: Familial correlations in the Québec Family Study. *Arteriosclerosis, Thrombosis and Vascular Biology, 17,* 3270-3277.

Pérusse, L., Rice, T., Province, M.A., Gagnon, J., Leon, A.S., Skinner, J.S., Wilmore, J.H., Rao, D.C., & Bouchard, S. (2000). Familial aggregation of amount and distribution of subcutaneous fat and their responses to exercise training in the HERITAGE family study. *Obesity Research, 8,* 140-150.

Poehlman, E.T., Després, J.P., Marcotte, M., Tremblay, A., Thériault, G., & Bouchard, C. (1986). Genotype dependency of adaptation in adipose tissue metabolism after short-term overfeeding. *American Journal of Physiology, 250,* E480-E485.

Poehlman, E.T., Tremblay, A., Marcotte, M., Pérusse, L., Thériault, G., & Bouchard, C. (1987). Heredity and changes in body composition and adipose tissue metabolism after short-term exercise training. *European Journal of Applied Physiology, 56,* 398-402.

Price, R.A., Cadoret, R.J., Stunkard, A.J., & Troughton, E. (1987). Genetic contributions to human fatness: An adoption study. *American Journal of Psychiatry, 144,* 1003-1008.

Price, R.A., & Gottesman, I.I. (1991). Body fat in identical twins reared apart: Roles for genes and environment. *Behavioral Genetics, 21,* 1-7.

Ramirez, M.E. (1993). Familial aggregation of subcutaneous fat deposits and the peripheral fat distribution pattern. *International Journal of Obesity, 17,* 63-68.

Rankinen, T., Perusse, L., Weisnagel, S.J., Snyder, E.E., Chagnon, Y.C., & Bouchard, C. (2002). The human obesity gene map: The 2001 update. *Obesity Research, 10,* 196-243.

Rao, D.C. (2001). Genetic dissection of complex traits: An overview. In D.C. Rao & M.A. Province (Eds.), *Genetic dissection of complex traits* (pp. 13-34), San Diego: Academic Press.

Rice, T., & Borecki, I.B. (2001). Familial resemblance and heritability. In D.C. Rao & M.A. Province (Eds.), *Genetic dissection of complex traits* (pp.35-44). San Diego: Academic Press.

Rice, T., Borecki, I.B., Bouchard, C., & Rao, D.C. (1993). Segregation analysis of fat mass and other body composition measures derived from underwater weighing. *American Journal of Human Genetics, 52,* 967-973.

Rice, T., Bouchard, C., Pérusse, L., & Rao, D.C. (1995). Familial clustering of multiple measures of adiposity and fat distribution in the Québec Family Study: A trivariate analysis of percent body fat, body mass index, and trunk-to-extremity skinfold ratio. *International Journal of Obesity, 19,* 902-908.

Rice, T., Chagnon, Y.C., Pérusse, L., Borecki, I.B., Ukkola, O., Rankinen, T., Gagnon, J., Leon, A.S., Skinner, J.S., Wilmore, J.H., Bouchard, C., & Rao, D.C. (2002). A genomewide linkage scan for abdominal subcutaneous and visceral fat in black and white families: The HERITAGE Family Study. *Diabetes, 51,* 848-855.

Rice, T., Daw, E.W., Gagnon, J., Bouchard, C., Leon, A.S., Skinner, J.S., Wilmore, J.H., & Rao, D.C. (1997a). Familial resemblance for body composition measures: The HERITAGE Family Study. *Obesity Research, 5,* 557-562.

Rice, T., Després, J.-P., Daw, E.W., Gagnon, J., Borecki, I.B., Pérusse, L., Leon, A.S., Skinner, J.S., Wilmore, J.H., Rao, D.C., & Bouchard, C. (1997b). Familial resemblance for abdominal visceral fat: The HERITAGE Family Study. *International Journal of Obesity 21,* 1024-1031.

Rice, T., Després, J.-P., Pérusse, L., Gagnon, J., Leon, A.S., Skinner, J.S., Wilmore, J.H., Rao, D.C., & Bouchard, C. (1997c). Segregation analysis of abdominal visceral fat: The HERITAGE Family Study. *Obesity Research 5,* 417-424.

Rice, T., Nadeau, A., Pérusse, L., Bouchard, C., & Rao, D.C. (1996a). Familial correlations in the Québec Family Study: Cross-trait familial resemblance for body fat with plasma glucose and insulin. *Diabetologia, 39,* 1357-1364.

Rice, T., Pérusse, L., Bouchard, C., & Rao, D.C. (1996b). Familial clustering of abdominal visceral fat and total fat mass: The Québec Family Study. *Obesity Research, 4,* 253-261.

Rice, T., Pérusse, L., Bouchard, C., & Rao, D.C. (1999). Familial aggregation of body mass index and subcutaneous fat measures in the longitudinal Québec Family Study. *Genetic Epidemiology, 16,* 316-334.

Rice, T., Province, M., Pérusse, L., Bouchard, C., & Rao, D.C. (1994). Cross-trait familial resemblance for body fat and blood pressure: Familial correlations in the Québec Family Study. *American Journal of Human Genetics, 55,* 1019-1029.

Roche, A.F. (1994). Sarcopenia: A critical review of its measurements and health-related significance in the middle-aged and elderly. *American Journal of Human Biology, 6,* 33-42.

Saltin, B., & Gollnick, P.D. (1983). Significance for metabolism and performance. In L.D. Peachy, R.H. Adrian, & S.R. Geiger (Eds.), *Skeletal muscle. Handbook of physiology, section 10* (pp. 555-631). Bethesda, MD: American Physiological Society.

Savard, R., & Bouchard, C. (1990). Genetic effects in the response of adipose tissue lipoprotein lipase activity to prolonged exercise: A twin study. *International Journal of Obesity, 14,* 771-777.

Schieken, R.M., Mosteller, M., Goble, M.M., Moskowitz, W.B., Hewitt, J.K., Eaves, L.J., & Nance, W.E. (1992). Multivariate genetic analysis of blood pressure and body size: The Medical College of Virginia Twin Study. *Circulation, 86,* 1780-1788.

Schork, N.J., Weder, A.B., Trevisan, M., & Laurenzi, M. (1994). The contribution of pleiotropy to blood pressure and body-mass index variation: The Gubbio Study. *American Journal of Human Genetics, 54,* 361-373.

Schulte, P.A., & Perera, F.P. (1993). Validation. In P.A. Schulte & F.P. Perera (Eds.), *Molecular epidemiology. Principles and practices* (pp. 81-107). San Diego: Academic Press.

Selby, J.V., Newman, B., Quesenberry, C.P. Jr., Fabsitz, R.R., King, M.C., & Meaney, J.M. (1989). Evidence of genetic influence on central body fat in middle-aged twins. *Human Biology, 61,* 179-193.

Simoneau, J.-A., & Bouchard, C. (1989). Human variation in skeletal muscle fiber-type proportion and enzyme activities. *American Journal of Physiology, 257,* E567-E572.

Simoneau, J.-A., & Bouchard, C. (1995). Genetic determinism of fiber type proportion in human skeletal muscle. *FASEB Journal, 9,* 1091-1095.

Simoneau, J.-A., Lortie, G., Boulay, M.R., Marcotte, M., Thibault, M.C., & Bouchard C. (1986). Repeatability of fiber type and enzyme activity measurements in human skeletal muscle. *Clinical Physiology, 6,* 347-356.

Sing, C.F., & Boerwinkle, E.A. (1987). Genetic architecture of inter-individual variability in apolipoprotein, lipoprotein and lipid phenotypes. In D. Weatherall (Ed.), *Molecular approaches to human polygenic disease* (pp. 99-127). New York: Wiley.

Sing, C.F., Boerwinkle, E., Moll, P.P., & Templeton, A.R. (1988). Characterization of genes affecting quantitative traits in humans. In B.S. Weir, E.J. Eisen, M.M. Goddman, & G. Namkoong (Eds.), *Proceedings of the Second International Conference on Quantitative Genetics* (pp. 250-269). Sunderland, MA: Sinauer Associates.

Sørensen, T.I.A., Holst, C., & Stunkard, A.J. (1992a). Childhood body mass index—genetic and familial environmental influences assessed in a longitudinal adoption study. *International Journal of Obesity, 16,* 705-714.

Sørensen, T.I.A., Holst, C., Stunkard, A.J., & Theil, L. (1992b). Correlations of body mass index of adult adoptees and their biological relatives. *International Journal of Obesity, 16,* 227-236.

Sørensen, T.I.A., Price, R.A., Stunkard, A.J., & Schulsinger, F. (1989). Genetics of obesity in adult adoptees and their biological siblings. *British Journal of Medicine, 298,* 87-90.

Stunkard, A.J., Foch, T.T., & Hrubec, Z. (1986a). A twin study of human obesity. *Journal of the American Medical Association, 256,* 51-54.

Stunkard, A.J., Harris, J.R., Pedersen, N.L., & McClearn, G.E. (1990). The body-mass index of twins who have been reared apart. *New England Journal of Medicine, 322,* 1483-1487.

Stunkard, A.J., Sorensen, T.I.A., Hannis, C., Teasdale, T.W., Chakraborty, R., Schull, W.J., & Schulsinger, F. (1986b). An adoption study of human obesity. *New England Journal of Medicine, 314,* 193-198.

Tambs, K., Moum, T., Eaves, L., Neale, M., Midthjell, K., Lund-Larsen, P.G., Naess, S., & Holmen, J. (1991). Genetic and environmental contributions to the variance of the body mass index in a Norwegian sample of first and second-degree relatives. *American Journal of Human Biology, 3,* 257-267.

Warden, C.H., Daluiski, A., & Lusis, A.J. (1992). Identification of new genes contributing to atherosclerosis: The mapping of genes contributing to complex disorders in animal models. In A.J. Lusis, J.I. Rotter, & R.S. Sparkes (Eds.), *Molecular genetics of coronary artery disease* (Monographs in Human Genetics, Vol. 14, pp. 419-441). Basel: Karger.

Chapter 17

Baumgartner, R.N. (2000). Body composition in healthy aging. *Annals of the New York Academy of Sciences 904:* 437-448.

Baumgartner, R.N., S.B. Heymsfield, S. Lichtman, J. Wang, & R.N. Pierson. (1991). Body composition in elderly people: effect of criterion estimates on predictive equations. *American Journal of Clinical Nutrition 53:* 1345-1349.

Baumgartner, R.N., K. Koehler, L. Romero, R. Lindeman, & P. Garry. (1998). Epidemiology of sarcopenia in elderly people in New Mexico. *American Journal of Epidemiology* 147: 744-763.

Baumgartner, R.N., R.L. Rhyne, P.J. Garry, & S.B. Heymsfield. (1993). Imaging techniques and anatomical body composition. *Journal of Nutrition* 123: 444-448.

Baumgartner, R.N., R.L. Rhyne, C. Troup, S. Wayne, & P.J. Garry. (1992). Appendicular skeletal muscle areas assessed by magnetic resonance imaging in older persons. *Journal of Gerontology: Medical Sciences* 47: M67-M72.

Baumgartner, R.N., A.F. Roche, S. Guo, T. Lohman, R. Boileau, & M. Slaughter. (1986). Adipose tissue distribution—the stability of principal components by sex, ethnicity, and maturation stage. *Human Biology* 58: 719-735.

Baumgartner, R.N., R.M. Siervogel, C. Chumlea, & A. F. Roche. (1989). Associations between plasma lipoprotein cholesterols, adiposity, and adipose tissue distribution during adolescence. *International Journal of Obesity* 13: 31-42.

Baumgartner, R.N., P.M. Stauber, D. McHugh, K. Koehler, & J.P. Garry. (1995). Cross-sectional age differences in body composition in persons 60+ years of age. *Journal of Gerontology: Medical Sciences* 50A: M307-M316.

Baumgartner, R.N., D.L. Waters, D. Gallagher, J. Morley, & P.J. Garry. (1999). Predictors of skeletal muscle mass in elderly men and women. *Mechanisms of Aging and Development* 107: 123-126.

Biolo, G., R. Antonione, R. Barrazoni, M. Zanetti, & G. Guarnieri. (2003). Mechanisms of altered protein turnover in chronic diseases: a review of human kinetic studies. *Current Opinion in Clinical Nutrition and Metabolic Care* 6: 55-63.

Blunt, B.A., M.R. Klauber, E. Barrett-Conor, & S. Edelstein. (1994). Sex differences in bone mineral density in 1653 men and women in the sixth through tenth decades of life: The Rancho Bernardo Study. *Journal of Bone and Mineral Research* 9: 1333-1338.

Boesch, C., and R. Kreis (2000). Observation of intramyocellular. *Annals of the New York Academy of Sciences* 904: 25-31.

Borkan, G.A., D.E. Hults, S.G. Gerzof, A.H. Robbins, & C.K. Silbert. (1983). Age changes in body composition revealed by computed tomography. *Journal of Gerontology* 38: 673-677.

Borkan, G.A., and A.H. Norris (1977). Fat redistribution and the changing body dimensions of the adult male. *Human Biology* 49: 495-514.

Bouchard, C., and F. E. Johnston (1988). *Fat distribution during growth and later health outcomes*. New York, Liss.

Butte, N., J.M. Hopkinson, W.W. Wong, E.O. Smith, & K.J. Ellis. (2000). Body composition during the first 2 years of life: an updated reference. *Pediatric Research* 47: 578-585.

Cameron, N., and E.W. Demerath (2001). Growth, maturation and the development of obesity. *Obesity, growth and development*. F.E. Johnston and G.D. Foster. London, Smith-Gordon, pp. 37-56.

Casey, V.A., J.H. Dwyer, C.S. Berkey, S.M. Bailey, K.A. Coleman, & I. Valadian. (1994). The distribution of body-fat from childhood to adulthood in a longitudinal-study population. *Annals of Human Biology* 21: 39-55.

Cheek, D.B. (1961). Extracellular volume: its structure and measurement and the influence of age and disease. *Journal of Pediatrics* 58: 103-125.

Chumlea, W.C., and R.N. Baumgartner (1989). Status of anthropometry and body composition data in elderly subjects. *American Journal of Clinical Nutrition* 50: 1158-1166.

Chumlea, W.C., S.S. Guo, R.J. Kucsmarski, K.M. Flegal, C.L. Johnson, S.B. Heymsfield, H.C. Jukaski, K. Friedl, & V.S. Hubbard. (2002). Body composition estimates from NHANES III bioelectric impedance data. *International Journal of Obesity* 26: 1596-1609.

Chumlea, W.C., S.S. Guo, C. Zeller, N.V. Reo, & R.M. Siervogel. (1999). Total body water data for white adults 18 to 64 years of age: The Fels Longitudinal Study. *Kidney International* 56: 244-252.

Cohn, S.H., A. Vaswani, I. Zanzi, J.F. Aloia, M.S. Roginsky, & K.J. Ellis. (1976). Changes in body chemical composition with age measured by total body neutron activation. *Metabolism* 25: 85-95.

Coppoletta, J.M., and S.B. Wolbach (1933). Body length and organ weights of infants and children: study of body lengths and normal weights of more important vital organs of body between birth and 12 years of age. *American Journal of Pathology* 9: 55-70.

Das, U.N. (2001). Is obesity an inflammatory condition? *Nutrition* 17: 953-966.

Dekaban, A.S., and D. Sadowsky (1978). Changes in brain weights during the span of human life: relation of brain weights to body heights and body weights. *Annals of Neurology* 4: 345-356.

Dietz, W.H. (1994). Critical periods in childhood for the development of obesity. *American Journal of Clinical Nutrition* 59: 955-959.

Elia, M. (1992). Organ and tissue contribution to metabolic rate. *Energy metabolism: tissue determinants and cellular corollaries*. J.M. Kinney and H.N. Tucker. New York, Raven Press, pp. 61-77.

Ellis, K.J. (1990). Reference man and woman more fully characterized: variations on the basis of body size, age, sex, and race. *Biological Trace Element Research* 26: 385-400.

Ellis, K.J., R.J. Shypailo, S.A. Abrams, & W.W. Wong. (2000). The reference child and adolescent models of body composition: a contemporary comparison. *Annals of the New York Academy of Sciences* 904: 374-382.

Eurich, R.E., and J. Linder (1984). Body weights, absolute and relative organ weights by maturation and aging (with sexual differences), and their importance as measures of reference for metabolic investigations. *Zeitschrift fur Gerontologie* 17: 60-68.

Evans, W. (1997). Functional and metabolic consequences of sarcopenia. *Journal of Nutrition* 127: 998S-1003S.

Figueroa-Colon, R., M.S. Mayo, M.S. Treuth, R.A. Aldridge, G.R. Hunter, L. Berland, M.I. Gorna, & R.I. Weinsier. (1998). Variability in abdominal adipose tissue measurements using computed tomography in prepubertal girls. *International Journal of Obesity* 22: 1019-1023.

Flynn, M.A., G.B. Nolph, A.S. Baker, W.M. Martin, & G. Krause. (1989). Total body potassium in aging humans: a longitudinal study. *American Journal of Clinical Nutrition* 48: 713-717.

Fomon, S.J., F. Haschke, E.E. Ziegler, & S.E. Nelson. (1982). Body composition of reference children from birth to age 10 years. *American Journal of Clinical Nutrition* 35: 1169-1175.

Forbes, G.B. (1986). Body composition in adolescence. *Human Growth*. F. Falkner and J. M. Tanner. New York, Plenum Press, pp. 119-145.

Forbes, G.B. (1987). *Human body composition: growth, aging, nutrition and activity.* New York, Springer Verlag.

Forbes, G.B. (1991). The companionship of lean and fat: some lessons from body composition studies. *New techniques in nutritional research.* R.G. Whitehead and A. Prentice. New York, Academic Press, pp. 318-330.

Forbes, G.B., and J.C. Reina (1970). Adult lean body mass declines with age: some longitudinal observations. *Metabolism* 19: 653-663.

Frantzell, A., and B.E. Ingelmark (1951). Occurence and distribution of fat in human muscles at various age levels: a morphologic and roentgenologic investigation. *Acta Societa Medica Upsalien* 56: 59-87.

Friis-Hansen, B. (1961). Body water compartments in children: changes during growth and related changes in body composition. *Pediatrics* 28: 169-181.

Frisch, R.E. (1985). Body fat, menarche, and reproductive ability. *Seminars in Reproductive Endocrinology* 3: 45-54.

Fulop, T., I. Worum, J. Csongor, G. Foris, & A. Leovey. (1985). Body composition in elderly people: determination of body composition by multiistope method and the elimination kinetics of these isotopes in healthy elderly subjects. *Gerontology* 31: 6-14.

Gallagher, D., A. Allen, Z.M. Wang, S.B. Heymsfield, & N. Krasnow. (2000). Smaller organ tissue mass in the elderly fails to explain lower resting metabolic rate. *Annals of the New York Academy of Sciences* 904: 449-455.

Gallagher, D., D. Belmonte, P. Deurenberg, Z-M. Wang, N. Krasnow, F.X. Pi-Sunyer, & S.B. Heymsfield. (1998). Organ tissue mass measurement allows modelling of resting energy expenditure and metabolically active tissue mass. *American Journal of Physiology* 275: E249-E258.

Gallagher, D., M. Visser, R.E. De Meersman, D. Sepulveda, R.N. Baumgartner, R.N. Pierson, T. Harris, & S.B. Heymsfield. (1997). Appendicular skeletal muscle mass: effects of age, gender and ethnicity. *Journal of Applied Physiology* 83: 229-239.

Gallagher, D., M. Visser, Z-M. Wang, T. Harris, R.N. Pierson, & S.B. Heymsfield. (1996). Metabolically active component of fat-free body mass: influences of age, adiposity, and gender. *Metabolism* 45: 992-997.

Gaylord, S.A., and M.E. Williams (1994). A brief history of the development of geriatric medicine. *Journal of the American Geriatrics Society* 42: 335-340.

Goodpaster, B.H., F.L. Thaete, & D.E. Kelley. (2000). Composition of skeletal muscle evaluated with computed tomography. *Annals of the New York Academy of Sciences* 904: 18-24.

Goodpaster, B.H., F.L. Thaete, et al. (1997). Subcutaneous abdominal fat and thigh muscle composition predict insulin sensitivity independently of visceral fat. *Diabetes* 46: 1579-1585.

Gower, B.A., T.R. Nagy, & M.I. Goran. (1999). Visceral fat, insulin sensitivity, and lipids in prepubertal children. *Diabetes* 48: 1515-1521.

Guo, S.S., W.C. Chumlea, A.F. Roche, & R.M. Siervogel. (1997). Age- and maturity-related changes in body composition during adolescence into adulthood: The Fels Longitudinal Study. *International Journal of Obesity* 21: 1167-1175.

Guo, S.S., C. Zeller, W.C. Chumlea, & R.M. Siervogel. (1999). Aging, body composition, and lifestyle: the Fels Longitudinal Study. *American Journal of Clinical Nutrition* 70: 405-411.

Haschke, F. (1989). Body composition during adolescence. *Body composition measurements in infants and children.* W.J. Klish and N. Kretchmer. Columbus, OH, Ross Laboratories, pp. 76-83.

He, Q., M. Heo, S. Heshka, J. Wang, R.N. Pierson, J. Albu, Z.M. Wang, S.B. Heymsfield, D. Gallagher. (2003). Total body potassium differs by sex and race across the adult age span. *American Journal of Clinical Nutrition* 78: 72-77.

Heitmann, B.L. (1991). Body fat in the adult Danish population aged 35 to 65 years: an epidemiologic study. *International Journal of Obesity* 15: 535-545.

Henry, C.J.K. (2000). Mechanisms of changes in basal metabolism during ageing. *European Journal of Clinical Nutrition* 54(Suppl): S77-S91.

Heymsfield, S.B., C. Arteaga, B.S. McManus, J. Smith, & S. Moffitt. (1983). Measurement of muscle mass in humans: validity of the 24-hour urinary creatinine method. *American Journal of Clinical Nutrition* 37: 478-494.

Heymsfield, S.B., C. Nunez, C. Testolin, & D. Gallagher. (2000). Anthropometry and methods of body composition measurement for research and field applications in the elderly. *European Journal of Clinical Nutrition* 54(Suppl 3): S26-S32.

Heymsfield, S.B., R. Smith, M. Aulet, B. Bensen, S. Lichtman, J. Wang, & R.N. Pierson. (1990). Appendicular skeletal muscle mass: measurement by dual-photon absorptiometry. *American Journal of Clinical Nutrution* 52: 214-218.

Heymsfield, S.B., Z.-M. Wang, R.N. Baumgartner, F.A. Dilmanian, R. Ma, & S. Yasumura. (1993). Body composition and aging: a study by in vivo neutron activation analysis. *Journal of Nutrition* 123: 432-437.

Hofbauer, K. (2002). Molecular pathways to obesity. *International Journal of Obesity* 26(Suppl 2): S18-S27.

Holliday, A.M. (1971). Metabolic rate and organ size during growth from infancy to maturity and during late gestation and early infancy. *Pediatrics* 47: 169-179.

Huang, T.T.K., M.S. Johnson, R. Figueroa-Colon, J.H. Dwyer, & M.I Goran. (2001). Growth of visceral fat, subcutaneous abdominal fat, and total body fat in children. *Obesity Research* 9: 283-289.

Janssen, I., S.B. Heymsfield, & R. Ross. (2002). Low relative skeletal muscle mass (sarcopenia) in older persons is associated with functional impairment and physical disability. *Journal of the American Geriatrics Society* 50: 889-896.

Janssen, I., S.B. Heymsfield, Z.-M. Wang, & R. Ross. (2000). Skeletal muscle mass and distribution in 468 men and women aged 18-88 yr. *Journal of Applied Physiology* 89: 81-88.

Janssen, I., D.S. Shepard, P.T. Katmarzyk, & R. Roubenoff. (2004). The healthcare costs of sarcopenia in the United States. *Journal of the American Geriatrics Society* 52: 80-85.

Johnston, F.E., and G.D. Foster (2001). *Obesity, growth and development.* London, Smith-Gordon.

Keyahias, J., M.F. Fiatarone, H. Zhuang, & R. Roubenoff. (1997). Total body potassium and body fat: relevance to aging. *American Journal of Clinical Nutrition* 66: 904-910.

Lesser, G.T., and J. Markofsky (1979). Body water compartments with human aging using fat-free mass as the reference standard. *American Journal of Physiology* 236: R215-R220.

Lev-Ran, A. (2001). Human obesity: an evoluationary approach to understanding our bulging waistline. *Diabetes-Metabolism and Research Reviews* 17: 347-362.

Mazariegos, M., Z.-M. Wang, D. Gallagher, R.N. Baumgartner, D.B. Allison, J. Wang, R.N. Pierson, & S.B. Heymsfield. (1994). Differences between young and old females in the five levels of body composition and their relevance to the two-compartment chemical model. *Journal of Gerontology: Medical Sciences* 49: M201-M208.

Melton, L.J., S. Khosla, C.S. Crowson, M.K. O'Connor, M. O'Fallon, & B.L. Riggs. (2000). Epidemiology of sarcopenia. *Journal of the American Geriatrics Society* 48: 625-630.

Metter, E.J., N. Lynch, R. conwit, R. Lindle, J. Tobin, & B. Hurley. (1999). Muscle quality and age: cross-sectional and longitudinal comparisons. *Journal of Gerontology: Biological Sciences* 54A: B207-B218.

Molarius, A., J.C. Seidell, S. Sans, J. Toumilehto, & K. Kuulasmaa. (1999). Waist and hip circumferences and waist/hip ratio in 19 populations of the WHO MONICA Project. *International Journal of Obesity* 23: 116-125.

Moore, F.D., K.H. Olesen, J.D. McMurrey, H.V. Parker, M.R. Ball, & C.M. Boyden. (1963). *The body cell mass and its supporting environment.* Philadelphia, Saunders.

Mora, S., and V. Gilsanz (2003). Establishment of peak bone mass. *Endocrinology and Metabolism Clinics of North America* 32: 39-40.

Morley, J.E., R.N. Baumgartner, R. Roubenoff, & K.S. Nair. (2001). Sarcopenia. *Journal of Laboratory and Clinical Medicine* 137: 231-243.

Mott, J.W., J. Wang, J.C. Thornton, D.B. Allison, S.B. Heymsfield, & R.N Pierson. (1999). Relation between body fat and age in 4 ethnic groups. *American Journal of Clinical Nutrition* 69: 1007-1013.

Mueller, W.H. (1982). The changes with age of the anatomical distribution of fat. *Social Science and Medicine* 16: 191-196.

Mueller, W.H., J.A. Grunbaum, & D.R. Labarthe. (2001). Anger expression, body fat, and blood pressure in adolescents: Project HeartBeat! *American Journal of Human Biology* 13: 531-538.

Mueller, W.H., and J. C. Wohlleb (1981). Anatomical distribution of subcutaneous fat and its distribution by multivariate methods: how valid are principal components? *American Journal of Physical Anthropology* 54: 25-35.

Novak, L.P. (1972). Aging, total body potassium, fat-free mass, and cell mass in males and females between 18 and 85 years. *Journal of Gerontology* 27: 438-443.

Pierson, R.N., D.H.Y. Lin, & R.A. Phillips. (1974). Total body potassium in health: effects of age, sex, height, and fat. *American Journal of Physiology* 226:206-212.

Poehlman, E.T., M.J. Toth, L.B. Bunyard, A.W. Gardner, K.E. Donaldson, E. Coleman, T. Fonong, & P.A. Ades. (1995). Physiological predictors of increasing total and central adiposity in aging men and women. *Archives of Internal Medicine* 155: 2443-2448.

Roche, A.F., R.N. Baumgartner, & S. Guo. (1986). Population methods: anthropometry or estimations. *Human body composition and fat distribution.* N.G. Norgan. Wageningen, Euro-Nut, pp. 31-47.

Rolland-Cachera, M.-F., M. Deheeger, F. Bellisle, M. Sempe, M. Guilloud-Batouille, & E. Patois. (1984). Adiposity rebound in children: a simple indicator for predicting obesity. *American Journal of Clinical Nutrition* 39: 129-135.

Ross, R., J. Rissanen, & R. Hudson. (1996). Sensitivity assocaited with the identification of visceral adipose tissue levels using waist circumference in men and women: effects of weight loss. *International Journal of Obesity* 20: 533-538.

Roubenoff, R. (2000). Sarcopenia and its implications for the elderly. *European Journal of Clinical Nutrition* 54(Suppl 3): S40-S47.

Roubenoff, R. (2000). Sarcopenic obesity: does muscle loss cause fat gain? Lessons from rheumatoid arthritis and osteoarthritis. *Annals of the New York Academy of Sciences* 904: 374-376.

Roubenoff, R., and J. Kehayias (1991). The meaning and measurement of lean body mass. *Nutrition Reviews* 49: 163-175.

Roubenoff, R., J. Kehayias, I.H. Rosenberg, S.B. Heymsfield, & J.G. Cannon. (1997). Standardization of nomenclature of body composition in weight loss. *American Journal of Clinical Nutrition* 66: 192-196.

Schoeller, D.A. (1989). Changes in total body water with age. *American Journal of Clinical Nutrition* 50(Suppl): 1176-1181.

Seidell, J.C., J.G.A.J. Hautvast, & P. Deurenberg. (1989). Overweight—fat distribution and health risks—epidemiological observations—a review. *Infusions Therapie und Transfusionmedizin* 16: 276-286.

Shimokata, H., D. Muller, & R. Andres. (1987). Waist hip ratio, age, sex—a longitudinal study of the effects of change in weight. *International Journal of Obesity* 11: A425.

Siervogel, R.M., W. Wisemandle, L. Maynard, S. Guo, A.A. Roche, W.C. Chumlea, & B. Towne. (1998). Serial changes in body composition throughout adulthood and their relationships to changes in lipid and lipoprotein levels: the Fels Longitudinal Study. *Arteriosclerosis Thrombosis Vascular Biology* 18: 1759-1764.

Snyder, W.S., M.J. Cook, E.S. Nasset, L.R. Karhausen, G.P. Howells, & I.H. Tipton. (1975). *Report of the task group on reference man.* Oxford, UK, Pergamon Press.

Steen, B. (1988). Body composition and aging. *Nutrition Reviews* 46: 45-52.

Steen, B., A. Bruce, B. Isaksson, T. Lewin, & A. Svanborg. (1977). Body composition in 70 year-old males and females in Gothenberg, Sweden. *Acta Medica Scandinavia* 611: 87-112.

Tanko, L.B., L. Movsesyan, U. Mouritzen, C. Christiansen, & O.L. Svendsen. (2002). Appendicular lean tissue mass and the prevalence of sarcopenia among healthy women. *Metabolism* 51: 69-74.

Toth, M.J., A. Tchernof, C.K. Sites, & E.T. Poehlman. (2000). Effect of menopause status on body composition and abdominal fat distribution. *International Journal of Obesity* 24: 226-231.

Visser, M., D. Gallagher, P. Deurenberg, J. Wang, R.N. Pierson, & S.B. Heymsfield. (1997). Density of fat-free body mass: relationship with race, age, and level of body fatness. *American Journal of Physiology: Endocrinology and Metabolism* 272: E781-E787.

Visser, M., M. Pahor, F. Tylavsky, S.B. Kritchevsky, J.A. Cauley, A.B. Newman, B.A. Blunt, & T. Harris. (2003). One- and two-year change in body composition as measured by DXA in a population-based cohort of older men and women. *Journal of Applied Physiology* 94: 2368-2374.

Waki, M., J.G. Kral, M. Mazariegos, J. Wang, R.N. Pierson, & S.B. Heymsfield. (1991). Relative expansion of extracellular fluid in obese vs non-obese women. *American Journal of Physiology: Endocrinology and Metabolism* 261: E199-E203.

Wanagat, J., Z.J. Cao, P. Pathare, & J. M. Aiken. (2001). Mitochondrial DNA deletion mutations colocalize with segmental electron transport system abnormalities, muscle fiber atrophy, fiber splitting, and oxidative damage in sarcopenia. *FASEB Journal* 15: 322-332.

Wang, Z.-M., P. Deurenberg, W. Wang, A. Pietrobelli, R.N. Baumgartner, & S.B. Heymsfield. (1999). Hydration of fat-free body mass: review and critique of a classic body-composition model. *American Journal of Clinical Nutrition* 69: 833-841.

Chapter 18

Adams, W.C.; Deck-Cote, K.; Winters, K.M. (1992) Anthropometric estimation of bone mineral content in young adult females. Am. J. Hum. Biol. 4:767-774.

Aloia, J.F.; Vaswani, A.; Ma, R.; Flaster, E. (1997) Comparison of body composition in Black and White premenopausal women. J. Lab. Clin. Med. 129:294-299.

Baumgartner, R.N.; Rhyne, R.L.; Troup, C.; Wayne, S.; Garry, P.J. (1992) Appendicular skeletal muscle areas assessed by magnetic resonance imaging in older persons. J. Gerontol. 47:M67-M72.

Baumgartner, R.N.; Roche, A.F.; Guo, S.; Lohman, T.G.; Boileau, R.A.; Slaughter, M.H. (1986) Adipose tissue distribution: The stability of principal components by sex, ethnicity and maturation stage. Hum. Biol. 58:719-735.

Bhudhikanok, G.S; Wang, M.-C.; Eckert, K.; Matkin, C.; Marcus, R.; Bachrach, L.K. (1996). Differences in bone mineral in young Asian and Caucasian Americans may reflect differences in bone size. J. Bone Min. Res. 11:1545-1556.

Bonjour, J-P.; Theintz, G.; Buchs, B.; Slosman, D.; Rizzoli, R. (1991) Critical years and stages of puberty for spinal and femoral bone mass accumulation during adolescence. J. Clin. Endocrinol. Metab. 73:555-563.

Borkan, G.A.; Hults, D.E.; Gerzof, S.G.; Robbins, A.H.; Silbert, C.K. (1983) Age changes in body composition revealed by computed tomography. J. Gerontol. 38:673-677.

Bouchard, C. (1994) Genetics of human obesities: Introductory notes. In Bouchard, C., ed. The Genetics of Obesity. Boca Raton, FL: CRC Press, pp. 1-15.

Bouchard, C.; Malina, R.M.; Perusse, L. (1997) Genetics of Fitness and Physical Performance. Champaign, IL: Human Kinetics.

Bouchard, C.; Wilmore, J.H. Unpublished data from the HERITAGE Family Study. See Wilmore et al. 1999 for a description of the study.

Brozek, J.; Grande, F.; Anderson, J.T.; Keys, A. (1963) Densitometric analysis of body composition: revision of some quantitative assumptions. Ann. N. Y. Acad. Sci. 110:113-140.

Chang, C.J.; Wu, C.H.; Chang, C.S.; Yao, W.J.; Yang, Y.C.; Wu, J.S.; Lu, F.H. (2003) Low body mass index but high percent fat in Taiwanese subjects: Implications of obesity cutoffs. Int. J. Obes. Relat. Metab. Disord. 27:253-259.

Chumlea, W.C.; Guo, S.S.; Kuczmarski, R.J.; Flegal, K.M.; Johnson, C.L.; Heymsfield, S.B.; Lukaski, H.C.; Friedl, K.; Hubbard, V.S. (2002) Body composition estimates from NHANES III bioelectrical impedance data. Int. J. Obes. 26:1596-1609.

Cowell, C.T.; Briody, J.; Lloyd-Jones, S.; Smith, C.; Moore, B.; Howman-Giles, R. (1997) Fat distribution in children and adolescents—The influence of sex and hormones. Horm. Res. 48 (Suppl. 5):93-100.

de Koning, F.L.; Binkhorst, R.A.; Kauer, J.M.G.; Thijssen, H.O.M. (1986) Accuracy of an anthropometric estimate of the muscle and bone area in a transversal cross-section of the arm. Int. J. Sports Med. 7:246-249.

de Ridder, C.M.; de Boer, R.W.; Seidell, J.C.; Nieuwenhoff, C.M.; Jeneson, J.A.L.; Bakker, C.J.G.; Zonderland, M.L.; Erich, W.B.M. (1992) Body fat distribution in pubertal girls quantified by magnetic resonance imaging. Int. J. Obes. 16:443-449.

Deurenberg, P.; Yap, M.; van Staveren, W.A. (1998) Body mass index and percent body fat: A meta-analysis among different ethnic groups. Int. J. Obes. 22:1164-1171.

Ellis, K.J. (1990) Reference man and woman more fully characterized: Variations on the basis of body size, age, sex, and race. In Schrauzer, G.N., ed. Biological Trace Element Research. Totowa, N.J.: Humana Press, pp. 385-400.

Ellis, K.J. (1997) Body composition of a young, multiethnic male population. Am. J. Clin. Nutr. 66:1323-1331.

Ellis, K.J.; Abrams, S.A.; Wong, W.W. (1997) Body composition of a young, multiethnic female population. Am. J. Clin. Nutr. 65:724-731.

Ellis, K.J.; Shypailo, R.J.; Abrams, S.A.; Wong, W.W. (2000). The reference child and adolescent models of body composition: A contemporary comparison. Ann. N. Y. Acad. Sci. 904:374-382.

Enzi, G.; Gasparo, M.; Biondetti, P.R.; Fiore, D.; Semisa, M.; Zurlo, F. (1986) Subcutaneous and visceral fat distribution according to sex, age, and overweight, evaluated by computed tomography. Am. J. Clin. Nutr. 44:739-746.

Faulkner, R.A.; Bailey, D.A.; Drinkwater, D.T.; Wilkinson, A.A., Houston, C.S.; McKay, H.A. (1993) Regional and total body bone mineral content, bone mineral density, and total body tissue composition in children 8-16 years of age. Calcif. Tissue Int. 53:7-12.

Fiatarone, M.A.; Marks, E.C.; Ryan, N.D.; Meredith, C.N.; Lipsitz, L.A.; Evans, W.J. (1990) High-intensity strength training in nonagenarians. J. Am. Med. Assoc. 263:3029-3034.

Fomon, S.J. (1966) Body composition of the infant. Part I. The male "reference infant." In Falkner, F., ed. Human Development. Philadelphia: Saunders, pp. 239-246.

Fomon, S.J.; Haschke, F.; Ziegler, E.E.; Nelson, S.E. (1982) Body composition of reference children from birth to age 10 years. Am. J. Clin. Nutr. 35:1169-1175.

Fox, K.; Peters, D.M.; Sharpe, P.; Bell, M. (2000) Assessment of abdominal fat deposition in young adolescents using magnetic resonance imaging. Int. J. Obes. 24:1653-1659.

Gallagher, D.; Visser, M.; de Meersman, R.E.; Sepulveda, D.; Baumgartner, R.N.; Pierson, R.N.; Harris, T.; Heymsfield, S.B. (1997) Appendicular skeletal muscle mass: Effects of age, gender, and ethnicity. J. Appl. Physiol. 83:229-239.

Gallagher, D.; Visser, M.; Sepulveda, D.; Pierson, R.N.; Harris, T.; Heymsfield, S.B. (1996) How useful is body mass index for comparison of body fatness across age, sex, and ethnic groups? Am. J. Epidemiol. 143:228-239.

Gasperino, J.A.; Wang, J.; Pierson, R.N.; Heymsfield, S.B. (1995) Age-related changes in musculoskeletal mass between Black and White women. Metabolism 44:30-34.

Gerace, L.; Aliprantis, A.; Russell, M.; Allison, D.B.; Buhl, K.M.; Wang, J.; Wang, Z-M., Pierson, R.N., Jr.; Heymsfield, S.B. (1994) Skeletal differences between Black and White men and their relevance to body composition estimates. Am. J. Hum. Biol. 6:255-262.

Geusens, P.; Cantatore, F.; Nijs, J.; Proesmans, W.; Emma, F.; Dequeker, J. (1991) Heterogeneity of growth of bone in children at the spine, radius and total skeleton. Growth Develop. Aging 55:249-256.

Glastre, C.; Braillon, P.; David, L.; Cochat, P.; Meunier, P.J.; Delmas, P.D. (1990) Measurement of bone mineral content of the lumbar spine by dual energy x-ray absorptiometry in normal children: Correlations with growth parameters. J. Clin. Endocrinol. Metab. 70:1330-1333.

Goldsmith, N.F.; Johnston, J.O.; Picetti, G.; Garcia, C. (1973) Bone mineral in the radius and vertebral osteoporosis in an insured population: A correlative study using ^{125}I photon absorption and miniature roentgenography. J. Bone Joint Surg. 55A:1276-1293.

Goran, M.I.; Kaskoun, M.; Shuman, W.P. (1995) Intra-abdominal adipose tissue in young children. Int. J. Obes. 19:279-283.

Goran, M.I.; Nagy, T.R.; Treuth, M.S.; Trowbridge, C.; Dezenberg, C.; McGloin, A.; Gower, B.A. (1997) Visceral fat in White and African American prepubertal children. Am. J. Clin. Nutr. 65:1703-1708.

Gordon, C.L.; Halton, J.M.; Atkinson, S.A.; Webber, C.E. (1991) The contributions of growth and puberty to peak bone mass. Growth Develop. Aging 55:257-262.

Goulding, A.; Taylor, R.W.; Gold, E.; Lewis-Barned, N.J. (1996) Regional body fat distribution in relation to pubertal stage: A dual-energy X-ray absorptiometry study of New Zealand girls and young women. Am. J. Clin. Nutr. 64:546-551.

Greig, C.A.; Botella, J.; Young, A. (1993) The quadriceps strength of healthy, elderly people remeasured after eight years. Muscle Nerve 16:6-10.

Guo, S.S.; Chumlea, W.C.; Roche, A.F.; Siervogel, R.M. (1997) Age- and maturity-related changes in body composition during adolescence into adulthood: The Fels Longitudinal Study. Int. J. Obes. 21:1167-1175.

Haschke, F. (1989) Body composition during adolescence. In Klish, W.J.; Kretchmer, N., eds. Body Composition Measurements in Infants and Children (Report of the 98th Ross Conference on Pediatric Research). Columbus, OH: Ross Laboratories, pp. 76-82.

Hattori, K. (1987) Subcutaneous fat distribution pattern in Japanese young adults. J. Anthrop. Soc. Nippon 95:353-359.

Hattori, K.; Numata, N.; Ikoma, M.; Matsuzaka, A.; Danielson, R.R. (1991) Sex differences in the distribution of subcutaneous and internal fat. Hum. Biol. 63:53-63.

Heitmann, B.L.; Swinburn, B.A.; Carmichael, H.; Rowley, K.; Plank, L.; McDermott, R.; Leonard, D.; O'Dea, K. (1997) Are there ethnic differences in the association between body weight and resistance, measured by bioelectrical impedance? Int. J. Obes. 21:1085-1092.

Horber, F.F.; Gruber, B.; Thomi, F.A.; Jensen, E.X.; Jaeger, P. (1997) Effect of sex and age on bone mass, body composition and fuel metabolism in humans. Nutrition 13:524-534.

Huang, T.T.-K.; Johnson, M.S.; Figueroa-Colon, R.; Dwyer, J.H.; Goran, M.I. (2001) Growth of visceral fat, subcutaneous abdominal fat, and total body fat in children. Obes. Res. 9:283-289.

Ishida, Y.; Kanehisa, H.; Fukunaga, T.; Pollock, M.L. (1992) A comparison of fat and muscle thickness in Japanese and American women. Ann. Physiol. Anthropol. 11:29-35.

Ishida, Y.; Kanehisa, H.; Kondo, M.; Fukunaga, T.; Carroll, J.F.; Pollock, M.L.; Graves, J.E.; Leggett, S.H. (1994) Body fat and muscle thickness in Japanese and Caucasian females. Am. J. Hum. Biol. 6:711-718.

Kin, K.; Lee, J.H.E.; Kushida, K.; Sartoris, D.J.; Ohmura, A.; Clopton, P.L., Inque, T. (1993) Bone density and body composition on the Pacific rim: A comparison between Japan-born and U.S.-born Japanese-American women. J. Bone Min. Res. 8:861-869.

Komiya, S.; Muraoka, Y.; Zhang, F-S.; Masuda, T. (1992) Age-related changes in body fat distribution in middle-aged and elderly Japanese. J. Anthropol. Soc. Nippon 100:161-169.

Lemieux, S.; Prud'homme, D.; Bouchard, C.; Tremblay, A.; Despres, J-P. (1993) Sex differences in the relation of visceral adipose tissue accumulation to total body fatness. Am. J. Clin. Nutr. 58:463-467.

Li, J-Y.; Specker, B.L.; Ho, M.L.; Tsang, R.C. (1989) Bone mineral content in Black and White children 1 to 6 years of age. Am. J. Dis. Child. 143:1346-1349.

Liel, Y.; Edwards, J.; Shary, J.; Spicer, K.M.; Gordon, L.; Bell, N.H. (1988) The effects of race and body habitus on bone mineral density of the radius, hip, and spine in premenopausal women. J. Clin. Endocrinol. Metab. 66:1247-1250.

Lohman, T.G. (1986) Applicability of body composition techniques and constants for children and youths. Exerc. Sport Sci. Rev. 14:325-357.

Luckey, M.M.; Meier, D.E.; Mandeli, J.P.; DaCosta, M.C.; Hubbard, M.L.; Goldsmith, S.J. (1989) Radial and vertebral bone density in White and Black women: Evidence for racial differences in premenopausal bone homeostasis. J. Clin. Endocrinol. Metab. 69:762-770.

Malina, R.M. (1969) Quantification of fat, muscle and bone in man. Clin. Orthop. Rel. Res. 65:9-38.

Malina, R.M. (1973) Biological substrata. In: Miller, K.S.; Dreger, R.M., eds. Comparative Studies of Blacks and Whites in the United States. New York: Seminar Press, pp. 53-123.

Malina, R.M. (1989) Growth and maturation: Normal variation and the effects of training. In Gisolfi, C.V.; Lamb, D.R., eds. Perspectives in Exercise Science and Sports Medicine, Vol. II, Youth, Exercise, and Sport. Indianapolis, IN: Benchmark Press, pp. 223-265.

Malina, R.M. (1996) Regional body composition: Age, sex, and ethnic variation. In Roche, A.F.; Heymsfield, S.B.; Lohman, T.G., eds. Human Body Composition. Champaign, IL: Human Kinetics, pp. 217-255.

Malina, R.M.; Bouchard, C. (1988) Subcutaneous fat distribution during growth. In: Bouchard, C.; Johnston, F.E., eds. Fat Distribution during Growth and Later Health Outcomes. New York: Plenum Press; pp. 63-84.

Malina, R.M.; Bouchard, C.; Bar-Or, O. (2004) Growth, Maturation, and Physical Activity, 2nd edition. Champaign, IL: Human Kinetics.

Malina, R.M.; Bouchard, C.; Beunen, G. (1988) Human growth: Selected aspects of current research on well nourished children. Ann. Rev. Anthropol. 17:187-219.

Malina, R.M.; Huang, Y.-C.; Brown, K.H. (1995) Subcutaneous adipose tissue distribution in adolescent girls of four ethnic groups. Int. J. Obes. 19:793-797.

Malina, R.M.; Little, B.B.; Stern, M.P.; Gaskill, S.P.; Hazuda, H.P. (1983) Ethnic and social class differences in selected anthropometric characteristics of Mexican American and Anglo adults. Hum. Biol. 55:867-883.

Maughan, R.J.; Watson, J.S.; Weir, J. (1984) The relative proportions of fat, muscle and bone in the normal human forearm as determined by computed tomography. Clin. Sci. 66:683-689.

Maynard, L.M.; Guo, S.S.; Chumlea, W.C.; Roche, A.F.; Wisemandle, W.A., Zeller, C.; Town, B.; Siervogel, R.M. (1998) Total body and regional bone mineral content and area bone mineral density in children aged 8-18 y: The Fels Longitudinal Study. Am. J. Clin. Nutr. 68:1111-1117.

Mazess, R.B.; Peppler, W.W.; Chesney, R.W.; Lange, T.A.; Lindgren, U.; Smith, E., Jr. (1984) Total body and regional bone mineral by dual-photon absorptiometry in metabolic bone disease. Calcif. Tissue Int. 36:8-13.

McCormick, D.P.; Ponder, S.W.; Fawcett, H.D.; Palmer, J.L. (1991) Spinal bone mineral density in 335 normal and obese children and adolescents: Evidence for ethnic and sex differences. J. Bone Min. Res. 6:507-513.

Moulton, C.R. (1923) Age and chemical development in mammals. J. Biol. Chem. 57:79-97.

Mueller, W.H. (1988) Ethnic differences in fat distribution during growth. In Bouchard, C.; Johnston, F.E., eds. Fat Distribution During Growth and Later Health Outcomes. New York: Plenum Press, pp. 127-145.

Mueller, W.H.; Shoup, R.F.; Malina, R.M. (1982) Fat patterning in athletes in relation to ethnic origin and sport. Ann. Hum. Biol. 9:371-376.

Nelson, D.A.; Feingold, M.; Bolin, F.; Parfitt, A.M. (1991) Principal components analysis of regional bone density in Black and White women: Relationship to body size and composition. Am. J. Phys. Anthropol. 86:507-514.

Nelson, D.A.; Kleerekoper, M.; Parfitt, A.M. (1988) Bone mass, skin color and body size among Black and White women. Bone Min. 4:257-264.

Norgan, N.G. (1994) Population differences in body composition in relation to the body mass index. Eur. J. Clin. Nutr. 48(Suppl. 3):S10-S27.

Ortiz, O.; Russell, M.; Daley, T.L.; Baumgartner, R.N.; Waki, M.; Lichtman, S.; Wang, J.; Pierson, R.N., Jr.; Heymsfield, S.B. (1992) Differences in skeletal muscle and bone mineral mass between Black and White females and their relevance to estimates of body composition. Am. J. Clin. Nutr. 55:8-13.

Park, Y-W.; Allison, D.B.; Heymsfield, S.B.; Gallagher, D. (2001) Larger amounts of visceral adipose tissue in Asian Americans. Obes. Res. 9:381-387.

Rice, C.L.; Cunningham, D.A.; Paterson, D.H.; Lefcoe, M.S. (1990) A comparison of anthropometry with computed tomography in limbs of young and aged men. J. Gerontol. 45:M174-M179.

Rico, H.; Revilla, M.; Hernandez, E.R.; Villa, L.F.; Alvarez de Buergo, M. (1992) Sex differences in the acquisition of total bone mineral mass peak assessed through dual-energy x-ray absorptiometry. Calcif. Tissue Int. 51:251-254.

Riggs, B.L.; Wahner, H.W.; Dunn, W.L.; Mazess, R.B.; Offord, K.P.; Melton, L.J. (1981) Differential changes in bone mineral density of the appendicular and axial skeleton with aging. J. Clin. Invest. 67:328-335.

Rolland-Cachera, M-F.; Bellisle, F.; Deheeger, M.; Pequignot, F.; Sempe, M. (1990) Influence of body fat distribution during childhood on body fat distribution in adulthood: A two-decade follow-up study. Int. J. Obes. 14:473-481.

Ross, P.D.; Orimo, H.; Wasnich, R.D.; Vogel, J.M.; MacLean, C.J.; Davis, J.W.; Nomura, A. (1989) Methodological issues in comparing genetic and environmental influences on bone mass. Bone Min. 7:67-77.

Russell-Aulet, M.; Wang, J.; Thornton, J.; Colt, E.W.D.; Pierson, R.N., Jr. (1991) Bone mineral density and mass by total-body dual-photon absorptiometry in normal White and Asian men. J. Bone Min. Res. 6:1109-1113.

Schantz, P.; Randall-Fox, E.; Hutchinson, W.; Tyden, A.; Åstrand, P-O. (1983) Muscle fiber type distribution, muscle cross-sectional area and maximal voluntary strength in humans. Acta Physiol. Scand. 117:219-226.

Seidell, J.C.; Oosterlee, A.; Deurenberg, P.; Hautvast, J.G.A.; Ruijs, J.H.J. (1988) Abdominal fat depots measured with computed tomography: Effects of degree of obesity, sex, and age. Eur. J. Clin. Nutr. 42:805-815.

Sipila, S.; Suominen, H. (1991) Ultrasound imaging of the quadriceps muscle in elderly athletes and untrained men. Muscle Nerve 14:527-533.

Sipila, S.; Suominen, H. (1993) Muscle ultrasonography and computed tomography in elderly trained and untrained women. Muscle Nerve 16:294-300.

Slaughter, M.H.; Lohman, T.G.; Boileau, R.A.; Crist, C.B.; Stillman, R.J. (1990) Differences in the subcomponents of fat-free body in relation to height between Black and White children. Am. J. Hum. Biol. 2:209-217.

Snyder, W.S.; Cook, M.J.; Nasset, E.S.; Karhausen, L.R.; Howells, G.P.; Tipton, T.H., eds. (1984) Report on the Task Group on Reference Man (ICRP 23). New York: Pergamon Press.

Sugimoto, T.; Tsutsumi, M.; Fujii, Y.; Kawakatsu, M.; Negishi, H.; Lee, M.C.; Tsai, K-S.; Fukase, M.; Fujita, T. (1992) Comparison of bone mineral content among Japanese, Koreans, and Taiwanese assessed by dual-photon absorptiometry. J. Bone Min. Res. 7:153-159.

Sun, S.S.; Chumlea, W.C.; Heymsfield, S.B.; Lukaski, H.C.; Schoeller, D.; Friedl, K.; Kuczmarski, R.J.; Flegal, K.M.; Johnson, C.L.; Hubbard, V.S. (2003) Development of bioelectrical impedance analysis prediction equations for body composition with the use of a multicomponent model for use in epidemiologic surveys. Am. J. Clin. Nutr. 77:331-340.

Tahara, Y.; Moji, K.; Aoyagi, K.; Nishizawa, S.; Yukawa, K.; Tsunawake, N.; Muraki, S.; Mascie-Taylor, C.G.N. (2002a) Age-related pattern of body density and body composition in Japanese males and females, 11 and 18 years of age. Am. J. Hum. Biol. 14:327-337.

Tahara, Y.; Moji, K.; Aoyagi, K.; Tsunawake, N.; Muraki, S.; Mascie-Taylor, C.G.N. (2002b) Age-related pattern of body density and body composition of Japanese men and women 18 to 59 years of age. Am. J. Hum. Biol. 14:743-752.

Theintz, G.; Buchs, B.; Rizzoli, R.; Slosman, D.; Clavien, H.; Sizonenko, P.C.; Bonjour, J-P. (1992) Longitudinal monitoring of bone mass accumulation in healthy adolescents: Evidence for a marked reduction after 16 years of age at the levels of lumbar spine and femoral neck in female subjects. J. Clin. Endocrinol. Metab. 75:1060-1065.

Trotter, M.; Broman, G.E.; Peterson, R.R. (1959) Density of cervical vertebrae and comparisons with densities of other bones. Am. J. Phys. Anthropol. 17:19-25.

Trotter, M.; Broman, G.E.; Peterson, R.R. (1960) Densities of bones of White and Negro skeletons. J. Bone Joint Surg. 42A:50-58.

Trotter, M.; Hixon, B.B. (1974) Sequential changes in weight, density, and percentage ash weight of human skeletons from an early fetal period through old age. Anat. Rec. 179:1-18.

Trotter, M.; Peterson, R.R. (1962) The relationship of ash weight and organic weight of human skeletons. J. Bone Joint Surg. 44A:669-681.

Trotter, M.; Peterson, R.R. (1970) The density of bones in the fetal skeleton. Growth 34:283-292.

Tsunenari, T.; Tsutsumi, M.; Ohno, K.; Yamamoto, Y.; Kawakatsu, M.; Shimogaki, K.; Negishi, H.; Sugimoto, T.; Fukase, M.; Fujita, T. (1993) Age- and gender-related changes in body composition in Japanese subjects. J. Bone Min. Res. 8:397-402.

Visser, M.; Gallagher, D.; Duerenberg, P.; Wang, J.; Pierson, R.N. Jr.; Heymsfield, S.B. (1997) Density of fat-free body mass: Relationship with race, age, and level of body fatness. Am. J. Physiol. 272 (Endocrinol. Metab. 35):E781-E787.

Wang, J.; Thornton, J.C.; Burastero, S.; Shen, J.; Tanenbaum, S.; Heymsfield, S.B.; Pierson, R.N. Jr. (1996) Comparisons for body mass index and body fat percent among Puerto Ricans, Blacks, Whites and Asians living in the New York City area. Obes. Res. 4:377-384.

Wang, J.; Thornton, J.C.; Russell, M.; Burastero, S.; Heymsfield, S.B.; Pierson, R.N. Jr. (1994) Asians have lower body mass index (BMI) but higher percent fat than do Whites: Comparisons of anthropometric measurements. Am. J. Clin. Nutr. 60:23-28.

Wang, M.-C.; Aguirre, M.; Bhudhikanok, G.S.; Kendall, C.G.; Kirsch, S.; Marcus, R.; Bachrach, L.K. (1997) Bone mass and hip axis length in healthy Asian, Black, Hispanic, and White American youths. J. Bone Min. Res. 12:1922-1935.

Weits, T.; van der Beek, E.J.; Wedel, M.; ter Haar Romeny, B.M. (1988) Computed tomography measurement of abdominal fat deposition in relation to anthropometry. Int. J. Obes. 12:217-225.

Wilmore, J.H.; Despres, J.-P.; Stanforth, P.R.; Mandel, S.; Rice, T.; Gagnon, J.; Leon, A.S.; Rao, D.C.; Skinner, J.S.; Bouchard, C. (1999) Alterations in body weight and composition consequent to 20 wk of endurance training: The HERITAGE Family Study. Am. J. Clin. Nutr. 70:346-352.

Yano, K.; Wasnich, R.D.; Vogel, J.M.; Heilbrun, L.K. (1984) Bone mineral measurements among middle-aged and elderly Japanese residents in Hawaii. Am. J. Epidemiol. 119:751-764.

Yanovski, J.A.; Yanovski, S.Z.; Filmer, K.M.; Hubbard, V.S.; Avila, N.; Lewis, B.; Reynolds, J.C.; Flood, M. (1996) Differences in body composition of Black and White girls. Am. J. Clin. Nutr. 64:833-839.

Young, A.; Stokes, M.; Crowe, M. (1984) Size and strength of the quadriceps muscles of old and young women. Eur. J. Clin. Invest. 14:282-287.

Young, A.; Stokes, M.; Crowe, M. (1985) The size and strength of the quadriceps muscles of old and young men. Clin. Physiol. 5:145-154.

Chapter 19

Briend, A. (1985). Do maternal energy reserves limit fetal growth? *Lancet, 1,* 38-40.

Brown, J.E. (1988). Weight gain during pregnancy: What is optimal? *Clinical Nutrition, 7,* 181-190.

Butte, N.F., Hopkinson, J.M., & Nicolson, M.A. (1997). Leptin in human reproduction: serum leptin levels in pregnant and lactating women. *Journal of Clinical Endocrinology and Metabolism, 82,* 585-589.

Catalano, P.M., Wong, W.W., Drago, N.M., & Amini, S.B. (1995). Estimating body composition in late gestation: a new hydration constant for body density and total body water. *American Journal of Physiology, 268,* E153-E158.

de Groot, L.C., Boekholt, H.A., Spaaij, C.K., van Raaij, J.M., Drijvers, J.J., van der Heijden, L.J., van der Heide, D., & Hautvast, J.G. (1994). Energy balances of healthy Dutch women before and during pregnancy: limited scope for metabolic adaptations in pregnancy. *American Journal of Clinical Nutrition, 59,* 827-832.

Duffus, G.M., MacGillivray, I., & Dennis, K.J. (1971). The relationship between baby weight and changes in maternal weight, total body water, plasma volume, electrolytes and proteins and urinary oestriol excretion. *Journal of Obstetrics and Gynaecology of the British Commonwealth, 78,* 97-104.

Durnin, J.V., & Womersley, J. (1974). Body fat assessed from total body density and its estimation from skinfold thickness: measurements on 481 men and women aged from 16 to 72 years. *British Journal of Nutrition, 32,* 77-97.

Emerson, K., Jr., Poindexter, E.L., & Kothari, M. (1975). Changes in total body composition during normal and diabetic pregnancy. Relation to oxygen consumption. *Obstetrics and Gynecology, 45,* 505-511.

Fidanza, F. (1987). The density of fat-free body mass during pregnancy. *International Journal of Vitamin and Nutrition Research, 57,* 104.

Forbes, G. (1987). Body composition of the fetus. In G. Forbes, *Human body composition* (pp. 101-126). New York: Springer-Verlag.

Forsum, E., Sadurskis, A., & Wager, J. (1988). Resting metabolic rate and body composition of healthy Swedish women during pregnancy. *American Journal of Clinical Nutrition, 47,* 942-947.

Forsum, E., Sadurskis, A., & Wager, J. (1989). Estimation of body fat in healthy Swedish women during pregnancy and lactation. *American Journal of Clinical Nutrition, 50,* 465-473.

Fuller, N.J., Jebb, S.A., Laskey, M.A., Coward, W.A., & Elia, M. (1992). Four-compartment model for the assessment of body composition in humans: comparison with alternative methods, and evaluation of the density and hydration of fat-free mass. *Clinical Science, 82,* 687-693.

Goldberg, G.R., Prentice, A.M., Coward, W.A., Davies, H.L., Murgatroyd, P.R., Wensing, C., Black, A.E., Harding, M., & Sawyer, M. (1993). Longitudinal assessment of energy expenditure in pregnancy by the doubly labeled water method. *American Journal of Clinical Nutrition, 57,* 493-505.

Gutersohn, A., Naber, C., Muller, N., Erbel, R., & Siffert, W. (2000) G protein β3 subunit 825TT genotype and post-pregnancy weight retention. *Lancet, 355,* 1240-1241.

Hediger, M.L., Scholl, T.O., Schall, J.I., Healey, M.F., & Fischer, R.L. (1994). Changes in maternal upper arm fat stores are predictors of variation in infant birth weight. *Journal of Nutrition, 124,* 24-30.

Highman, T.J., Friedman, J.E., Huston, L.P., Wong, W.W., & Catalano, P.M. (1998). Longitudinal changes in maternal serum leptin concentrations, body composition, and resting metabolic rate in pregnancy. *American Journal of Obstetrics and Gynecology, 178,* 1010-1015.

Hocher, B., Slowinski, T., Stolze, T., Pleschka, A., Neumayer, H-H., & Halle, H. (2000). Association of maternal G protein β3 subunit 825TT allele with low birthweight. *Lancet, 355,* 1241-1242.

Hopkinson, J.M., Butte, N.F., Ellis, K.J., Wong, W.W., Puyau, M.R., & Smith, E.O. (1997). Body fat estimation in late pregnancy and early postpartum: comparison of two-, three-, and four-component models. *American Journal of Clinical Nutrition, 65,* 432-438.

Hytten, F.E., Thomson, A.M., & Taggart, N. (1966). Total body water in normal pregnancy. *Journal of Obstetrics and Gynaecology of the British Commonwealth, 73,* 553-561.

Hytten, F.E., & Leitch, I. (1964). *The physiology of human pregnancy.* Oxford, UK: Blackwell.

Hytten, F., & Chamberlain, G. (Eds.) (1980). *Clinical physiology in obstetrics*. Oxford, UK: Blackwell Scientific.

IOM (Institute of Medicine), National Academy of Sciences. (1990). *Nutrition during pregnancy*. Washington, DC: National Academy Press.

Jaque-Fortunato, S.V., Khodiguian, N., Artal, R., & Wiswell, R.A. (1996). Body composition in pregnancy. *Seminars in Perinatology, 20,* 340-342.

Kopp-Hoolihan, L.E., van Loan, M.D., Wong, W.W., & King, J.C. (1999). Fat mass deposition during pregnancy using a four-component model. *Journal of Applied Physiology, 87,* 196-202.

Langhoff-Roos, J., Lindmark, G., & Gebre-Medhin, M. (1987). Maternal fat stores and fat accretion during pregnancy in relation to infant birth weight. *British Journal of Obstetrics and Gynaecology, 94,* 1170-1177.

Lederman, S.A., Paxton, A., Heymsfield, S.B., Wang, J., Thornton, J., & Pierson, R.N., Jr. (1997). Body fat and water changes during pregnancy in women with different body weight and weight gain. *Obstetrics and Gynecology, 90,* 483-488.

Lederman, S.A., Paxton, A., Heymsfield, S.B., Wang, J., Thornton, J.C., & Pierson, R.N., Jr. (1999). Maternal fat and water gain during pregnancy: Do they raise infant birth weight? *American Journal of Obstetrics and Gynecology, 180,* 235-240.

Lederman, S.A., Pierson, R.N., Jr., Wang, J., Paxton, A., Thornton, J., Wendel, J., & Heymsfield, S.B. (1993). Body composition measurements during pregnancy. *Basic Life Sciences, 60,* 193-195.

Lukaski, H.C., Siders, W.A., Nielsen, E.J., & Hall, C.B. (1994). Total body water in pregnancy: assessment by using bioelectrical impedance. *American Journal of Clinical Nutrition, 59,* 578-585.

Masuda, K., Osada, H., Iitsuka, Y., Seki, K., & Sekiya, S. (2002). Positive association of maternal G protein beta3 subunit 825T allele with reduced head circumference at birth. *Pediatric Research, 52,* 687-691.

Paxton, A., Lederman, S.A., Heymsfield, S.B., Wang, J., Thornton, J., & Pierson, R.N., Jr. (1998). Anthropometric equations for studying body fat in pregnant women. *American Journal of Clinical Nutrition, 67,* 104-110.

Pedersen, S., Gotfredson, A., & Knudsen, F.U. (1989). Total body bone mineral in light-for-gestational-age infants and appropriate-for-gestational-age infants. *Acta Paediatrica Scandinavia, 78,* 347-350.

Pipe, N.G., Smith, T., Halliday, D., Edmonds, C.J, Williams, C., & Coltart, T.M. (1979). Changes in fat, fat-free mass and body water in human normal pregnancy. *British Journal of Obstetrics and Gynaecology, 86,* 929-940.

Prentice, A.M., Goldberg, G.R., Davies, H.L., Murgatroyd, P.R., & Scott, W. (1989) Energy-sparing adaptations in human pregnancy assessed by whole-body calorimetry. *British Journal of Nutrition, 62,* 5-22.

Rees, J.M., Lederman, S.A., & Kiely, J.L. (1996). Birth weight associated with lowest neonatal mortality: infants of adolescent and adult mothers. *Pediatrics, 98,* 1161-1166.

Ritchie, L.D., Fung, E.B., Halloran, B.P., Turnlund, J.R., Van Loan, M.D., Cann, C.E., & King, J.C. (1998). A longitudinal study of calcium homeostasis during human pregnancy and lactation and after resumption of menses. *American Journal of Clinical Nutrition, 67,* 693-701.

Rosso, P., Donoso, E., Braun, S., Espinoza, R., Fernandez, C., & Salas, S.P. (1993). Maternal hemodynamic adjustments in idiopathic fetal growth retardation. *Gynecologic and Obstetric Investigation, 35,* 162-165.

Seeds, J.W., & Peng, T.C. (2000). Does augmented growth impose an increased risk of fetal death? *American Journal of Obstetrics and Gynecology, 183,* 316-322.

Seitchik, J., Alper, C., & Szutka, A. (1963). Changes in body composition during pregnancy. *Annals of the New York Academy of Sciences, 110,* 821-829.

Selinger, A. (1977). *The body as a three component system*. Ann Arbor, MI: University Microfilms International.

Siri, W.E. (1961). Body composition from fluid spaces and density: analysis of methods. In J. Brozek & A. Henschel (Eds.), *Techniques for measuring body composition* (pp. 223-244). Washington, DC: National Academy of Sciences.

Sowers, M., Crutchfield, M., Jannausch, M., Updike, S., & Corton, G. (1991). A prospective evaluation of bone mineral change in pregnancy. *Obstetrics and Gynecology, 77,* 841-845.

Valensise, H., Andreoli, A., Lello, S., Magnani, F., Romanini, C., & De Lorenzo, A. (2000). Multifrequency bioelectrical impedance analysis in women with a normal and hypertensive pregnancy. *American Journal of Clinical Nutrition, 72,* 780-783.

van Raaij, J.M., Peek, M.E., Vermaat-Miedema, S.H., Schonk, C.M., & Hautvast, J.G. (1988). New equations for estimating body fat mass in pregnancy from body density or total body water. *American Journal of Clinical Nutrition, 48,* 24-29.

Villar, J., Cogswell, M., Kestler, E., Castillo, P., Menendez, R., & Repke, J.T. (1992). Effect of fat and fat-free mass deposition during pregnancy on birth weight. *American Journal of Obstetrics and Gynecology, 167,* 1344-1352.

Ziegler, E.E., O'Donnell, A.M., Nelson, S.E., & Fomon, S.J. (1976). Body composition of the reference fetus. *Growth, 40,* 329-341.

Chapter 20

American College of Sports Medicine. (1998). Position stand: Exercise and physical activity for older adults. *Medicine and Science in Sports and Exercise,* 30:975-991.

Bailey, D. (2000). Is anyone out there listening? *Quest,* 52: 344-350.

Bailey, D.A., Faulkner, R.A., McKay, H.A (1996). Growth, physical activity, and bone mineral acquisition. *Exercise and Sport Sciences Reviews,* 24:233-267.

Bailey, D.A., McKay, H.A., Mirwald, R.L., Crocker, P.R.E., Faulkner, R.A. (1999). A six-year longitudinal study of the relationship of physical activity to bone mineral accrual in growing children: the University of Saskatchewan bone mineral accrual study. *Journal of Bone and Mineral Research,* 14:1672-1679.

Ballor, D.L. (1996). Exercise training and body composition changes. In A.F. Roche, S.B. Heymsfield, & T.G. Lohman (Eds.), *Human Body Composition* (pp. 287-304). Champaign, IL: Human Kinetics.

Barbeau, P., Gutin, B., Litaker, M., Owens, S., Riggs, S., Okuyama, T. (1999). Correlates of individual differences in body-composition changes resulting from physical training in obese children. *American Journal of Clinical Nutrition,* 69: 705-711.

Barengolts, E.I., Lathon, P.V., Curry, D.J., Kukreja, S.C. (1994). Effects of endurance exercise on bone histomorphometric parameters in intact and ovariectomized rats. *Bone Mineral* 26:133-140.

Bass, S., Pearce, G., Bradney, M., Hendrich, E., Delmas, P.D., Harding, A., Seeman, E. (1998). Exercise before puberty may confer residual benefits in bone density in adulthood: Studies in active prepubertal and retired female gymnasts. *Journal of Bone and Mineral Research*, 13: 500-507.

Bassey, E.J., Rothwell, M.C., Littlewood, J.J., Pye, D.W. (1998). Pre- and postmenopausal women have different bone mineral density responses to the same high-impact exercise. *Journal of Bone and Mineral Research*, 13:1805-1813.

Baumgartner, R.N., Koehler, K.M., Gallagher, D., Romero, L., Heymsfield, S.B., Ross, R.R., Garry, P.J., Lindeman, R.D. (1998). Epidemiology of sarcopenia among the elderly in New Mexico. *American Journal of Epidemiology*, 147: 755-763.

Blair, S.N., Cheng, Y., Holder, J.S. (2001). Is physical activity or physical fitness more important in defining health benefits? *Medicine and Science in Sports and Exercise*, 33: S379-399.

Block, J.E. (1997). Interpreting studies of exercise and osteoporosis: a call for rigor. *Controlled Clinical Trials*, 18:54-57.

Blundell, J.E., King, N.A. (1999). Physical activity and regulation of food intake: Current evidence. *Medicine and Science in Sports and Exercise*, 31:S573-S583.

Bouchard, C., & Rankinen, T. (2001). Individual differences in response to regular physical activity. *Medicine and Science in Sports and Exercise*, 33:S446-S451.

Bradney, M., Pearce, G., Naughton, G., Sullivan, C., Bass, S., Beck, T., Carlson, J., Seeman, E. (1998). Moderate exercise during growth in prepubertal boys: Changes in bone mass, size, volumetric density, and bone strength: A controlled prospective study. *Journal of Bone and Mineral Research*, 13:1814-1821.

Brooks, G.A., Falney, T.D., White, T.P., Baldwin, K.M. (2000). *Exercise Physiology: Human Bioenergetics and its Applications*. Mountain View, CA: Mayfield.

Cussler, E.C., Lohman, T.G., Going, S.B., Houtkooper, L.B., Metcalfe, L.L., Flint Wagner, H.G., Harris, R.B. Teixeira, P.J. (2003). Weight lifted in strength training predicts bone change in postmenopausal women. *Medicine and Science in Sports and Exercise*, 35:10-17.

Dalean, N. Laftman, P., Ohlsen, H., Stromberg, L. (1985). The effect of athletic activity on the bone mass in human diaphyseal bone. *Orthopedics*, 8:1139-1141.

Dalsky, G.P., Stocke, K.S., Ehsani, A.A., Slatopolsky, E., Lee, W.C. Birge, S.J. (1988). Weight bearing exercise training and lumbar bone mineral content in postmenopausal women. *Annals of Internal Medicine*, 108:824-828.

Dengel, D.R., Brown, M.D., Ferrell, R.E., Reynolds, T.H., Supiano, M.A. (2002). Exercise-induced changes in insulin action are associated with ACE gene polymorphisms in older adults. *Physiological Genomics*, 11:73-80.

Dietz, W.H. (1998). Health consequences of obesity in youth: Childhood predictors of adult disease. *Pediatrics*, 101: S518-S525.

DiPietro, L. (1999). Physical activity in the prevention of obesity: Current evidence and research issues. *Medicine and Science in Sports and Exercise*, 31:S542-S546.

Drinkwater, B.L. (1994). Physical activity, fitness and osteoporosis. In C. Bouchard, R.J. Shephard, & Stephens (Eds), *Physical Activity, Fitness and Health. International Proceedings and Consensus Statement* (pp. 724-736). Champaign, IL: Human Kinetics.

Esposito, K., Pontillo, A., Di Palo, C., Giugliano, G., Masella, M., Marfella, R., Giugliano, D. (2003). Effect of weight loss and lifestyle changes on vascular inflammatory markers in obese women. *Journal of the American Medical Association*, 289:1799-1804.

Evans, E.M., Van Pelt, R.E., Binder, E.F., Williams, D.B., Ehsani, A.A., Kohrt, W.M. (2001). Effects of HRT and exercise training on insulin action, glucose tolerance, and body composition in older women. *Journal of Applied Physiology*, 90: 2033-2040.

Fiatarone, M.A., Marks, E.C., Ryan, N.D., Meredith, C.N., Lipsitz, L.A., Evans, W.J. (1990). High-intensity strength training in nonagenarians. Effects on skeletal muscle. *Journal of the American Medical Association*, 263:3029-3034.

Fiatarone, M.A., O'Neill, E.F., Ryan, N.D., Clements, K.M., Solares, G.R., Nelson, M.E., Roberts, S.B., Kehayias, J.J., Lipsitz, LA., Evans, W.J. (1994). Exercise training and nutritional supplementation for physical frailty in very elderly people. *New England Journal of Medicine*, 330:1769-1775.

Flegal, K.M., Carroll, M.D., Ogden, C.L, Johnson, C.L. (2002). Prevalence and trends in obesity among US adults, 1999-2000. *Journal of the American Medical Association*, 288: 1723-1727.

Forbes, G.B. (2000). Body fat content influences the body composition response to nutrition and exercise. *Annals of the New York Academy of Science*, 904:359-365.

Forwood, M.R. (2001). Mechanical effects on the skeleton: Are there clinical implications? *Osteoporosis International*, 12: 77-83.

Forwood, M., Burr, D. (1993). Physical activity and bone mass: Exercise in futility? *Bone Mineral*, 21:89-112.

Fried, S.K., Leibel, R.L., Edens, N.K, Kral, J.G. (1993). Lipolysis in intraabdominal adipose tissue of obese women and men. *Obesity Research*, 1:443-448.

Friedlander, A.L., Genant, H.K., Sadowsky, S., Byl, N.N, Gluer, C.C. (1995). A two-year program of aerobics and weight training enhances bone mineral density of young women. *Journal of Bone and Mineral Research*, 10:574-585.

Frost, H.M. (1987). The mechanostat: A proposed pathogenic mechanism of osteoporosis and the bone mass effects of mechanical and nonmechanical agents. *Bone Mineral* 2: 73-85.

Fuchs, R.K, Bauer, J.J., Snow, C.M. (2001). Jumping improves hip and lumbar spine bone mass in prepubescent children: A randomized controlled trial. *Journal of Bone and Mineral Research*, 16:148-156.

Garenc, C., Perusse, L., Bergeron, J., Gagnon, J., Chagnon, Y.C., Borecki, I.B., Leon, A.S., Skinner, J.S., Wilmore, J.H., Rao, D.C., Bouchard, C. (2001). Evidence of LPL gene-exercise interaction for body fat and LPL activity: The HERITAGE family study. *Journal of Applied Physiology*, 91:1334-1340.

Gleeson, P.B., Protas, E.J., LeBlanc, A.D., Schneider, V.S. Evans, H.J. (1990). Effects of weight lifting on bone mineral density in premenopausal women. *Journal of Bone and Mineral Research* 5:153-158.

Going, S.B., Lohman, T.L., Houtkooper, L.B., Metcalfe, L., Flint-Wagner, H., Blew, R., Stanford, V., Cussler, E., Weber, J. (2003). Effects of exercise on bone mineral density in calcium-supplemented postmenopausal women with and without hormone replacement therapy. *Osteoporosis International*, 14:637-643.

Goran, M.I., Poehlman, E.T. (1992). Endurance training does not enhance total energy expenditure in healthy elderly persons. *American Journal of Physiology,* 263:E950-E957.

Haapsalo, H., Sievanen, H., Kannus, P., Heinonen, A., Oja, P., & Vouri I. (1996). Dimensions and estimated mechanical characteristics of the humerus after long-term tennis loading. *Journal of Bone and Mineral Research,* 11:864-872.

Halliday, D., Hesp, R., Stalley, S.F., Warwick, P., Altman, D.G., Garrow, J.S. (1979). Resting metabolic rate, weight, surface area and body composition in obese women. *International Journal of Obesity,* 3:1-6.

Hardman, A.E. (2001). Issues of fractionization of exercise (short vs. long bouts). *Medicine and Science in Sports and Exercise,* 33:S421-S427.

Haskell, W.L. (1994). Health consequences of physical activity: Understanding and challenges regarding dose-response. *Medicine and Science in Sports and Exercise,* 26:649-660.

Haskell, W.L. (2001). What to look for in assessing responsiveness to exercise in a health context. *Medicine and Science in Sports and Exercise,* 33:S454-S458.

Hass, C.J., Feigenbaum, M.S., & Franklin, B.A. (2001). Prescription of resistance training for healthy populations. *Sports Medicine,* 31:953-964.

Heaney, R.P., Abrams, S., Dawson-Hughes, B., Looker, A., Marcus, R., Matkovic, V., Weaver, C. (2000). Peak bone mass. *Osteoposis,* 11:985-1009.

Heikkinen, J., Kurttila-Matero, E., Kyllonen, E., Vuori, J., Takala, T., Vaananen, H.K. (1991).Moderate exercise does not enhance the positive effect of estrogen on bone mineral density in postmenopausal women. *Calcified Tissue International,* 49:S83-S84.

Heinonen, A., Oja, P., Kannus, P., Sievanen, H., Haapasalo, H., Manttari, A. Vuori, I. (1995). Bone mineral density in female athletes representing sports with different loading characteristics of the skeleton. *Bone,* 17:197-203.

Heinrich, C.H., Going, S.B., Pamenter, R.W., Perry C.D., Boyden, T.W., Lohman, T.G. (1990). Bone mineral content of cyclically menstruating female resistance and endurance trained athletes. *Medicine and Science in Sports and Exercise,* 22: 558-563.

Hesselink, M.K.C., Mensink, M., Schrauwen P (2003). Human uncoupling protein-3 and obesity: An update. *Obesity Research,* 11:1429-1443.

Hickey, M.S., Calsbeek, D.J. (2001). Plasma leptin and exercise. *Sports Medicine,* 31:583-589.

Huddleston A.L., Rockwell. D., Kulund, D.N., & Harrison, R.B. (1980). Bone mass in lifetime tennis athletes. *Journal of the American Medical Association,* 244:1107-1109.

Hunter, G.R., Bryan, D.R., Wetzstein, C.J., Zuckerman, P.A., Bamman, M.M. (2002). Resistance training and intra-abdominal adipose tissue in older men and women. *Medicine and Science in Sports and Exercise,* 34:1023-1028.

Irwin, M.L., Yasui, Y., Ulrich, C.M., Bowen, D., Rudolph, R.E., Schwartz, R.S., Yukawa, M., Aiello, E., Potter, J.D., McTiernan, A. (2003). Effect of exercise on total and intra-abdominal body fat in postmenopausal women: A randomized controlled trial. *Journal of the American Medical Association,* 289:323-330.

Jakicic, J.M., Clark, K., Coleman, E., Donnelly, J.E. Foreyt, J., Melanson, E., Volek, J., Vope, S.L, American College of Sports Medicine. (2001). American College of Sports Medicine position stand. Appropriate intervention strategies for weight loss and prevention of weight regain for adults. *Medicine and Science in Sports and Exercise,* 33:2145-2156.

Karlsson, M.K., Linden, C., Karlsson, C., Johnell, O., Obrant, K. Seeman, E. (2000). Exercise during growth and bone mineral density and fractures in old age. *Lancet,* 355:469-470.

Kelley, G.A. (1998a). Aerobic exercise and bone density at the hip in postmenopausal women: A meta-analysis. *Preventive Medicine* 27:798-807.

Kelley, G.A. (1998b). Exercise and regional bone mineral density in postmenopausal women: A meta-analytic review of randomized trials. *American Journal of Physical Medicine and Rehabilitation,* 77: 76-87.

Kerr, D., Morton, A., Dick, I., & Prince, R. (1996). Exercise effects on bone mass in postmenopausal women are site-specific and load-dependent. *Journal of Bone and Mineral Research,* 11: 212-225.

Khan, K.M., McKay, H.A., Haapsalo, H., Bennell, K.L., Forwood, M.R., Kannus, P., Wark, J.D. (2000). Does childhood and adolescence provide a unique opportunity for exercise to strengthen the skeleton? *Journal of Science and Medicine in Sports,* 3:150-164.

Kiens, B., Lithell, H. (1989). Lipoprotein metabolism influenced by training-induced changes in human skeletal muscle. *Journal of Clinical Investigation,* 83:558-564.

Kohrt, W.M., Ehsani, A.A., Birge, S.J. (1997).Effects of exercise involving predominantly either joint-reaction or ground-reaction forces on bone mineral density in older women. *Journal of Bone and Mineral Research,* 12:1253-1261.

Kohrt, W.M., Snead, D.B., Slatopolsky, E., Birge, S.J., Jr. (1995). Additive effects of weight-bearing exercise and estrogen on bone mineral density in older women. *Journal of Bone and Mineral Research,* 10:1303-1311.

Kraemer, W.J., Adams, K., Cafarelli, E., Dudley, G.A., Dooly, C., Feigenbaum, M.S., et al. (2002). American College of Sports Medicine position stand. Progression models in resistance training for healthy adults. *Medicine and Science in Sports and Exercise,* 34:364-380.

Kraemer, W.J., Fry, A.C., Frykman, P.N., Conroy, B., Hoffman, J. (1989). Resistance training and youth. *Pediatric Exercise Science* 1:336-345.

Lamarche, B., Despres, J.P., Moorjani, S., Nadeau, A., Lupien, P.J., Tremblay, A., Theriault, G., Bouchard, C. (1993). Evidence for a role of insulin in the regulation of abdominal adipose tissue lipoprotein lipase response to exercise training in obese women. *International Journal of Obesity and Related Metabolic Disorders,* 17:255-261.

Lanouette, C.M., Chagnon, Y.C., Rice, T., Perusse, L, Muzzin, P., Giacobino, J.P., Gagnon, J., Wilmore, J.H., Leon, A.S., Skinner, J.S., Rao, D.C., Bouchard, C. (2002). Uncoupling protein 3 gene is associated with body composition changes with training in HERITAGE study. *Journal of Applied Physiology,* 92:1111-1118.

Lanyon, L.E., Rubin, C.T. (1984). Static vs dynamic loads as an influence on bone remodeling. *Journal of Biomechanics,* 17:897-905.

LeMura, L.M., & Maziekas, M.T. (2002). Factors that alter body fat, body mass, and fat-free mass in pediatric obesity. *Medicine and Science in Sports and Exercise,* 34:487-496.

Lin, B.Y., Jee, W.S., Chen, M.M., Ma, Y.F., Ke, H.Z., Li, X.J. (1994). Mechanical loading modifies ovariectomy-induced cancellous bone loss. *Bone Mineral,* 25:199-210.

Lohman, T.G. (1995). Exercise training and bone mineral density. *Quest,* 47:354-361.

Lohman, T., Going, S., Pamenter, R., Hall, M., Boyden T., Houtkooper L., Ritenbaugh, C., Bare, L., Hill, A., Aickin, M. (1995). Effects of resistance training on regional and total bone mineral density in premenopausal women: A randomized prospective study. *Journal of Bone and Mineral Research,* 10:1015-1024.

MacKelvie, K.J., McKay, H.A., Khan, K.M., Crocker, P.R. (2001). A school-based exercise intervention augments bone mineral accrual in early pubertal girls. *Journal of Pediatrics,* 139:501-508.

Marcus, R., Cann, C., Madvig, P., Minkoff, J., Goddard, M., Bayer, M., Martin, M., Gaudiani, L., Haskell, W., Genant, H. (1985). Menstrual function and bone mass in elite women distance runners. Endocrine and metabolic features. *Annals of Internal Medicine,* 102:158-163.

Marcus, R. (2002). Mechanisms of exercise effects on bone. In: *Principles of Bone Biology,* Second Edition. New York: Academic Press.

Marshall, D.O., Johnell, O., Wedel, H. (1996). Meta-analysis of how measures of BMD predict occurrence of osteoporotic fractures. *British Medical Journal,* 312:1254-1259.

Matsuzawa, Y., Shimomura, I., Nakamura, T., Keno, Y., Kotani, K., Tokunaga, K. (1995). Pathophysiology and pathogenesis of visceral fat obesity. *Obesity Research,* 3:S187-S194.

McCarty, M.F. (2001). Modulation of adipocyte lipoprotein lipase expression as a strategy for preventing or treating visceral obesity. *Medical Hypotheses,* 57:192-200.

McKay, H.A., Petit, M.A., Schutz, R.W., Prior, J.C., Barr, S.I., Khan, K.M. (2000). Augmented trochanteric bone mineral density after modified physical education classes: A randomized school-based exercise intervention study in prepubescent and early pubescent children. *Journal of Pediatrics,* 136: 156-162.

Miller, W.C., Koceha, D.M., Hamilton, E.J. (1997). A meta-analysis of the past 25 years of weight loss research using diet, exercise or diet plus exercise intervention. *International Journal of Obesity,* 21:941-947.

Montgomery, H., Clarkson, P., Barnard, M., Bell, J., Brynes, A., Dollery, C., et al. (1999). Angiotensin-converting-enzyme gene insertion/deletion polymorphism and response to physical training. *Lancet,* 353:541-545.

Mora, S., Pessin, J.E. (2002). An adipocentric view of signaling and intracellular trafficking. *Diabetes/Metabolism Research and Reviews,* 18:345-356.

Morris, F.L., Naughton, G.A., Gibbs, J.L., Carlson, J.S., Wark, J.D. (1997). Prospective ten month exercise intervention in premenarcheal girls: Positive effects on bone and lean mass. *Journal of Bone and Mineral Research,* 12:1453-1462.

National Institues of Health: National Heart, Lung, and Blood Institute. (1998). Clinical guidelines on the identification, evaluation and treatment of overweight and obesity in adults: The evidence report. *Obesity Research,* 6(Suppl 2): S51-S210.

Nelson, M.E., Fiatarone, M.A., Morganti, C.M., Trice, I., Greenberg, R.A., Evans, W.J. (1994). Effects of high-intensity strength training on multiple risk factors for osteoporotic fractures. A randomized controlled trial. *Journal of the American Medical Association,* 272:1909-1914.

Nelson, M.E., Meredith, C.N., Dawson-Hughes, B., Evans, W.J. (1988). Hormone and bone mineral status in endurance-trained and sedentary postmenopausal women. *Journal of Clinical Endocrinology and Metabolism,* 66: 927-933.

Newhall, K.M., Rodnick, K.J., van der Meulen, M.C., Carter, D.R., Marcus, R. (1991). Effects of voluntary exercise on bone mineral content in rats. *Journal of Bone and Mineral Research,* 6:289-296.

Notelovitz, M., Martin, D., Tesar, R., Khan, F.Y., Probart, C., Fields, C., McKenzie, L. (1991). Estrogen therapy and variable-resistance weight training increase bone mineral in surgically menopausal women. *Journal of Bone and Mineral Research,* 6:583-590.

O'Connor, J.A., Lanyon, L.E., MacFie, H. (1982). The influence of strain rate on adaptive bone remodelling. *Journal of Biomechanics,* 15:767-781.

Ogden, C.L., Flegal, K.M., Carroll, M.D., Johnson, C.L. (2002). Prevalence and trends in overweight among US children and adolescents, 1999-2000. *Journal of the American Medical Association,* 288:1728-1732.

Owens, S., Gutin, B., Allison, J., Riggs, S., Ferguson, M., Litaker, M., Thompson, W. (1999). Effect of physical training on total and visceral fat in obese children. *Medicine and Science in Sports and Exercise,* 31:143-148.

Petit, M.A., McKay, H.A., MacKelvie, K.J., Heinonen, A., Khan, K.M., Beck, T.J. (2002). A randomized school-based jumping intervention confers site and maturity-specific benefits on bone structural properties in girls: A hip structural analysis study. *Journal of Bone and Mineral Research,* 17: 363-372.

Pikosky, M., Faigenbaum, A., Westcott, W., Rodriguez, N. (2002). Effects of resistance training on protein utilization in healthy children. *Medicine and Science in Sports and Exercise,* 34: 820-827.

Poehlman, E.T., Dvorak, R.V., Denino, W.F., Brochu, M., Ades, P.A. (2000). Effects of resistance training and endurance training on insulin sensitivity in nonobese, young women: A controlled randomized trial. *Journal of Clinical Endrocrinology and Metabolism,* 85:2463-2468.

Powers, S.K., Howley, E.T., Cox, R.H. (1982). A differential catecholamine response during prolonged exercise and passive heating. *Medicine and Science in Sports and Exercise,* 14:435-439.

Pradhan, A.D., Ridker, P.M. (2002). Do atherosclerosis and type 2 diabetes share a common inflammatory basis? *European Heart Journal,* 23:831-834.

Prince, R., Devine, A., Dick, I., Criddle, A., Kerr, D., Kent, N., Price, R., Randall, A. (1995). The effects of calcium supplementation (milk powder or tablets) and exercise on bone density in postmenopausal women. *Journal of Bone and Mineral Research,* 10:1068-1075.

Pruitt, L.A., Jackson, R.D., Bartels, R.L., Lehnhard, H.J. (1992). Weight-training effects on bone mineral density in early postmenopausal women. *Journal of Bone and Mineral Research,* 7:179-185.

Remes, T., Vaisanen, S.B., Mahonen, A., Huuskonen, J., Kroger, H., Jurvelin, J.S., Penttila, I.M., Rauramaa, R. (2003). Aerobic exercise and bone mineral density in middle-aged Finnish men: A controlled randomized trial with reference to androgen receptor aromatase, and estrogen receptor α gene polymorphisms. *Bone,* 32:412-420.

Robinson, T.L., Snow-Harter, C., Taaffe, D.R., Gillis, D., Shaw J., Marcus, R. (1995). Gymnasts exhibit higher bone mass than runners despite similar prevalence of amenorrhea and oligomenorrhea. *Journal of Bone and Mineral Research,* 10: 26-35.

Ross, R., Dagnone, D., Jones, P.J., Smith, H., Paddogs, A., Hudson, R., Janssen, I. (2000). Reduction in obesity and related comorbid conditions after diet-induced weight loss or exercise-induced weight loss in men. A randomized, controlled trial. *Annals of Internal Medicine,* 133:92-103.

Ross, R., Janssen, I. (1999). Is abdominal fat preferentially reduced in response to exercise-induced weight loss? *Medicine and Science in Sports and Exercise,* 31:S568-S572.

Ross, R., Janssen, I. (2001). Physical activity, total and regional obesity: Dose-response considerations. *Medicine and Science in Sports and Exercise,* 33:S521-S527.

Roth, S.M., Ivey, F.M., Martel, G.F., Lemmer, J.T., Hurlbut, D.E., Siegel, E.L., Metter, E.J., Fleg, J.L., Fozard, J.L., Kostek, M.C., Wernick, D.M., Hurley, B.F. (2001). Muscle size responses to strength training in young and older men and women. *Journal of the American Geriatrics Society,* 49:1428-1433.

Rubin, C.T., Lanyon, L.E.(1985). Regulation of bone mass by mechanical strain magnitude. *Calcified Tissue International,* 37:411-417.

Schrauwen, P., Hesselink, M (2003). Uncoupling protein 3 and physical activity: the role of uncoupling protein 3 in energy metabolism revisited. *Proceedings of the Nutrition Society,* 62:635-643.

Schwingshandl, J., Borkenstein, M. (1995). Changes in lean body mass in obese children during a weight reduction program: Effect on short term and long term outcome. *International Journal of Obesity and Related Metabolic Disorders,* 19:752-755.

Shephard, R.J. (2001). Absolute versus relative intensity of physical activity in a dose-response context. *Medicine and Science in Sports and Exercise,* 33:S400-S418.

Skerry, T.M. (1997). Mechanical loading and bone: What sort of exercise is beneficial to the skeleton? *Bone* 20:179-181.

Snow, C.M., Williams, D.P., LaRiviere, J., Fuchs, R.K., Robinson, T.L (2001). Bone gains and losses follow seasonal training and detraining in gymnasts. *Calcified Tissue International,* 69:7-12.

Snow-Harter, C., Bouxsein, M.L., Lewis, BT., Carter, D.R., Marcus, R. (1992). Effects of resistance and endurance exercise on bone mineral status of young women: A randomized exercise intervention trial. *Journal of Bone and Mineral Research,* 7: 761-769.

Sun, G., Gagnon, J., Chagnon, Y.C., Perusse, L., Despres, J.P., Leon, A.S., Wilmore, J.H., Skinner, J.S., Borecki, I., Rao, D.C., Bouchard, C. (1999). Association and linkage between an insulin-like growth factor-1 gene polymorphism and fat free mass in the HERITAGE Family Study. *International Journal of Obesity and Related Metabolic Disorders,* 23:929-935.

Sung, R.Y., Yu, C.W., Chang, S.K., Mo, S.W., Woo, K.S., Lam, C.W. (2002). Effects of dietary intervention and strength training on blood lipid level in obese children. *Archives of Disease in Childhood,* 86:407-410.

Swezey, R.L. (1996). Exercise for osteoporosis—Is walking enough? The case for site specificity and resistive exercise. *Spine,* 21:2809-2813.

Taaffe, D.R,. Robinson, T.L., Snow, C.M., Marcus, R. (1997). High-impact exercise promotes bone gain in well-trained female athletes. *Journal of Bone and Mineral Research,* 12: 255-260.

Taaffe, D.R., Snow-Harter, C., Connolly, D.A., Robinson, T.L., Brown, M.D., Marcus, R. (1995). Differential effects of swimming versus weight-bearing activity on bone mineral status of eumenorrheic athletes. *Journal of Bone and Mineral Research,* 10:586-593.

Teixeira, P.J., Going, S.B., Houtkooper, L.B., Metcalfe, L.L., Blew, R.M., Flint-Wagner, H.G., Cussler, E.C., Sardihna, L.B., Lohman, T.G. (2003). Resistance training in postmenopausal women with and without hormone therapy. *Medicine and Science in Sports and Exercise,* 35:555-562.

Toth, M.J., Beckett, T., Poehlman, E.T. (1999). Physical activity and the progressive change in body composition with aging: Current evidence and research issues. *Medicine and Science in Sports and Exercise,* 31:S590-S596.

Treuth, M.S., Hunter, G.R., Figueroa-Colon, R., & Goran, M. I. (1998). Effects of strength training on intra-abdominal adipose tissue in obese prepubertal girls. *Medicine and Science in Sports and Exercise,* 30:1738-1743.

Turner, C.H. (1998). Three rules for bone adaptation to mechanical stimuli. *Bone,* 23:399-407.

Turner, C.H., Robling, A.G. (2003). Designing exercise regimens to increase bone strength. *Exercise and Sport Sciences Reviews,* 31:45-50.

Turner, C.H., Takano, Y., Owan, I. (1995). Aging changes mechanical loading thresholds for bone formation in rats. *Journal of Bone and Mineral Research* 10:1544-1549.

Ukkola, O., Rankinen, T., Rice, T., Gagnon, J., Leon, A.S., Skinner, J.S., Wilmore, J.H., Rao, D.C., Bouchard, C., HERITAGE family study. (2003). Interactions among the β2- and β3-adrenergic receptor genes and total body fat and abdominal fat level in the HERITAGE family study. *International Journal of Obesity and Related Metabolic Disorders,* 27:389-393.

U. S. Department of Health and Human Services. (1996). *Physical Activity and Health: A Report of the Surgeon General.* Atlanta, GA: USDHHS, Centers for Disease Control and Prevention.

Van Aggel-Leijssen, D.P.C.V., Saris, W.H, Wagenmakers, A., Senden, J.M., Van Baak, M.A. (2002). Effect of exercise training at different intensities on fat metabolism of obese men. *Journal of Applied Physiology,* 92:1300-1309.

van der Wiel, H.E., Lips, P., Graafmans, W.C., Danielsen, C.C., Nauta, J., van Lingen A., Mosekilde, L. (1995). Additional weight-bearing during exercise is more important than duration of exercise for anabolic stimulus of bone: A study of running exercise in female rats. *Bone* 16:73-80.

Whalen, R.T., Carter, D.R., Steele, C.R. (1988). Influence of physical activity on the regulation of bone density. *Journal of Biomechanics,* 21:825-837.

Williams, P.T. (2001). Health effects resulting from exercise versus those from body fat loss. *Medicine and Science in Sports and Exercise,* 33:S611-S621.

Wilmore, J.H., Despres, J.P., Stanforth, P.R., Mandel, S., Rice, T., Gagnon, J., Leon, A.S., Rao, D., Skinner, J.S., Bouchard, C. (1999). Alterations in body weight and composition consequent to 20 wk of endurance training: the HERITAGE Family Study. *American Journal of Clinical Nutrition,* 70: 346-352.

Winters, K.M., Snow, C.M. (2000). Detraining reverses positive effects of exercise on the musculoskeletal system in premenopausal women. *Journal of Bone and Mineral Research,* 15:2495-2503.

Witzke, K.A., Snow, C.M. (2000). Effects of plyometric jump training on bone mass in adolescent girls. *Medicine and Science in Sports and Exercise* 32:1051-1057.

Wolff, I., van Croonenborg, J.J., Kemper, H.C., Kostense, P.J., Twisk, J.W. (1999). The effect of exercise training programs on bone mass: A meta-analysis of published controlled trials in pre- and postmenopausal women. *Osteoporosis International,* 9:1-12.

Wood, P.D., Stefanick, M.L., Dreon, D.M., Frey-Hewitt, B., Garay, S.C., Williams, P.T., Superko, H.R., Fortmann, S.P., Albers, J.J., Vranizan, K.M. (1988). Changes in plasma lipids and lipoproteins in overweight men during weight loss through dieting as compared with exercise. *New England Journal of Medicine,* 319:1173-1179.

World Health Organization. (1994). Assessment of fracture risk and its application to screening for postmenopausal osteoporosis. *WHO Technical Report Series,* No. 843. Geneva: World Health Organization.

Yeh, J.K., Aloia, J.F., Barilla, ML (1994). Effects of 17-beta-estradiol replacement and treadmill exercise on vertebral and femoral bones of the ovariectomized rat. *Bone Mineral,* 24: 223-234.

Chapter 21

Abate, N., Haffner, S.M., Garg, A., Peshock, R.M., & Grundy, S.M. (2002). Sex steroid hormones, upper body obesity, and insulin resistance. Journal of Clinical Endocrinology and Metabolism, 87, 4522-4527.

Alen, M., Hakkinen, A., & Komi, P. (1994). Changes in neuromuscular performance and muscle fiber characteristics of elite power athletes self-administering androgen and anabolic steroids. Acta Physiologica Scandinavica, 122, 535-544.

Altman, J. (2002). Weight in the balance. Neuroendocrinology, 76, 131-136.

Andersson, B., Marin, P., Lissner, L., Vermeulen, A., & Bjorntorp, P. (1994). Testosterone concentrations in women and men with non-insulin dependent diabetes mellitus. Diabetes Care, 17, 405-411.

Appel, B., & Fried, S. (1992). Effects of insulin and dexamethasone on lipoprotein lipase in human adipose tissue. American Journal of Physiology, 262, E695-E699.

Asplin, C.M., Faria, A.C., Carlsen, E.M., Vaccaro, V.A., Barr, R.E., Iranmanesh, A., Lee, M.M., Veldhuis, J.D., & Evans, W.S. (1989). Alterations in the pulsatile mode of growth hormone release in men and women with insulin-dependent diabetes mellitus. Journal of Clinical Endocrinology and Metabolism, 69, 239-245.

Bengtsson, B.A., Brummer, R., Eden, S., & Bosaeus, I. (1989). Body composition in acromegaly. Clinical Endocrinology, 30, 121-130.

Bengtsson, B.A., Eden, S., Lonn, L., Kvist, H., Stokland, A., Lindstedt, G., Bosaeus, I., Tolli, J., Sjostrom, L., & Isaksson, O.G. (1993). Treatment of adults with growth hormone deficiency with recombinant human growth hormone. Journal of Clinical Endocrinology and Metabolism, 76, 309-317.

Bjorntorp, P. (1993). Androgens, the metabolic syndrome and non-insulin dependent diabetes mellitus. Annals of the New York Academy of Science, 676, 242-252.

Bjorntorp, P. (1999). Neuroendocrine perturbations as a cause of insulin resistance. Diabetes/Metabolism Research and Reviews, 15, 427-441.

Bjorntorp, P., & Ostman, J. (1971). Human adipose tissue: Dynamics and regulation. Advances in Metabolic Disorders, 5, 277-327.

Bjorntorp, P., & Rosmond, R. (1999). Hypothalamic origin of the metabolic syndrome X. Annals of the New York Academy of Science, 892, 297-307.

Blackman, M.R., Sorkin, J.D., Munzer, T., Bellantoni, M.F., Busby-Whitehead, J., Stevens, T.E., Jayme, J., O'Connor, K.G., Christmas, C., Tobin, J.D., Stewart, K.J., Cottrell, E., St. Clair, C., Pabst, K.M., & Harman, S.M. (2002). Growth hormone and sex steroid administration in healthy aged women and men. Journal of the American Medical Association, 288, 2282-2292.

Bonnet, F., Vanderscheuren-Lodeweyckx, M., Beckel, R., & Malvaux, P. (1974). Subcutaneous adipose tissue and lipids in blood in growth hormone deficiency before and after treatment with human growth hormone. Pediatric Research, 8, 800-805.

Brill, K.T., Weltman, A.L., Gentili, A., Patrie, J.T., Fryburg, D.A., Hanks, J.B., Urban, R.J., & Veldhuis, J.D. (2002). Single and combined effects of growth hormone and testosterone administration on measures of body composition, physical performance, mood, sexual function, bone turnover, and muscle gene expression in healthy older men. Journal of Clinical Endocrinology and Metabolism, 87, 5649-5657.

Carson-Jurica, M., Schrader, T., & O'Malley, B. (1990). Steroid receptor family: Structure and functions. Endocrine Reviews, 11, 201-220.

Carter-Su, C., & Okamoto, K. (1987). Effect of insulin and glucocorticoids on glucose transporters in rat adipocytes. American Journal of Physiology, 252, E441-E453.

Cheek, D., & Hill, D. (1974). Effect of growth hormone on cell and somatic growth. In R. Greep & E. Astwood (Eds.), Handbook of Physiology (pp. 159-185). Washington, DC: American Physiological Society.

Cigolini, M., & Smith, U. (1979). Human adipose tissue in culture, VIII: Studies on the insulin-antagonistic effect of glucocorticoids. Metabolism, 28, 502-510.

Cnop, M., Havel, P.J., Utzschneider, K.M., Carr, D.B., Sinha, M.K., Boyko, E.J., Retzlaff, B.M., Knopp, R.H., Brunzell, J.D., & Kahn, S.E. (2003). Relationship of adiponectin to body fat distribution, insulin sensitivity and plasma lipoproteins: Evidence for independent roles of age and sex. Diabetologia, Apr. 10 [epub ahead of print].

Cohen, J.C., & Hickman, R. (1987). Insulin resistance and diminished glucose tolerance in power lifters ingesting anabolic steroids. Journal of Clinical Endocrinology and Metabolism, 64, 960-963.

Corbould, A.M., Bawden, M.J., Lavranos, T.C., Rodgers, R.J., & Judd, S.J. (2002). The effect of obesity on the ratio of type 3 17β-hydroxysteroid dehydrogenase mRNA to cytochrome P450 aromatase mRNA in subcutaneous abdominal and intra-abdominal adipose tissue of women. International Journal of Obesity and Related Metabolic Disorders, 26, 165-175.

De Gasquet, P., Pequignot-Planche, E., Tomm, N., & Diaby, F. (1975). Effect of glucocorticoids on lipoprotein lipase activity in rat heart and adipose tissue. Hormone and Metabolic Research, 7, 152-157.

De Pergola, G., Holmang, A., Svedberg, J., et al. (1990). Testosterone treatment of ovariectomized rats: Effects on lipolysis regulation in adipocytes. Acta Endocrinologica (Copenhagen), 123, 61-66.

Dietz, J., & Schwartz, J. (1991). Growth hormone alters lipolysis and hormone sensitive lipase activity in 3T3-F442A adipocytes. Metabolism, 40, 800-806.

DiGirolamo, M., Eden, S., Enberg, G., Isaksson, O., Lonnroth, P., Hall, K., & Smith, U. (1986). Specific binding of human growth hormone but not insulin-like growth factors by human adipocytes. FEBS Letters: Federation of European Biochemical Societies, 205, 15-19.

Divertie, G., Jensen, M., & Miles, J. (1991). Stimulation of lipolysis in humans by physiological hypercortisolemia. Diabetes, 40, 1228-1232.

Ebert, K.M., Low, M.J., Overstrom, E.W., Overstrom, E.W., Buonomo, F.C., Baile, C.A., Roberts, T.M., Lee, A., Mandel, G., & Goodman, R.H. (1988). A moloney MLV-rat somatotropin fusion gene produces biologically active somatotropin in a transgenic pig. Molecular Endocrinology, 2, 277-283.

Feldman, D., & Loose, G. (1977). Glucocorticoid receptors in adipose tissue. Endocrinology, 100, 389-405.

Friedl, K. (1990). Reappraisal of the health risks associated with the use of high doses of oral and injectable androgenic steroids. In G. Lin & L. Erinoff (Eds.), Anabolic steroid abuse (pp. 142-177). Washington, DC: Department of Health and Human Services.

Friedl, K., Hannan, C., Jones, R., & Plymate, S. (1990). High-density lipoprotein cholesterol is not decreased if an aromatizable androgen is administered. Metabolism, 39, 69-74.

Gale, S.M., Castracane, V.D., & Mantzoros, C.S. (2004). Energy homeostasis, obesity, and eating disorders: Recent advances in endocrinology. Journal of Nutrition, 134, 295-298.

Gause, I., & Eden, S. (1985). Hormonal regulation of growth hormone binding and responsiveness in adipose tissue and adipocytes of hypophysectomized rats. Journal of Endocrinology, 105, 331-337.

Gregory, J.W., Greene, S.A., Thompson, J., Scrimgeour, C.M., & Rennie, M.J. (1992). Effects of oral testosterone undecanoate on growth, body composition, strength and energy expenditure of adolescent boys. Journal of Clinical Endocrinology, 37, 207-213.

Guillaume-Gentil, C., Assimacopoulos-Jeannet, F., & Jeanrenaud, B. (1993). Involvement of non-esterified fatty acid oxidation in glucocorticoid-induced peripheral insulin resistance in vivo in rats. Diabetologia, 36, 899-906.

Haarbo, J., Marskew, U., Gottfredsen, A., & Christiansen, C. (1991). Postmenopausal hormone replacement therapy prevents central distribution of body fat after menopause. Metabolism, 40, 323-326.

Haffner, S.H., Waldez, R.A., Stern, M.P., & Katz, M.S. (1993). Obesity, body fat distribution and sex hormones in men. International Journal of Obesity, 17, 643-650.

Hartman, M.L., Veldhuis, J.D., Johnson, M.L., Lee, M.M., Alberti, K.G., Samojlik, E., & Thorner, M.O. (1992). Augmented growth hormone (GH) secretory burst frequency and amplitude mediate enhanced GH secretion during a two-day fast in normal men. Journal of Clinical Endocrinology and Metabolism, 74, 757-765.

Hauner, H. (1992). Physiology of the fat cell, with emphasis on the role of growth hormone. Acta Paediatrica Scandinavica, 383 (Suppl.), 47-51.

Hervey, G.R., Knibbs, A.V., Burkinshaw, L., Morgan, D.B., Jones, P.R., Chettle, D.R., & Vartsky, K. (1981). Effects of methandienone on the performance and body composition of men undergoing athletic training. Clinical Science, 60, 457-461.

Heymsfield, S.B., Greenberg, A.S., Fujioka, K., Dixon, R.M., Kushner, R., Hunt, T., Lubina, J.A., Patane, J., Self, B., Hunt, P., & McCamish, M. (1999). Recombinant leptin for weight loss in obese and lean adults. Journal of the American Medical Association, 282, 1568-1575.

Holmang, A., & Bjorntorp, P. (1992). The effects of testosterone on insulin sensitivity in male rats. Acta Physiologica Scandinavica, 146, 505-510.

Horlick, M.B., Rosenbaum, M., Nicolson, M., Levine, L.S., Fedun, B., Wang, J., Pierson, R.N. Jr., & Leibel, R.L. (2000). Effect of puberty on the relationship between circulating leptin and body composition. Journal of Clinical Endocrinology and Metabolism, 85, 2509-2518.

Iranmanesh, A., Lizarralde, G., & Veldhuis, J. (1991). Age and relative adiposity are specific negative determinants of the frequency and amplitude of growth hormone (GH) secretory bursts and the half-life of endogenous GH in healthy men. Journal of Clinical Endocrinology and Metabolism, 73, 1081-1088.

Isidori, A.M., Strollo, F., Mori, M., Caprio, M., Aversa, A., Moretti, C., Frajese, G., Riondino, G., & Fabbri, A. (2000). Leptin and aging: correlation with endocrine changes in male and female healthy adult populations of different body weights. Journal of Clinical Endocrinology and Metabolism, 85, 1954-1962.

Jorgensen, J., Pedersen, S., Laurberg, P., Weeke, J., Skakkebaek, N.E., & Christiansen, J.S. (1989a). Effects of growth hormone therapy on thyroid function of growth hormone-deficient adults with and without concomitant thyroxine-substituted central hypothyroidism. Journal of Clinical Endocrinology and Metabolism, 69, 1127-1132.

Jorgensen, J.O., Pedersen, S.A., Thuesen, L., Jorgensen, J., Ingemann-Hansen, T., Skakkebaek, N.E., & Christiansen, J.S. (1989b). Beneficial effects of growth hormone treatment in GH-deficient adults. Lancet, 2, 1221-1225.

Khaw, K., Chir, B., & Barrett-Connor, E. (1992). Lower endogenous androgens predict central adiposity in men. American Journal of Epidemiology, 2, 675-682.

Krotkiewski, M., Bjorntorp, P., & Smith, U. (1976). The effect of long-term dexamethasone treatment on lipoprotein lipase activity in rat fat cells. Hormone and Metabolic Research, 8, 245-246.

Kvist, H., Chowdhury, B., Grangard, U., Tylen, U., Sjostrom, L. (1988). Total and visceral adipose-tissue volumes derived from measurements with computed tomography in adult men and women: Predictive equations. American Journal of Clinical Nutrition, 48, 1351-1361.

Lam, P., Jimenez, M., Zhaung, T.N., Celermajer, D.S., Conway, A.J., & Handelsman, D.J. (2001). A double-blind, placebo controlled, randomized clinical trial of transdermal dihydrotestosterone gel on muscular strength, mobility, and quality of life in older men with partial androgen deficiency. Journal of Clinical Endocrinology and Metabolism, 86, 4078-4088.

Li, M., & Bjorntorp, P. (1995). Estrogens and AR density in adipose tissue of female rats. Unpublished raw data.

Marin, P., Darin, N., Amemiya, T., Andersson, B., Jern, S., & Bjorntorp, P. (1992). Cortisol secretion in relation to body fat distribution in obese premenopausal women. Metabolism, 41, 882-886.

Marin, P., Gustafsson, C., Oden, B., & Bjorntorp, P. (1995). Assimilation and mobilization of triglycerides in subcutaneous abdominal and femoral adipose tissue in vivo in men: Effects of androgens. Journal of Clinical Endocrinology and Metabolism, 80, 239-243.

Marin, P., Holmang, S., Gustafsson, C., Jonsson, L., Kvist, H., Elander, A., Eldh, J., Sjostrom, L., Holm, G., & Bjorntorp, P. (1993a). Androgen treatment of abdominally obese men. Obesity Research, 1, 245-251.

Marin, P., Kvist, H., Lindstedt, G., et al. (1993b). Low concentrations of insulin-like growth factor I in abdominal obesity. International Journal of Obesity, 17, 83-89.

Marin, P., Lonn, L., Andersson, B., et al. (1996). Assimilation of triglycerides in subcutaneous and intraabdominal adipose tissue in vivo in men: Effects of testosterone. Journal of Clinical Endocrinology and Metabolism, 81, 1018-1022.

McDonald, A., & Goldfine, J. (1988). Glucocorticoid regulation of insulin receptor gene transcription in IM-9 cultured lymphocytes. Journal of Clinical Investigation, 81, 499-504.

Morikawa, M., Nixon, T., & Green, H. (1982). Growth hormone and the adipose conversion of 3T3 cells. Cell, 29, 783-789.

Nass, R., & Thorner, M.O. (2002). Impact of the GH-cortisol ratio on the age-dependent changes in body composition. Growth Hormone and IGF Research, 12, 147-161.

Nielsen, J. (1982). Effects of growth hormone, prolactin, and placental lactogen on insulin content and release, and deoxyribonucleic acid synthesis in cultured pancreatic islets. Endocrinology, 110, 600-606.

Niswender, K.D., & Schwartz, M.W. (2003) Insulin and leptin revisited: Adiposity signals with overlapping physiological and intracellular signaling capabilities. Frontiers in Neuroendocrinology, 24, 1-10.

Olsen, D., & Ferin, M. (1987). Corticotropin-releasing hormone inhibits gonadotropin secretion in ovariectomized Rhesus monkey. Journal of Clinical Endocrinology and Metabolism, 65, 262-267.

O'Sullivan, A.J., Crampton, L.J., Freund, J., & Ho, K.K. (1998). The route of estrogen replacement therapy confers divergent effects on substrate oxidation and body composition in postmenopausal women. Journal of Clinical Investigations, 102, 1035-1040.

Ottosson, M., & Bjorntorp, P. (1994). [Cortisol-induced stimulation of LPL and T]. Unpublished raw data.

Ottosson, M., Lonnroth, P., Bjorntorp, P., & Eden, S. (2000). Effects of cortisol and growth hormones on lipolysis in human adipose tissue. Journal of Clinical Endocrinology and Metabolism, 85, 799-803.

Ottosson, M., Vikman-Adolfsson, K., Enerback, S., Olivecrona, G., & Bjorntorp, P. (1994). The effects of cortisol on the regulation of lipoprotein lipase activity in human adipose tissue. Journal of Clinical Endocrinology and Metabolism, 79, 820-825.

Pannacciulli, N., Vettor, R., Milan, G., Granzotto, M., Catucci, A., Federspil, G., De Giacomo, P., Giorgino, R., & De Pergola, G. (2003). Anorexia nervosa is characterized by increased adiponectin plasma levels and reduced nonoxidative glucose metabolism. Journal of Clinical Endocrinology and Metabolism, 88, 1748-1752.

Parra, A., Argote, R.M., Garcia, G., Cervantes, C., Alatorres, S., & Perez-Pasten, E. (1979). Body composition in hypopituitary dwarfs before and during human growth hormone therapy. Metabolism, 28, 851-857.

Peeke, P., Oldfield, E., Alexander, R., et al. (1993). Glucocorticoid receptor protein in omental and subcutaneous adipose tissue of patients with Cushing syndrome prior to and after surgical cure. Obesity Research, 1 (Suppl. 2), Abstract No. 49.

Rebuffe-Scrive, M. (1991). Neuroregulation of adipose tissue: Molecular and hormonal mechanisms. International Journal of Obesity, 15, 83-86.

Rebuffe-Scrive, M., Andersson, B., Olbe, L., & Bjorntorp, P. (1989). Metabolism of adipose tissue in intraabdominal depots of non obese men and women. Metabolism, 38, 453-458.

Rebuffe-Scrive, M., Bronnegard, M., Nilsson, A., Eldh, J., Gustafsson, J.A., & Bjorntorp, P. (1990). Steroid hormone receptors in human adipose tissues. Journal of Clinical Endocrinology and Metabolism, 71, 1215-1219.

Rebuffe-Scrive, M., Eldh, J., Hafstrom, L.-O., & Bjorntorp, P. (1986). Metabolism of mammary, abdominal and femoral adipocytes in women before and after menopause. Metabolism, 35, 792-797.

Rebuffe-Scrive, M., Enk, L., Crona, N., Lonnroth, P., Abrahamsson, L., Smith, U., & Bjorntorp, P.. (1985a). Fat cell metabolism in different regions in women: Effect of menstrual cycle, pregnancy, and lactation. Journal of Clinical Investigation, 75, 1973-1976.

Rebuffe-Scrive, M., Krotkiewski, M., Elfverson, J., & Bjorntorp, P. (1988). Muscle and adipose tissue morphology and metabolism in Cushing's syndrome. Journal of Clinical Endocrinology and Metabolism, 67, 1122-1128.

Rebuffe-Scrive, M., Lundholm, K., & Bjorntorp, P. (1985b). Glucocorticoid hormone binding to human adipose tissue. European Journal of Clinical Investigation, 15, 267-272.

Richelsen, B., Pedersen, S.B., Borglum, J.D., Moller-Pedersen, T., Jorgensen, J., & Jorgensen, J.O. (1994). Growth hormone treatment of obese women for 5 wk: Effect on body composition and adipose tissue LPL activity. American Journal of Physiology, 266, E211-E216.

Rosell, S., & Belfrage, E. (1979). Blood circulation in adipose tissue. Physiological Reviews, 59, 1078-1104.

Rosen, T., Bosaeus, I., Tolli, J., Heymsfield, S.B., Gallagher, D., Chu, F., & Leibel., R.L. (1993). Increased body fat and decreased extracellular fluid volume in adults with growth hormone deficiency. Clinical Endocrinology, 38, 63-71.

Rosenbaum, M., Nicolson, M., Hirsch, J., et al. (1996). Effects of gender, body composition, and menopause on plasma concentrations of leptin. Journal of Clinical Endocrinology and Metabolism, 81, 3424-3427.

Rosenbaum, M., Nicolson, M., Hirsch, J., Murphy, E., Chu, F., & Leibel, R.L. (1997). Effects of weight change on plasma leptin concentrations and energy expenditure. Journal of Clinical Endocrinology and Metabolism, 82, 3647-3654.

Salans, L.B., Cushman, S.W., & Weismann, R.E. (1973). Studies of human adipose tissue: Adipose cell size and number in non-obese and obese patients. Metabolism, 52, 929-941.

Salomon, F., Cuneo, R., Hesp, R., & Sonksen, P. (1989). The effects of treatment with recombinant human growth hormone on body composition and metabolism in adults with growth hormone deficiency. New England Journal of Medicine, 321, 1797-1803.

Schwartz, J., Foster, C., & Satin, M. (1985). Growth hormone and insulin-like growth factors-I and II produce distinct alterations in glucose metabolism in STS-F442A adipocytes. Proceedings of the National Academy of Science, U.S.A., 82, 8724-8728.

Seidell, J.C., Bjorntorp, P., Sjostrom, L., Kvist, H., & Sannerstedt, R.. (1990). Visceral fat accumulation in men is positively associated with insulin, glucose and C-peptide levels, but negatively with testosterone levels. Metabolism, 39, 897-901.

Silverman, M.S., Mynarcik, D.C., Corin, R.E., Roger, M., Saint-Paul, M., Nahoul, K., & Papoz, L. (1989). Antagonism by growth hormone of insulin sensitive hexose transport in 3T3-F442A adipocytes. Endocrinology, 125, 2600-2604.

Simon, D., Preziosi, P., Barrett-Connor, E., Roger, M., Saint-Paul, M., Nahoul, K., & Papoz, L. (1992). Interrelation between plasma testosterone and plasma insulin in healthy adult men: The Telecom Study. Diabetologia, 35, 173-177.

Sjogren, J., Li, M., & Bjorntorp, P. (1995). Androgen hormone binding to adipose tissue in rats. Biochimica Biophysica Acta, 1244, 117-120.

Snyder, P.J., Peachey, H., Hannoush, P., Berlin, J.A., Loh, L., Lenrow, D.A., Holmes, J.H., Dlewati, A., Santanna, J., Rosen, C.J., & Strom, B.L. (1999). Effect of testosterone treatment on body composition and muscle strength in men over 65 years of age. Journal of Clinical Endocrinology and Metabolism, 84, 2647-2653.

Stefan, N., Brent, J.C., Salbe, A.D., Funahashi, T., Matsuzawa, Y., & Tataranni, P.A. (2002). Plasma adiponectin concentrations in children: Relationships with obesity and insulinemia. Journal of Clinical Endocrinology and Metabolism, 87, 4652-4656.

Tanner, J.M., & Whitehouse, R.H. (1967). The effect of human growth hormone on subcutaneous fat thickness in hyposomatotrophic and panhypopituitary dwarfs. Journal of Endocrinology, 39, 263-275.

Tenover, J. (1992). Effects of testosterone supplementation in the aging male. Journal of Clinical Endocrinology and Metabolism, 75, 1092-1098.

Tomlinson, J.W., Crabtree, N., Clark, P.M.S., Holder, G., Toogood, A.A., Shackleton, C.H., & Stewart, P.M. (2003). Low-dose growth hormone inhibits 11 β-hydroxysteroid dehydrogenase type 1 but has no effect upon fat mass in patients with simple obesity. Journal of Clinical Endocrinology and Metabolism, 88, 2113-2118.

Udden, J., Bjorntorp, P., Arner, P., Barkeling, B., Meurling, L., & Rossner, S. (2003). Effects of glucocorticoids on leptin levels and eating behaviour in women. Journal of Internal Medicine, 253, 225-231.

Vernon, R., & Flint, D. (1989). Role of growth hormone in the regulation of adipocyte growth and function. In R. Heap, C. Prosser, & G. Lamming (Eds.), Biotechnology in growth regulation (pp. 57-71). London: Butterworths.

Vikman, K., Carlsson, B., Billig, H., & Eden, S. (1991). Expression and regulation of growth hormone (GH) receptor messenger ribonucleic acid (mRNA) in rat adipose tissue, adipocytes and adipose precursor cells: GH regulation of GH receptor mRNA. Endocrinology, 129, 1155-1161.

Wade, G.N., & Gray, J.M. (1979). Gonadal effects on food intake and adiposity: A metabolic hypothesis. Physiology and Behavior, 22, 583-593.

Wade, N. (1972). Anabolic steroids: Doctors denounce them, but athletes aren't listening. Science, 176, 1399-1403.

Weiss, R., Dufour, S., Groszmann, A., Petersen, K., Dzuira, J., Taksali, S.E., Shulman, G., & Caprio, S. (2003). Low adiponectin levels in adolescent obesity: A marker of increased intramyocellular lipid accumulation. Journal of Clinical Endocrinology and Metabolism, 88, 2014-2018.

West, D.B., Prinz, W.A., & Greenwood, M.R.C. (1989). Regional changes in adipose tissue, blood flow and metabolism in rats after a meal. American Journal of Physiology, 257, R711-R716.

Weyer, C., Funahashi, T., Tanaka, S., Hotta, K., Matsuzawa, Y., Pratley, R.E., & Tataranni, P.A. (2001). Hypoadiponectinemia in obesity and type 2 diabetes: Close association with insulin resistance and hyperinsulinemia. Journal of Clinical Endocrinology and Metabolism, 86, 1930-1935.

Wiedemann, E. (1981). Adrenal and gonadal steroids. In W.H. Daughaday (Ed.), Endocrine control of growth (pp. 67-119). New York: Elsevier Science.

Xu, X., & Bjorntorp, P. (1994). [Activity of LPL and other enzymes for triglyceride synthesis]. Unpublished raw data.

Yamauchi, T., Kamon, J., Waki, H., Murakami, K., Motojima, K., Komeda, K., Ide, T., Kubota, N., Terauchi, Y., Tobe, K., Miki, H., Tsuchida, A., Akanuma, Y., Nagai, R., Kimura, S., & Kadowaki, T. (2001). The fat-derived hormone adiponectin reverses insulin resistance associated with both lipoatrophy and obesity. Nature Medicine, 7, 941-946.

Yang, S., Xu, X., Bjorntorp, P., & Eden, S. (1995). Additive effects of growth hormone and testosterone on lipolysis in adipocytes of hypophysectomized rats. Journal of Endocrinology, 147, 147-152.

Chapter 22

Allison, D.B., Faith, M.S., Heo, M., & Kotler, D.P. (1997). Hypothesis concerning the u-shaped relation between body mass index and mortality. *Am J Epidemiol* 146, 339-49.

Allison, D.B., Zannolli, R., Faith, M.S., Heo, M., Pietrobelli, A., van Itallie, T.B., Pi-Sunyer, F.X., & Heymsfield, S.B. (1999). Weight loss increases and fat loss decreases all-cause mortality rate: results from two independent cohort studies. *Int J Obesity* 23, 603-11.

Allison, D.B., Faith, M.S., Heo, M., Townsend-Butterworth, D., & Williamson, D. (1999). Meta-analysis of the effect of excluding early deaths on the estimated relationship between body mass index and mortality. *Obes Res* 7, 417-9.

Andres, R., Elahi, D., Tobin, J.D., Muller, D.C., & Brant, L. (1985). Impact of age on weight goals. *Ann Intern Med* 103, 1030-3.

Blair, S.N., & Brodney, S. (1999). Effects of physical inactivity and obesity on morbidity and mortality: current evidence and research issues. *Med Sci Sports Exerc* 31 (suppl 11), S646-62.

Borkan, G.A., Hults, D.E., Gerzof, S.G., & Robbins, A.H. (1985). Comparison of body composition in middle-aged and elderly males using computed tomography. *Am J Anthropol* 66, 289-95.

Bjorntorp, P. (1990). "Portal" adipose tissue as a generator of risk factors for cardiovascular disease and diabetes. *Arteriosclerosis* 10, 493-6.

Chowdhury, B., Lantz, H., & Sjöström, L. (1996). Computed tomography-determined body composition in relation to cardiovascular risk factors in Indian and matched Swedish males. *Metabolism* 45, 634-44.

Deurenberg, P., Weststrate, J.A., & Seidell, J.C. (1991). Body mass index as a measure of body fatness: age- and sex-specific prediction formulas. *Br J Nutr* 65, 105-114.

Deurenberg, P., Yap, M., & van Staveren, W.A. (1998). Body mass index and percent body fat: a meta analysis among different ethnic groups. *Int J Obesity* 22, 1164-71.

Deurenberg, P., & Deurenberg-Yap, M. (2001). Differences in body-composition assumptions across ethnic groups: practical consequences. *Curr Opin Clin Nutr Metab Care* 4, 377-83.

Fontaine, K.R., Redden, D.T., Wang, C., Westfall, A.O., & Allison, D.B. (2003). Years of life lost due to obesity. *JAMA* 289, 187-93.

Frayn, K.N., Samra, J.S., & Summers, L.K. (1997). Visceral fat in relation to health: is it a major culprit or simply an innocent bystander? *In J Obesity* 21, 1191-2.

Harris, T.B., Launer, L.J., Madans, J., & Feldman, J.J. (1997). Cohort study of effect of being overweight and change in weight on risk of coronary heart disease in old age. *BMJ* 314, 1791-4.

Heitmann, B.L., Erikson, H., Ellsinger, B.M., Mikkelsen, K.L., & Larsson, B. (2000). Mortality associated with body fat, fat-free mass and body mass index among 60-year-old Swedish men: a 22-year follow-up. The study of men born in 1913. *Int J Obesity* 24, 33-7.

Lahmann, P.H., Lissner, L., Gullberg, B., & Berglund, G. (2002). A prospective study of adiposity and all-cause mortality: the Malmö Diet and Cancer Study. *Obes Res* 10, 361-9.

Launer, L.J., Harris, T., Rumpel, C., & Madans, J. (1994). Body mass index, weight change, and risk of mobility disability in middle-aged and older women. The epidemiologic follow-up study of NHANES I. *JAMA* 271, 1093-8.

Launer, L.J., Barendregt, J.J., & Harris, T. (1995). Shift in body mass index distributions due to height loss. *Epidemiology* 6, 98-9.

Lee, C.D., Blair, S.N., & Jackson, A.S. (1999). Cardiorespiratory fitness, body composition, and all-cause and cardiovascular disease mortality in men. *Am J Clin Nutr* 69, 373-80.

Lindstedt, K.D., & Singh, P.N. (1997). Body mass and 26-year risk of mortality among women who never smoked: finding from the Adventist mortality study. *Am J Epidemiol* 146, 1-11.

Lissner, L., Björkelund, C., Heitmann, B.L., Seidell, J.C., & Bengtsson, C. (2001). Larger hip circumference independently predicts health and longevity in a Swedish female cohort. *Obes Res* 9, 644-6.

Ostman, J., Efendic, S., & Arner, P. (1969). Catecholamines and metabolism of human adipose tissue I. Comparison between in vitro effects of noradrenaline, adrenaline and theophylline on lipolysis in omental adipose tissue. *Acta Med Scand* 186, 241-6.

Peeters, A., Barendregt, J.J., Willekens, F., Mackenbach, J.P., Al Mamun, A., & Bonneux, L. (2003). Obesity in adults and its consequences for life expectancy: a life-table analysis. *Ann Intern Med* 138, 24-32.

Peters, E.Th.J., Seidell, J.C., Menotti, A., Aravanis, C., Dontas, A., Fidanza, F., Karvonen, M., Nedeljkovic, S., Nissinen, A., Buzina, R., Bloemberg, B., & Kromhout, D. (1995). Changes in body weight in relation to mortality in 6441 European middle-aged men: the Seven Countries Study. *Int J Obesity* 19, 862-8.

Poehlman, E.T., Toth, M.J., Fishman, P.S., Vaitkevicius, P., Gottlieb, S.S., Fisher, M.L., & Fonong, T. (1995). Sarcopenia in aging humans: the impact of menopause and disease. *J Gerontol A* 50, 73-7.

Roubenoff, R. (2000). Sarcopenia and its implications for the elderly. *Eur J Clin Nutr* 54, S40-7.

Rowland, M. (1990). Self-reported weight and height. *Am J Clin Nutr* 52, 1125-33.

Seidell, J.C., Oosterlee, A., Deurenberg, P., Hautvast, J.G.A.J., & Ruijs, J.H.J. (1988). Abdominal fat depots measured with computed tomography: effects of degree of obesity, sex and age. *Eur J Clin Nutr* 42, 805-15.

Seidell, J.C., Cigolini, M., Charzewska, J., Ellsinger, B.M., DiBiase, G., Björntorp, P., Hautvast, J.G.A.J., Contaldo, F., Szostak, V., Scuro, L.A. (1989). Indicators of fat distribution, serum lipids, and blood pressure in European women born in 1948—the European fat distribution study. *Am J Epidemiol* 130, 53-66.

Seidell, J.C., Andres, R., Sorkin, J.D., & Muller, D.C. (1994). The sagittal waist diameter and mortality in men: the Baltimore Longitudinal Study on Aging. *Int J Obesity* 18, 61-7.

Seidell, J.C., & Bouchard, C. (1997). Visceral fat in relation to health: is it a major culprit or simply an innocent bystander? *Int J Obesity* 21, 626-31.

Seidell, J.C., Han, T.S., Feskens, E.J.M., & Lean, M.E.J. (1997). Narrow hips and broad waist circumferences independently contribute to increased risk of non-insulin-dependent diabetes mellitus. *J Int Med* 242, 401-6.

Seidell, J.C., & Visscher, T.L.S. (2000). Body weight and weight change and their health implications for the elderly. *Eur J Clin Nutr* 54 (suppl 3), S33-9.

Seidell, J.C., Visscher, T.L.S., & Hoogeveen, R.T. (1999). Overweight and obesity in the mortality rate data: current evidence and research issues. *Med Sci Sports Exerc* 31 (11 suppl): S597-601.

Sjöström, L.V. (1992). Mortality of severely obese subjects. *Am J Clin Nutr* 55, 516s-23s.

Snijder, M.B., Dekker, J.M., Visser, M., et al. (2003). Larger thigh and hip circumferences are associated with better glucose tolerance: the Hoorn study. *Obes Res* 11, 104-11.

Snijder, M.B., Dekker, J.M., Visser, M., Bouter, L.M., Stehouwer, C.D., Yudkin, J.S., Heine, R.J., Nijpels, G., & Seidell, J.C. (2004). Trunk fat and leg fat have independent and opposite associations with fasting and postload glucose levels: the Hoorn Study. *Diabetes Care* 27, 372-7.

Tankó, L.B., Bagger, Y.Z., Alexandersen, P., Larsen, P.J., & Christansen, C. (2003). Peripheral adiposity exhibits an independent dominant antiatherogenic effect in elderly women. *Circulation* 107, 1626-31.

Tayback, M., Kumanyika, S., & Chee, E. (1990). Body weight as a risk factor in the elderly. *Arch Intern Med* 150, 1065-72.

Visser, M., van den Heuvel, E., & Deurenberg, P. (1994). Prediction equations for the estimation of body composition in the elderly using anthropometric data. *Br J Nutr* 71, 823-33.

Visser, M., Harris, T.B., Langlois, J., Hannan, M.T., Roubenoff, R., Felson, D.T., Wilson, P.W.F., & Kiel, D.P. (1998a). Body fat and skeletal muscle mass in relation to physical disability in very old men and women of the Framingham Heart Study. *J Gerontol* 53A, M214-21.

Visser, M., Langlois, J., Guralnik, J.M., Kronmal, R.A., Robbins, J., Williamson, J.D., & Harris, T.B. (1998b). High body fatness, but not low fat-free mass, predicts disability in older men and women: the Cardiovascular Health Study. *Am J Clin Nutr* 68, 584-90.

Visscher, T.L.S., Seidell, J.C., Menotti, A., Blackburn, H., Nissinen, A., Feskens, E.J.M., & Kromhout, D. (2000). Under and overweight in relation to mortality among men aged 40-59 and 50-69: the seven countries study. *Am J Epidemiol* 151, 660-6.

Visscher, T.L.S., Seidell, J.C., Molarius, A., van der Kuip, D., Hofman, A., & Witteman, J.C. (2001). A comparison of body mass index, waist-hip ratio and waist circumference as predictors of all-cause mortality among the elderly: the Rotterdam Study. *Int J Obesity*, 25, 18-24.

Willett, W.C., Stampfer, M., Manson, J., & Van Itallie, T. (1991). New weight guidelines for Americans: justified or injudicious? *Am J Clin Nutr* 53, 1102-3.

Woo, J., Ho, S.C., Yu, A.L., & Sham, A. (2002). Is waist circumference a useful measure in predicting health outcomes in the elderly? *Int J Obesity* 26, 1349-55.

Chapter 23

Ali, P.A., al-Ghorabie, F.H., Evans, C.J., el-Sharkawi, A.M., and Hancock, D.A. (1998). Body composition measurements using DXA and other techniques in tamoxifen-treated patients. *Applied Radiation and Isotopes* 49, 643-5.

Almendingen, K., Hofstad, B., and Vatn, M.H. (2001). Does high body fatness increase the risk of presence and growth of colorectal adenomas followed up in situ for 3 years? *American Journal of Gastroenterology* 96, 2238-46.

Anonymous. (2002a). International Agency for Research on Cancer Biennial Report 2000/2001. *WHO Report 175001*, pp. 163. Geneva: World Health Organization.

Anonymous. (2002b). Prostate Cancer Trends 1973-1995 Survival. SEER program, *Internet Sources:* http://seer.cancer.gov/Publication/ProstMono/survival.pdf. National Cancer Institute.

Arikoski, P., Komulainen, J., Riikonen, P., Parvainen, M., Jurvelin, J.S., Voutilainen, R., and Kroger, H. (1999). Impaired development of bone mineral density during chemotherapy: a prospective analysis of 46 children newly diagnosed with cancer. *Journal of Bone and Mineral Research.* 14, 2002-9.

Aslani, A., Smith, R.C., Allen, B.J., Pavlakis, N., and Levi, J.A. (1999). Changes in body composition during breast cancer chemotherapy with the CMF-regimen. *Breast Cancer Research and Treatment* 57, 285-90.

Aslani, A., Smith, R.C., Allen, B.J., Pavlakis, N., and Levi, J.A. (2000). The predictive value of body protein for chemotherapy-induced toxicity. *Cancer* 88, 796-803.

Ballard-Barbash, R., Schatzkin, A., Taylor, P.R., and Kahle, L.L. (1990). Association of change in body mass with breast cancer. *Cancer Research* 50, 2152-5.

Ballard-Barbash, R., and Swanson, C.A. (1996). Body weight: estimation of risk for breast and endometrical cancer. *American Journal of Clinical Nutrition* 63, 437S-41S.

Banfi, A., Podesta, M., Fazzuoli, L., Sertoli, M.R., Venturini, M., Santini, G., Cancedda, R., and Quarto, R. (2001). High-dose chemotherapy shows a dose-dependent toxicity to bone marrow osteoprogenitors: a mechanism for post-bone marrow transplantation osteopenia. *Cancer* 92, 2419-28.

Barber, M.D., McMillan, D.C., Preston, T., Ross, J.A., and Fearon, K.C. (2000). Metabolic response to feeding in weight-losing pancreatic cancer patients and its modulation by a fish-oil-enriched nutritional supplement. *Clinical Science* 98, 389-99.

Barni, S., Lissoni, P., Tancini, G., Ardizzoia, A., and Cazzaniga, M. (1996). Effects of one-year adjuvant treatment with tamoxifen on bone mineral density in postmenopausal breast cancer women. *Tumori* 82, 65-7.

Basaria, S., Lieb, J., II, Tang, A.M., DeWeese, T., Carducci, M., Eisenberger, M., and Dobs, A.S. (2002). Long-term effects of androgen deprivation therapy in prostate cancer patients. *Clinical Endocrinology* 56, 779-86.

Bergstrom, A., Pisani, P., Tenet, V., Wolk, A., and Adami, H.O. (2001a). Overweight as an avoidable cause of cancer in Europe. *International Journal of Cancer* 91, 421-30.

Bergstrom, A., Hsieh, C.C., Lindblad, P., Lu, C.M., Cook, N.R., and Wolk, A. (2001b). Obesity and renal cell cancer—a quantitative review. *British Journal of Cancer* 85, 984-90.

Berruti, A., Dogliotti, L., Osella, G., Cerutti, S., Reimondo, G., Martino, A., Gorzegno, G., Catolla, R., and Angeli, A. (2000). Evaluation by dual energy X-ray absorptiometry of changed bone density in metastatic bone sites as a consequence of systemic treatment. *Oncology Reports* 7, 777-81.

Berruti, A., Dogliotti, L., Terrone, C., Cerutti, S., Isaia, G., Tarabuzzi, R., Reimondo, G., Mari, M., Ardissone, P., De Luca, S., Fasolis, G., Fontana, D., Rossetti, S.R., Angeli, A., and Gruppo Onco Urologico Piemontese, R.O.P. (2002). Changes in bone mineral density, lean body mass and fat content as measured by dual energy x-ray absorptiometry in patients with prostate cancer without apparent bone metastases given androgen deprivation therapy. *Journal of Urology* 167, 2361-7.

Bird, C.L., Frankl, H.D., Lee, E.R., and Haile, R.W. (1998). Obesity, weight gain, large weight changes, and adenomatous polyps of the left colon and rectum. *American Journal of Epidemiology* 147, 670-80.

Bruning, P.F., Bonfrer, J.M., Hart, A.A., van Noord, P.A., van der Hoeven, H., Collette, H.J., Battermann, J.J., de Jong-Bakker, M., Nooijen, W.J., and de Waard, F. (1992). Body measurements, estrogen availability and the risk of human breast cancer: a case-control study. *International Journal of Cancer* 51, 14-9.

Buist, D.S., LaCroix, A.Z., Barlow, W.E., White, E., Cauley, J.A., Bauer, D.C., and Weiss, N.S. (2001a). Bone mineral density and endogenous hormones and risk of breast cancer in postmenopausal women (United States). *Cancer Causes and Control* 12, 213-22.

Buist, D.S., LaCroix, A.Z., Barlow, W.E., White, E., and Weiss, N.S. (2001b). Bone mineral density and breast cancer risk in postmenopausal women. *Journal of Clinical Epidemiology* 54, 417-22.

Caan, B.J., Coates, A.O., Slattery, M.L., Potter, J.D., Quesenberry, C.P., Jr., and Edwards, S.M. (1998). Body size and the risk of colon cancer in a large case-control study. *International Journal of Obesity and Related Metabolic Disorders: Journal of the International Association for the Study of Obesity* 22, 178-84.

Calle, E.E., Rodriguez, C., Walker-Thurmond, K., and Thun, M.J. (2003). Overweight, obesity, and mortality from cancer in a prospectively studied cohort of U.S. adults. *New England Journal of Medicine* 348, 1625-38.

Cauley, J.A., Gutai, J.P., Kuller, L.H., LeDonne, D., and Powell, J.G. (1989). The epidemiology of serum sex hormones in postmenopausal women. *American Journal of Epidemiology* 129, 1120-31.

Cauley, J.A., Lucas, F.L., Kuller, L.H., Vogt, M.T., Browner, W.S., and Cummings, S.R. (1996). Bone mineral density and risk of breast cancer in older women: the study of osteoporotic fractures. Study of Osteoporotic Fractures Research Group. *JAMA* 276, 1404-8.

Cerhan, J.R., Torner, J.C., Lynch, C.F., Rubenstein, L.M., Lemke, J.H., Cohen, M.B., Lubaroff, D.M., and Wallace, R.B. (1997). Association of smoking, body mass, and physical activity with risk of prostate cancer in the Iowa 65+ Rural Health Study (United States). *Cancer Causes and Control* 8, 229-38.

Chang, S., Hursting, S.D., Contois, J.H., Strom, S.S., Yamamura, Y., Babaian, R.J., Troncoso, P., Scardino, P.S., Wheeler, T.M., Amos, C.I., and Spitz, M.R. (2001). Leptin and prostate cancer. *Prostate* 46, 62-7.

Chen, H.H., Lee, B.F., Guo, H.R., Su, W.R., and Chiu, N.T. (2002a). Changes in bone mineral density of lumbar spine after pelvic radiotherapy. *Radiotherapy and Oncology* 62, 239-42.

Chen, Z., Maricic, M., Nguyen, P., Ahmann, F.R., Bruhn, R., and Dalkin, B.L. (2002b). Low bone density and high percentage of body fat among men who were treated with androgen deprivation therapy for prostate carcinoma. *Cancer* 95, 2136-44.

Chyou, P.H., Nomura, A.M., and Stemmermann, G.N. (1994). A prospective study of weight, body mass index and other anthropometric measurements in relation to site-specific cancers. *International Journal of Cancer* 57, 313-7.

Coates, R.J., Uhler, R.J., Hall, H.I., Potischman, N., Brinton, L.A., Ballard-Barbash, R., Gammon, M.D., Brogan, D.R., Daling, J.R., Malone, K.E., Schoenberg, J.B., and Swanson, C.A. (1999). Risk of breast cancer in young women in relation to body size and weight gain in adolescence and early adulthood. *British Journal of Cancer* 81, 167-74.

Demark-Wahnefried, W., Conaway, M.R., Robertson, C.N., Mathias, B.J., Anderson, E.E., and Paulson, D.F. (1997a). Anthropometric risk factors for prostate cancer. *Nutrition and Cancer* 28, 302-7.

Demark-Wahnefried, W., Peterson, B.L., Winer, E.P., Marks, L., Aziz, N., Marcom, P.K., Blackwell, K., and Rimer, B.K. (2001). Changes in weight, body composition, and factors influencing energy balance among premenopausal breast cancer patients receiving adjuvant chemotherapy. *Journal of Clinical Oncology* 19, 2381-9.

Demark-Wahnefried, W., Rimer, B.K., and Winer, E.P. (1997b). Weight gain in women diagnosed with breast cancer. *Journal of the American Dietetic Association* 97, 519-26.

Demark-Wahnefried, W., Winer, E.P., and Rimer, B.K. (1993). Why women gain weight with adjuvant chemotherapy for breast cancer. *Journal of Clinical Oncology* 11, 1418-29.

den Broeder, E., Lippens, R.J., van't Hof, M.A., Tolboom, J.J., Sengers, R.C., van den Berg, A.M., van Houdt, N.B., Hofman, Z., and van Staveren, W.A. (2000). Nasogastric tube feeding in children with cancer: the effect of two different formulas on weight, body composition, and serum protein concentrations. *JPEN: Journal of Parenteral and Enteral Nutrition* 24, 351-60.

den Tonkelaar, I., de Waard, F., Seidell, J.C., and Fracheboud, J. (1995). Obesity and subcutaneous fat patterning in relation to survival of postmenopausal breast cancer patients participating in the DOM-project. *Breast Cancer Research and Treatment* 34, 129-37.

den Tonkelaar, I., Seidell, J.C., Collette, H.J., and de Waard, F. (1994). A prospective study on obesity and subcutaneous fat patterning in relation to breast cancer in post-menopausal women participating in the DOM project. *British Journal of Cancer* 69, 352-7.

Dixon, J.K., Moritz, D.A., and Baker, F.L. (1978). Breast cancer and weight gain: an unexpected finding. *Oncology Nursing Forum* 5, 5-7.

Douchi, T., Ijuin, H., Nakamura, S., Oki, T., Maruta, K., and Nagata, Y. (1997a). Correlation of body fat distribution with grade of endometrial cancer. *Gynecologic Oncology* 65, 138-42.

Douchi, T., Kosha, S., Kan, R., Nakamura, S., Oki, T., and Nagata, Y. (1997b). Predictors of bone mineral loss in patients with ovarian cancer treated with anticancer agents. *Obstetrics and Gynecology* 90, 12-5.

Douchi, T., Tsuji, T., Tokuhisa, T., Fujie, Y., Katanozaka, M., Yoshinaga, M., and Nagata, Y. (2000). Relation of estrogen receptor expression in endometrial cancer specimens to bone mineral density. *Acta Obstetricia et Gynecologica Scandinavica* 79, 1011-4.

Douchi, T., Yamamoto, S., Nakamura, S., Oki, T., Maruta, K., and Nagata, Y. (1999). Bone mineral density in postmenopausal women with endometrial cancer. *Maturitas* 31, 165-70.

Drinkard, C.R., Sellers, T.A., Potter, J.D., Zheng, W., Bostick, R.M., Nelson, C.L., and Folsom, A.R. (1995). Association of body mass index and body fat distribution with risk of lung cancer in older women. *American Journal of Epidemiology* 142, 600-7.

Fairfield, K.M., Willett, W.C., Rosner, B.A., Manson, J.E., Speizer, F.E., and Hankinson, S.E. (2002). Obesity, weight gain, and ovarian cancer. *Obstetrics and Gynecology* 100, 288-96.

Folsom, A.R., Kaye, S.A., Potter, J.D., and Prineas, R.J. (1989). Association of incident carcinoma of the endometrium with body weight and fat distribution in older women: early findings of the Iowa Women's Health Study. *Cancer Research* 49, 6828-31.

Folsom, A.R., Kaye, S.A., Prineas, R.J., Potter, J.D., Gapstur, S.M., and Wallace, R.B. (1990). Increased incidence of carcinoma of the breast associated with abdominal adiposity in postmenopausal women. *American Journal of Epidemiology* 131, 794-803.

Ford, E.S. (1999). Body mass index and colon cancer in a national sample of adult US men and women. *American Journal of Epidemiology* 150, 390-8.

Furuya, Y., Akakura, K., Akimoto, S., and Ito, H. (1998). Prognosis of patients with prostate carcinoma presenting as nonregional lymph node metastases. *Urologia Internationalis* 61, 17-21.

Gallus, S., La Vecchia, C., Levi, F., Simonato, L., Dal Maso, L., and Franceschi, S. (2001). Leanness and squamous cell oesophageal cancer. *Annals of Oncology* 12, 975-9.

Ganz, P.A., Schag, C.C., Polinsky, M.L., Heinrich, R.L., and Flack, V.F. (1987). Rehabilitation needs and breast cancer: the first month after primary therapy. *Breast Cancer Research and Treatment* 10, 243-53.

Garfinkel, L. (1985). Overweight and cancer. *Annals of Internal Medicine* 103, 1034-6.

Giovannucci, E., Ascherio, A., Rimm, E.B., Colditz, G.A., Stampfer, M.J., and Willett, W.C. (1995). Physical activity, obesity, and risk for colon cancer and adenoma in men. *Annals of Internal Medicine* 122, 327-34.

Gluck, O., and Maricic, M. (2002). Raloxifene: recent information on skeletal and non-skeletal effects. *Current Opinion in Rheumatology* 14, 429-32.

Goodwin, P.J., Ennis, M., Pritchard, K.I., McCready, D., Koo, J., Sidlofsky, S., Trudeau, M., Hood, N., and Redwood, S. (1999). Adjuvant treatment and onset of menopause predict weight gain after breast cancer diagnosis. *Journal of Clinical Oncology* 17, 120-9.

Gredmark, T., Kvint, S., Havel, G., and Mattsson, L.A. (1999). Adipose tissue distribution in postmenopausal women with adenomatous hyperplasia of the endometrium. *Gynecologic Oncology* 72, 138-42.

Hankinson, S.E., Willett, W.C., Manson, J.E., Hunter, D.J., Colditz, G.A., Stampfer, M.J., Longcope, C., and Speizer, F.E. (1995). Alcohol, height, and adiposity in relation to estrogen and prolactin levels in postmenopausal women. *Journal of the National Cancer Institute* 87, 1297-302.

Hardwick, J.C., Van Den Brink, G.R., Offerhaus, G.J., Van Deventer, S.J., and Peppelenbosch, M.P. (2001). Leptin is a growth factor for colonic epithelial cells. *Gastroenterology* 121, 79-90.

Harvie, M., Hooper, L., and Howell, A.H. (2003). Central obesity and breast cancer risk: a systematic review. *Obesity Review* 4, 157-173.

Headley, J.A., Theriault, R.L., LeBlanc, A.D., Vassilopoulou-Sellin, R., and Hortobagyi, G.N. (1998). Pilot study of bone mineral density in breast cancer patients treated with adjuvant chemotherapy. *Cancer Investigation* 16, 6-11.

Heasman, K.Z., Sutherland, H.J., Campbell, J.A., Elhakim, T., and Boyd, N.F. (1985). Weight gain during adjuvant chemotherapy for breast cancer. *Breast Cancer Research and Treatment* 5, 195-200.

Helzlsouer, K.J., Alberg, A.J., Gordon, G.B., Longcope, C., Bush, T.L., Hoffman, S.C., and Comstock, G.W. (1995). Serum gonadotropins and steroid hormones and the development of ovarian cancer. *Journal of the American Medicine Association* 274, 1926-30.

Henderson, R.C., Madsen, C.D., Davis, C., and Gold, S.H. (1998). Longitudinal evaluation of bone mineral density in children receiving chemotherapy. *Journal of Pediatric Hematology/Oncology* 20, 322-6.

Hesseling, P.B., Hough, S.F., Nel, E.D., van Riet, F.A., Beneke, T., and Wessels, G. (1998). Bone mineral density in long-term survivors of childhood cancer. *International Journal of Cancer Supplement* 11, 44-7.

Hsing, A.W., Deng, J., Sesterhenn, I.A., Mostofi, F.K., Stanczyk, F.Z., Benichou, J., Xie, T., and Gao, Y.T. (2000). Body size and prostate cancer: a population-based case-control study in China. *Cancer Epidemiology, Biomarkers and Prevention* 9, 1335-41.

Huang, Z., Hankinson, S.E., Colditz, G.A., Stampfer, M.J., Hunter, D.J., Manson, J.E., Hennekens, C.H., Rosner, B., Speizer, F.E., and Willett, W.C. (1997). Dual effects of weight and weight gain on breast cancer risk. *Journal of the American Medicine Association* 278, 1407-11.

Hung, Y.C., Yeh, L.S., Chang, W.C., Lin, C.C., and Kao, C.H. (2002). Prospective study of decreased bone mineral density in patients with cervical cancer without bone metastases: a preliminary report. *Japanese Journal of Clinical Oncology* 32, 422-4.

Iemura, A., Douchi, T., Yamamoto, S., Yoshimitsu, N., and Nagata, Y. (2000). Body fat distribution as a risk factor of endometrial cancer. *Journal of Obstetrics and Gynaecology Research* 26, 421-5.

Jain, M.G., Rohan, T.E., Howe, G.R., and Miller, A.B. (2000). A cohort study of nutritional factors and endometrial cancer. *European Journal of Epidemiology* 16, 899-905.

Jemal, A., Murray, T., Ward, E., Samuels, A., Tiwari, R.C., Ghafoon, A., Feuer, E.J., and Thum, M.J. (2005). Cancer statistics, 2005. *CA Cancer J Clin* 55:10-30.

Jernstrom, H., and Barrett-Connor, E. (1999). Obesity, weight change, fasting insulin, proinsulin, C-peptide, and insulin-like growth factor-1 levels in women with and without breast cancer: the Rancho Bernardo Study. *Journal of Women's Health and Gender-Based Medicine* 8, 1265-72.

Kaaks, R., and Lukanova, A. (2002). Effects of weight control and physical activity in cancer prevention: role of endogenous hormone metabolism. *Annals of the New York Academy of Sciences* 963, 268-81.

Kaaks, R., Van Noord, P.A., Den Tonkelaar, I., Peeters, P.J., Riboli, E., and Grobbee, D.E. (1998). Breast-cancer incidence in relation to height, weight and body-fat distribution in the Dutch "DOM" cohort. *International Journal of Cancer* 76, 647-51.

Kadar, L., Albertsson, M., Areberg, J., Landberg, T., and Mattsson, S. (2000). The prognostic value of body protein in patients with lung cancer. *Annals of the New York Academy of Sciences* 904, 584-91.

Kato, I., Severson, R.K., and Schwartz, A.G. (2001). Conditional median survival of patients with advanced carcinoma: surveillance, epidemiology, and end results data. *Cancer* 92, 2211-9.

Kono, S., Handa, K., Hayabuchi, H., Kiyohara, C., Inoue, H., Marugame, T., Shinomiya, S., Hamada, H., Onuma, K., and Koga, H. (1999). Obesity, weight gain and risk of colon adenomas in Japanese men. *Japanese Journal of Cancer Research* 90, 805-11.

Kristensen, B., Ejlertsen, B., Dalgaard, P., Larsen, L., Holmegaard, S.N., Transbol, I., and Mouridsen, H.T. (1994). Tamoxifen and bone metabolism in postmenopausal low-risk breast cancer patients: a randomized study. *Journal of Clinical Oncology* 12, 992-7.

Kumar, N.B., Cantor, A., Allen, K., and Cox, C.E. (2000). Android obesity at diagnosis and breast carcinoma survival: Evaluation of the effects of anthropometric variables at diagnosis, including body composition and body fat distribution and weight gain during life span, and survival from breast carcinoma. *Cancer* 88, 2751-7.

Kumar, N.B., Lyman, G.H., Allen, K., Cox, C.E., and Schapira, D.V. (1995). Timing of weight gain and breast cancer risk. *Cancer* 76, 243-9.

Kutynec, C.L., McCargar, L., Barr, S.I., and Hislop, T.G. (1999). Energy balance in women with breast cancer during adjuvant treatment. *Journal of the American Dietetic Association* 99, 1222-7.

Le Marchand, L., Wilkens, L.R., Hankin, J.H., Kolonel, L.N., and Lyu, L.C. (1997). A case-control study of diet and colorectal cancer in a multiethnic population in Hawaii (United States): lipids and foods of animal origin. *Cancer Causes and Control* 8, 637-48.

Lee, I.M., and Paffenbarger, R.S., Jr. (1992). Quetelet's index and risk of colon cancer in college alumni. *Journal of the National Cancer Institute* 84, 1326-31.

Lee, I.M., Sesso, H.D., and Paffenbarger, R.S., Jr. (2001). A prospective cohort study of physical activity and body size in relation to prostate cancer risk (United States). *Cancer Causes and Control* 12, 187-93.

Lew, E.A., and Garfinkel, L. (1979). Variations in mortality by weight among 750,000 men and women. *Journal of Chronic Diseases* 32, 563-76.

Li, S.D., and Mobarhan, S. (2000). Association between body mass index and adenocarcinoma of the esophagus and gastric cardia. *Nutrition Reviews* 58, 54-6.

Liedman, B. (1999). Symptoms after total gastrectomy on food intake, body composition, bone metabolism, and quality of life in gastric cancer patients—is reconstruction with a reservoir worthwhile? *Nutrition* 15, 677-82.

Liedman, B., Henningsson, A., Mellstrom, D., and Lundell, L. (2000). Changes in bone metabolism and body composition after total gastrectomy: results of a longitudinal study. *Digestive Diseases and Sciences* 45, 819-24.

Liedman, B., Svedlund, J., Sullivan, M., Larsson, L., and Lundell, L. (2001). Symptom control may improve food intake, body composition, and aspects of quality of life after gastrectomy in cancer patients. *Digestive Diseases and Sciences* 46, 2673-80.

Lucas, F.L., Cauley, J.A., Stone, R.A., Cummings, S.R., Vogt, M.T., Weissfeld, J.L., and Kuller, L.H. (1998). Bone mineral density and risk of breast cancer: differences by family history of breast cancer. Study of Osteoporotic Fractures Research Group. *American Journal of Epidemiology* 148, 22-9.

MacInnis, R.J., English, D.R., Gertig, D.M., Hopper, J.L., and Giles, G.G. (2003). Body size and composition and prostate cancer risk. *Cancer Epidemiology, Biomarkers, and Prevention* 12, 1417-1421.

May, P.E., Barber, A., D'Olimpio, J.T., Hourihane, A., and Abumrad, N.N. (2002). Reversal of cancer-related wasting using oral supplementation with a combination of beta-hydroxy-beta-methylbutyrate, arginine, and glutamine. *American Journal of Surgery* 183, 471-9.

Mayne, S.T., and Navarro, S.A. (2002). Diet, obesity and reflux in the etiology of adenocarcinomas of the esophagus and gastric cardia in humans. *Journal of Nutrition* 132, 3467S-70S.

McMillan, D.C., Watson, W.S., Preston, T., and McArdle, C.S. (2000). Lean body mass changes in cancer patients with weight loss. *Clinical Nutrition* 19, 403-6.

Merrill, R.M. (2001). Partitioned prostate cancer prevalence estimates: an informative measure of the disease burden. *Journal of Epidemiology and Community Health* 55, 191-7.

Mink, P.J., Folsom, A.R., Sellers, T.A., and Kushi, L.H. (1996). Physical activity, waist-to-hip ratio, and other risk factors for ovarian cancer: a follow-up study of older women. *Epidemiology* 7, 38-45.

Moller, H., Mellemgaard, A., Lindvig, K., and Olsen, J. H. (1994). Obesity and cancer risk: a Danish record-linkage study. *European Journal of Cancer* 30A, 344-50.

Moore, J.W., Key, T.J., Clark, G.M., Hoare, S.A., Allen, D.S., and Wang, D.Y. (1987). Sex-hormone-binding globulin and breast cancer risk. *Anticancer Research* 7, 1039-47.

Morimoto, L.M., White, E., Chen, Z., Chlebowski, R.T., Hays, J., Kuller, L., Lopez, A.M., Manson, J., Margolis, K.L., Muti, P.C., Stefanick, M.L., and McTiernan, A. (2002). Obesity, body size, and risk of postmenopausal breast cancer: the Women's Health Initiative (United States). *Cancer Causes and Control* 13, 741-51.

Moyad, M.A. (2001). Obesity, interrelated mechanisms, and exposures and kidney cancer. *Seminars in Urologic Oncology* 19, 270-9.

Murphy, T.K., Calle, E.E., Rodriguez, C., Kahn, H.S., and Thun, M.J. (2000). Body mass index and colon cancer mortality in a large prospective study. *American Journal of Epidemiology* 152, 847-54.

Murray, R.D., Brennan, B.M., Rahim, A., and Shalet, S.M. (1999). Survivors of childhood cancer: long-term endocrine and metabolic problems dwarf the growth disturbance. *Acta Paediatrica Supplement* 88, 5-12.

Newcomb, P.A., Trentham-Dietz, A., Egan, K.M., Titus-Ernstoff, L., Baron, J.A., Storer, B.E., Willett, W.C., and Stampfer, M.J. (2001). Fracture history and risk of breast and endometrial cancer. *American Journal of Epidemiology* 153, 1071-8.

Nguyen, M.C., Stewart, R.B., Banerji, M.A., Gordon, D.H., and Kral, J.G. (2001). Relationships between tamoxifen use, liver fat and body fat distribution in women with breast cancer. *International Journal of Obesity and Related Metabolic Disorders: Journal of the International Association for the Study of Obesity* 25, 296-8.

Nguyen, T.V., Center, J.R., and Eisman, J.A. (2000). Association between breast cancer and bone mineral density: the Dubbo Osteoporosis Epidemiology Study. *Maturitas* 36, 27-34.

Nysom, K., Holm, K., Fleischer Michaelsen, K., Hertz, H., Muller, J., and Molgaard, C. (2001). Bone mass after treatment of malignant lymphoma in childhood. *Medical and Pediatric Oncology* 37, 518-24.

Nysom, K., Molgaard, C., Holm, K., Hertz, H., and Michaelsen, K. F. (1998). Bone mass and body composition after cessation of therapy for childhood cancer. *International Journal of Cancer Supplement* 11, 40-3.

O'Brien, K., and Caballero, B. (1997). High bone mass as a marker for breast cancer risk. *Nutrition Reviews* 55, 284-6.

Olso, J.E., Anderson, K.E., Cerhan, J.R., Follsom, A.R., and Sellers, T.A. (2000). An investigation of the biological basis of an interaction of abdominal fat distribution and family history of breast cancer. A nested study of sisters in the Iowa Women's Health Study (United States). *Cancer Causes and Control* 11, 941-54.

Oppert, J.M., and Charles, M.A. (2002). Anthropometric estimates of muscle and fat mass in relation to cardiac and cancer mortality in men: the Paris Prospective Study. *American Journal of Clinical Nutrition* 75, 1107-13.

Putnam, S.D., Cerhan, J.R., Parker, A.S., Bianchi, G.D., Wallace, R.B., Cantor, K.P., and Lynch, C.F. (2000). Lifestyle and anthropometric risk factors for prostate cancer in a cohort of Iowa men. *Annals of Epidemiology* 10, 361-9.

Quan, M.L., Pasieka, J.L., and Rorstad, O. (2002). Bone mineral density in well-differentiated thyroid cancer patients treated with suppressive thyroxine: a systematic overview of the literature. *Journal of Surgical Oncology.* 79, 62-70.

Resch, A., Biber, E., Seifert, M., and Resch, H. (1998). Evidence that tamoxifen preserves bone density in late postmenopausal women with breast cancer. *Acta Oncologica* 37, 661-4.

Rodriguez, C., Calle, E.E., Fakhrabadi-Shokoohi, D., Jacobs, E.J., and Thun, M.J. (2002). Body mass index, height, and the risk of ovarian cancer mortality in a prospective cohort of postmenopausal women. *Cancer Epidemiology, Biomarkers and Prevention* 11, 822-8.

Rodriguez, C., Patel, A.V., Calle, E.E., Jacobs, E.J., Chao, A., and Thun, M.J. (2001). Body mass index, height, and prostate cancer mortality in two large cohorts of adult men in the United States. *Cancer Epidemiology, Biomarkers and Prevention* 10, 345-53.

Saarto, T., Blomqvist, C., Valimaki, M., Makela, P., Sarna, S., and Elomaa, I. (1997). Clodronate improves bone mineral density in post-menopausal breast cancer patients treated with adjuvant antioestrogens. *British Journal of Cancer* 75, 602-5.

Salazar-Martinez, E., Lazcano-Ponce, E.C., Lira-Lira, G.G., Escudero-De los Rios, P., Salmeron-Castro, J., Larrea, F., and Hernandez-Avila, M. (2000). Case-control study of diabetes, obesity, physical activity and risk of endometrial cancer among Mexican women. *Cancer Causes and Control* 11, 707-11.

Sarhill, N., Walsh, D., Nelson, K., Homsi, J., and Komurcu, S. (2000). Bioelectrical impedance, cancer nutritional assessment, and ascites. *Supportive Care in Cancer* 8, 341-3.

Schapira, D.V., Clark, R.A., Wolff, P.A., Jarrett, A.R., Kumar, N.B., and Aziz, N.M. (1994). Visceral obesity and breast cancer risk. *Cancer* 74, 632-9.

Schapira, D.V., Kumar, N.B., and Lyman, G.H. (1993). Variation in body fat distribution and breast cancer risk in the families of patients with breast cancer and control families. *Cancer* 71, 2764-8.

Schapira, D.V., Kumar, N.B., Lyman, G.H., Cavanagh, D., Roberts, W.S., and LaPolla, J. (1991a). Upper-body fat distribution and endometrial cancer risk. *Journal of the American Medical Association* 266, 1808-11.

Schapira, D.V., Kumar, N.B., Lyman, G.H., and Cox, C.E. (1990). Abdominal obesity and breast cancer risk. *Annals of Internal Medicine* 112, 182-6. (erratum appears in *Annals of Internal Medicine* 1990; 112:798).

Schapira, D.V., Kumar, N.B., Lyman, G.H., and Cox, C.E. (1991b). Obesity and body fat distribution and breast cancer prognosis. *Cancer* 67, 523-8.

Schuurman, A.G., Goldbohm, R.A., Dorant, E., and van den Brandt, P.A. (2000). Anthropometry in relation to prostate cancer risk in the Netherlands Cohort Study. *American Journal of Epidemiology* 151, 541-9.

Sellers, T.A., Davis, J., Cerhan, J.R., Vierkant, R.A., Olson, J.E., Pankratz, V.S., Potter, J.D., and Folsom, A.R. (2002). Interaction of waist/hip ratio and family history on the risk of hormone receptor-defined breast cancer in a prospective study of postmenopausal women. *American Journal of Epidemiology* 155, 225-33.

Sellers, T.A., Gapstur, S.M., Potter, J.D., Kushi, L.H., Bostick, R.M., and Folsom, A.R. (1993). Association of body fat distribution and family histories of breast and ovarian cancer with risk of postmenopausal breast cancer. *American Journal of Epidemiology* 138, 799-803.

Sellers, T.A., Kushi, L.H., Potter, J.D., Kaye, S.A., Nelson, C.L., McGovern, P.G., and Folsom, A.R. (1992). Effect of family history, body-fat distribution, and reproductive factors on the risk of postmenopausal breast cancer. *New England Journal of Medicine* 326, 1323-9. (erratum appears in *New England Journal of Medicine* 1992; 327:1612)

Shapiro, C.L., Keating, J., Angell, J.E., Janicek, M., Gelman, R., Hayes, D., and LeBoff, M.S. (1999). Monitoring therapeutic response in skeletal metastases using dual-energy x-ray absorptiometry: a prospective feasibility study in breast cancer patients. *Cancer Investigation* 17, 566-74.

Shapiro, C.L., Manola, J., and Leboff, M. (2001). Ovarian failure after adjuvant chemotherapy is associated with rapid bone loss in women with early-stage breast cancer. *Journal of Clinical Oncology* 19, 3306-11.

Sherman, B.M., and Korenman, S.G. (1974). Inadequate corpus luteum function: a pathophysiological interpretation of human breast cancer epidemiology. *Cancer* 33, 1306-12.

Shoff, S.M., and Newcomb, P.A. (1998). Diabetes, body size, and risk of endometrial cancer. *American Journal of Epidemiology* 148, 234-40.

Shu, X.O., Brinton, L.A., Zheng, W., Swanson, C.A., Hatch, M.C., Gao, Y.T., and Fraumeni, J.F., Jr. (1992). Relation of obesity and body fat distribution to endometrial cancer in Shanghai, China. *Cancer Research* 52, 3865-70.

Simons, J.P., Schols, A.M., Buurman, W.A., and Wouters, E.F. (1999a). Weight loss and low body cell mass in males with lung cancer: relationship with systemic inflammation, acute-phase response, resting energy expenditure, and catabolic and anabolic hormones. *Clinical Science* 97, 215-23.

Simons, J.P., Schols, A.M., Westerterp, K.R., Ten Velde, G.P., and Wouters, E.F. (1999b). Bioelectrical impedance analysis to assess changes in total body water in patients with cancer. *Clinical Nutrition.* 18, 35-9.

Smith, M.R. (2002). Osteoporosis during androgen deprivation therapy for prostate cancer. *Urology* 60, 79-86.

Smith, M.R., Finkelstein, J.S., McGovern, F.J., Zietman, A.L., Fallon, M.A., Schoenfeld, D.A., and Kantoff, P.W. (2002). Changes in body composition during androgen deprivation

therapy for prostate cancer. *Journal of Clinical Endocrinology and Metabolism* 87, 599-603.

Sonnenschein, E., Toniolo, P., Terry, M.B., Bruning, P.F., Kato, I., Koenig, K.L., and Shore, R.E. (1999). Body fat distribution and obesity in pre- and postmenopausal breast cancer. *International Journal of Epidemiology* 28, 1026-31.

Stoch, S.A., Parker, R.A., Chen, L., Bubley, G., Ko, Y.J., Vincelette, A., and Greenspan, S.L. (2001). Bone loss in men with prostate cancer treated with gonadotropin-releasing hormone agonists. *Journal of Clinical Endocrinology and Metabolism* 86, 2787-91.

Stoll, B.A. (1994). Breast cancer: the obesity connection. *British Journal of Cancer* 69, 799-801.

Stoll, B.A. (1995). Timing of weight gain in relation to breast cancer risk. *Annals of Oncology* 6, 245-8.

Stoll, B.A. (1998). Teenage obesity in relation to breast cancer risk. *International Journal of Obesity and Related Metabolic Disorders: Journal of the International Association for the Study of Obesity* 22, 1035-40.

Stoll, B.A. (2000). Adiposity as a risk determinant for postmenopausal breast cancer. *International Journal of Obesity and Related Metabolic Disorders* 24, 527-33.

Stoll, B.A. (2002). Upper abdominal obesity, insulin resistance and breast cancer risk. *International Journal of Obesity and Related Metabolic Disorders: Journal of the International Association for the Study of Obesity* 26, 747-53.

Strom, B.L., Soloway, R.D., Rios-Dalenz, J., Rodriguez-Martinez, H.A., West, S.L., Kinman, J.L., Crowther, R.S., Taylor, D., Polansky, M., and Berlin, J.A. (1996). Biochemical epidemiology of gallbladder cancer. *Hepatology* 23, 1402-11.

Swanson, C.A., Coates, R.J., Schoenberg, J.B., Malone, K.E., Gammon, M.D., Stanford, J.L., Shorr, I.J., Potischman, N.A., and Brinton, L.A. (1997). Body size and breast cancer risk among women under age of 45 years. *American Journal of Epidemiology* 145, 669-70.

Swanson, C.A., Potischman, N., Wilbanks, G.D., Twiggs, L.B., Mortel, R., Berman, M.L., Barrett, R.J., Baumgartner, R.N., and Brinton, L.A. (1993). Relation of endometrial cancer risk to past and contemporary body size and body fat distribution. *Cancer Epidemiology, Biomarkers and Prevention* 2, 321-7.

Tisdale, M.J. (1999). Wasting in cancer. *Journal of Nutrition* 129, 243S-6S.

Titus-Ernstoff, L., Egan, K.M., Newcomb, P.A., Ding, J., Trentham-Dietz, A., Greenberg, E.R., Baron, J.A., Trichopoulos, D., and Willett, W.C. (2002). Early life factors in relation to breast cancer risk in postmenopausal women. *Cancer Epidemiology, Biomarkers and Prevention* 11, 207-10.

Tornberg, S.A., and Carstensen, J.M. (1994). Relationship between Quetelet's index and cancer of breast and female genital tract in 47,000 women followed for 25 years. *British Journal of Cancer* 69, 358-61.

Toso, S., Piccoli, A., Gusella, M., Menon, D., Bononi, A., Crepaldi, G., and Ferrazzi, E. (2000). Altered tissue electric properties in lung cancer patients as detected by bioelectric impedance vector analysis. *Nutrition* 16, 120-4.

Trentham-Dietz, A., Newcomb, P.A., Egan, K.M., Titus-Ernstoff, L., Baron, J.A., Storer, B.E., Stampfer, M., and Willett, W.C. (2000). Weight change and risk of postmenopausal breast cancer (United States). *Cancer Causes and Control* 11, 533-42.

Trentham-Dietz, A., Newcomb, P.A., Storer, B.E., Longnecker, M.P., Baron, J., Greenberg, E.R., and Willett, W.C. (1997).

Body size and risk of breast cancer. *American Journal of Epidemiology* 145, 1011-9.

Trichopoulos, D., and Lipman, R.D. (1992). Mammary gland mass and breast cancer risk. *Epidemiology*. 3, 523-6.

Twiss, J.J., Waltman, N., Ott, C.D., Gross, G.J., Lindsey, A.M., and Moore, T.E. (2001). Bone mineral density in postmenopausal breast cancer survivors. *Journal of the American Academy of Nurse Practitioners* 13, 276-84.

van den Brandt, P.A., Spiegelman, D., Yaun, S., Adami, H., Beeson, L., Folsom, A.R., Fraser, G., Goldbohm, R.A., Graham, S., Kushi, L., Marshall, J.R., Miller, A.B., Rohan, T., Smith-Warner, S.A., Speizer, F.E., Willett, W.C., Wolk, A., and Hunter, D.J. (2000). Pooled analysis of prospective cohort studies on height, weight, and breast cancer risk. *American Journal of Epidemiology* 152, 514-527.

van Leeuwen, B.L., Kamps, W.A., Jansen, H.W., and Hoekstra, H.J. (2000). The effect of chemotherapy on the growing skeleton. *Cancer Treatment Review* 26, 363-76.

Vassilopoulou-Sellin, R., Brosnan, P., Delpassand, A., Zietz, H., Klein, M.J., and Jaffe, N. (1999). Osteopenia in young adult survivors of childhood cancer. *Medical and Pediatric Oncology* 32, 272-8.

Vehmanen, L., Saarto, T., Elomaa, I., Makela, P., Valimaki, M., and Blomqvist, C. (2001). Long-term impact of chemotherapy-induced ovarian failure on bone mineral density (BMD) in premenopausal breast cancer patients. The effect of adjuvant clodronate treatment. *European Journal of Cancer* 37, 2373-8.

Warner, J.T., Bell, W., Webb, D.K., and Gregory, J.W. (1997). Relationship between cardiopulmonary response to exercise and adiposity in survivors of childhood malignancy. *Archives of Disease in Childhood* 76, 298-303.

Warner, J.T., Evans, W.D., Webb, D.K., Bell, W., and Gregory, J.W. (1999). Relative osteopenia after treatment for acute lymphoblastic leukemia. *Pediatric Research* 45, 544-51.

Weiderpass, E., Persson, I., Adami, H.O., Magnusson, C., Lindgren, A., and Baron, J.A. (2000). Body size in different periods of life, diabetes mellitus, hypertension, and risk of postmenopausal endometrial cancer (Sweden). *Cancer Causes and Control* 11, 185-92.

Yoneda, K., Tanji, Y., Ikeda, N., Miyoshi, Y., Taguchi, T., Tamaki, Y., and Noguchi, S. (2002). Influence of adjuvant tamoxifen treatment on bone mineral density and bone turnover markers in postmenopausal breast cancer patients in Japan. *Cancer Letters* 186, 223-30.

Zatonski, W.A., Lowenfels, A.B., Boyle, P., Maisonneuve, P., Bueno de Mesquita, H.B., Ghadirian, P., Jain, M., Przewozniak, K., Baghurst, P., Moerman, C.J., Simard, A., Howe, G.R., McMichael, A.J., Hsieh, C.C., and Walker, A.M. (1997). Epidemiologic aspects of gallbladder cancer: a case-control study of the SEARCH Program of the International Agency for Research on Cancer. *Journal of the National Cancer Institute* 89, 1132-8.

Zhang, S., Folsom, A.R., Sellers, T.A., Kushi, L.H., and Potter, J.D. (1995). Better breast cancer survival for postmenopausal women who are less overweight and eat less fat. *The Iowa Women's Health Study Cancer* 76, 275-83.

Zhang, Y., Felson, D.T., Ellison, R.C., Kreger, B.E., Schatzkin, A., Dorgan, J.F., Cupples, L.A., Levy, D., and Kiel, D.P. (2001). Bone mass and the risk of colon cancer among postmenopausal women: the Framingham study. *American Journal of Epidemiology* 153, 31-7.

Zhang, Y., Kiel, D.P., Kreger, B.E., Cupples, L.A., Ellison, R.C., Dorgan, J.F., Schatzkin, A., Levy, D., and Felson, D.T. (1997). Bone mass and the risk of breast cancer among postmenopausal women. *New England Journal of Medicine* 336, 611-7.

Zmuda, J.M., Cauley, J.A., Ljung, B.M., Bauer, D.C., Cummings, S.R., Kuller, L.H., and The Study of Osteoporotic Fractures Research Group. (2001). Bone mass and breast cancer risk in older women: differences by stage at diagnosis. *Journal of the National Cancer Institute* 93, 930-6.

Chapter 24

Abate, N., Garg, A., Peshock, R.M., Stray-Gundersen, J., Adams-Huet, B., & Grundy, S.M. (1996). Relationship of generalized and regional adiposity to insulin sensitivity in men with NIDDM. *Diabetes* 45, 1684-1693.

Abate, N., Garg, A., Peshock, R.M., Stray-Gundersen, J., & Grundy, S.M. (1995). Relationships of generalized and regional adiposity to insulin sensitivity in men. *Journal of Clinical Investigation* 96, 88-98.

Albu, J.B., Murphy, L., Frager, D.H., Johnson, J.A., & Pi-Sunyer, F.X. (1997). Visceral fat and race-dependent health risks in obese nondiabetic premenopausal women. *Diabetes* 46, 456-62.

Anonymous. (2000). Obesity: preventing and managing the global epidemic. Report of a WHO consultation. *World Health Organization Technical Report Series* 894, i-xii, 1-253.

Arslanian, S.A. (2000). Type 2 diabetes mellitus in children: pathophysiology and risk factors. *Journal of Pediatric Endocrinology and Metabolism* 13(Suppl 6), 1385-1394.

Banerji, M., Chaiken, R., Gordon, D., & Lebowitz, H. (1995). Does intra-abdominal adipose tissue in black men determine whether NIDDM is insulin resistant or insulin sensitive? *Diabetes* 44, 141-146.

Banerji, M., Lebowitz, J., Chaiken, R., Gordon, D., Kral, J., & Lebovitz, H. (1997). Relationship of visceral adipose tissue and glucose disposal is independent of sex in black NIDDM subjects. *American Journal of Physiology—Endocrinology and Metabolism* 273, E425-E432.

Banerji, M.A., Faridi, N., Atluri, R., Chaiken, R.L., & Lebovitz, H.E. (1999). Body composition, visceral fat, leptin, and insulin resistance in Asian Indian men. *Journal of Clinical Endocrinology and Metabolism* 84, 137-144.

Boden, G., Chen, X., Ruiz, J., White, J., & Rossetti, L. (1994). Mechanisms of fatty acid-induced inhibition of glucose uptake. *Journal of Clinical Investigation* 93, 2438-2446.

Boden, G., Lebed, B., Schatz, M., Homko, C., & Lemieux, S. (2001). Effects of acute changes of plasma free fatty acids on intramyocellular fat content and insulin resistance in healthy subjects. *Diabetes* 50, 1612-1617.

Boesch, C., Slotboom, J., Hoppeler, H., & Kreis, R. (1997). In vivo determination of intra-myocellular lipids in human muscle by means of localized 1H-MR-spectroscopy. *Magnetic Resonance in Medicine* 37, 484-493.

Bogardus, C., Lillioja, S., Mott, D., Hollenbeck, C., & Reaven, G. (1985). Relationship between degree of obesity and in vivo insulin action in man. *American Journal of Physiology—Endocrinology and Metabolism* 248, E286-E291.

Brochu, M., Tchernof, A., Dionne, I.J., Sites, C.K., Eltabbakh, G.H., Sims, E.A.. & Poehlman, E.T. (2001). What are the physical characteristics associated with a normal metabolic profile despite a high level of obesity in postmenopausal women? *Journal of Clinical Endocrinology and Metabolism* 86, 1020-5.

Campbell, P.J., & Carlson, M.G. (1993). Impact of obesity on insulin action in NIDDM. *Diabetes* 42, 405-410.

Carey, D., Jenkins, A., Campbell, L., Freund, J., & Chisholm, D. (1996). Abdominal fat and insulin resistance in normal and overweight women. *Diabetes* 45, 633-638.

Caro, J.F., Dohm, L.G., Pories, W.J., & Sinha, M.K. (1989). Cellular alterations in liver, skeletal muscle, and adipose tissue responsible for insulin resistance in obesity and type II diabetes. *Diabetes—Metabolism Reviews* 5, 665-689.

Chitturi, S., & Farrell, G.C. (2001). Etiopathogenesis of nonalcoholic steatohepatitis. *Seminars in Liver Disease* 21, 27-41.

Clark, J.M., & Diehl, A.M. (2002). Hepatic steatosis and type 2 diabetes mellitus. *Current Diabetes Reports* 2, 210-215.

Coggan, A.R., Spina, R.J., King, D.S., Rogers, M.A., Brown, M., Nemeth, P.M., & Holloszy, J.O. (1992). Histochemical and enzymatic comparison of the gastrocnemius muscle of young and elderly men and women. *Journal of Gerontology* 47, B71-B76.

Colberg, S.R., Simoneau, J.-A., Thaete, F.L., & Kelley, D.E. (1995). Skeletal muscle utilization of FFA in women with visceral obesity. *Journal of Clinical Investigation* 95, 1846-1853.

Colditz, G.A., Willett, W.C., Rotnitzky, A., & Manson, J.E. (1995). Weight gain as a risk factor for clinical diabetes mellitus in women. *Annals of Internal Medicine* 122, 481-486.

Cruz, M.L., Bergman, R.N., & Goran, M.I. (2002). Unique effect of visceral fat on insulin sensitivity in obese Hispanic children with a family history of type 2 diabetes. *Diabetes Care* 25, 1631-1636.

Day, C., & Saksena, S. (2002). Non-alcoholic steatohepatitis: definitions and pathogenesis. *Journal of Gastroenterology and Hepatology* 17, S377-S384.

Despres, J. (1998). The insulin resistance-dyslipidemic syndrome of visceral obesity: effect on patients' risk. *Obesity Research* 6 (Suppl 1), 8S-17S.

Eckel, R.H. (1992). Insulin resistance: an adaptation for weight maintenance. *Lancet* 340, 1452-1453.

Evan, D., Murray, R., & Kissebah, A. (1984). Relationship between skeletal muscle insulin resistance, insulin-mediated glucose disposal, and insulin binding: effects of obesity and body fat topography. *Journal of Clinical Investigation* 74, 1515-1525.

Evans, W.J., & Campbell, W.W. (1993). Sarcopenia and age-related changes in body composition and functional capacity. *Journal of Nutrition* 123(2 Suppl), 465-8.

Falck-Ytter, Y., Younossi, Z.M., Marchesini, G., & McCullough, A.J. (2001). Clinical features and natural history of nonalcoholic steatosis syndromes. *Seminars in Liver Disease* 21, 17-26.

Felig, P., Wahren, J., Hendler, R., & Brundin, T. (1974). Splanchnic glucose and amino acid metabolism in obesity. *Journal of Clinical Investigation* 53, 582-590.

Ferland, M., Despres, J.-P., Tremblay, A., Pinault, S., Nadeau, A., Moorjani, S., Lupien, P., Theriault, G., & Bouchard, C. (1989). Assessment of adipose tissue distribution by computed axial tomography in obese women: association with body density and anthropometric measurements. *British Journal of Nutrition* 61, 139-148.

Fink, R.I., Wallace, P., & Olefsky, J.M. (1986). Effects of aging on glucose-mediated glucose disposal and glucose transport. *Journal of Clinical Investigation* 77, 2034-2041.

Forouhi, N.G., Jenkinson, G., Thomas, E.L., Mullick, S., Mierisova, S., Bhonsle, U., McKeigue, P.M., & Bell, J.D. (1999). Relation of triglyceride stores in skeletal muscle cells to central obesity and insulin sensitivity in European and South Asian men. *Diabetologia* 42, 932-935.

Gabriely, I., Ma, X.H., Yang, X.M., Atzmon, G., Rajala, M.W., Berg, A.H., Scherer, P., Rossetti, L., & Barzilai, N. (2002). Removal of visceral fat prevents insulin resistance and glucose intolerance of aging: an adipokine-mediated process? *Diabetes* 51, 2951-2958.

Gallagher, D., Visser, M., Sepulveda, D., Pierson, R.N., Harris, T., & Heymsfield, S.B. (1996). How useful is body mass index for comparison of body fatness across age, sex, and ethnic groups? *American Journal of Epidemiology* 143, 228-239.

Gavrilova, O., Marcus-Samuels, B., Graham, D., Kim, J.K., Shulman, G.I., Castle, A.L., Vinson, C., Eckhaus, M., & Reitman, M.L. (2000). Surgical implantation of adipose tissue reverses diabetes in lipoatrophic mice. *Journal of Clinical Investigation* 105, 271-278.

Goodpaster, B., He, J., Watkins, S., & Kelley, D. (2001). Skeletal muscle lipid content and insulin resistance: evidence for a paradox in endurance-trained athletes. *Journal of Clinical Endocrinology and Metabolism* 86, 5755-5761.

Goodpaster, B.H., Kelley, D.E., Thaete, F.L., He, J., & Ross, R. (2000a). Skeletal muscle attenuation determined by computed tomography is associated with skeletal muscle lipid content. *Journal of Applied Physiology* 89, 104-110.

Goodpaster, B.H., Kelley, D.E., Wing, R.R., Meier, A., & Thaete, F.L. (1999). Effects of weight loss on regional fat distribution and insulin sensitivity in obesity. *Diabetes* 48, 839-847.

Goodpaster, B., Krishnaswami, S., Resnick, H., Kelley, D., Haggerty, C., Harris, T., Schwartz, A., Kritchevsky, S., & Newman, A. (2003). Association between regional adipose tissue distribution and both type 2 diabetes and impaired glucose tolerance in elderly men and women. *Diabetes Care* 26, 372-379.

Goodpaster, B., Stenger, A., Boada, F., McKolanis, T., Davis, D., Ross, R., & Kelley, D. (2004). Skeletal muscle lipid concentration quantified by MRI. *American Journal of Clinical Nutrition,* 79, 748-754.

Goodpaster, B., Thaete, F., & Kelley, D. (2000b). Thigh adipose tissue distribution is associated with insulin resistance in obesity and type 2 diabetes mellitus. *American Journal of Clinical Nutrition* 71, 885-892.

Goodpaster, B.H., Thaete, F.L., Simoneau, J.-A., & Kelley, D.E. (1997). Subcutaneous abdominal fat and thigh muscle composition predict insulin sensitivity independently of visceral fat. *Diabetes* 46, 1579-1585.

Goodpaster, B.H., Theriault, R., Watkins, S.C., & Kelley, D.E. (2000c). Intramuscular lipid content is increased in obesity and decreased by weight loss. *Metabolism* 49, 467-472.

Gray, R.E., Tanner, C.J., Pories, W.J., MacDonald, K.G., & Houmard, J.A. (2003). Effect of weight loss on muscle lipid content in morbidly obese subjects. *American Journal of Physiology—Endocrinology and Metabolism* 284, E726-E732.

Greco, A.V., Mingrone, G., Giancaterini, A., Manco, M., Morroni, M., Cinti, S., Granzotto, M., Vettor, R., Camastra, S., & Ferrannini, E. (2002). Insulin resistance in morbid obesity: reversal with intramyocellular fat depletion. *Diabetes* 51, 144-151.

Harris, M.I., Flegal, K.M., Cowie, C.C., Eberhardt, M.S., Goldstein, D.E., Little, R.R., Wiedmeyer, H.M., & Byrd-Holt, D.D. (1998). Prevalence of diabetes, impaired fasting glucose, and impaired glucose tolerance in US adults: the Third

National Health and Nutrition Examination Survey. *Diabetes Care* 21, 518-524.

He, J., Watkins, S., & Kelley, D. (2001). Skeletal muscle lipid content and oxidative enzyme activity in relation to muscle fiber type in type 2 diabetes and obesity. *Diabetes* 50, 817-823.

Hollenbeck, C., Chen, Y.-D., & Reaven, G. (1984). A comparison of the relative effects of obesity and non-insulin-dependent diabetes mellitus on in vivo insulin stimulated glucose utilization. *Diabetes* 33, 622-626.

Hotamisligil, G.S., & Spiegelman, B.M. (1994). Tumor necrosis factor alpha: a key component of the obesity-diabetes link. *Diabetes* 43, 1271-1278.

Huang, T.T., Johnson, M.S., Gower, B.A., & Goran, M.I. (2002). Effect of changes in fat distribution on the rates of change of insulin response in children. *Obesity Research* 10, 978-984.

Hulver, M.W., Berggren, J.R., Cortright, R.N., Dudek, R.W., Thompson, R.P., Pories, W.J., MacDonald, K.G., Cline, G.W., Shulman, G.I., Dohm, G.L., & Houmard, J.A. (2003). Skeletal muscle lipid metabolism with obesity. *American Journal of Physiology—Endocrinology and Metabolism* 284, E741-E747.

Jackson, A.S., Stanforth, P.R., Gagnon, J., Rankinen, T., Leon, A.S., Rao, D.C., Skinner, J.S., Bouchard, C., & Wilmore, J.H. (2002). The effect of sex, age and race on estimating percentage body fat from body mass index: The Heritage Family Study. *International Journal of Obesity and Related Metabolic Disorders: Journal of the International Association for the Study of Obesity* 26, 789-796.

Jacob, S., Machann, J., Rett, K., Brechtel, K., Volk, A., Renn, W., Maerker, E., Matthaei, S., Schick, F., Claussen, C.D., & Haring, H.U. (1999). Association of increased intramyocellular lipid content with insulin resistance in lean nondiabetic offspring of type 2 diabetic subjects. *Diabetes* 48, 1113-1119.

Johnson, M.S., Figueroa-Colon, R., Huang, T.T., Dwyer, J.H., & Goran, M.I. (2001). Longitudinal changes in body fat in African American and Caucasian children: influence of fasting insulin and insulin sensitivity. *Journal of Clinical Endocrinology and Metabolism* 86, 3182-3187.

Kelley, D., Goodpaster, B., Wing, R., & Simoneau, J. (1999). Skeletal muscle fatty acid metabolism in association with insulin resistance, obesity and weight loss. *American Journal of Physiology—Endocrinology and Metabolism* 277, E1130-E1141.

Kelley, D., Kuller, L., McKolanis, T., Harper, P., Mancino, J., & Kalhan, S. (2004). Effects of moderate weight loss and orlistat on insulin resistance, regional adiposity and fatty acids in type 2 diabetes mellitus. *Diabetes Care* 27, 33-40.

Kelley, D., McKolanis, T., Hegazi, R., Kuller, L., & Kalhan, S. (2003). Fatty liver in type 2 diabetes mellitus: relation to regional adiposity, fatty acids, and insulin resistance. *American Journal of Physiology—Endocrinology and Metabolism* 285, E906-E916.

Kelley, D.E., Mokan, M., & Mandarino, L. (1993a). Metabolic pathways of glucose in skeletal muscle of lean NIDDM patients. *Diabetes Care* 16, 1158-1166.

Kelley, D.E., Slasky, B.S., & Janosky, J. (1991). Skeletal muscle density: effects of obesity and non-insulin-dependent diabetes mellitus. *American Journal of Clinical Nutrition* 54, 509-515.

Kelley, D., Troost, F., Huwe, T., Thaete, F., & Goodpaster, B. (2000). Subdivisions of subcutaneous abdominal adipose tissue and insulin resistance. *American Journal of Physiology—Endocrinology and Metabolism* 278, E941-E948.

Kelley, D., Wing, R.R., Buonocore, C., Fitzsimmons, M., Sturis, J., & Polonsky, K. (1993b). Relative effects of calorie restriction and weight loss in noninsulin dependent diabetes mellitus. *Journal of Clinical Endocrinology and Metabolism* 77, 1287-1293.

Kelley, D.E., Williams, K., Price, J., McKolanis, T., Goodpaster, B., & Thaete, F. (2001). Plasma fatty acids, adiposity, and variance of skeletal muscle insulin resistance in type 2 diabetes mellitus. *Journal of Clinical Endocrinology and Metabolism* 86, 5412-5419.

Kim, J., Kim, Y.-J., Fillmore, J., Chen, Y., Moore, I., Lee, J., Yuan, M., Li, Z.W., Karin, M., Perret, P., Shoelson, S., & Shulman, G. (2001). Prevention of fat-induced insulin resistance by salicylate. *Journal of Clinical Investigation* 108, 437-446.

Kim, J.Y., Hickner, R.C., Cortright, R.L., Dohm, G.L., & Houmard, J.A. (2000). Lipid oxidation is reduced in obese human skeletal muscle. *American Journal of Physiology—Endocrinology and Metabolism* 279, E1039-E1044.

Kissebah, A.H., & Peiris, A.N. (1989). Biology of regional fat distribution: relationship to non-insulin-dependent diabetes mellitus. *Diabetes-Metabolism Reviews* 5, 83-109.

Knowler, W.C., Barrett-Connor, E., Fowler, S.E., Hamman, R.F., Lachin, J.M., Walker, E.A., Nathan, D.M., & Diabetes Prevention Program Research, G. (2002). Reduction in the incidence of type 2 diabetes with lifestyle intervention or metformin. *New England Journal of Medicine* 346, 393-403.

Kohrt, W.M., & Holloszy, J.O. (1995). Loss of skeletal muscle mass with aging: effect on glucose tolerance. *Journals of Gerontology Series A, Biological Sciences and Medical Sciences. (50 Spec No)*, 68-72.

Kriketos, A.D., Pan, D.A., Lillioja, S., Cooney, G.J., Baur, L.A., Milner, M.R., Sutton, J.R., Jenkins, A.B., Bogardus, C., & Storlien, L.H. (1996). Interrelationships between muscle morphology, insulin action, and adiposity. *American Journal of Physiology* 270, R1332-R1339.

Krotkiewski, M., Seidell, J., & Björntorp, P. (1990). Glucose tolerance and hyperinsulinaemia in obese women: role of adipose tissue distribution, muscle fibre characteristics and androgens. *Journal of Investigative Medicine* 228, 385-392.

Krssak, M., Falk, P.K., Dresner, A., DiPietro, L., Vogel, S.M., Rothman, D.L., Shulman, G.I., & Roden, M. (1999). Intramyocellular lipid concentrations are correlated with insulin sensitivity in humans: a 1H NMR spectroscopy study. *Diabetologia* 42, 113-116.

Lillioja, S., Mott, D.M., Howard, B.V., Bennett, P.H., Yki-Jarvinen, H., Freymond, D., Nyomba, B.L., Zurlo, F., Swinburn, B., & Bogardus, C. (1988). Impaired glucose tolerance as a disorder of insulin action. Longitudinal and cross-sectional studies in Pima Indians. *New England Journal of Medicine* 318, 1217-1225.

Lillioja, S., Young, A.A., Culter, C., Ivy, J.L., Abbot, W., Zawadzki, J.K., Yki-Järvinen, H., Christin, L., Secomb, T.W., & Bogardus, C. (1987). Skeletal muscle capillary density and fiber type are possible determinants of in vivo insulin resistance in man. *Journal of Clinical Investigation* 80, 415-424.

Long, S., O'Brien, K., MacDonald, K., Jr., Leggett-Frazier, N., Swanson, M., Pories, W., & Caro, J. (1994). Weight loss in severely obese subjects prevents the progression of impaired glucose tolerance to type II diabetes. A longitudinal interventional study. *Diabetes Care* 17, 372-375.

Ludvik, B., Nolan, J.J., Baloga, J., Sacks, D., & Olefsky, J. (1995). Effect of obesity on insulin resistance in normal subjects and patients with NIDDM. *Diabetes* 44, 1121-1125.

Marchesini, G., Brizi, M., Morselli-Labate, A.M., Bianchi, G., Bugianesi, E., McCullough, A.J., Forlani, G., & Melchionda, N. (1999). Association of nonalcoholic fatty liver disease with insulin resistance. *American Journal of Medicine* 107, 450-455.

McTigue, K.M., Garrett, J.M., & Popkin, B.M. (2002). The natural history of the development of obesity in a cohort of young US adults between 1981 and 1998. *Annals of Internal Medicine* 136, 857-864.

Miyazaki, Y., Glass, L., Triplitt, C., Wajcberg, E., Mandarino, L.J., & DeFronzo, R.A. (2002). Abdominal fat distribution and peripheral and hepatic insulin resistance in type 2 diabetes mellitus. *American Journal of Physiology Endocrinology and Metabolism* 283, E1135-E1143.

Mokdad, A.H., Ford, E.S., Bowman, B.A., Nelson, D., Engelgau, M.M., Vinicor, F., & Marks, J.S. (2000). Diabetes trends in the US: 1990-1998. *Diabetes Care* 23, 1125-1133.

Montague, C.T., & O'Rahilly, S. (2000). The perils of portliness: causes and consequences of visceral adiposity. *Diabetes* 49, 883-888.

Mott, D., Lillioja, S., & Bogardus, C. (1986). Overnutrition induced decrease in insulin action for glucose storage: in vivo and in vitro in man. *Metabolism* 35, 160-165.

O'Rahilly, S. (1997). Science, medicine, and the future. Non-insulin dependent diabetes mellitus: the gathering storm. *British Medical Journal* 314, 955-959.

Olefsky, J. (1976). Decreased insulin binding to adipocytes and monocytes from obese subjects. *Journal of Clinical Investigation* 57, 1165-1172.

Pan, D.A., Lillioja, S., Kriketos, A.D., Milner, M.R., Baur, L.A., Bogardus, C., Jenkins, A.B., & Storlien, L.H. (1997). Skeletal muscle triglyceride levels are inversely related to insulin action. *Diabetes* 46, 983-988.

Perseghin, G., Scifo, P., DeCobelli, F., Pagliato, E., Arcelloni, C., Vanzulli, A., Testolin, G., Pozza, G., Del Maschio, A., & Luzi, L. (1999). Intramyocellular triglyceride content is a determinant of in vivo insulin resistance in humans: a 1H-13C nuclear magnetic resonance spectroscopy assessment in offspring of type 2 diabetic patients. *Diabetes* 48, 1600-1606.

Petersen, K.F., West, A.B., Reuben, A., Rothman, D.L., & Shulman, G.I. (1996). Noninvasive assessment of hepatic triglyceride content in humans with 13C nuclear magnetic resonance spectroscopy. *Hepatology* 24, 114-7.

Phillips, D., Caddy, S., Llic, V., Frayn, K., Borthwick, A., & Taylor, R. (1996). Intramuscular triglycerides and muscle insulin sensitivity: evidence for a relationship in nondiabetic subjects. *Metabolism* 45, 947-950.

Pories, W.J., Swanson, M.S., MacDonald, K.G., Long, S.B., Morris, P.G., Brown, B.M., Barakat, H.A., deRamon, R.A., Israel, G., & Dolezal, J.M. (1995). Who would have thought it ? An operation proves to be the most effective therapy for adult-onset diabetes mellitus. *Annals of Surgery* 222, 339-352.

Reaven, G.M. (1988). Banting lecture 1988. Role of insulin resistance in human disease. *Diabetes* 37, 1595-1607.

Reitman, M.L., Arioglu, E., Gavrilova, O., & Taylor, S.I. (2000). Lipoatrophy revisited. *Trends in Endocrinology and Metabolism* 11, 410-416.

Resnick, H.E., Harris, M.I., Brock, D.B., & Harris, T.B. (2000). American Diabetes Association diabetes diagnostic criteria, advancing age, and cardiovascular disease risk profiles: results from the Third National Health and Nutrition Examination Survey. *Diabetes Care* 23, 176-180.

Ross, R., Aru, J., Freeman, J., Hudson, R., & Janssen, I. (2002). Abdominal adiposity and insulin resistance in obese men. *American Journal of Physiology—Endocrinology and Metabolism* 282, E657-E663.

Ross, R., Fortier, L., & Hudson, R. (1996). Separate associations between visceral and subcutaneous adipose tissue distribution, insulin and glucose levels in obese women. *Diabetes Care* 19, 1404-1411.

Ruderman, N.B., Schneider, S.H., & Berchtold, P. (1981). The "metabolically-obese," normal-weight individual. *American Journal of Clinical Nutrition* 34, 1617-1621.

Ryysy, L., Hakkinen, S.M., Goto, T., Vehkavaara, S., Westerbacka, J., Halavaara, J., & Yki-Jarvinen, H. (2000). Hepatic fat content and insulin action on free fatty acids and glucose metabolism rather than insulin absorption are associated with insulin requirements during insulin therapy in type 2 diabetic patients. *Diabetes* 49, 749-758.

Schick, F., Machman, J., Bretchtel, K., Strempfer, A., Klumpp, B., Stein, D., & Jacob, S. (2002). MRI of muscular fat. *Magnetic Resonance in Medicine* 47, 720-727.

Schmitz-Peiffer, C., Craig, D., & Biden, T. (1999). Ceramide generation is sufficient to account for the inhibition of the insulin-stimulated PKB pathway in C2C12 skeletal muscle cells pretreated with palmitate. *Journal of Biological Chemistry* 274, 24202-24210.

Seppala-Lindroos, A., Satu, V., Hakkinen, A.-M., Goto, T., Westerbacka, J., Sovijarvi, A., Juha, H., & Yki-Jarvinen, H. (2002). Fat accumulation in the liver is associated with defects in insulin suppression of glucose production and serum fatty acids independent of obesity in normal men. *Journal of Clinical Endocrinology and Metabolism* 87, 3023-3028.

Sial, S., Coggan, A.R., Carroll, R., Goodwin, J., & Klein, S. (1996). Fat and carbohydrate metabolism during exercise in elderly and young subjects. *American Journal of Physiology* 271, E983-E989.

Simoneau, J., & Kelley, D. (1997). Altered glycolytic and oxidative capacities of skeletal muscle contribute to insulin resistance in NIDDM. *Journal of Applied Physiology* 83, 166-171.

Simoneau, J.-A., Veerkamp, J., Turcotte, L., & Kelley, D. (1999). Markers of capacity to utilize fatty acids in human skeletal muscle: relation to insulin resistance and obesity and effects of weight loss. *FASEB Journal* 13, 2051-2060.

Sinha, R., Dufour, S., Peterson, K.F., LeBon, V., Enoksson, S., Ma, Y.Z., Savoye, M., Rothman, D.L., Shulman, G.I., & Caprio, S. (2002). Assessment of skeletal muscle triglyceride content by (1)H nuclear magnetic resonance spectroscopy in lean and obese adolescents: relationships to insulin sensitivity, total body fat, and central adiposity. *Diabetes* 51, 1022-1027.

Sjostrom, C.D., Peltonen, M., Wedel, H., & Sjostrom, L. (2000). Differentiated long-term effects of intentional weight loss on diabetes and hypertension. *Hypertension* 36, 20-25.

Szcepaniak, L., Babcock, E., Schick, F., Dobbins, R., Garg, A., Burns, D., McGarry, J., & Stein, D. (1999). Measurement of intracellular triglyceride stores by H spectroscopy: validation in vivo. *American Journal of Physiology—Endocrinology and Metabolism* 276, E977-E989.

Tanner, C.J., Barakat, H.A., Dohm, G.L., Pories, W.J., MacDonald, K.G., Cunningham, P.R., Swanson, M.S., & Houmard, J.A. (2002). Muscle fiber type is associated with obesity and weight loss. *American Journal of Physiology—Endocrinology and Metabolism* 282, E1191-E1196.

Travers, S.H., Jeffers, B.W., & Eckel, R.H. (2002). Insulin resistance during puberty and future fat accumulation. *Journal of Clinical Endocrinology and Metabolism* 87, 3814-3818.

Vague, J. (1956). The degree of masculine differentiation of obesity: a factor determining predisposition to diabetes, atherosclerosis, gout, and uric calculous disease. *American Journal of Clinical Nutrition* 4, 20-34.

Warram, J.H., Martin, B.C., Krolewski, A.S., Soeldner, J.S., & Kahn, C.R. (1990). Slow glucose removal rate and hyperinsulinemia precede the development of type II diabetes in the offspring of diabetic parents. *Annals of Internal Medicine* 113, 909-915.

Weyer, C., Foley, J.E., Bogardus, C., Tataranni, P.A., & Pratley, R.E. (2000). Enlarged subcutaneous abdominal adipocyte size, but not obesity itself, predicts type 2 diabetes independent of insulin resistance. *Diabetologia* 43, 1498-1506.

Williams, K., and Kelley, D. (2000). Metabolic consequences of weight loss on glucose metabolism and insulin action in Type 2 diabetes mellitus. *Diabetes, Obesity and Metabolism* 2, 121-129.

Wing, R., Blair, E., Bononi, P., Marcus, M., Watanabe, R., & Bergman, R. (1994). Caloric restriction per se is a significant factor in improvements in glycemic control and insulin sensitivity during weight loss in obese NIDDM patients. *Diabetes Care* 17, 30-36.

Yki-Järvinen, H., & Koivisto, V. (1983). Effects of body composition on insulin sensitivity. *Diabetes* 32, 765-969.

Yu, C., Chen, Y., Cline, G.W., Zhang, D., Zong, H., Wang, Y., Bergeron, R., Kim, J.K., Cushman, S.W., Cooney, G.J., Atcheson, B., White, M.F., Kraegen, E.W., & Shulman, G.I. (2002). Mechanism by which fatty acids inhibit insulin activation of insulin receptor substrate-1 (IRS-1)-associated phosphatidylinositol 3-kinase activity in muscle. *Journal of Biological Chemistry* 277, 50230-50236.

Chapter 25

Agin D, Gallagher D, Wang J, Heymsfield SB, Pierson RN, Kotler DP. Effects of whey protein and resistance exercise on body cell mass, muscle strength and quality of life in women with HIV. AIDS 2001, 15:2431-2440.

Arpadi SM, Cuff PA, Horlick M, Wang J, Kotler DP. Lipodystrophy in HIV-infected children is associated with high viral load and low CD4+ lymphocyte count and CD4+ lymphocyte percent age at baseline and use of protease inhibitors and stavudine. J Acquir Immunodefic Syndr 2001;27:30-4.

Babameto G, Kotler DP, Burastero S, Wang J, Pierson RN. Alterations in hydration in HIV-infected individuals. (Abstract) Clin Res 1994;42:279.

Babameto G, Kotler DP. Malnutrition in HIV infection. GI Clin North America 1997;26:393-417.

Bhasin S, Storer TW, Javanbakht M, Yarasheski K E, Phillips J, Dike M, Sinha-Hikim I, Shen R, Hays RD, Beall G. Testosterone Replacement and Resistance Exercise in HIV-infected Men with Weight Loss and Low Testosterone Level. JAMA, 2000, 6:763-77.

Burkinshaw L, Hill GL, Morgan DB. Assessment of the distribution of protein in the human body by in vivo neutron activation analysis. International Symposium on Nuclear Activation Techniques in the Life Sciences. Vienna: International Atomic Energy Association. 1978.

Carr A, Samaras K, Burton S, Law M, Freund J, Chisholm DJ, Cooper DA. A syndrome of peripheral lipodystrophy, hyperlipidaemia and insulin resistance in patients receiving HIV protease inhibitors. AIDS. 1998;12:F51-F58.

Carr A, Workman C, Smith DE, Hoy J, Hudson J, Doong N, Martin A, Amin J, Freund J, Law M, Cooper DA, MITOX Study Group. Abacavir substitution for nucleoside analogs in patients with HIV lipoatrophy: a randomized trial. JAMA. 2002;288:207-15.

Cohn SH, Sawitsky A, Vartsky D, Yasumura S, Zanzi I, Ellis KJ. Compartmental body composition of cancer patients by measurement of total body nitrogen, potassium and water. Metabolism 1980;30:222-9.

Dong K, Bausserman LL, Flynn MM, Dickinson BP, Flanigan TP, Mileno MD, Tashima KT, Carpenter CC. Changes in body habitus and serum lipid abnormalities in HIV-positive women on highly active antiretroviral therapy. J Acquir Immunodefic Syndr 1999;23:107-12.

Engelson ES, Tierney AR, Pi-Sunyer FX, Kotler DP. Effects of megestrol acetate upon body composition and circulating testosterone in patients with AIDS. J Acquir Immunodefic Syndr 1995;9:1107-8.

Engelson ES, Rabkin JG, Rabkin R, Kotler DP. Effects of testosterone upon body composition. Letter In: J Acquir Immune Defic Syndr and Human Retrovir. 1996;11:510-1.

Engelson EE, Kotler DP, Tan YX, Agin D, Wang J, Pierson RN, Heymsfield SB. Fat distribution in HIV-infected patients reporting truncal enlargement quantified by whole-body magnetic resonance imaging. Am J Clin Nutr. 1999;69:1162-9.

Engelson ES, Glesby MJ, Mendez D, Albu JB, Wang J, Heymsfield SB, Kotler DP. Effect of recombinant human growth hormone in the treatment of visceral fat accumulation in HIV infection. J Acquir Immunodefic Syndr. 2002;30:379-91.

Gibert CL, Wheeler DA, Collins G, Madans M, Muurahainen N, Raghavan SS, Bartsch G. Randomized, Controlled Trial of Caloric Supplements in HIV infection. J Acquir Immune Defic Syndr 1999;22:253-9.

Gold J, High HA, Li Y, Michelmore H, Bodsworth NJ, Finlayson R, Furner VL, Allen BJ, Oliver CJ. Safety and efficacy of nandrolone decanoate for treatment of wasting in patients with HIV infection. AIDS 1996;10:745-51.

Grinspoon S, Corcoran C, Parlman K, Costello M, Rosenthal D, Anderson E, Stanley T, Schoenfeld D, Burrows B, Hayden D, Basgoz N, Klibanski A. Effects of Testosterone and Progressive Resistance Training in Eugonadal Men with AIDS Wasting. Ann Intern Med 2000;133:348-55.

Grunfeld C, Kotler DP, Hamadeh R, Tierney A, Wang J, Pierson RN. Hypertriglyceridemia in the acquired immunodeficiency syndrome. Am J Med 1989;86:27-31.

Grunfeld C, Kotler DP, Shigenaga JK, Doerrler W, Tierney A, Wang J, Pierson RN Jr, Feingold KR. Circulating interferon alpha levels and hypertriglyceridemia in the acquired immunodeficiency syndrome. Am J Med 1991;90:154-7.

Hellerstein MK, Wu K, McGrath M, Faix D, George D, Shackleton CH, Horn W, Hoh R, Neese RA. Effects of Dietary n-3 Fatty Acids Supplementation in Men with Weight Loss Associated with the Acquired Immune Deficiency Syndrome: Relation to Indices of Cytokine Production. J Acquir Immune Defic Syndr 1996;11:258-270.

Hommes MJT, Romijn JA, Endert E, Sauerwein HP. Resting energy expenditure and substrate oxidation in human immunodeficiency virus (HIV)-infected asymptomatic men: HIV affects host metabolism in the early asymptomatic stage. Am J Clin Nutr 1991;54:311-5.

Johannsson G, Marin P, Lonn L, Ottoson K, Stenlof P, Bjorntorp P, Sjostrom L, Bengtsson B. Growth hormone treatment of abdominally obese men reduces abdominal fat mass, improves glucose and lipoprotein metabolism, and reduces diastolic blood pressure. J Clin Endocrinol Metab. 1997;82:727-34.

Kotler DP, Gaetz HP, Klein EB, Lange M, Holt PR. Enteropathy associated with the acquired immunodeficiency syndrome. Ann Intern Med. 1984;101:421-8.

Kotler DP, Wang J, Pierson R. Studies of body composition in-patients with the acquired immunodeficiency Syndrome. Am J Clin Nutr. 1985;42:1255-65

Kotler DP, Tierney AR, Altilio D, Wang J, Pierson RN Jr. Body mass repletion during ganciclovir therapy of cytomegalovirus infections in patients with the acquired immunodeficiency syndrome. Arch Int Med. 1989a;149:901-5.

Kotler DP, Tierney AR, Francisco A, Wang J, Pierson RN Jr. The magnitude of body cell mass depletion determines the timing of death from wasting in AIDS. Am J Clin Nutr 1989b;50:444-7.

Kotler DP, Tierney AR, Brenner SK, Couture S, Wang J, Pierson RN Jr. Preservation of short-term energy balance in clinically stable patients with AIDS. Am J Clin Nutr 1990a;57:7-13.

Kotler DP, Tierney AR, Wang J, Pierson RN Jr. Effect of home total parenteral nutrition upon body composition in AIDS. J Parenter Enteral Nutr 1990b;14:454-8.

Kotler DP, Tierney AR, Ferraro R, Cuff P, Wang J, Pierson RN, Heymsfield S. Effect of enteral feeding upon body cell mass in AIDS. Am J Clin Nutr 1991a;53:149-54.

Kotler DP, Tierney AR, Dilmanian FA, Kamen Y, Wang J, Pierson RN Jr, Weber D. Correlation between total body potassium and total body nitrogen in-patients with acquired immunodeficiency syndrome. (Abstract) Clin Res 1991b;39:649.

Kotler DP, Fogleman L, Tierney AR. Comparison of total parenteral nutrition and oral, semielemental diet upon body composition and quality of life in AIDS patients with malabsorption. J Parent Enteral Nutr. 1998;22:120-6.

Kotler DP, Heymsfield SB. Human Immunodeficiency virus infection: a model chronic illness for studying wasting disorders. Editorial. Am J Clin Nutr 1998;68:519-20.

Kotler DP, Thea DM, Heo M, Allison DB, Engelson ES, Wang J, Pierson RNJr, St. Louis M, Keusch GT. Relative influences of race, sex, environment, and HIV infection upon body composition in adults. Am J Clin Nutr. 1999;69:432-9.

Kotler DP, Rosenbaum K, Wang J, Pierson RN. Studies of body composition and fat distribution in HIV-infected and control subjects. J Acquir Immunodefic Syndr. 1999;20:228-37.

Lichtenstein KA, Ward DJ, Moorman AC, Delaney KM, Young B, Palella FJ Jr, Rhodes PH, Wood KC, Holmberg SD, HIV Outpatient Study Investigators. Clinical assessment of HIV-associated lipodystrophy in an ambulatory population. AIDS 2001;15:1389-98.

Lo JC, Mulligan K, Tai VW, Algren H, Schambelan M. Buffalo hump in men with HIV-1 infection. Lancet. 1998;351:867-70.

Lo JC, Mulligan K, Noor MA, Schwarz JM, Halvorsen RA, Grunfeld C, Schambelan M. The effects of recombinant human growth hormone on body composition and glucose metabolism in HIV-infected patients with fat accumulation. J Clin Endocrinol Metab. 2001;86:3480-7

Martinez E, Conget I, Lozano L, Casamitjana R, Gatell JM. Reversion of metabolic abnormalities after switching from HIV protease inhibitors to nevirapine. AIDS. 1999;13:805-10.

Melchior JC, Raguin G, Boulier A, Bouvet E, Rigaud D, Matheron S, Casalino E, Vilde JL, Vachon F, Coulaud JP. Resting energy expenditure in human immunodeficiency virus-infected patients: comparison between patients with and without secondary infections. Am J Clin Nutr 1993;57:614-19.

Oster MH, Enders SP, Samuels SJ, Cone LA, Hooton TM, Browder HP, Flynn NM. Megestrol acetate in patients with AIDS and cachexia. Ann Intern Med 1994;121:400-4.

Palella FJ Jr, Delaney KM, Moorman AC, Loveless MO, Fuhrer J, Satten GA, Aschman DJ, Holmberg SD. Declining morbidity and mortality among patients with advanced human immunodeficiency virus infection. HIV Outpatient Study Investigators. N Engl J Med. 1998;338:853-860.

Palenicek JG, Graham NMH, He YH, Hoover DA, Oishi J, Kingsley L, Saah AJ, and the MACS. Weight loss prior to clinical AIDS as a predictor of survival. J Acquir Immune Defic Syndr 1995;10:366-73.

Pichard C, Sudre P, Karsegard V, Yerly S, Slosman DO, Delley V, Perrin L, Hirschel B. A randomized double-blind controlled study of 6 months of oral nutritional supplementation with arginine and n-3 fatty acids in HIV-infected patients. AIDS 1998;12:53-63.

Reyes-Teran G, Sierra-Madero JG, Martinez del Cerro V, Arroyo-Figueroa H, Pasquetti A, Calva JJ, Ruiz-Palacios G. Effects of Thalidomide on HIV-associated wasting syndrome: a randomized, double-blind, placebo-controlled trial. AIDS 1996, 10:1501-7.

Roubenoff R, Weiss L, McDermott A, Heflin T, Cloutier GJ, Wood M, Gorbach S. A pilot study of exercise training to reduce trunk fat in adults with HIV-associated fat redistribution. AIDS. 1999;13:1373-5.

Saint-Marc T, Touraine JL. The effects of discontinuing stavudine therapy on clinical and metabolic abnormalities in patients suffering from lipodystrophy. AIDS. 1999;13:2188-9.

Schambelan M, Mulligan K, Grunfeld C, Daar ES, LaMarca A, Kotler DP, Wang J, Bozzette SA, Breitmeyer JB. Recombinant human growth hormone in patients with HIV-associated wasting. Ann Intern Med. 1996;125:873-8.

Silva M, Skolnik PR, Gorbach SL, Spiegelman D, Wilson IB, Knox TA. The effect of protease inhibitors on weight and body composition in HIV-infected patients. AIDS 1998;12:1645-51

Siwek RA, Wales JK, Swaminathan R, Burkinshaw L, Oxby CB. Body composition of fasting obese patients measured by in vivo neutron activation analysis and isotope dilution. Clin Phys Physiol Meas 1987;8:271-82.

Strawford A, Barbieri T, Van Loan M, Parks E, Catlin D, Barton N, Neese R, Christiansen M, King J, Hellerstein MK. Resistance exercise and supraphysiologic androgen therapy in eugonadal men with HIV-related weight loss. JAMA 1999;281:1282-90.

Suttmann U, Ockenga O, Selber O, Hoogestraat L, Deicher H, Muller MJ. Incidence and prognostic value of malnutrition and wasting in human immunodeficiency virus-infected outpatients. J Acquir Immunodef Syndr 1995;8:239-45.

Torres RA, Unger KW, Cadman JA, Kassous JY. Recombinant human growth hormone improves truncal adiposity and 'buffalo humps' in HIV-positive patients on HAART. AIDS. 1999;13:2479-81.

Turner J, Muurahainen N, Terrell C, Graeber C, Kotler DP. Nutritional status and quality of life. Tenth International Conference on AIDS. 1994;2:35.

Vaswani AN, Vartsky D, Ellis KJ, Yasumura S, Cohn SH. Effects of caloric restriction on body composition and total body nitrogen as measured by neutron activation. Metabolism 1983;32:185-8.

Von Roenn JH, Armstrong D, Kotler D, Cohn DL, Klimas NG, Tchekmedyian NS, Cone L, Brennan PJ, Weitzman SA. Megestrol acetate in patients with AIDS-related cachexia. Ann Intern Med 1994;121:393-9.

Wang ZM, Visser M, Ma R, Baumgartner RN, Kotler DP, Gallagher D, Heymsfield SB. Skeletal muscle mass: validation of neutron activation and dual energy x-ray absorptiometry methods by computerized tomography. J Appl Physiol. 1996;80:824-831.

Wilson IB, Cleary PD. Clinical predictors of declines in physical functioning in persons with AIDS: results of a longitudinal study. J Acquir Immunodefic Syndr 1997;16:343-9.

Chapter 26

Abitbol, V., Roux, C., Chaussade, S., Guillemant, S., Kolta, S., Dougados, M., Couturier, D., and Amor, B. (1995): Metabolic bone assessment in patients with inflammatory bowel disease. *Gastroenterology* 108, 417-22.

ACCP/AACVPR Pulmonary Rehabilitation Guidelines Panel (1997): Pulmonary rehabilitation: joint ACCP/AACVPR evidence-based guidelines. *Chest* 112, 1363-96.

Al-Jaouni, R., Hebuterne, X., Pouget, I., and Rampal, P. (2000): Energy metabolism and substrate oxidation in patients with Crohn's disease. *Nutrition* 16, 173-8.

Anker, S.D., Chua, T.P., Ponikowski, P., Harrington, D., Swan, J.W., Kox, W.J., Poole-Wilson, P.A., and Coats, A.J. (1997a): Hormonal changes and catabolic/anabolic imbalance in chronic heart failure and their importance for cardiac cachexia. *Circulation* 96, 526-34.

Anker, S.D., Clark, A.L., Teixeira, M.M., Hellewell, P.G., and Coats, A.J. (1999a): Loss of bone mineral in patients with cachexia due to chronic heart failure. *Am J Cardiol* 83, 612-5, A10.

Anker, S.D., and Coats, A.J. (1999): Cardiac cachexia: a syndrome with impaired survival and immune and neuroendocrine activation. *Chest* 115, 836-47.

Anker, S.D., Ponikowski, P.P., Clark, A.L., Leyva, F., Rauchhaus, M., Kemp, M., Teixeira, M.M., Hellewell, P.G., Hooper, J., Poole-Wilson, P.A., and Coats, A.J. (1999b): Cytokines and neurohormones relating to body composition alterations in the wasting syndrome of chronic heart failure. *Eur Heart J* 20, 683-93.

Anker, S.D., Swan, J.W., Volterrani, M., Chua, T.P., Clark, A.L., Poole-Wilson, P.A., and Coats, A.J. (1997b): The influence of muscle mass, strength, fatigability and blood flow on exercise capacity in cachectic and non-cachectic patients with chronic heart failure. *Eur Heart J* 18, 259-69.

Arden, N.K., Nevitt, M.C., Lane, N.E., Gore, L.R., Hochberg, M.C., Scott, J.C., Pressman, A.R., and Cummings, S.R. (1999): Osteoarthritis and risk of falls, rates of bone loss, and osteoporotic fractures. Study of Osteoporotic Fractures Research Group. *Arthritis Rheum* 42, 1378-85.

Aris, R.M., Renner, J.B., Winders, A.D., Buell, H.E., Riggs, D.B., Lester, G.E., and Ontjes, D.A. (1998): Increased rate of fractures and severe kyphosis: sequelae of living into adulthood with cystic fibrosis. *Ann Intern Med* 128, 186-93.

Azcue, M., Rashid, M., Griffiths, A., and Pencharz, P.B. (1997): Energy expenditure and body composition in children with Crohn's disease: effect of enteral nutrition and treatment with prednisolone. *Gut* 41, 203-8.

Bacon, M.C., White, P.H., Raiten, D.J., Craft, N., Margolis, S., Levander, O.A., Taylor, M.L., Lipnick, R.N., and Sami, S. (1990): Nutritional status and growth in juvenile rheumatoid arthritis. *Semin Arthritis Rheum* 20, 97-106.

Baxter, J.D., and Forsham, P.H. (1972): Tissue effects of glucocorticoids. *Am J Med* 53, 573-89.

Bernard, S., LeBlanc, P., Whittom, F., Carrier, G., Jobin, J., Belleau, R., and Maltais, F. (1998): Peripheral muscle weakness in patients with chronic obstructive pulmonary disease. *Am J Respir Crit Care Med* 158, 629-34.

Bernard, S., Whittom, F., Leblanc, P., Jobin, J., Belleau, R., Berube, C., Carrier, G., and Maltais, F. (1999): Aerobic and strength training in patients with chronic obstructive pulmonary disease. *Am J Respir Crit Care Med* 159, 896-901.

Bernstein, B.H., Stobie, D., Singsen, B.H., Koster-King, K., Kornreich, H.K., and Hanson, V. (1977): Growth retardation in juvenile rheumatoid arthritis (JRA). *Arthritis Rheum* 20, 212-6.

Bhudhikanok, G.S., Lim, J., Marcus, R., Harkins, A., Moss, R.B., and Bachrach, L.K. (1996): Correlates of osteopenia in patients with cystic fibrosis. *Pediatrics* 97, 103-11.

Blum, A., and Miller, H. (2001): Pathophysiological role of cytokines in congestive heart failure. *Annu Rev Med* 52, 15-27.

Bonfield, T.L., Panuska, J.R., Konstan, M.W., Hilliard, K.A., Hilliard, J.B., Ghnaim, H., and Berger, M. (1995): Inflammatory cytokines in cystic fibrosis lungs. *Am J Respir Crit Care Med* 152, 2111-8.

Boot, A.M., Bouquet, J., Krenning, E.P., and de Muinck Keizer-Schrama, S.M. (1998): Bone mineral density and nutritional status in children with chronic inflammatory bowel disease. *Gut* 42, 188-94.

Borovnicar, D.J., Stroud, D.B., Bines, J.E., Haslam, R.H., and Strauss, B.J. (2000): Comparison of total body chlorine, potassium, and water measurements in children with cystic fibrosis. *Am J Clin Nutr* 71, 36-43.

Buchanan, N., Cane, R.D., Kinsley, R., and Eyberg, C.D. (1977): Gastrointestinal absorption studies in cardiac cachexia. *Intensive Care Med* 3, 89-91.

Bucuvalas, J.C., Chernausek, S.D., Alfaro, M.P., Krug, S.K., Ritschel, W., and Wilmott, R.W. (2001): Effect of insulinlike growth factor-1 treatment in children with cystic fibrosis. *J Pediatr Gastroenterol Nutr* 33, 576-81.

Burger, H., van Daele, P.L., Odding, E., Valkenburg, H.A., Hofman, A., Grobbee, D.E., Schutte, H.E., Birkenhager, J.C., and Pols, H.A. (1996): Association of radiographically evident osteoarthritis with higher bone mineral density and increased bone loss with age. The Rotterdam Study. *Arthritis Rheum* 39, 81-6.

Capristo, E., Addolorato, G., Mingrone, G., Greco, A.V., and Gasbarrini, G. (1998a): Effect of disease localization on the anthropometric and metabolic features of Crohn's disease. *Am J Gastroenterol* 93, 2411-9.

Capristo, E., Mingrone, G., Addolorato, G., Greco, A.V., and Gasbarrini, G. (1998b): Metabolic features of inflammatory bowel disease in a remission phase of the disease activity. *J Intern Med* 243, 339-47.

Chan, J.M., Rimm, E.B., Colditz, G.A., Stampfer, M.J., and Willett, W.C. (1994): Obesity, fat distribution, and weight gain as risk factors for clinical diabetes in men. *Diabetes Care* 17, 961-9.

Chandra, S., and Chandra, R.K. (1986): Nutrition, immune response, and outcome. *Prog Food Nutr Sci* 10, 1-65.

Coats, A.J., Adamopoulos, S., Meyer, T.E., Conway, J., and Sleight, P. (1990): Effects of physical training in chronic heart failure. *Lancet* 335, 63-6.

Coats, A.J., Adamopoulos, S., Radaelli, A., McCance, A., Meyer, T.E., Bernardi, L., Solda, P.L., Davey, P., Ormerod, O., and Forfar, C. (1992): Controlled trial of physical training in chronic heart failure. Exercise performance, hemodynamics, ventilation, and autonomic function. *Circulation* 85, 2119-31.

Coats, A.J., Clark, A.L., Piepoli, M., Volterrani, M., and Poole-Wilson, P.A. (1994): Symptoms and quality of life in heart failure: the muscle hypothesis. *Br Heart J* 72, S36-9.

Cowan, F.J., Warner, J.T., Lowes, L.M., Riberio, J.P., and Gregory, J.W. (1997): Auditing paediatric diabetes care and the impact of a specialist nurse trained in paediatric diabetes. *Arch Dis Child* 77, 109-14.

Cystic Fibrosis Foundation (1998): Cystic Fibrosis Foundation, Patient Registry Annual Data Report, Cystic Fibrosis Foundation, Bethesda, MD.

Davis, R.C., Hobbs, F.D., and Lip, G.Y. (2000): ABC of heart failure. History and epidemiology. *BMJ* 320, 39-42.

De Boer, W.I. (2002): Cytokines and therapy in COPD: a promising combination? *Chest* 121, 209S-18S.

de Meer, K., Gulmans, V.A., Westerterp, K.R., Houwen, R.H., and Berger, R. (1999): Skinfold measurements in children with cystic fibrosis: monitoring fat-free mass and exercise effects. *Eur J Pediatr* 158, 800-6.

Dean, G.S., Tyrrell-Price, J., Crawley, E., and Isenberg, D.A. (2000): Cytokines and systemic lupus erythematosus. *Ann Rheum Dis* 59, 243-51.

Elders, M.J. (2000): The increasing impact of arthritis on public health. *J Rheumatol Suppl* 60, 6-8.

Elsasser, U., Wilkins, B., Hesp, R., Thurnham, D.I., Reeve, J., and Ansell, B.M. (1982): Bone rarefaction and crush fractures in juvenile chronic arthritis. *Arch Dis Child* 57, 377-80.

Engelen, M.P., Schols, A.M., Baken, W.C., Wesseling, G.J., and Wouters, E.F. (1994): Nutritional depletion in relation to respiratory and peripheral skeletal muscle function in out-patients with COPD. *Eur Respir J* 7, 1793-7.

Engelen, M.P., Schols, A.M., Heidendal, G.A., and Wouters, E.F. (1998): Dual-energy X-ray absorptiometry in the clinical evaluation of body composition and bone mineral density in patients with chronic obstructive pulmonary disease. *Am J Clin Nutr* 68, 1298-303.

Eubanks, V., Koppersmith, N., Wooldridge, N., Clancy, J.P., Lyrene, R., Arani, R.B., Lee, J., Moldawer, L., Atchison, J., Sorscher, E.J., and Makris, C.M. (2002): Effects of megestrol acetate on weight gain, body composition, and pulmonary function in patients with cystic fibrosis. *J Pediatr* 140, 439-44.

Fisher, N.M., Pendergast, D.R., Gresham, G.E., and Calkins, E. (1991): Muscle rehabilitation: its effect on muscular and functional performance of patients with knee osteoarthritis. *Arch Phys Med Rehabil* 72, 367-74.

FitzSimmons, S.C. (1993): The changing epidemiology of cystic fibrosis. *J Pediatr* 122, 1-9.

Gulmans, V.A., de Meer, K., Binkhorst, R.A., Helders, P.J., and Saris, W.H. (1997): Reference values for maximum work capacity in relation to body composition in healthy Dutch children. *Eur Respir J* 10, 94-7.

Hambrecht, R., Gielen, S., Linke, A., Fiehn, E., Yu, J., Walther, C., Schoene, N., and Schuler, G. (2000): Effects of exercise training on left ventricular function and peripheral resistance in patients with chronic heart failure: a randomized trial. *JAMA* 283, 3095-101.

Hamilton, A.L., Killian, K.J., Summers, E., and Jones, N.L. (1995): Muscle strength, symptom intensity, and exercise capacity in patients with cardiorespiratory disorders. *Am J Respir Crit Care Med* 152, 2021-31.

Hansen, M., Florescu, A., Stoltenberg, M., Podenphant, J., Pedersen-Zbinden, B., Horslev-Petersen, K., Hyldstrup, L., and Lorenzen, I. (1996): Bone loss in rheumatoid arthritis. Influence of disease activity, duration of the disease, functional capacity, and corticosteroid treatment. *Scand J Rheumatol* 25, 367-76.

Hardin, D.S., Ellis, K.J., Dyson, M., Rice, J., McConnell, R., and Seilheimer, D.K. (2001): Growth hormone improves clinical status in prepubertal children with cystic fibrosis: results of a randomized controlled trial. *J Pediatr* 139, 636-42.

Harrington, D., Anker, S.D., Chua, T.P., Webb-Peploe, K.M., Ponikowski, P.P., Poole-Wilson, P.A., and Coats, A.J. (1997): Skeletal muscle function and its relation to exercise tolerance in chronic heart failure. *J Am Coll Cardiol* 30, 1758-64.

Hata, J.S., and Fick, R.B., Jr. (1988): Pseudomonas aeruginosa and the airways disease of cystic fibrosis. *Clin Chest Med* 9, 679-89.

Haugen, M.A., Hoyeraal, H.M., Larsen, S., Gilboe, I.M., and Trygg, K. (1992): Nutrient intake and nutritional status in children with juvenile chronic arthritis. *Scand J Rheumatol* 21, 165-70.

Henderson, R.C., and Madsen, C.D. (1999): Bone mineral content and body composition in children and young adults with cystic fibrosis. *Pediatr Pulmonol* 27, 80-4.

Henderson, R.C., and Specter, B.B. (1994): Kyphosis and fractures in children and young adults with cystic fibrosis. *J Pediatr* 125, 208-12.

Hill, G.L. (1992): Jonathan E. Rhoads Lecture. Body composition research: implications for the practice of clinical nutrition. *JPEN J Parenter Enteral Nutr* 16, 197-218.

Hochberg, M.C., Lethbridge-Cejku, M., Plato, C.C., Wigley, F.M., and Tobin, J.D. (1991): Factors associated with osteoarthritis of the hand in males: data from the Baltimore Longitudinal Study of Aging. *Am J Epidemiol* 134, 1121-7.

Hochberg, M.C., Lethbridge-Cejku, M., Scott, W.W., Jr., Reichle, R., Plato, C.C., and Tobin, J.D. (1995): The association of body weight, body fatness and body fat distribution with osteoarthritis of the knee: data from the Baltimore Longitudinal Study of Aging. *J Rheumatol* 22, 488-93.

Holl, R.W., Wolf, A., Thon, A., Bernhard, M., Buck, C., Missel, M., Heinze, E., von der Hardt, H., and Teller, W.M. (1997): Insulin resistance with altered secretory kinetics and reduced proinsulin in cystic fibrosis patients. *J Pediatr Gastroenterol Nutr* 25, 188-93.

Ionescu, A.A., Chatham, K., Davies, C.A., Nixon, L.S., Enright, S., and Shale, D.J. (1998): Inspiratory muscle function and body composition in cystic fibrosis. *Am J Respir Crit Care Med* 158, 1271-6.

Issenman, R.M., Atkinson, S.A., Radoja, C., and Fraher, L. (1993): Longitudinal assessment of growth, mineral metabolism, and bone mass in pediatric Crohn's disease. *J Pediatr Gastroenterol Nutr* 17, 401-6.

Janssen, I., Heymsfield, S.B., and Ross, R. (2002): Low relative skeletal muscle mass (sarcopenia) in older persons is associated with functional impairment and physical disability. *J Am Geriatr Soc* 50, 889-96.

Kanof, M.E., Lake, A.M., and Bayless, T.M. (1988): Decreased height velocity in children and adolescents before the diagnosis of Crohn's disease. *Gastroenterology* 95, 1523-7.

Keller, C., Hafstrom, I., and Svensson, B. (2001): Bone mineral density in women and men with early rheumatoid arthritis. *Scand J Rheumatol* 30, 213-20.

Kelsey, J.L., and Hochberg, M.C. (1986): Epidemiology and prevention of musculoskeletal disorders, pp. 1277-95. In J.M. Last (Ed.): *Public Health and Preventive Medicine*, Appleton-Century-Crofts, Norwalk, CT.

Kent-Braun, J.A., and Ng, A.V. (2000): Skeletal muscle oxidative capacity in young and older women and men. *J Appl Physiol* 89, 1072-8.

Kern, P.A., Saghizadeh, M., Ong, J.M., Bosch, R.J., Deem, R., and Simsolo, R.B. (1995): The expression of tumor necrosis factor in human adipose tissue. Regulation by obesity, weight loss, and relationship to lipoprotein lipase. *J Clin Invest* 95, 2111-9.

Khoshoo, V., Reifen, R., Neuman, M.G., Griffiths, A., and Pencharz, P.B. (1996): Effect of low- and high-fat, peptide-based diets on body composition and disease activity in adolescents with active Crohn's disease. *JPEN J Parenter Enteral Nutr* 20, 401-5.

Kipen, Y., Briganti, E.M., Strauss, B.J., Littlejohn, G.O., and Morand, E.F. (1999): Three year follow-up of body composition changes in pre-menopausal women with systemic lupus erythematosus. *Rheumatology (Oxford)* 38, 59-65.

Kipen, Y., Strauss, B.J., and Morand, E.F. (1998): Body composition in systemic lupus erythematosus. *Br J Rheumatol* 37, 514-9.

Kirschner, B.S., Klich, J.R., Kalman, S.S., deFavaro, M.V., and Rosenberg, I.H. (1981): Reversal of growth retardation in Crohn's disease with therapy emphasizing oral nutritional restitution. *Gastroenterology* 80, 10-5.

Kotler, D.P., Tierney, A.R., Wang, J., and Pierson, R.N., Jr. (1989): Magnitude of body-cell-mass depletion and the timing of death from wasting in AIDS. *Am J Clin Nutr* 50, 444-7.

Kotler, D.P., Wang, J., and Pierson, R.N. (1985): Body composition studies in patients with the acquired immunodeficiency syndrome. *Am J Clin Nutr* 42, 1255-65.

Kunnamo, I., Kallio, P., and Pelkonen, P. (1986): Incidence of arthritis in urban Finnish children. A prospective study. *Arthritis Rheum* 29, 1232-8.

Kushner, I. (1993): Regulation of the acute phase response by cytokines. *Perspect Biol Med* 36, 611-22.

Laan, R.F., van Riel, P.L., van de Putte, L.B., van Erning, L.J., van't Hof, M.A., and Lemmens, J.A. (1993): Low-dose prednisone induces rapid reversible axial bone loss in patients with rheumatoid arthritis. A randomized, controlled study. *Ann Intern Med* 119, 963-8.

Lin, C.H., Lerner, A., Rossi, T.M., Feld, L.G., Riddlesberger, M.M., and Lebenthal, E. (1989): Effects of parenteral nutrition on whole body and extremity composition in children and adolescents with active inflammatory bowel disease. *JPEN J Parenter Enteral Nutr* 13, 366-71.

Mancini, D.M., Walter, G., Reichek, N., Lenkinski, R., McCully, K.K., Mullen, J.L., and Wilson, J.R. (1992): Contribution of skeletal muscle atrophy to exercise intolerance and altered muscle metabolism in heart failure. *Circulation* 85, 1364-73.

Mangge, H., and Schauenstein, K. (1998): Cytokines in juvenile rheumatoid arthritis (JRA). *Cytokine* 10, 471-80.

Marchand, V., Baker, S.S., Stark, T.J., and Baker, R.D. (2000): Randomized, double-blind, placebo-controlled pilot trial of megestrol acetate in malnourished children with cystic fibrosis. *J Pediatr Gastroenterol Nutr* 31, 264-9.

Markowitz, J., Grancher, K., Rosa, J., Aiges, H., and Daum, F. (1993): Growth failure in pediatric inflammatory bowel disease. *J Pediatr Gastroenterol Nutr* 16, 373-80.

Mauras, N., George, D., Evans, J., Milov, D., Abrams, S., Rini, A., Welch, S., and Haymond, M. W. (2002): Growth hormone has anabolic effects in glucocorticosteroid-dependent children with inflammatory bowel disease: a pilot study. *Metabolism* 51, 127-35.

McGoey, B.V., Deitel, M., Saplys, R.J., and Kliman, M.E. (1990): Effect of weight loss on musculoskeletal pain in the morbidly obese. *J Bone Joint Surg Br* 72, 322-3.

Meyer, K. (2001): Exercise training in heart failure: recommendations based on current research. *Med Sci Sports Exerc* 33, 525-531.

Miller, W.C., Koceja, D.M., and Hamilton, E.J. (1997): A meta-analysis of the past 25 years of weight loss research using diet, exercise or diet plus exercise intervention. *Int J Obes Relat Metab Disord* 21, 941-7.

Mingrone, G., Greco, A.V., Benedetti, G., Capristo, E., Semeraro, R., Zoli, G., and Gasbarrini, G. (1996): Increased resting lipid oxidation in Crohn's disease. *Dig Dis Sci* 41, 72-6.

Morrow, W.J.W., Nelson, L., Watts, R., and Isenberg, D.A. (1999): *Autoimmune rheumatic disease*. Oxford University Press, Oxford, UK.

Motil, K.J., Grand, R.J., Davis-Kraft, L., Ferlic, L.L., and Smith, E.O. (1993): Growth failure in children with inflammatory bowel disease: a prospective study. *Gastroenterology* 105, 681-91.

Nair, K.S., Ford, G.C., Ekberg, K., Fernqvist-Forbes, E., and Wahren, J. (1995): Protein dynamics in whole body and in splanchnic and leg tissues in type I diabetic patients. *J Clin Invest* 95, 2926-37.

Nicholls, M.G., Espiner, E.A., Donald, R.A., and Hughes, H. (1974): Aldosterone and its regulation during diuresis in patients with gross congestive heart failure. *Clin Sci Mol Med* 47, 301-15.

Nixon, P.A., Orenstein, D.M., Kelsey, S.F., and Doershuk, C.F. (1992): The prognostic value of exercise testing in patients with cystic fibrosis. *N Engl J Med* 327, 1785-8.

O'Reilly, S.C., Muir, K.R., and Doherty, M. (1999): Effectiveness of home exercise on pain and disability from osteoarthritis of the knee: a randomised controlled trial. *Ann Rheum Dis* 58, 15-9.

Piepoli, M.F., Scott, A.C., Capucci, A., and Coats, A.J. (2001): Skeletal muscle training in chronic heart failure. *Acta Physiol Scand* 171, 295-303.

Pomposelli, J.J., Flores, E.A., and Bistrian, B.R. (1988): Role of biochemical mediators in clinical nutrition and surgical metabolism. *JPEN J Parenter Enteral Nutr* 12, 212-8.

Radaelli, A., Coats, A.J., Leuzzi, S., Piepoli, M., Meyer, T.E., Calciati, A., Finardi, G., Bernardi, L., and Sleight, P. (1996): Physical training enhances sympathetic and parasympathetic control of heart rate and peripheral vessels in chronic heart failure. *Clin Sci (Lond)* 91 Suppl, 92-4.

Rall, L.C., Meydani, S.N., Kehayias, J.J., Dawson-Hughes, B., and Roubenoff, R. (1996a): The effect of progressive resistance training in rheumatoid arthritis. Increased strength without changes in energy balance or body composition. *Arthritis Rheum* 39, 415-26.

Rall, L.C., Rosen, C.J., Dolnikowski, G., Hartman, W.J., Lundgren, N., Abad, L.W., Dinarello, C.A., and Roubenoff, R. (1996b): Protein metabolism in rheumatoid arthritis and aging. Effects of muscle strength training and tumor necrosis factor alpha. *Arthritis Rheum* 39, 1115-24.

Rall, L.C., and Roubenoff, R. (1996): Body composition, metabolism, and resistance exercise in patients with rheumatoid arthritis. *Arthritis Care Res* 9, 151-6.

Reid, I.R. (1989): Pathogenesis and treatment of steroid osteoporosis. *Clin Endocrinol (Oxford)* 30, 83-103.

Reilly, J.J., Jr., Hull, S.F., Albert, N., Waller, A., and Bringardener, S. (1988): Economic impact of malnutrition: a model system for hospitalized patients. *JPEN J Parenter Enteral Nutr* 12, 371-6.

Rexrode, K.M., Carey, V.J., Hennekens, C.H., Walters, E.E., Colditz, G.A., Stampfer, M.J., Willett, W.C., and Manson, J.E. (1998): Abdominal adiposity and coronary heart disease in women. *JAMA* 280, 1843-8.

Riordan, J.R., Rommens, J.M., Kerem, B., Alon, N., Rozmahel, R., Grzelczak, Z., Zielenski, J., Lok, S., Plavsic, N., and Chou, J.L. (1989): Identification of the cystic fibrosis gene: cloning and characterization of complementary DNA. *Science* 245, 1066-73.

Rogler, G., and Andus, T. (1998): Cytokines in inflammatory bowel disease. *World J Surg* 22, 382-9.

Roubenoff, R., and Hughes, V.A. (2000): Sarcopenia: current concepts. *J Gerontol A Biol Sci Med Sci* 55, M716-24.

Roubenoff, R., Roubenoff, R.A., Cannon, J.G., Kehayias, J.J., Zhuang, H., Dawson-Hughes, B., Dinarello, C.A., and Rosenberg, I.H. (1994): Rheumatoid cachexia: cytokine-driven hypermetabolism accompanying reduced body cell mass in chronic inflammation. *J Clin Invest* 93, 2379-86.

Roubenoff, R., Roubenoff, R.A., Ward, L.M., Holland, S.M., and Hellmann, D.B. (1992): Rheumatoid cachexia: depletion of lean body mass in rheumatoid arthritis. Possible association with tumor necrosis factor. *J Rheumatol* 19, 1505-10.

Roubenoff, R., Roubenoff, R.A., Ward, L.M., and Stevens, M.B. (1990): Catabolic effects of high-dose corticosteroids persist despite therapeutic benefit in rheumatoid arthritis. *Am J Clin Nutr* 52, 1113-7.

Roubenoff, R., Walsmith, J., Lundgren, N., Snydman, L., Dolnikowski, G., and Roberts, S. (2002): Low physical activity reduces total energy expenditure in women with rheumatoid arthritis: implications for dietary intake recommendations. *Am J Clin Nutr* 76, 774-9.

Royall, D., Greenberg, G.R., Allard, J.P., Baker, J.P., and Jeejeebhoy, K.N. (1995): Total enteral nutrition support improves body composition of patients with active Crohn's disease. *JPEN J Parenter Enteral Nutr* 19, 95-9.

Royall, D., Jeejeebhoy, K.N., Baker, J.P., Allard, J.P., Habal, F.M., Cunnane, S.C., and Greenberg, G.R. (1994): Comparison of amino acid v peptide based enteral diets in active Crohn's disease: clinical and nutritional outcome. *Gut* 35, 783-7.

Sambrook, P.N., Eisman, J.A., Champion, G.D., Yeates, M.G., Pocock, N.A., and Eberl, S. (1987): Determinants of axial bone loss in rheumatoid arthritis. *Arthritis Rheum* 30, 721-8.

Sasayama, S., Matsumori, A., and Kihara, Y. (1999): New insights into the pathophysiological role for cytokines in heart failure. *Cardiovasc Res* 42, 557-64.

Schols, A.M., Mostert, R., Soeters, P.B., Greve, L.H., and Wouters, E.F. (1989): Nutritional state and exercise performance in patients with chronic obstructive lung disease. *Thorax* 44, 937-41.

Selvadurai, H.C., Blimkie, C.J., Meyers, N., Mellis, C.M., Cooper, P.J., and Van Asperen, P.P. (2002): Randomized controlled study of in-hospital exercise training programs in children with cystic fibrosis. *Pediatr Pulmonol* 33, 194-200.

Sentongo, T.A., Semeao, E.J., Piccoli, D.A., Stallings, V.A., and Zemel, B.S. (2000): Growth, body composition, and nutritional status in children and adolescents with Crohn's disease. *J Pediatr Gastroenterol Nutr* 31, 33-40.

Shepherd, R. (2002): Achieving genetic potential for nutrition and growth in cystic fibrosis. *J Pediatr* 140, 393-5.

Sidossis, L.S., Coggan, A.R., Gastaldelli, A., and Wolfe, R.R. (1995): Pathway of free fatty acid oxidation in human subjects. Implications for tracer studies. *J Clin Invest* 95, 278-84.

Simon, D., Prieur, A., and Czernichow, P. (2000): Treatment of juvenile rheumatoid arthritis with growth hormone. *Horm Res* 53 Suppl 1, 82-6.

Simon, D., Touati, G., Prieur, A.M., Ruiz, J.C., and Czernichow, P. (1999): Growth hormone treatment of short stature and metabolic dysfunction in juvenile chronic arthritis. *Acta Paediatr Suppl* 88, 100-5.

Sowers, M., Lachance, L., Jamadar, D., Hochberg, M.C., Hollis, B., Crutchfield, M., and Jannausch, M.L. (1999): The associations of bone mineral density and bone turnover markers with osteoarthritis of the hand and knee in pre- and peri-menopausal women. *Arthritis Rheum* 42, 483-9.

Staun, M., Tjellesen, L., Thale, M., Schaadt, O., and Jarnum, S. (1997): Bone mineral content in patients with Crohn's disease. A longitudinal study in patients with bowel resections. *Scand J Gastroenterol* 32, 226-32.

Stettler, N., Kawchak, D.A., Boyle, L.L., Propert, K.J., Scanlin, T.F., Stallings, V.A., and Zemel, B.S. (2000): Prospective evaluation of growth, nutritional status, and body composition in children with cystic fibrosis. *Am J Clin Nutr* 72, 407-13.

Stokes, M.A., and Hill, G.L. (1993): Total energy expenditure in patients with Crohn's disease: measurement by the combined body scan technique. *JPEN J Parenter Enteral Nutr* 17, 3-7.

Sturge, R.A., Beardwell, C., Hartog, M., Wright, D., and Ansell, B.M. (1970): Cortisol and growth hormone secretion in relation to linear growth: patients with Still's disease on different therapeutic regimens. *Br Med J* 3, 547-51.

Sullivan, D.B., Cassidy, J.T., and Petty, R.E. (1975): Pathogenic implications of age of onset in juvenile rheumatoid arthritis. *Arthritis Rheum* 18, 251-5.

Svenson, K.L., Lundqvist, G., Wide, L., and Hallgren, R. (1987): Impaired glucose handling in active rheumatoid arthritis: relationship to the secretion of insulin and counter-regulatory hormones. *Metabolism* 36, 940-3.

Swan, J.W., Walton, C., Godsland, I.F., Clark, A.L., Coats, A.J., and Oliver, M.F. (1994): Insulin resistance in chronic heart failure. *Eur Heart J* 15, 1528-32.

Taylor, A.M., Bush, A., Thomson, A., Oades, P.J., Marchant, J.L., Bruce-Morgan, C., Holly, J., Ahmed, L., and Dunger, D.B. (1997): Relation between insulin-like growth factor-I, body mass index, and clinical status in cystic fibrosis. *Arch Dis Child* 76, 304-9.

Tellado, J.M., Garcia-Sabrido, J.L., Hanley, J.A., Shizgal, H.M., and Christou, N.V. (1989): Predicting mortality based on body composition analysis. *Ann Surg* 209, 81-7.

Thomas, R.D., Silverton, N.P., Burkinshaw, L., and Morgan, D.B. (1979): Potassium depletion and tissue loss in chronic heart-disease. *Lancet* 2, 9-11.

Tjellesen, L., Nielsen, P.K., and Staun, M. (1998): Body composition by dual-energy X-ray absorptiometry in patients with Crohn's disease. *Scand J Gastroenterol* 33, 956-60.

Toda, Y. (2001): The effect of energy restriction, walking, and exercise on lower extremity lean body mass in obese women with osteoarthritis of the knee. *J Orthop Sci* 6, 148-54.

Toda, Y., Segal, N., Toda, T., Kato, A., and Toda, F. (2000): A decline in lower extremity lean body mass per body weight is characteristic of women with early phase osteoarthritis of the knee. *J Rheumatol* 27, 2449-54.

Toth, M.J., Gottlieb, S.S., Goran, M.I., Fisher, M.L., and Poehlman, E.T. (1997): Daily energy expenditure in free-living heart failure patients. *Am J Physiol* 272, E469-75.

Touati, G., Prieur, A.M., Ruiz, J.C., Noel, M., and Czernichow, P. (1998): Beneficial effects of one-year growth hormone administration to children with juvenile chronic arthritis on chronic steroid therapy. I. Effects on growth velocity and body composition. *J Clin Endocrinol Metab* 83, 403-9.

Ulivieri, F.M., Lisciandrano, D., Ranzi, T., Taioli, E., Cermesoni, L., Piodi, L.P., Nava, M.C., Vezzoli, M., and Bianchi, P.A. (2000): Bone mineral density and body composition in patients with ulcerative colitis. *Am J Gastroenterol* 95, 1491-4.

Ulivieri, F.M., Piodi, L.P., Taioli, E., Lisciandrano, D., Ranzi, T., Vezzoli, M., Cermesoni, L., and Bianchi, P. (2001): Bone mineral density and body composition in ulcerative colitis: a six-year follow-up. *Osteoporos Int* 12, 343-8.

van den Berg, W.B. (1999): The role of cytokines and growth factors in cartilage destruction in osteoarthritis and rheumatoid arthritis. *Z Rheumatol* 58, 136-141.

Varonos, S., Ansell, B.M., and Reeve, J. (1987): Vertebral collapse in juvenile chronic arthritis: its relationship with glucocorticoid therapy. *Calcif Tissue Int* 41, 75-8.

Visser, M., Harris, T.B., Langlois, J., Hannan, M.T., Roubenoff, R., Felson, D.T., Wilson, P.W., and Kiel, D.P. (1998a): Body fat and skeletal muscle mass in relation to physical disability in very old men and women of the Framingham Heart Study. *J Gerontol A Biol Sci Med Sci* 53, M214-21.

Visser, M., Langlois, J., Guralnik, J.M., Cauley, J.A., Kronmal, R.A., Robbins, J., Williamson, J.D., and Harris, T.B. (1998b): High body fatness, but not low fat-free mass, predicts disability in older men and women: the Cardiovascular Health Study. *Am J Clin Nutr* 68, 584-90.

Walsmith, J., and Roubenoff, R. (2002): Cachexia in rheumatoid arthritis. *Int J Cardiol* 85, 89.

Westacott, C.I., and Sharif, M. (1996): Cytokines in osteoarthritis: mediators or markers of joint destruction? *Semin Arthritis Rheum* 25, 254-72.

Westhovens, R., Nijs, J., Taelman, V., and Dequeker, J. (1997): Body composition in rheumatoid arthritis. *Br J Rheumatol* 36, 444-8.

Winick, M.E. (1979): *Hunger Disease: Studies by Jewish Physicians in the Warsaw Ghetto.* Wiley, New York.

Yilmaz, M., Kendirli, S.G., Altintas, D., Bingol, G., and Antmen, B. (2001): Cytokine levels in serum of patients with juvenile rheumatoid arthritis. *Clin Rheumatol* 20, 30-5.

Zhang, Y., Hannan, M.T., Chaisson, C.E., McAlindon, T.E., Evans, S.R., Aliabadi, P., Levy, D., and Felson, D.T. (2000): Bone mineral density and risk of incident and progressive radiographic knee osteoarthritis in women: the Framingham Study. *J Rheumatol* 27, 1032-7.

Appendix

Butte NF, Hopkinson JM, Wong WW, Smith EO, Ellis KJ. Body composition during the first 2 years of life: an undated reference. Pediatr. Res. 2000;47:578-585.

Elia M. Organ and tissue contribution to metabolic rate. In: Kinney JM, Tucker HN (eds.) Energy metabolism: Tissue determinants and cellular corollaries. New York: Raven Press, pp. 19-60; 1992.

Fomon SJ, Haschke F, Ziegler EZ, Nelson SE. Body composition of reference children from birth to age 10 years. Am. J. Clin. Nutr. 1982;35:1169-1175.

Fomon SJ, Nelson SE. Body composition of the male and female reference infants. Annu. Rev. Nutr. 2002;22:1-17.

Snyder WS, Cook MJ, Nasset ES, Karhausen LR, Howells GP, Tipton IH. Report of the task group on reference Man. Oxford, UK: Pergamon Press, 1975.

Ziegler EE, O'Donnell AM, Nelson SE, Fomon SJ. Body composition of the reference fetus. Growth, 1976;40:329-341.

Index

Page numbers followed by an italicized *f* or *t* indicate a figure or table, respectively.

About the Editors

Steven B. Heymsfield, MD, brings a broad range of experience in research to the writing of this book. He is trained in physics, chemistry, biology, and medicine, all of which relate to the measurement and study of body composition. Dr. Heymsfield is currently the executive director of clinical studies, metabolism at Merck in Rahway, New Jersey, where he oversees Merck's Clinical Obesity Research Program. Additionally, he is a visiting scientist at the Obesity Research Center at St. Luke's–Roosevelt Hospital. Prior to his current position, Dr. Heymsfield was a professor of medicine at the Columbia University, New York, College of Physicians and Surgeons.

Dr. Heymsfield has conducted national and international presentations and made many contributions to publications in the field. He serves on the editorial boards of the *Journal of Parenteral and Enteral Nutrition, American Journal of Clinical Nutrition, International Journal of Body Composition Research, Age & Nutrition, Nutrition Reviews, Clinical Nutrition,* and *Adipocytes.* He is also an active member of the North American Association for the Study of Obesity and past president of both the American Society of Clinical Nutrition and the American Society of Parenteral and Enteral Nutrition.

Timothy G. Lohman, PhD, is a leading scientist in the field of body composition assessment. A respected researcher, he explores body composition methodology and changes in body composition with growth and development, exercise, and aging. His leadership in standardization of body composition methodology is well recognized.

Dr. Lohman is a professor in the department of exercise science at the University of Arizona. He is also a fellow of the American Academy of Physical Education, a member of the American College of Sports Medicine, and a member of the Youth Fitness Advisory Committee of the Cooper Institute for Aerobics Research in Dallas, Texas. He is the author of many research articles and an editor of the *Anthropometric Standardization Reference Manual,* published by Human Kinetics. Dr. Lohman is also author of *Advances in Body Composition Assessment.*

ZiMian Wang, PhD, is an associate research scientist at Columbia University College and research associate in the Obesity Research Center of St. Luke's–Roosevelt Hospital. Trained in biology, biochemistry, physiology, and chemistry, he has published more than 100 research papers on the topic of human body composition.

Scott B. Going, PhD, is an associate professor in the department of nutritional sciences at the University of Arizona. An expert in body composition models and methods, he has more than 20 years of related teaching experience and research experience in this field.